Bridges London

Mathematics, Music, Art, Architecture, Culture

2006

BRIDGES
Mathematical Connections in Art, Music, and Science

Conference Proceedings
2006

Reza Sarhangi and John Sharp, Editors

tarquin publications

Bridges London, Mathematics, Music, Art, Architecture, Culture
Conference Proceedings, 2006

Editors:

Reza Sarhangi
Department of Mathematics
Towson University
Towson, Maryland, USA

John Sharp
London Knowledge Lab
Institute of Education, University of London, London, UK

ISBN: 0-9665201-7-3
ISSN: 1099-6702

Printed in the UK by Print Solutions Partnership

Distributed by MATHARTFUN.COM (http://mathartfun.com) and Tarquin Books (www.tarquinbooks.com)

Cover layout: John Sharp
Front cover design: *Poincaré lace* by Carlo Séquin
Back Cover: Puzzle: "The root of the proportion", clockwise from top left:
Pentagonal Interference by John Sharp, *Zome Atomium* by Samuel Verbiese and Scott Vorthmann, *Bacteriophage* by Richard Ahrens, *SpiralSurface* by RinusRoelofs, *Dodecaspidroball* by Dániel Erdély, Marc Pelletier andWalt van Ballegooijen,

Bridges London

Mathematics, Music, Art, Architecture, Culture

Organizers

Reza Sarhangi
Department of Mathematics
Towson University
Towson, Maryland, USA

Phillip Kent
London Knowledge Lab
Institute of Education, University of London
London, UK

John Sharp
London Knowledge Lab
Institute of Education, University of London
London, UK

Conference Advisors

Michael Field
Department of Mathematics
University of Houston,
Texas, USA

Robert Moody
Department of Mathematics and Statistics
University of Victoria
Victoria, British Columbia, Canada

Richard Noss
Director, London Knowledge Lab
Institute of Education, University of London
London, UK

Michael Reiss
Head, School of Mathematics,
Science and Technology Education
Institute of Education, University of London
London, UK

Carlo Séquin
Computer Science Division, EECS Department
University of California
California, Berkeley, USA

Bridges for Teachers, Teachers for Bridges

Mara Alagic
Department of Curriculum and Instruction
Wichita State University
Wichita, Kansas, USA

Paul Gailiunas
Gosforth High School
Newcastle, UK

Bridges Art exhibit

Robert Fathauer
Curator
Tesselations Company
Phoenix, Arizona, USA

Anne Burns
Juror, Web and CD Designer
Department of Mathematics
Long Island University,
Brookeville, New York, USA

Nat Friedman
Juror
Department of Mathematics
University at Albany,
Albany, New York, USA

Contents

Bridges for Teachers, Teachers for Bridges

Preface

When a man is tired of London, he is tired of life; for there is in London all that life can afford.
— Samuel Johnson 1777

In 1998 one man from the UK made a trip, all alone, half-way around the world to the small town of Winfield in the plains of Kansas, to the first Bridges Conference. On the first night there was a terrific thunderstorm with sheet lightening he had never seen before, which made him wonder what he had let himself in for. He does wonder, because now this man, John Sharp, is one of the organizers for what promises to be the largest Bridges so far. With many international faces, Bridges itself has travelled half-way around the world, visiting Spain and Canada, to be in London, a city centred on the river Thames which has many real bridges.

For the conference logo, Phillip Kent, the young and enthusiastic co-organizer, whose involvement has enriched the conference greatly, cleverly drew some lines on a photograph of the bridge most commonly seen in tourist photographs: Tower Bridge. The logo is a set of five lines, which is a play on the diagram of Euclid's Book I, Proposition 5, commonly known as the *pons asinorum* or "Bridge of Asses". The logo can also be suggestive of construction lines for drawing in linear perspective. Apart from its obvious symbolism as a bridge, the story goes that only if you could pass over this bridge (understand the proof) was there hope for you as a mathematician. Perhaps because it is a humorous allusion only geometers will understand, no one seems to have noticed it! Maybe Bridges participants do not study Euclid any more. They obviously do study modern geometries as is evident by many of the papers in this book.

The range of topics in connections among mathematics and the arts in this book brings hope for a rich exchange of ideas during the conference and aftermath. The Bridges logo subtitle is "Mathematical Connections in Art, Music and Science". Over the years there has not been much science. Some papers in previous years have been very much at the pure mathematics end of the spectrum. This year we have made a conscious effort to pull them back from the edge if they did not have an artistic element. This is not to say that there are papers at the extreme both ways. However, this is not where the cross fertilization occurs best. Artists need to be made aware of mathematical possibilities; artists whose work has a mathematical element which is "intuitive" often are not able to include detailed mathematics. There are many papers this year that are collaborative efforts in the true spirit of Bridges. There are many mathematicians who are expressing their work in an artistic way. There are historical papers, and there are cutting edge descriptions in architecture. Analyzing the papers yields a tangled web of interactions that would take many pages to describe. The *Bridges Proceedings* are one of those books which one can come back to again and again, dipping in and finding something new every time.

Since the first Bridges, the size of the Proceedings has more than doubled with a wider variety of topics than ever before. The increase in submissions, of course, caused more work for the referees and editors. No words can express our appreciation for the work the referees have done to enable publication and also to those who helped some authors prepare and format their papers. In order to keep the book a reasonably manageable size, some authors were requested to shorten their papers. The complexities of bringing together so many ideas and the amount of work that goes on behind the scenes is described by John Sharp in the following pages. Because of the varieties of subject and author experience, Bridges is unique in the way it accepts such a large proportion of papers. In effect this changes the attitudes of the people submitting the papers even before they are published and the conference then strengthens the bridges as well as building them.

The front cover is by Carlo Séquin who has been a valued advisor to Bridges since the beginning. He is very much an "ideal" member of the Bridges community in the way he worked with artists like Brent Collins both to learn from them and to collaborate. His efforts symbolize the way art and mathematics is interconnected.

The *Bridges Visual Art Exhibit* is the result of Robert Fathauer's hard work in communicating with a large number of artists in order to carefully select the artwork and properly set them up for the exhibit. His job (together with fellow jurors Nat Friedman and Anne Burns) has been harder this year because there have been a record number of submissions. The website pages assembled by Anne Burns, which are also on the CD that accompanies this book, record the images presented at the *Bridges Visual Art Exhibit*.

Although we do not include the details of the conference in the Proceedings, it is worth mentioning that Bridges London has taken some new directions this year. We have more excursions since we are in one of the world's major cultural centers. We are also aiming this year to reach out to the general public with two free and open events. We will have a concert, or Musical Event, mixing music and mathematics by Bridges participants and well-known UK mathematicians. The workshop papers at the end of the Proceedings show how *Bridges for Teachers - Teachers for Bridges* helps the educational specialists. This has been organized by Mara Alagic and Paul Gailiunas.

We are privileged to have been able to collaborate with the Royal Institution of Great Britain. This organization has been a major player in scientific progress in the last 200 years, with Sir Humphrey Davy, Michael Faraday and James Dewar amongst its scientific members and it is famous for its annual Christmas Lectures on scientific topics delivered to school students. It is noteworthy the 1978 Lectures were titled 'Mathematics into pictures' and were given by Professor Sir Christopher Zeeman, FRS, famous for his work on catastrophe theory. Building on these lectures, in 1981 he launched the Royal Institution Mathematics Masterclass program, which aims to mathematically engage and inspire 12 to 14 year olds. We are delighted to celebrate the 25th anniversary of these classes at Bridges by having him and three other expert masterclass presenters as part of a Bridges Family Day to be held on the last day of the conference. This again is open to the public and includes a Maths Fair which allows Bridges participants to get involved in small workshops, either as contributors or participants.

Special thanks must also be given to the staff at the Institute of Education Conference Office and the School of Mathematics, Science and Technology and the London Knowledge Lab. We also gratefully acknowledge financial support of the conference by the London Mathematical Society, Sibelius Software, the Institute of Education and the London Knowledge Lab. We also thank Aida Jones for her time and effort to assist in the reviewing and registration process, and Barbara Kaiser at Southwestern College in Winfield, Kansas who generously spent days to deal with the registration process again this year and everyone else who has helped in their own way, not least the participants.

How Bridges Proceedings come together

This book is product of many people's time and creativity. The results are obvious once it is completed. The path to those results is long and tortuous in many ways, not all of them obvious. The following is an attempt to describe how this takes place in a way which shows why Bridges is different from most other conferences. Although I have reviewed and edited papers in previous years, I have never been involved to the level I have this year and it has been illuminating to see how much work is necessary.

Looking through the papers, it is not easy to define where art ends and mathematics begins, and vice versa. There is also the question of "what is art" which is why Bridges is about the connections between mathematics and the arts in all its forms, from the visual and the decorative to music, architecture and many more. Pablo Picasso said that "great artists don't borrow, they steal". Artists certainly copy and some look through mathematics books to gain ideas. This book has a wealth of material to inspire, copy and steal for both the artist and mathematician. Why do I say all of this? Because it shows the problems presented in bringing Bridges together. Because defining Bridges other than through the people who make it up is very difficult.

A Bridges conference starts in earnest as soon as the previous one finishes. There is publicity and announcements about when and where and what to submit. When the papers start rolling in the next job is to review them for suitability. These reviews are carried out by a group of anonymous (to those submitting the papers and all but the editors) reviewers. The choice of reviewers for a paper (usually three, but often two this year with so many papers) is based on the experience of the editors and is not easy. Sometimes it is not easy for the reviewer. Because a reviewer might not be as expert in the field as the writer, not all errors may be picked up. Some reviewers are not necessarily aware of how they should approach the job. They may give opinion rather than a check for accuracy. Some write a few lines and some another paper as their review with many more references, but on the whole they write a set of points that need to be fixed, which the editors can then pass back to the author before the paper is published. In some cases, though not many, the reviewers conflict with one saying accept and one reject. In some cases both suggest rejection and rejections can be quite strongly worded. Then editorial diplomacy comes into play.

Regular conference participants will recognise this as fairly typical, but there are other aspects which make Bridges different. Many participants (usually at the mathematical end of the spectrum) are academics used to conferences and often relying on getting papers published to further their careers and thus are used to preparing and formatting papers. There is a strong contrast with artists and others who have never been to a conference before or written a paper. The latter are equally likely to be unfamiliar with the finer points of formatting in a Word Processor or presenting their ideas in a structured way, good though their ideas may be. This is very evident with artists who speak visually, and are not used to working with words and they might not know the correct technical words, or how to use them in the right way. Mostly this is picked up in the review process, but even so the editors have had to explain and edit some texts before they were suitable for publication. There is also the issue of someone who might be good at creating a picture but not technically aware of how to prepare it for publication. This means we have had a few pictures which do not have enough resolution and so appear blurred because we have not been able to get them improved, usually because of time constraints.

So while the editorial job has often been one of reformatting or re-writing a paper there is a more important situation where changes have been required which makes Bridges different from other conferences. Many people submitted papers which were generally aimed at the idea of making connections between art and mathematics but did not make the link as strongly as they could have done.

Some did not even do so at all, even though the author was aware of the link, otherwise they would not have submitted the paper. This is further complicated by the need for mathematicians to say "here are some ideas which I think artists can use, but I am not an artist so I can't write about that side". At the other end of the scale artists can be mathematically intuitive and see things before the mathematicians have found them. They can use mathematics in a way which is too "free" for some mathematicians. Then there are authors who come from what one might call "the fringes" and their original submissions were not of the required standard, especially when compared with the more academic papers. So the editors have had to perform the role that a conductor of an orchestra would and mould many papers to fit the conference. Rather than reject a paper, there has been much discussion with some authors who were not in tune with Bridges, most (but not all) of whom have not been to a previous conference. Moreover, since Bridges is unique in bringing so many disciplines together, the authors may not have been exposed to an audience that ranges so widely as Bridges. So while they might be distinguished mathematicians or artists, and used to communicating in their own peer group, their work has to be focussed in a different way. With so many papers and so many new faces, the orchestration has been a greater task than in previous years and a full-time job in parts of the process.

Many conferences publish rejections rates. It is not Bridges policy to do this. The main reason is because this would invite comparison of situations which are not equivalent. Bridges, by its very nature needs to spend time with the author, to coax them if necessary. So Bridges is unusual in not rejecting outright in the first instance. Whereas a reviewer might suggest rejection many papers that might have been rejected have been "brought into line" to a certain extent. Authors have been very accommodating in this respect and in some cases produced papers which are excellent examples of building the kind of bridges that the conference is aiming at. There are papers here which might have been more suitable for a pure mathematics conference. There are papers which mathematical readers will find incorrect or even confused and we do not have the luxury of many iterations to get a paper correct; with a journal this can take years and we only have a couple of months. Mathematicians can be very polarized (binary) in their outlook and often intolerant of what does not conform to the rigour of their subject. So if you are reading with a supercritical eye remember that an artist is not necessarily a mathematician and vice versa. Before being too critical ask yourself if the author has seen something that you haven't. After all that is what we have had to do when we have brought these papers together for Bridges and with such a short time to do so, how Bridges comes together sometimes feels like a miracle.

Bridges always produces the Proceedings in time for the conference. This benefits both the person giving the paper and the participants listening to the talk. Many people read the papers they are especially interested in before going to the talk. Participants can really get involved and not meet the ideas for the first time out of the blue and the paper comes more alive at the conference. Since there are only a few months to accomplish this, the method has its pros and cons. The ideas might not be polished, but then the author gains a chance to iron out any points that are not clear before, possibly, taking it further for journal publication. Bridges has many gems; often they are uncut and unpolished; many will get cut into diamonds which later appear in journals. There may be some ideas akin to common stones as well as gems, but even those have their uses and they might not see the light of day with another approach. The success and reputation of Bridges needs variety above all.

Bridges London

The Papers

Dodecaspidroball by Dániel Erdély, Marc Pelletier andWalt van Ballegooijen,

Collaboration on the Integration of Sculpture and Architecture in The Eden Project

Peter Randall-Page
Veet Mill Farm
Crockernwell
Exeter
Devon
EX6 6NL
E-mail: contact@peterrandall-page.com

Abstract

This paper and talk document my collaboration with Jolyon Brewis of Grimshaw Architects on the design of a new education building for the Eden Project, Cornwall. The roof structure of the building is based on plant geometry in the form of spiral phyllotaxis and incorporates a granite sculpture which will be sited in it's own specially designed chamber at the centre of the building.

This very large sculpture is based on the same growth pattern as the roof and has involved collaboration with professionals from many disciplines including quarrymen, stone masons, engineers and computer experts.

My first involvement with the Eden project was in April 2001 when I was asked to consider the idea of making an 'iconic' sculpture for the site. This was not an easy brief as the biomes themselves are such an iconic and sculptural image, and most of the landscape design had already been finalised.

Figure 1 The biomes at the Eden Project, Cornwall

I wanted to draw attention to the way in which the world of plants has always nourished the human spirit and imagination as well as providing for our physical needs. In particular I wanted to make something which would graphically illustrate how botanical imagery has been incorporated into architecture.

I thought of making a series of free standing architectural columns drawing attention to the idea of columns being symbolic trees with their capitals as stylised foliage. This idea was initially met with enthusiasm by the Eden team, but for various reasons it proved impractical and inappropriate for the existing site.

Subsequently, Jolyon Brewis from the architects Grimshaw, had been commissioned to design a new education building for Eden and Eden's Creative Director, Peter Hampel and CEO, Tim Smit had the idea of establishing a collaboration between me and the architects at the start of the design process. The brief was to contribute ideas that could influence the form and fabric of the new building and also to identify opportunities where artists could enhance the overall scheme, demonstrate the power of human imagination and illustrate the inspiration that can be drawn from natural and organic forms, and in particular plant life.

In March 2003, I was appointed as artist in residence to work with Jolyon on the concept and design of the new building and the appropriate integration of artwork. When Jolyon and I had our first meeting, at my studio, to discuss the building, Jolyon had only received the brief a week or so before.

The brief was that the building should be like a tree. Trees are the largest living things on the planet and are a graphic example of our dependence on plants for oxygen, food, fuel and numerous vital materials from paper to timber, coal to charcoal, fruit to rubber.

Jolyon's initial response to this unusual brief was the broad concept of a lattice roof structure (the canopy) emanating from a central hollow core (the trunk). This lattice produced a pattern of diamond shapes set on a grid of opposing spirals based on concentric circles and radial lines, the diagonals of which created a symmetrical pattern with equal numbers of spirals of the same curvature in both directions.

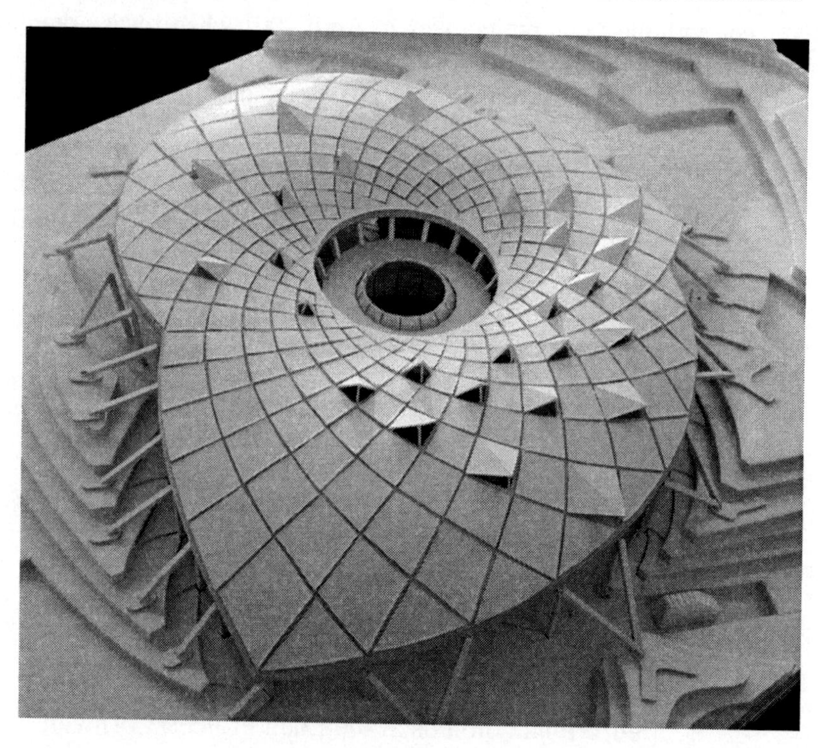

Figure 2 Scale model of The Core, the new education resource centre for The Eden Project

This concept was enthusiastically received by both the client and the Millennium Commission, but there was a problem - the engineers calculated that the structure would require extremely thick (2,000mm) timber members to support the roof at the outer edge.

My own practice has always been informed by observation of living things in general and botanical form in particular. My work has taken me on a journey from relatively literal interpretations of specific forms in nature to an enquiry into the underlying geometric principles which determine the myriad forms that nature produces.

Figure 3 Peter Randall-Page, 'Cone', 1998,
Forest of Dean stone, 61 x 124 x 70 cm

For me the challenge for this building, and any associated artwork, was how to incorporate botanical imagery in a genuinely contemporary and meaningful way. Architecture of almost all periods and all cultures is redolent of plant allusion and imagery. The lotus flower in the far East, the Acanthus in Greece, European medieval stiff leaf carving, the list goes on.

One of the major areas of inquiry in my own work has been the Fibonacci sequence and the golden proportion and the way in which plant growth is determined by these fundamental mathematical principals.

Figure 4 Photograph of a sunflower head

Jolyon's latticed roof structure had an organic feel but was unlike nature in the symmetry of it's radial grid structure. We talked about how nature's love of economy results in the kind of patterns one finds in flowers, cones and seed pods and how these patterns can be rationalised mathematically in terms of the golden angle and the Fibonacci sequence. The golden proportion has been understood and used in art and architecture for millennia, but it is only in recent decades that it's relevance to phyllotaxis (the study of the geometry of plant growth) has been fully appreciated.

Figure 5 Phyllotaxis pattern

We started to speculate about how the roof structure might be based on the kind of spiral phyllotaxis seen in many plant forms and Jolyon passed this concept onto the engineers (Anthony Hunt Associates). The response from the engineers was very encouraging; based on the spiral phyllotaxis pattern found in plants the structure was far more viable, reducing the depth of timbers at the perimeter of the structure from 2,000 mm to nearer 800 mm. In a way this is not so surprising, the sunflower (for example) wants to pack as many seeds of a similar size into a given circle. Taking the gaps between the seeds as the roof supports was bound to result in an efficient structural geometry.

Figure 6 The building of The Core

The design that Jolyon then produced had a genuine connection with plant growth and was also viable as a roof structure combining botanical imagery with structural efficiency. The roof will be light and elegant and, unlike the geodesic biomes, has a definite centre, in botanical terms the apex from which the primordia emanate.

I had long wanted to make a massive, volumetric sculpture to be contained within a chamber with carefully controlled lighting. I showed Jolyon drawings of these ideas and we began to think of this central space as a chamber to house a massive symbolic seed at the kernel of the building; a distillation of the structural principals of the roof.

Figure 7 The inside roof structure of The Core

Figure 8 Model of the sculpture and inner chamber for The Core

We thought of this central chamber as a space designed specifically for the sculpture, echoing it's shape like a giant seed pod. We also wanted to carefully control the transition that visitors would experience when passing from the main body of the building into this inner chamber. Jolyon designed the central core with a double skin incorporating a circular passageway with low light and dampened sound to maximise the drama of moving from the hustle and bustle of the main exhibits hall to the tranquility of the central space.

Figure 9 Excavating the stone for the sculpture

The sculpture itself will be carved from a massive piece of granite measuring 4 x 3 x 3m and weighing approximately 80 tonnes. The stone is being specially quarried in Cornwall and it's surface will be carved in relief with a pattern based on the same spiral phyllotaxis geometry as the roof structure. This geometry will be reflected in various ways throughout the building, from the arrangement of the windows to the division of the internal spaces.

Figure 10 Roughing out the overall form for the sculpture and marking out

Much of the success of this collaboration comes down to the personal chemistry between Jolyon and myself and in particular Jolyon's generosity and openness to my ideas, but also to an awareness by the Eden project of the potential in such a collaboration in the first place. The result is a unique example of collaboration where concept and form, object and structure, architecture and art are inextricably linked.

Figure 11 Projecting on the pattern

The Work of Foster and Partners Specialist Modelling Group

Brady Peters and Xavier DeKestellier
Foster and Partners
Architects and Designers
Riverside Three
22 Hester Road
London, UK
SW11 4AN

Abstract

The following paper is a brief introduction to Foster and Partners and the work of its Specialist Modelling Group (SMG). The SMG was formed in 1997 and has been involved in over 100 projects. The SMG expertise encompasses architecture, art, math and geometry, environmental analysis, geography, programming and computation, urban planning, and rapid prototyping. The SMG brief is to carry out project-driven research and development. The group consults in the area of project workflow, advanced three-dimensional modelling techniques, and the creation of custom digital tools. The specialists in the team are a new breed of architectural designer, requiring an education based in design, math, geometry, computing, and analysis.

1. Foster and Partners

Foster and Partners is an international studio for architecture, planning and design led by Norman Foster and a group of Senior Partners. Norman Foster's philosophy of integration can be seen in the way the practice's London design studio works; it is essentially one large open space, shared equally by everyone, and free of subdivisions to encourage good communication between the many people who come together there. The practice's work ranges in scale from the largest construction project on the planet, Beijing International airport to its smallest commission, a range of door furniture. The scope of its work includes masterplans for cities, the design of buildings, interior and product design, graphics and exhibitions. These can be found throughout the world, from Britain, Europe and Scandinavia to the United States, Hong Kong, Japan, China, Malaysia, Saudi Arabia and Australia.

In developing and communicating the design concept, the project teams are supported by a broad spectrum of in-house disciplines. The advent of digital technologies has allowed the practice to design and build structures with complex geometric forms that would not have been feasible as little as twenty years ago. The practice's in-house Specialist Modelling Group (SMG) has introduced a highly advanced three-dimensional computer modelling capability that allows architects both to explore design solutions rapidly and to communicate data to consultants and contractors

2. The Specialist Modelling Group

The Specialist Modelling Group (SMG) was formed in 1997 and is lead by Hugh Whitehead, a Partner at Foster and Partners. Some of the over 100 projects that the SMG has made a contribution to include the Swiss Re Headquarters, the Sage Gateshead Music Centre, London City Hall, Albion Riverside residences, the Chesa Futura apartment building, and the new Beijing International Airport.

Figure 1: Swiss Re Headquarters, London City Hall, Chesa Futura Residences

The Specialist Modelling Group currently has seven members and has within it expertise in complex geometry, environmental simulation, parametric design, computer programming, and rapid prototyping. The SMG brief is to carry out project-driven research and development in the intense design environment of the Foster and Partners office. The group consults in the areas of project workflow, digital techniques, and the creation of custom CAD tools. These specialists work with project teams on either a short or long term basis and are involved with projects from concept design through to fabrication.

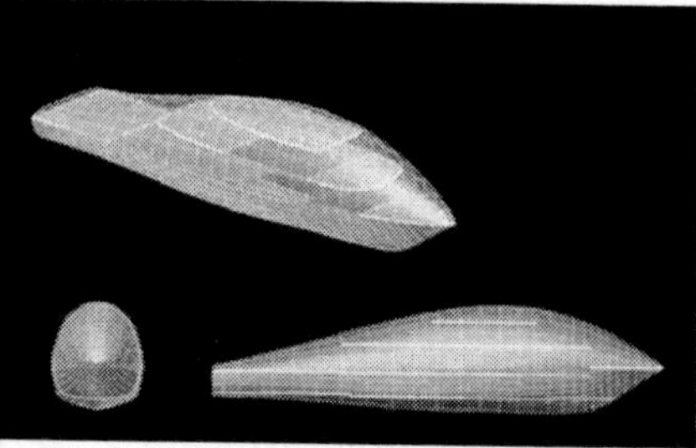

Figure 2: Parametric Design

One of the primary goals of the Specialist Modelling Group is to develop control mechanisms that drive geometry in response to relationships. These control mechanisms can be parametric models or custom programmed scripts. The CAD geometry that these mechanisms are driving responds to the constraints acting on the architectural design. These digital design tools must be flexible to allow designs to make dramatic shifts as the solution space is explored. Through the use of these techniques the designer is presented with feedback about the design through dynamic reporting, analysis, and evaluation. This system can rapidly generate comparative options. The evolution of the design concept starts with open-ended exploration and progresses towards a specific and detailed geometric definition. The linking of the parametric model to simulation and analysis tools is necessary to produce a design that responds to performance criteria. The use of rapid prototyping technology closes the loop in the digital design process by recognizing the fact that key decisions are still made by the designer from the study of physical models.

Complex arrangements of three dimensional surfaces and solids are generated using parametric tools and generative programmed scripts. The controlling variables of these tools respect design constraints and proportional relationships. Parametric plans and sections are produced as templates specific to each project, for use by the design teams, while programmed scripts are produced to either perform a specific task, or as a general tool that can be used on many projects. For example, custom in-house software can produce mathematical surfaces while another can populate these surfaces with structural and cladding components. These panels can be coded and referenced in the 3D model so that they can be automatically laid out, unfolded and scheduled. Designs are rationalised as a precursor to communication with consultants and fabricators. The setting out geometry of a design is specified in a method statement, which describes a sequence of geometric procedures. The method statement is to be used by consultants and subcontractors and provides the framework for the validation of all consultants and subcontractors work.

Figure 3: Generative Design: Mathematical Surfaces Populated with Structure and Cladding

3. The SmartGeometry Group

The SmartGeometry Group is an independent organisation whose aim is to further advanced education and research in the area of advanced 3D CAD applications. The group is dedicated to educating the construction professions in the new skills which will be required to use these new systems effectively. The SmartGeometry Group includes Lars Hesselgren (KPF), Hugh Whitehead (Foster and Partners), J Parrish (Arup Sport) and Robert Aish (Bentley). They are all pioneers of parametric modelling and digital technologies as applied to architecture.

Currently SmartGeometry's development efforts are focused on the new Generative Components technology being created by Robert Aish at Bentley Systems. The group has been running series of schools and seminars where this new technology is being explored both in the context of highly experienced professionals, and in advanced educational institutions. The SmartGeometry conferences bring together highly skilled professionals and educators from all over the world. The students and tutors participate in lectures and workshops focussing on advanced 3D parametric modelling and the programming of custom generative scripts. These workshops are one of the few opportunities for professionals to freely share ideas, tools, and techniques that relate geometry to architecture.

4. The Emergence of the Digital Design Specialist in Architecture

Architectural education must respond to the changing nature of the profession to accommodate the emergence of the Digital Design Specialist. Architects are not trained in the necessary skills and, as a result, one of the roles of the SMG is the education of architects in the use of digital techniques and the role of geometry in design of buildings. While many of these tools and techniques are new to architecture, they are currently used in other disciplines such as product design, aircraft manufacture, mathematics, and computer science.

The architect must understand the constraints that can act upon an architectural design and be able to make inspired and creative decisions that respond to these constraints. A successful way to explore a new design concept is the translation of the sketch into a parametric model. The parametric model should capture the designer's intent and involves the careful consideration of the controlling parameters and variables. The architect must understand of how geometric entities (linear elements, surfaces, solids) are created and controlled.

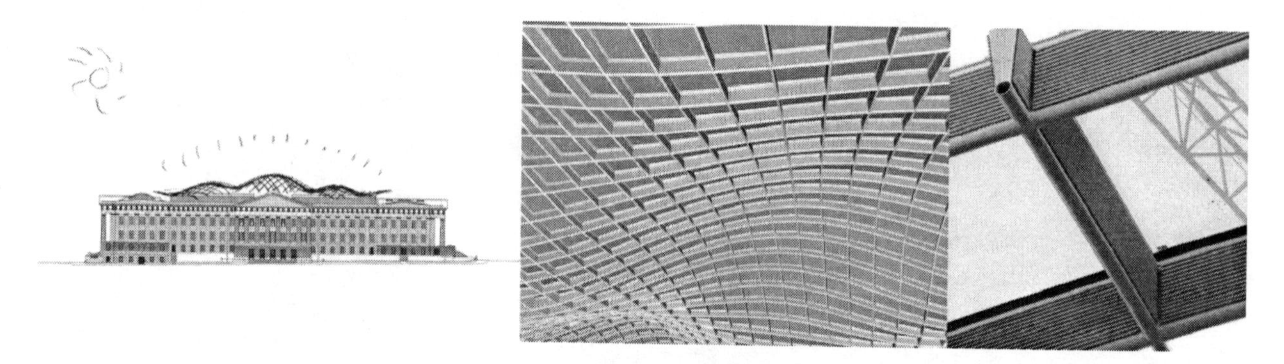

Figure 4: Sketch to Digital Design Model to Building: Smithsonian Courtyard Enclosure

Computer programming is becoming an increasingly valuable tool for these design specialists. The design logic for a project can be used as the basis of a computer program that can rapidly generate many options and large numbers of elements. In order to create a performance based model the digital design specialist needs to understand the analysis method. This can require one of many different skills depending on what aspect of building performance is being investigated. The designer must work with a broad range of consultants and have knowledge of specific disciplines previously considered non-architectural such as environmental analysis and form finding.

The new digital design specialists use geometry as a means of description and as a way to harness the complexity of a building design. Geometry is used to rationalise designs instead of being used to create more elaborate formal propositions. Increases in computational power and advances in digital fabrication have introduced the potential for mass-customisation in architecture. In order to properly take advantage of these new technologies, the designer must be able to design and generate the adaptive components, understand the method by which they will be fabricated, and have the ability to communicate the design information.

There is a new range of skills that is increasingly relevant to the practice of architecture. Digital tools and techniques are now used to solve the problems faced by architects and designers in the building industry. The new digital design specialist must have the ability to use science in an artistic manner. These designers must be able to imaginatively combine knowledge of building performance and fabrication methods with geometry and computer science to create beautiful architecture that responds intelligently to its environment.

The Borromean Rings - A Tripartite Topological Relationship

Louis H. Kauffman
Department of Mathematics, Statistics, and Computer Science
University of Illinois at Chicago
Chicago, Illinois, USA
E-Mail: Kauffman@ uic.edu

1. Introduction

The following dialogue is between Bertie and George. Bertie and George are following an evolutionary path upward from the Void. In this episode they encounter a frighteningly beautiful ternary relation. The following is an account of their conversation and the epistemological issues that emanate from this experience.

The reader will note that one might interpret Bertie as a precursor to the twentieth century philosopher Bertrand Russell and George as a precursor to the twentieth century maverick mathematician George Spencer-Brown. Such interpretations should be taken lightly.

2. An Encounter with a Ternary Relation

Bubbling up from the void, you never know what sorts of structures are likely to come into view. Old Bertrand here thinks that it should all happen in completely orderly fashion, with the emergence of binary relations as the basis of everything.

Bertie, I said, why do you think that binary relations are the primary generator of all form?

He says to me, well look here George, distinctions are made. Distinctions are made within and without the spaces of the distinctions that have already been made. Everything comes out of that process of distinguishing, and distinguishing distinguishing. A distinction connotes a binary relation between its parts. One part is dominant (marked) to the other. There is an ordered binary relationship between the two sides. This leads to higher forms and all of Art and Mathematics.

Amazing fellow this Bertie, that he could make a speech like that when he was still an amoeba but he did, waving a primordial flagellum from time to time. We were all together at one point when the argument started, and that argument went on, hot and heavy, while we underwent uncountable transformations of form in our journey out of the void. But this time I figured I had him. There were new forms underfoot. I said:

Bertie!! Look sharp! There! In red, blue and green!!!!

And there she was indeed, in all her curvaceous beauty, *a topological tripartite relation in full bloom.*

Bertie, I says, look at her. She's autopoetically constructed from three colored toroids, red, blue and green. If you remove any one toroid, she'll just float apart and drift away. There's binary relations in there, but she's a true integral tripartite relation. Look at her structure:

Red surrounds blue.
Blue surrounds green.
Green surrounds red.

Figure 1. *Borommean Rings*

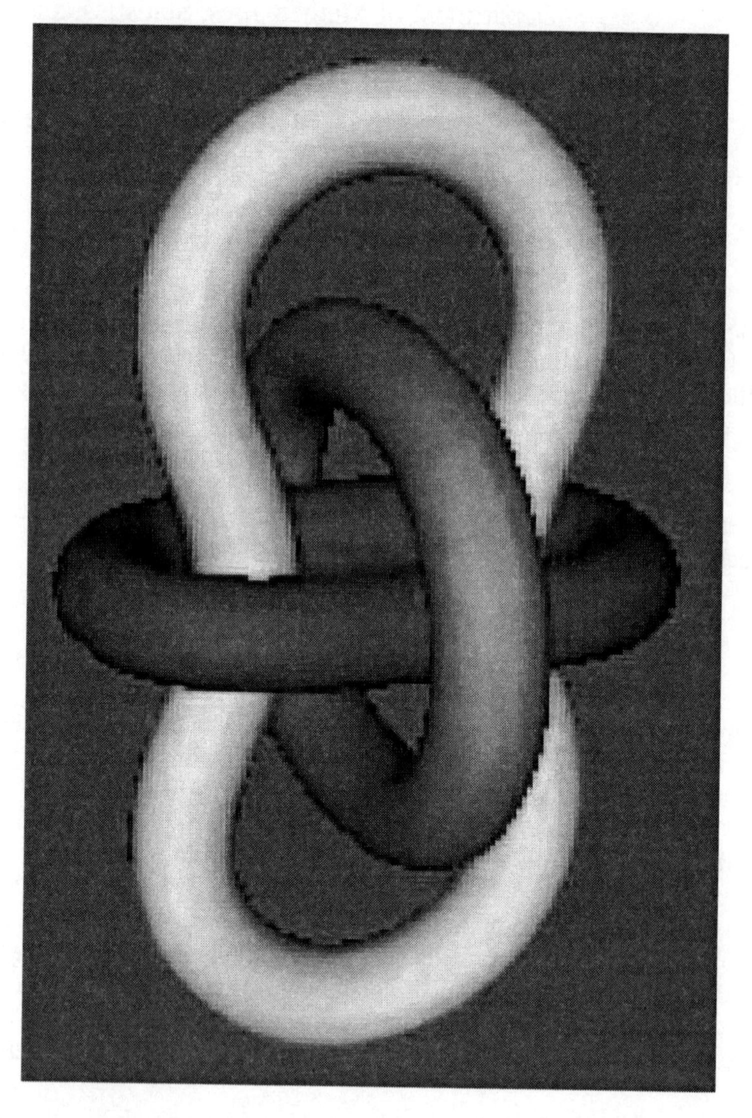

Bertie, if I were to ask you to make a circularity like that in your binary logic, why you'd probably blow a gasket and send off a packet of paradice (paradoxes, paradoxen? (does anybody know the plural of paradox?)). She's held together by a circularity. Floating out there in the void of topological space girdled in glorious triparticity. One and yet Three. Made from Three, All One. A Trinity in the pregeometry prior to space and time.

I paused and Bertie started up. He says: Good Lord George she is beautiful and triparitous indeed. A vision of the Trinity, an angel of circularity. Almost enough to make me a believer. You would have me believe that she's an *elemental form.* You would have me take her and my beloved empty set , and put them in the same category. But I shall have none of that! She is a *composite.* Mark my words and I shall prove it to you.

Wait Bertie! (I says.) Before you go into an interminable analysis of this situation, think! You agree that the Borommean Rings (That's what we called them then in our amoebic and paramecium states respectively. I t was not until millennia later that the Borommeo Family in Italy (who would have guessed "Italy" way back there at the beginning?) used the Rings as a coat of arms. But somehow, we did call them by that appellation. Time is an illusion.) are a single entity and yet composed of three entities, the individual rings. So of course they are composite. Your binary relations are composite as well, being composed of the two sides of the distinction that gives rise to them. See Figures 1 and 2 [PC].

Figure 2. (Left) *From the Capella Rucellai in the Church of San Pancrazio in Florence* (1467). *Designed by Leon Batista Alberti, the rings are said to be a symbol of the Medici Family.*

Figure 3. (Right) *Odin's Triangle, or the Walknot, used by the Norse people of Scandinavia.*

So George, you regard these Borommean Rings as equally fundamental with the first distinction? Says Bertie. I do, says I. You can analyze them every which way, and you'll come back and realize that there is fundamental topology occurring at the bottom of the world. Those rings were co-created along with your empty set and your notion of binary distinction and binary relation.

Bertie goes on: A little set theoretic magic, a few equations and the Borommean Rings come into existence with no more than a continuum of binary distinctions! They are a beautiful example of a composite form unfolding in the (not yet written) 137th volume of Principia Mathematica.
There goes Bertie again, only an amoeba but going on about the 137th volume of Principia Mathematica, his great future work on the emanation of the World and Mathematics from pure Logic! I say to him. Bertie, I do not doubt it. But look, how about the empty set, isn't that also a composite?

Bertie: The empty set stands for a distinction in the void. The empty set itself is distinct from the void. Its contents are void. The empty set is a composite of nothing and the first something. It is a true composite. Without the empty set there would be nothing at all.

George: That reminds me of a riddle: What is better, eternal happiness or a ham sandwich?

Bertie: Well, nothing is better than eternal happiness, and a ham sandwich is better than nothing. Therefore, a ham sandwich is better than eternal happiness.

George: That's Logic for you.

Bertie: But look here, you do agree that the Rings are complex!

George: I agree, but they are not so complex as you might think.
I am going to tell you about knot set theory [KL], and how a generalization of it, captures the Rings.

Bertie: You'd best do that in the next section.

George: Well yes, but first let me just draw a diagram.

Bertie: You mean the diagram below?

A over B
B over C
C over A

George: Well, now its above us. They keep shifting our sentences down the page, and we are only located in those sentences after all.

Bertie: We used to be an amoeba and a paramecium, and now we are just a pair of alternating disembodied bits of text.

George: Well I am still a sign of myself!

Bertie: Yes, yes. But what about the diagram above?

George: Well it is an example of a link diagram, and it represents the Borommean Rings.

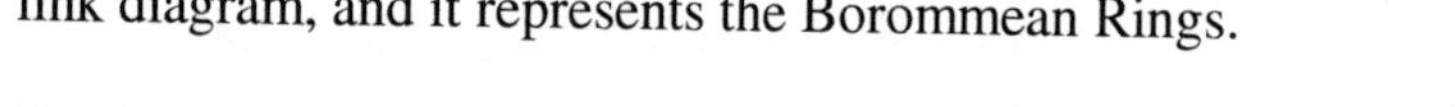

Bertie: I see that. I suppose you are now going to give the Rings purely syntactic existence inside a language of diagrams.

George: Of course. Wouldn't you like to exist in a language of diagrams?

Bertie: And be tied in reference to some particular form of diagram? That's not for me. I will take my chances in these sentences. Sentences are based ultimately on the binary relation of marked/unmarked. Nothing unremarkable there, and I feel quite safe. Your diagrams make me nervous.

George: Oh Bertie, take a leap. Read the next section.

Bertie: That would indeed be a leap. How is a text supposed to read another text? Am I to become interpretant as well as sign and signifier?

George: You interpret all the time. Let's go to the next section.

3. Knot Sets, Ordered Knot Sets and the Borommean Rings

We shall use knot and link diagrams to represent sets. More about this point of view can be found in the author's paper "Knot Logic" [KL]. Diagrams were first used in this way by Flatlanders before the invention of the third dimension. After that, it turned out that the diagrams represented knotted and linked curves in space, a concept far beyond the ken of those original flatlanders.

Set theory is about an asymmetric relation called *membership.* We write **a** $\in$ **S** to say that **a** is a member of the set **S.** In this section we shall diagram the membership relation as follows:

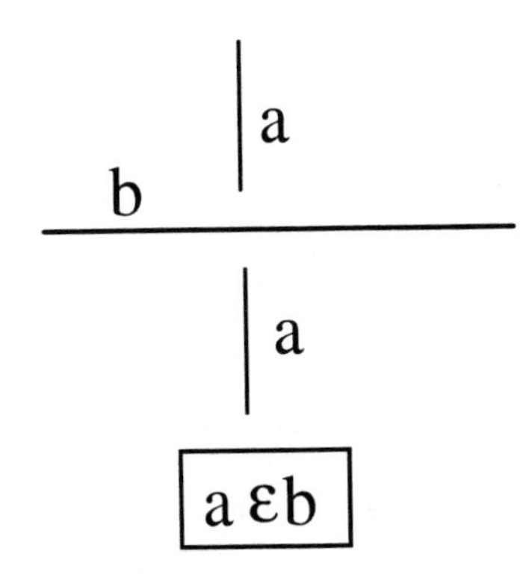

This is *knot-set notation.*

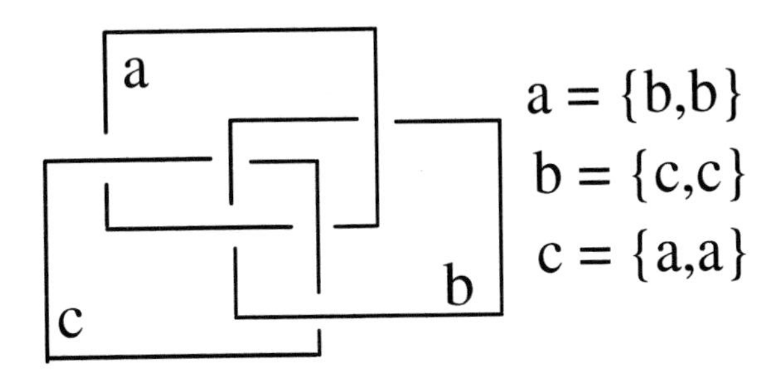

Here again are the *Borommean Rings.* The Rings have the property that if you remove any one of them, then the other two are topologically unlinked. They form a topological tripartite relation. Their knot-set is described by the three equations in the diagram. Thus we see that this representative knot-set is a "scissors-paper-stone" pattern. Each component of the Rings lies over one other component, in a cyclic pattern.

4. Commentary

Bertie. Interesting manuscript. Where did you get it?

George. I told you. It is the third section of this paper.

Bertie. No no! I mean where did you find it before it was included in this paper. Surely that document has a different source than our conversation.

George. I don't know about that, but you are right. The document was found written on parchment and attached to an old bronze copy of the Borommean Rings. It could be centuries old.

Bertie. How could that be? It refers to a paper written by Kauffman in 1994!

George. You are naive as always. This is a result of time travel.

Bertie. Time travel you say?

George. Yes. The old breed of mathematicians, before the twentieth century were prone to travel forward in time, sometimes stealing the work of future mathematicians, sometimes just referring obscurely to their papers. In this case the author of this paper on knot-sets has borrowed from Kauffman's 1994 paper on Knot Logic, but he or she has been kind enough to give a reference. Other time-travelers could then find the paper and read it.

Bertie. What happened? Why don't we time travel anymore?

George. Actually, Bertie, it is your fault. Some mathematicians from the 19th century went forward and read your now-famous Russell Paradox. They were so shocked, that they quit time travel. Of course others had gone farther forward in time, but there was some singularity associated with the Russell paradox. It induced a one-way blindness that kept mathematicians then on from traveling forward into the future and taking future results from our contemporaries.

Bertie. You're pulling my leg again George. In my opinion, this manuscript proves my point. It constructs the Borommean knot set. All this occurs in a formal system where all the language is built from binary relations. Why I warrant that if Whitehead were born yet, the two of us could build a mathematical system based on logic and binary relations so that these Rings would be one of the simpler constructions of the system. Why I wager I will call this system Principia Mathematica!

George. Well you will have to evolve a bit for that Bertie. It is deucedly difficult for an amoeba to write a book. I am sure that any such endeavor will be incomplete, and you will always have to ask yourself whether the Rings in Principia are really the same as the beautiful vision of colored rings surrounding one another that we have seen today. Perhaps I shall be able to reduce the noise level of your work after you have done it. That will be a good future for a paramecium.

References

[KL] L. H. Kauffman. Knot Logic. In *Knots and Applications* ed. by L. Kauffman, World Scientific Pub. (1994) pp. 1-110.

[PC] Peter Cromwell, University of Wales, <http://www.liv.ac.uk/~spmr02/>.

[GSB] G. Spencer-Brown, "Laws of Form", Julian Press, New York (1969).

Cultural Insights from Symmetry Studies

Dorothy K. Washburn
Laboratory of Anthropology
Museum of New Mexico
Santa Fe, New Mexico 87504
dkwashburn@worldnet.att.net

Donald W. Crowe
Department of Mathematics
University of Wisconsin
Madison, Wisconsin 53706
crowe@math.wisc.edu

Abstract

Washburn and Crowe have published texts and studies documenting the procedure for and application of the use of plane pattern symmetries to classify cultural patterns [8, 9]. This paper contrasts the difference in cultural insights gained between pattern studies that simply describe patterns by motif type and shape and those that describe the way motifs are repeated by plane pattern symmetries.

Culturally produced patterns can be described in many ways, each useful for different purposes. We describe how early pattern studies aimed at designers of textiles and wallpapers created classificatory groupings that were descriptively idiosyncratic, grouping patterns by motif similarities that are arranged by very different symmetries. We then cite several recent studies that illustrate how a symmetry rather than a motif similarity grouping reveals new insights from continuities, changes and preferential symmetry use that can enhance our understanding and interpretation of the material.

One of the best-known pattern studies is that of Archibald Christie, Pattern Design [1]. This 1929 edition is a revised version of the original 1910 study entitled Traditional Methods of Pattern Designing. The revised edition has been reprinted in its entirety by Dover Publications and so remains a very available example of an early comprehensive pattern study. Christie focused on illustrating how the rhythmic movement of element repetition, both naturalistic and geometric, has pervaded pattern from the earliest times. More recent improvements in the technology of pattern production, such as mechanized looms for woven textiles, have enabled the generation of an endless succession of different patterns.

Figure 1 *Left: Egyptian border, 16th century. Right: Persian rug border, 16th century. From* [1, *Figures* 183, 185 *respectively*]

Christie reduced the units that comprise pattern to two main types--*isolated* units (spots) and *continuous* units (stripes)--and showed how they have been used to expand designs into patterns by types of repetitions he labels as "powdering, striping, interlocking, interlacing, branching, and counterchanging" (reversing colors). It is essentially a historical survey that differentiates pattern structures only by general descriptive terms, such as stripes, borders, waves and chevrons and cross bands. While he showed how patterns repeated by the same symmetry can be created with many different motifs (Figure 1), it is not clear what we are to do with this recognition of common symmetries among designs from many different cultures and many different periods except, of course, appreciate them for their beautiful rhythmic appearance.

The pattern book Abstract Design: A Practical Manual on the Making of Patterns for the Use of Students Teachers Designers and Craftsmen by Amor Fenn [2] is, in contrast, dedicated to the explicit instruction of designers about the basics of pattern construction. However, although Fenn understands that repeating patterns are constructed on a geometric basis, he does not differentiate patterns by their generating symmetries but rather by the angles by which "enclosed shapes"—squares, circles, polygons—are juxtaposed and repeated. He considers the range of pattern arrangements on two kinds of layouts: borders and textiles. His borders generally correspond to one-dimensional band designs, and his textiles correspond to two-dimensional overall or wallpaper patterns. For designers he has provided line diagrams of numerous border and textile patterns, showing how simple units can be recombined and elaborated into very complex, decorative patterns.

However, Fenn describes his units in such a way that border bands called frets include examples generated by different symmetries. For example, No. 67 is a fret generated by bifold rotation and vertical reflection, *pma2*, while No. 68 is a fret generated only by bifold rotation, *p112*, and No. 79 is a fret generated by vertical and horizontal mirror reflections and bifold rotation, *pmm2* (Figure 2).

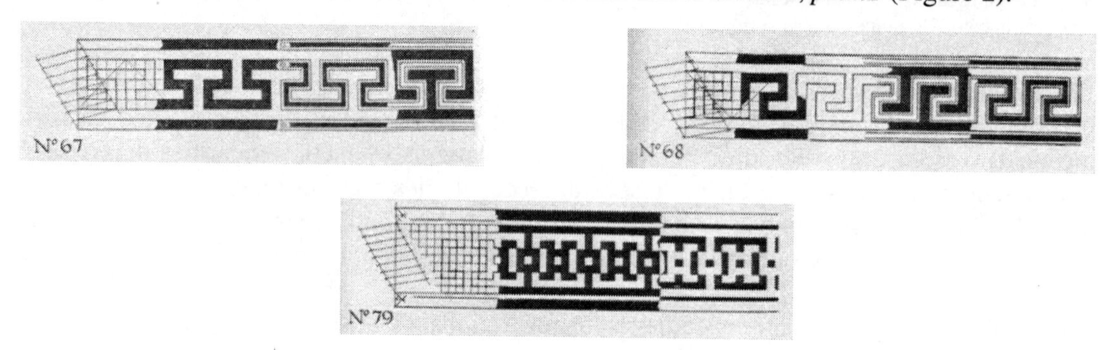

Figure 2 *Nos.* 67, *pma2*; ,68, *p112*; 79, *pmm2*. *From* [2]

In another example, interlacings can be arranged by several of the seven one-dimensional symmetries, such as by simple translation, *p111*, as on No. 106, and by bifold rotation, *p112*, as on No. 104 (Figure 3).

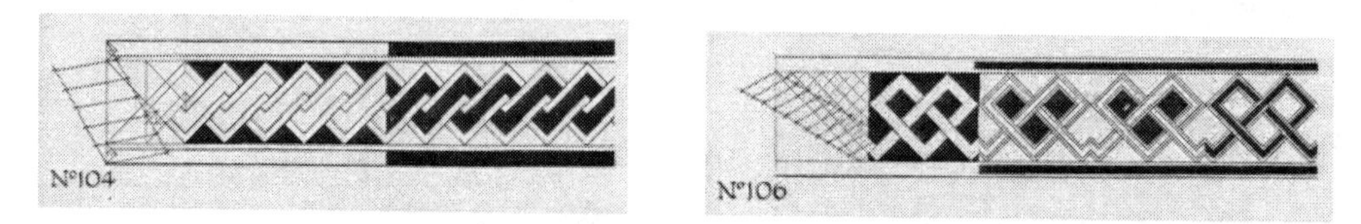

Figure 3 *Nos.* 104, *p112*; 106, *p111*. *From* [2]

For overall wallpaper and textile patterns Fenn bases his discussion on the repeated unit, such as squares, hexagons, and undulate lines. He shows how these are arranged so that, in some cases, the units completely cover space, as in drop patterns such as No. 296. In other cases, they appear to cover the surface as closely spaced or intertwined leaf and flower elements, as in No. 358. Although these two examples render quite different decorative effects, both Nos. 296 and 358 are organized and repeated by symmetry *cmm* (Figure 4).

The issue is: In what way(s) is a symmetry classification of pattern superior to the descriptive ones based on motif shape and/or shared rhythmic repetition configurations as exemplified above? The descriptive classifications group pattern by similarity in motif appearance regardless of underlying differences in symmetrical structure. This approach focuses on the decorative features of design. In contrast, we have set aside interest in the way pattern *decorates*, an issue that may well be a Western preoccupation, and instead queried how pattern *informs*. We have found that a focus on symmetrical structure rather than motif enables us to explore how cultures without writing systems use pattern in different kinds of information transmitting capacities. We present here several examples of studies that demonstrate how symmetry differences in pattern correspond to important ethnic differences and geographical interaction patterns as well as to environmental changes that stimulated major social adaptations.

Figure 4 *Nos.* 296, 358 *have cmm symmetry. From* [2]

We begin with an archaeological study of decorated pottery made during the Neolithic on sites throughout mainland Greece [3]. Art historians have often described the red/on/cream designs on the Early Neolithic ceramics and the incised designs on Late Neolithic ceramics in terms of four motifs: flames, triangles, zigzags, or nets (Figure 5).

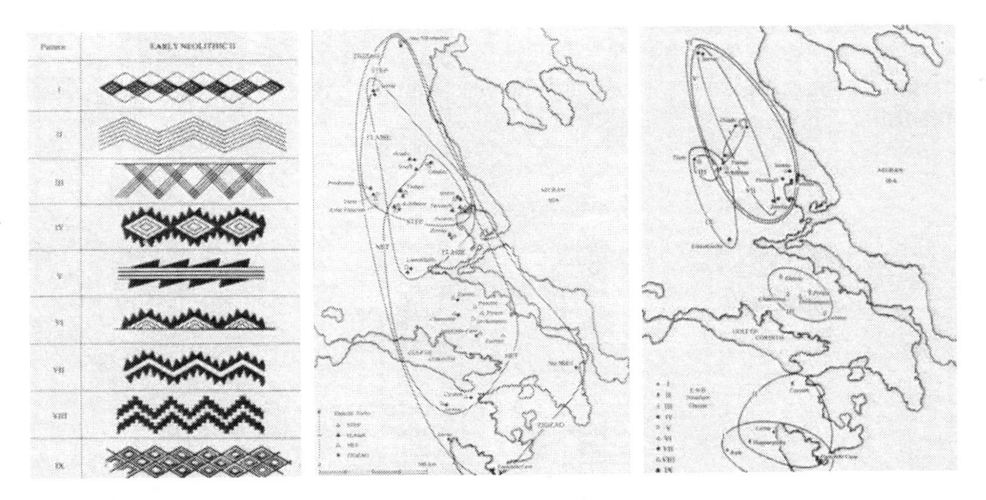

Figure 5 *(Left) The Four Motifs in Nine Patterns from the Early Neolithic, Greece. From* [3]
Figure 6 *(Center) Distribution of the Four Motif Styles in the Early Neolithic, Greece. From* [3]
Figure 7 *(Right) Distribution of the Nine Symmetry Configurations of the Four Motifs. From* [3].

A distributional study of these motifs shows them to occur on pottery from every area where Greece was occupied (Figure 6). However, if the motifs are described by the symmetries used to configure them into patterns, then an entirely different geographical distribution appears (Figure 7). That is, for example, if we trace the occurrence of all patterns composed of triangles during the Early Neolithic that have been typically called "flame" patterns by Classical archaeologists, we find that they are present throughout occupied Greece. In contrast if we use symmetry to distinguish among the configurations of these flame patterns we find that there are mutually exclusive, geographically separate enclaves of different flame patterns, each characterized by a different structural symmetry. Notably these enclaves are separated by mountain ranges or bodies of water, suggesting that geographic factors impeded free movement and interchange during this early period.

Interestingly, if we apply this same methodology to incised patterns from the Late Neolithic, we find a new distribution. Now, a centrally located site appears to have a number of patterns and symmetries, while the surrounding sites display only one or two patterns and symmetries. This new distribution correlates precisely with the beginnings of trade in the Aegean, suggesting that the central site with the greatest variety of patterned pottery was a market place as well as a bulking center where goods from outlying sites were brought in to be sold and later shipped throughout the Aegean. It would appear that here is an excellent example of the power of symmetry analysis to quickly highlight cultural spheres and changes in interaction routes over time and space.

We next examine an ethnographic study of designs on twined baskets from three Indian tribes in northern California, the Yurok, Karok, and Hupa [4]. While these peoples are all salmon fishers living in large villages along the swiftly flowing northern rivers of the Sierras, they speak mutually exclusive languages. Nevertheless, their common lifestyle has resulted in basket forms and designs that are difficult to differentiate in technique and pattern motif. In the 1930s Lila O'Neale of the University of California, Berkeley, visited a number of weavers with the object of studying the aesthetic principles that guided them as they produced their baskets [5]. She discovered a dichotomy between baskets that were said to be "good" and thus would be worn for tribal ceremonies and those that were "bad" that were made for sale to non-Indians, whether dealers, tourists or anthropologists. Notably, both the good and bad baskets were made with the *same* twined technology, raw materials, care in execution, and design motifs. The ONLY feature that distinguished the good and bad baskets was the difference in the symmetries that they used to make good and bad basket designs--the good designs being constructed exclusively by *p112* and *pma2* symmetries (Figure 8) and the bad designs being constructed by other symmetries (Figure 9). A blind sorting of basket hat images by the symmetries used to repeat the designs on them resulted in a perfect separation of the good basket hats made for traditional home use and the bad basket hats made to sell outside the tribal sphere.

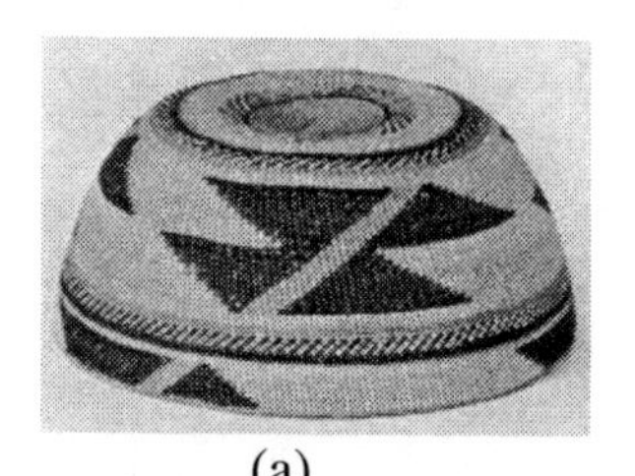

(a)

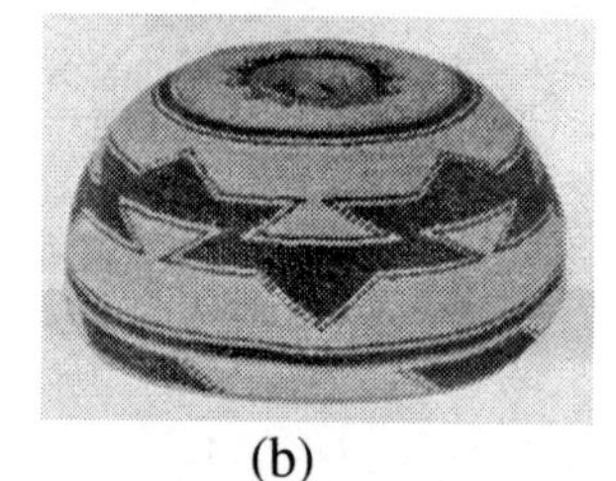

(b)

Figure 9

Figure 8 *a: Hat with "good" p112 design From* [5, *Figure* 24a], b: *Hat with "good" pma2 design. From* [5, *Figure* 21a]

Figure 9 *Hat made for sale with "bad" pla1 design. From* [5, *Figure* 22a]

It would appear that this case is an example of a deliberate decision by basket makers to use pattern structure, rather than motif, as a way to differentiate objects that carry designs appropriate for internal use versus those made to satisfy the desires of outside buyers for "traditional" crafts. Non-indigenous buyers, unfamiliar with the structural requirements for appropriate pattern, willingly buy any baskets that *appear* traditional on visible grounds—technique, materials, design elements---even though the configuration of the designs elements has no ethnic authenticity.

The final example illustrates a correlation between environmental change and pattern structure change that seems to reflect the kinds of social configurations and changes therein that best organize communities of different size and subsistence regimes. The data comes from an extensive ongoing study of ceramic designs made by the prehistoric puebloan peoples known as the Anasazi in the American Southwest. Between AD 600 and 1600 these corn agriculturalists decorated their pottery with geometric designs. Figure 10 charts the changing use frequencies of five different symmetries over this 1000-year period, revealing clear shifts in the AD 800-900 period from *C2* and *D2* to *p112* and then in the post-AD 1175 period, from *p112* to designs that are asymmetric *C1* or have simple translational *p111* arrangements, and finally to designs that have mirror reflection arrangements, both in a finite *D2* arrangement or in banded *pm11* configurations.

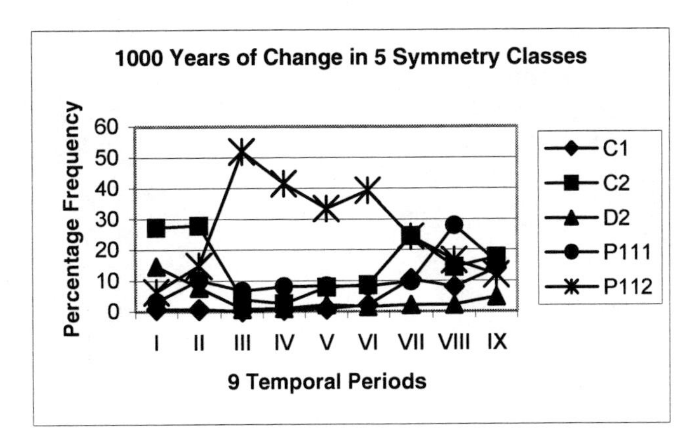

Figure 10 *Prevalence of five symmetries on Anasazi ceramics during nine periods. Period* I=AD625-795; II+825-890; III=920-1025; IV=1025-1100; V=1100-1175; VI=1175-1285; VII=1287-1400; VIII=1400-1480; IX=1490-1650.

Both shifts correspond to periods of environmental change severe enough to require changes in subsistence practices, habitation type and location, and social organization. During the first shift, cooler temperatures and lower rainfall forced people to abandon large pithouse villages on mesa tops whose inhabitants decorated pottery with C2 and D2 designs (Figures 11,12). They moved into smaller masonry pueblo units scattered adjacent to valley bottoms, floodplains or arroyo fans that could better capture the rainfall from the summer thunderstorm rainfall regime. A lengthy 400-year period of conditions generally favorable for this dry-farming corn agriculture lifeway ensued that enabled the spread of these small unit pueblo villages throughout the Four Corners area. The prevailing *p112* symmetry on the ceramic designs of this period (Figure 13) appears to be a structural metaphor of the simple reciprocities at all levels of society that maintained these small farming villages.

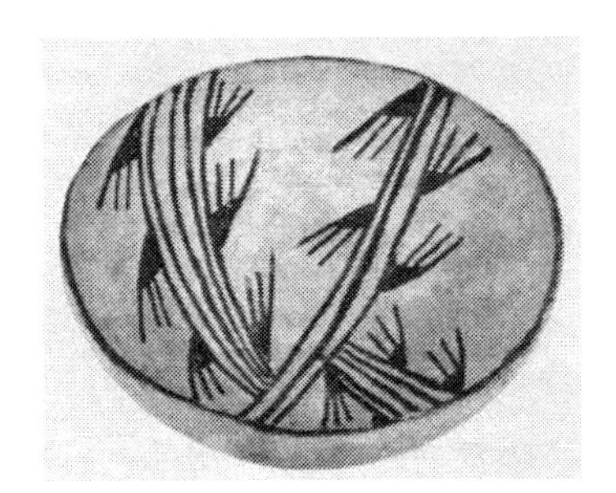

Figure 11 *(Left) C2 ceramic design. From* [6, *Figure* 17, #9622]
Figure 12 *(Right) D2 Ceramic design. From* [6, *Figure* 5, *center*]

But by the late 12th century successively longer droughts defeated even the best storage arrangements and massive areas were depopulated. In New Mexico people moved to the permanently watered tributaries of the northern Rio Grande where many of their descendents continue to live today. In Arizona, people moved to sites along the Little Colorado River and its tributaries as well as to sites adjacent to permanent springs, such as on the southern edge of Black Mesa where the Hopi live in 12 villages today. These locational shifts were accompanied by the development of new agricultural techniques. The mulched gravel fields and irrigated plots in sites along the Rio Grande enabled the growth of large populations and thus necessitated the development of new forms of social organization to organize these larger groups. We suspect that the rise of *D2* and other mirror reflection symmetries, such as the *pm11* design on the jar in Figure 14, reflects the development of the two-part moiety divisions of the pueblos.

Figure 13 *(Left) p112 ceramic design. From* [6, *Figure* 18, *top*]
Figure 14 *(Right) pm11 ceramic design. From* [7, *Plate* XXVII]

These studies have revealed clear correlations between design symmetry and patterns of interaction and trade, ethnic identity, environmental change and social organization. They suggest that the structural symmetries underlying designs and patterns may be more than compositional vehicles for creating pleasing decoration. Not only do the consistencies and changes in design symmetries appear to mirror correlations between key factors in the environmental and social domains, but also the symmetries themselves may have functioned in the past for their makers and users as visual displays of socially important information.

References

[1] Archibald H. Christie, *Pattern Design. An Introduction to the Study of Formal Ornament.* 1929 2nd edition, republished by Dover, NY, 1969.

[2] Amor Fenn, *Abstract Design. A Practical Manual on the Making of Patterns for the Use of Students Teachers Designers and Craftsmen.* Scribner's Sons, NY, 1920.

[3] Dorothy K. Washburn, Symmetry Analysis of Ceramic Design: two tests of the method on Neolithic material from Greece and the Aegean. In *Structure and Cognition in Art*, ed. D. Washburn, pp. 138-164. Cambridge University Press, Cambridge, 1983.

[4] Dorothy K. Washburn, Symmetry Analysis of Yurok, Karok, and Hupa Indian Basket Designs. *Empirical Studies of the Arts* 4:19-45, 1986.

[5] Lila M. O'Neale, *Yurok-Karok Basket Weavers.* University of California Publications in American Archaeology and Ethnology Vol.32, No. 1, 1932.

[6] Robert H. Lister and Florence C. Lister, *The Earl H. Morris Memorial Pottery Collection.* University of Colorado Studies, Series in Anthropology, No. 16, 1969.

[7] H.P. Mera, *Style Trends of Pueblo Pottery 1500-1840.* Memoirs, Laboratory of Anthropology, Vol. 3, Museum of New Mexico, Santa Fe, 1939.

[8] Dorothy K. Washburn and Donald W. Crowe, *Symmetries of Culture: Theory and Practice of Plane Pattern Analysis.* University of Washington Press, Seattle, 1988.

[9] Dorothy K. Washburn and Donald W. Crowe, eds. *Symmetry Comes of Age: The Role of Pattern in Culture.* University of Washington Press, Seattle, 2004.

Non-Euclidean Symmetry and Indra's Pearls

Caroline Series
Mathematics Institute, University of Warwick
Coventry CV4 7AL, UK
E-mail: cms@maths.warwick.ac.uk
David Wright
Department of Mathematics, Oklahoma State University
Stillwater, Oklahoma, 74078 USA
E-mail: wright@math.okstate.edu

Abstract

Escher's well known picture of devils and angels is an example of a symmetrical tiling of two dimensional hyperbolic space. We discuss similar symmetries of three dimensional hyperbolic space, modelled as the inside of a solid ball. The 'shadows' of the solid tiles on the boundary of the ball themselves form patterns governed by a new kind of symmetry, that of Möbius maps on the complex plane. All aspects of such pictures, together with instructions for making them, are explored in the authors' book *Indra's Pearls*. We give examples of beautiful fractal patterns created in this way.

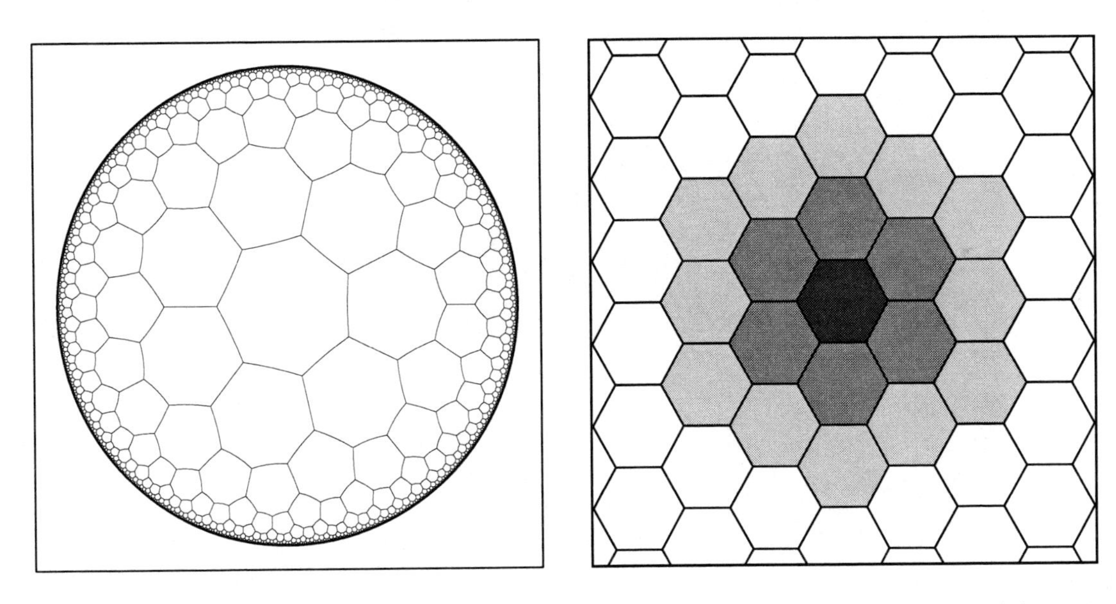

Figure 1: (a) Left: A non-Euclidean tiling of the disk by 'regular' heptagons. (b) Right: A Euclidean tiling of the plane by regular hexagons.

Many people will be familiar with Escher's famous picture of devils and angels. An image with similar symmetry is shown in Figure 1(a). The symmetries are those of hyperbolic, or non-Euclidean geometry. In this geometry, things behave in unexpected ways. For example, the circumference of a circle is proportional not to its radius, but to e^{radius}. This means that to fit into ordinary Euclidean space, a big hyperbolic disk has to crinkle up round its edges like a kale leaf. Once you start looking for it, you see this type of growth throughout the natural world.

The same exponential growth law is manifested in Figure 1(a), in which the tiles are arranged in 'layers' around the centre. If you count carefully, you will find that the nth layer contains roughly 3.1×2.6^n tiles (or exactly 7 times the $2n$-th Fibonacci number). This is in marked contrast with

Euclidean tilings, where the growth law is linear. For example, the honeycomb tiling in Figure 1(b) has exactly $6n$ hexagons in the nth layer.

Despite appearances, in the world of hyperbolic geometry the tiles in Figure 1(a) all have the same size and shape. To fit them into a Euclidean picture, we have to shrink their apparent size as we move away from the centre, so that to our Euclidean glasses the tiles look smaller and smaller as they pile up near the edge of the disk. Since you can fit infinitely many layers of congruent tiles between the centre of the disk and its boundary, the boundary must be infinitely far away from the centre. In this strange geometry, the diameter of the disk is infinite. For this reason, the boundary circle is called the 'circle at infinity'. All the points in the boundary circle are infinitely far away from the centre.

Now imagine a similar geometry in 3-dimensions. Solid tiles or crystals of the same hyperbolic size will be fitted together to fill up 3-dimensional hyperbolic space. This hyperbolic universe can be enclosed in a Euclidean sphere whose boundary is infinitely far, in hyperbolic terms, away from its centre.

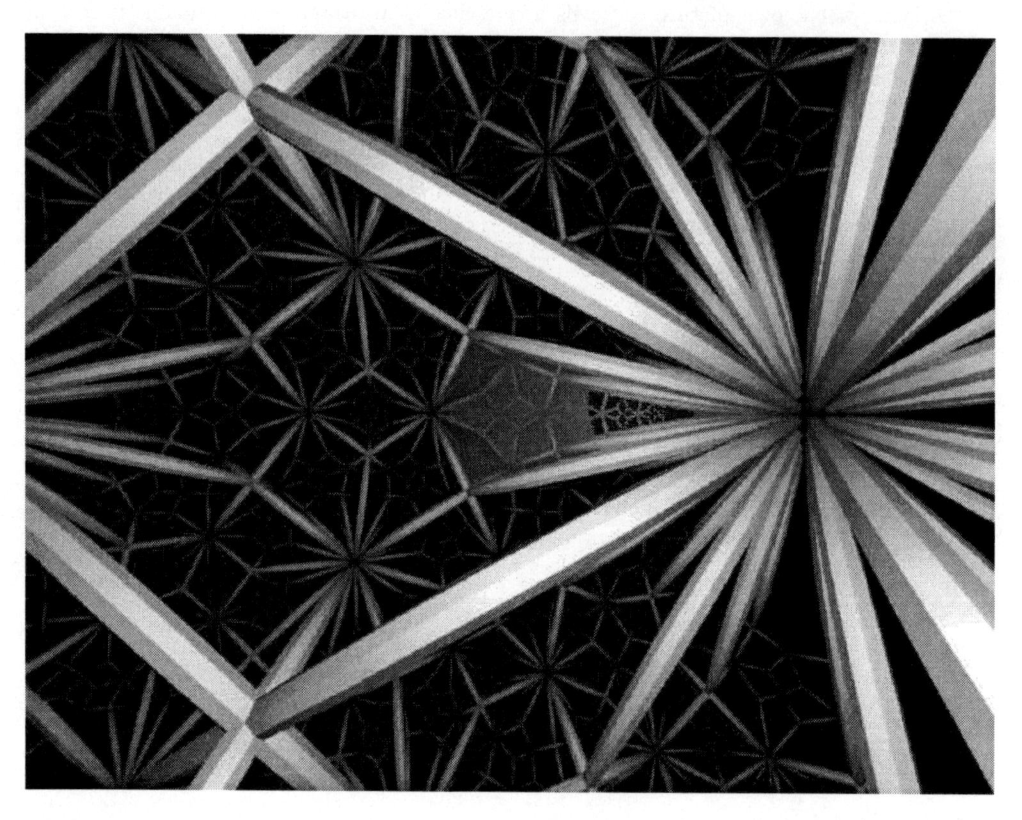

Figure 2: Non-Euclidean tiling of three-dimensional space. Still from *Not Knot!*

Tilings of hyperbolic 3-space are rather hard to draw, although some remarkable pictures have been made by Charles Gunn at the Geometry Center of the University of Minnesota, for the film *Not Knot!* published by A K Peters, Ltd. As an easier substitute, mathematicians usually study what they see on the boundary of the sphere. Patterns seen here, being two dimensional, can be flattened out by projecting onto a plane.

Figure 3: Non-Euclidean tiling of the disk by a polygon with some sides at infinity.

In the two dimensional analogue Figure 1(a) this is not very interesting, since tiles pile up all the way round the circle at infinity. But suppose that the original tile, still a polygon with a finite number of sides, stretched all the way out to the boundary, as in Figure 3. Each copy also meets the circle in 4 circular arcs. Although the tiles fill up all of hyperbolic 2-space (the interior of the disk), the totality of the arcs do not fill up the whole circle. The set of omitted points has interesting properties, for example it has a 'fractal dimension'. It is an example of what is called mathematically a *Cantor set*, more colloquially, a 'fractal dust'.

Returning to 3-dimensions, the analogue is a polyhedron with a finite number of faces, some of which reach all the way out to the sphere at infinity. Outside observers will see these faces as 'shadows' where the polyhedron meets the sphere. Figure 4 shows the shadows of the tiles (in this case triangular) piling up in remarkable patterns on the boundary of the sphere, like noses pressed against a window pane.

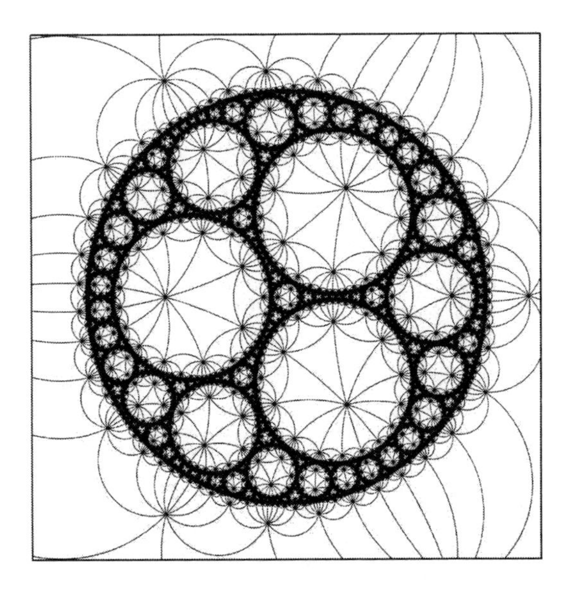

Figure 4: Faces of polyhedra in a 3-dimensional hyperbolic tiling pressed against the sphere at infinity.

Patterns like this were among the novel symmetries studied by the German mathematician Felix Klein (1849 – 1925). In the 1980's, David Mumford realised they were a natural target for computer exploration. With David Wright, he embarked on a systematic study which eventually resulted, not only in inspiring new mathematics, but also in our book *Indra's Pearls* [1].

The book shows off some of the remarkable pictures which resulted. We wanted to explain them with the minimum of mathematical baggage, but with enough detail for the mathematically inclined to follow the reasoning and for the computationally inclined to make their own pictures. By writing down mathematical formulae which describe the symmetries, we can discover what happens to a basic tile as the symmetries move it around.

All rigid motions of Euclidean space can be obtained by repeated reflections in lines or planes. The same is true in hyperbolic space, as long as we remember to replace 'reflection' by 'hyperbolic reflection'. Without thinking about that too hard, remember we are interested in what happens on the window pane at infinity. In Figure 1, two dimensional hyperbolic lines appear as circular arcs. Similarly in 3-dimensions, hyperbolic planes look like pieces of spheres. Such a spherical shell inside hyperbolic space meets infinity in a circle. So the motion we want to pin down could be described as 'reflecting in circles'. There is a nice bit of elementary mathematics which implements exactly this: 'reflecting' in a plane inside hyperbolic space translates to 'inverting' in the circle in which the plane hits infinity. Figure 5 shows a straight-laced stick figure inverted in a circle into slightly curvier figure.

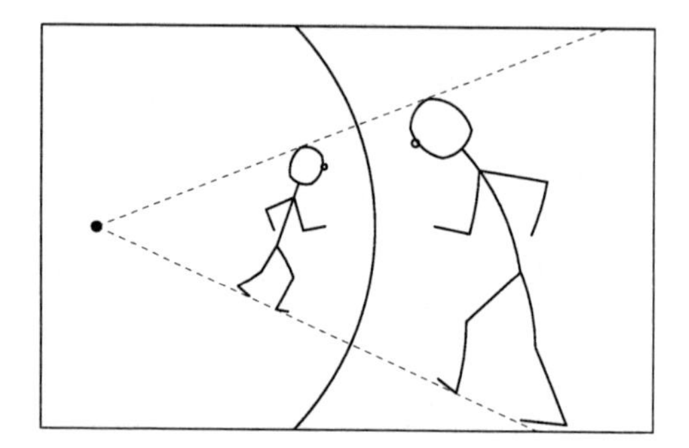

Figure 5: Inversion of a stick figure in a circle.

There is an important difference with the familiar symmetries of Euclidean space, which preserve Euclidean distance. Inside hyperbolic space, the symmetries preserve *hyperbolic* distance. On the window pane at infinity, however, it is impossible to find a way to measure distance which is preserved. All is not lost: it turns out that the transformations we need are exactly those which transform circles into circles, with changes of radius being allowed. Such transformations are called *Möbius maps*, after the German mathematician August Möbius (1790 – 1868).

For mathematicians, what we do is flatten out the sphere by stereographic projection and view it as the Riemann sphere or extended complex plane. Möbius maps are exactly the maps which send the complex number z (possibly including ∞) to the new complex number

$$\frac{az+b}{cz+b},$$

where a, b, c, d are fixed complex numbers. (The shape of the formula gives Möbius maps their other common name *linear fractional transformations.*) This enables us to study Klein's new symmetries using the algebra of two-by-two matrices $\begin{pmatrix} a & b \\ c & d \end{pmatrix}$.

Indra's Pearls begins with a review of the language of symmetry and complex numbers, before going into the detailed effects of maps like these. What we really want to study is this. Take a pair of matrices such as for example

$$\begin{pmatrix} 1 & 0 \\ -2i & 1 \end{pmatrix} \text{ and } \begin{pmatrix} 1-i & 1 \\ 1 & 1+i \end{pmatrix}.$$

Then take a basic figure, say a stick man, and apply both of these transformations to the figure over and over again. As you can see in Figure 6, the images get small and are distorted as the level of the repetition increases. If you choose the starting matrices cleverly, the patterns which emerge when the images pile up are can be amazingly beautiful.

Figure 6: Images of a man piling up on the limit set.

To see more clearly what is going on, we often drop the original figure altogether and just look at the region where its smaller and smaller images pile up. This is called the *limit set* or *chaotic set* of the iteration, because in this part of the pictures, the symmetry group acts in a chaotic way. (Though to a mathematician, the chaos is very controlled.)

Figure 7: The same limit set with the man taken away to get a better view.

By choosing the initial symmetries with enough care, we can create limit sets with intricate patterns of tangent circles. The two matrices written down above produce the famous Apollonian Gasket shown in Figure 8.

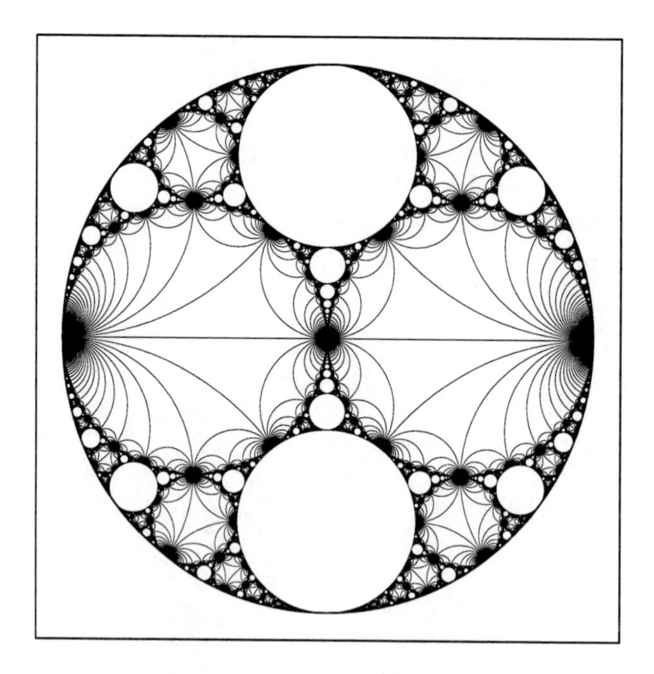

Figure 8: The Apollonian gasket.

In other examples like those in Figures 9 and 10, the tangent circles spiral in beautiful patterns. How and why this happens is explored in great detail in the book, which also contains instructions for making such images.

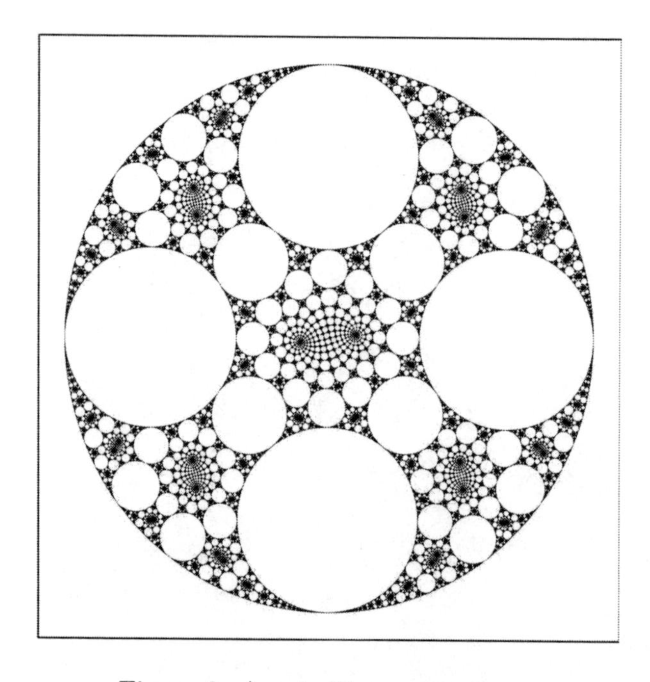

Figure 9: A spiralling cusp group.

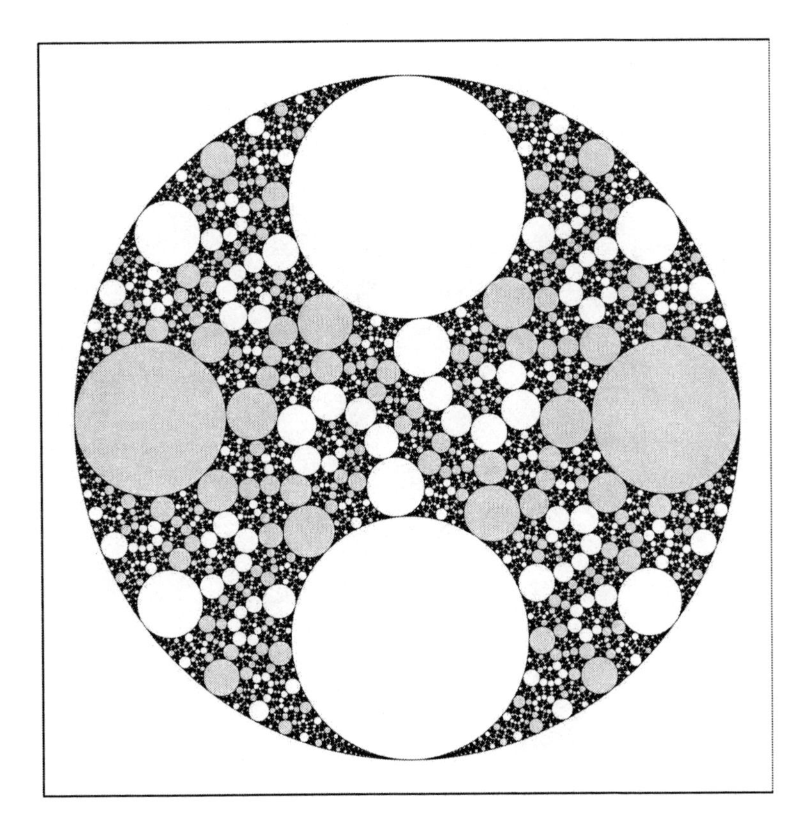

Figure 10: A more intricate cusp group with one side of circles shaded.

Why did we call our book Indra's Pearls? In western thought, the infinite is conceived as counting without end, a flock of sheep going through the gate forever: one, two, three, .. one hundred and one, one hundred and two, ... But there are other ways of getting to infinity. Remember the man with seven wives. Kits, cats, sacks and wives, quite a lot of traffic enroute to St Ives! In fact the traffic increases exponentially with the number of levels (kits, cats,...).

Such exponential growth reminds us of hyperbolic tilings, typically formed by a similar repetitive process. In Figure 11, we start with six disjoint circles. Suppose we reflect in one of these circles C. The five other initial circles which are outside C get reflected into five smaller circles inside C. The repeat operation produces five more small circles inside each of these second level circles: $5^2 = 25$ circles in all. At the next level, we will have $5^3 = 125$ tiny circles. And so on.

In many eastern philosophies, especially Buddhist, this idea of the infinite appearing from copies within copies is pervasive: *"In a single atom, great and small lands, as many as atoms."* This concept was so exactly reflected in the mathematics of our pictures that it inspired our title, taken from the ancient Buddhist myth of Indra's Web:

In the heaven of the great god Indra is said to be a vast and shimmering net, finer than a spider's web, stretching to the outermost reaches of space. Strung at the each intersection of its diaphanous threads is a reflecting pearl. Since the net is infinite in extent, the pearls are infinite in number. In the glistening surface of each pearl are reflected all the other pearls, even those in the furthest corners of the heavens. In each reflection, again are reflected all the infinitely many other pearls, so that by this process, reflections of reflections continue without end.

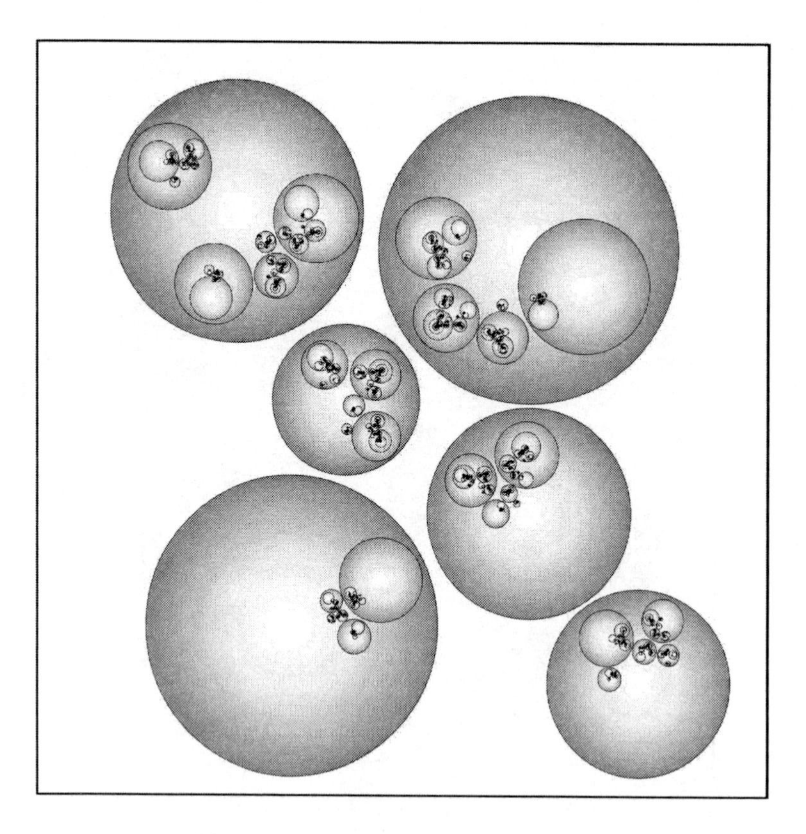

Figure 11: Worlds within worlds.

References

[1] *Indra's Pearls, The Vision of Felix Klein,* D. Mumford, C. Series and D. Wright, Cambridge University Press, 2002.

Love, Understanding, and Soap Bubbles

Simon Thomas
Lantivet Reach
Lanteglos by Fowey
Cornwall PL23 1NP
England
e-mail: sileethom@yahoo.co.uk
www.simonthomas-sculpture.com

Abstract

As an artist my interest in mathematics has evolved through a love of nature and a desire to better understand the "nature of things". An evolving interest in natural efficiencies has recently led to a thorough investigation of soap bubble foam, where I have found the relationship between pressure differentials and geometric organisation of particular interest. Through this study I have developed a physical modelling system, which is the foundation of my latest Artwork(s).

Development

Fascination focused on an all embracing notion of nature remains the motivation behind my artistic activities. This fascination is nourished by that revealed from observations and investigations into the ways of nature, a process which more often than not will include a mathematical analysis of patterns, structures, and functions, found all around. Simultaneously love has developed a voice, expressed when successful as "poetry of space" within the marriage of spirit and matter. Over the years any distinction between these developments has become more and more blurred. I have come to consider mathematics not only as a language to describe the weave of life, but also as the thread of its fabric.

Figure 1 "Orb" by Simon Thomas.

For someone who learned more on the riverbank than in the classroom my connection with mathematics has evolved along less orthodox lines than some. My first brush with mathematical magic was at primary school, when with a compass I drew my first "flower".

Many years later whilst a post-graduate student at the Royal College of Art I rediscovered geometry's flowery gateway, and this time armed with compass, pencil, and ruler, embarked on a journey into the wonderland of divided space. In the early days I employed my new found knowledge and skills to create practical devices such as templates and the like, but soon enough I was seduced by the process itself, investigating the properties of various proportions, sequences, and symmetries.

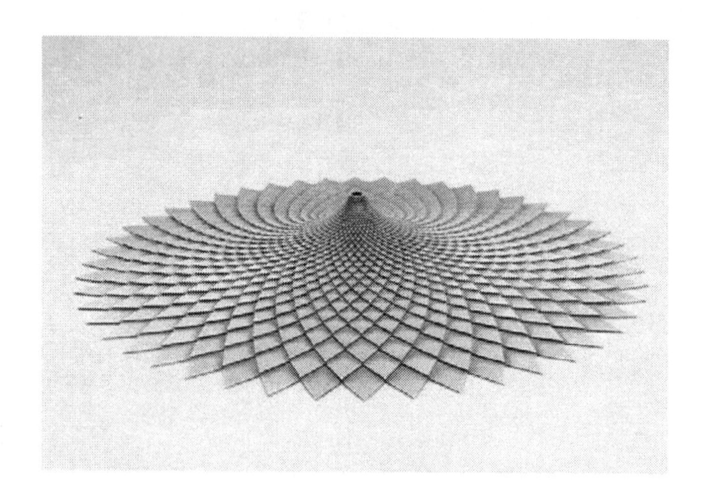

Figure 2 "Plane Liner" by Simon Thomas.

Humanity's sense of beauty has long been a topic of debate, thought of by some as purely subjective, whilst others prefer the "reflection of God notion". It is my belief that there is a strong relationship between natural efficiencies and our perception of what we call beautiful. Why are we predisposed to appreciate beauty in such things as the design and fabric of a dragonfly's wing? Or, what about the dreadful and awesome sight of a sardine "bait-ball" under attack? Could it be that these things are fascinating because throughout human evolution we have learned to recognize "efficiencies" as a vital part of our survival mechanism; an appreciation of how to gain most through least possible effort?

Figure 3 "Eye" by Simon Thomas.

Figure 4 "Small Worlds" by Simon Thomas.

Studies in Soap Bubble Foam

Introduction

Throughout my study of soap bubble foam geometry I have been transported by a sense of wonder in the beauty of this optimized fluid structure. It is surely an unsurpassed expression of energy conservation, the qualities of efficiency and beauty inexorably interwoven. It is an example of a dynamic system where the parameters of geometric possibilities and the laws of physics co-exist in an ephemeral harmony.

Closed and open celled foams seem to be found across the scale range of existence, in organic and inorganic structures. From quantum foam at Planck length, right through to that state suggested in the research of cosmologist Margaret Geller, where the distribution of galaxies within the cosmos appear to be located within a foam structure.

Soap bubble foam is a randomly arranged congregation of air pockets varying in volume, encapsulated by films of detergent/water. Detergent lowers the amount of surface tension experienced by the water it is mixed with. If air is forced through the soapy water it will surface as a bubble. When this process is repeated many times, these bubbles find themselves positioned within a colony of neighboring bubbles, and through this coming together the original sphere bubbles transmute into polyhedral cells. The structure of this water matrix has long been a muse, and the numerous attempts at understanding its math and physics are testament to its charm.

Studying the history of this subject with its experiments and hypotheses has been totally captivating, inspiring me to join the chase in hunting down this particular grail. Rigorous observation in tandem with the relevant reading matter has been accompanied by various experiments to form my own research.

Within the apparent chaos of closed cell foam there are certain constants. Plateau's laws tell us about the shapes and connections of soap films:

(1) Films can only meet three at a time and they do so symmetrically, so that the angles between them are 120 degrees.
(2) The lines along which they meet are themselves joined in vertices at which only four lines (or six films) can meet. Again they are symmetric, so that the angle between the lines is 109 degrees (the tetrahedral, or Maraldi, angle).
(3) The films and the lines are curved in general: the average amount by which the films are bowed in or out is determined by the difference in pressure between the gas on either side (Laplace's law).
Note that the third law does not dictate that a zero pressure difference implies a flat film. Saddle-shaped surfaces can have zero mean curvature.

Now, you may ask, what is the role of an Artist in all of this? Well, I am in the process of creating an artwork which aspires to reflect the geometric conditions of this water matrix, and aims to celebrate the mesmeric relationship between beauty and natural efficiencies. In re-presenting experience the Artist, to a greater or lesser degree edits, orders, and interprets perceptions, entertaining certain compromises such as the use of a particular medium, or "freezing" a dynamic system. These compromises are an intrinsic part of the creative process, setting parameters, and where appropriate should be seen as opportunities. I also have to consider the various modelling techniques available to me, always with an eye on the practicality of any given process, its durability, and how they might impact on the overall "feel" I am aiming at.

Experimentation

To enable the creation of a credible Artwork concerning the structure of foam, the Artist should, at least, have a partial understanding of the subject. As mentioned earlier, I have sought to build my knowledge through observation, reading, and experimentation.

The most fruitful experiments I've conducted to date have been achieved with the help of a particular molecular modelling node. As I came to understand the structure of foam a little more, I realised that this node along with the use of flexible joining rods of differing lengths, could enable a very handy modelling technique, one I have subsequently found to behave in a very similar way to the edges of real foam.

In my past experience of working alongside Mathematicians, one trick learned in understanding difficult spatial problems is to whenever possible reduce the issues involved by spatial dimensions. So, for instance, when investigating the properties of a sphere, where possible simplify things by applying the question to a circle. When appropriate this approach can enable an understanding where previous attempts proved overly complex.

The benefits of dimensional "downsizing" (by which I reduce the foam to only edges and corners), allied with the development of a simple modelling technique where I have utilised a particular molecular node, has been invaluable. The node I am referring to is that associated with the fourfold carbon bonds found in diamond, the one where the four equally spaced valency pegs would fit into the four corners of a regular tetrahedron. The angle between pegs is approximately 109 degrees and is known as the "Maraldi". This is the same symmetry we find at the very heart of the four edged corners of foam.

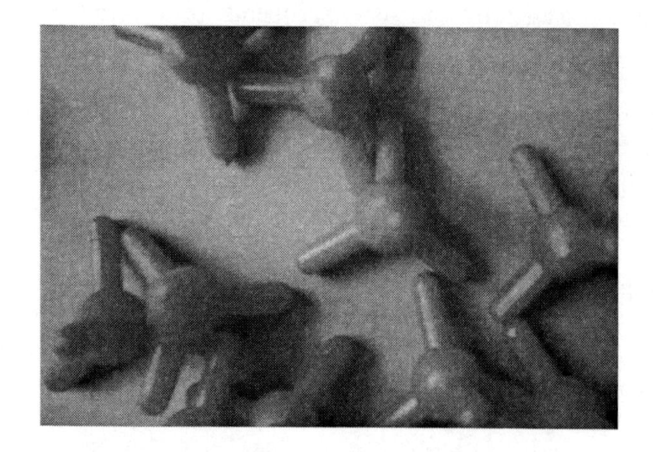

Figure 5. Carbon atom nodes with "Maraldi" angles.

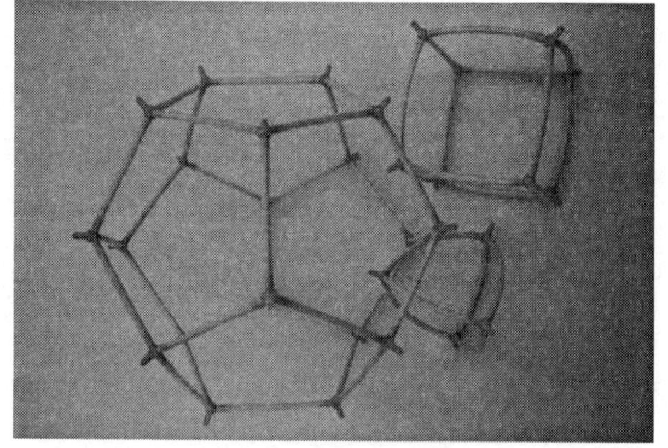

Figure 6. Polyhedra indicating relative edge curvatures.

While working with this modelling method I have developed an interest concerning the role of geometric parameters in fashioning the structure. The nodes, in combination with flexible bond rods of various lengths, create a network where the rods are obliged to deform through curvature, seemingly just as the edges of foam cells do. The positions and alignments of the nodes, and the way rods set in certain arrangements correspond to certain curvatures, is striking in its resemblance of the cell edges and corners of real foam.

There is a general scaling associated with the characteristic curvature and geometric arrays in relation to cell sizes, and this is tied up with the effects of surface tension. The larger cells have proportionally less pressure than smaller ones, and therefore are spatially invaded by the latter, causing a concaved face on the larger cell. This state of affairs is due to the surface tension being weaker over a larger surface area. The surface tension therefore exerts proportionally less constrictive force on the larger encapsulated air pockets or cells.

Larger individual cells can have convexed, concaved, flat and saddle shaped faces, all part of the same cell. This indicates a rule that wherever a cell is larger in volume than a neighbour, the face it shares will always be concaved into the larger one.

The relationship between cell size/pressure, and its geometric expression, is very much influenced by the Maraldi angle sat in the cell corners. The smaller cells of higher pressure create the most convexed curvature. Development of this modelling system with it's 109 degree angle, and through observations of real foam, have informed me that the maximum convexed curvature in rods (or edges) is witnessed when they form part of triangular faces as found on tetrahedra. A little less convexed edge curvature is found as part of a square face, approximately no edge curvature with pentagonal faces, and edges gradually increase in negative curvature (concaved) for faces with more than five edges. Of note here is that an edge is not exclusive to one face, or "ring" of edges. With each edge sharing three faces, and the faces having an angle of 120 degrees set between them, it is likely that the curvature of one edge is the mean of these three sets of force.

Figure 7. Clustered "Rings of edges" showing positive, neutral, and negative curvature.

When modelling in this way and as a foam bank is evolving, decisions on comparative rod lengths become more critical and choices are taken to eradicate any compound curvature in the rods, which seem to seldom occur in real foam. It is an intuitive process where the "trial and error" approach eventually pays off. The high degree of freedom in this modelling technique apparently corresponds to the character of cell edges and corners in real foam, where countless combinations of cell shapes and cell sizes will meet.

Worth bearing in mind here is that as with real foam, the "cells" on the periphery of the colony are different from those within the bulk of the complex. In the modelling system, cell edges surrounded by neighbours have full influence from the three "rings of edges" which they are part of (rings of face edges in real foam) whereas those cell edges not totally enclosed have rods of a more "relaxed" nature. Also of note is that where soap cells meet the internal surface of a containing vessel, cell edges linking that 2-D "vessel surface design" to the foam within, inwardly project at tangents of approximately ninety degrees. This is where the transition of spatial rules between three and two dimensions is taking place.

My modelling process is always accompanied by a large glass vessel filled with soap foam, so that I may compare as I progress.

Summary of findings

Foam expresses the dynamic equilibrium formed when the forces of pressurised air pockets co-exist with the surface tension of water, forcing an accommodation of each other's energy. So, in simple terms we see a dual system of air pressure versus water tension, although in reality they are inextricably involved in each other's status; and other lesser forces are involved.

Understanding the transition from a wet foam to a dry foam, is essential for a clear view of the geometric characteristics of a closed cell dry foam.

Imagine the moment when a wet foam is being formed, many spheres of air are surrounded by adequate water so that no contact is made with neighbouring air bubbles. As gravity drains the excess water from around the bubbles they will soon come to make contact. The first thing to happen is that they will kiss, that is to say that they touch at a point. These points rapidly open out laterally on the surfaces of neighbouring bubbles, forming a disk of contact. The disks enlarge until other disks similarly generated around the bubbles meet. This is the moment that edges and ultimately corners are created, changing the bubbles into polyhedral cells. Surface tension is the cause of this constriction, and the process ceases when the pressure within the air pockets is strong enough to resist further progress.

What were once independent bubbles are now a colony of interdependent cells. The characteristics of corners, edges, and films, are arranged by default in accordance with underlying geometrical efficiencies necessary to sustain a structural integrity. The most interesting aspects of this geometric default is the consistency of Maraldi angles (approximately 109 degrees) at the corners of cells, and the 120 degree angles emanating between film faces "spline-like" along the curved edges.

These particular angles are present owing to their fundamental nature. The Maraldi, or tetrahedral angle, is the first symmetrical division of 3-space when emanating from a point. It has four directions equally spaced, the least possible. The three directional symmetry associated with film co-angles, is a consequence of this corner angle, the cell walls could not be arranged in any other way.

These constant corner angles, where four edges and six films meet within a random colonisation of various sized air pockets, cannot join up in straight lines (edges). Curvature along the edges in between these corners is how this problem is overcome, and similar forces replicate these curves in the modelling system. By subtracting the cell faces from the equation, leaving only corners and edges, the difficulty of linking these angles is taken up by flexibility in the rods.

Whilst understanding that the films (faces) are remnants of original independent bubble surfaces, in the context of a cell complex I prefer to see films as simply spanning "rings" of edges. These films are minimal surfaces, indicators of pressure differences yes, but also subject to the powers of geometric default in the corners and edges of that network.

The Artwork

Foam structure is an eminent expression of energy conservation, and in my opinion beautiful because of this. This mysterious structure where these qualities of efficiency and beauty are interwoven is a real wonder, and fruitful meditation on this subject will draw the viewer in contact with elemental geometric rules, and their physical consequences.

Now, for me the science of the subject is only half of the story, I also have to evolve an artwork, which above all has presence and meaning. The reader may appreciate that building a durable structure which is closely related to soap foam is not going to be an easy ride. It seems likely to be fraught with endless technical problems, inhibiting geometric parameters, and visual flaws; and so it is! I have worked through quite a few ideas concerning the look and build-technique of such a sculpture, and within the particular circumstances of my current commission I have decided to more or less stick with uncomplicated look of the modelling system, a design I feel is both technically plausible, and visually arresting. One thing I have done in order to fortify the visual impact of the piece is to exponentially increase the cell sizes in the vertical direction, adding a sense of the foam "billowing forth".

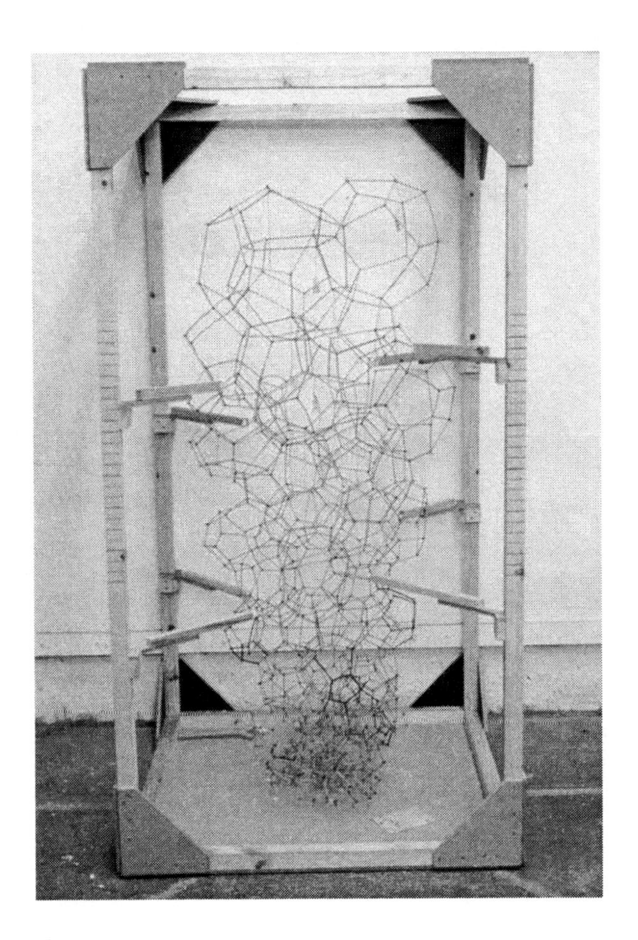

Figure 8. Modelled foam complex in frame.

Figure 9. "Cell" being cast.

In this way the Artwork will describe the edges and corners only, of a foam complex. Fabricated in 6mm stainless steel rod, every weld will be carved /ground back to form an "oily" minimal surface look. The feel will be one of a continuous homogeneous surface over the whole network, not lots of spars "spot welded" at their ends.

In order to transfer accurate co-ordinate information from the plastic model to the sculpture itself I have taken a cast from every one of the 71 "cells". These are composed of a heat resistant plaster based material (developed on the job) and will be used to weld upon. The rods will be rolled to the correct curvatures before being incorporated into the final design. Finally, the frame within which the plastic model has been located will be utilised to offer datum points, ensuring a faithful representation of the forces freely expressed by the tensile plastic model.

Figure 10. Seventy one "cells/ formers", the edges of which are to be used for welding upon.

Creating Penrose-type Islamic Interlacing Patterns

John Rigby
Cardiff School of Mathematics
Cardiff University
Senghennydd Road
Cardiff CF24 4AG, Wales UK
E-mail: rigbycathedralcourt@tiscali.co.uk

Abstract

Some of the most interesting Islamic interlacing patterns involve ten-pointed stars or ten-petalled rosettes. These motifs have local ten-fold symmetry, yet they are often included as part of a plane periodic pattern, which can have no overall five- or ten-fold symmetries. Instead of using these motifs in periodic patterns, can we incorporate them in patterns based in some way on Penrose tilings (which have many local five-fold symmetries)?

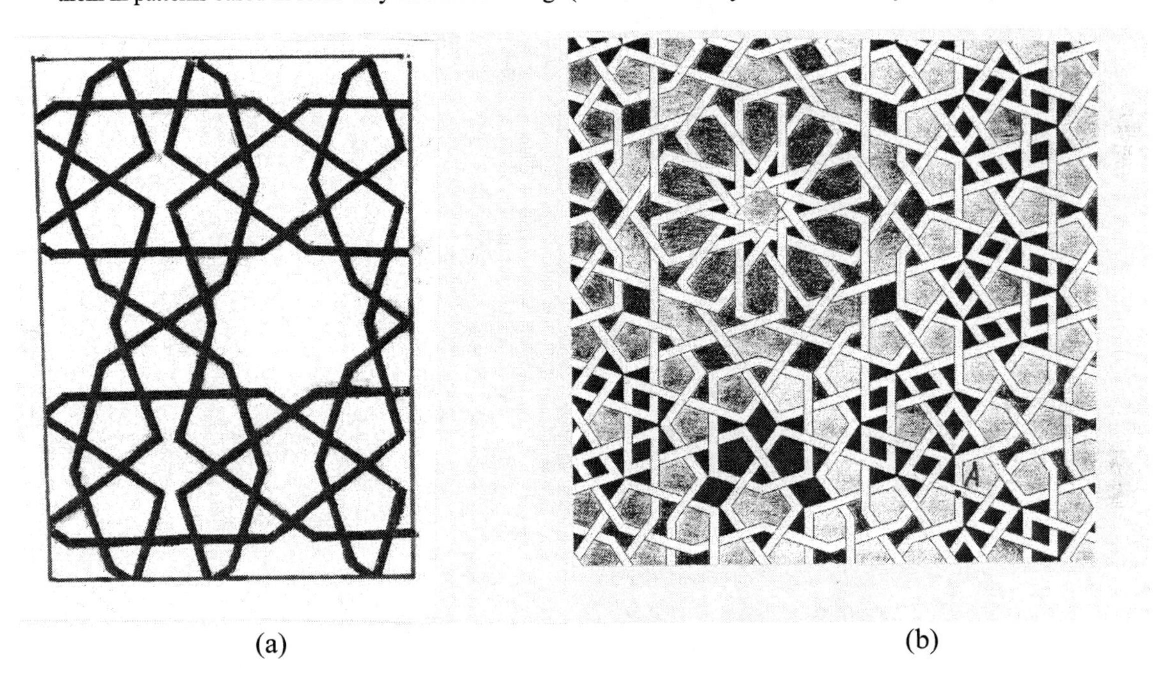

(a) (b)

Figure1. *Interlacing patterns from Afghanistan and Turkey.*

1. Interlacing Patterns and Penrose Tilings

1.1. Interlacing Patterns. Ten-pointed stars (Figure 1a) and ten-petalled rosettes (Figure 1b) occur in Islamic interlacing patterns from many countries. Figure 1b shows part of a very elaborate pattern from the Karatay Medrese at Konya in Turkey; horizontal and vertical lines through the centre of the rosette are mirror lines of the pattern if we disregard the under-over interlacing, and the point marked A is a 2-fold rotational centre. Although the rosettes and other small motifs in this pattern have local five- or ten-fold symmetry, it is well known that periodic patterns in a plane cannot have any overall five-fold symmetry. One situation in which five-fold symmetry can extend over a complete periodic pattern is on a dodecahedron or on a sphere. This idea does not seem to have been much employed in architecture, but an alcove in the Prince's Room in the Reales Alcázares, Seville, Spain, has a polyhedral domed ceiling containing eight trapezium-shaped planes with 72° and 108° angles, into which ten-petalled rosettes fit neatly.

After giving a brief description of Penrose tilings, we shall show how to create patterns in an Islamic style that, instead of being periodic, share some of the geometrical properties of Penrose tilings.

1.2. Penrose Tilings. Penrose kite-and-dart tilings are aperiodic – they do not repeat regularly. An account of such tilings can be found in [5, Chap. 10], and Figure 2 shows an example; we shall simply state some of their basic properties without proof. The kite and dart tiles have side-lengths in the ratio τ :1 where $\tau = (1 + \sqrt{5})/2$ is the golden ratio, and their angles are multiples of 36°. One way of ensuring aperiodicity is to colour the corners of the tiles alternately black and white: the 144° angle (or corner) of each kite, and the 36° angles (or corners) of each dart, are black, and the corners of the tiles meeting at a vertex of the tiling must be either all black or all white. This extra requirement ensures that the vertex-coloured kites and darts can only be put together in an aperiodic manner, as described in [5].

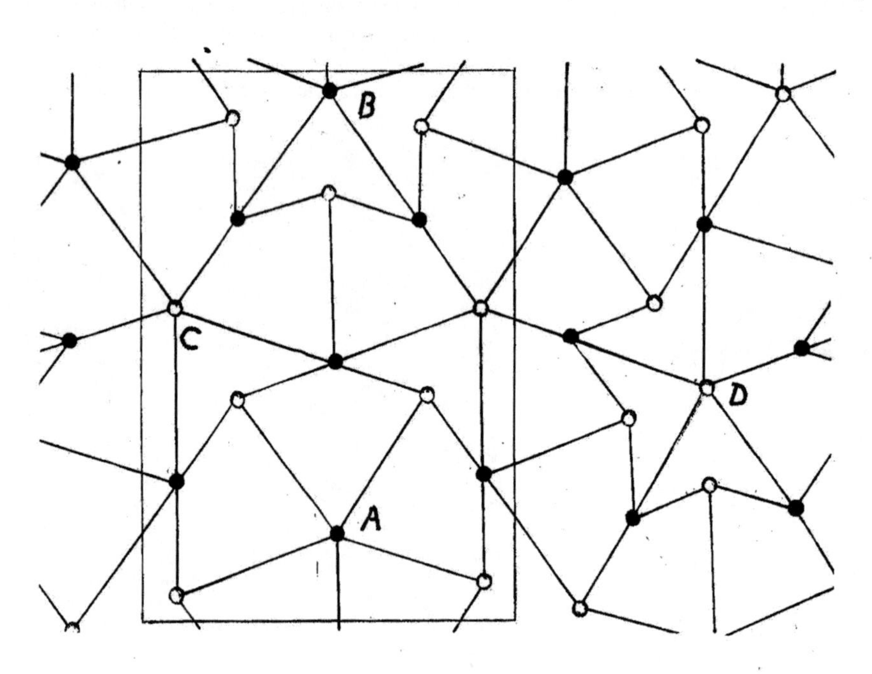

Figure 2. *A Penrose tiling of kites and darts.*

1.3. Penrose-type Patterns. There are two properties that make Penrose tilings of interest to us here: although they are aperiodic, they contain arbitrarily large finite regions with five-fold rotational symmetry, and any finite region is repeated infinitely often in the tiling. So, instead of producing periodic Islamic-style patterns associated with the plane symmetry groups, it would be nice to be able to produce Islamic patterns, especially patterns containing star or rosette motifs, that are in some way associated with Penrose tilings. We shall proceed to do this in an exploratory way, and shall wait until the end before attempting to give a definition of the term "Penrose-type pattern". The Karatay pattern in Figure 1b, and a Penrose-type pattern in Karatay style, will form the beginning and the end of our investigation.

2. Creating Penrose-type Patterns

2.1. A Method of Creating Penrose-type Patterns. The obvious way to associate a pattern with a Penrose tiling is to draw part of a pattern on a kite, and part on a dart, in such a way that when a copies of these partial patterns are used on each kite and each dart of a Penrose tiling, (a) the partial patterns fit together neatly whenever two tiles meet edge to edge, and (b) the resulting overall pattern looks Islamic in style. In other words, we use copies of a patterned kite and a patterned dart to create an overall Penrose-type pattern. Satisfying the criteria (a) and (b) is not as easy as it sounds, and (b) involves a subjective judgment.

2.2. Skeletal Islamic Patterns. First we must get used to studying and creating Islamic patterns. When studying the structure and the geometry of an interlacing pattern, it is convenient to use a *skeletal form* of the pattern, in which the interlaced braiding is reduced to a single line; the background shapes in the original pattern now become tiles in a tiling, and the braids become the edges of the tiles. The skeletal forms of the star motif from Figure 1a, and the lion head and five-diamond motifs from Figure 1b, are shown in Figure 3, and the rosette from Figure 1b is shown in Figure.4 The term "skeletal pattern" is convenient, but perhaps unfair, since such patterns often occur as ceramic wall patterns in their own right, not just as simplified drawings of braided patterns. (The artist M.C.Escher copied a pattern in the Alhambra which occurs in both skeletal and braided form with only minor variations; [1, pp.41, 53].)

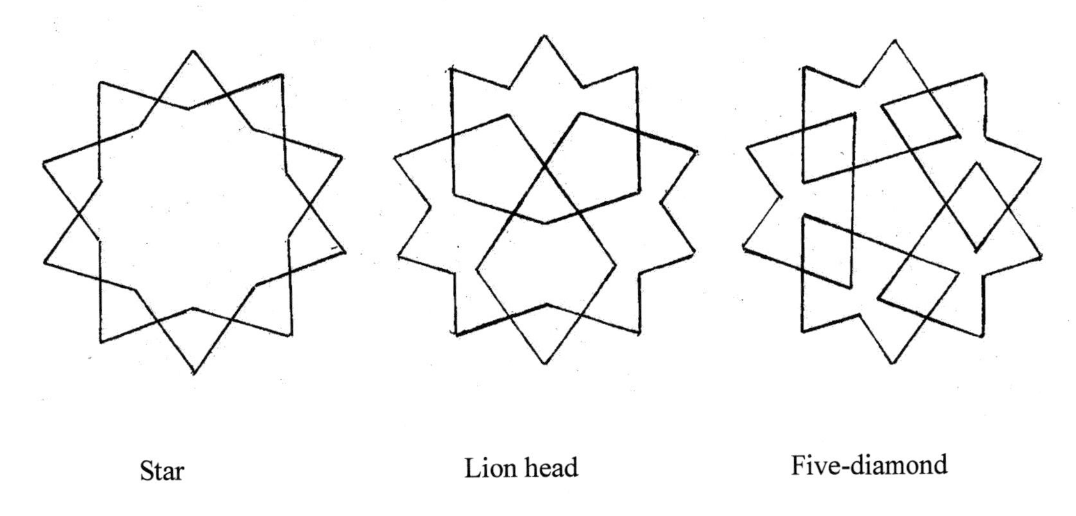

Figure 3. *Skeletal forms of motifs from Figure 1.*

2.3. A Penrose-type Pattern Using a Restricted Set of Tiles. The lion head motif in Figure 3 is made up of shapes of tile that occur frequently in patterns from many sources: a regular pentagon with side-length 1 unit, a non-convex 10-gon with sides of length 1, and a non-convex 8-gon that is always combined with two small kites to form a rhombus with side-length 2; I shall call these *the four tiles*. Many interesting patterns can be created using copies of just the four tiles, but such patterns do not seem to occur in Islamic architecture. However, examples in which the four tiles are combined with the star motif (Figure 3) do occur architecturally and in collections of patterns [2, 4, 6]. Figure 1a shows one such example. Whilst collaborating with a colleague in the creation of four-tile patterns – a form of artistic and geometrical relaxation! [8] – I considered the possibility of producing a Penrose-type four-tile pattern; to make a start I decided to introduce stars also, and I was then able to devise the patterned kite and dart shown in Figure 5. There is a partial star centred at each corner of the kite and dart, and the half-rhombuses and partial stars on the individual kites and darts fit together to form complete rhombuses and stars. When the patterns of Figure 5 are applied to the portion of Penrose tiling enclosed within the rectangle in Figure 3, the resulting pattern is shown in Figure 6.

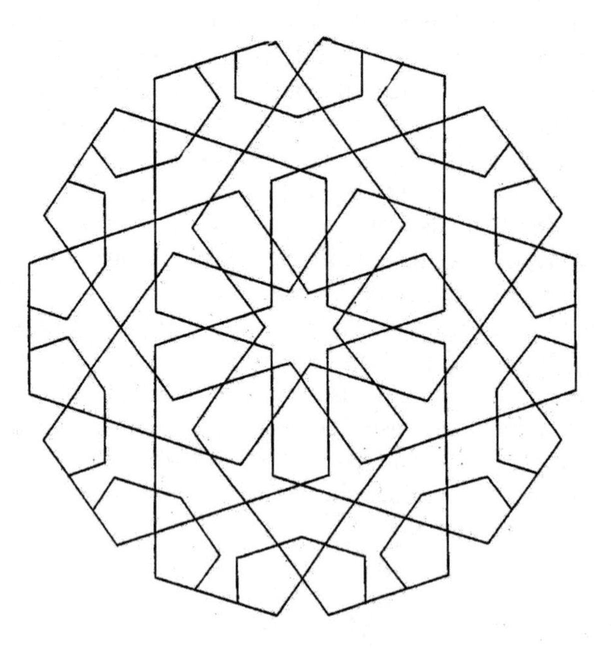

Figure 4. *The skeletal form of the rosette motif.*

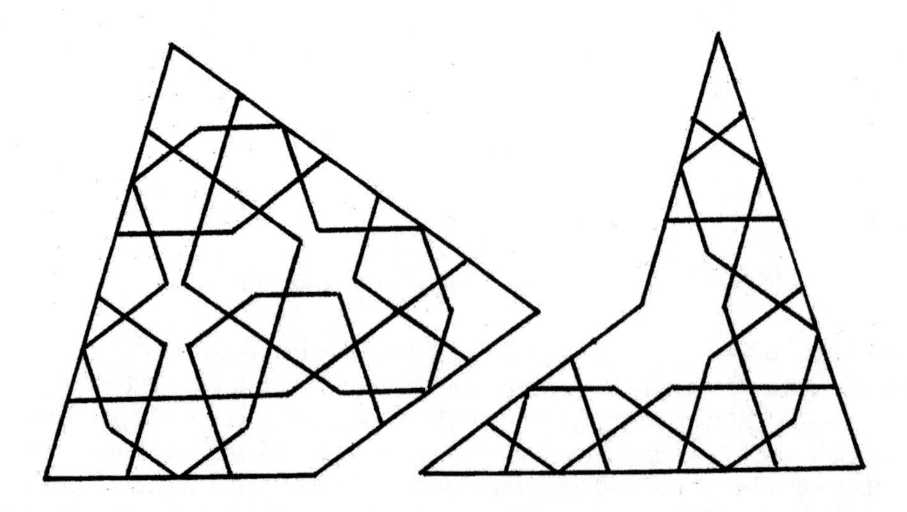

Figure 5. *A patterned kite and dart.*

2.4 A Penrose-type Karatay Pattern. Figure 6 is itself a perfectly good Penrose-type pattern, but at this stage I was still trying to produce a Penrose-type four-tile pattern. My original idea was to replace each star in Figure 6 by a lion head motif, which we can do because the two motifs have the same size and shape, but this process destroys much of the local 5-fold rotational symmetry of the Penrose tiling, so the result can hardly be called Penrose-type. However, having laid out Figure 6 using cardboard tiles, it occurred to me that we can transform this pattern into a pattern having many of the characteristics of the Karatay pattern in Figure 1b. The star marked *A* in the figure is surrounded by a ring of ten pentagons, ten rhombuses and ten 10-gons, then by a ring of twenty pentagons. We notice that the rosette motif in Figure 4 is surrounded in exactly the same way by twenty pentagons, so the star marked *A*, and its immediate surroundings, can be replaced by a rosette motif. At the same time, let us replace the stars closest to star *B* (three of these are shown in the figure, marked *X*, *Y*, *Z*) by lion heads all facing in towards the centre of *B*. The result is shown in Figure 7.

We find that star *B* is now surrounded by twenty pentagons, so it also (with its immediate surroundings) can be replaced by a rosette. Star *C* can be treated in much the same way as star *B*; the steps are not quite the same, but again we finish up with a rosette, as in Figure 8. Finally, since the Karatay pattern contains no stars, we can replace the remaining stars by five-diamond motifs and lion head motifs in some suitable manner.

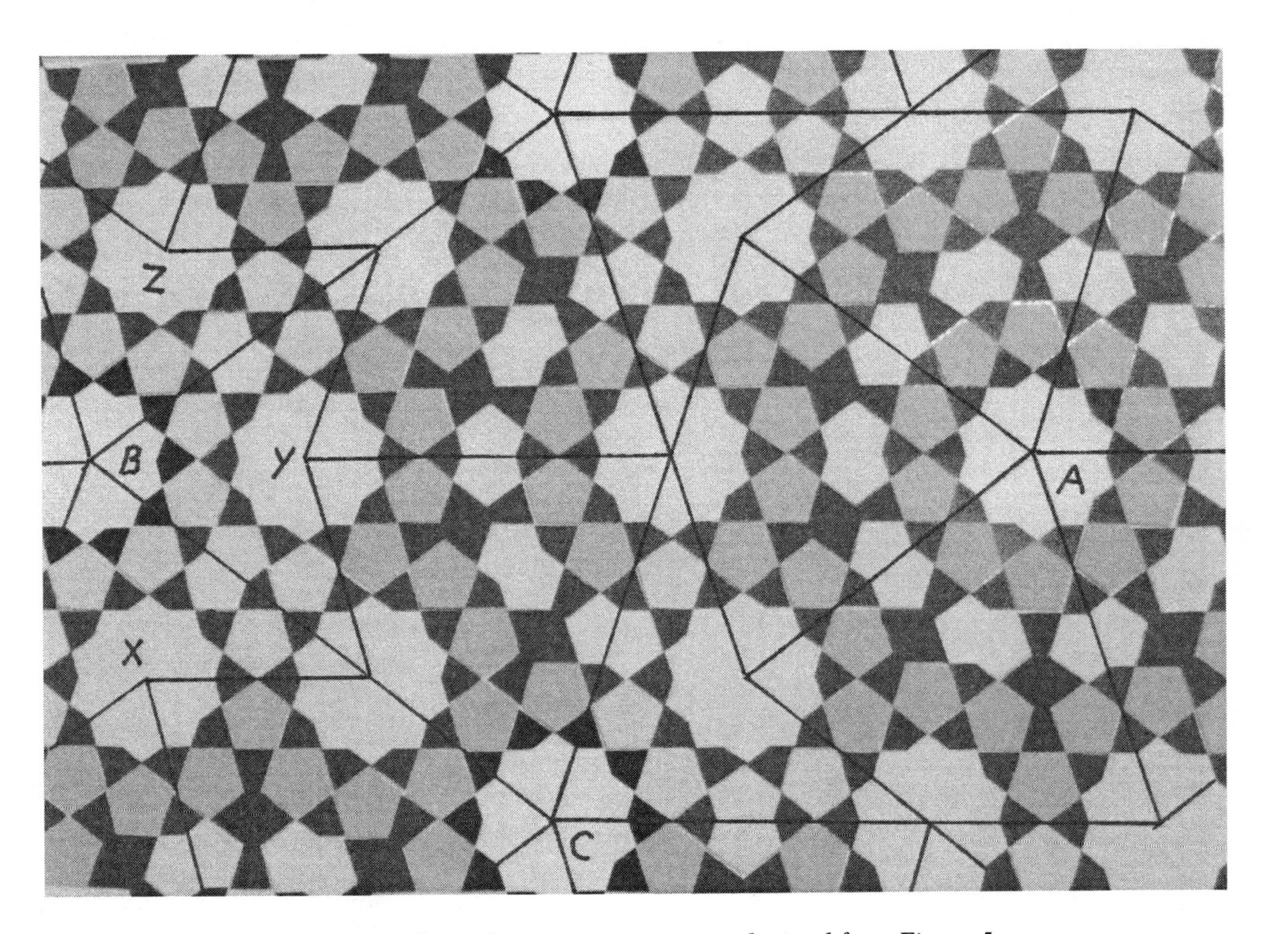

Figure 6. *A basic Penrose-type pattern obtained from Figure 5.*

Note that Figure 6 shows merely a small part of an infinite pattern. The centre of star *A* lies at a point of the underlying tiling (Figure 2) where five kites meet; we subject all stars with this property in the infinite pattern to the same treatment as *A*, replacing them and their surroundings by rosettes. The centre of star *B* lies at a point of the underlying tiling where five darts meet; we subject all stars with this same property to the same treatment as *B*. Star *C* lies at a point of the underlying tiling where four kites and one dart meet; where possible we subject all stars such as *C* to the same treatment as *C*, but this cannot be done if they lie too close to an *A*-star or *B*-star. So, Figures 6 – 8 show small finite portions of infinite patterns;

unfortunately these patterns only begin to exhibit their Penrose quality in a visual manner when we can view a much larger portion than has been possible here.

2.5 Thoughts on a Definition of "Penrose-type". What is meant in the last phrase of the previous paragraph but one by "suitable manner"? What modifications can we make to a pattern whilst still describing it as "Penrose-type"? Here are some thought on the matter, but they fall short of providing a precise definition. Two patches or regions in a pattern or tiling, P and Q say, are *congruent* if there is an isometry (a distance-preserving transformation) transforming the portion of the pattern on P to the portion of the pattern on Q.

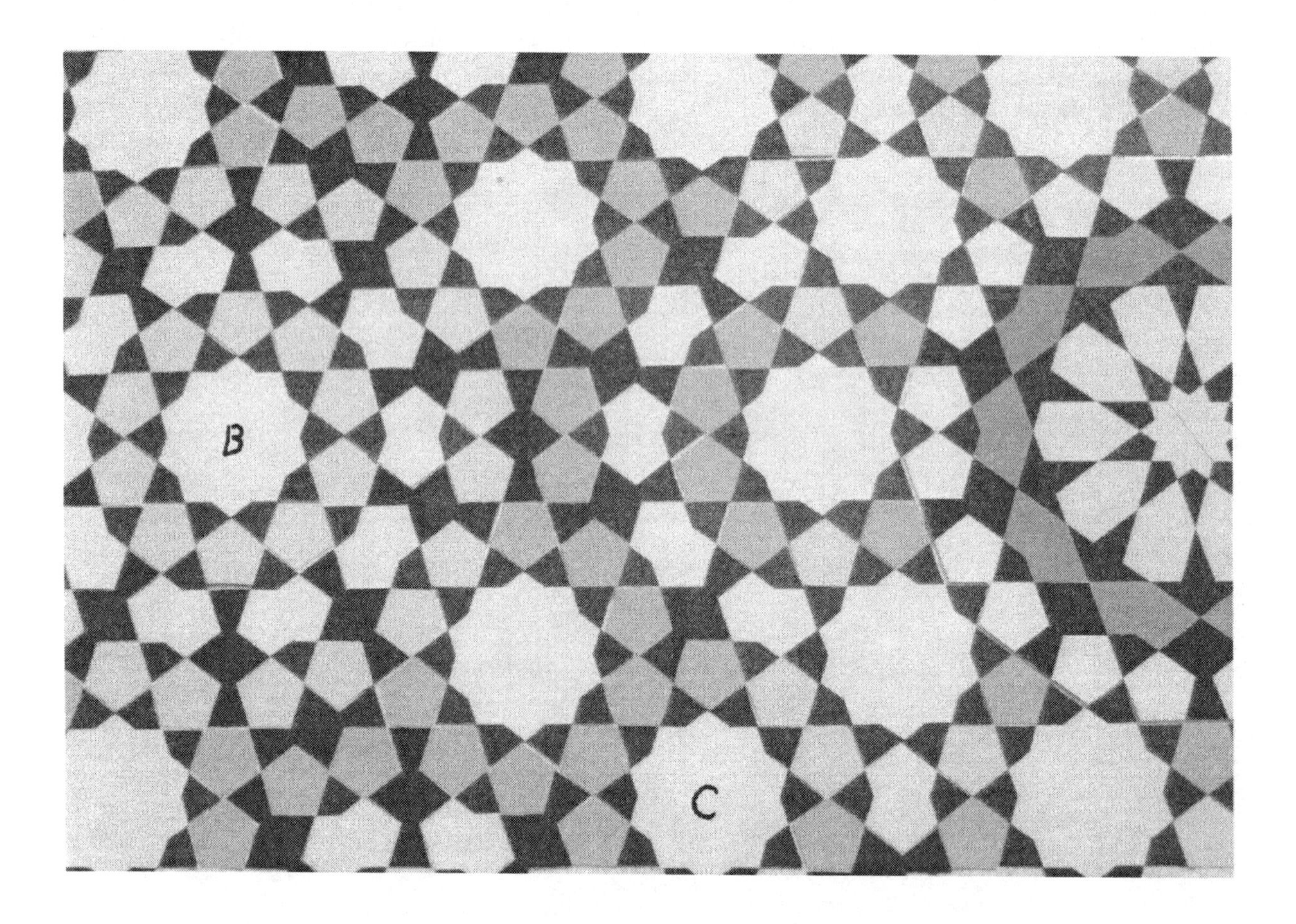

Figure 7. *A modification of the pattern in Figure 6.*

If we start with a Penrose tiling **T**, and replace its kites and darts by patterned kites and darts, as described in 2.1, **T** is transformed into a pattern **T***; let us agree to say that **T*** is a Penrose-type pattern, or more precisely a *basic* **T**-*type pattern*. To obtain a *modified* **T**-*type pattern*, we replace a patch of tiles P in **T*** by another patch P′ (where the boundaries of P and P′ have the same size and shape) and at the same time we replace *every other patch* Q *congruent to* P by a patch Q′ congruent to P′. (For instance, in 2.4 we replaced all stars of a certain type, with their surrounding tiles, by rosettes.) Any number of such modifications can be carried out.

With this terminology, Figure 6 shows a basic Penrose-type pattern; Figures 7 and 8 show modified patterns obtained from Figure 6.

3. Other Penrose-type Patterns

3.1. A Pattern Similar to 2.4. The centre of every star in Figure 5 lies at a vertex of the underlying tiling shown in Figure 3. There are only seven types of vertex in a kite-and-dart tiling [5, p.561] (two vertices are of the same type if they are surrounded in the same way by their adjacent kites and darts), and a brief investigation shows that any star can be subjected to a treatment similar to that given to C in 2.4, replacing adjacent stars by lion head motifs and then replacing the star and its surroundings by a rosette. Two stars can be subjected to this treatment simultaneously as long as the distance between their centres is not less than $s + t$, where s and t are the lengths of the short and long edges of the kites and darts.

Figure 8. *Further modifications of Figures 6 and 7.*

In 2.4 we replaced all A-type and B-type stars, and as many C-type stars as possible, by rosettes. Another type of vertex in Figure 2 is the one labelled D. It is easily verified that if we consider all vertices of types A, B and D, the distance between any two of these vertices is at least $s + t$, so we can replace all the corresponding stars by rosettes to obtain a Penrose-type pattern different from the one in 2.4.

We cannot carry out a similar process on the remaining types of star – there will always be pairs of stars somewhere in the plane that are too close together. But if we are interested merely in creating a pleasing pattern in a small region of the plane, it may be possible to find other stars that can be replaced.

3.2 Other Patterns. It is possible to create Penrose-type patterns using only the four tiles as described in 2.3. As patterns of shapes they are unexciting, but suitable colouring brings them to life. J.-M. Castéra [3, p.287] describes a different method of creating Penrose-type patterns based on Penrose rhombs. There is not space here to describe and illustrate these ideas, but it may be possible to include them in my presentation. We conclude with another pattern, which seems more elaborate because we can see more of it in Figure 9. The centre of the rosette in the middle of the top edge is a 5-fold rotational centre of the complete pattern

Figure 9. *Another Penrose-type pattern obtained by a similar process.*

References

[1] F. H. Bool *et al.*, Escher, with a Complete Catalogue of the Graphic Works, Thames and Hudson, 1982.

[2] J. Bourgoin, Arabic Geometrical Pattern and Design, Dover, New York, 1971.

[3] Jean-Marc Castéra, Arabesques: Decorative Art in Morocco, ACR Édition, Paris, 1999.

[4] Issam El-Said and Ayşe Parman, Geometric Concepts in Islamic Art, Scorpion Publishing and World of Islam Festival Trust, 1976.

[5] Branko Grünbaum & G. C. Shephard, Tilings and Patterns, W. H. Freeman and Co., 1987.[6] G. Necipoglu, The Topkapi Scroll – Geometry and Ornament in Islamic Art, The Getty Centre for the History of Art and the Humanities, 1995.

[7] John Rigby, *A Turkish Interlacing Pattern and the Golden Ratio,* Mathematics in School, Vol. 34 No. 1. pp. 16 – 24, 2005.

[8] John Rigby and Brian Wichmann, *Some Patterns Using Specific Tiles*. Not yet published.

Steve Reich's *Clapping Music* and the *Yoruba* Bell Timeline

Justin Colannino* Francisco Gómez† Godfried T. Toussaint‡

Abstract

Steve Reich's *Clapping Music* consists of a rhythmic pattern played by two performers each clapping the rhythm with their hands. One performer repeats the pattern unchangingly throughout the piece, while the other shifts the pattern by one unit of time after a certain fixed number of repetitions. This shifting continues until the the performers are once again playing in unison, which signals the end of the piece. Two intriguing questions in the past have been: how did Steve Reich select his pattern in the first place, and what kinds of explanations can be given for its success in what it does. Here we compare the *Clapping Music* rhythmic pattern to an almost identical *Yoruba* bell timeline of West Africa, which strongly influenced Reich. Reich added only one note to the *Yoruba* pattern. The two patterns are compared using two mathematical measures as a function of time as the piece is performed. One measure is a dissimilarity measure between the two patterns as they are being played, and the other is a measure of syncopation computed on both patterns, also as they are played. The analysis reveals that the pattern selected by Reich has greater rhythmic changes and a larger variety of changes as the piece progresses. Furthermore, a phylogenetic graph computed with the dissimilarity matrix yields additional insights into the salience of the pattern selected by Reich.

1 Introduction

The history of music is often the history of humanity's reactions to it. A good example of this may be observed in Minimalism. Since the Second World War, mainstream classical music has been dominated by composers such as Boulez, Berio, Cage, Ligeti, and Stockhausen, among others. These composers represent postwar Modernism, either through postserialism, Boulez being its most prominent figure, or through indeterminacy, where Cage is its most notable figure. Although the term Minimalism was originally used for the visual arts, it was later applied to a style of music characterized by an intentionally simplified rhythmic, melodic and harmonic vocabulary (see [19]). Its main representatives are LaMonte Young, Philip Glass, Terry Riley and Steve Reich. Their music and ideas become the major reaction to the Modernism epitomized by the aforementioned composers. Indeed, whereas Modernism is decisively atonal, Minimalism is clearly modal or tonal; whereas Modernism is aperiodic and fragmented, Minimalism is characterized by great rhythmic regularity; and whereas Modernism is structurally and texturally complex, Minimalism is simply transparent.

Minimalism has different materializations depending upon the particular composer, but minimalist works share concern for non-functional tonality and reiteration of musical phrases, often small motifs or cells, which evolve gradually. For example, while Young uses sustained drones for long periods of time, Glass chooses recurrent chord arpeggios, and Riley and Reich incorporate repeated melodies and quick pulsating harmonies. No less significant is the fact that minimalist music possesses almost

*School of Computer Science, McGill University. jcolan@cs.mcgill.ca

†Departamento de Matemática Aplicada, Universidad Politécnica de Madrid. fmartin@eui.upm.es

‡School of Computer Science, McGill University. godfried@cs.mcgill.ca

none of the main features of Western music (at least since the time of the Romantic period) that is, harmonic movement, key modulation, thematic development, complex textures, or musical forms with well-designed structures. On the contrary, this music deliberately skirts around any sense or awareness of climax or development, and seems to ignore the dialectic of tension and release, at least as it is usually posited in the classical music tradition.

It is probably Reich, the minimalist composer who most unhesitatingly repudiates the Western classical tradition. Reich objects to both European serialism and American indeterminacy because in these traditions the processes by which the music is constructed cannot be heard and discerned clearly by the listener. This rejection, formulated not only by Reich but by other minimalist composers as well, may be the reason why minimalist music has been so incomprehensibly ignored by many critics and scholars. In his essay *Music as a Gradual Process*, included in [20], Reich states his principles as follows: "I am interested in perceptible processes. I want to be able to hear the process happening throughout the sounding music." For such processes to be accessible to the listener, they must flow in an extremely gradual manner. The process itself must be related to the idea of shifting phases. First, a melody is played by two or more players, and after a while one of them gradually shifts phase. At the beginning of the phasing a kind or rippling broken chord is produced; later, as the process moves forward, the second melody is at a distance of an eighth note, and a new interlocking melody arises. The process continues until the two melodies are in phase again.

These ideas are fulfilled in many of Reich's works composed between 1965 and 1973. This experimentation starts with *It's Gonna Rain* and *Come out* (both composed in 1966), where he uses phasing on tape music; it continues with *Piano Phase* (1967), and *Violin Phase* (1967), where he experiments within an instrumental context (no electrical devices); and finally, Reich reaches the highest development in *Drumming* (1970-71), *Clapping Music* (1972), and *Music for Mallet Instruments, Voices and Organ* (1973), where he incorporates gradual changes of timbre and rhythmic augmentation, among other musical resources. By the end of 1972, he abandoned the gradual phase shifting processes, because "it was time for something new" [20].

This paper is concerned with a mathematical comparative analysis of *Clapping Music* and the bell rhythmic timelines of West African *Yoruba* music. This relation is not as distant as it might seem at first glance. During the summer of 1970 Reich traveled to Ghana where he studied African drumming. He learned *Gahu, Agdabza* and other musical styles, which influenced his music (later he also studied Gamelan music). Such an influence can be perceived in works such as *Drumming* and *Clapping Music*, where the phasing is discrete rather than continuous, as in his previous works. More specifically, we study *Clapping Music* with regards to syncopation, inasmuch as it forms an essential part of that piece.

2 *Clapping Music*

Clapping Music is a phase piece for two performers clapping the same pattern throughout the duration of the piece. The phasing is discrete, with one performer advancing an eighth note after several repetitions of the pattern, while the other imperturbably remains playing the pattern without shifting. See Figure 1 for further details. In the following, variations produced by shifting are numbered in ascending order as $\{V_0, V_1, \cdots, V_{11}, V_{12}\}$, where $V_0 = V_{12}$ indicates that the two performers play the pattern in unison.

This piece, in spite of its apparent simplicity, does not lack musical interest. First of all, *Clapping Music* constitutes a synthesis and a refinement of Reich's ideas by means of a piece with few but well combined elements. Secondly, *Clapping Music* enjoys a profound metrical ambiguity (something common to Reich's pieces) as well as a great deal of interlocking rhythmic patterns. The analysis

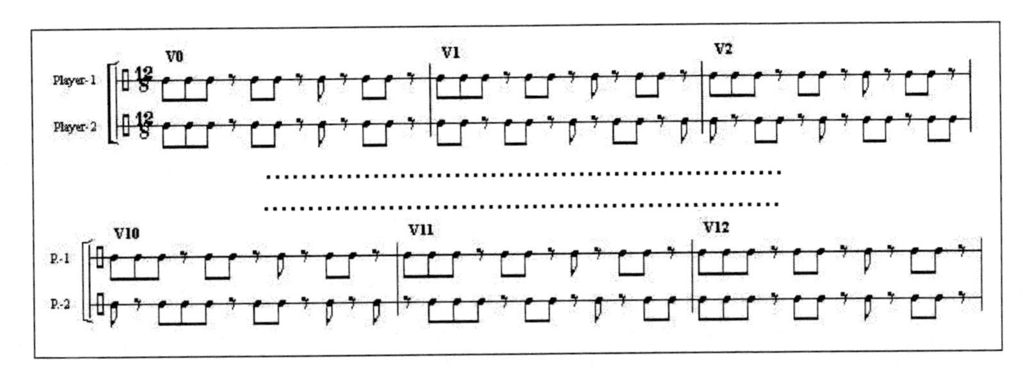

Figure 1: The first and last few bars of *Clapping Music.*

of those interlocking rhythmic patterns and its distribution along the piece is quite elucidating on its musical structure, as we will see later. Paradoxically enough, in spite of this rich structure, *Clapping Music* has received little *musical* analysis, in sharp contrast to the much more attention received from a *mathematical* point of view [8, 4, 9, 21].

3 Measuring Features of *Clapping Music*

When one listens to *Clapping Music*, a question that arises naturally is how Reich came to select that particular pattern. As the pattern shifts, a series of interlocking rhythms emerge, creating great variety. Furthermore, there is a sense of balance in the whole piece, between the resulting variations, as they create and release rhythmic tension. Once the pattern is defined however, the rulebook does not allow us to change it. Therefore, the pattern must be carefully chosen to begin with.

Phylogenetic graphs have been already used to analyze musical rhythms. In [21], a phylogenetic analysis of binary *claves* from Brazil, Cuba and some parts of Africa was carried out. *Claves* are rhythmic patterns repeated throughout a piece whose main functions include rhythmic stabilization as well as the organization of phrasing [16]. Subsequently such an analysis was extended to ternary claves taken from the African tradition, and to the hand-clapping metric rhythms of Flamenco music ([22] and [5], respectively). In all cases, worthwhile conclusions were drawn from the phylogenetic analyses. In this paper we use phylogenetic graphs to both, analyze the structure of *Clapping Music* itself, and to compare it to the *Yoruba* rhythmic timeline.

The key mystery in *Clapping Music* is how Reich was able to find a pattern that would work so well within the constraints of process music. We believe that his inspiration for the pattern came from his study of African drumming. In particular, we note an extraordinary resemblance between the pattern of *Clapping Music* and a clave bell pattern used by the *Yoruba* people of West Africa: only one additional seemingly inconsequential note has been added by Reich! The two patterns are shown in Figure 2.

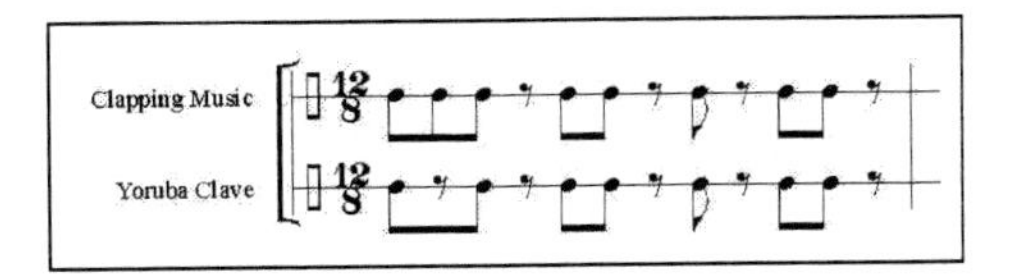

Figure 2: The pattern of *Clapping Music* and the *Yoruba* clave.

The *Yoruba* people live on the west coast of Africa, mainly in Nigeria, although they can be found also in the eastern Republic of Benin and Togo. Because most of the slaves were taken from West

Africa, a diaspora took place and the descendants of the *Yoruba* people can also be found in Brazil, Cuba, The Caribbean, the United States and the United Kingdom. They are one of the largest cultural groups in Africa, and musically speaking are of great relevance. *Yoruba* music has exerted much influence on the music of the surrounding countries.

The clave considered here is widely employed as a timeline in the sacred music among the *Yoruba* people [18]. Bettermann [3] calls this rhythm the *Omele.* It is also found in Cuba, where it is used in several styles [17] like the *Columbia.*

It might be argued that there could likewise exist other clave patterns very close to that of *Clapping Music.* This is not the case, as we will see in the next section. Once we formally introduce the measure of distance between a pair of rhythms, we will verify that the *Yoruba* clave is the rhythm closest to the *Clapping Music* pattern among a great number of African ternary claves. These reasons constitute our primary motivation for comparing these two patterns. In fact, an intriguing natural question is whether the *Yoruba* clave itself would work just as well as the pattern employed in *Clapping Music.*

The musical effectiveness of *Clapping Music* is partly due to the way in which syncopation is dealt with. The problem of defining a mathematical measure of syncopation has not been addressed until recently. In [7], Gómez et al. reviewed several measures of syncopation, and proposed a new measure, the so-called *weighted note-to-beat distance* measure (*WNBD* measure from here on). That measure will be used here to analyze the syncopation of *Clapping Music* and the *Yoruba* clave pattern.

4 Phylogenetic Analysis

Phylogenetic graphs were originally used in Biology to determine the proximity and evolution of species. Biologists measure the degree of proximity between two species by comparing their genes. In our context, rhythmic patterns take the place of genes, and as a consequence, we define a new measure of proximity (similarity) between rhythmic patterns (measures used in Biology are not appropriate in the context of Music). The question of how to define similarity measures for rhythms has already been considered [21, 22, 24, 5]. Among the many existing similarity measures (Euclidean interval vector distance, interval-difference vector distance, swap distance, etc.), the most satisfactory one in these studies has been the directed-swap distance, first introduced in [5]. For further information on measures of rhythmic similarity the reader is referred to [24].

The directed-swap distance is a generalization of the swap distance to handle comparison of rhythms that do not have the same number of onsets. Let P and Q be two rhythms such that P has more onsets than Q. Positions of the rhythm that contain an onset will be referred to as occupied positions. Thus, the directed-swap distance is the minimum number of swaps required to convert P to Q according to the following constraints: (1) Each onset in P must move to some occupied position of Q; (2) All occupied positions of Q must receive at least one onset from P; (3) No onset may travel across the boundary between the first and the last position in the rhythm.

For example, the directed-swap distance between Player-1 and Player-2 in variation V_1 is 4, since we have to perform 4 swaps in Player-1 at positions $3, 6, 8$ and 11 to convert Player-1 to Player-2; see Figure 1. In our case, since all rhythms have the same number of onsets, computing the directed-swap distance is easy because it reduces to the computation of a sum of a linear number of terms [5, 24].

As mentioned before, one of the reasons for comparing the pattern of *Clapping Music* to the *Yoruba* clave is that this clave is the rhythm closest to it. The distance between them was measured with the directed-swap distance, and the set of claves used in the comparison was taken mainly from well-established African musical traditions. To obtain precise details about these claves consult [22] and the references therein.

The distance matrix corresponding to the directed-swap distance is shown in Figure 3. Box notation is used for the variations of *Clapping Music*. For each rhythm the bottom of each column indicates the sum of the swap distances to all other rhythms. Surprisingly, the sums take on only two values, 48 and 74, where V_0, V_3, V_6 and V_9 are the variations that obtain the highest score. Also noteworthy is the diagonal below the zeros in Figure 3, that is, $\{4,4,4,8,4,4,8,4,8,4,4\}$. This diagonal gives the directed-swap distances between consecutive variations. It takes on only two values, 4 and 8, but these values differ considerably, and change 6 times over a total of 12 variations.

Variations	V_0	V_1	V_2	V_3	V_4	V_5	V_6	V_7	V_8	V_9	V_{10}	V_{11}
V_0=xxx.xx.x.xx.	0											
V_1=xx.xx.x.xx.x	4	0										
V_2=x.xx.x.xx.xx	8	4	0									
V_3=.xx.x.xx.xxx	12	8	4	0								
V_4=xx.x.xx.xxx.	4	2	4	8	0							
V_5=x.x.xx.xxx.x	8	4	2	4	4	0						
V_6=.x.xx.xxx.xx	12	8	4	2	8	4	0					
V_7=x.xx.xxx.xx.	4	4	4	8	2	4	8	0				
V_8=.xx.xxx.xx.x	8	4	4	4	4	2	4	4	0			
V_9=xx.xxx.xx.x.	2	4	8	12	4	8	12	4	8	0		
V_{10}=x.xxx.xx.x.x	4	2	4	8	4	4	8	2	4	4	0	
V_{11}=.xxx.xx.x.xx	8	4	2	4	4	4	4	4	2	8	4	0
$\sum$	74	48	48	74	48	48	74	48	48	74	48	48

Figure 3: The directed-swap distance matrix of the *Clapping Music* pattern.

In Figure 4 the phylogenetic graph associated with the above matrix is depicted. The graph returns the degree of fit with the data, listed as a percentage. If the percentage reaches 100%, then the minimum distances between nodes in the graph correspond exactly to the entries in the matrix. The algorithm initially tries to impose a tree structure on the distance matrix. If this is not possible, it introduces extra nodes in order to maintain a high fit value. See [11] for further details on the construction and properties of these graphs. Black dots correspond to the variations, while the rest of the nodes are without dots; the central node is labeled as A.

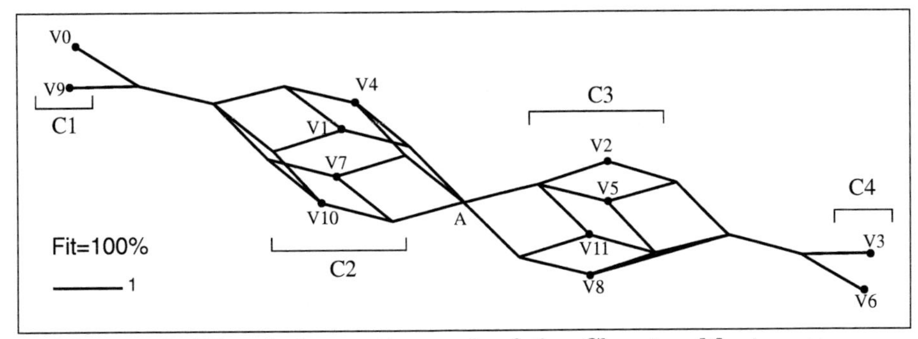

Figure 4: The phylogenetic graph of the *Clapping Music* pattern.

This phylogenetic graph provides valuable information about the structure of *Clapping Music*. For the moment, we consider the central node A. Such a node corresponds to an "ancestral" rhythm, and is also the center of the graph (i.e., it is the vertex that minimizes the maximum distance to any other vertex in the graph). Therefore, it seems that this central node plays a key role in the piece. There is as yet no known algorithm to compute the "ancestral" nodes for rhythms in phylogenetic graphs constructed with our distance measure. However, in this case, given the small number of rather short rhythms involved, the "ancestral" rhythm can be reconstructed by hand without much difficulty. It turns out to be the rhythm in Figure 5:

Figure 5: The "ancestral" rhythm of *Clapping Music.*

This fundamental rhythmic pattern is none other than a group of trochees. A trochee is a rhythmic grouping consisting of a long note followed by a short note. This "ancestral" rhythm has a strong metric time-keeping character. The trochee, expressed in box notation as [x . x], is a common Afro-Cuban drum pattern, also found in disparate areas of the globe. For example, it is the conga rhythm of the (6/8)-time *Swing Tumbao* [14]. It is common in Latin American music, as for example in the Chilean *Cueca* [15], and the Cuban *coros de clave* [6]. It is common in Arab music, as for example in the *Al Táer* rhythm of Nubia [10]. It is also a rhythmic pattern of the Drum Dance of the *Slavey* Indians of Northern Canada [2]. Furthermore, the entire pattern of 8 onsets in a time span of 12 pulses shown in Figure 5 is also the *Euclidean* rhythm $E(8,12)$ which distributes the onsets as evenly as possible [25].

The phylogenetic graph has four distinguishable clusters $C1, C2, C3$ and $C4$, that can be easily seen in Figure 4. When *Clapping Music* is performed these clusters appear in the order given in the following table in Figure 6:

Clusters	C1	C2	C3	C4
	V_0	V_1	V_2	V_3
		V_4	V_5	V_6
		V_7	V_8	
	V_9	V_{10}	V_{11}	
	V_{12}			

Figure 6: Clustering in *Clapping Music.*

From this sequence of clusters we may observe the evolution of the variations through time. There is a first section formed by variations V_0 to V_3; in it, each variation moves further away from V_0. In a second section, which goes from V_4 to V_6, variations are still kept away from V_0. In the third section, variations V_7 and V_8 remain around the center of the graph, and represent a turning-point after which the subsequent variations move towards V_0. The variations in section four, consisting of V_9, V_{10} and V_{11}, tend towards V_0. Lastly, *Clapping Music* closes by coming back to the main pattern ($V_0 = V_{12}$) played in unison. This evolution may be detected, although less visually, on the diagonal below the zeros in the directed-swap distance.

Let us now compare the *Clapping Music* pattern to the *Yoruba* clave. To start with, we present the swap distance matrix for the *Yoruba* bell pattern in Figure 7, and its corresponding phylogenetic graph in Figure 8. By looking at the bottom row of the matrix we see that the sums of the distances take many different values. The most similar rhythm is V_5 and the most different one is V_4. That is not surprising: V_5 is the so-called *Bembé*, a very popular ternary rhythm. Toussaint [22] already proved that it is one of the most similar of an important family of ternary claves. Here we note that the *Bembé* is the most similar rhythm in its own wheel (the wheel of a rhythm consists of those rhythms obtained by its rotations that begin on an onset). The diagonal below the zeros is $\{5,7,5,7,5,5,7,5,7,5,5\}$. The difference between two consecutive variations is smaller than in the case of *Clapping Music.* Changes are more frequent in the case of the *Yoruba* clave.

Variations	V_0	V_1	V_2	V_3	V_4	V_5	V_6	V_7	V_8	V_9	V_{10}	V_{11}
V_0=x.x.xx.x.xx.	0											
V_1=.x.xx.x.xx.x	5	0										
V_2=x.xx.x.xx.x.	2	7	0									
V_3=.xx.x.xx.x.x	3	2	5	0								
V_4=xx.x.xx.x.x.	4	9	2	7	0							
V_5=x.x.xx.x.x.x	1	4	3	2	5	0						
V_6=.x.xx.x.x.xx	6	1	8	3	10	5	0					
V_7=x.xx.x.x.xx.	1	6	1	4	3	2	7	0				
V_8=.xx.x.x.xx.x	4	1	6	1	8	3	2	5	0			
V_9 =xx.x.x.xx.x.	3	8	1	6	1	4	9	2	7	0		
V_{10}=x.x.x.xx.x.x	2	3	4	1	6	1	4	3	2	5	0	
V_{11}=.x.x.xx.x.xx	7	2	9	4	11	6	1	8	4	9	5	0
$\sum$	38	48	48	38	66	36	56	42	43	55	41	66

Figure 7: The directed-swap distance matrix of the *Yoruba* clave.

The graph is a chain with a rather disappointing fit of 89%, with no ancestral nodes. Take into account that if the fit is not 100%, reasoning on the graph does not accurately reflect reasoning on the distance matrix, and accordingly, neither on the rhythms. For example, on the graph the distance from V_0 to V_2 is 2.5, but in the matrix it is actually 2. The role of A could be played by variation V_5 in this phylogenetic graph. It is the center of the graph, and as before, variations alternately go from left to right and from right to left around V_5. Nevertheless, this does not seem to yield anything in particular about the structure of the *Yoruba* clave. Hence, there is no remarkable clustering analysis to be discussed. In addition, the graph does not exhibit special symmetries or regularities of musical significance either. In reality, when the same musical process as *Clapping Music* is carried out, a rather awkward result is obtained.

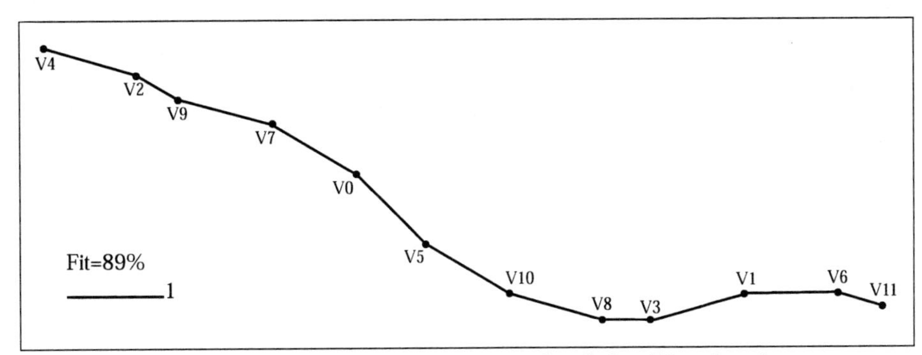

Figure 8: The phylogenetic graph of the *Yoruba clave.*

5 Syncopation Analysis of *Clapping Music*

The definition of syncopation includes a momentary contradiction of the prevailing meter [19]. According to this fact, the *WNBD* measure is based on the durations of notes and how they cross over the strong beats of the meter. This measure has a different approach than others, like Keith's measure [13], based on the structure of metrical levels, or Toussaint's off-beatness measure [23, 26], based on the underlying polyrhythmic structure of the meter. The *WNBD* measure has proven to be more flexible and precise than the others [7].

Now we introduce the *WNBD* measure, which will enable us to analyse the syncopation of *Clapping Music*. We assume that a note ends where the next note begins. Let e_i, e_{i+1} be two consecutive strong beats in the meter. Also, let x denote a note that starts after or on the strong beat e_i

but before the strong beat e_{i+1}. We define $T(x) = \min\{d(x, e_i), d(x, e_{i+1})\}$, where d denotes the distance between notes in terms of durations. Here the distance between two adjacent strong beats is taken as the unit, and therefore, the distance d is always a fraction (see Figure 9 (a)).

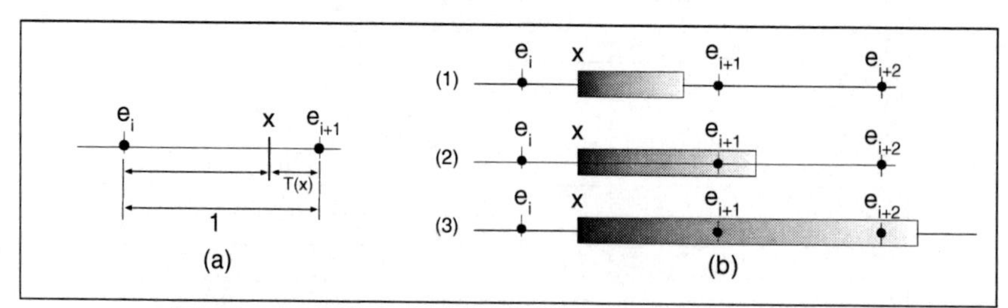

Figure 9: Definition of the *WNBD* measure.

The *WNBD* measure $D(x)$ of a note x is then defined according to the following cases: (1)$D(x) = 0$, if $x = e_i$; $D(x) = \frac{1}{T(x)}$, if note $x \neq e_i$ ends before or at e_{i+1}; (2) $D(x) = \frac{2}{T(x)}$, if note $x \neq e_i$ ends after e_{i+1} but before or at e_{i+2}; and (3) $D(x) = \frac{1}{T(x)}$, if note $x \neq e_i$ ends after e_{i+2}. See Figure 9 (b) for an illustration of this definition. Now, let n denote the number of notes of a rhythm. Then, the *WNBD* measure of a rhythm is the sum of all $D(x)$, for all notes x in the rhythm, divided by n.

Figure 10 plots the *WNBD* measure of the variations in *Clapping Music* with respect to a 12/8 meter. The measure produces only three different values, namely, $24/8, 21/8$ and $12/8$, but the graph is quite revealing about how syncopation works in *Clapping Music*. Variation V_0 by itself has a high value of syncopation. Two identical ascending-descending cycles, $V_1 - V_2 - V_3 - V_4$ and $V_4 - V_5 - V_6 - V_7$, follow after V_0. From V_7, we find a symmetric cycle with respect to the previous cycle, namely, $V_7 - V_8 - V_9 - V_{10}$. Finally, we discover an ascending path to V_{12} (which is a half of the previous cycle). If variation V_1 were moved after V_{12}, then the resulting graph would have a perfect symmetry about V_7. Therefore, a strong symmetry in musical form is evident in *Clapping Music*.

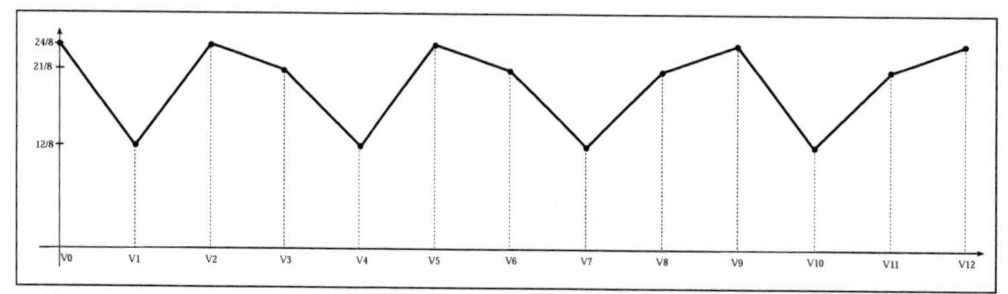

Figure 10: The graph of the syncopation measure of *Clapping Music*.

On the other hand, when looking at the graph of the *WNBD* measure for the *Yoruba* clave, depicted in Figure 11, several differences come out. As in *Clapping Music*, the measure only takes three values, namely, $21/7, 18/7$ and $15/7$. The range of the syncopation values is much smaller than in the case of *Clapping Music*, and consequently, so is its rhythmic variety. The smaller the range of the measure is, the less interesting the rhythms are from the syncopation standpoint. The graph of the *Yoruba* clave exhibits the same quasi-symmetry about V_7.

6 Concluding Remarks

Although phylogenetic graphs have been already used for analyzing families of rhythms, in this paper we use them for studying the musical structure of process music, in particular Reich's *Clapping Music*. The resulting graph allows us to explore a variety of musical properties of the piece, such

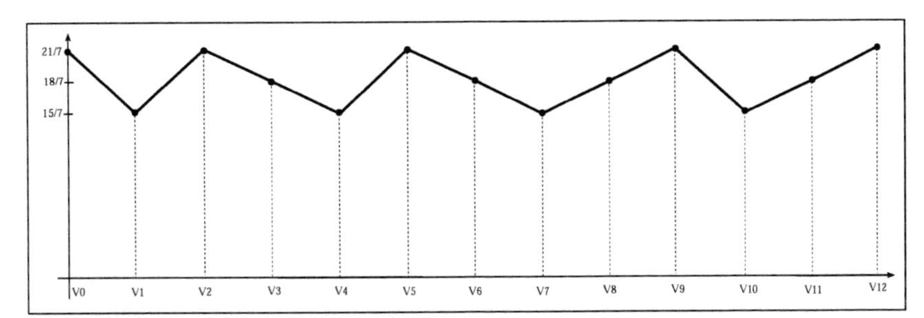

Figure 11: The graph of the syncopation measure of the *Yoruba* clave.

as the classification, evolution, and transformation of variations, or the structure of the musical form. We advocate the use of phylogenetic graphs as a useful tool for musical analysis in general, especially for musical styles as characteristic as Minimalism. Reich's other pieces in this style, such as *Music for Pieces of Wood*, may also be analyzed in this way.

We have compared *Clapping Music* to the *Yoruba* bell timeline since they are almost identical. One would expect that changing only one note could not make such a significant difference in musical terms. Quite to the contrary, as we have seen, the phylogenetic graph of *Clapping Music* has a richer structure than that of the *Yoruba* clave. This is a consequence of its inherent musical structure.

The *WNBD* measure produces interesting conclusions as well, mainly that the rhythmic variety, at least from a syncopation standpoint, can be measured in terms of the range and distribution of the syncopation values.

Previously, Haack [8] proved that *Clapping Music* is unique from a combinatorial point of view. In this paper we have added arguments of a geometrical nature to support this conclusion.

Finally, the properties exhibited by the phylogenetic graphs and the *WNBD* measure pose several open problems. For example, one may ask which rhythms yield nice graphs, that is, graphs with good properties (symmetry, clustering, 100% fit, etc.) The *WNBD* measure gives poor results when it is used to measure the interlocking melodies (for that we just use the *Clapping Music* pattern as the meter, and recompute the measure). Therefore, finding a function that measures the overall complexity of two rhythms is an interesting open problem. Also open is the question of which other rhythmic patterns would work just as well as the pattern used in *Clapping Music.*

References

[1] Amira, J. and Label, S.C. *The Music Of Santería – Traditional Rhythms Of The Batá Drums: The Oru Del Igbodu.* White Cliffs Media , 1999.

[2] Asch, M. I. Social context and the musical analysis of Slavey drum dance songs. *Ethnomusicology,* 19:2:245–257, 1975.

[3] Bettermann H.; Amponsah, D.; Cysarz, D.; and Van Leeuwen, P. Musical rhythms in heart period dynamics: a cross-cultural and interdisciplinary approach to cardiac rhythms. *Proceedings of the American Physiological Society,* pp. H1762–H1770, 1999.

[4] Cohn, R. Transpositional Combination of Beat-Class Sets in Steve Reich's Phase-Shifting Music *Perspectives of New Music,* 30:2:146-176, 1992.

[5] Díaz-Báñez, J. M.; Farigu, G.; Gómez, F.; Rappaport, D.; Toussaint, G. T. El Compás Flamenco: A Phylogenetic Analysis. In *Proceedings of BRIDGES: Mathematical Connections in Art, Music, and Science,* Winfield, Kansas, 61-70, July, 2004.

[6] Eli Rodríguez, V. and alt. *Instrumentos de la música folclórico-popular de Cuba.* Centro de Investigación y Desarrollo de la Música Cubana, 1997.

[7] Gómez, F.; Melvin, A.; Rappaport, D.; Toussaint, G. T. Mathematical Measures of Syncopation In *Proceedings of BRIDGES: Mathematical Connections in Art, Music, and Science*, Banff, Canada, pp. 73-84, August, 2005.

[8] Haack, J. K. Clapping Music – a Combinatorial Problem. *The College Mathematical Journal*, 22:224-227, May, 1991.

[9] Haack, J. K.; "Mathematics of Steve Reich's *Clapping Music*. In *Proceedings of BRIDGES: Mathematical Connections in Art, Music and Science*, pp. 87-92, Winfield, Kansas, 1998.

[10] Hagoel, K. *The Art of Middle Eastern Rhythm.*, OR-TAV Music Publications, Kfar Sava, Israel, 2003.

[11] Huson, D. H. SplitsTree: Analyzing and visualizing evolutionary data. *Bioinformatics*, 14:68-73, 1998.

[12] Stone, R.M. (editor) *The Garland Encyclopedia of World Music.* Vol. 1, Garland Publishing, 1998.

[13] Keith, M. *From Polychords to Pólya: Adventures in Music Combinatorics.* Vinculum Press, Princeton, 1991.

[14] Klőwer, T. *The Joy of Drumming: Drums and Percussion Instruments from Around the World.* Binkey Kok Publications, Diever, Holland, 1997.

[15] van der Lee, P. H. Zarabanda: esquemas rítmicos de acompañamiento en 6/8. *Latin American Music Review*, 16:2:199-220, 1995.

[16] Ortiz, F. *La Clave.* Editorial Letras Cubanas. La Habana, Cuba, 1995.

[17] Ortiz, F. *Los Instrumentos de la Música Cubana.* Dirección de Cultura del Ministerio de Educación, La Habana, Cuba, 1952-55. Republished by Editorial Música Mundana, Madrid, 1998.

[18] Pressing, J. Cognitive isomorphisms between pitch and rhythm in world musics: West Africa, the Balkans and Western tonality. *Studies in Music*, vol. 17, 38–61, 1983.

[19] Randel, D. (editor). *The New Grove Dictionary of Music and Musicians.* Akal, 1986.

[20] Reich, S. *Writings about Music.* The Press of the Nova Scotia College of Art and Design, New York, 1974.

[21] Toussaint, G. T. A Mathematical Analysis of African, Brazilian, and Cuban *Clave* Rhythms. In *Proceedings of BRIDGES: Mathematical Connections in Art, Music and Science*, pp. 157-168, Towson University, Towson, MD, 2002.

[22] Toussaint, G. T. Classification and Phylogenetic Analysis of African Ternary Rhythm Timelines. In *Proceedings of BRIDGES: Mathematical Connections in Art, Music and Science*, pp. 25-36, Universidad de Granada, Granada, 2003.

[23] Toussaint, G. T. A Mathematical Measure of Preference in African Rhythm. In *Abstracts of Papers Presented to the American Mathematical Society*, volumen 25, pp. 248, Phoenix, Arizona, January, 2004. American Mathematical Society.

[24] Toussaint, G.T. A Comparison of Rhythmic Similarity Measures. In *Proceedings of the Fifth International Conference on Music Information Retrieval*, pages 10-14, Barcelona, Spain, October, 2004.

[25] Toussaint, G. T. The Euclidean Algorithm Generates Traditional Musical Rhythms. In *Proceedings of BRIDGES: Mathematical Connections in Art, Music and Science*, pp. 47-56, Banff, Canada, July, 2005.

[26] Toussaint, G. T. Mathematical Features for Recognizing Preference in Sub-Saharan African Traditional Rhythm Timelines. In *Proceedings of the 3rd International Conference on Advances in Pattern Recognition*, pp. 18-27, University of Bath, United Kingdom, August, 2005.

Illuminating Chaos – Art on Average

Mike Field
Department of Mathematics
University of Houston
Houston
TX 772004-3008
E-mail: MikeField@gmail.com

Abstract

At first sight, chaos and structure seem antithetical. Yet there is an intimate connection between randomness and structure. In this talk we explain some of the ideas we have used for creative artistic design that depend on results from the study of chaotic dynamics. Our intention is to avoid the Platonistic perspective that the role of the mathematician is to dig out and discover the beauty hidden within the mathematics. Our view will be more that of an engineer. How can we use mathematics in a creative way to produce aesthetically pleasing art? (as opposed to 'pretty patterns'.) How can we achieve the effects we want to emphasize in a particular design? We illustrate the talk with examples of (symmetric) designs, many of which have appeared in art exhibitions in the Americas and Europe. As well we give some visual demonstrations and explanations of chaos and, if there is time, indicate some practical applications of these ideas to teaching art students (some mathematics) and mathematics teachers (some art).

Figure 1: A chaotic frieze pattern of type **pm′m′2**

1 Background notes on the presentation

Associations between art and mathematics are most often tied to geometry and symmetry. This is seen in the way geometry, especially symmetry, is used in design, and in the intrinsic beauty of many geometric objects in mathematics ranging from Platonic solids, through minimal surfaces (soap bubbles), representations of singular algebraic surfaces and more recently a whole new range of visually attractive mathematical objects, such as Julia sets and fractals. However, there is a feeling among many mathematicians that their subject is one of discovery rather than creation. This viewpoint seems antithetical to that of an artist or painter and perhaps this difference in view explains some of the misunderstandings. G H Hardy[1] is quite direct in his views

> "A mathematician, like a painter or a poet, is a maker of patterns. If his patterns are more permanent than theirs, it is because they are made with *ideas*. A painter makes patterns with shapes and colours, a poet with words. A painting may embody an idea but the idea is usually commonplace and unimportant..."

[1] *A Mathematician's Apology*, Cambridge University Press, Cambridge Paperbacks, 1993.

Much has changed since Hardy wrote these words in 1940 – indeed, one interest of the essay is the many statements it contains that are just flat out wrong. However, if there is to be – at some level – a synthesis (or even a dialog) between mathematics and art then I believe it important to develop new perspectives on the way mathematics can contribute to art (and conversely).

Mathematics is not just an extension of the work of the Greeks on geometry and number theory. The last two hundred years have seen the rise of statistics and probability theory in mathematics and science. It is quite remarkable how many of the basic laws of physics are framed in statistical terms. The same is true of biology, through the laws of genetics, and evolution, through the effects of random mutations. Although it sometimes seems intensely paradoxical (not to mention disturbing), the enormous variety of shape and form we see around us appears to be a consequence of statistical irregularities rather than in spite of them. Without the statistics we would probably be living in Laplace's deterministic and inevitably dull clockwork universe[2]. Statistics and probability are often thought of as the ugly ducklings of mathematics. Look closer and you will see the swans.

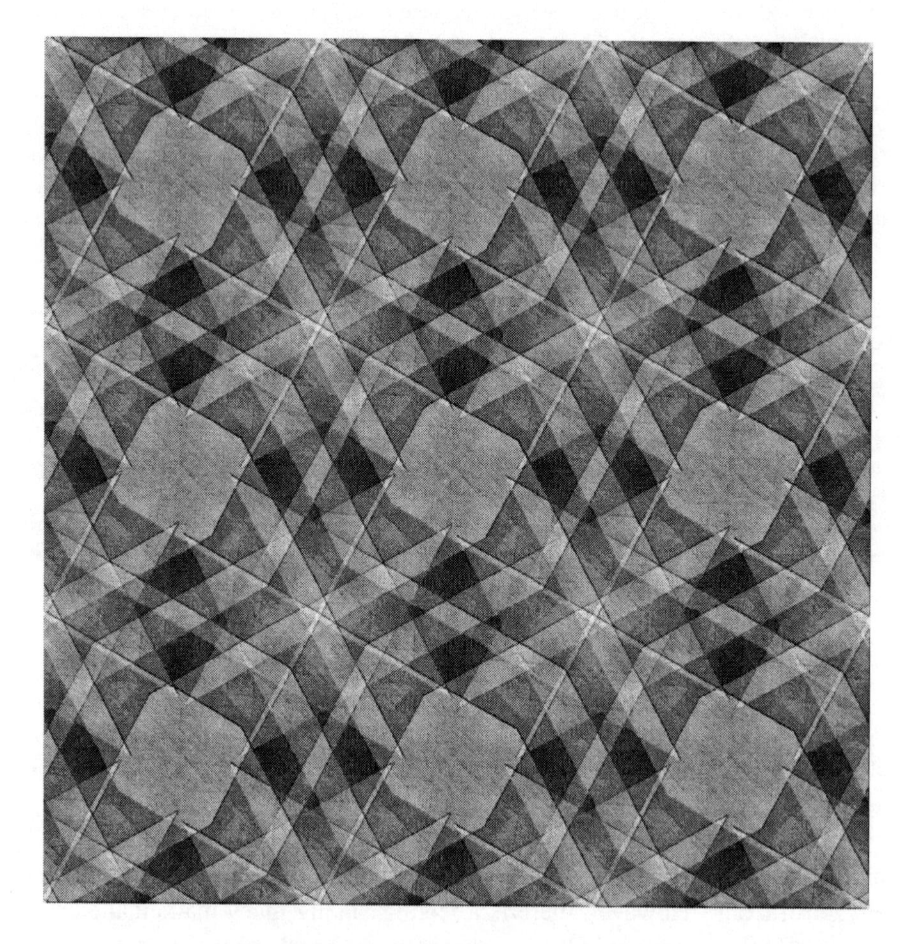

Figure 2: A chaotic pattern of type **p4**

[2]Of course, this is grossly unfair to Laplace, the founder of probability theory. Although responsible for the famous quote on determinism, he also observed that "It is remarkable that a science which began with the consideration of games of chance should have become the most important object of human knowledge" (Théorie Analytique des Probabilitiés (1812))

Bridging the gap - a search for a braid language

Jacqui Carey
Summercourt
Ridgeway
Ottery St Mary
Devon, EX11 1DT
England.
e-mail: jacqui@careycompany.com

Abstract

As a braidmaker, my work encompasses both maths and art. However, language can be a bridge, or a barrier, between different disciplines and without a 'mathematical language' it has been difficult for me to access work done in this field. This paper describes my search for a visual language that provides me with a practical and theoretical way of comparing and analysing braid structure. From this comes the means of discovering all possible braid structures for a set of given constraints. Although braids have been made for millennia, they tend to be limited to certain types of structure. These have usually evolved from the characteristics found within the methods of production. Approaching the subject from a mathematical viewpoint, enables me to find new structures from the wealth of possibilities that have yet to be explored.

I label myself as neither mathematician nor artist, preferring to place myself on the periphery of many subjects. The advantages of cross-discipline communication are becoming better understood and appreciated, and braiding is an ideal medium for traversing into different realms. It is an ancient technique that can be found in many forms, all over the world. Its chameleon-like ability means it can be found in seemingly diverse spheres: from fashion to warfare, surgery to mechanics, and sports to cuisine. It can, quite literally, include the kitchen sink (with its braided metal pipes). It can be incorporated into the study of disparate disciplines: from maths to art, history to religion, and anthropology to physics. However, in order to bring these worlds together, common ground and more importantly, a common language must be found. Even a simple concept cannot be understood if it is explained in a foreign language. So in order to bridge the gap between disciplines, there is a need to find new ways of communicating. Not only will this give access to new territory, but also the actual search for a new language can widen and enrich our understanding of areas of specialization.

Finding a format.

It was the issue of language that led to the search for a visual way of representing braid structure. A quest to find a simple method that would enable braids to be analysed and compared, and new structures discovered. The problem was finding a format that was universal for all braids. Carey [1] established a grid system that provides a means of working out all of the pattern possibilities on one braid structure . Now the challenge was to find a way to calculate all possible braid structures.

Flat braids are easy to represent, although problems in uniformity can be found, even with some basic, well-known braids (see figure 1). Three-dimensional structures presented more complex problems. Ashley [2] uses a simple cross-section to give a sense of the braid structure (see figure 2). Speiser [3] takes this a step further with her track plans. Although these are intended as a visual representation of technique (the method of creating) rather than structure (the finished result), they do offer a certain sense of the type of braid being produced. However, once again, uniformity is a problem, because track plans cannot be made for all braids.

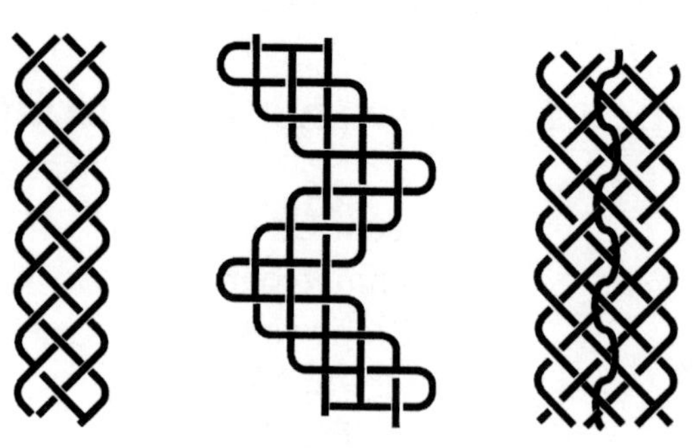

Figure 1: *Three different flat braids: an 'oblique interlacing' (left), a 'zigzag interlacing' (centre), and a 'triaxle interlacing' (right).*

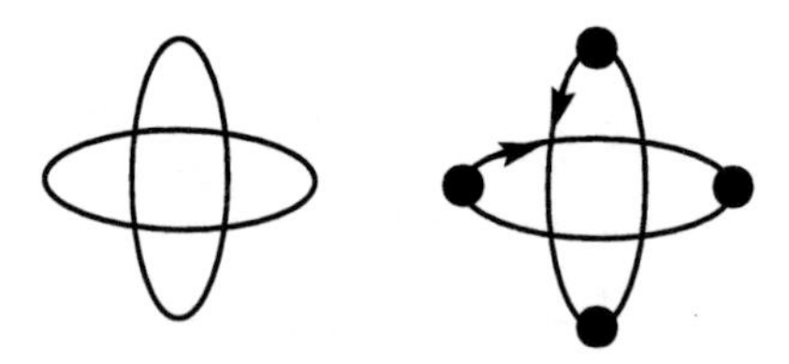

Figure 2: *A simple 4-element tubular braid represented in the style of Ashley (left) and Speiser (right).*

There are also inherent problems in attempting to look at braid possibilities through technique. There are many different ways of making braids, such as plaiting, loop-manipulation, and stand and bobbin techniques, such as kumihimo. These various methods can be used to create identical structures. It is even possible to make the same braid structure using a variation within the same technique. This has caused a certain amount of confusion as can be seen in Owen [4]. Here, two routes are used to create two braids. One braid is described as square, whilst the other is said to be round, when in fact, they are both identical. Basically, there are many ways to arrive at one particular structure. Each method has idiosyncratic features that may effect the visual outcome but the underlying structure remains the same. This is best illustrated with a comparison to plain weave - the 'under one, over one' structure that is common to many fabrics. It can be warp-faced, weft-faced, or even-weave, but always maintains the 'under one, over one' structure (see figure 3).

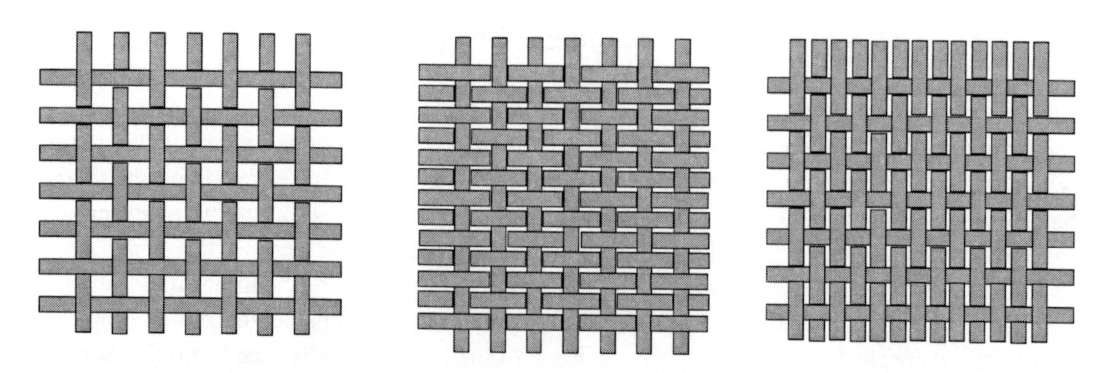

Figure 3: *Even weave, weft-faced and warp-faced are all plain weave structures.*

In fact, it was by making a comparison to weaving, that a breakthrough was made. The drafting of weave patterns can be done on squared paper, and there is a magical moment when the drafting results not in a flat single cloth, but in a 'double cloth' (see figure 4). These simple 3-dimensional structures consists of two interconnected layers of weaving. The drafting 'flattens' the structure with each layer represented by alternate rows of interlacing. The same idea could be applied to braids. By following the basic rules of drawing flat braids on squared paper, the 3-dimensional aspects could be incorporated by elongating the braid structure so that the different layers became integrated in adjacent rows. The final problems were resolved with the addition of a 'no-intersection', which made the system work for all braid structures. Although it is difficult to visualize the more complex braids, the fact that they can be reduced to a simple three-digit code provides a useful tool for theoretic discovery.

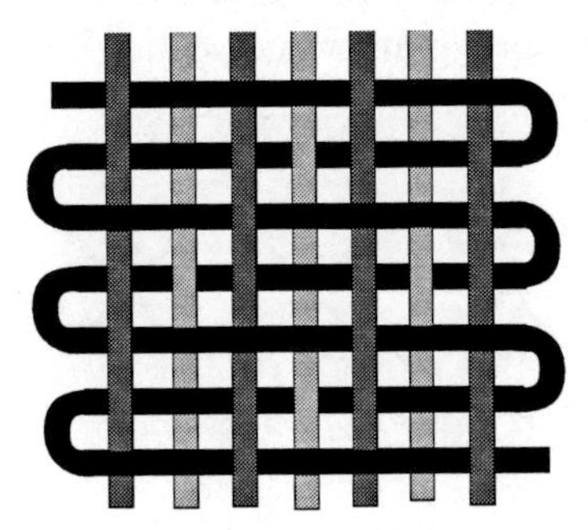

Figure 4: *An example of 'double cloth'.*
This particular version will form a tubular plain weave structure.

The basic rules

The diagrams are generated on squared paper worked at a 45-degree angle. But, before we begin, it is worth clarifying some of the terms used in the explanation. In this context, element refers to one of the working units from which the braid is made (for example, the common hair plait is a 3-element braid). The diagram is the visual representation of the braid, whilst the grid and lines refer to the squared paper on which it is drawn. The point at which the lines on the grid meet will be called an intersection.

The braid diagram must follow the lines on the squared paper, with one element travelling along each line. The number of braid elements determines the width of the diagram. So, if the number of elements is known, boundaries can be drawn vertically on the grid. For example, a 6-element braid will use six lines and have the following boundaries.

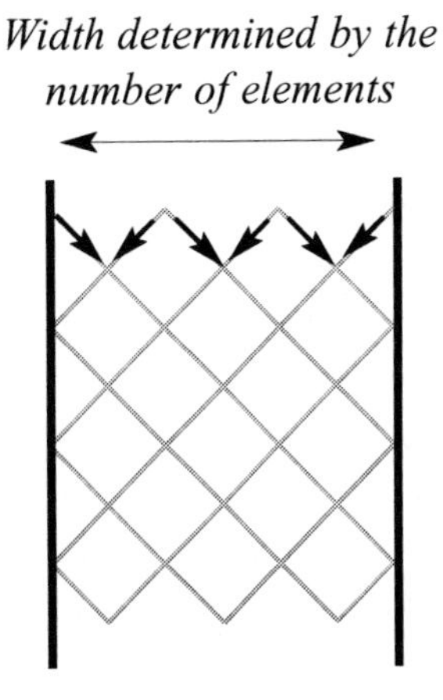

Figure 5: *Diagram showing the boundaries for a 6-element braid.*

At each point where the grid lines meet, the elements will intersect, with one element arriving from the top left-hand side, and the other from the top right-hand side. After they intersect one element will travel down to the bottom left-hand side, and the other to the bottom right-hand side.

Figure 6: *Elements arriving and departing from an intersection.*

There are three types of intersection (referred to as S, Z and O). Either the top left-hand side element goes over the top right-hand side element (making an S intersection) or vice versa (making a Z intersection). The third option (O) is for neither element to cross, they simply meet, turn and travel downward.

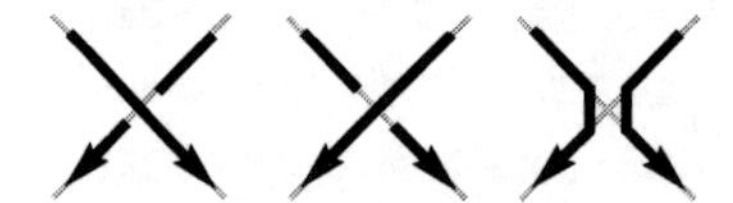

Figure 7: *Three types of intersection: S, Z, and O.*

The braid elements work their way obliquely down the grid, following the lines and making intersection as they go. Whenever an element reaches a boundary, it turns 90 degrees back into the diagram, as if making half an O intersection.

The width of the diagram determines the number of elements in the braid, whilst the length/depth of the diagram shows the intersections required to make the braid structure. The rows of intersections have to work in pairs, referred to as a set of intersections.The braid is formed when these sets are repeated. Note that if a turn at the boundary is considered half an intersection, then the number of intersections in a set will always equal the number of elements.

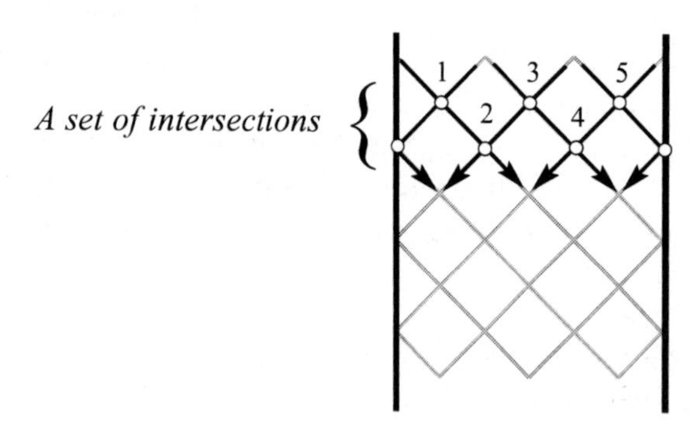

Figure 8: *For a 6-element braid, there are six intersections in the set - five complete ones, and two halves at the boundaries.*

Using these rules, it is now possible to diagrammatically illustrate all known braid structures. They can also be written in a verbal language by translating the information into a simple code based on the three letters S, Z and O. Here, each letter describes the intersections in a set, reading from left to right. Note that as the half intersections at the boundaries are not included, the set is written as a group of letters that is one less than the number of elements in the braid. The diagram, or its code, can now be used to discover all of the outcomes for given constraints. For example, there are nine possible outcomes for three elements, working repeats after one set of intersections (see figure 9).

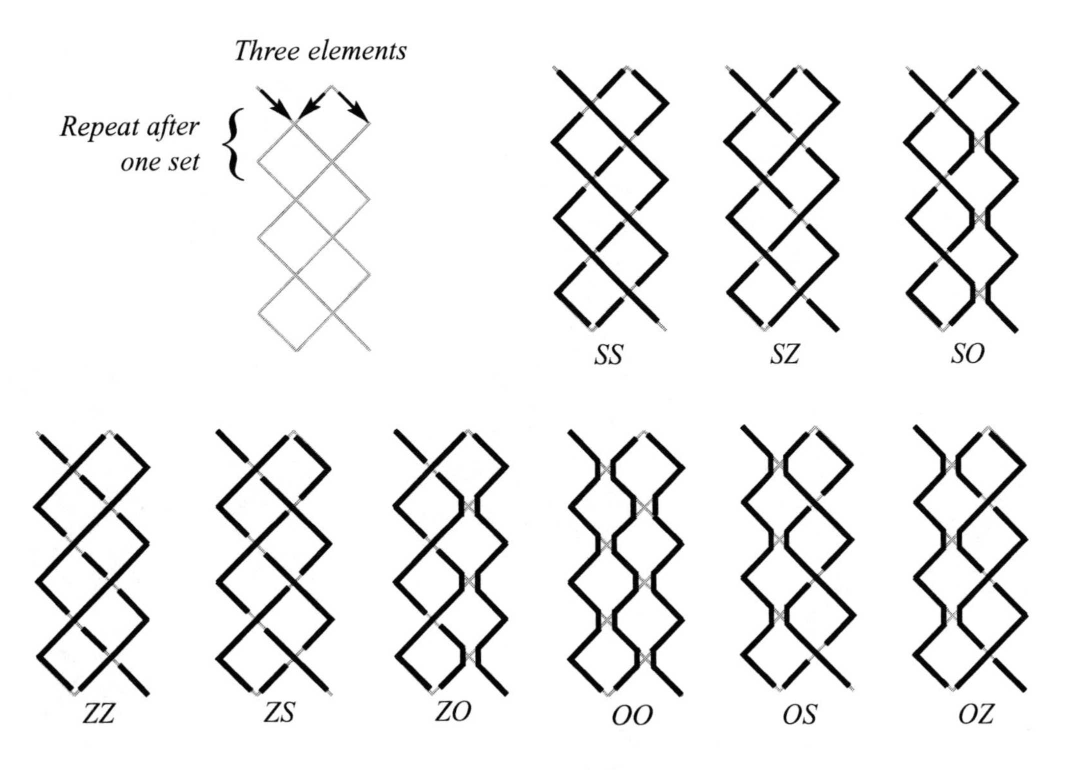

Figure 9: *The constraints and the nine outcomes.*

However, not all of the outcomes shown in figure 9 actually produce a braid - and here we enter the rather contentious issue of what actual constitutes a braid. None the less, it has ensured that all options have been considered, and that no solution is left undiscovered. Of course, the process can be simplified through elimination. For example, all pairs of diagrams that have rotational symmetry of 180 degrees are, by their nature, the same braid viewed upside down. So the examples in figure 9 can now be narrowed down to six solutions. For those who like a full analysis, these are the precise results:

SS = a 3-ply, S-twist cord
SZ = ZS = the common 3-element braid.
SO = OO = a 2-ply, S-twist cord and a single element.
ZZ = a 3-ply, Z-twist cord.
ZO = OZ = a 2-ply, Z-twist cord and a single element.
OO = three single straight elements.

Of course, expanding the constraints could increase these results: either by increasing the width of the diagram (for example to four elements), or by increasing the depth of the repeat (for example 3-elements working repeats after two sets of intersections).

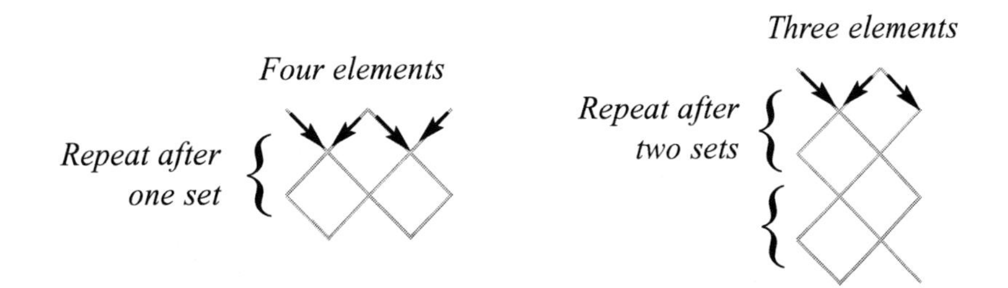

Figure 10: *Two different sets of constraints.*

As the complexity increases, so does the need for refining the search for 'same solutions'. This is inevitable as braid sequences can be repeated from any point, and 3-dimensional structures can be 'flattened' from any face. Unfortunately, the 'flattening' can make it difficult to visualize some of the structures, especially the more complex ones. However, the fact that all braids can be 'translated' into the same format provides a useful tool for comparing and analysing braids. Furthermore, patterns of behaviour within the diagram, or code, can be studied and compared with actual braid samples. This provides a fascinating realm for research and material for further development.

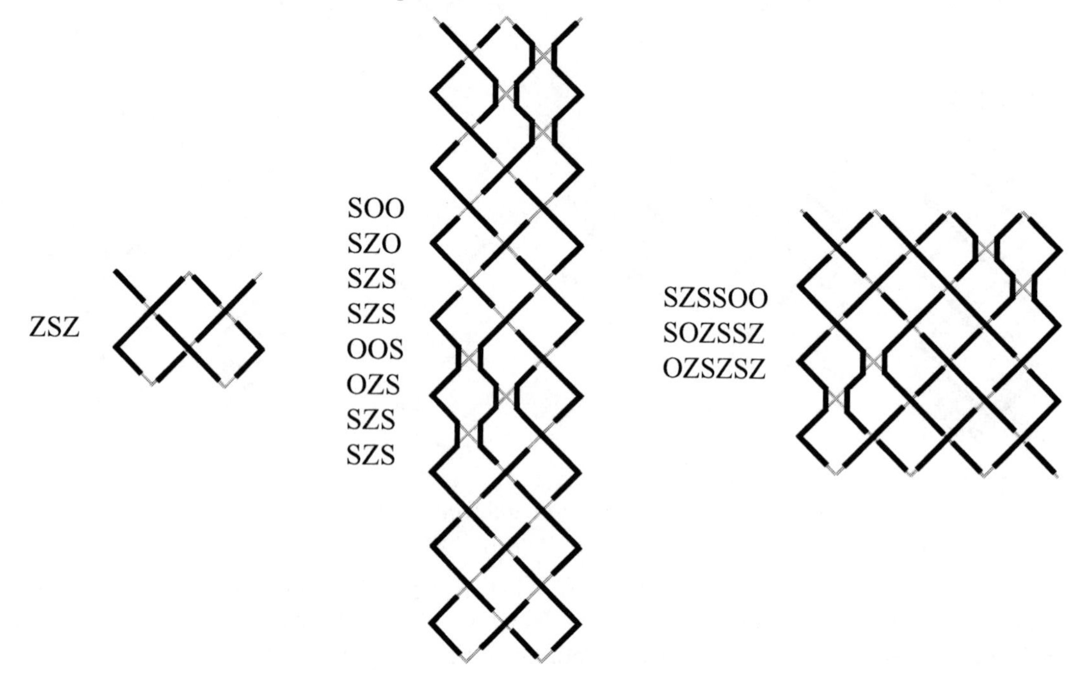

Figure 11: *The three braids, shown in figures 1, can now be represented in the same format.*

Figure 12: *3-dimensional braids, such as the one shown in figure 2, can also be represented in the same manner.*

Beyond the basics

Ultimately, this 'language' is just a tool for understanding and exploring the underlying construction of braid structures. Braid design takes on a whole new meaning when areas such as scale, material, colour and tension are explored. All of these need to be considered and understood if they are to be manipulated with control. However, the real mystery and challenge is to assimilate this with other disciplines, creating a sensual and intellectual union - to find a beauty that combines both underlying and outward aesthetics, quite literally, interlacing all aspects together in harmony.

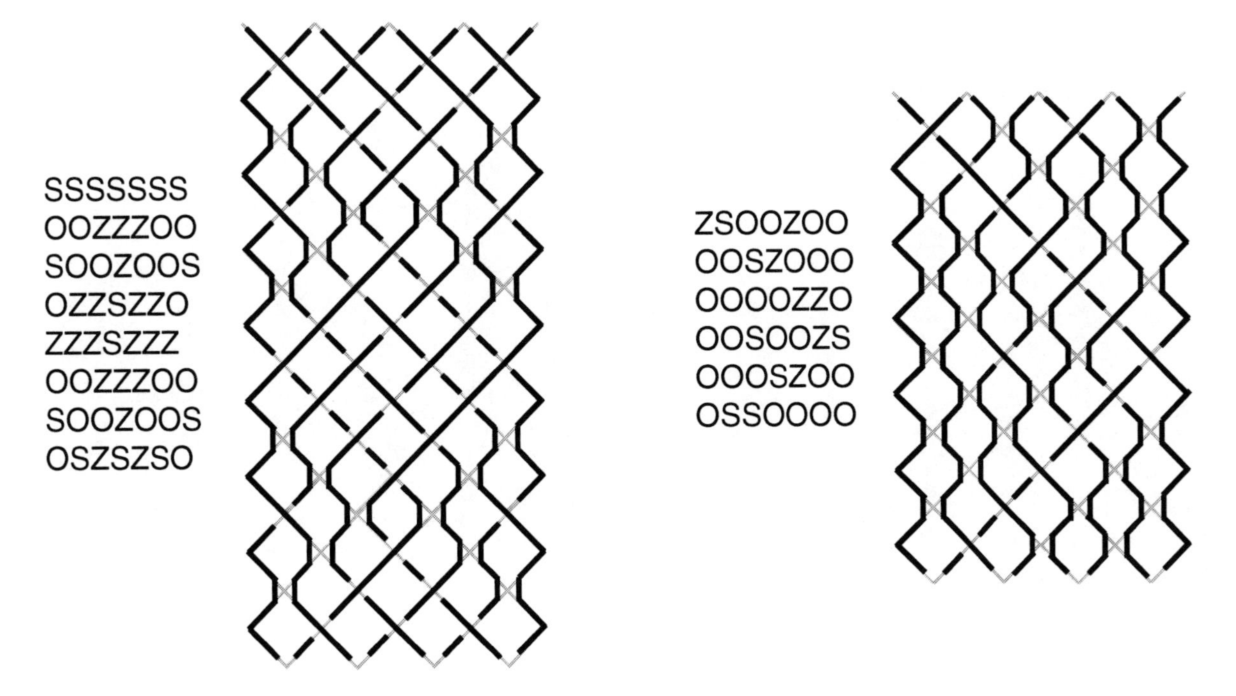

Figure 13: *The diagram and code for more complex braids. Left: a known braid that has been difficult to represent (top photograph). Right: a 'newly discovered' structure (bottom photograph).*

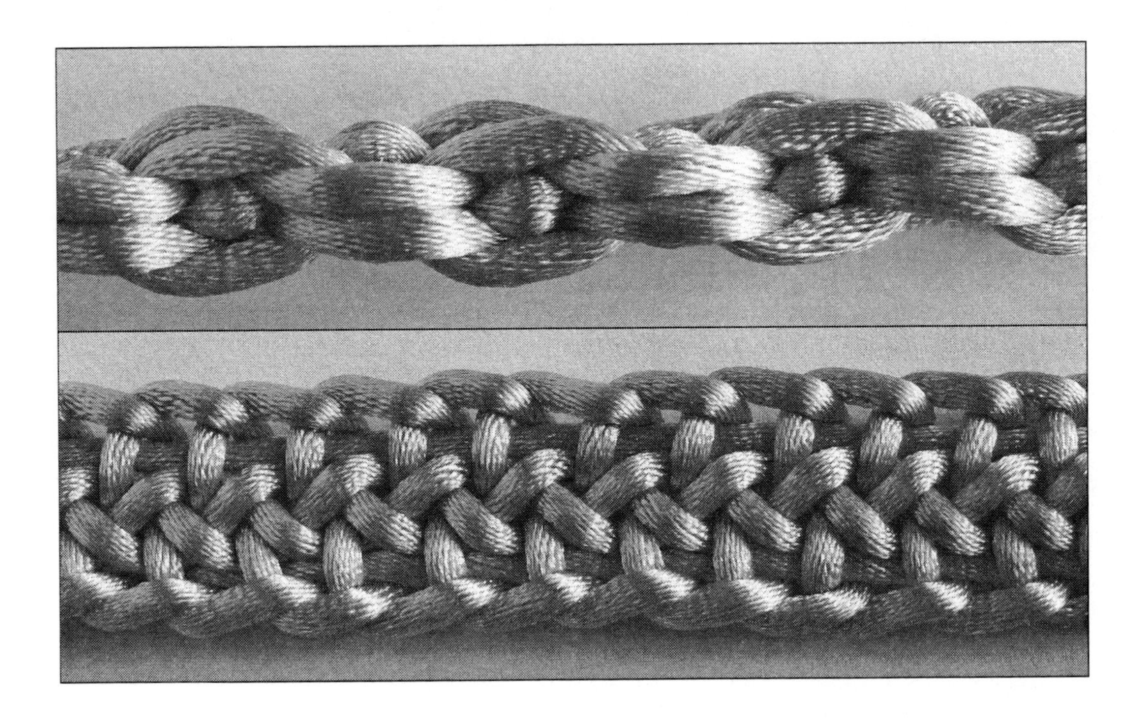

Figure 14: *A range of different patterns created on the same braid structure.These can be designed and analysed using the 'grid' system (Carey 1994).*

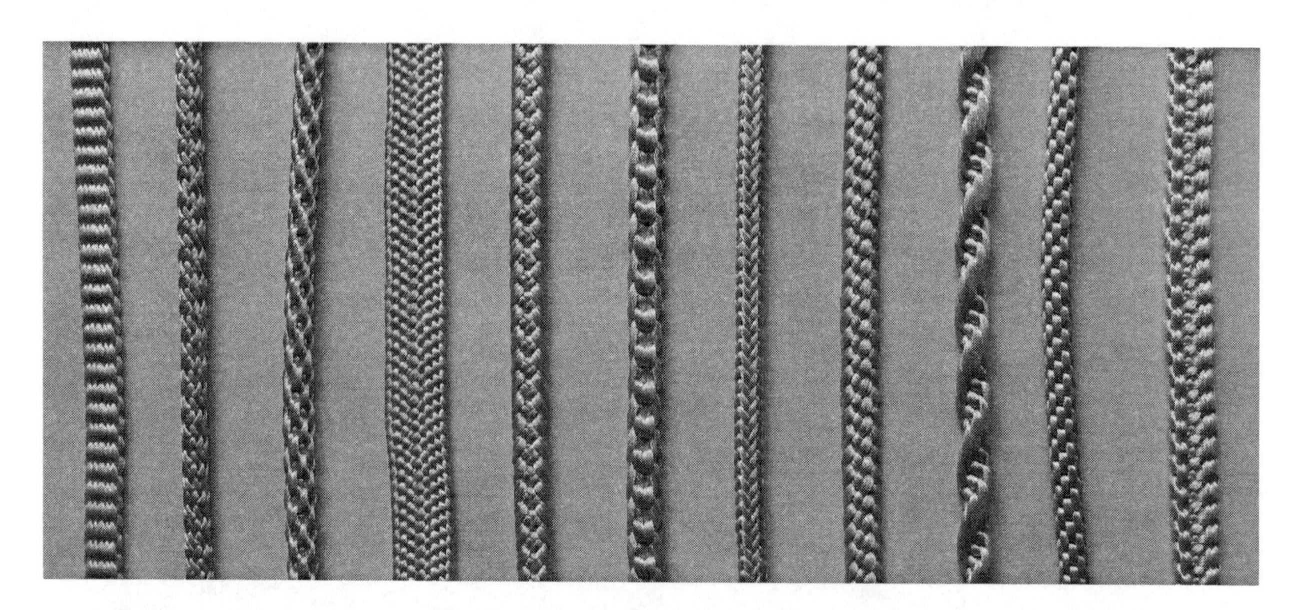

Figure 14: *A range of different braid structures that can now be designed and analysed using the code and diagrams.*

Bibliography

[1] J. Carey, *Creative Kumihimo* Carey Company, 1994.
[2] C. Ashley, *The Ashley Book of Knots* Faber and Faber, 1993.
[3] N. Speiser, *The Manual of Braiding* Self published, 1983.
[4] R. Owen, *The Big Book of Sling and Rope Braids* Cassell, pp53, 60. 1995.

Magic Stars and Their Components

Sergei Zagny
Svyatoozerskaya dom 11, kv. 49
111625 Moscow
Russia
E-mail: sergei-zagny@mtu-net.ru

Abstract

Magic Stars is the title of a musical work based on mathematical objects of the same name. Six six-pointed magic stars provide six two-dimensional 12-tone structures, which constitute the building blocks of the work. These structures are subjected to analysis, transformations, disintegration and recombination of their components. The parts of the score, which is richly visual, look like tables rather than traditional musical pieces. While pitches ('space') are fixed, time is not, giving ultimate freedom to the performer, who may find out his own time and thereby meet quite mathematical and objective things in a very personal and intimate way.

1. Magic Stars as Mathematical Objects

The six-pointed magic star is a figure having the following appearance:

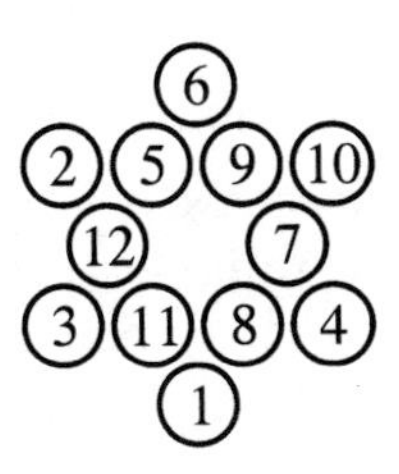

Figure 1: *An example of a Magic Star.*

Twelve non-recurrent numbers, from 1 to 12, are located on the vertices of the **main triangles** that form the star, and on the points of their intersection. The vertices of the main triangles (1, 2, 10 and 3, 6, 4 in the example) are called the **outer vertices**, and the points of intersection of the triangles (11, 12, 5, 9, 7 and 8 in the example) are called the **inner vertices**. The **side** of a star – which is also the side of one of the main triangles – is a segment, and four vertices are placed on it. The numbers are arranged so that each side can produce the same sum. For example, 1+11+12+2 = 3+12+5+6 = 2+5+9+10 = 6+9+7+4 = 10+7+8+1 = 4+8+11+3. The sum of the six outer vertices also has to be the same, for example, 1+3+2+6+10+4. There are a total of six different magic stars, disregarding variants produced by mirror images or rotation.

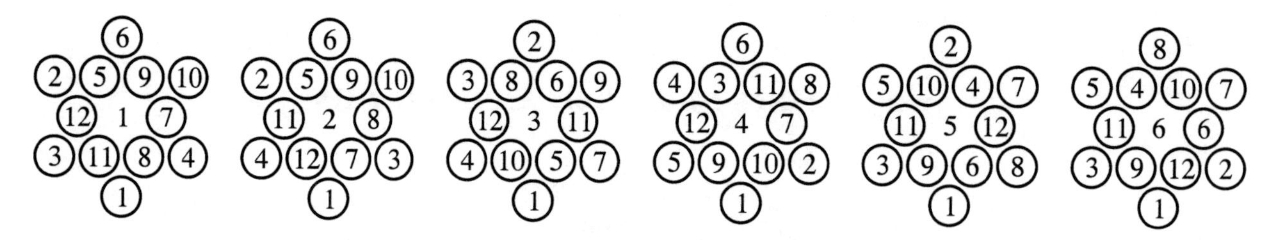

Figure 2: *All six Magic Stars in their original form.*

One can distinguish the following components in the stars: vertices (outer and inner), sides, complements, hexagons (outer and inner), triangles (outer and inner), enneagons (main triangles) and dodecagons, and also inner equilateral triangles.

There are twelve **vertices** in each star: six outer and six inner ones. Here the expression 'sum of vertices' means the sum of the numbers that lie on these vertices.

A **side** is four vertices placed on one segment. On the ends of the segment there are outer vertices and between them there are inner vertices. The sum of the vertices of any side of a six-pointed magic star is always 26.

Opposite (or **parallel**) **sides** of a star are sides that have no common vertices. Each side has exactly one opposite side. Parallel sides are formed by eight vertices, their sum being $26\times2=52$. A group formed by the remaining four vertices is called **complement**. The sum of the complement vertices is always 26.

A **hexagon** is six outer vertices (**outer hexagon**), or six inner vertices (**inner hexagon**). The sum of the vertices of an outer hexagon is 26. The sum of the vertices of an inner hexagon is $26\times2=52$.

A **triangle** is three outer or three inner vertices, which form a particular sum. An **outer triangle** is formed by three outer (corner) vertices, which are the points of the main triangle of a six-pointed star. The sum of the vertices of an outer triangle is 13. An **inner triangle** is a set of three inner vertices that make the sum of 26. For each six-pointed star, two outer triangles and two inner triangles are always clearly defined. One can add the following distinction. The **first outer triangle** is one which has the vertex '1' (vertex which contains the number 1; one of the two outer triangles always has such a vertex). The **first inner triangle** is one which has the vertex '12' (one of the two inner triangles always has such a vertex as well). The **second outer triangle** or the **second inner triangle** is one which has no vertex '1' or '12', respectively.

An **equilateral triangle** (unlike a triangle) is three outer or three inner vertices which form a particular geometrical figure, namely that of an equilateral triangle. An **outer equilateral triangle** always coincides with an outer triangle. An **inner equilateral triangle** never coincides with an inner triangle. The sum of the vertices of an inner equilateral triangle, while being different in different cases, is never 26.

An **enneagon** (or **main triangle**) is three outer vertices of one of the two outer triangles and all six inner vertices. In other words, it is a set of all the vertices of a main triangle of a star (of all the vertices of three adjacent sides of a star, which form such a triangle). The sum of the vertices of an enneagon is $13+52=65$.

A **dodecagon** is a figure formed by all the vertices of a star. The sum of the vertices of a dodecagon is $26+52=78$.

In all cases (with the partial exception of complements), fundamental importance attaches not only to which elements (vertices) form this or that component of a star, but the order of these elements as well. Thus, for example, we distinguish the side '1, 11, 12, 2' from the side '2, 12, 11, 2' (reverse order) or from '1, 12, 11, 2'. For the same reason, one can say that for the first star there exists the dodecagon '1, 11, 3, 12, 2, 5, 6, 9, 10, 7, 4, 8' or '8, 4, 7, 10, 9, 6, 5, 2, 12, 3, 11, 1', but not '1, 3, 11, 12, 2, 5, 6, 9, 10, 7, 4, 8'.

One can transform magic star also by 'inside-out/outside-in' operation. As the result, outer vertices of an original star take now inner positions and vice versa. A star transformed in such a way is called an **Antistar**. In antistars, hexagons of equal type will still have an equal sum (now, 52 for outer hexagons and 26 for inner), while sides lose their quantitative sameness.

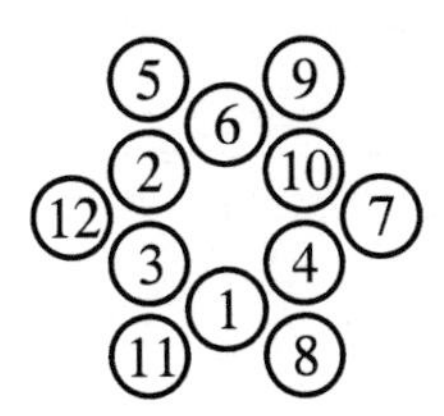

Figure 3: *The first Antistar (compare with Figure 1).*

How can stars be turned into music? For example, like this: let 1 be E, 2 be F and so on, up the chromatic scale... Now one can examine stars and their components, in different combinations and correlations, by means of music.

2. Magic Stars as Musical Work

The full title of the work is **Magic Stars // Tables for Piano or Other Appropriate Instruments**. There are a number of tables – more than two hundred and the work is still in progress. The tables are grouped in a following way:

Stars, this section includes Vertical Stars, Vertical Stars Expanded, Vertical Stars Compressed, Horizontal Stars, Horizontal Stars Sloped, Horizontal Stars Extra Sloped, Horizontal Stars Compressed, Horizontal Stars Degenerated (8 tables on 36 pages in total in this section).

Components, this section includes Sides and Complements (in different forms, sorting and combinations, 44 tables on 40 pages), Triangles (12 tables on 12 pages), Hexagons (30 tables on 20 pages), Enneagons (18 tables on 26 pages), Dodecagons (18 tables on 24 pages), Opposite Vertices (4 tables on 4 pages) (126 tables on 126 pages in total in this section).

Ordered Components, this section includes Ordered Sides (20 tables on 20 pages), Ordered Complements (19 tables on 20 pages), Ordered Triangles (16 tables on 16 pages), Ordered Hexagons (26 tables on 26 pages), Ordered Enneagons (6 tables on 12 pages), Ordered Dodecagons (6 tables on 12 pages) (93 tables on 106 pages in total in this section).

Stars in Base-n Mode, this section includes Stars represented in different numerical systems, from Base-2 to Base-13 Mode (12 tables on 12 pages in this section).

Antistars, this section includes Vertical Antistars, Horizontal Antistars, Horizontal Antistars Expanded (3 tables on 12 pages in this section).

(So, by now there are 242 tables on 292 pages in total.)

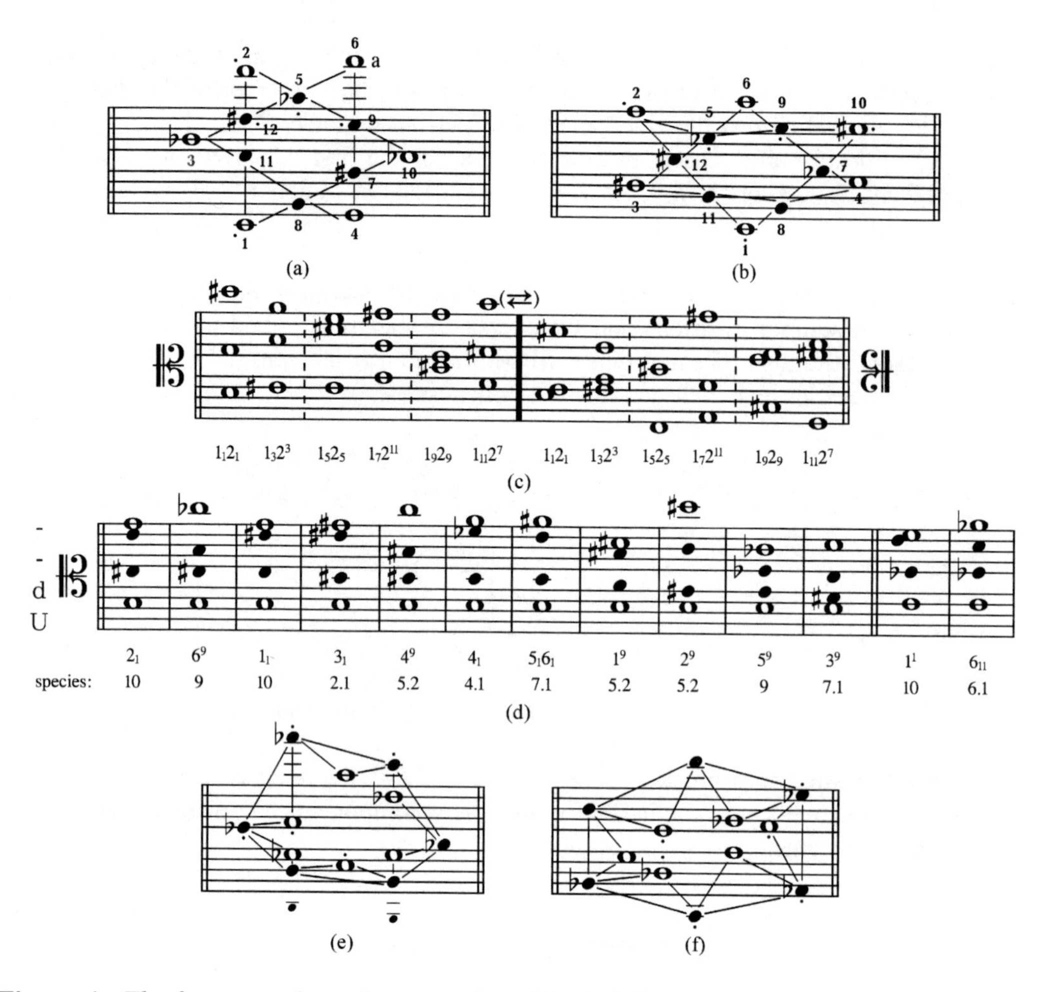

Figure 4: *The fragments from the score: from* Vertical Stars (a); *from* Horizontal Stars (b); *from* Outer and Inner Triangles 1 (c); *from* Ordered Sides 1 (d); *from* Vertical Antistars (e); *from* Horizontal Antistars (f).

Stars and their components are sorted, classified and distributed in the score in a systematic way (often followed by index marks of different types). In terms of comparison, the score is a dictionary rather than a novel. Any traces of 'subjectivity' and 'expression' have, as far as possible, been eliminated. Conversely, the performer's subjectivity is entirely permissible: any sort of expression may be generated by him and 'added' to the music. While note pitches are fixed in the score, note durations and their time positions are not. Each table or its fragment may be played from left to right, from right to left, from top to bottom, at any angle 'diagonally', etc. If desired, the notes may follow one after another in any other order, with reiterations and omissions. (For example, instead of playing a three- or four-note chord progression, the performer can play only the *soprano* or the *alto* part.) This principle also applies generally: for any particular performance any number of tables can be selected, which may follow one another in any order, with reiterations, etc. When playing, one can follow and emphasise – or hide and break – a logical structure in any desired way.

The score in its entirety and any of its components can also be compared to a landscape or a park. While the spatial structure is well defined, the routes are not, but are determined by the walker's moods, wishes and decisions, either spontaneous or planned ahead, which may be different each time. One can experience the things of the outer world in a deeply personal and intimate way. The music in itself is 'emotionally void'. However, when communicated with, it may be richly 'coloured' and 'filled up' with human emotions and feelings, like a sunset, a river or a starry sky...

Introducing the Precious Tangram Family

Stanley Spencer
The Sycamores
Queens Road
Hodthorpe
Worksop
Nottinghamshire
England
S80 4UT
pythagoras@bcs.org.uk
www.pythagoras.org.uk

Abstract

The Author of this paper has developed a family of Precious Tangrams based upon dissections of the first six regular polygons. Each set of tiles has similar properties to that of the regular tangram. In particular the property called Preciousness. It includes a discussion of some of the mathematical aspects of the dissections with examples of non periodic tessellating patterns. It continues with examples of the unique way in which they can produce an infinite number of designs. It explains the iterative nature of the process as applied to designs for mosaics, quilts and animation.

1. Introduction

This work follows from previous work on Precious Triangles and Polygons [1],[2],[3],[4], [5]. The idea has been expanded to include dissections of the n sided regular polygons for n=3 up to n=8. Each dissection results in tiles that have angles which are multiples of p,n. Each dissection contains triangles, quadrilaterals and, as one of its tiles, a smaller version of the original regular polygon. Finally each set of tiles displays the property of preciousness.

2. What are Precious Polygons?

Precious Polygons are sets of different polygons that can be used to form other sets of similar polygons. The necessary conditions for preciousness are that a larger version of each polygon can be produced using only the original polygons, secondly, the enlargement factor in each case must be the same and finally, all the elements of the nth power of the Precious Matrix must be non zero, where n is the number of different tiles [1]. This characteristic ensures that all designs, even a single tile, are expandable to infinity. This process is similar to that of Solomon Golomb's Rep-tiles [7],[8] but involving sets of polygons rather than a single polygon. Self similarity was also a feature of 14th and 15th Century Islamic Geometry [9] as well much earlier 1st Century Celtic Art.

3. Precious Dissection of Regular Polygons.

The following diagrams show dissections of the regular polygons with the corresponding Precious scheme including the Precious Matrix. It is left to the reader to show that all the elements of the nth power of the matrix are non zero. In addition the enlargement factor is given. This must be a constant for the dissection to be Precious and is known as the Precious Ratio. Once we have a Precious set then we can take any design made from the original tangram shapes and produce a larger version using the schemes in figures 1 to 6. Since we end up with a design using only the original shapes, we can repeat the process ad infinitum. Each successive design is larger than the previous one by a factor of P, the Precious Ratio.

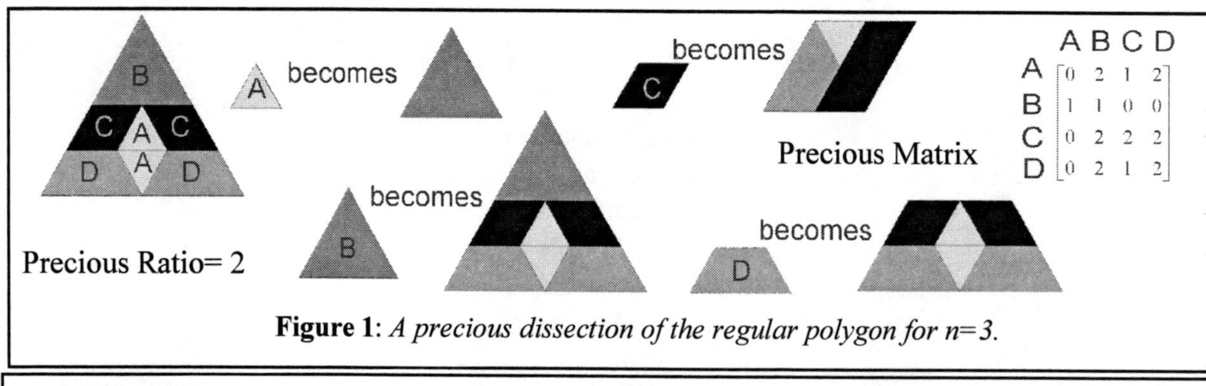

Figure 1: *A precious dissection of the regular polygon for n=3.*

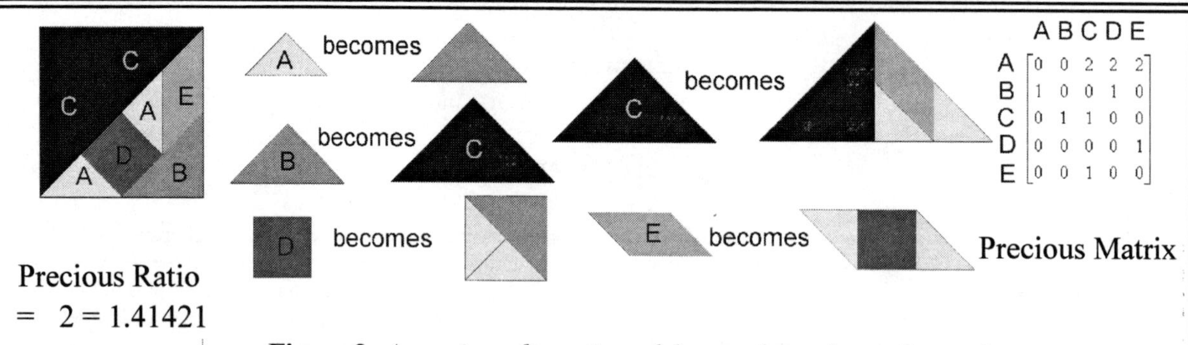

Figure 2: *A precious dissection of the regular polygon for n=4.*

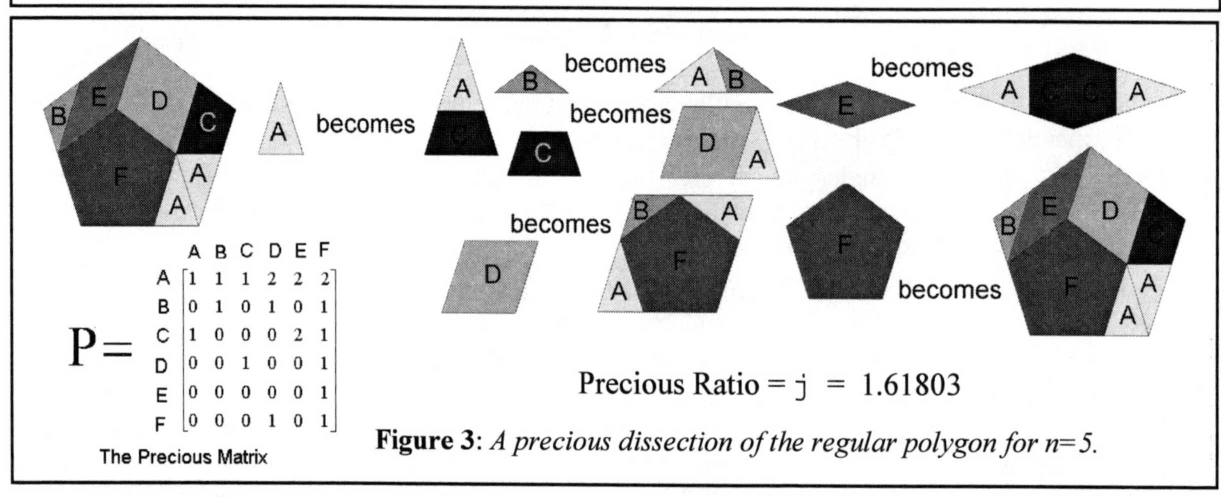

Figure 3: *A precious dissection of the regular polygon for n=5.*

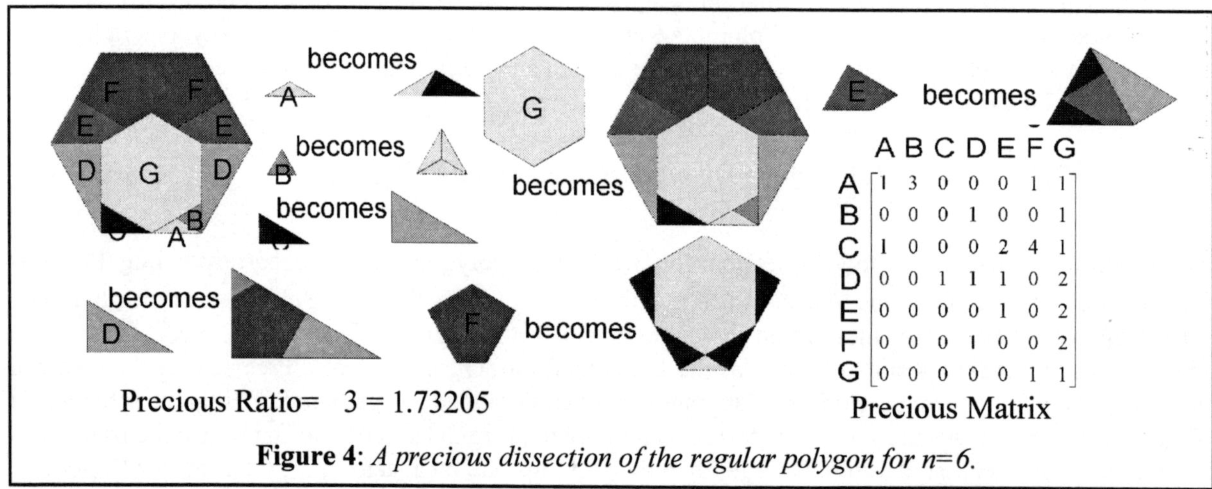

Figure 4: *A precious dissection of the regular polygon for n=6.*

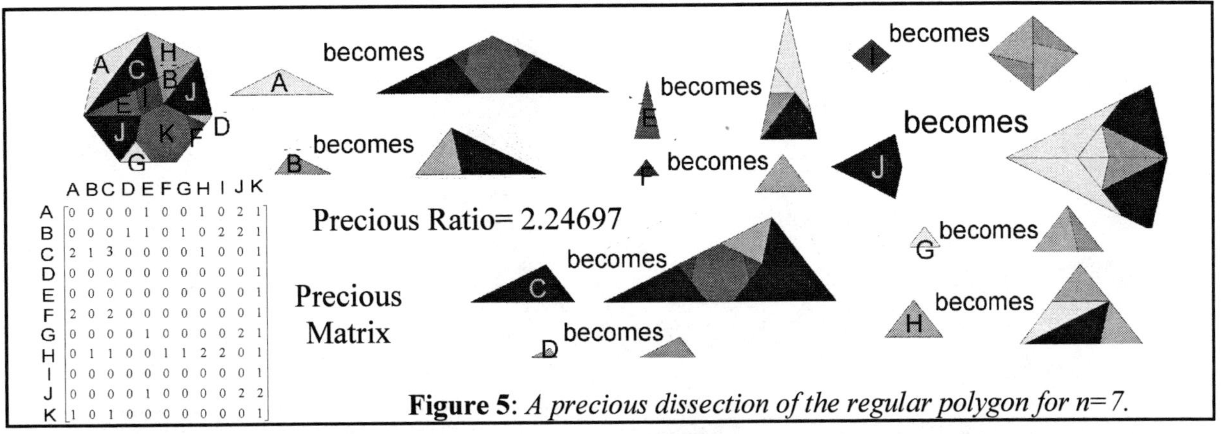

	A	B	C	D	E	F	G	H	I	J	K
A	0	0	0	0	1	0	0	1	0	2	1
B	0	0	0	1	1	0	1	0	2	2	1
C	2	1	3	0	0	0	0	1	0	0	1
D	0	0	0	0	0	0	0	0	0	0	1
E	0	0	0	0	0	0	0	0	0	0	1
F	2	0	2	0	0	0	0	0	0	0	1
G	0	0	0	0	1	0	0	0	0	2	1
H	0	1	1	0	0	1	1	2	2	0	1
I	0	0	0	0	0	0	0	0	0	0	1
J	0	0	0	0	1	0	0	0	0	2	2
K	1	0	1	0	0	0	0	0	0	0	1

Figure 5: *A precious dissection of the regular polygon for n= 7.*

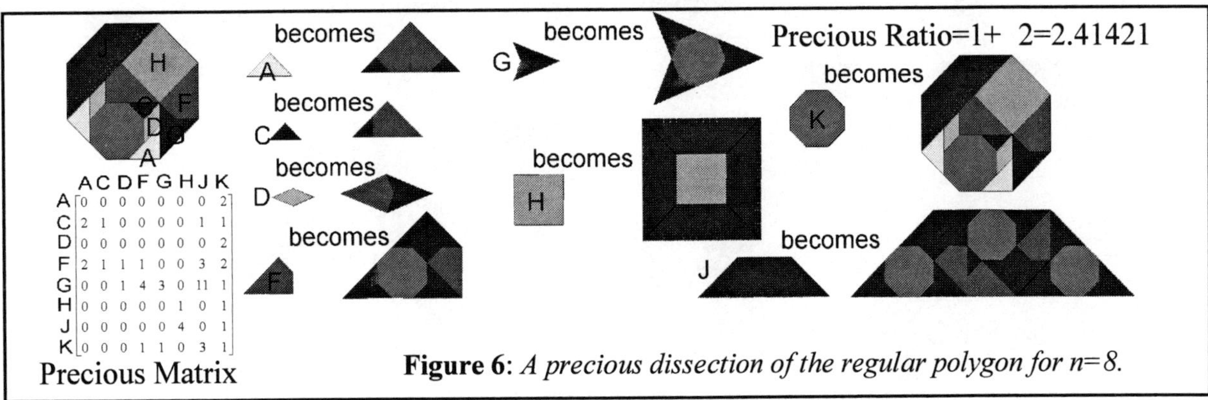

	A	C	D	F	G	H	J	K
A	0	0	0	0	0	0	0	2
C	2	1	0	0	0	0	1	1
D	0	0	0	0	0	0	0	2
F	2	1	1	1	0	0	3	2
G	0	0	1	4	3	0	11	1
H	0	0	0	0	0	1	0	1
J	0	0	0	0	0	4	0	1
K	0	0	0	1	1	0	3	1

Figure 6: *A precious dissection of the regular polygon for n= 8.*

4. Examples of simple Designs.

Figures 7, 8 and 9 are examples of simple designs using the dissection of the heptagon and octagon. They also show the first few of an infinite number of non periodic tessellations using the Precious properties of the tiles. Further examples for the square and pentagon can be seen at [1],[2].

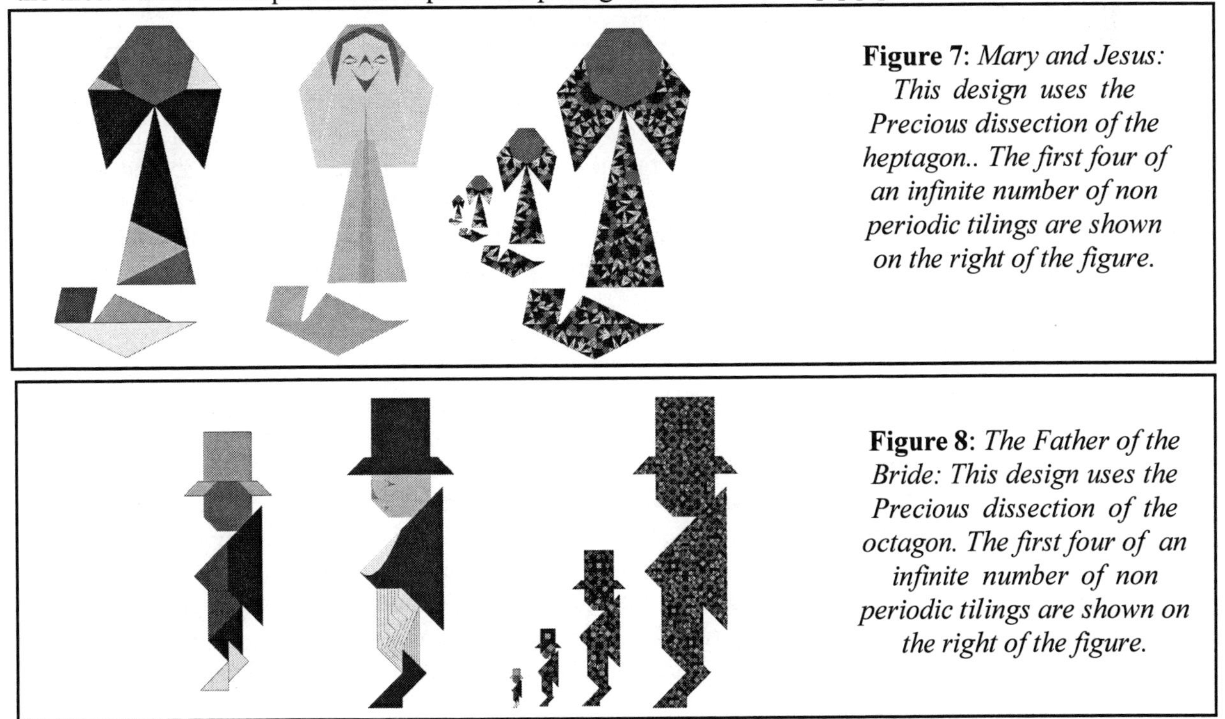

Figure 7: *Mary and Jesus: This design uses the Precious dissection of the heptagon.. The first four of an infinite number of non periodic tilings are shown on the right of the figure.*

Figure 8: *The Father of the Bride: This design uses the Precious dissection of the octagon. The first four of an infinite number of non periodic tilings are shown on the right of the figure.*

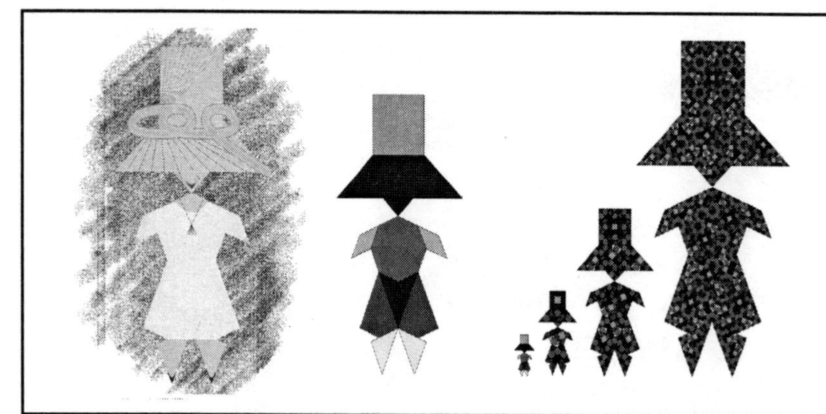

Figure 9: *The Mother of the Groom: This design uses the Precious dissection of the octagon. The first four of an infinite number of non periodic tilings are shown on the right of the figure.*

5 The Creation of Mosaics from the Precious Tangram

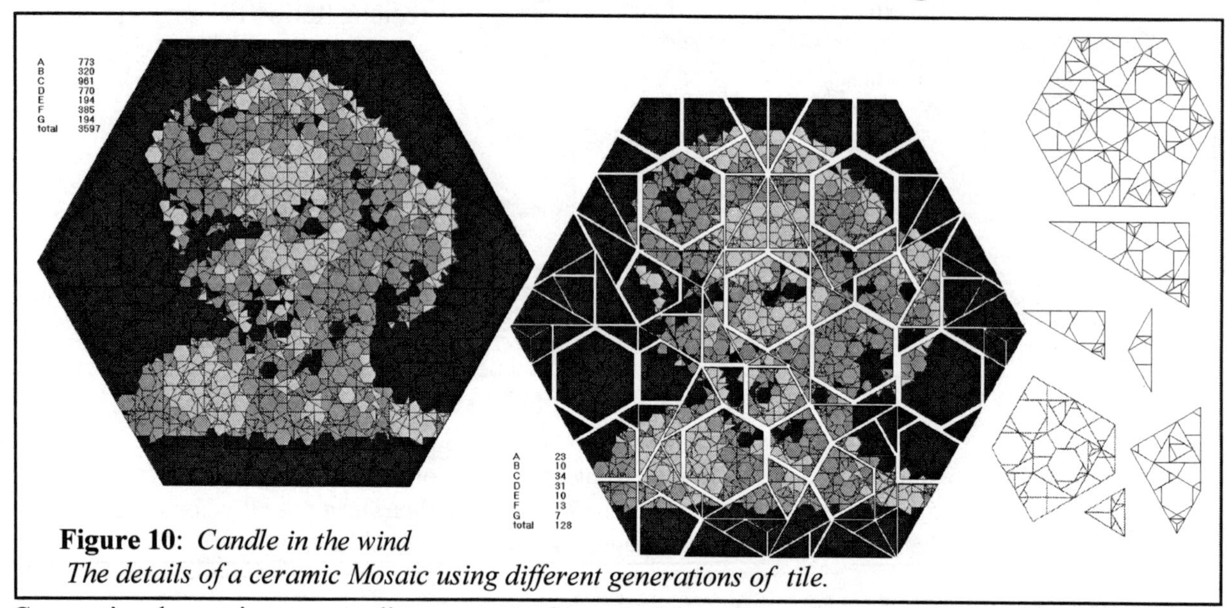

Figure 10: *Candle in the wind*
The details of a ceramic Mosaic using different generations of tile.

Conventional mosaics use small tesserae to form a picture. The picture at the left in Figure 10 is constructed from several thousand tesserae, each being one of the six Precious Shapes. The picture on the right shows how the picture can be created from 162 higher order tiles. Each one being one of the same original 6 shapes, but larger. To maintain the detail, the large tiles would need to be embossed with the original smaller tiles. The outlines of the 6 moulds can be seen in the top, right of the picture. The software, written by the Author, used to create the pictures in this paper also produces a disc that controls a CNC milling machine that makes the moulds from which the high order ceramic tiles can be accurately produced.

6 The Creation of Patchwork from the Precious Tangram

Many of the Precious schemes lend themselves to patchwork. That is, the schemes are quiltable. This means that they can be assembled by sewing in straight lines. Some preplanning of the sewing sequence is necessary to avoid sewing around a corner. To date such schemes have been developed for the equilateral triangle, the square, a subset of the pentagon [3] and a subset of the heptagon. Schemes are being developed for a subset of the hexagon and octagon. The designs in Figures 11 and 12 are based upon the equilateral triangle dissection. It is easy to see that the Precious scheme is quiltable in these cases. From a practical point of view it is necessary to be accurate with the sewing if the geometry is to work. This attention to accuracy starts with the cutting of the material. To this end a series of templates were made each with an extra border of 0.25 inches. A rotary cutter was used to cut out the various shapes. These were then sewn together with a hem of exactly 0.25 inches.

Figure 11: *The details of patchwork designs using different generations of tile.*

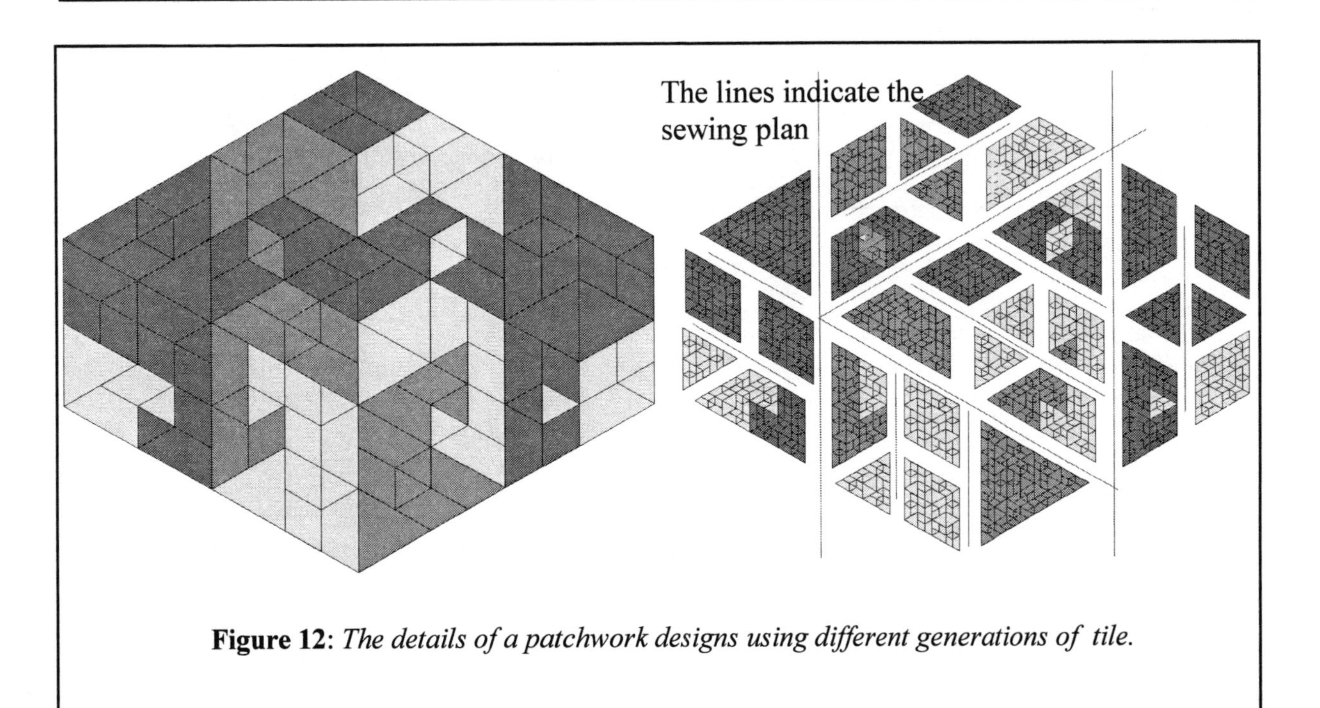
The lines indicate the sewing plan

Figure 12: *The details of a patchwork designs using different generations of tile.*

7 The creation of Animation from the Precious Tangram

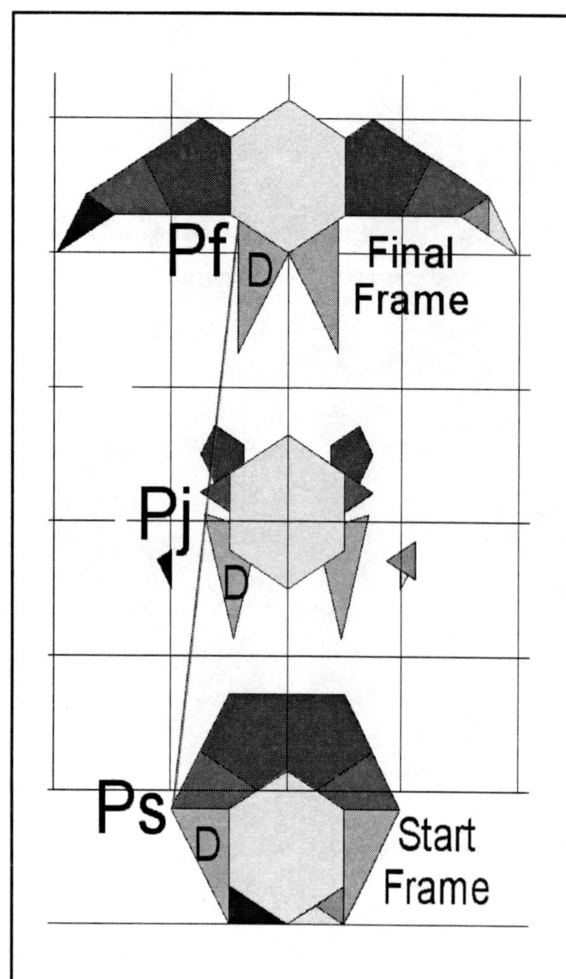

Figure 13: *The hatching of the swallow is an example of an animated Tangram design.*

To produce animated clips it is necessary to create a number of frames, each being a snapshot of the required movement. The method adopted by the Author was to produce two pictures from a set of tangram shapes. These represent the starting and finishing positions of the animated clip. The example in Figure 13 shows a hexagon and a swallow as the first and last frames of the sequence. It is important to note that the two designs use the same set of tiles. The intermediate frames can be produced by considering a line joining corresponding points on the start and final frames. If these points are called Ps and Pf, the number of frames n and the number of the frame being considered j then the distance PsPj is PsPf x j , n.
If this calculation is repeated for each corner of each tile then an intermediate frame can be produced. Figure 13 has the start, finish and jth frames. Normally these are separate frames but have been assembled together for illustration.
The frames were printed and assembled into a booklet that can be flicked through and given the illusion of movement. An alternative was to assemble the frames into an animated .gif file. The software used to create the individual frames was written by the Author and the JASC Animation Shop2 was used to assemble the individual frames into an Animated gif. The size of the individual frames and the number of frames both have an effect on the quality and speed of the animated clip. A few examples of animated gifs will be presented at the conference.

8 References

[1] Spencer, Stanley J, *The Tangram Route to Infinity* ISBN 141202917-1
[2] *Mathematical Connections in Art, Music and Science*. Bridges Conference 2004 ISBN 0-9665201-5-7.
[3] *Mathematical Connections in Art, Music and Science*. Bridges Conference 2005 ISBN 0-9665201-6-5.
[4] Spencer, Stanley J (Accessed 6.12.2003)<http://pythagoras.org.uk>
[5] *Meeting Alhambra. ISAMA-Bridges 2003 Conference Proceedings* ISBN 84-930669-1-5
[6] Fu Traing Wang and Chuan-Chih Hsiung *A Theorem on the Tangram*. American Mathematical Monthly, vol. 49, 1942
[7] Steven Dutch *Rep-Tiles* (Accessed 1.4.2004) http://www.uwgb.edu/dutchs/symmetry/reptile1.htm
[8] Solomon W Golomb, *Polynomials Puzzles, Patterns, Problems, and Packings,* pg. 8, Appendix C, Pg 148, Princeton University Press, 2nd edition, 1994.
[9] Jay Bonner, *Three Traditions of Self Similarity in 14th and 15th Century Islamic Geometric Ornament,* Meeting Alhambra, ISAMA Bridges Conference Proceedings 2003, ISBN 84-930669-1-5.

Sand Drawings and Gaussian Graphs

Erik D. Demaine Martin L. Demaine
Computer Science and AI Laboratory
Massachusetts Institute of Technology
{edemaine,mdemaine}@mit.edu

Perouz Taslakian
SOCS
McGill University
perouz@cs.mcgill.ca

Godfried T. Toussaint
SOCS & CIRMMT
McGill University
godfried@cs.mcgill.ca

Abstract

Sand drawings form a part of many cultural artistic traditions. Depending on the part of the world in which they occur, such drawings have different names such as *sona*, *kolam*, and *Malekula* drawings. *Gaussian graphs* are mathematical objects studied in the disciplines of graph theory and topology. We uncover a bridge between sand drawings and Gaussian graphs, leading to a variety of new mathematical problems related to sand drawings. In particular, we analyze sand drawings from combinatorial, graph-theoretical, and geometric points of view. Many new mathematical open problems are illuminated and listed.

1. Introduction

Different cultures around the world contain mathematics in one form or another. Not all of these mathematical ideas have developed out of necessity, like counting and calculation: some mathematical ideas are developed in the context of cultural arts. The study of this mathematical art within and without its cultural context constitutes the field of *ethnomathematics* [2, 3, 12, 22].

In this paper, we explore the mathematics and geometry found in a particular kind of visual art that seems to have developed independently in different forms in disparate cultures. In its basic form, the artist draws dots and one surrounding continuous loop, which crosses itself repeatedly, in the sand or on a floor sprinkled with powder. Collectively, we refer to these practices as *sand drawings*, though each practicing culture has its own name for the visual art.

For example, the women in Tamil Nadu (South India) create geometric designs using rice flower, called *kolam*, at the entrances of their homes [3]. One type of kolam, called *pulli kolam*, consists of first drawing a grid of dots (the *pulli*), and then drawing a continuous closed curve that partitions the planar space into as many bounded regions as there are dots, such that each bounded region contains exactly one dot. Each kolam drawing has a name. Figure 1 illustrates two such drawings.

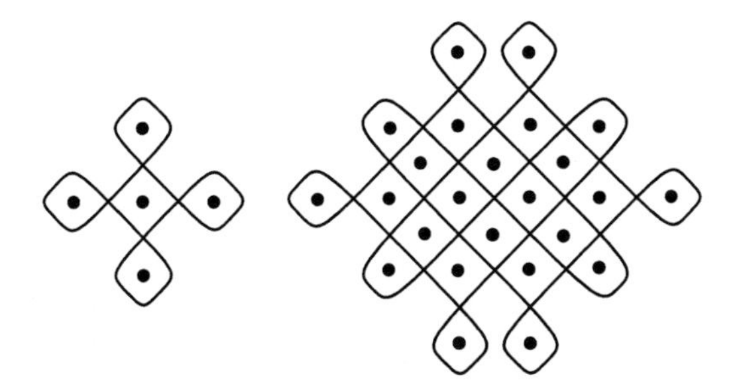

Figure 1: *Two examples of* pulli kolam*: the* Anklet of Krishna *(left) and* The Ring *(right).*

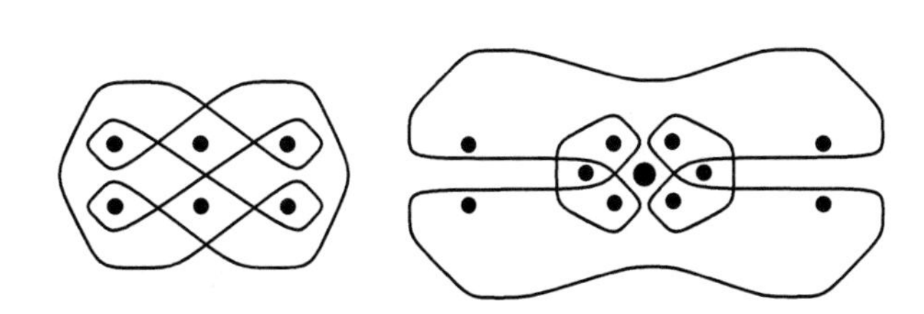

Figure 2: *Two examples of* sona*: the* Antelope's Paw *(left) and the* Spider *(right).*

The *Tshokwe* people of the West Central Bantu area of Africa make similar drawings called *sona*. In these drawings, it may happen that some regions contain more than one dot, or that some regions are empty [2, 12]. Figure 2 shows two such drawings: the drawing on the left has two empty bounded regions, while the drawing on the right has one bounded region with five dots. Some sona drawings do not have dots at all. However, in most sona drawings, there is exactly one dot per region.

Paulus Gerdes [12] has developed several geometric algorithms for constructing some families of *sona* drawings. One such algorithm uses Euclid's algorithm for computing the greatest common divisor of two natural numbers. It is interesting that the Euclidean algorithm [9] not only generates traditional drawings in visual art, but also traditional rhythms in music [19].

In other traditions, people make drawings on the sand without first placing a grid of dots. The *Malekula* sand drawings are one such example. The Malekula people live on the island of Vanuatu in the South Pacific. Marcia Ascher [4] has analysed these drawings from the graph theoretic, geometric, and topological points of view.

From a geometric perspective, a sand drawing can be viewed as a self-intersecting closed curve drawn in the two-dimensional plane. Translating this definition of sand drawings into graph theory, the intersection points of the curve map to vertices and the portions of the curve connecting these vertices map to edges of a planar map. We are also interested in how dots can be placed so that exactly one dot lies in each region of the planar map (except for the infinite outside face). Although not all sand drawings have this property, this seems to be the most natural mathematical abstraction of the majority of sand drawings.

In this paper we unveil the bridge between sand drawings and a class of graphs known as Gaussian graphs. We pose a variety of basic questions about sand drawings and Gaussian graphs, solve some of them, and leave the reader with many interesting mathematical problems to consider.

2. Gaussian Graphs

Gaussian graphs implicitly go back to an observation of Carl Gauss around 1830 [11] that was proved by Julius v. Sz. Nagy almost a hundred years later [21]. The formal notion of Gaussian graphs was introduced more recently by Michael Gargano and John Kennedy [10], and then generalized by John Kennedy and Brigitte and Herman Servatius [14]. We follow the latter, more general, definition here, although our main interest is in the original definition of Gargano and Kennedy.

First we need some basic terminology from graph theory. A *graph* is a collection of vertices (or points) together with a collection of edges (or lines) connecting pairs of vertices. The *degree* of a vertex is the number of edges incident to it. A *circuit* in a graph is a cyclic sequence alternating between vertices and edges such that the two endpoints of each edge are the two adjacent vertices in the sequence. Thus a circuit may repeat vertices and/or edges. A graph is *Eulerian* if it has a circuit visiting every edge exactly once. It is well known that a graph is Eulerian if and only if it is connected and every vertex has even degree. A special family of Eulerian graphs is 4-*regular* graphs, in which every vertex has degree exactly 4. A *planar map* is a graph together with a *planar embedding*, a placement of the vertices as points in the plane and a drawing of the edges as curves that intersect each other only at common endpoints. When such a planar embedding exists, the graph itself is also called *planar*. A planar graph can have several different planar embeddings. We call two planar embeddings *(combinatorially) equivalent* if they define the same clockwise cyclic order of edges around each vertex.

Intuitively, a *Gaussian graph* [14] is a planar map that can be drawn by a single closed curve that "goes straight" at each vertex. More precisely, two edges incident to a common vertex v are *parallel* if these edges partition the clockwise order of edges around v into two pieces with an equal number of edges. In other words, if we label the edges incident to v by $1, 2, \ldots, \text{degree}(v)$ in clockwise order, then two of these edges are parallel precisely if they have the same label modulo $\text{degree}(v)/2$. A *transverse circuit* of a planar map is a circuit whose successive edges are parallel. A planar map is Eulerian if and only if it is connected and its edges decompose into one or more transverse circuits. A *Gaussian map* is an Eulerian map that decomposes into exactly one transverse circuit.

Intuitively, a *sand drawing* or *sona drawing* is a closed curve drawn in the plane such that no more than two pieces of the curve intersect at the same point and such that the curve does not "touch" itself without crossing itself. More precisely, if we label the four portions of the curve around an intersection point v by $1, 2, 3, 4$ in clockwise order, then the curve in a sona drawing should connect portion 1 to portion 3, connect portion 2 to 4, and connect no other pairs. In other words, a sona drawing is a 4-regular Gaussian map, so we also use the term *sona map* to make clear the underlying graph and embedding structure. Kennedy et al. [14, Theorem 2] show that, for any 4-regular planar graph, the number of its transverse circuits is independent of its embedding. Thus, all embeddings of the underlying graph in a sona map are sona maps. We can therefore define a *sona graph* to be a planar 4-regular graph whose planar embeddings (all) create sona maps.

There are efficient (linear-time) algorithms to determine whether a given graph or map are sona. To decide whether a planar map is sona, check 4-regularity and check whether all edges are visited by starting on an arbitrary edge e_0 in an arbitrary orientation and, upon entering a vertex v along an edge e, exiting v along the unique edge incident to v that is parallel to e. Kennedy et al. [14] give an incremental characterization of all sona maps, which also leads to an efficient algorithm for deciding whether a planar map is sona. To decide whether a graph is sona, find a planar embedding using, e.g., the algorithm of [6], and upon success, test whether the resulting planar map is sona as above.

Throughout this paper, n denotes the number of bounded regions or *faces* of a sona map, or the number of such regions that a sona graph would have if it were embedded into the plane. This count n corresponds to the number of dots that would be placed, exactly one per bounded face, in many cultural practices. We therefore sometimes speak of "a sona map on n dots" or "a sona graph on n dots".

Lemma 1 *A sona graph on n dots has $n-1$ vertices and $2(n-1)$ edges.*

Proof: The number E of edges is half the total degree of the V vertices. Because every vertex has degree 4, $E = 2V$. By Euler's Formula, $V - E + F = 2$ where $F = n + 1$ is the number of faces including the infinite outside face. Substituting E and F, we obtain $V - 2V + n + 1 = 2$, i.e., $V = n - 1$. Therefore, $E = 2V = 2(n-1)$. □

In addition to their connection to Gaussian graphs in graph theory, sona maps are equivalent to generic closed curves (immersions of the unit circle into the plane) in the field of topology. By *generic*, we mean that the curve is not tangent to itself anywhere and that no more than two portions of the curve cross at any point. Arnol′d [1], for example, proves several topological invariants about such curves. Craveiro [7] considers the curves that are "maximally looped". Ozawa [17] considers the number of bitangents, shared tangents between different points on the curve.

3. Combinatorics of Sand Drawings

In this section, we analyze the number of different sona drawings. There are two main different objects we can count: sona graphs and sona maps. Each sona graph has one or more associated sona maps, so in general there are more maps than graphs.

3.1. Enumeration and Drawing. We have developed a software program that generates all sona graphs and sona maps on n dots, for small n. More precisely, the program generates all distinct sona graphs on n dots, and all distinct sona maps on n dots, incrementally for $n = 2, 3, 4, \ldots$. It uses the incremental characterization of sona maps by Kennedy et al. [14] to generate, for each sona map on $n-1$ dots, a list of $O(n^2)$ sona maps on n dots. The combined list of all such sona maps on n dots includes all possible sona maps on n dots, but it lists the same sona map more than once if it can be generated in multiple ways. The time consuming part of the computation is to then remove duplicates from the resulting list, first according to combinatorial equivalence of sona maps, and second according to isomorphism of sona graphs. The program simply tests each pair of sona maps for equivalence of either type, by testing all possible bijections between the vertices of the two maps, and removes duplicates.

Table 1 shows the computed number of sona graphs and sona maps on n dots for n between 1 and 8.

n dots	sona graphs	sona maps
$n = 1$	1	1
$n = 2$	1	2
$n = 3$	1	5
$n = 4$	3	21
$n = 5$	5	102
$n = 6$	13	639
$n = 7$	38	4,492
$n = 8$	133	34,032

Table 1: *Number of sona graphs and sona maps on n dots for small n.*

The program also draws a polygonal planar embedding of each sona map, where each edge is represented by a chain of up to three line segments, by triangulating and applying a theorem of Tutte [20]. Unfortunately, these drawings make poor use of area and are barely visible without zooming in extremely close. We redrew each sona map for n between 1 and 4 using a combination of circular arcs and straight lines, joined at common tangents, so that the resulting curve always crosses itself at right angles, and to illustrate all symmetries in the map. Figure 3 shows these drawings of all sona maps for n between 1 and 4.

We distinguish sona maps from their reflections, but the only sona map in Figure 3 that lacks reflectional symmetry is the second and fourth maps in the second row of $n = 4$, which resemble a treble clef.

3.2. Combinatorial Complexity. One measure of the *combinatorial complexity* of a sona map on n dots is the sum of the depths of all its faces; the *depth* of a face is the minimum number of edges that a point within the face needs to cross in order to reach the outer face of the map. For $n = 4$, the sona map with the highest complexity is the third map on the second row of $n = 4$ in Figure 3, It consists of a loop nested within double edges and a loop, resembles a rose, and has complexity 10. The *face depth sequence* of a sona map is the ordered sequence of the depths of all its bounded faces. The face depth sequence does not uniquely define a sona map. For example, the two rightmost maps on the first row of $n = 4$ in Figure 3 have the same face depth sequence $1, 1, 2, 2$.

3.3. Sona Maps. The general combinatorics of sona maps remains open:

Open Problem 1 *How many different sona maps are there on n dots?*

This question has been studied before for small n by Arnol'd [1]. The sequence is in OEIS [18, A008981]. The rightmost column of Table 1 shows the sequence for small n, computed by Arnol'd [1] and verified exhaustively by our software. To our knowledge, however, no closed-form solution or asymptotic bounds have been published previously. Again we can provide asymptotic upper and lower bounds of $2^{\Theta(n)}$:

Theorem 2 *There are at least $\Omega(4^n/n^{3/2})$ different sona maps on n dots.*

Proof: As defined by Arnol'd [1], a sona map is *tree-like* if cutting any vertex disconnects the map into two components. For example, the top-right map of $n = 4$ in Figure 3 is not tree-like, while the one to its left is tree-like. Every rooted ordered tree can be converted into a different tree-like sona map, so the number of such sona maps is at least the number of rooted ordered trees. The number of such trees on n nodes is the nth Catalan number: $\binom{2n}{n}/(n+1) = \frac{1}{\sqrt{\pi}} 4^n/n^{3/2} - O(4^n/n^{5/2})$. □

Theorem 3 *There are at most $16^{n-O(\lg n)}$ different sona maps on n dots.*

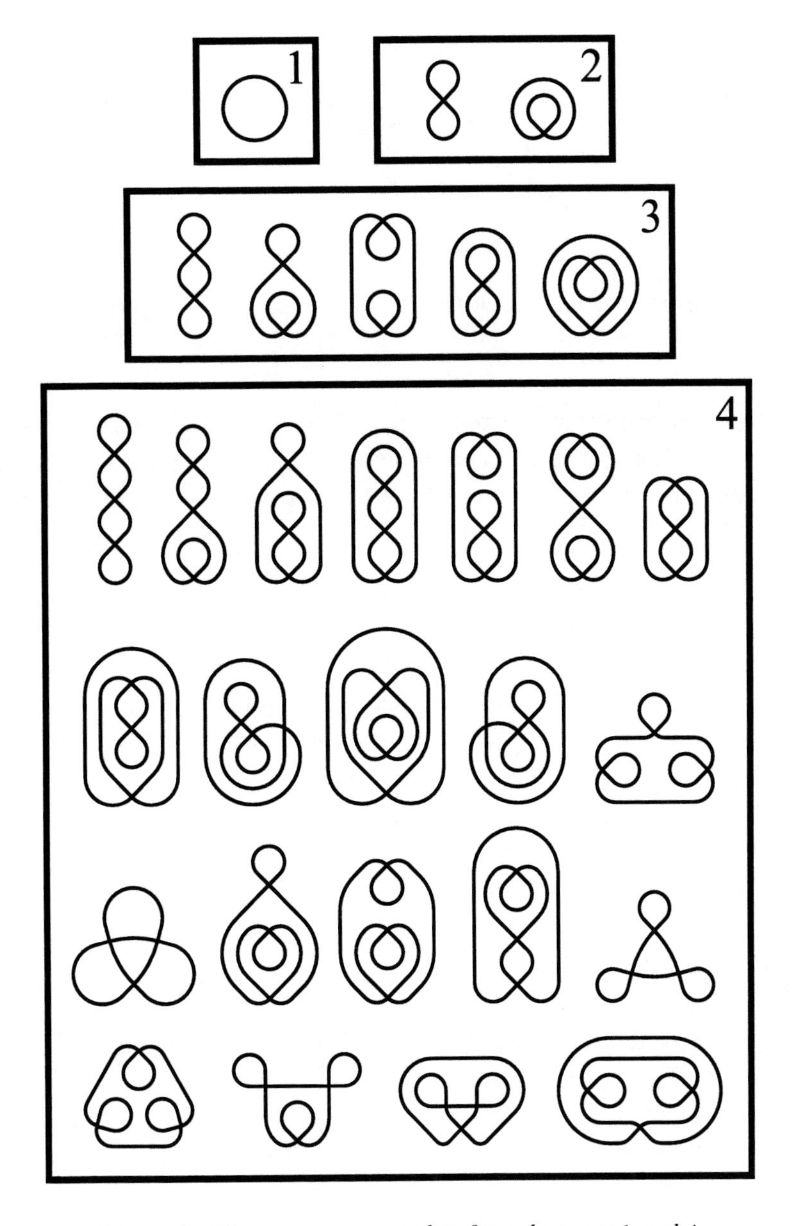

Figure 3: *All sona maps on* $\mathbf{n}$ *dots for* $\mathbf{n}$ *between* 1 *and* 4.

Proof: The number of sona maps on n dots is at most the number of Eulerian planar maps on $n' = 2(n-1)$ edges. This sequence is in the OEIS [18, A069727]. Liskovets and Walsh [15] give an explicit formula for this sequence. Also they prove that the number of Eulerian planar maps on n' edges is asymptotically equal to the number of rooted Eulerian planar maps on n' edges divided by $2n'$, while the number of rooted Eulerian planar maps on n' edges is

$$\frac{3 \cdot 2^{n'-1}}{(n+1)(n+2)}\binom{2n}{n} = 8^{n'-O(\lg n')}.$$

Therefore, the number of sona maps on n dots is asymptotically at most $16^{n-O(\lg n)}$. □

3.4. Sona Graphs. The general combinatorics of sona graphs also remains open:

Open Problem 2 *How many different sona graphs are there on* n *dots?*

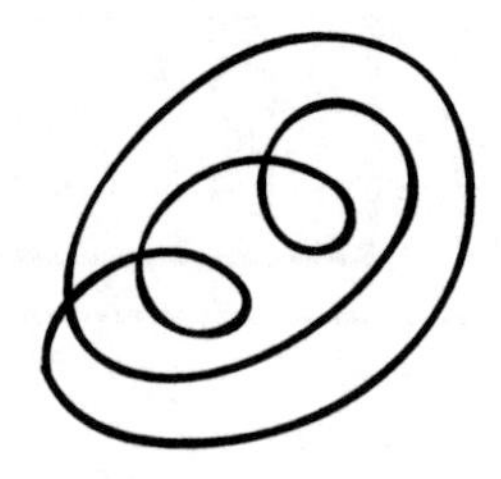

Figure 4: *Two different sona maps with the same face degree sequence:* $1, 1, 4, 5$.

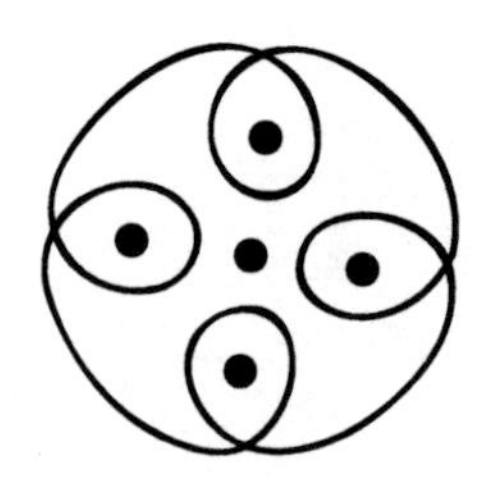

Figure 5: *A sona map in which all but two faces have degree 1.*

The middle column of Table 1 shows the answer for small n, found via exhaustive search. This sequence is not currently listed in the *On-line Encyclopedia of Integer Sequeneces* (OEIS) [18], and we plan to submit it. While the exact counts remain open for general n, we can provide asymptotic upper and lower bounds of $2^{\Theta(n)}$:

Theorem 4 *There are at least $\Omega(2.95^n)$ different sona graphs on n dots.*

Proof: As in the proof of Theorem 2, we consider tree-like sona maps. Every unrooted unordered tree can be converted into a different tree-like sona graph, so the number of such sona graphs is at least the number of unrooted unordered trees. The latter sequence is in the OEIS [18, A000055]. Otter [16] computes the asymptotic number of unrooted unordered trees on n vertices to be $\frac{\beta^3}{3\alpha^{9/2}} \left|\binom{3/2}{n}\right| \alpha^n + O(\alpha^n/n^{7/3}) \approx 0.53479\, \alpha^n/n^{5/2} + O(\alpha^n/n^{7/2})$, where $\alpha \approx 2.955765$ and $\beta \approx 7.924$. □

Our best upper bound on the number of sona graphs follows trivially from Theorem 3's upper bound on the number of sona maps:

Corollary 5 *There are at most $16^{n-O(\lg n)}$ different sona graphs on n dots.*

We are not aware of any better upper bounds on the number of Eulerian planar graphs compared to what we used in Theorem 3 about Eulerian planar maps. For example, the best known asymptotic upper bound on the number of unlabeled planar graphs with n vertices is given by Bonichon et al. [5]: $2^{\alpha n+O(\log n)}$ where $\alpha \approx 4.9098$. This result implies an upper bound of $O(30.1^n)$, which is strictly worse than Corollary 5.

4. Face Degrees of Sand Drawings

This section investigates the "face degree sequence" of sona maps. The *degree* of a face in a planar map is the number of edges that bound the face. The *face degree sequence* of a planar map is the sequence of the degrees of all bounded faces (excluding the infinite outside face), sorted by degree. There can be multiple sona maps with the same face degree sequence. Figure 4 shows one such example.

We omit the degree of the outside face from the face degree sequence because it is determined by n and the rest of the sequence. The total degree of all faces, including the outside face, is twice the number of edges. By Lemma 1, the degree of the outside face is $4(n-1)$ minus the total degree of all bounded faces, i.e., the sum of the values in the face degree sequence.

4.1. Equal Face Degrees.

We begin by investigating the extent to which most of the degrees in the face degree sequence can be equal. First we prove that most of the face degrees are equal, they must be at most 4:

Lemma 6 *The average face degree of any sona map, counting the outside face, is $4(n-1)/(n+1) = 4 - 8/n + O(1/n^2)$.*

Figure 6: *A sona map in which all but three faces have degree exactly* **2**.

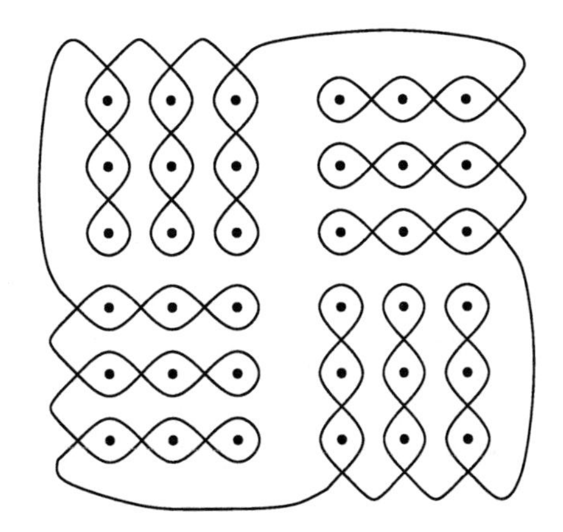

Figure 7: *"Men-lions that, stealthily, plan their intrigues" [12].*

Proof: Let D denote the average face degree. The number of edges is half the total face degree: $E = (DF)/2$. By Lemma 1, $2(n-1) = (D(n+1))/2$, so $D = 4(n-1)/(n+1)$. Asymptotically, $(n-1)/(n+1) = 1 - 2/n + 2/n^2 - O(1/n^3)$. □

Lemma 7 *For any n, there is a sona map on n dots with $n-1$ faces of degree exactly* 1.

Proof: See Figure 5. The construction loops around each of the n dots in turn. For the last dot, however, it just circles around it and connects to the starting point. □

Lemma 8 *For any n, there is a sona map on n dots with $n-2$ faces of degree exactly 2, and with two faces of degree exactly* 1.

Proof: See Figure 6. The construction takes a loop of string and twists the two strands (reversing their order) in between passing over each dot. □

Interestingly, this construction appears as an element in real-world sona drawings; see Figure 7.

Lemma 9 *For any n, there is a sona map on $3n+1$ dots with $3n-2$ faces of degree exactly 3, and with two faces of degree exactly* 2.

Proof: See Figure 8. Take the construction from Figure 6 and, just before closing the loop, continue the drawing by crossing each face, dividing each face into two; then return to the original point to close the loop. □

Lemma 10 *For any n, there is a sona map on n dots with $n - O(\sqrt{n})$ faces of degree exactly* 4.

Proof: See Figure 9. Consider the "grid" sona map where all faces except for $O(\sqrt{n})$ on the boundary have degree exactly 4. Now, for any such grid sona map with n faces, we can draw a grid sona map with a higher value of n by adding two rows or two columns, whichever is smaller. The total number of faces added to the new drawing is at most $2\sqrt{n}$. Thus, the gap in the number of faces between any two consecutive grid sona maps is $O(\sqrt{n})$. To construct a sona map whose number of faces is between n and $n + 2\sqrt{n}$, we can simply pick a non-degree-4 face and add loops to the face until we obtain the number of faces we want. This modification changes the degree of one face and will add up to $O(\sqrt{n})$ new faces, thus keeping the number of faces not of degree 4 bounded by $O(\sqrt{n})$. □

This construction also appears in real-world sona drawings; see Figure 10. A similar grid-like construction appears in the pulli kolam of Figure 1.

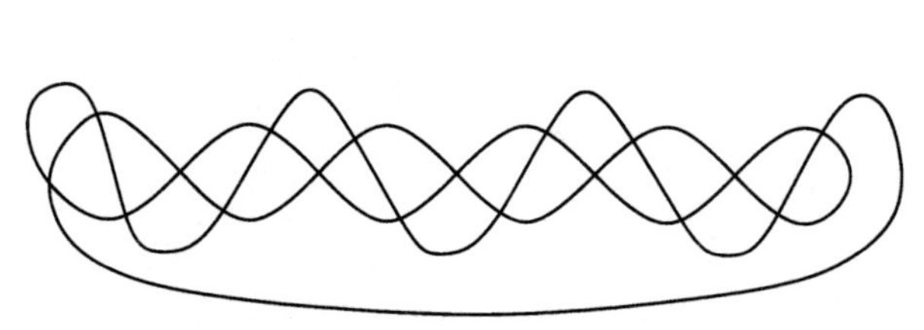

Figure 8: *A sona map in which all but four faces have degree exactly* **3**.

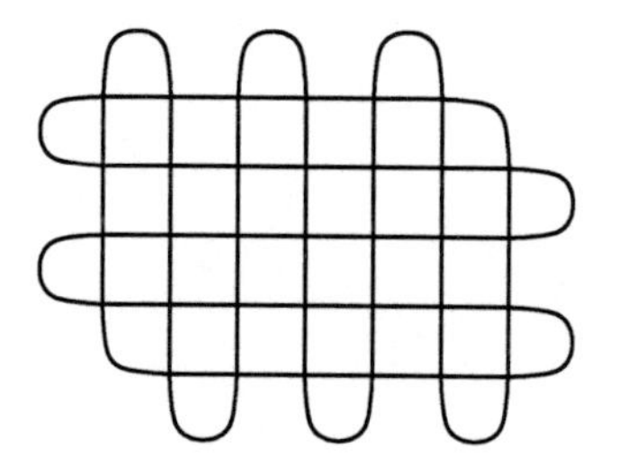

Figure 9: *A sona map with* $n - O(\sqrt{n})$ *faces of degree exactly* **4**.

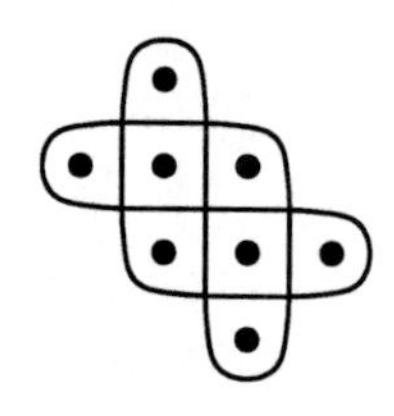

Figure 10: *"A small animal that lives in a tree hole and pierces the intestines" [2].*

4.2. Characterizing Face Degree Vectors.

Open Problem 3 *What is the complexity of deciding whether a given finite nondecreasing sequence of positive integers is the face degree sequence of some sona map? When it exists, can we construct such a sona map?*

An obvious necessary condition for a finite sequence of positive integers to be the (vertex or face) degree sequence of a graph is that the sum must be even (twice the number of edges). Hakimi [13] proved that every such sequence is the vertex degree sequence of some graph, possibly with loops and multiple edges. Note that this problem is substantially easier than characterizing the degree sequences of simple graphs (with no loops or multiple edges), considered by Erdős and Gallai [8] and Hakimi [13].

Naturally, this universality no longer holds when we add the restrictions of planarity, Eulerian, Gaussian, and/or sona. One specific related problem, which we are not aware of being studied before, is the following:

Open Problem 4 *What is the complexity of deciding whether a given finite nondecreasing sequence of positive integers is the (vertex or face) degree sequence of some planar map? When it exists, can we construct such a planar map?*

Face and vertex degree sequences are closely related to face depth sequences defined in Section 3.2: all three are integer sequences that do not uniquely define a sona map. Thus, the open problems of this section can also be asked for face depth sequences.

5. Sand Drawings on Given Dots

In this section, we consider a family of geometric sand-drawing problems: given n dots in the plane, find a sona map with exactly one of the given dots in each face (except the outside face) subject to some constraints or optimizing some objective function. This extra constraint captures how sand drawings are often performed in practice.

5.1. Minimum-Length Sona Maps. A natural geometric objective, albeit of dubious artistic merit, is to minimize the Euclidean length of the curve (or equivalently, the total Euclidean length of edges in the map). This minimum length is in fact an infimum, and may not be exactly achievable.

Open Problem 5 *What is the complexity of finding the minimum-length sona map for a given set of dots in the plane?*

A natural value to compare to is the minimum length of a Euclidean Traveling Salesman (TSP) tour, i.e., the shortest closed tour that visits every dot. We refer to the length of this tour as TSP.

Lemma 11 *For any $\varepsilon > 0$, every set of dots has a sona map of length* $\mathrm{TSP} + \varepsilon$.

Proof: As in the proof of Lemma 7 and Figure 5, we loop around each dot in turn, now with very small loops, connecting the loops together with the TSP tour. We do not loop around the last dot, but we do visit it to ensure that it is enclosed by the high-degree bounded face. □

Lemma 12 *There is a set of four dots on which the minimum-length sona map has length* TSP/c_0 *where* $c_0 = 2/3 + 2\sqrt{3}/9 > 1.05$.

Proof: Consider four dots, three dots forming an equilateral triangle with side length 1 and the fourth dot in the center of the triangle (Figure 11). The optimal TSP tour has length $2 + 2\sqrt{3}/3 > 3.15$, while the optimal solution visits only the corners of the triangle, at a cost of $3 + \varepsilon$ for some small positive ε. □

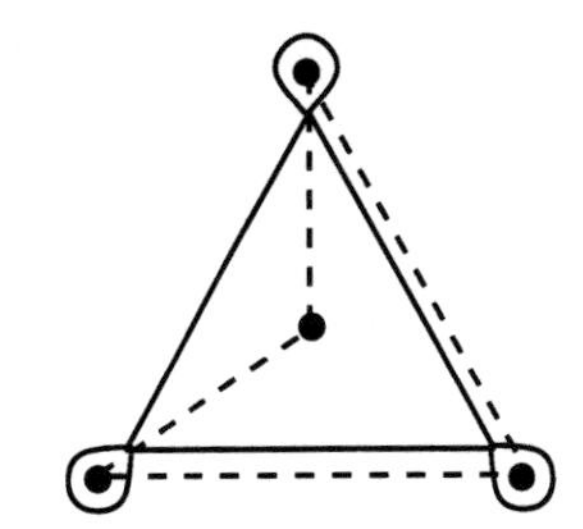

Figure 11: *The TSP tour (dashed lines) is slightly longer than the sona map whose faces encircle the given dots.*

Open Problem 6 *Does every sona map have length at least* TSP/c *for some constant* $c > 0$*?*

5.2. Other Objectives. Geometrically, it is natural to consider optimizing objectives other than total Euclidean length. It remains to determine whether these objectives are artistically more or less interesting.

Open Problem 7 *What is the complexity of finding a sona map on a given set of dots, where each edge is drawn as a polygonal chain of links, that minimizes the total number of links?*

Open Problem 8 *What is the complexity of finding a sona map on a given set of dots that minimizes the total absolute turn angle along the transverse circuit?*

5.3. Two Coloring. A more difficult family of problems arises when each dot has one of two colors, and the face 2-coloring of the sona map must match the specified coloring of dots. We can extend Lemma 11 to this case:

Lemma 13 *For any $\varepsilon > 0$, every 2-colored set of dots has a sona map of length* $\mathrm{TSP} + \varepsilon$.

Proof: See Figure 12. We follow the proof of Lemma 11, but loop right or loop left around each dot according to color. The last dot is in the high-degree face; if it has the wrong color, invert all of the other dots' orientations to fix it. □

5.4. Clockwise Turning.

Open Problem 9 *Given a sona map, can we decide in polynomial time whether it can be drawn using only clockwise turns? Characterize such sona maps. How many are there on n dots?*

It is not the case that every face degree vector of a sona map is the face degree vector of a clockwise-turning sona map. A simple example is the sona drawing illustrated in Figure 13 whose face degree vector has nine 1's and one 10. Another interesting question is whether the family of sona maps with face degree vector $1, 1, \ldots, 1, k + 1$, where the number of 1's is $k > 1$, constitutes a family of sona maps that cannot be drawn with only clockwise turns.

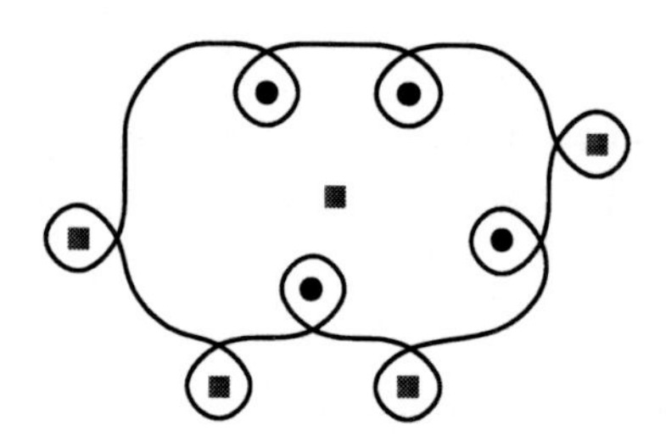

Figure 12: *Given dots colored as either circles or squares, we can always find a sona map in which no two faces containing elements of the same color share an edge.*

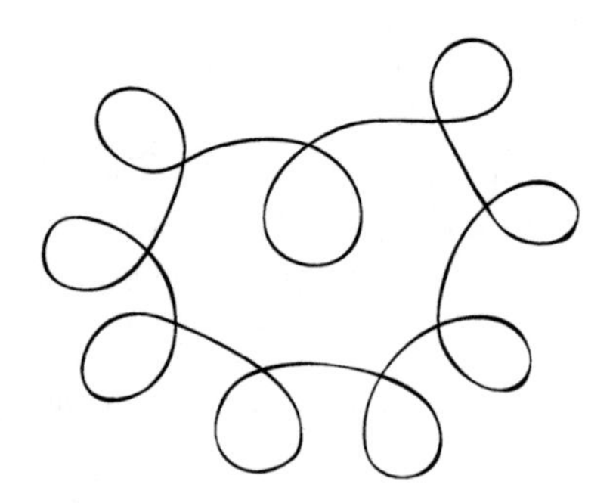

Figure 13: *This sona map with face degree vector* $1, 1, 1, 1, 1, 1, 1, 1, 1, 10$ *cannot be drawn with only clockwise turns.*

References

[1] V. I. Arnol′d. *Topological Invariants of Plane Curves and Caustics*. American Math. Soc., 1994.

[2] M. Ascher. *Ethnomathematics: A Multicultural View of Mathematical Ideas*. Chapman and Hall, 1998.

[3] M. Ascher. *Mathematics Elsewhere: An Exploration of Ideas Across Cultures*. Princeton, 2002.

[4] M. Ascher. Malekula sand tracings: A case in ethnomathematics. In *Proceedings of BRIDGES: Mathematical Connections in Art, Music and Science*, pages 57–64, Banff, Canada, 2005.

[5] N. Bonichon, C. Gavoille, N. Hanusse, D. Poulalhon, and G. Schaeffer. Planar graphs, via well-orderly maps and trees. In *Proc. 30th Int. Workshop Graph Th. Concepts Comp. Sci.*, pp. 270–284, 2004.

[6] J. M. Boyer and W. J. Myrvold. On the cutting edge: simplified $O(n)$ planarity by edge addition. *Journal of Graph Algorithms and Applications*, 8(3):241–273, 2004.

[7] F. J. C. de Carvalho. Characterizing maximally looped closed curves. *Amer. Math. Monthly*, 92(3):202–207, 1985.

[8] P. Erdős and T. Gallai. Graphs with prescribed degrees of vertices. *Mat. Lapok*, 11:264–274, 1960.

[9] P. Franklin. The Euclidean algorithm. *Amer. Math. Monthly*, 63(9):663–664, 1956.

[10] M. L. Gargano and J. W. Kennedy. Gaussian graphs and digraphs. In *Proc. 25th Southeastern Int. Conf. Combinatorics, Graph Theory and Computing, Congressus Numerantium* 101:161–170, 1994.

[11] C. F. Gauß. Zur Geometrie der Lage für zwei Raumdimensionen. In *Werke (Collected Works)*, volume 8, pages 282–286. Springer-Verlag, 1990.

[12] P. Gerdes. *Geometry From Africa: Mathematical and Educational Explorations*. MAA, 1999.

[13] S. L. Hakimi. On realizability of a set of integers as degrees of the vertices of a linear graph. I. *SIAM Journal on Applied Mathematics*, 10:496–506, 1962.

[14] J. Kennedy, B. Servatius, and H. Servatius. A note on gaussian graphs. *Congressus Numerantium*, 112:112–117, 1995.

[15] V. A. Liskovets and T. R. S. Walsh. Enumeration of eulerian and unicursal planar maps. *Discrete Math.*, 282(1–3):209–221, May 2004.

[16] R. Otter. The number of trees. *Annals of Mathematics*, 49:583–599, 1948.

[17] T. Ozawa. On Halpern's conjecture for closed plane curves. *Proc. AMS*, 92(4):554–560, 1984.

[18] N. Sloane. On-line encyclopedia of integer sequences. http://www.research.att.com/~njas/sequences/.

[19] G. T. Toussaint. The Euclidean algorithm generates traditional musical rhythms. In *Proceedings of BRIDGES: Mathematical Connections in Art, Music and Science*, pages 47–56, Banff, Canada, 2005.

[20] W. T. Tutte. How to draw a graph. *Proc. London Math. Soc., Series 3*, 13:743–767, 1963.

[21] J. v. Sz. Nagy. Über ein topologisches Problem von Gauß. *Math. Zeit.*, 26(1):579–592, 1927.

[22] C. Zaslavsky. *Africa Counts: Number and Pattern in African Cultures*. Lawrence Hill Books, 1999.

Symmetric Characteristics of Traditional Hawaiian Patterns: a Computer Model

Tony Cao
School of Architecture
University of Hawaii at Manoa
Honolulu, HI 96822, U.S.A.
E-mail: tcao@hawaii.edu

Jin-Ho Park
Department of Architecture
Inha University
Incheon, 402-751 Korea
Email: jinhopark@inha.ac.kr

Abstract

Most of Hawaiian quilts, fabrics and traditional handicrafts are lavishly decorated with patterns. Reflecting the culture of Hawaii, Hawaiian flora and fauna find their creation in a fabric of symmetrical patterns. Although this exotic and highly balanced symmetry is an essential component of many traditional handicrafts in Hawaiian patterns, the symmetric principles of Hawaiian patterns have rarely been discussed. To provide insight into the creation of Hawaiian patterns, this article analyzes the symmetric characteristics of the traditional Hawaiian patterns. In addition, the article presents a computer model using a java applet that has been developed to generate an exponential number of different Hawaiian patterns.

1. Introduction

Traditional Hawaiian patterns appearing on quilt and bark cloth, known as *kapa and tapa*, have long been appreciated for their beauty, due to their unique design patterns. Hawaiians began to decorate their fabrics not only to represent Hawaii's natural beauty, but also Hawaii's culture. Little is known about the actual emergence of the patterns and about the thought process behind the patterns. The original designers of these patterns kept the techniques and their meanings a guarded secret. Little or no access to the design was allowed until the quilt was completed. In addition, Hawaiians did not have written language until the missionaries from New England arrived in 1820 to record the techniques. Nevertheless, as the centuries passed, this practice has been developed into a system of sophisticated ornamentation that spread into other areas of Polynesia.

Many attempts have been made to explain how Hawaiian patterns were designed. One such attempt was to decipher the traditional Hawaiian patterns using the "snowflake". For example, Poakalani and John Serrao (1997), Stewart (1986) analyzed the Hawaiian quilt with the paper folding and cutting technique. They dissected the Hawaiian patterns, and then used the process to create a larger collection of traditional Hawaiian quilt patterns. In doing so, they found that the pivotal part of the pattern was its center, for it was the center that allowed creativity to branch out within the borders. Similarly, the Hawaiians believed that a spiritual and physical center provided the foundation for growth. This snowflake method closely resembles the method for creating the appliquéd designs.

Although a variety of explanations is possible in analyzing Hawaiian patterns, fundamental to all these designs is the principle of symmetry. Yet, little attention has been given to the principles of symmetry in the traditional Hawaiian patterns. Moreover, a systematic analysis of Hawaiian patterns with regard to symmetry principles is almost nonexistent, even when literature suggests that, interesting enough, most of the Hawaiian patterns are based on what we call "point group symmetry", and the "frieze and wallpaper group symmetry". It is compelling, then to discuss the patterns with regard to symmetry principles.

In this article, the historic backgrounds and evolutions of Hawaiian patterns are discussed first. Second, the characteristics of Hawaiian quilts are described. Third, a procedural traditional technique for constructing the patterns is explained. Fourth, the patterns with regard to symmetry principles that underlie composing the patterns are detailed. Finally, given the symmetric principles and Hawaiian motifs, an interactive computer model using a Java applet to generate a number of variations of Hawaiian patterns is presented. The computer model is a procedure for constructing a myriad of new Hawaiian patterns. Examples of patterns generated with a software implementation of the technique are provided.

2. History and Evolutions of Hawaiian Patterns

Even before the missionaries arrived on the islands, Hawaiian women used beautifully textured bark cloth (*tapa*) made from the inner bark of the paper mulberry tree (*wauke)* for clothing and other utilitarian and ceremonial purposes (Titcomb, 1975).

The making of *tapa* was a time-consuming and labor intensive process. The outer bark was stripped away and the inner bark was soaked in water to make it soft. Artistically carved round wooden mallets were used to pound the strips of the inner bark into meshed fiber to form sheets of various sizes, thicknesses, and textures. Then, the *tapa* was colored by native dyes and decorated with block prints (Arthur, 2002). The beauty of the tapa creations were found in fine bedding, consisting of six sheets of *tapa*: four white sheets, a dyed sheet, and a decorated top sheet (Schleck, 1986). Women took pride and care in developing their own designs in the mallets, and tapa stamps.

New England missionaries arrived in 1820 and taught high-ranking Hawaiian women to make American-style patchwork quilts. On April 3, 1820, seven young New England missionary women held the first "Sewing Circle" aboard the brig Thaddeus. The high-ranking Hawaiian women were Kalakua, mother of King Liholiho, her sister Namahan, and two wives of Chief Kalanimoku. The missionary wives were Lucy Thurston, Lucia Holman, Sybil Bingham, Nancy Ruggles, Mercy Whitney, Jerusha Chamberlin, and Elisha Loomis. The missionaries supplied the scissors and furnished the native women with calico patchwork to sew. Sewing calico piecework was new to the ranked Hawaiian women, but the geometric and symmetrical patterns associated with quilting were not (Stewart, 1986). The Hawaiian women were taught both types of quilt making. The pieced variety technique was used to make geometric repeated blocks, and the appliqué style used the "snowflakes" method (Stewart, 1986).

While the missionaries had fabric scraps left over after cutting out their garment making, Hawaiian's garments left few scraps of fabric. The Hawaiians saw no point in cutting up the whole piece of fabric into small pieces only to be sewn back together again. Quilting began to change as the Hawaiian women created their own methods of constructing and designing a quilt, which is known today as the traditional Hawaiian quilt. The patterns on the quilts reflected their culture, heritage, and beliefs; the beauty of the islands; and historical events (Poakalani and J. Serrao, 1997).

Figure 1: Historic quilts used in a room decoration (Courtesy from the Mission Houses Museum, Honolulu, Hawaii)

By late 1970, while the traditional Hawaiian quilts still existed, a new form of quilting known as the contemporary Hawaiian quilt began to emerge. Contemporary Hawaiian quilt designs had no rules. The designs were often visual representations of Hawaii, often realistic and pictorial, and easily understood by the viewer. This new quilt evolution was possibly due to rapid changes society and the integration of different cultures in Hawaii (Arthur, 2002).

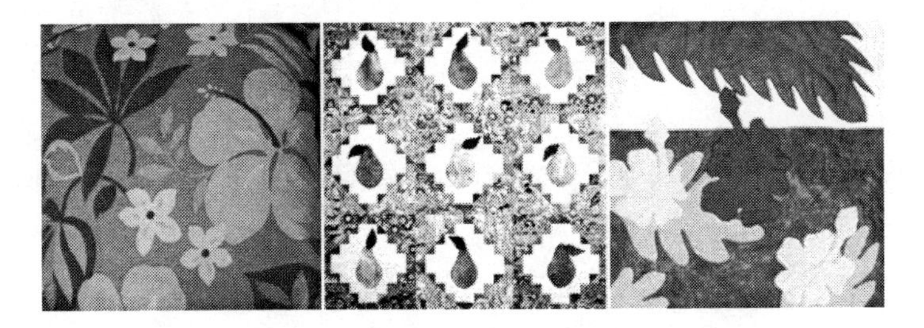

Figure 2: Three examples of contemporary Hawaiian quilts

3. The Symmetry of the Hawaiian Quilt Designs

This paper focuses on the symmetry of the Hawaiian quilt designs. Since ancient times, designs on Hawaiian quilts were created by using techniques of symmetry. Early designs were produced by carving designs onto bamboo strips called 'ohe kapala strips" (Figure 3a). These bamboo strips could be arranged into groups side by side, end to end, or in other unique patterns. Multiple groups of bamboo strips could also be formed to create motifs (Kaeppler, 1980). Examples of these patterns are shown in Figure 2b & 2c. These patterns were then used to make impressions on tapa, or cloth made of tree bark.

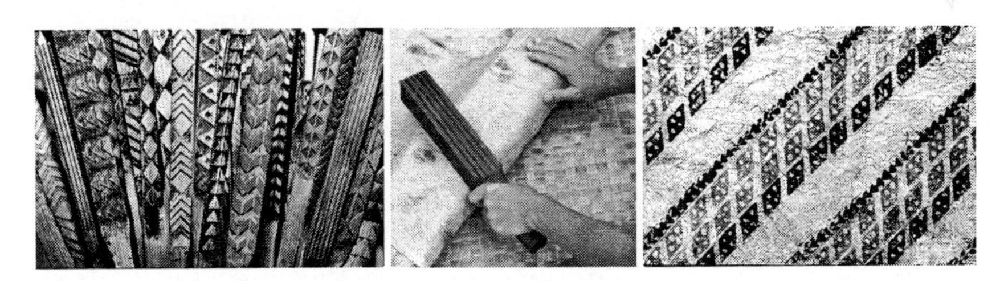

Figure 3: a. Different types of 'ohe kapala strips; b. Beating Kapa; c. Repetative Kapa pattern

Tapa can be characterized by its texture and thickness. The thin and soft-textured tapa had lighter and smaller designs created by pressing it onto carved patterns on the surface of bamboo strips. Thick tapa had bold designs created by pressing it into the grooves of the narrow channels in the patterns of the bamboo strips (Kaeppler, 1980).

Over time, one technique that dominated Hawaiian quilt designs was the symmetrical design in the square. This popular design utilizes the "snowflake" method to create a symmetrical pattern in a square. The use of the square finds its roots in Hawaiian religious beliefs (Poakalani and John Serrao, 1997).

The "snowflake" method begins with a square paper of the size that will be appliqu‎éd to produce the quilt. The square paper can be folded into halves, fourths, sixths, or eighths. This produces perfect symmetry in both simple and complex designs (Root, 1989). For example, folding a square in sixths can create a circular symmetrical configuration. Folding a square into eighths, the most common technique, extended designs to the corners of the square (Figure 4). Because the folds are either on a bias (diagonal), horizontal, or vertical, the patterns can be aligned with the grain of the fabric. The "snowflake" design is cut from a single piece of fabric, then appliquéd onto a background (top) square of fabric, and finally quilted with the batting and the bottom fabric backing.

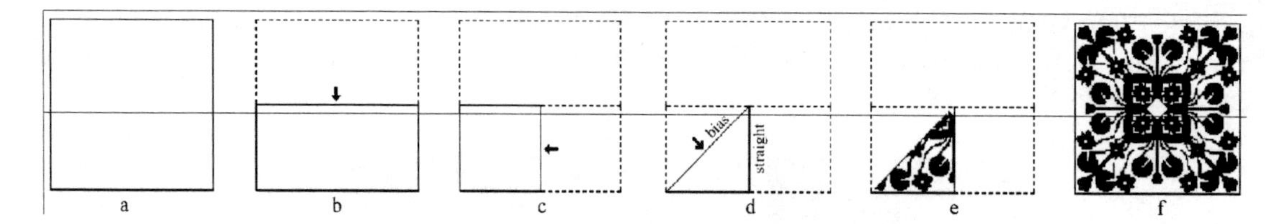

Figure 4: Paper folding technique with a square: a. Square flat material; b. Folding the material in half; c. Folding the material in half again; d. Folding the material diagonally; e. Asymmetric motif to folded square and cut out design; f. final pattern unfolded

Figure 5: a: Typical Hawaiian quilt pattern; b: Contemporary asymmetric quilt pattern (Redrawn by the authors)

The symmetry of the quilt patterns captured the interest of the authors of this article because they were intrigued with the possibility that the designs could be replicated using computer technology. The symmetrical techniques of the designs were, thus, analyzed. The findings of this analysis are presented next.

4. Analysis of the Symmetrical Techniques Used in the Quilt Designs

Two dimensional plane symmetry designs can be categorized into two groups: finite (or point symmetry) and infinite. Spatial transformations take place in a fixed point or line. The transformations involve rotation about the point and reflection along the lines, or the combination of both. In the point symmetry group, no translation takes place.

In the infinite symmetry group, spatial transformations occur where the basic movement is either a translation or a glide reflection. In this group, designs which are invariant under one directional translation are called the frieze group, and designs under two directional translations are called the wallpaper group (March and Steadman, 1971; Park, 2004).

Finite or Point Symmetry: Many variations of the symmetrical designs in the square can be found (Hammond, 1986). See Figure 6. The typical "snowflake" eight-fold design is shown in Figure 6a. Figure 6b illustrates a vertical reflective symmetry. Figure 6c, "My Beloved Flag," is a renowned emblem of an overthrown monarchy. Although the central symbol is asymmetrical, the flags that surround the symbol are produced by four quarter-turn cyclical symmetry. Figure 6d represents a quilt pattern based on four quarter-turns without reflections. Figure 6e illustrates a motif in which four quarter-turns, 10-fold, and 18-fold rotations are respectively superimposed. Stewart (1986) provides other designs that rely on symmetry for their creation (Figure 7). Interestingly enough, a half turn rotation design could not be found in the research conducted for this paper.

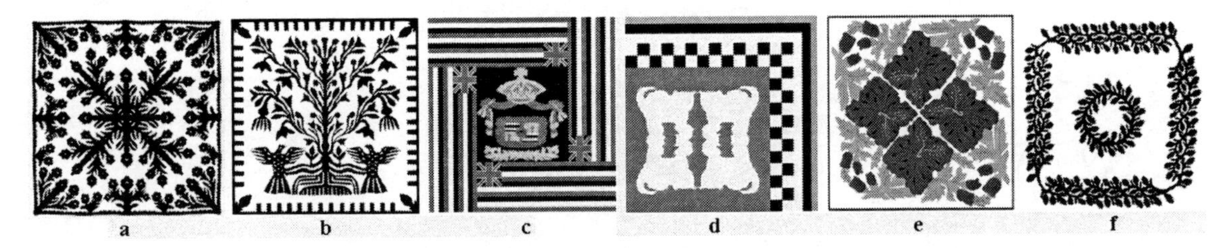

Figure 6: Six Hawaiian Quilts examples (Redrawn by the authors)

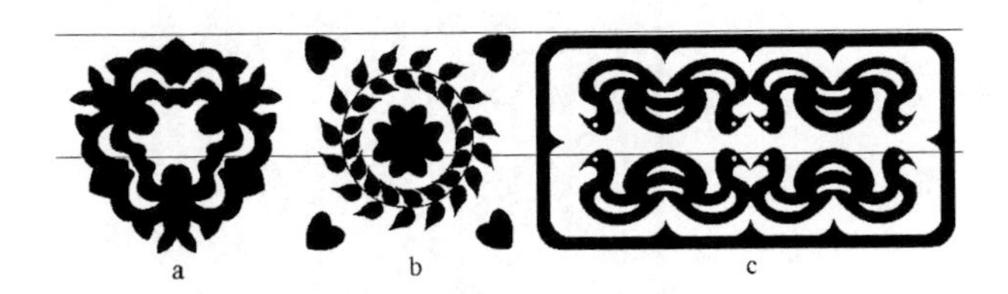

Figure 7: Three Hawaiian Quilts that show different symmetry: a, triangular symmetry; b. superimposition of the symmetry of a square and a 16-fold rotation; c. rectangular symmetry (Redrawn by the authors)

Infinite Symmetry: Frieze and wallpaper symmetry reflect the use of bamboo stamps (Figure 8) used long ago to create designs. Geometric designs such as lines, triangles, circles, and other symbolic motifs carved into bamboo were used to repeat patterns. Symmetrical patterns created by the repetitions show patterns that were reflected, rotated, and translated along a line (Figure 9). The prolific repetition of the patterns on a single item created rich designs. Opportunities to use each design to produce other designs are infinite.

Wallpaper patterns (Figure 10) are similar to frieze patterns. Distributions of patterns are of equal distance horizontally and/or vertically.

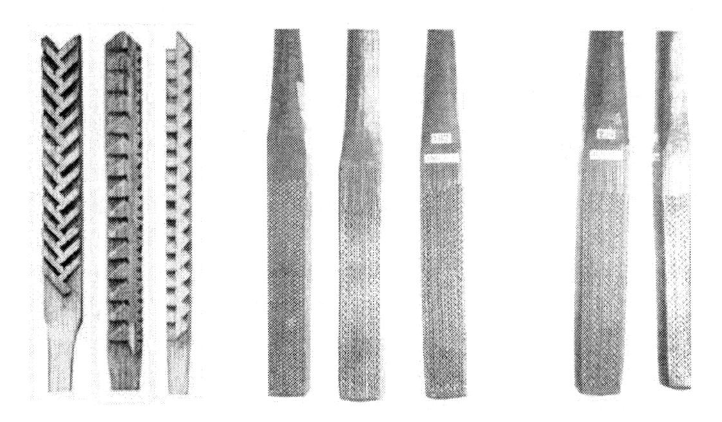

Figure 8: Contemporary (a) and traditional (b) examples of various bamboo stamps

Figure 9: Left. Various Frieze patterns found in Hawaiian ornamentation

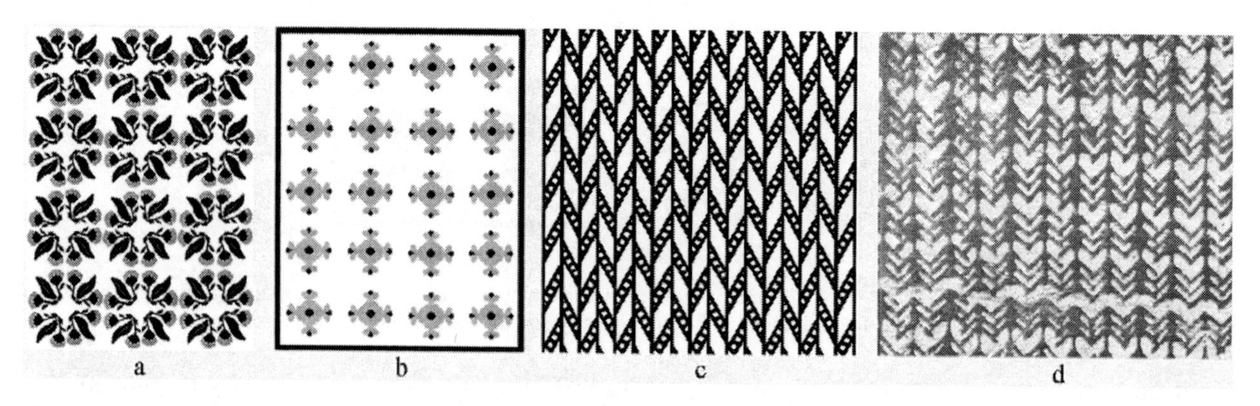

Figure 10: Four Wallpaper patterns found in Hawaiian patchwork (a and b) and *kapa* (c and d). The way that symmetry was used is simple and straightforward.

5. Creating Hawaiian Design Symmetry on the Computer

Computer technology can not only be used to copy Hawaiian design symmetry, but to create an infinite number of patterns by duplicating, reflecting, rotating, superimposing, attaching, and shrinking or reducing single designs. The authors saw this potential and developed a Java applet model that includes point, line, and plane group symmetry.

Using the Java applet application, motifs are viewed as lines and/or curves that can be created by plotting points (x and y coordinates) with mathematical lines and third cubical Spline equations. X and y coordinates of points in rows represent a line or a curve segment of the motif. The Java program then generates a motif path that can be filled with color when displayed on the computer screen.

To create a motif, a Java input file must be created. First, the motif image is scanned into a digital format. Second, the digital image application is used to read the scanned digital file. Third, the x and y coordinates of point in the motif are recorded. Fourth, the coordinates of the points for each line or curve segment are entered. Then, the process is repeated until the entire motif is completed. Line and curve segments can be made continuous from their end points.

Figure 11 shows snapshots of the implementation of a Java-based interface for dynamic retrieval of a set of pattern designs. Initially, the simplest common denominator of a design is created using x and y coordinates. The reason for using the simplest possible motif is that, on one hand, the hierarchical structure of the whole pattern is clearly revealed; on the other hand, developments in complexity can occur rapidly by adding other symmetric motifs, one-by-one (Budden, 1972).

One can design a quilt by clicking on buttons in the design where motif transformation is desired. The Java program ensures that the design is always transformed according to symmetry principles.

Applying the Java motif files created for this paper, the user can create various patterns by selecting a basic motif from three categories of motifs. Then, the user who wishes to create a design according to the point group symmetry can choose the reflect or rotate button. Otherwise, if the user wants to create a design based on the line or plan group symmetry, the user can select from 7 to 17 different types of symmetry patterns. In essence, this powerful applet model can generate a myriad of Hawaiian designs using point, line, and plane group symmetry with the chosen motif that is limited only by one's imagination.

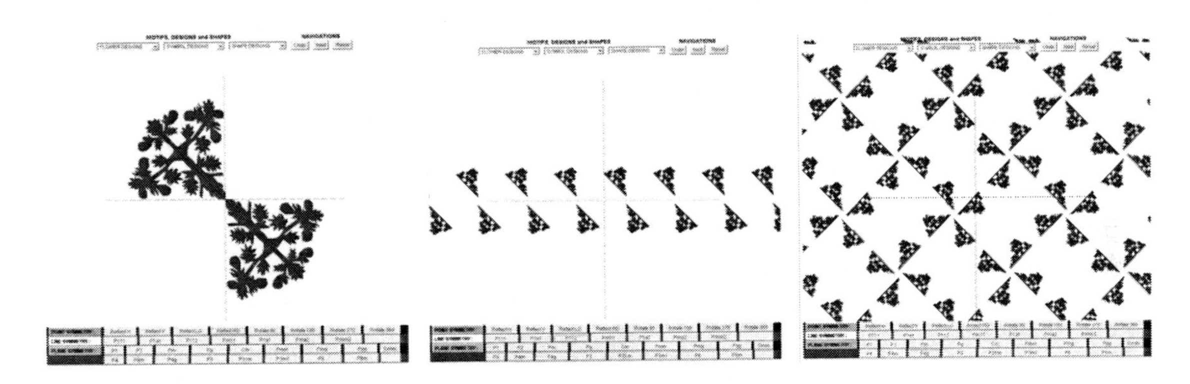

Figure 11: Java applet design examples of the point, line, and plane group symmetry

Conclusion

This paper has surveyed and analyzed traditional Hawaiian patterns with regard to symmetry. At a glance, Hawaiian quilt patterns seem to vary greatly as if disengaged from a systematized design. However, this study identifies an underlying carefully ordered and balanced symmetry in the Hawaiian patterns. Without the help of mathematical models, Hawaiians seemed to have understood the notion of geometric order as they intuitively constructed patterns base on countless experiences.

With the Java applet model, however, the possibilities of creating symmetrical Hawaiian patterns are endless. The most striking implication of this Java model is that it can generate bountiful numbers of new Hawaiian patterns based on existing motifs, thus enriching Hawaiian designs on various products.

References

[1] L. March and P. Steadman, The Geometry of Environment, RIBA Publications Limited, London, 1971.
[2] F. J. Budden, The Fascination of Groups, London: Cambridge University Press, 1972.
[3] C. J. Stewart, Snowflakes in the Sun: a How-To-Guide To Hawaiian Quilting, Lombard, Ill: Wallace-Homestead Book Co., 1986.
[4] Poakalani and J. Serrao, The Hawaiian Quilt, A Spiritual Experience, Honolulu: Mutual Publishing, 1997.

[5] R. M. Brandon and L. Woodard, Hawaiian Quilts: Tradition and Transition, Honolulu: Honolulu Academy of Arts, 2004
[6] M. Titcomb, The ancient Hawaiians: how they clothed themselves, Honolulu: Hogarth Press, 1975.
[7] L. Arthur, At the Cutting Edge – Contemporary Hawaiian Quilting, Honolulu: Island Heritage Publishing, 2002.
[8] A. Kaeppler, Kapa: Hawaiian bark cloth, Hilo Bay, Hawaii: Boom Books, 1980.
[9] E. Root, Hawaiian Quilting, Instructions and Full-Size Patterns for 20 Blocks, New York: Dover Publications, 1989.
[10] R. Schleck, The Wilcox Quilts in Hawaii, Kauai: Grove Farm Homestead & Waioli Mission House, 1986.
[11] J. D. Hammond, Tifaifai and Quilts of Polynesia, Honolulu: University of Hawaii Press, 1986.
[12] J.-H. Park, "Symmetry and Subsymmetry as Characteristic of Form-making - The Schindler Shelter," Journal of Architectural and Planning Research, 21(1): 24-37, 2004.

Circle Folded Helices

Bradford Hansen-Smith
4606 N. Elston #3
Chicago, IL, 60630, USA
wholemovement@sbcglobal.net
www.wholemovement.com

Abstract

Helices are explored as functions of circle reformation using observations that the circle functions as both Whole and parts in ways no other shape or form demonstrates. The generalization of tubes and cones, parallel surface and non-parallel surface, is fundamental to reforming the circle reveling countless variations in the helix and conical helices. The circle can generate forms that in multiples will model natural growth systems revealing a dynamic process reflecting the interrelated nature of universe order. The helix and conical helix are uniquely demonstrated in the first right angle movement of the circle to itself and fundamental to all subsequent folding of the circle.

Curving the circle

Draw a circle and cut it from the paper plane. The circle has three surfaces and two edges. The compression of the sphere shows the same properties as the cutout circle; the only form that demonstrates the concept of a singular self-referencing spherical Whole. [1].
Draw a diameter on the circle (line AB in fig.1). Curve the boundary around so the opposite points on the diameter touch (point AB fig.1). This forms a tube with a circumference formed from diameter AB, reconfiguring the flat surface to a parallel surface. Theoretically this can be carried out indefinitely as a right angle function of the self-referencing circle diameter.

The movement of line AB to point AB is perpendicular to the diameter CD on the opposite side of the curved surface. This is a right angle function of movement AB to line CD forming a triangle plane of points AB,C,D. By rolling the circumference on itself the connection point AB separates forming a tetrahedral pattern of four points A,B,C,D. The surface is now sloping away from and into itself reconfiguring the tube into a cone. The point of connection has moved away from the diameter which continues to be at right angle to line CD. An end view will show concentric openings of circles as the connection point moves away from point AB. The cone is a generalized movement towards an infinitely remote outer boundary. Infinite movement out without boundary suggests infinite movement in without center. In this regard the circle has no center point beyond the limitations of the material.

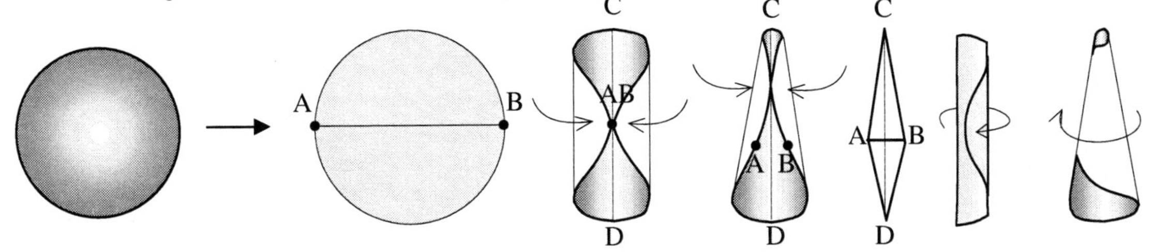

Figure 1. The sphere compressed to a circle with diameter curved end-to-end forms a tube. The tube moves into a cone as the circumference rolls. Overlapping the surface rolls the tube and cone tighter reducing diameters. The cone's open ends extend infinitely large out from, and infinitely small in to.

There are more combinations of touching points on the circumference than there are endless diameters in a circle. That suggests a preponderance of helix formations; a small part observable in natural growth forms. The potential in cone variations can be demonstrated in the numerous spiral formations that arise from folding and joining circles, in the same way spirals are observed in cosmic systems, flow forms, shells, molecular arrangements, and endless scaling universally throughout.

The four points of the open diameter AB, Fig.1 have six relationships between them forming the pattern of an irregular tetrahedron. The tetrahedron is structural pattern giving consistency necessary for reformation, transformations, generation and sustained growth. The tetrahedron/helices are observed everywhere as right angle functions of scaling patterns of movement.

Touching any two points on the circumference and creasing the circle will form two more points perpendicular and half way between the touching points. When the circle is folded in half a tetrahedron pattern is formed. Four points in space is tetrahedral. There are only six relationships between four points. The number ten describes the tetrahedron. This is observed in both curving and creasing the circle. This first movement of the circle is tetrahedral and it must follow that touching any two points on any surface and folding is a tetrahedron function.

Tetrahelix

The helix is reflected in the tetrahedron, the form is not obvious. The six edges of the tetrahedron can be divided into two sets of three, a right hand and a left hand path. There are no parallel edges showing a helix and no scale change to indicate a spiral. The tetrahedron "solid" is traditionally formed as an isolated object, an abstraction separated from any context. The context for the tetrahedron is spherical pattern, thus the importance of the circle/sphere.

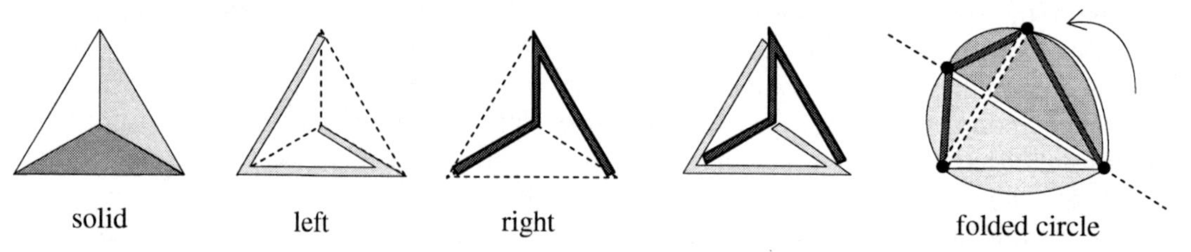

Figure 2. The dual right and left hand helix individually describe all four points of the tetrahedron and show the relationship of six edges. This is observable in the first fold of the circle, principle to all folding.

The tetrahedral packing of spheres (fig.3a) shows the tetrahedron pattern to be regular and at minimum four tetrahedra containing the octahedron interval and multiple relationships of bisectors. The unity of four spheres is a 2-frequency tetrahedron fundamental to all spatial formation. (2-frequency is an edge length in 2 equal parts.)The full helix unit of six segments, 3 sets of 2 each forms parallel edges (fig.3b). The two helix forms show 4 spheres and the 6 points of connection in spherical packing.

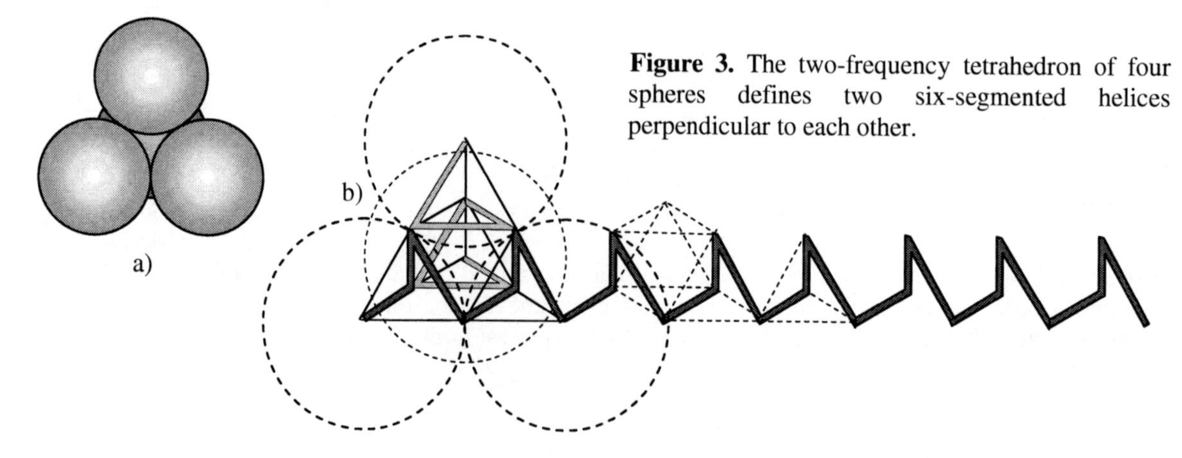

Figure 3. The two-frequency tetrahedron of four spheres defines two six-segmented helices perpendicular to each other.

By connecting the two ends of three edge tetrahedron through the center location of a single tetrahedron the helix becomes a loop connecting the inside and outside. Three open segments are now a five-segment path and a closed helix system connecting the center location and the outside boundary.

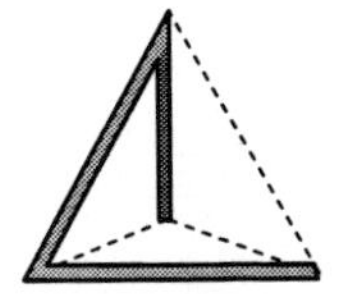
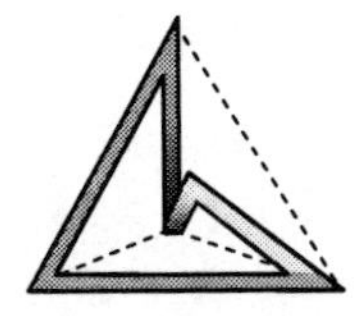
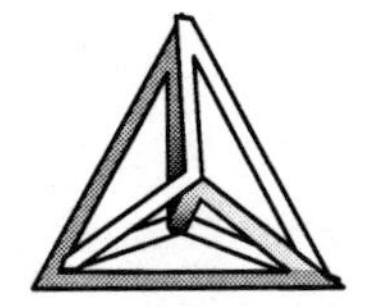

Figure 4. A one-frequency tetrahedron showing two associated loops of five segments each, where two of each five segments are shorter.

a)
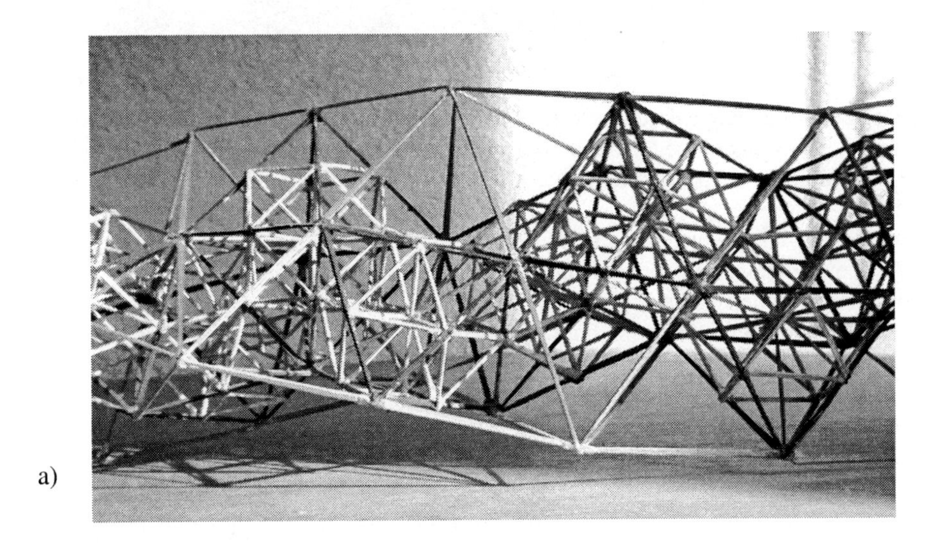

b)
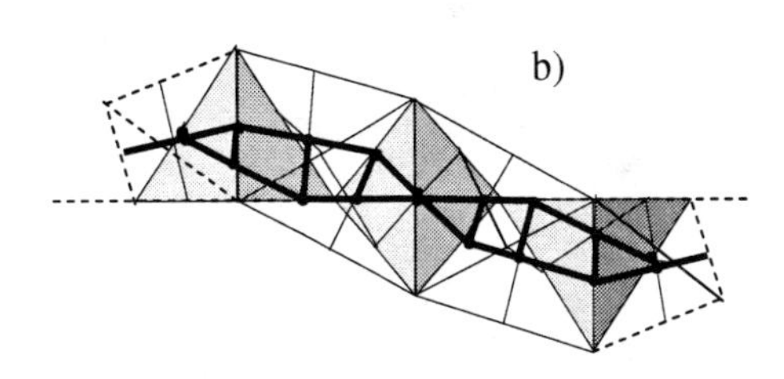

Figure 5. a) A stick model of a two-frequency tetrahelix showing the octahedron and vector equilibrium (traditionally called a cuboctahedron) helix forms. The relationship of these forms reveal 13 individual arrangements of primary helices that run through the two-frequency tetrahelix. b) This drawing shows two tetrahedral intervals between each of two tetrahedra touching on the points. The connection of one half of the three axis of each octahedron (3 of 6 radii at right angles) runs through the center of each tetrahedron and models the DNA double helix ladder. The direction of twisting of the double helix is opposite to the tetrahelix.

Circle Reformation to Helix

Forming a tetrahedron using the circle reveals nine folds [2]. By reconfiguring these nine creases many different forms of the circle can be created. Any number of these reconfigurations when produced in multiples of the same diameter circle will when consistently joined in a line create a helix formation. The most fundamental is in the form of the tetrahedron/octahedron combination reflected in the helix function of the two-frequency tetrahedron (fig.3).

The nine tetrahedron folds are a primary subset of folds in the 24 creases of a folded equilateral triangle grid (fig.6). Three diameters are perpendicularly divided into eight equal parts forming this grid [3]. The parallel creases of this grid allow the circle to reform into the fundamental helix forms. Folding the circle into an 8-frequency diameter grid is similar to an octave in music. From that primary differentiation of fundamentals and the intervals between there is endless potential of reformation.

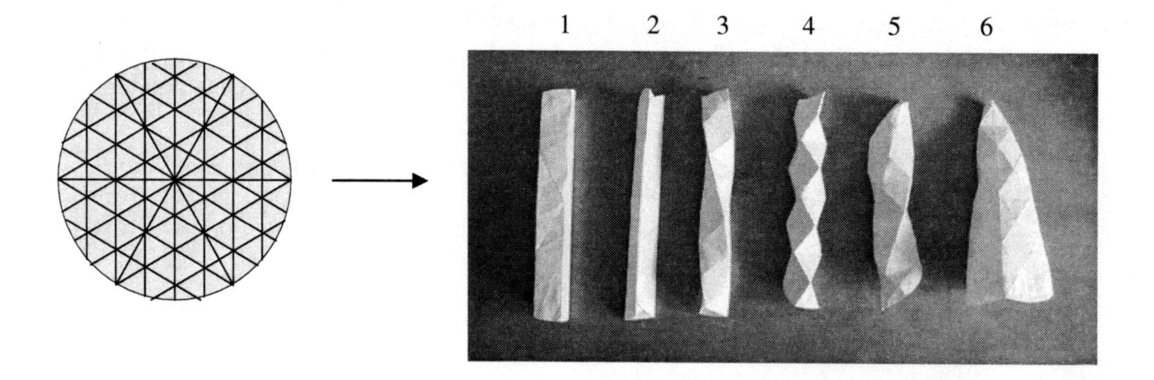

Figure 6. Folding the circle into an eight-frequency grid allows reforming the circle into the individual reconfigurations pictured above. The square tube (1) transforms into a triangular tube (2), into a tetrahelix (3), into an octahelix (4), then to a four-sided helix (5), and then opens to a conical helix (6). The triangle tube and helix forms are triangulated and rigid. The conical helix breaks the parallel edges and is less rigid. The square tube is without triangulation and collapses. The diameter in the triangle tube is joined to itself perpendicular to the tube length as shown in Fig. 1. The tetrahelix (3) comes from opening the diameter and connecting with the next parallel cord. The direction of opening determines right or left handedness. The diameter when opened further and keeping edges parallel forms the octahelix (4) with both right and left hand twisting edges. The square and triangle tubes are prisms in design and the twisting of parallel planes are anti-prisms.

Transformational Nature of the Tetrahelix

The tetrahelix modeled as a static object has properties but is without function (fig.6, #3). Using individual tetrahedra with hinge-joining edge connections between them rather than rigid face-to-face joining reveals some interesting changes. Fig. 7 shows the edge-to-edge joining of tetrahedra that can be considered a helix pattern since corresponding edge positions are parallel and twist around in a continuous path. There are two right hand and two left hand edge paths, four of the six edges. The top view shows a square image where each side is an edge path and diagonals are joining edges.

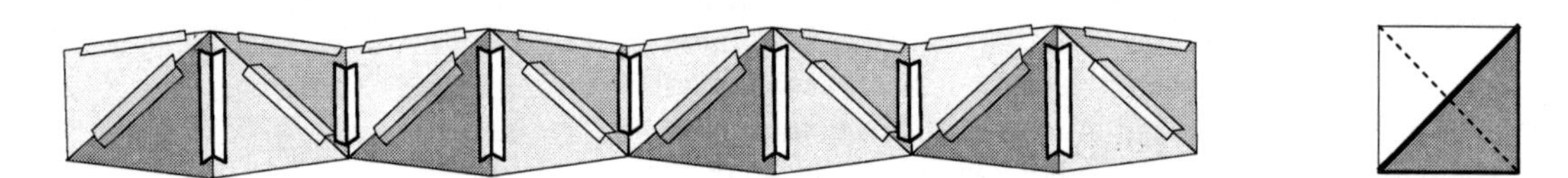

Figure 7. Eight tetrahedra hinge-joined on opposite edges. The square shows the tetrahedron, top view of the tetrahelix, where the diagonals are connecting edges perpendicular to each other. This right angle hinge-joining shows many variations in helix reformation. (By connecting the ends together with a tape hinge the helix will form a rotating Torus ring.)

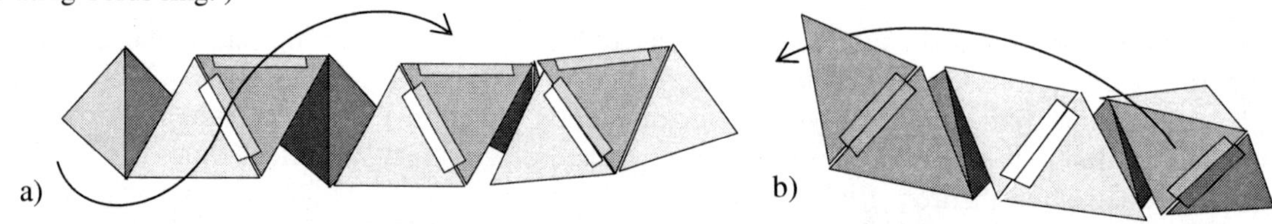

Figure 8. a) Eight tetrahedra twisted in one direction will reform to a tetrahelix joined surface-to-surface. Twisting in the opposite direction exposes the inside surfaces, hiding the outside to the inside. The direction of the twisting planes dose not change. b) Twisting to either direction, with alternating two triangle sides out and two triangle sides in, will change the direction of the three twisting planes from a right to a left hand helix. (Notice the dark and light differentiation between triangle surfaces.)

Figure 9. The tetrahelix with tetrahedra joined surface-to-surface

Figure 10. Using hinge-joining at intervals between sets of two tetrahedra the helix can be reformed to a tighter twisting.

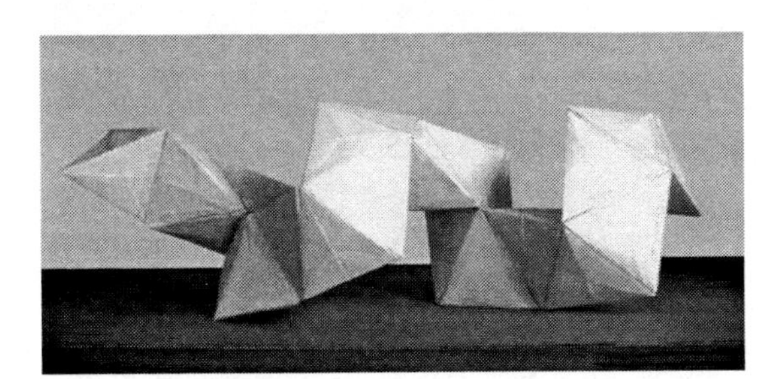

Figure 11. Here the tetrahelix is hinged in sets of six and hinge-joined together.

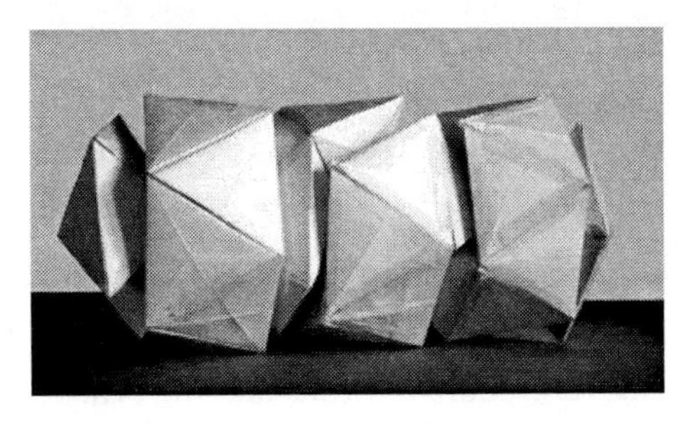

Figure 12. Another variation of the tetrahelix wound tighter through hinging.

Modular Growth of Conical Helices

Diminishing or enlarging the diameter of the circles used for a helix segments will turn it into a conical function. Much like curving the circle and separating the diameter points AB (fig.1) making multiple levels of parallel circle planes in different sizes, Each segment in a helix changes scale depending on how much the diameter for each circle folded is reduction or increased. The circle configurations are folded the same, the diameters differ corresponding to scale differential in individual stages of growth (fig.13a,b). There are many options in forming conical helices. The degree of open and closedness of the helix is correlated to the proportional difference between diameters and specific reformations being used. Generally the closer the intervals between diameters the number of units used the tighter it is. These differences can be generalized whereas the controlling development appears to be conditional, dependent on configuration interaction within the environment towards specific growth function. Natural growth forms are regulated by larger systems inherent to pattern. In modeling conical helices all variations and diversity of forms are inherent to the circle, otherwise they could not be formed. The possibilities lie between the generalizations of mathematical functions and the forming interactions of parts inherent to life systems. This seems to apply to both the circle and to growth forms in nature.

Figure 13 Two conical helices folded from different reforming of circles. a) 24 circles with a printed design have been folded, each fitting into the other. b) 14 circles folded into tetrahedra forming an open pentagon.

a)

b)

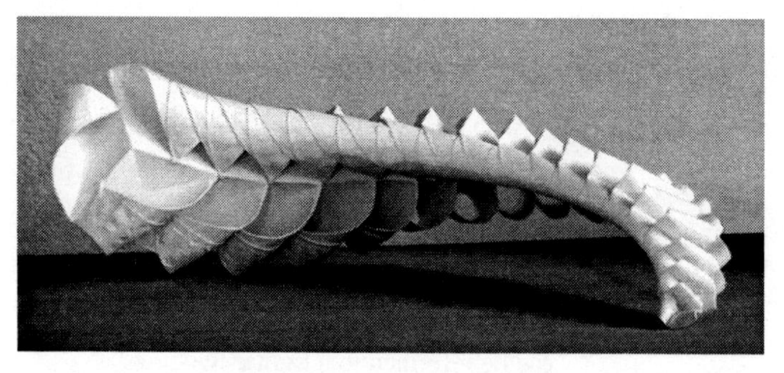

Figure 14. 18 paper plate circles. The first unit shows the form of the reconfigured, triangle folded circle. The folded segment is consistently the same with diminishing diameters.

Figure 15. 12 reformed paper plates held together with hair pins.

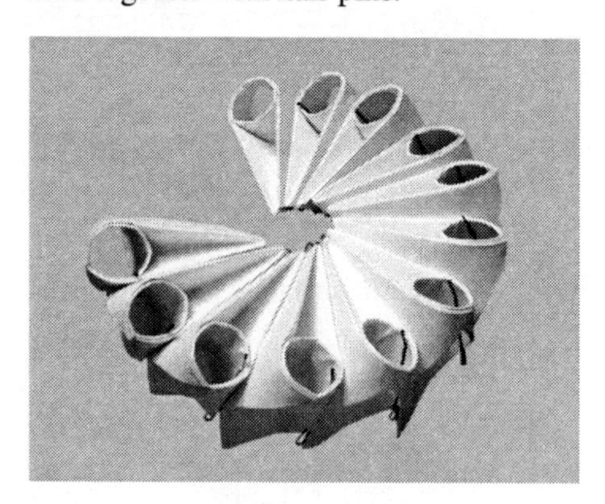

Variations in Conical Helices

There are endless variations in forming conical helices. This is obvious in nature and in folding and joining circles. The circumference can be folded in, out, and in combinations; only the circle can do that. The changing diameters, the number of units, sequencing of segmented sets, and open/closed forming of units are all variables that can be creased, curved, or both. When working with triangular units there are three primary positions that can change direction of orientation. The sequencing of set numbers and the rotational positions of each unit will change the angle and direction of the movement in relationship to the possible ways of joining. This is similar to the articulation between musical notes.

Figure 16. Double end conical helix using paper with printed image in gradation. Angles of direction changed by rotation of units.

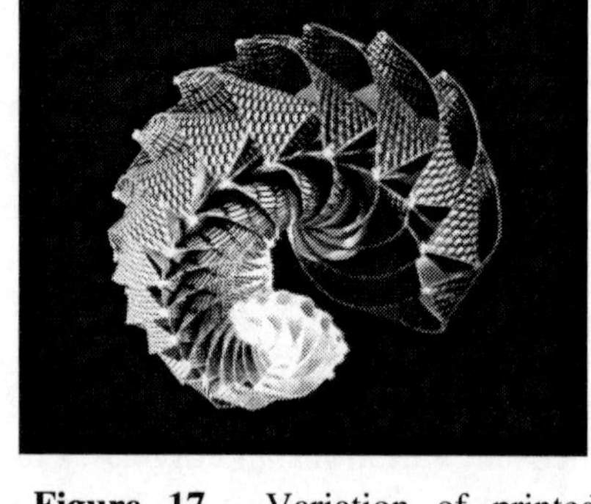

Figure 17. Variation of printed units used in fig.16.

a) 

b) 

Figure 18. a) View of 14 circles in tight twist. b) Opposite view shows how the circles have been formed where each circle is folded into an equilateral triangle and two ends twisted into a mobius surface.

a)

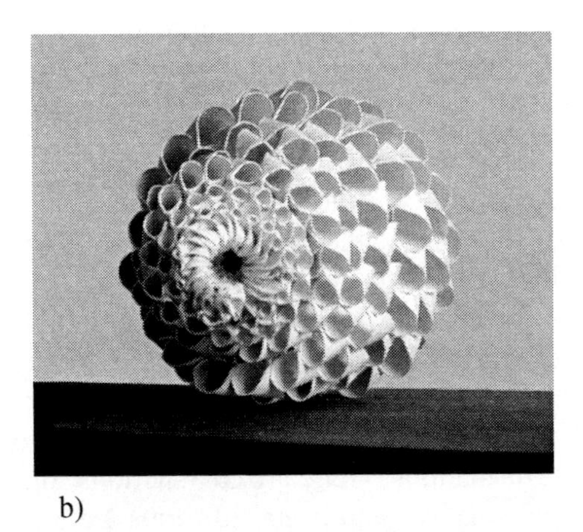

b)

Figure 20. a) This model is made using 162 circles all folded the same. It is a variation from the unit used in Fig. 14. These circles are folded tighter and joined closer together making a more compact growth path. Each revolution touches itself. The units decrease in sets of eight until they get to the end where the numbers of units in a set diminishes. b) The end view shows the axial hole open all the way through both ends that is unobservable from the side.

Figure 21. One rotation using 15 reconfigured circles.

Figure 22. The units used here are a variation to those in fig.15. They are arranged differently showing 1 ½ turns.

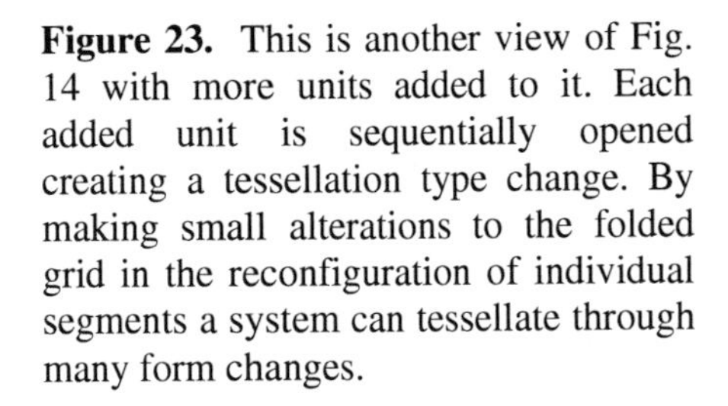

Figure 23. This is another view of Fig. 14 with more units added to it. Each added unit is sequentially opened creating a tessellation type change. By making small alterations to the folded grid in the reconfiguration of individual segments a system can tessellate through many form changes.

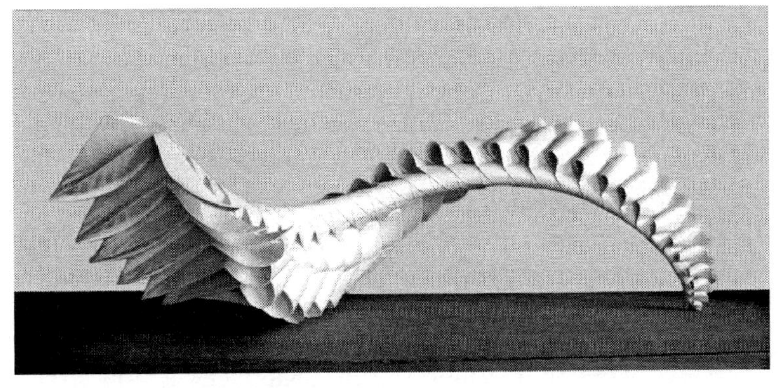

There are great advantages for using circles to model the geometry in nature. The greatest advantage is that it is far more comprehensive than any other form of modeling and will generate forms and designs not possible in other ways. The circle is not restricted by limited number of sides or scale. It can form all traditional planer configurations as well as those things that can not be done by other means.

Nothing is added or taken away. There is simply the movement of the circle to itself demonstrating the process of transformation and change which these pictures of models indicate. With the circle everything becomes informational, nothing is ambiguous or arbitrary. The circle reveals a principled process of movement in the context of endless forming systems through folded reformations. Generalized functions, discrete information, and beautiful forms are modeled that reflect patterns of growth in the natural world. These models show but a few forms of the endless possibilities of the helix growth pattern.

Figure 25. Detail of a helix branching system of tightly folded circles in conical form showing a fractal like growth.

Figure 24. The units for this system are the same used in Fig.16,17, and 20. In each case the fit of one segment into another has been modified changing the angulation. Different diameter intervals change the angle of rotation making each model a uniquely different form.

Conclusion

Folding circles expands the information of traditional geometry and generalized mathematical functions, revealing beautiful and dynamic forms in a rather poetic way, that allows exploration of the richness in diverse forms observed in nature, particularly evident with helical growth. The forms tend to show process and structural design rather than reproducing outward appearances. This is a small part of what is an enormous benefit in folding circles.

References

[1] Bradford Hansen-Smith, *The Geometry of Wholemovement: folding the circle for information.* Wholemovement publications, 2000

[2] Demonstrated on website; www.wholemovement.com

[3]Bridges Mathematical Connections in Art, Music, and Science. Conference Proceedings, 2004. Reza Sarhangi & Carlo Sequin, Editors. *Folding the circle, both Whole and Part,* B.Hansen-Smith, pp.87-94

The images in **Figures, 1, 2, 4, 5b,7 8a,b, 9-12, 13b, 15, 19b, 21, 24**, are from the author's book *Folding Circle Tetrahedra: truth in the geometry of wholemovement.* Wholemovement Publications, 2005

The Taming of Roelofs Polyhedra

Frits Göbel
Schubertlaan 28
7522 JS Enschede
Netherlands
E-mail: a.gobel@wxs.nl

Abstract

Roelofs polyhedra form a vast collection of polyhedra containing many interesting solids and including very irregular ones. The purpose of this paper is to consider two special subsets: polyhedra with the symmetry of the prism and polyhedra with just two different types of vertices. Beside the figures in the paper PowerPoint pictures, all made by Rinus Roelofs, will be presented.

1. Introduction

Some time ago Rinus Roelofs defined the following class of polyhedra:
a) The faces are convex regular polygons;
b) Some of the faces are intersecting.

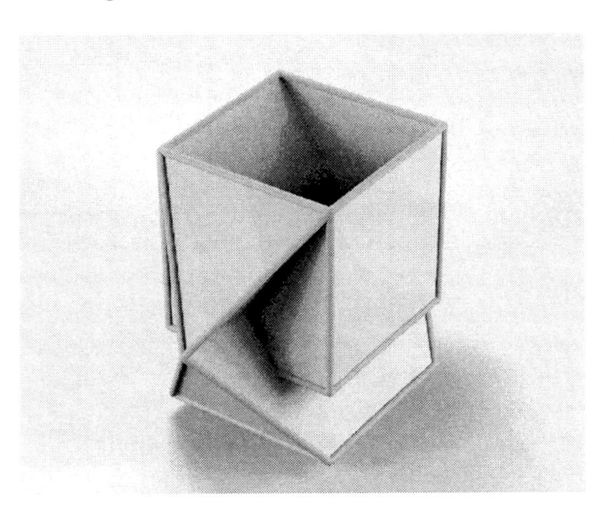

Figure 1: *Example of a Roelofs polyhedron.*

An example is shown in Figure.. These solids are now generally known as Roelofs polyhedra. From the definition it will be clear that there exists overlap with other categories. For example, all non-convex uniform polyhedra without star polygons belong to the class of Roelofs polyhedra. However, the latter contains a great deal more, including many irregular, not to say wild polyhedra. It is true that many of these irregular solids have interesting properties. Nevertheless, the subject of this paper to present a restricted subset of the Roelofs polyhedra with stress on simplicity, beauty, and symmetry.

In the next section we consider prism-like polyhedra, and in Section 3 we examine polyhedra with two types of vertices.

2. Prism-like Roelofs polyhedra

Archimedean prisms and antiprisms form two infinite series among the convex Archimedean solids. They are quite simple in their symmetry and are generally considered as the ugly ducklings. With the Roelofs polyhedra, several other series with the same symmetry arise, which makes them more interesting. Consider the following example.

Figure 2a, 2b, 2c: *Building a prism-like Roelofs polyhedron.*

Start with a four-sided antiprism on top of a cube; see Fig. 2a. The square half-way up is not one of the faces. Now push the cube upward; the faces of the two solids will intersect. The result is the 14-hedron shown in Fig. 2b. Repeating the process with an antiprism on the opposite face gives the solid in Fig. 2c; this one has the full prism symmetry D4. Please note that the solids in Fig. 2a, b, and c are not compounds.

The trick also works for n-sided prisms and antiprisms. And it can be repeated: form a pile of prisms and antiprisms, now and then applying intersections as above. Star prisms and star antiprisms (grammic or grammic-crossed) can be used provided the two extremal stars are covered by star-shaped pyramids. In the next section, where we consider a restricted subset, conditions on the stars will be given.

An alternative variation of the prisms is obtained when the lateral faces are used as bases for other polyhedra.

3. Two types of vertices

The uniform polyhedra have, by definition, only one type of vertex; they are vertex-transitive. The polyhedra we consider in this section are *next-to-uniform* in that they have two types of vertices.

As a first category we consider a subset of the prism-like solids mentioned in the previous section. The polyhedron in Fig. 2b has three types of vertices. After adding the second antiprism, as in Fig. 2c, the number of types drops to two. When allowing star prisms and antiprisms, the star faces must be covered by a star pyramid. Of course, the side length of the star has to be sufficiently large to make this possible. In detail: for an (n,k)-star (see Fig. 3) we must have gcd(n,k)=1, k>1 and moreover 2k<n<6k for a prism and 2k<n<3k for an antiprism.

Figure 3: *the 7/3 star.*

When working this out, nine infinite series of next-to-uniform solids with prism symmetry are obtained.

Other types of symmetry appear when starting with the Platonic solids. The simplest example is the tetrahedron with on each face a prism pushed inside. Applying the same trick on the cube, it turns out that the number of new solids need not be 6; also with 2 or 4 of them, appropriately placed, a next-to-uniform solid is obtained. The other Platonic solids have similar properties. An interesting example is the following. On the icosahedron choose four faces such that twelve vertices are all different. (This can be done, in fact in only one way up to isomorphism.) On each of these triangles push a tetrahedron inside. The result is a Roelofs polyhedron with 28 triangular faces with "microscopic" intersections.

Finally we consider Roelofs polyhedra with a small number of faces. Roelofs investigated the possible cases with 6 to 12 faces. For 12 faces he found about 60 solids, with no claim for completeness! From his list, eleven next-to-uniform cases can be extracted. These will be shown in the power-point presentation. The unique one with 11 faces is shown in Fig. 4.

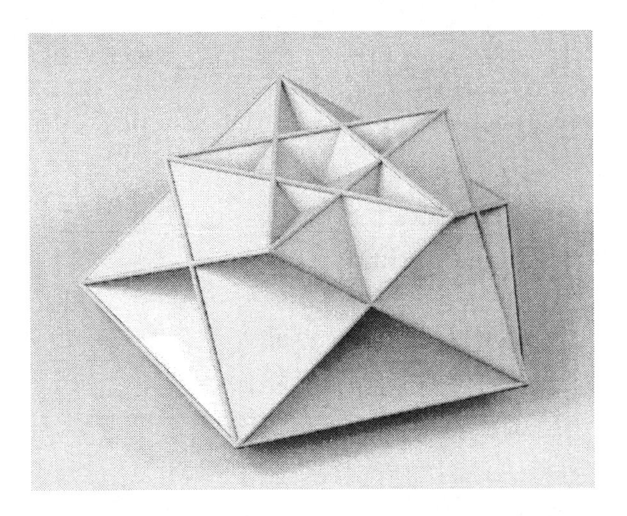

Figure 4: *the next-to-uniform 11-hedron.*

Roelofs also found a 30-hedron bounded by six dodecagons and 24 squares; see Figure 5. Johan van de Konijnenberg noted that this solid still has only two types of vertices when the dodecagons are replaced by 16-gons; the number of squares then increases to 36. These two polyhedra are the start of a remarkable infinite series with cube-symmetry. However, the number of types of vertices increases.

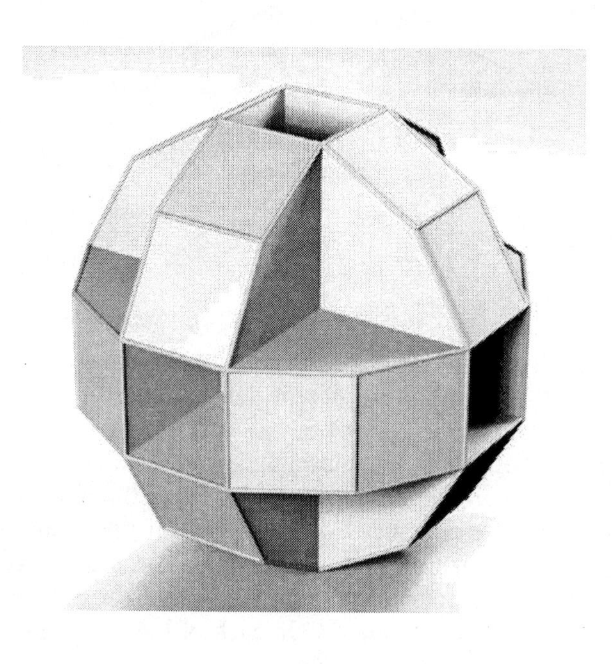

Figure 5: *the 12(6)4(24)-hedron.*

4. Conclusion

You have seen attempts to define subsets of Roelofs polyhedra which contain fewer irregular specimen. The restriction to next-to-uniform solids seems promising. Further research is necessary to find out whether certain characteristics of the irregular solids, such as invisible faces, will occur in the restricted set.

Acknowledgement

I am grateful to Rinus Roelofs who made all figures in the text as well as all power point pictures..

A Program to Interpolate (and Extrapolate) Between Turtle Programs

Ken Kahn
London Knowledge Lab
Institute of Education, University of London
23-29 Emerald Street
London, WC1N 3QS, UK
kenkahn@toontalk.com

Abstract

People have been creating geometric figures with computer programs consisting of turtle commands such as forward and right since the late 1960s [1]. Here I describe a program that takes in two such programs and produces a new program capable of producing both figures and all the intermediate figures. It can produce a figure that is one third circle and two thirds triangle or one that is half star and half pentagon. The program produced by interpolating, say, a square and a circle program takes in a number between zero and one and produces a figure between a square and a circle. If, however, it is given a number greater than one, or a negative number, it will produce an extrapolation between a square and circle.

Interpolated programs can be the basis of playful aesthetic explorations. The intermediate forms can be drawn on the same image. Or animations can be generated where the figures morph into (and beyond) each other. Colours and other attributes of the turtle pen can also be interpolated. Unlike conventional morphing programs, we are interpolating between computational processes rather than static images.

Interpolating and Extrapolating Geometric Figures

Imagine one had a program that could draw a blue square, a red triangle, and any intermediate shape. One could use it to produce the following image by calling the program with inputs 0.0, 0.01, 0.02, …, 0.98. 0.99, and 1.0:

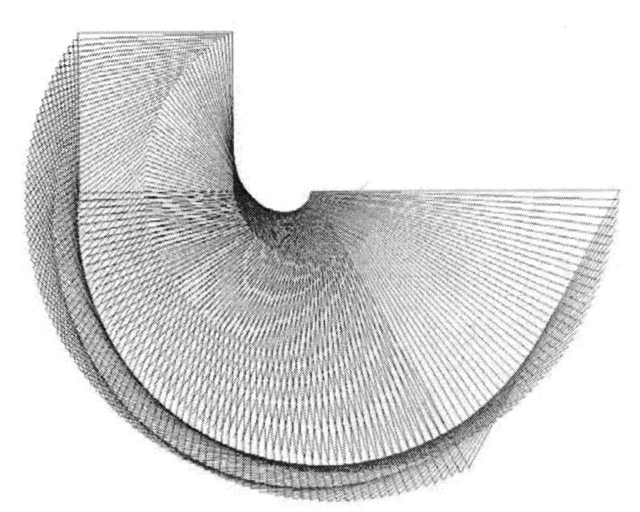

Figure 1: *Interpolation between a square and triangle*

If the input to the procedure is 0 it draws a square, if it is 1 it draws a triangle, and if it is 0.5 then it draws something that is half square and half triangle (an *interpolation*). But what if the input is 2? Can it draw something that is beyond triangle when starting from a square? Or if it is -1 can it draw something that is before square in the transition to triangle? Such figures are *extrapolations* between a square and triangle.

Here are 100 figures extrapolating between -0.5 to 1.5 for a circle (really a 360-side polygon) and a triangle. The original circle is at 9 o'clock in the image and the triangle is at 3 o'clock.

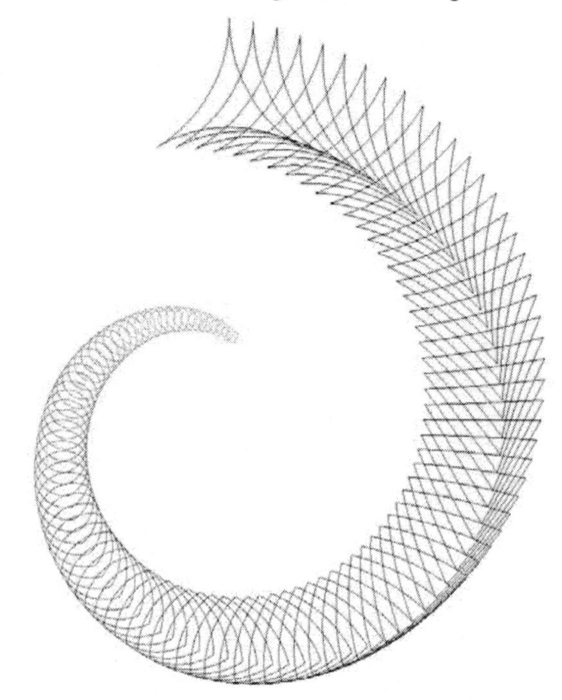

Figure 2: *Circle to triangle from -0.5 to 1.5 in 0.02 increments*

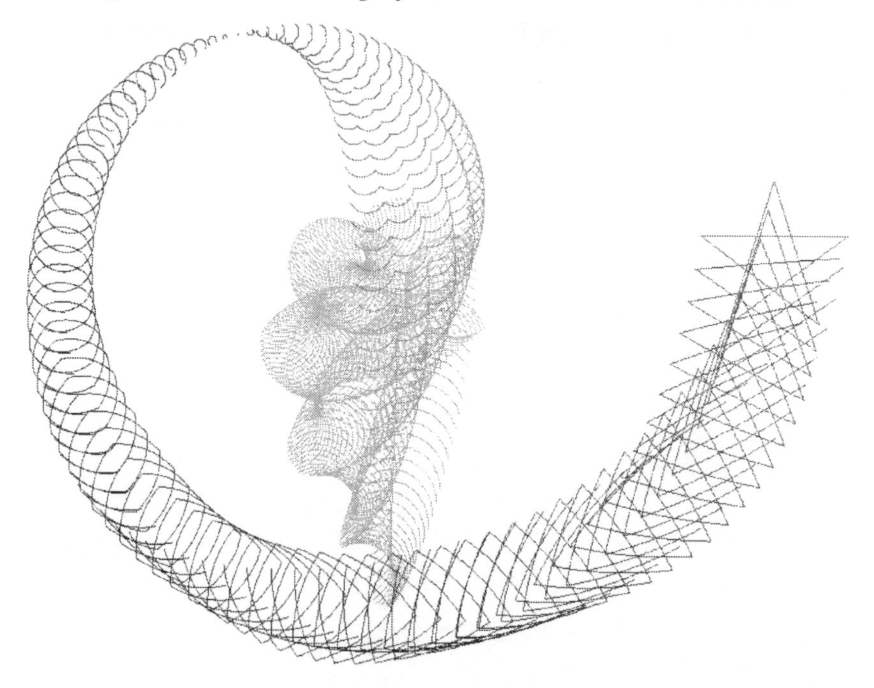

Figure 3: *A circle to a five-pointed star from -2 to 1 incremented by 0.02*

More extrapolations are illustrated in Figures 4 to 8.

How to interpolate turtle programs

Rather than building a special program for each pair of figures I wanted extrapolations, I built a program in Logo that takes as input two turtle Logo programs [1] and produces a new program. Consider the very simple example where the two programs draw lines:

```
to short
 forward 40
end
to long
 forward 100
end
```

The resulting interpolated program produced by my program is:

```
to short_to_long :x
 forward interpolate :x 40 100
end
```

Where *interpolate* is defined as:

```
to interpolate :x :a :b
 output :a + :x * (:b - :a)
end
```

When *short_to_long* is called with 0 it draws a line 40 units long and with 1, it is 100. When called with 0.5 the line is 70 units long, the average value. When called with 2 it is 160 units long, an extrapolation.

A challenging task is to produce an interpolation between a triangle drawn by **`repeat 3 [forward 100 right 120]`** and a square drawn by **`repeat 4 [forward 100 right 90]`**. The two programs need to be placed in a *canonical form* before the interpolation can be generated. My program does this by first computing the least common multiple of the repeat counts. The triangle program can be transformed to make 4 calls to *forward* repeated 3 times while the square program can be transformed to make 3 calls to *forward* repeated 4 times. A single call to *forward* is automatically transformed into many calls to *forward* by going forward a fraction of the original distance each time. However, for interpolation we need a standard form so the two programs "line" up. My program interpolator uses sequences of calls to *forward* and *right*. The triangle program is transformed to

```
repeat 3 [forward 100/4 right 0
          forward 100/4 right 0
          forward 100/4 right 0
          forward 100/4 right 120]
```

And the square program becomes

```
repeat 4 [forward 100/3 right 0
          forward 100/3 right 0
          forward 100/3 right 90]
```

Clearly the drawings produced by these programs will be identical to the originals. The next step is to "open code" the *repeat* statements so that we can interpolate corresponding commands. In the following table corresponding commands from the triangle and square programs are interpolated. When the values are the same there is no need to generate interpolation code since the value will be constant.

to triangle	to square	to triangle_to_square :x
forward 100/4	forward 100/3	forward interpolate :x 100/4 100/3
right 0	right 0	right 0
forward 100/4	forward 100/3	forward interpolate :x 100/4 100/3

right 0	right 0	right 0
forward 100/4	forward 100/3	forward interpolate :x 100/4 100/3
right 0	right 90	right interpolate :x 0 90
forward 100/4	forward 100/3	forward interpolate :x 100/4 100/3
right 120	right 0	right interpolate :x 120 0
forward 100/4	forward 100/3	forward interpolate :x 100/4 100/3
right 0	right 0	right 0
forward 100/4	forward 100/3	forward interpolate :x 100/4 100/3
right 0	right 90	right interpolate :x 0 90
forward 100/4	forward 100/3	forward interpolate :x 100/4 100/3
right 0	right 0	right 0
forward 100/4	forward 100/3	forward interpolate :x 100/4 100/3
right 120	right 0	right interpolate :x 120 0
forward 100/4	forward 100/3	forward interpolate :x 100/4 100/3
right 0	right 90	right interpolate :x 0 90
forward 100/4	forward 100/3	forward interpolate :x 100/4 100/3
right 0	right 0	right 0
forward 100/4	forward 100/3	forward interpolate :x 100/4 100/3
right 0	right 0	right 0
forward 100/4	forward 100/3	forward interpolate :x 100/4 100/3
right 120	right 90	right interpolate :x 120 90
end	end	end

Table 1: *The interpolation of two expanded turtle programs*

After adding a few commands to the triangle and circle programs to set the pen colour and initial position Figure 1 was produced by 100 calls to *triangle_to_square* with arguments ranging from 0 to 1.

The following figures were generated by extrapolating between a 9-pointed star, **`repeat 9 [forward 100 right 80]`**, and a doubly-drawn polygon with 360 sides, **`repeat 360 [forward 1 right 2]`**. One obtains very different (but also appealing) results if a normal circle program is used instead. Notice how "nine-ness" pervades all the figures. Both figures are drawn with two alternating colours.

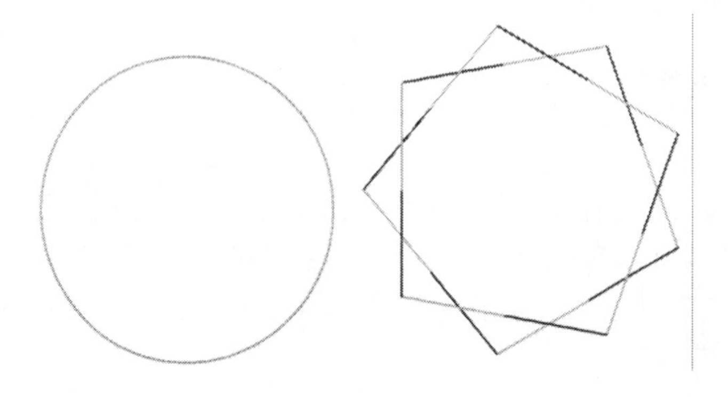

Figure 4: *The double circle (x = 0.0) and 9-pointed star (x = 1.0)*

Figure 5: *Circle to Star Extrapolations, x = 0.5, x = 2, x = -1*

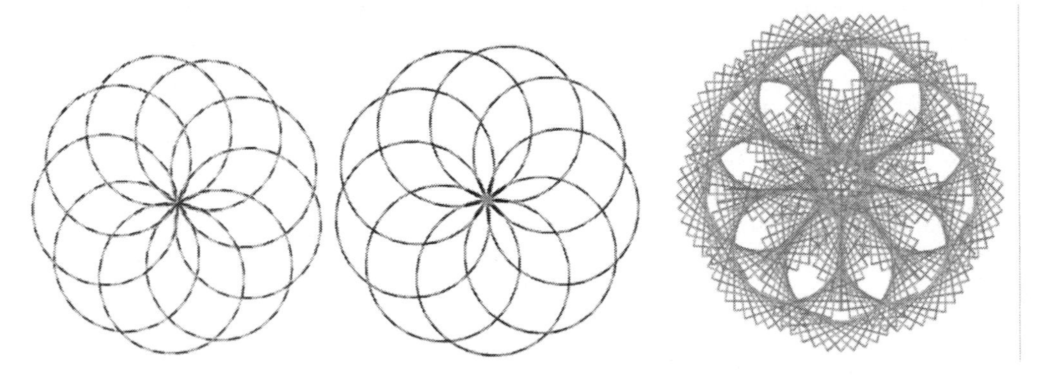

Figure 6: *Circle to Star Extrapolations, x = 10, x = -10, x = 100*

Figure 7: *Circle to Star Extrapolations, x = -100, x = 3.14159, x = -3.14159*

Isn't this "just" morphing?

Films and music videos have been morphing images and shapes to produce intermediate images and shapes for decades. How is this different? My program interpolates between other *programs* – not images or shapes. It automates the process of finding corresponding elements needed in morphing.

A program for a shape contains more information than the shape itself. A circle drawn once or twice looks the same but interpolates differently; as does a circle drawn clockwise instead of counter-clockwise. The *processes* for drawing two figures are what is being interpolated – not the figures themselves.

What does it really mean to extrapolate programs?

It is hard to get an intuition for program extrapolation in general. Some simple cases are intuitive. A positive extrapolation from a short line to a long line will be lines that are even longer. A negative extrapolation will be lines that are even shorter until a zero-length line is reached. Further negative extrapolation produces lines that start at the same location but grow in the other direction. The extrapolation from a small figure to a similar larger figure will be the figure even larger. (By "similar" I mean the sense in geometry that the two figures have identical angles and the same ratio of lengths of sides.) The extrapolation between a straight line and two lines that form a right angle will be angles greater than 90 degrees. What is hard to think about, for example, is the extrapolation of a circle to a triangle. How can something be more of a triangle than a triangle?

Are there always cycles in the extrapolations?

Corresponding turtle commands are matched with an interpolation function between the two values. We can see that each interpolated *right* command will cycle since for every unit increase in the interpolation/extrapolation parameter the angle is increased by the difference between the two angles in the original programs. So the values will cycle with a period of the greatest common divisor of the angle difference and 360. The full cycle is the least common multiple of the cycle length of each interpolated call to *right*. Similarly, the red, green, and blue colour components will cycle since they each have 256 distinct values.

Hence the sequence of the extrapolation between any two turtle programs whose sides have the same ratios will cycle except for the scale of the figures. If the canonicalised form of the two programs has identical calls to *forward* then the scale will remain constant.

Can all turtle programs be interpolated?

My program interpolator currently only works with programs built using the following primitives: *forward*, *right*, *repeat*, *setPenColour*, *penUp*, and *penDown*. This list could be greatly expanded. It is an open research question as to how to deal with conditionals, recursion, state variables, and calls to *repeat* with a non-constant repeat count. Note, however, that the current set of primitives is adequate for expressing all *fixed instruction turtle programs*. Only figures produced by infinite processes (e.g., fractals or spirals) are missing.

The history and future of the program interpolator

I first built the program interpolator in 1978. I used it an "improvisational" computer animated film called *Two Nights on a Computer* that was shown in the Boston area and in some film festivals. I recreated the program interpolator in 2006 in Imagine Logo. I have no plans to develop it further (though I may just for its recreational value). Anyone who wants to explore it further (either aesthetically or computationally) is welcome to the source code.

References

[1] Harold Abelson and Andrea A. diSessa. *Turtle Geometry, the Computer as a Medium for Exploring Mathematics*. The MIT Press, Cambridge, MA, 1981.

The Programmer as Poet

Russell Jay Hendel
Department of Mathematics, Towson University
Towson, MD, 21252, USA
E-mail: RHendel@Towson.edu

Abstract

In Tennyson's *Now Sleeps the Crimson Petal,* the poet requests from his lover, "*...slip into my bosom and be lost in me."* This theme is poetically developed by seeking an oneness with nature: The poet reviews many natural events which have a cycle of energetic wakefulness followed by a state of relative rest. A similar method of poetic development, by analogy with several other domains, occurs in other poetic passages: for example, Job wishes death by metaphorically seeking that his day of birth be *lost, stained, unlit, not allowed to come to the calendar,...* Computer scientists will immediately recognize this technique of poetic development as resembling **polymorphism**, which allows the naming of an abstract concept by its instantiation in one particular domain. This paper explores use of computer concepts to classify poetic technique; it also advocates enriching computer science curriculum with the teaching of poetic technique.

1. Goals

1.1. Overview. We first summarize 2 poems from different periods, with distinct styles, authors and topics. Our exploration of a common structure in these 2 poems naturally uses modern computer concepts. This suggests a computer-poetry classification project from which both disciplines can benefit.

1.2. Now Sleeps the Crimson Petal [1] [2]. The climax of Tennyson's poem, Now Sleeps the Crimson Petal is a request for the lover, after consummation, to slip into the poet's bosom: *So fold thyself, my dearest, thou, and slip into my bosom and be lost in me.* This theme is poetically developed by seeking an oneness with nature. The poet reviews many natural processes that similarly start in a more energetic state followed by a more restful state: (a) a *meteor* coming to rest, (b) *day* turning to night, (c) *lilies* nightly folding into a lake, etc.

1.3. The Death Wish, (Job, Old Testament, 3:3-11). In the dramatic opening passages of the poetic section of the book of Job, the poet wishes death: *Why did I not die from the womb? Why did I not perish when I came out of the belly?* This death wish is poetically developed by applying a variety of non-existence concepts from other domains to the date of birth of the poet. Examples of requested non-existence in the poem include (a) *losing* the date of birth, (b) *staining* it, (c) not letting it *come* to the calendar, (d) *darkening* it, etc.

1.4. Polymorphism. To define the commonality of poetic technique in the above passages we briefly review the computer concept of **polymorphism** [3]. A standard example is illustrated with a **class**, *OrchestraInstruments*. Suppose the only instrument you know is the *Violin*, which **extends** the **parent class** *OrchestraInstrument*. You know that violins are *fiddled* but at the time of writing the program you don't know the names of any other *OrchestraInstruments* or how they are *played*. Proper programming technique would write an *OrchestraInstrument* **method,** *fiddle()*. Then, every particular instrument could **inherit** this method but adapt its definition to that particular instrument. Thus, the program would speak about *flute.fiddle(), drum.fiddle()* etc. The word *fiddle* now represents the abstract concept of *playing* an instrument. The programmer finds useful the indefiniteness of the *fiddle* **method** since it can be defined in each new situation as needed [3]. Thus, the programmer has used metaphoric technique to write a program. Modern compilers can compile such a program even though at the time the program is written no knowledge is possessed of other instruments and other playing methods. Computer theory calls this **polymorphism**. **Polymorphism** enables us to articulate the commonality in the Tennyson and Job passages both of which poetically develop a single theme by polymorphically / metaphorically using similar concepts in other domains.

1.5. The Symmetry Classification Project. [4] describes a joint classification project by anthropologists and group theorists in which the group of symmetries of the patterns in the clothing, pottery and blankets of several distinct cultures and tribes were analyzed. The resulting analysis surprisingly uncovered mathematical categorizations of symmetries of different cultures. Based on the analysis presented in the preceding sections, we would suggest a similar project between poets and computer scientists: that is, advanced computer science concepts could be used to provide classifications of the types of poetic development employed in the poems of different poets or of different periods. This computer-poetry classification project could uncover mathematical categorizations of poetic development useful to poets.

1.6. Poetry vs. Mechanism. In this subsection we discuss the paradox that computer science with its classical emphasis on objectivity, precision and mechanism should be useful to poetry with its romantic emphasis on feeling and freedom from precise laws [5]. A possible resolution to this paradox was presented in section 1.4. There we showed how computer science, within its own mechanistic borders, now sees a need for concepts of an *indefinite* nature. To the poet metaphors create a "mood of musical like *indefiniteness"*[5]; to the programmer concepts that are *indefinite* are useful. It follows that the traditional classical-romantic distinction must be reexamined. Neurological studies have similarly uncovered a need for intelligence tests of higher order rationality not covered in standard intelligence tests[6].

1.7. Computer Pedagogy Enrichment. The preceding subsections dealt with computer theory aiding the poet. But the converse approach should also be fruitful. If computer programs are now using poetic concepts then perhaps knowledge of basic poetic technique will enhance the performance of the programmer. Anecdotally, all computer science instructors know that a basic pedagogic problem in introductory courses is an over-emphasis on mechanism, objectivity and correctness. It is equally important to emphasize overall structure and use of analogy. A re-examination of the computer curriculum with an emphasis on the poetic, would therefore be welcome.

References

[1] Helen Gardner, ed., *The New Oxford Book of English Verse*, Oxford: Oxford UP, pg. 648. 1972

[2] Gerald J. Steen, *Identifying Metaphor in Language: A Cognitive Approach*, Style, Vol. 36.3, pp. 386-407. 2002.

[3] Bruce Eckel, *Thinking in Java, 3rd edition*, Prentice Hall, 2003.

[4] D. K. Washburn and D. W. Crowe, *Symmetries of Culture: Theory and Practice of Plane Pattern Analysis*, Seattle, Washington: University of Washington Press, 1988.

[5] Edmund Wilson, *Axel's Castle*, N.Y.: Charles Scribner's Sons, pp. 1-25. 1959.

[6] Antonio R. Damasio, *Descartes' Error*, N.Y.: Grosset / Putnam Book, 1994.

Minkowski Sums and Spherical Duals

John M. Sullivan
Technische Universität Berlin
Institut für Mathematik, MA 3–2
Straße des 17. Juni 136
DE–10623 Berlin
Email: jms@isama.org

Abstract

At Bridges 2001, Zongker and Hart [8] gave a construction for "blending" two polyhedra using an overlay of dual spherical nets. The resulting blend, they noted, is the Minkowski sum of the original polyhedra. They considered only a restricted class of polyhedra, with all edges tangent to some common sphere. This note defines spherical duals of general convex polyhedra and proves that the Zongker/Hart construction is always valid. It can be used visually, for instance, to "morph" from any polyhedron to any other.

Polyhedra and their spherical duals

The notion of dual convex polyhedra, like the cube and octahedron, or the dodecahedron and icosahedron, is familiar. The faces of the polyhedron P correspond to vertices of its dual and vice versa. The combinatorics are thus clear, but in general (moving away from the Platonic solids) it is not always clear what geometry to give the dual. Indeed, most useful for us will be a dual which is itself not a convex polyhedron, but instead a network drawn on the surface of a sphere. We still consider it a dual of the original polyhedron P because it does have the dual combinatorics: a node for each face of P, an arc for each edge of P, and a region on the sphere for each vertex.

The nodes of this spherical dual are easy to find: each face f of P has an outward unit normal vector ν_f. Since ν_f is a unit vector in space, it can be viewed as a point on the unit sphere. We can think of this correspondance as follows: put a flashlight down on P such that its base rests flat on f. Its beam will then shine outwards in the normal direction, and it will hit the celestial sphere in the point ν_f.

Now consider an edge e of P. It lies between some pair of faces f and f' of P. If our flashlight is on f, and we tip it slowly across the edge e towards f', the beam will trace out an arc in the celestial sphere. This will be the geodesic or great-circle arc η_e from ν_f to $\nu_{f'}$. Suppose d_e is the edge vector of e (the difference between the endpoints of e). Then the arc η_e lies exactly in the great circle perpendicular to d_e. (We can check this as follows: the face normals ν_f and $\nu_{f'}$, being the endpoints of η_e, both lie on this circle, but they are also both perpendicular to d_e.) The length of the arc η_e is exactly the exterior dihedral angle of P along e (the angle between ν_f and $\nu_{f'}$).

The nodes and arcs we have described form the network on the sphere that we call the *spherical dual* $\widetilde{P}$ of P. (By analogy to smooth surfaces, where the normal vector ν is given by the so-called Gauss map, this spherical dual is sometimes also called the Gauss image of P.) The network $\widetilde{P}$ cuts the sphere into regions corresponding to the vertices v of P. Indeed, the region ρ_v associated to a vertex v is the one bounded by the arcs η_e corresponding to the edges e incident to v.

Let us recall the definition of *supporting plane*. A supporting plane through a point p on P is a plane such that all of P lies to one side (or in the plane). At a point within a face f, the unique supporting plane is the one with normal ν_f. At any point along an edge e, there is a one-parameter family of supporting planes; their normals are the points along the arc η_e. At a vertex v there is a two-parameter family of supporting planes: holding our flashlight at v, we can tip it to be perpendicular to any of these planes. The normal directions (in which the flashlight then shines) fill out the region ρ_v. The area of this region ρ_v of $\widetilde{P}$ is what one might call the *exterior solid angle* of P at v or the *Gauss curvature* of P at v.

Note that, by the combinatorial duality, an n-gon face f of P corresponds to a n-valent node ν_f in $\widetilde{P}$: indeed there are exactly n ways to tip our flashlight off f. Similarly, if k edges e_i meet at the vertex v of P, the the region ρ_v on the sphere has k sides, namely the k arcs η_{e_i}.

We can consider two ways in which a polyhedron can degenerate to be lower-dimensional. First, a planar n-gon in space, with its two sides thought of as two faces (with opposite normals), forms a *dihedron*; its spherical dual has two antipodal nodes $\pm\nu_f$, connected by n arcs. Second, a line segment in space has no faces but has an edge with vector, say, d_e; its spherical dual has no nodes, but consists of the entire great circle perpendicular to d_e.

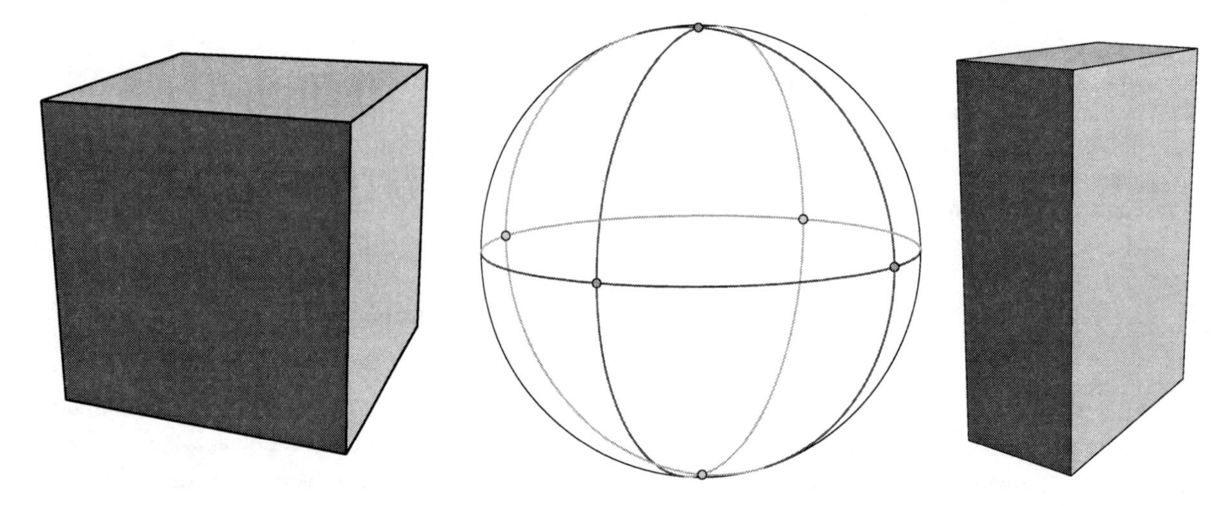

Figure 1: The spherical dual of a cube (left) is a network of three perpendicular great circles (center) which could be called a spherical octahedron. (Our spherical figures were drawn with the program "Spherical Easel" [1].) The six nodes of the network are the normals to the six cube faces; the twelve arcs correspond to the cube edges; these cut the sphere into eight congruent triangular regions corresponding to the cube vertices. The same network is also the dual of any rectangular parallelepiped (right). We can introduce labels on the arcs of the dual, recording the edge lengths of the original polyhedron, to distinguish these possibilities.

Recovering a polyhedron from its spherical dual

As a very simple example, the cube C and its dual $\widetilde{C}$ are shown in Figure 1. The dual consists of three great circles on the sphere, in the coordinate planes. These meet at six nodes (the north and south poles plus four along the equator) corresponding to the the cube's face normals. Just as the usual dual of a cube is an octahedron, this network $\widetilde{C}$ could be called as a spherical octahedron; indeed it is the radial projection of a regular octahedron to its circumscribed sphere.

Given this dual $\widetilde{C}$, can we reconstruct the original cube C? We know that if any polyhedron P has dual $\widetilde{C}$, then P must have six faces, with opposite pairs parallel to the coordinate planes. But any (axis-aligned) rectangular parallelepiped satisfies this conditions, and will indeed have this same dual network $\widetilde{C}$.

In general, given a spherical network N with convex regions, we can look for polyhedra P with dual N. Any such polyhedron will be the intersection of halfspaces H_ν, one for each node ν of N. We know that H_ν will be bounded by a plane p_ν with unit normal ν, but the location of this plane is not usually uniquely determined. Given a polyhedron P, we can move its faces inwards or outwards a bit, keeping them parallel. As long as we don't move any face so far that the combinatorics of P changes, the result is new polyhedron with the same spherical dual $\widetilde{P}$.

Note, however, that there are also rigid examples, like the octahedron O of Figure 2. It is uniquely determined (up to homothety) by its spherical dual $\widetilde{O}$, since its combinatorics would change immediately, no matter how little we moved any face.

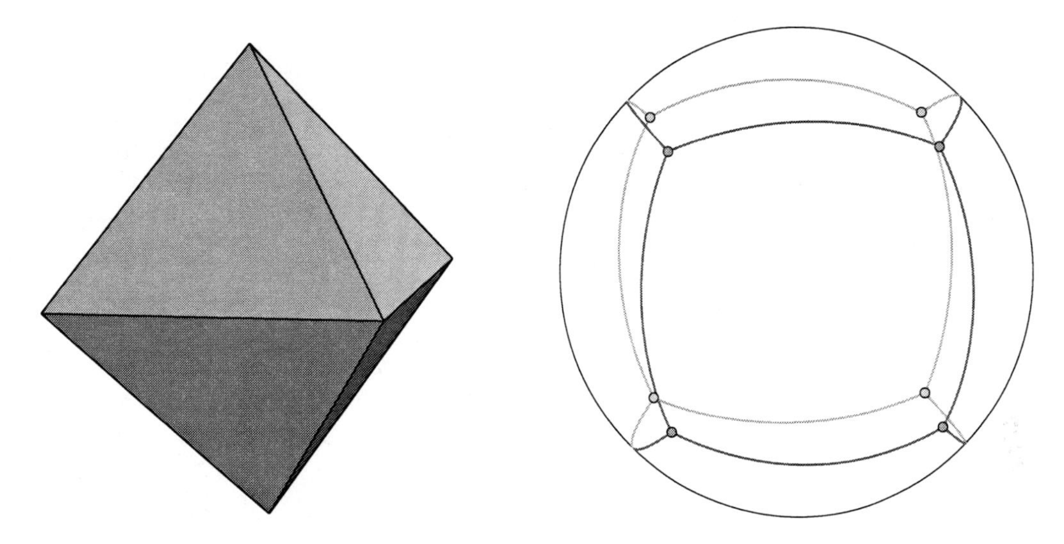

Figure 2: The octahedron (left) is uniquely determined by its spherical dual (right), since moving any face inwards or outwards would change the combinatorics. The dual $\widetilde{O}$, a spherical cube, has eight nodes (in the body-diagonal directions) and twelve arcs, dividing the sphere into six congruent square regions. Here, the only legal edge-labelings use equal labels on all twelve edges.

In general, though, to recover P from $\widetilde{P}$ we need some further information. One possibility would be to record, for each node ν_f, the distance from the origin to the plane of the face f. This would make the spherical dual carry exactly the information of the so-called *support function* of P. That is, for any direction we would know the distance from the origin to the supporting plane to P in that direction. It is well-known that any convex body is determined by its support function. (See for instance [7].) Given a spherical network, the various polyhedra with that dual correspond to the different labelings of the nodes. An n-faced polyhedron P is part of a large family with dual $\widetilde{P}$; any small adjustment of the n labels on $\widetilde{P}$ would lead to a particular member of this family.

Recent work of Fogel and Halperin [3] develops an exact algorithm for computing Minkowski sums based on spherical duals. (Actually, for computational efficiency, they project the dual radially onto a cube.) In order to be able to recover polyhedra from the dual networks, they store quite redundant information: namely, for each region of the dual, the three coordinates of the original vertex in space.

Zongker and Hart [8] suggest yet another way to encode the extra information needed to determine P from $\widetilde{P}$. They record, for each arc η_e, the length ℓ_e of the corresponding edge e. Since a polyhedron has more edges than it has faces, this leaves us with more labels than we need. That is, not every edge-labeling on $\widetilde{P}$ corresponds to a polyhedron; instead there are necessary conditions on these labels. However, we will show that certain constructions (in particular the overlay of two labeled dual nets) always do give correct labelings that correspond to polyhedra.

To understand the conditions on the labels, think first about a single arc η_e from ν_f to $\nu_{f'}$, with label ℓ_e. It must correspond to an edge e of length ℓ_e in the direction perpendicular to ν_f and $\nu_{f'}$. That is, the edge vector d_e is known to be the vector of length ℓ_e in the direction of the cross product $\nu_f \times \nu_{f'}$. Note that this direction lies tangent to the sphere at ν_f, perpendicular to the direction in which the arc η_e leaves the node ν_f.

Of course, because the face f is a closed polygon, the edge vectors d_e around f must sum to 0. Equivalently, think of the weights ℓ_e as tensions in the arcs η_e. The closure condition around f becomes the following condition at ν_f: the tensions in all the arcs η_e sum to 0. (Since this is a vector equation in the tangent plane at ν_f, it really represents two linear conditions on the incident labels there.)

Minkowski sums

Given two polyhedra A and B, their Minkowski sum $A + B := \{a + b : a \in A, b \in B\}$ is again a polyhedron. It has faces parallel to the faces of the original polyhedra, and additional faces which are parallelograms generated by one edge from A and one from B.

As above, we will allow a segment in space to count as a degenerate polyhedron. Then the Minkowski sum of two segments is a parallelogram, the sum of three (linearly independent) segments is a parallelepiped, and in general the sum of k segments is a special kind of polyhedron called a zonohedron. Unless two of the segments are collinear, each edge of the zonohedron is parallel and equal in length to one of the original segments. Unless three of the segments are coplanar, the faces of the zonohedron are parallelograms. Figure 3 shows two zonohedra which have the same duals as certain Archimedean solids.

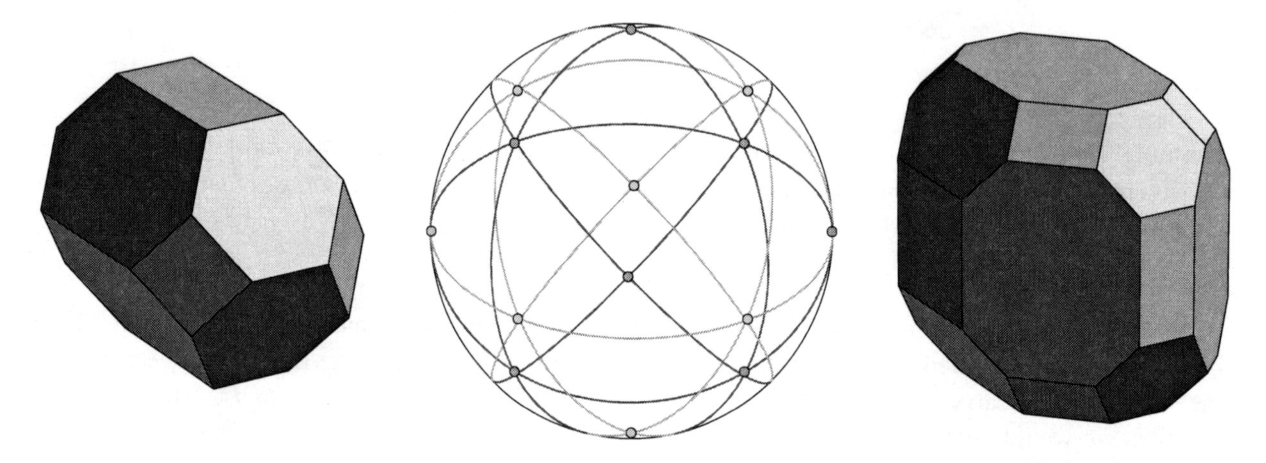

Figure 3: The zonohedron T (left) is a "stretched" truncated octahedron. Its spherical dual (center) is obtained by extending the arcs of the spherical cube $\widetilde{O}$ in Figure 2 until we have six full great circles. The truncated octahedron would have equal weights on all dual arcs; we obtain the stretched version T by varying the weights. The stretched truncated cuboctahedron (right), also a zonohedron, is the Minkowski sum of T with the box of Figure 1(right). Its dual (not shown) is obtained simply by overlaying their two duals.

We have noted that the spherical dual of a segment should be taken as the great circle perpendicular to this segment. If a zonohedron Z is the sum of k segments s_i, then its dual is the great-circle arrangement on the sphere consisting of the corresponding k great circles c_i. Each great circle in this arrangement is divided into a number of subarcs by its intersections with the other circles; these arcs correspond to a family of parallel edges (called a zone) on Z. To recover Z from the dual, we just need to know the length of the edges in each zone. But that is the same as the length ℓ_i of the original segment s_i. Thus, to properly label the dual of the Minkowski sum, we put the label ℓ_i on each segment of the great circle c_i.

The observation of Zongker and Hart [8] was that this overlay method works to produce Minkowski sums in general. Suppose A and B are two polyhedra, with labeled spherical duals $\widetilde{A}$ and $\widetilde{B}$. Then we construct a spherical network $\widetilde{P}$ by overlaying $\widetilde{A}$ and $\widetilde{B}$, simply drawing them both on the same sphere, inserting a new node wherever arcs cross. Each arc η of $\widetilde{P}$ is part of an arc from either $\widetilde{A}$ or $\widetilde{B}$, and inherits the label of that arc. In excpetional cases, η will lie along parts of arcs from both $\widetilde{A}$ and $\widetilde{B}$, in which case it gets the sum of those two labels.

The analysis of this construction in [8] was limited to the special case where A and B were each midscribed around some sphere (that is, had all edges tangent to that sphere). Here we show that, in fact, it works in complete generality.

Generically, the nodes in the overlay network $\widetilde{P}$ either will be nodes from $\widetilde{A}$ or $\widetilde{B}$ or will arise where an edge of $\widetilde{A}$ crosses one of $\widetilde{B}$. In the first case, the node and its incident arcs and their labels are exactly the same as seen in $\widetilde{A}$ or $\widetilde{B}$, so the closure condition is satisfied. In the second case, the node ν has four incident arcs. They come in two opposite pairs, with equal labels ℓ_i on either pair. Clearly, this node is the dual of a parallelogram, and the closure condition is satisfied by symmetry.

In special cases, a node of $\widetilde{A}$ may lie exactly on an arc or node of $\widetilde{B}$. But then the closure condition in $\widetilde{P}$ is just the sum of the closure conditions from these two overlapping nodes. The case where an arc of $\widetilde{A}$ (partially) overlaps an arc of $\widetilde{B}$ also causes no problems, as long as we have used the sum of the original labels on the overlap, as specified above.

Since the closure conditions are satisfied everywhere, the labeled network $\widetilde{P}$ does correspond to a polyhedron P, the Minkowski sum of A and B.

Connections and applications

This construction could be used to morph between any two convex polyhedra A and B. For time t ranging from 0 to 1, we would use the weighted sum $P_t := (1-t)A + tB$. These intermediate polyhedra P_t all have the same spherical dual $\widetilde{P}$, obtained by overlaying the duals $\widetilde{A}$ and $\widetilde{B}$. All we need to do as t varies is to linearly interpolate the labels.

As an example, consider a symmetric pattern of great circles drawn on the sphere as in Figure 4, each tilted slightly from the equator. It is the dual to a so-called "polar zonohedron" (see [2, 6]). Suppose we

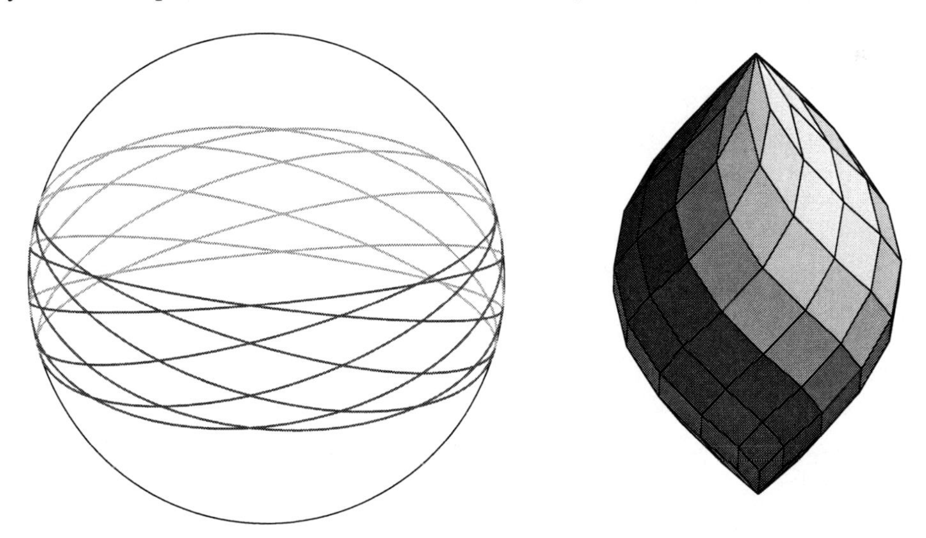

Figure 4: The symmetric pattern of great circles (left) is the spherical dual of a polar zonohedron like the one shown (right). Such a zonohedron is by definition generated by segments equally spaced around a cone.

place two different polar zonohedra in space with their axes perpendicular. Morphing between them the results in the sequence shown in Figure 5.

Aside from zonohedra, another interesting class to consider for this duality construction is deltahedra. A deltahedron is a polyhedron with equilateral-triangle faces. Its dual is then a network of great-circle arcs meeting in threes at equal (120°) angles. Such a network is reminiscient of a two-dimensional bubble cluster,

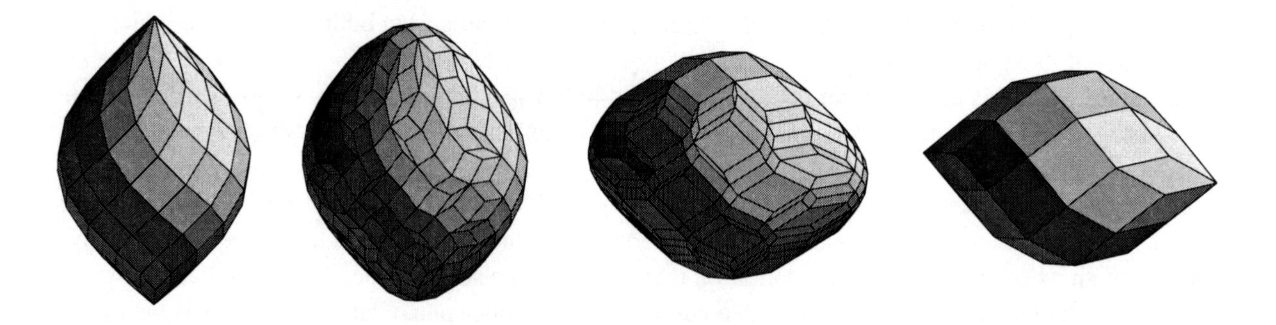

Figure 5: The polar zonohedra at left and right are generated by segments equally spaced around a cone. They have been placed with perpendicular axes. In the middle we see two intermediate stages of the morph between them.

and indeed the eight possible such networks describe the only candidate singularities for three-dimensional bubble clusters. These ideas generalize to arbitrary higher dimensions [5] where the classification of such soap-film singularities is not yet complete.

From a more abstract point of view, our spherical dual $\widetilde{P}$ with arcs labeled by edge lengths is simply the generalized mean curvature measure of the polyhedron P in the sense of Minkowski mixed volumes. (See for instance [4].) The overlay construction is then simply explained by the known fact that such measures are additive under Minkowski sum. It seems to be an open problem in general, however, to decide which measures on the sphere can arise as the generalized mean curvature of some convex body.

Acknowledgements

I wish to thank George Hart for helpful comments on a draft of this paper.

References

[1] David Austin and Will Dickinson. *Spherical Easel.* Software, 2002–2004. `http://merganser.math.gvsu.edu/easel/`

[2] David Eppstein. *Zonohedra and Zonotopes.* Mathematica in Education and Research **5**:4, pp. 15–21, 1996. `http://www.ics.uci.edu/~eppstein/junkyard/ukraine/ukraine.html`

[3] Efi Fogel and Dan Halperin. *Exact and Efficient Construction of Minkowski Sums of Convex Polyhedra with Applications.* Proc. ALENEX 2006, to appear. `http://www.math.tau.ac.il/~halperin/papers/exact_mink_3d.pdf`

[4] Jane R. Sangwine-Yager. *Mixed Volumes.* In *Handbook of Convex Geometry*, Gruber and Wills, eds., pp. 43–71, 1993.

[5] John M. Sullivan. *Convex Deltatopes in all Dimensions, and Polyhedral Soap Films.* Preprint, 1995. `http://torus.math.uiuc.edu/jms/Papers/`

[6] Russell Towle. *Graphics Gallery: Polar Zonohedra.* The Mathematica J. **6**:2, pp. 8–12, 1996.

[7] Günter M. Ziegler. *Lectures on Polytopes.* Springer GTM **152**, 1997.

[8] Douglas Zongker and George W. Hart. *Blending Polyhedra with Overlays.* Bridges 2001 Proceedings, Southwestern Coll., Kansas, pp. 167–174.

Polygon Foldups in 3D

Kate Mackrell
Faculty of Education
Queen's University
Kingston, ON K7L 3N6, Canada
E-mail: katemackrell@sympatico.ca

Abstract

The software Cabri 3D allows the nets of polyhedra to be constructed using one or more sets of connected polygons where the angle between all connected polygons is the same. These collections can be folded into the polyhedron by dragging a point controlling the angle between the polygons. Viewed from above, the polygons act as a kaleidoscope as the angle changes, and when the angle is decreased so that polygons intersect, surprisingly beautiful symmetric figures emerge, which can be constructed as physical artifacts or experienced as dynamic computer animations.

1. Introduction: an unusual dodecahedron construction

At the T^3 conference in February 2005 I watched Jean-Marie Laborde perform an extraordinary dodecahedron construction using the relatively new software Cabri 3D [1], based on the work of Schumann [2]. The outline of this construction is given in Figure 1 below. See [3] for a reference to a website containing a movie which demonstrates the construction in detail.

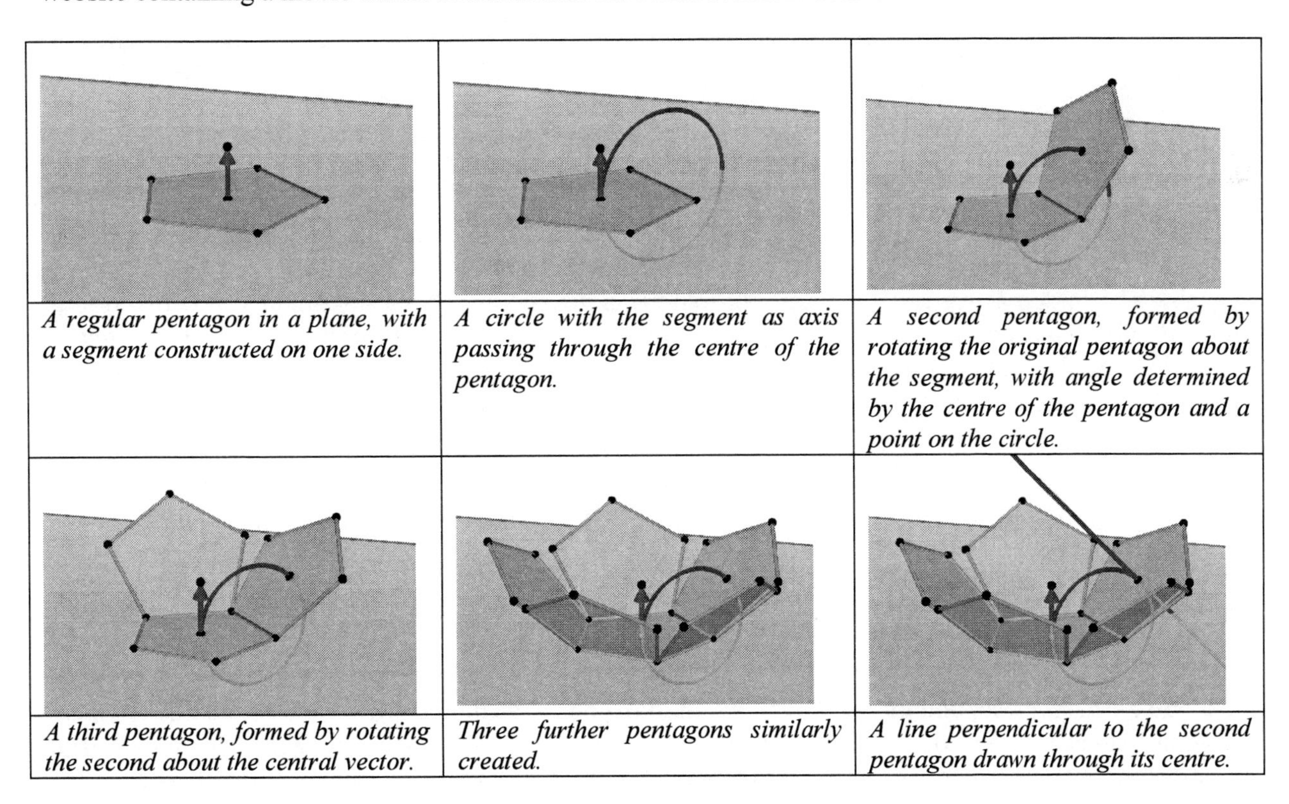

A regular pentagon in a plane, with a segment constructed on one side.	*A circle with the segment as axis passing through the centre of the pentagon.*	*A second pentagon, formed by rotating the original pentagon about the segment, with angle determined by the centre of the pentagon and a point on the circle.*
A third pentagon, formed by rotating the second about the central vector.	*Three further pentagons similarly created.*	*A line perpendicular to the second pentagon drawn through its centre.*

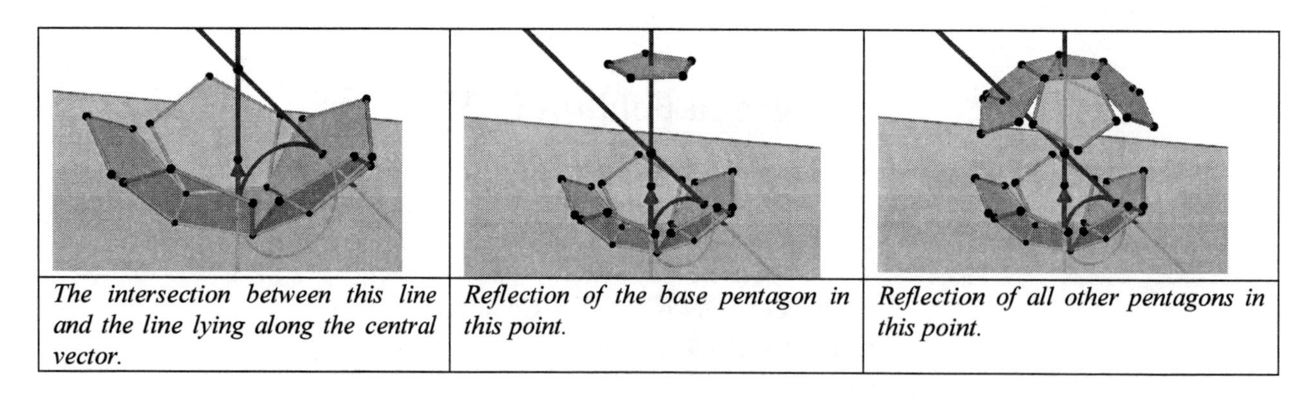

The intersection between this line and the line lying along the central vector.	*Reflection of the base pentagon in this point.*	*Reflection of all other pentagons in this point.*

Figure 1: *Construction of a dodecahedron net.*

This construction is pleasing in itself, both mathematically in that it uses the central symmetry of the dodecahedron, and artistically in that a net with high symmetry results. However, the most intriguing aspect of this net is its behaviour when "folded" by dragging the point on the circle as shown in figure 2 below:

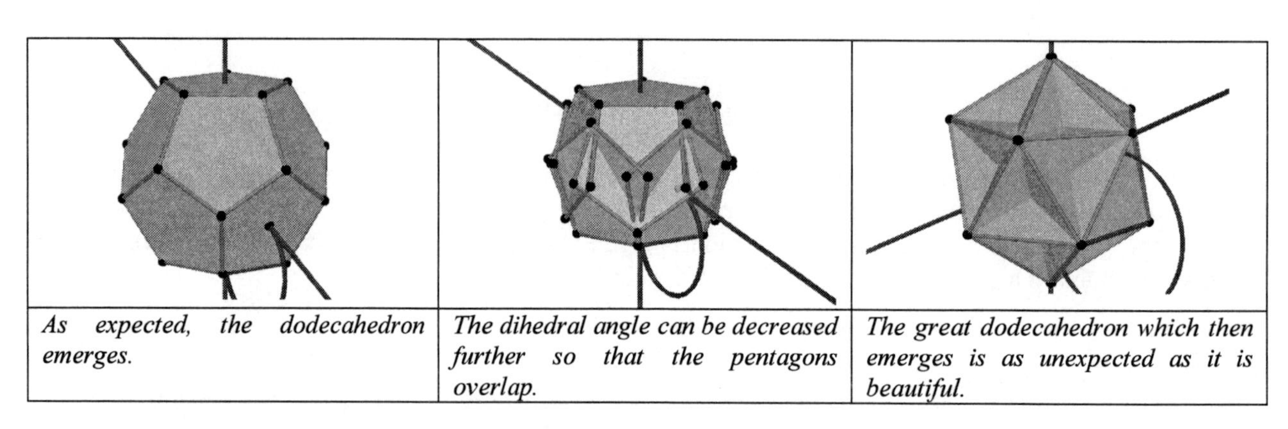

As expected, the dodecahedron emerges.	*The dihedral angle can be decreased further so that the pentagons overlap.*	*The great dodecahedron which then emerges is as unexpected as it is beautiful.*

Figure 2: *Folding the net.*

The collection of pentagons shown above is an example of what I will call a *foldup* which is defined to be a collection of connected polygons in which all dihedral angles between pairs of joined polygons are equal. The foldup above has two parts, but foldups in general may have any number of parts up to the number of polygons used. A *gathering* is defined to be a special state of a foldup in which the dihedral angle is such that edges and/or vertices of some polygons which are not directly joined coincide. The dodecahedron and great dodecahedron shown above are both gatherings.

2. Questions

Some of the mathematical questions arising out of this construction will be explored very briefly in this section. There is scope for much more extensive exploration: there are many more polygons or collections of polygons to explore and many more questions which arise.

2.1. Do gatherings depend on the way in which polygons are connected? There are a large number of ways in which nets can be configured. For example, the cube, with only six faces has more than ten possible nets consisting only of squares. An alternative construction of a dodecahedron net (starting from step 7 of the construction above) is given below:

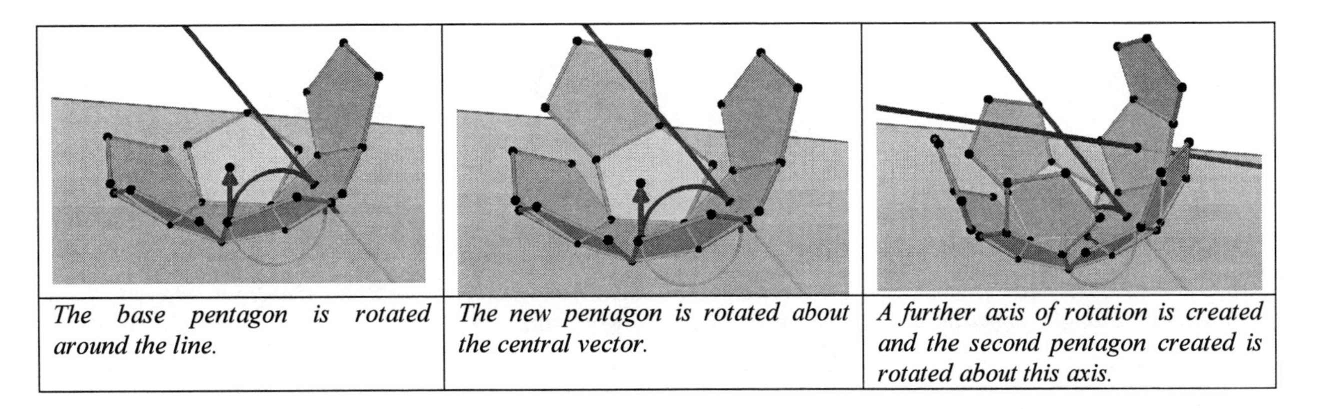

The base pentagon is rotated around the line.	*The new pentagon is rotated about the central vector.*	*A further axis of rotation is created and the second pentagon created is rotated about this axis.*

Figure 3: *An alternative dodecahedron net.*

This foldup does not share all the gatherings of the first pentagon foldup:

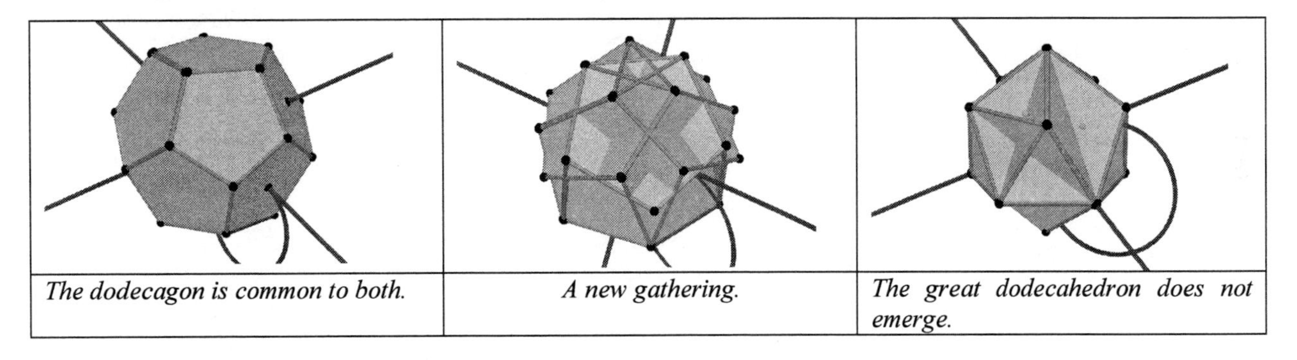

The dodecagon is common to both.	*A new gathering.*	*The great dodecahedron does not emerge.*

Figure 4: *Gatherings of the alternative dodecahedron net.*

2.2. What happens when the nets of other Platonic solids are folded? An icosahedron net can be constructed using a combination of the techniques for the two pentagon foldups shown above. This net, together with two of its gatherings is shown in Figure 5 below.

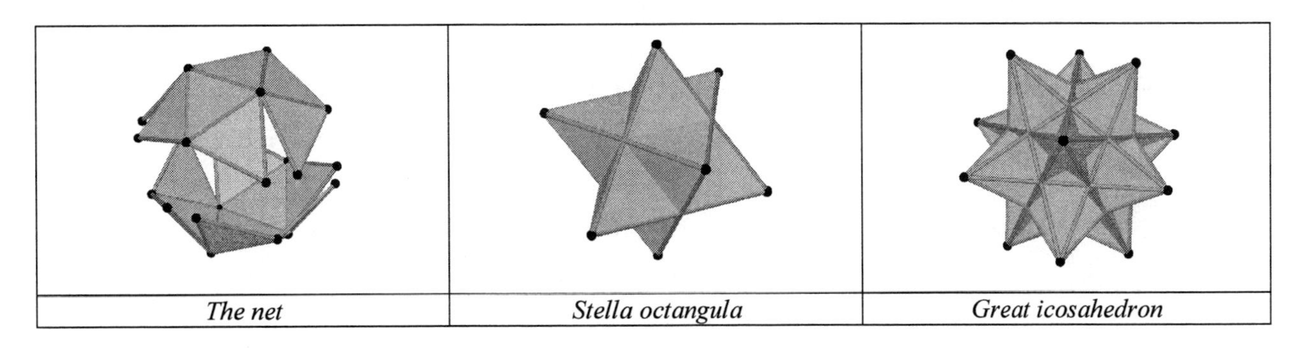

The net	*Stella octangula*	*Great icosahedron*

Figure 5: Some of the *shapes that emerge as an icosahedron net is folded.*

2.3. What happens when the nets of non-regular polyhedra are folded? Figure 6 on the next page shows one way to construct the net of a truncated dodecahedron. In this construction, in order to keep dihedral angles equal, no two triangles or decagons can have adjoining edges.

:

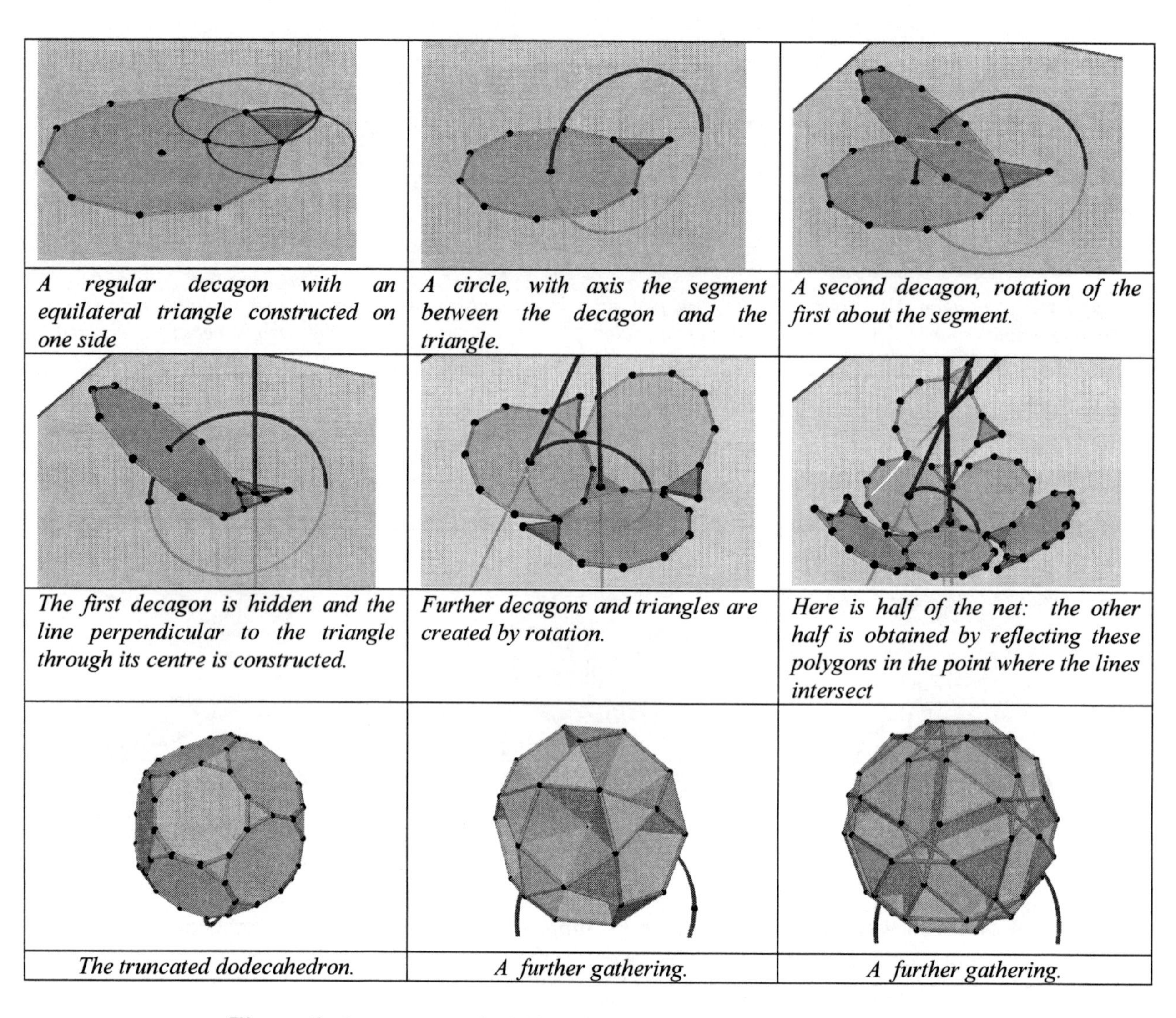

A regular decagon with an equilateral triangle constructed on one side	*A circle, with axis the segment between the decagon and the triangle.*	*A second decagon, rotation of the first about the segment.*
The first decagon is hidden and the line perpendicular to the triangle through its centre is constructed.	*Further decagons and triangles are created by rotation.*	*Here is half of the net: the other half is obtained by reflecting these polygons in the point where the lines intersect*
The truncated dodecahedron.	*A further gathering.*	*A further gathering.*

Figure 6: *Construction of a foldup which forms a truncated dodecahedron.*

2.4. What happens if a foldup is not the net of a polyhedron? An example follows.

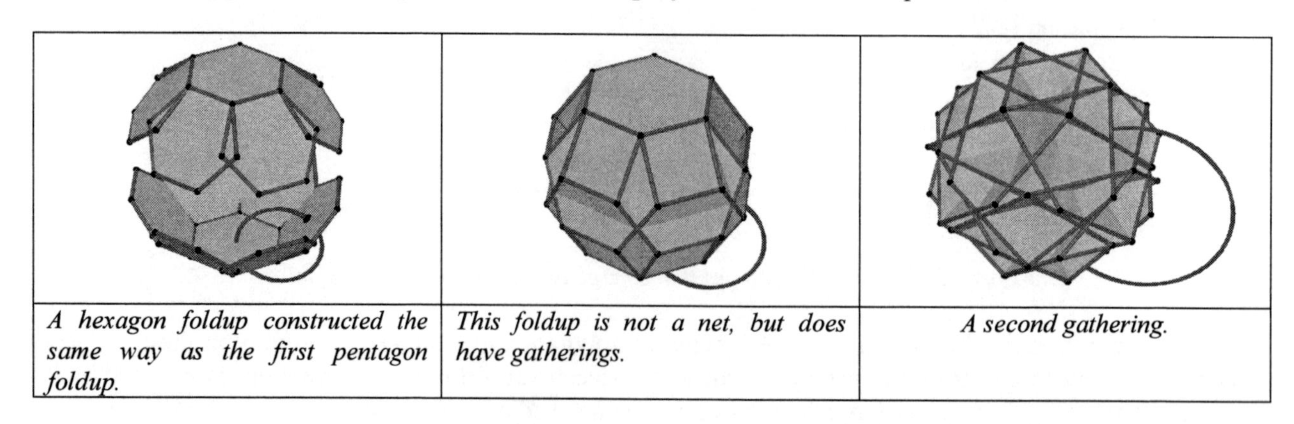

A hexagon foldup constructed the same way as the first pentagon foldup.	*This foldup is not a net, but does have gatherings.*	*A second gathering.*

Figure 7: *Gatherings of a hexagonal foldup that is not a net.*

2.5. Why do gatherings occur? The diagrams below in Figure 8 show the progressive folding of a foldup consisting of five regular nonagons arranged in a ring. These hint at the reasons that particular gatherings occur.

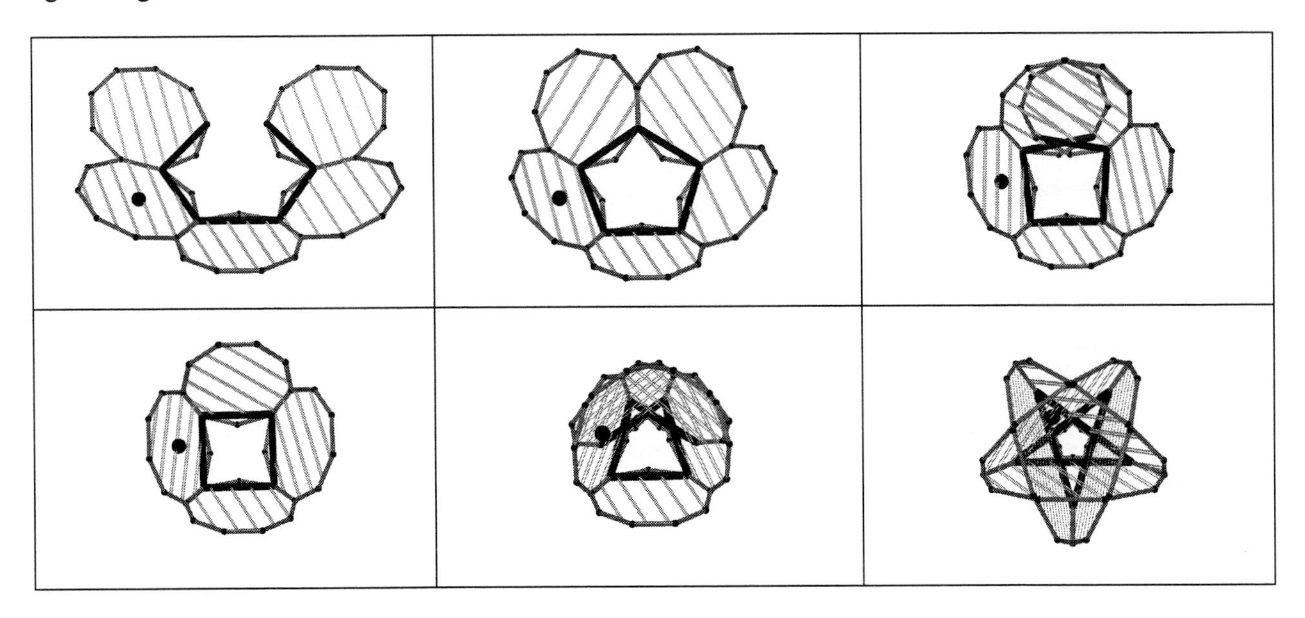

Figure 8: *Some indication of why gatherings occur.*

2.6. Which gatherings form polyhedra? Some gatherings are polyhedra, such as the great dodecahedron formed by the first pentagon foldup. Many other gatherings, however, appear to be polyhedra until examined closely, when, as with the gathering of the truncated icosahedron in figure 9 below, it is clear that not all edges meet other edges. A further question is whether the foldup can be changed in any way so that gatherings become polyhedra.

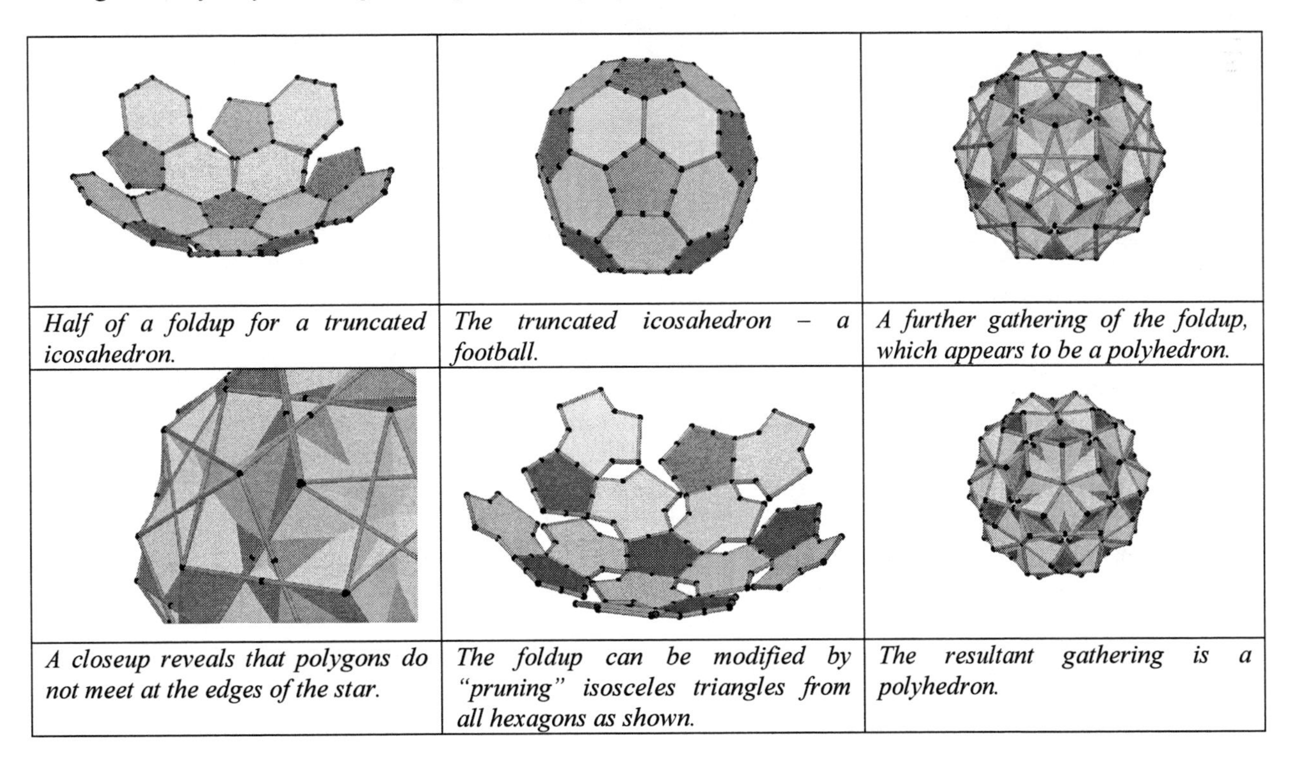

Half of a foldup for a truncated icosahedron.
The truncated icosahedron – a football.
A further gathering of the foldup, which appears to be a polyhedron.
A closeup reveals that polygons do not meet at the edges of the star.
The foldup can be modified by "pruning" isosceles triangles from all hexagons as shown.
The resultant gathering is a polyhedron.

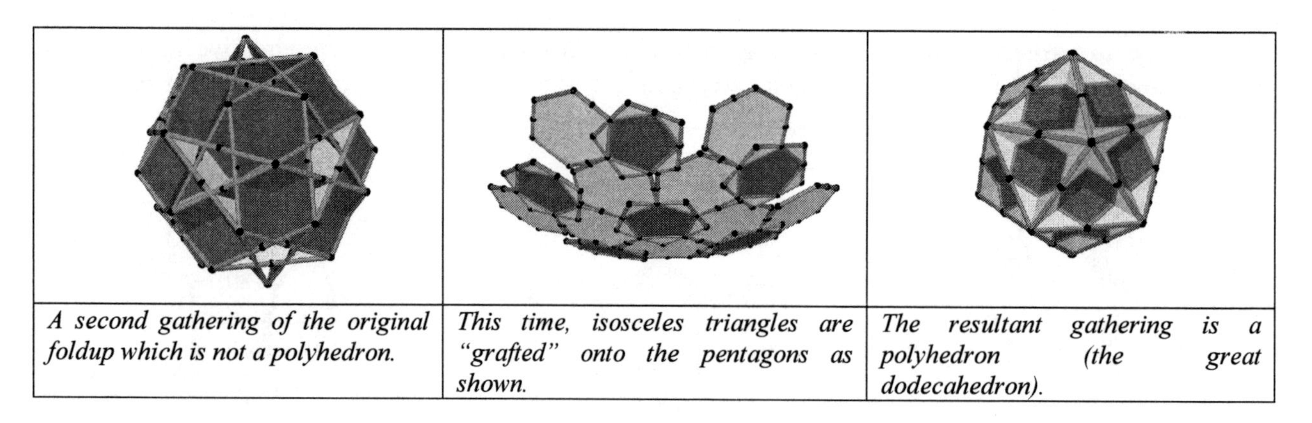

A second gathering of the original foldup which is not a polyhedron.	*This time, isosceles triangles are "grafted" onto the pentagons as shown.*	*The resultant gathering is a polyhedron (the great dodecahedron).*

Figure 9: *Gatherings and polyhedra.*

2.7. What happens when a foldup contains irregular polygons? One among many possibilities to explore is Erdely's [3] spidron, which consists of two infinite sequences of equilateral and isosceles triangles as shown below. The process by which the spidron is constructed has been used to create a foldup which is the beginning of a fractal.

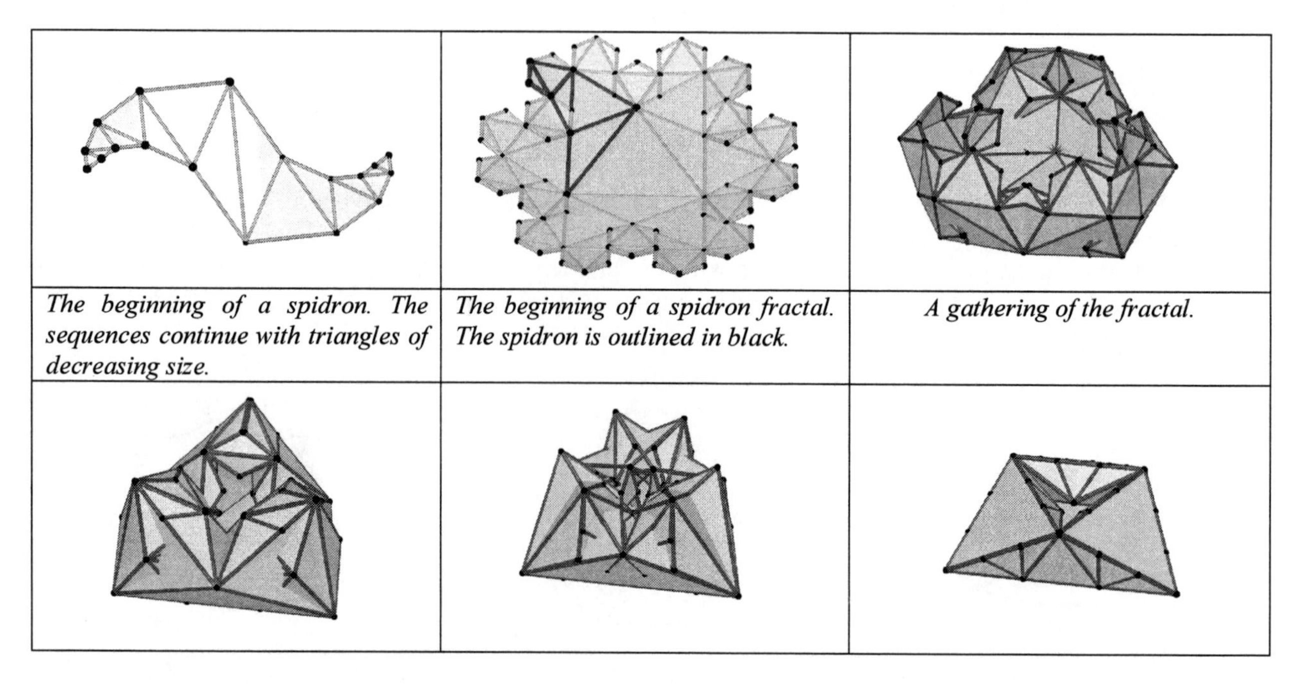

The beginning of a spidron. The sequences continue with triangles of decreasing size.	*The beginning of a spidron fractal. The spidron is outlined in black.*	*A gathering of the fractal.*

Figure 10: *A spidron "fractal" with gatherings.*

2.8. What happens when foldups are created to form a stellation of a polyhedron? As the gatherings of the dodecahedron net and icosahedron net include a stellation of the dodecahedron and of the icosahedron, I decided to find out what would happen with a foldup which was deliberately designed to fold to create the great dodecahedron. This foldup is formed from a number of pentagons with pentagrams cut out as shown in figure 11 on the next page. An interesting further gathering, resembling the great dodecahedron turned "inside out" resulted.

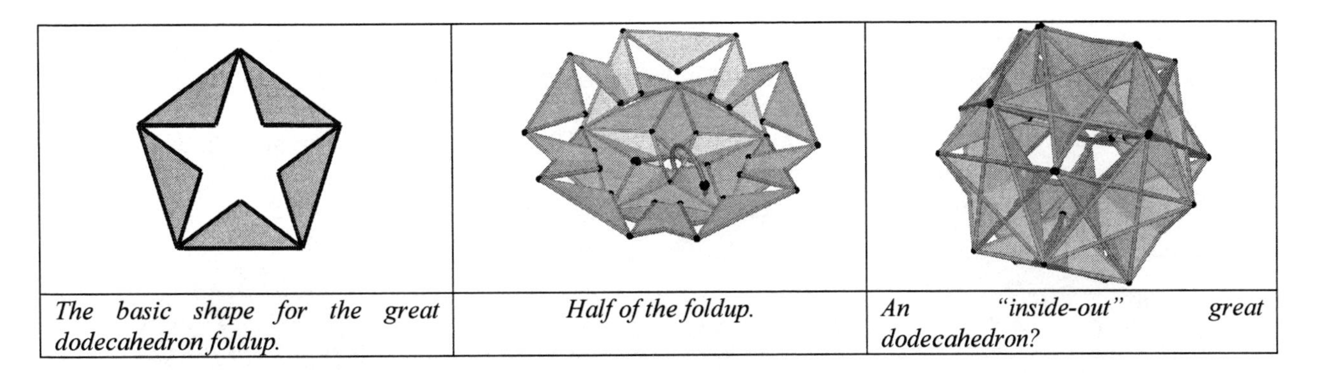

The basic shape for the great dodecahedron foldup.	*Half of the foldup.*	*An "inside-out" great dodecahedron?*

Figure 11: *A net for the great dodecahedron with a further gathering.*

2.9. Further possibilities. Here are a few further ideas to explore.

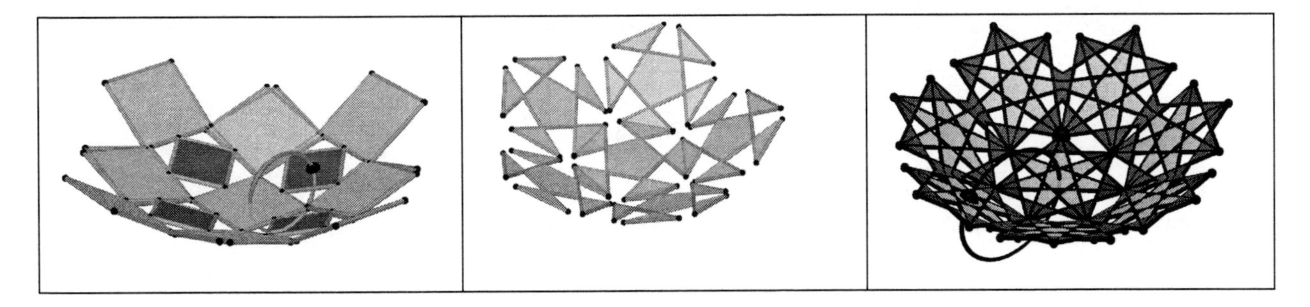

Figure 12: *Further possibilities.*

Pruning and grafting, introduced in section 2.6, also give rise to further complex and beautiful shapes. In figure 13 below, the truncated icosahedron foldup has first been set to a gathering and then the points controlling the degree of pruning and grafting have been dragged.

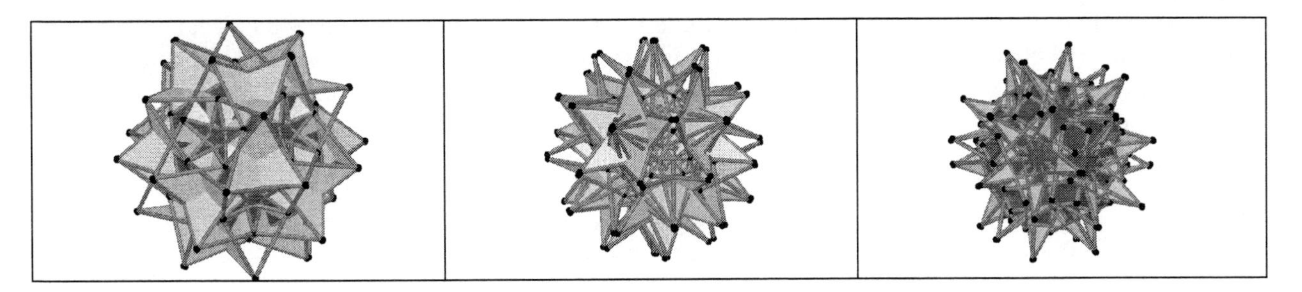

Figure 13: *Shapes formed by pruning and grafting a truncated icosahedron foldup.*

3. Polygon foldups as art

Hopefully the diagrams here speak for themselves: gatherings give rise to attractive visual images of objects which could be reproduced using physical materials.

The screenshots do not, however, capture the dynamic nature of polygon foldups: the point controlling the dihedral angle may be animated and as this angle changes, the entire configuration changes. Polygons are in constant motion, with little symmetry in the overall figure– until suddenly and unexpectedly a gathering with a high order of symmetry emerges and then disappears. Several dynamic foldups may be experienced at http://educ.queensu.ca/~mackrelk/Cabri3D/polygonfoldups.htm

Foldups may also be viewed looking directly down on the xy plane, and form attractive 2D objects with a high degree of symmetry whether or not the foldup forms a gathering. The pictures below show a number of foldups (none of which are in a gathering) viewed from above:

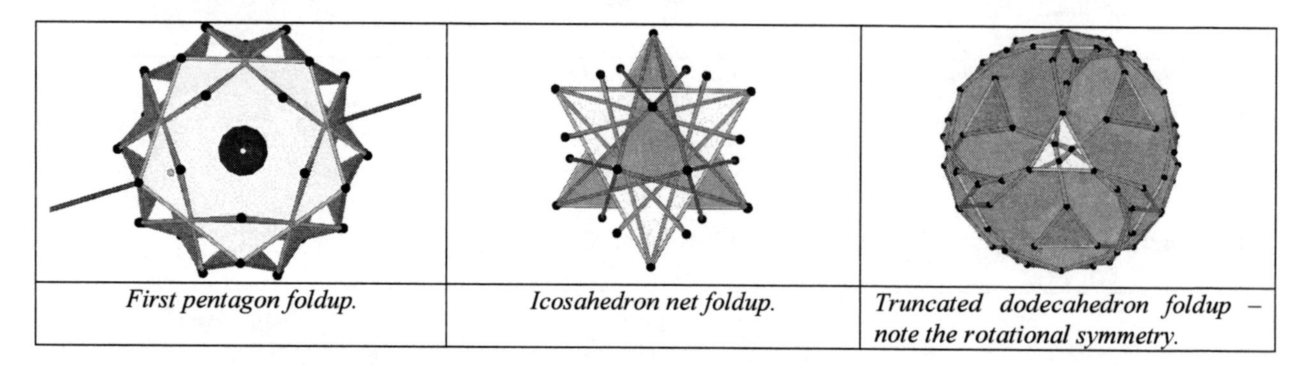

First pentagon foldup.	*Icosahedron net foldup.*	*Truncated dodecahedron foldup – note the rotational symmetry.*

Figure 14: *Foldups viewed from above.*

If the point controlling the dihedral angle is now animated the effect is of a kaleidoscope in which geometric figures are continually changing and symmetry is constantly preserved.

I would also stress that this paper illustrates a very few of the huge number of possible directions for exploration. Most of the shapes used above have been very simple – but the process can be applied to almost any geometric shape, however complex, giving great scope for individual creativity.

4. Conclusion

I would like to make a plea regarding polygon foldups. As far as I am aware, this is a new and potentially rich area of mathematical exploration. This is also an area of mathematics which is both accessible to school students and visually attractive, and this combination is very rare indeed. Could the further exploration of this area hence be left to school students in order that some students have the opportunity to do truly original work in mathematics?

Note: A free 30 day trial version of Cabri 3D can be downloaded from http://www.chartwellyorke.com/cabri3d/demo.html

References

[1] Cabri 3D (Version 1.1) (2005) [Computer Software]. Grenoble, France: Cabrilog

[2] H. Schumann, *Konstruktion von Polyedermodellen mit Cabri 3D im Umfeld der platonischen Körper,* Beiträge zum Computereinsatz in der Schule, number 2, pp. 3-48. 2004.

[3] http://www.ph-weingarten.de/homepage/lehrende/schumann/geometrie-seite/videoclips.html.

[4] D. Erdely, *Spidron System: A flexible space-filling structure.* Retrieved October 12, 2005 from http://www.szinhaz.hu/spidron/

Portraits of Groups

Jay Zimmerman
Mathematics Department
Towson University
8000 York Road
Towson, MD 21252, USA
E-mail: jzimmerman@towson.edu

Abstract

This paper represents some small finite groups as groups of transformations of a compact surface of small genus. In particular, we start with a designated pair of regions of this surface and each region is labeled with the group element, which transforms the designated region into it. This gives a portrait of that finite group. These surfaces and the regions corresponding to the group elements are shown in this paper. William Burnside first gave a simple example of such a portrait in his 1911 book, "Theory of Groups of Finite Order".

Introduction and Historical Perspective

There are many ways to draw a picture of a finite group. One possibility is to let the group elements be represented by one to one transformations of the points of a surface. This idea was developed by Dyck [3] and elaborated further in Burnside [1]. Burnside started with circles in the plane and the transformation was inversion in the circle. Inversion in a circle can be defined in a Euclidean plane with a "point at infinity" appended. The plane with a "point at infinity" can be identified with the Riemann sphere, Σ. It can be shown that inversion in circle C is given by the equation $I_C(z) = -\frac{\bar{b}\cdot\bar{z}+c}{a\cdot\bar{z}+b}$, where C has equation $az\bar{z}+bz+\bar{b}\bar{z}+c=0$ with a and c real and b complex. This map is an anti-automorphism of the Riemann sphere (Jones and Singerman [4], p. 29).

The group generated by these transformations is determined by the relationship between the circles. For example, starting with a circle and a straight line tangent to it (a circle of infinite radius), the set of inversions of the circles in each other gives the diagram given in Figure 1 (Burnside [1], p. 377). The transformation, S is given by composing first a reflection in the line and then an inversion in the circle. The plane is divided into black and white regions as in figure 1 and each transformation takes the white regions into themselves and the black regions into themselves. If we start with a white region labeled E for the identity, then the region into which E is transformed by S^n can be labeled by that group element. This gives a nice graphical picture of the integers as a group of transformations.

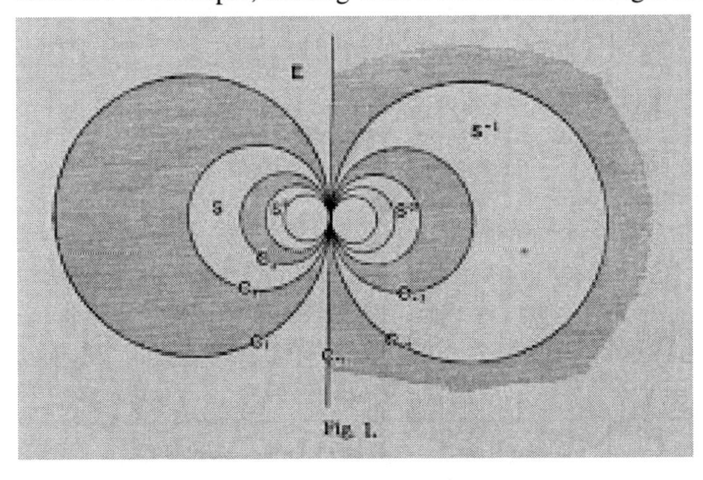

Figure 1, Burnside [1], Page 377.

The same ideas are used in Burnside [1, p. 379] to construct a free group on n generators, F_n. This construction fills up a unit disk with black and white regions and the transformations are given in the same way. We have used Geometer's SketchPad to reconstruct part of this portrait of a free group on two generators (Figure 2). This figure is very similar to the figure in Burnside [1], Page 380. Each "triangle" is bounded by arcs colored red, blue or black in our sketch. Inversion in any single arc will take a shaded region into a non-shaded region and vice versa. Therefore, each group action is represented by the composite of two such inversions. Inversion through first a red arc and then a blue arc corresponds to multiplying on the left by the generator S. Multiplying on the left by the generator T corresponds to inversion through black and then red. Multiplying on the left by ST corresponds to inversion through black and then blue. If we considered inversion through a black arc first and then a blue arc as the inverse of a single generator, R, then we could interpret this picture as a portrait of a group with presentation $\langle r,s,t \mid rst=1 \rangle$.

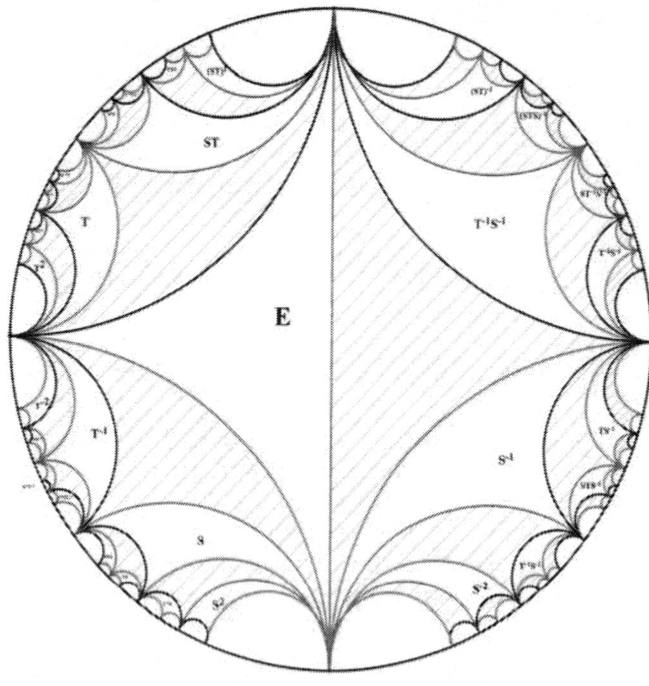

Figure 2, Portrait of a Free Group

Now suppose that we have a finite group, G, generated by n generators. This group is the image of F_n by a normal subgroup, N. After associating an element of F_n to each region, the final step is to identify all regions with labels from the subgroup, N. After this identification, we have the finite group, G, represented as a group of transformations on a surface of some genus. For n = 3, this is really the image of a quotient of a triangle group, $\Gamma(u,v,w) = \langle r,s,t \mid r^u = s^v = t^w = rst = 1 \rangle$. The transformation of inversion in a circle is an anti-analytic transformation of the Riemann sphere into itself. Therefore, any group represented in this way acts on a Riemann surface in an orientation-preserving way. The next section will attempt to give some portraits of small groups.

Group Portraits

Every compact Riemann surface with genus g is topologically equivalent to a sphere with g handles or equivalently, a sphere with g holes in it. This surface may be drawn and colored with white and black regions that represent a finite group of transformations, which act on the surface. Figure 3 gives Burnside's example of the picture of the group of quaternions, Q. The quaternions are the smallest group of strong symmetric genus 2. They also have a presentation as an image of the triangle group $\Gamma(4,4,4)$. This is a very symmetric presentation and that makes it easy to construct the portrait. The other groups of strong symmetric genus 2 are the dicyclic group, DC_3, the quasidihedral group, QD_4, the group $(4,6 \mid 2,2)$, using notation from Coxeter and Moser [2], $SL(2,3)$ and $GL(2,3)$ (See May and Zimmerman [5]). We consider the dicyclic group, DC_3 next.

Figure 3, Burnside [1], Page 396.

The dicyclic group of order 12 has presentation $\langle x, y \mid x^6 = 1, x^3 = y^2, y^{-1}xy = x^{-1} \rangle$. Its genus action is given by its presentation $\langle s,t \mid s^4 = t^4 = (st)^3 = 1, st = (ts)^2 \rangle$ as the image of the triangle

Figure 4 – Portrait of the Dicyclic Group of Order 12

group $\Gamma(3,4,4)$. The portrait of this group on a surface of genus 2 is shown in Figure 4. A red, a blue and a black arc of a circle bound each "triangle". The portrait consists of 12 white and 12 black triangles and has 10 vertices. The triangular regions that meet at a vertex are labeled in such a way that each white region is related to the adjoining white regions by multiplication on the left by either S, T or ST or its inverse. Therefore, each vertex could be classified as an S-vertex, a T-vertex or an ST-vertex depending on the labeling of its bounding regions. Since S and T have order 4, each S or T vertex has degree 8. Since the product ST has order 3, each ST vertex has degree 6. There are 3 S-vertices and the edges incident to them are blue and red. Similarly, there are 3 T-vertices and the edges incident to them are black and red. Finally, there are 4 ST-vertices and the edges incident to them are blue and black. This gives 10 vertices, 36 edges and 24 faces and so the Euler characteristic is –2.

Finally, we construct the portrait of the "quasiabelian" group, $QA_4 =< 2,2 \mid 2 >$ of order 16. This is a group of strong symmetric genus 3; it is also the rotation group of a regular map of genus 3. This group has presentation $< x, y \mid x^8 = y^2 = 1,\ yxy = x^5 >$ and it is the image of the triangle group $\Gamma(2,8,8)$. Its presentation as an image of $\Gamma(2,8,8)$ is $< s,t \mid s^8 = t^8 = (st)^2 = 1,\ s^2 = t^2 >$. The portrait consists of 16 white and 16 black triangles and has 12 vertices. Since S and T have order 8, each S or T vertex has degree 16. Since the product ST has order 2, each ST vertex has degree 4. There are 2 S-vertices, 2 T-vertices and 8 ST-vertices. Each ST-vertex connects only to the S and the T vertices. Each S or T vertex connects 4 times to the same S vertex, 4 times to the same T vertex and to each one of the ST vertices. This results in the portrait in Figure 5.

Figure 5 – Portrait of the Quasiabelian Group of Order 16

This gives 12 vertices, 48 edges and 32 faces and so the Euler characteristic is –4. Therefore, it may be drawn on a surface of genus 3. A surface of genus 3 may be constructed in many topologically equivalent ways. I have chosen a way where the symmetry of the construction complements the structure of the group.

References

1. W. Burnside, *Theory of groups of finite order*, (Cambridge University Press 1911).
2. H.S.M. Coxeter and W. Moser, *Generators and relations for discrete groups, 4th Ed.*, (Springer-Verlag, 1957).
3. Dyck, Gruppentheoretische Studien, Math. Ann., Vol. (1882), pp. 1 – 44.
4. G. Jones and D. Singerman, *Complex Functions, An algebraic and geometric viewpoint*, (Cambridge University Press 1987).
5. C. May and J. Zimmerman, Groups of small strong symmetric genus, J. Group Theory 3(2000), p. 233 – 245.

A New Use of the Basic Mathematical Idea of Twelve-Tone Music

Ward Douglas Maurer
Computer Science Department
George Washington University
Washington, DC 20052
202-994-5921
maurer@gwu.edu

Abstract

We here briefly describe a collection of pieces which we have written, and which have been performed for large audiences, in which mathematics is used. Specifically, every one of the 12 major chords, and every one of the 12 minor chords, appears in each of these pieces. We argue that this is a more pleasing use of the number 12 in music than the twelve-tone system of Schönberg.

The Twelve-Tone System

One of the most obvious connections between mathematics and music is that of twelve-tone music, pioneered by Schönberg and also used by Anton Webern, Alban Berg, and Hanns Eisler (see [1, 2, 3]). A piece of music is made up of tone rows, each of which consists of the 12 notes of the chromatic scale in some order. Mathematical transformations of a tone row may be used; for example, it may be transposed into a different key, or it may be played backwards. Also, it may be inverted, meaning that downward and upward intervals are interchanged. Any combination of these three transformations may be used. Also, adjacent notes in a tone row may be played together, as a chord. The number of possible variations of a tone row depends on that tone row; as a simple example, a chromatic scale (which is a tone row), inverted and then played backwards, gives the original scale, so it is not a different variation.

In academic music departments in the US, twelve-tone music was popular for a long time. Today, however, it seems to have fallen on hard times. Much of the blame for this has been placed on Schönberg's own initial insistence that the twelve-tone method, rather than the classical ideas of musical harmony, should be the basic principle behind composition: "As originally designed by Schönberg, the [twelve-tone] method was intended to preclude tonality" [4]. This was fascinating for a long time but has generally lost its appeal. Contrary to popular belief, however, not all twelve-tone music is atonal; both Berg and Schönberg wrote twelve-tone music in which some semblance of tonal harmony was preserved (also see [4]).

Our Alternate Use Of Twelve Tones

Our own interest in the mathematical ideas of twelve-tone music was an outgrowth of our interest in the composition of satirical songs. Not that we had any idea of satirizing the twelve-tone system itself (although Leonard Bernstein has done this). Rather, we were interested in expanding the range of tonal music, which normally uses only a tiny fraction of the available chord changes. Our interest was drawn, in this connection, to Bach's Well-Tempered Clavier, containing 24 preludes and fugues, one in each of the 12 major and 12 minor keys. However, we wanted to write, not 24 pieces, but a single piece. We were also fascinated by the Symphony in D Minor by Cesar Franck, an early exponent of chromaticism (later greatly expanded by Schönberg). Franck, although his music was unabashedly tonal, used sequences of chord changes which had never been heard before.

Could it be possible, we wondered, to write a long piece in the style of Cesar Franck, which used, at some point in the piece, every one of the 12 major chords and every one of the 12 minor chords? This would preserve the connection between mathematics and music that was inspired by Schönberg, without introducing atonality, which has little appeal to audiences for satirical songs. The result was a series of songs, written in this way, for the annual Hexagon Revue, which runs for over 15 performances each year in the city of Washington. These were called "Election Jeer" (about the election of 1996), an opener for the 2001 show (about the election of 2000), and "Wee Puns of Mass Distraction" (about the invasion of Iraq). Space here does not permit the reproduction of the lyrics of these songs. The following table, however, shows the measure numbers of typical measures in which these chords occur, for each song:

Chord	Election Jeer	Opener 2001	Wee Puns of Mass Distraction
C major	16, 66, 69	98, 99, 100	18, 20, 31
Db major	25, 28, 104	21, 22, 23	60, 136, 138
D major	55	63, 64, 105	67, 154
Eb major	7, 34, 38	25, 26, 115	12, 14, 24
E major	20, 91, 163	186, 187, 188	55, 56, 148
F major	32, 68	76, 95, 96	51, 58, 59
Gb major	30, 59, 102	7, 11, 39, 40	61, 62
G major	63, 65, 159	73, 74	10, 38
Ab major	1, 2, 4-6	27, 28, 44	145, 166
A major	12, 22, 24	161, 181	64, 134
Bb major	33, 48, 57	42, 67	52, 53, 83
B major	19, 31, 40	195	54, 148
C minor	115	25, 83, 231	4, 6
C# minor	138	213	43, 144, 147
D minor	18, 158	162	26, 150
Eb minor	58, 106, 107	43	85, 142, 159
E minor	62, 67, 169	249, 251	33, 101
F minor	53, 114	245	75, 140
F# minor	171	215, 253	160
G minor	54	55, 56, 57	46
G# minor	60, 89	5, 6, 8, 10	70, 92, 153
A minor	9, 10	250	29, 124
Bb minor	187	107	91
B minor	21	216	65

The full music and lyrics are available from the author.

References

[1] Perle, G., *Serial Composition and Atonality: An Introduction to the Music of Schoenberg, Berg, and Webern*, University of California Press, 1962.
[2] Schoenberg, A., *Fundamentals of Musical Composition*, Faber and Faber, London, 1967.
[3] Betz, A., *Hanns Eisler: Political Musician*, Cambridge University Press, Cambridge, 1983.
[4] Sadie, S., ed., *The Norton/Grove Concise Encyclopedia of Music*, W. W. Norton, New York, 1994.

A Braided Effort:
A Mathematical Analysis of Compositional Options

James Mai
School of Art
Campus Box 5620
Illinois State University
Normal, IL 61790-5620, USA
E-mail: jlmai@ilstu.edu

Daylene Zielinski
Mathematics Department
Bellarmine University
2001 Newburg Road
Louisville, KY 40205, USA
E-mail: dzielinski@bellarmine.edu

Abstract:

Artist James Mai created a system of forms in the developmental stages of his work *Epicycles*. This system offered mathematician Daylene Zielinski opportunities to provide mathematical analysis and to contribute to the final compositional organization of *Epicycles*. A set of eight new permutational forms are developed from a revised interrogation of a previously developed system of eighteen forms. The new set of forms lends itself to a variety of compositional arrangements including, with contributions from Zielinski, a "braided" ordering that creates a coherent sequence of the forms in the final work. This paper not only explicates the system of forms used in the resulting work, but it also illustrates the benefits and insights gained from interdisciplinary interactions between an artist and a mathematician during the development of a mathematically based work of art.

1. From Antecedent To Subsequent Forms

For artist James Mai, a system of forms previously developed for one painting may lend itself to later elaborations for new and different paintings. In these cases, subsets or entirely new sets of forms may be discovered by refining the rules that gave rise to the original set. This paper examines how such a new set of forms is derived from the system used in an earlier painting, *Permutations: Earthly*, which includes eighteen distinct forms created from permutations of upward and downward arcing semi-circles [1]. The new set of forms opens fresh considerations of visual characteristics and organization. Both aesthetic and mathematical analyses of these new forms have culminated in the digital composition, *Epicycles*, shown in Figure 5.

The motivation for the new set of forms is to convert the earlier closed shapes, used in *Permutations: Earthly*, to open-ended curves.[1] These curves are comprised of three semi-circular line-segments that connect four nodes distributed along a line. The original eighteen closed forms were composed of semi-circular segments from the parent figure, shown at the top of Figure 1. Each closed form yields four 3-segment paths, giving a total of 72 new forms. After eliminating rotational and reflective symmetric duplicates, there remain 28 visually distinct 3-segment forms, shown in Figure 1. Eight of these, shown in the last row of Figure 1, consist of an upward arc followed by a downward arc, then completed by an upward arc. This subset of eight curves is the formal basis of *Epicycles* and the subject of the discussion that follows.

[1] These eighteen closed shapes were derived from a set of four permutational forms developed by Victor Flach. See [1] for details.

Figure 1: *Mai's set of 3-segment forms.*

2. Visually Understandable Characteristics

Mai employs a series of aesthetic strategies to make the permutational characteristics of his forms visually recognizable. The underlying logical relationships connecting the forms are converted into visual relationships of shape, size, location, and color so that they become available to the eye. For Mai, this translation of the conceptual to the perceptual is critical to the success of a mathematically based work of art. While a work of art may benefit from subsequent interpretation, Mai endeavors to give these logical relationships a fully visual existence in a "self-revealing" artwork that transcends dependence on verbal explanation.

The initial aesthetic strategies for that visualization included holding all forms to the same scale and stacking them vertically so that the eye may directly compare the different orderings and sizes of the component arcs. In so doing, the viewer can recognize that each form is both similar to the others in its construction from arcs and nodes and distinct from the others as a unique configuration of arcs. The vertical alignment also makes apparent the varying locations of endpoints among the eight forms, which suggest the next level of information to be visualized.

If we identify the beginning node of each form with the number 1, the next node along the curvilinear path with 2, and so on, then a left-to-right reading of the node numbers, along with the restriction that the semi-circles must follow the up-down-up pattern, uniquely determines each form. In this manner, each of the curves is an expression of a distinct permutation of the set {1, 2, 3, 4}. For instance, the first form in the last row of Figure 1 represents 1 2 3 4, while the second form represents 1 2 4 3, and so on. It should be noted that without the up-down-up pattern Mai has forced on each form, a permutation of {1, 2, 3, 4} does not translate into a unique visual form. All of the forms in the first column of Figure 1 can be seen as a visual expression of the permutation 1 2 3 4, and all forms in the second column are associated with 1 2 4 3. In fact, each column of forms displayed in Figure 1 is associated with a distinct permutation of {1, 2, 3, 4}, while each row is a distinct permutation of three directional choices from the set {up, down}.

Since there are 24 distinct permutations of the set {1, 2, 3, 4}, one might wonder why the final row of Figure 1 contains only eight forms representing the permutations 1 2 3 4, 1 2 4 3, 1 3 2 4, 1 3 4 2, 1 4 2 3, 1 4 3 2, 2 1 4 3, and 2 4 1 3. This is because all other permutations of {1, 2, 3, 4} correspond to forms that are merely rotations or reflections of one of the eight forms already appearing in this row of Figure 1. So, these particular configurations of the eight forms should be thought of as archetypes, where each of the eight forms represents all rotated or reflected versions of that configuration. Mathematically, these eight forms are a set of equivalence class representatives from the larger set of 24 forms where the equivalence is "is symmetric to." Interestingly, these equivalence classes are not all of the same size because each of the eight forms does not appear exactly three times in the full set of 24 forms. Each of the four symmetric forms appears exactly twice, while each of the four asymmetric forms appears four times.

When stacking the forms, Mai wanted to establish a vertical distribution of nodes that was equal to the horizontal distribution, creating a square grid of nodes upon which the forms would be arrayed. This resulted in severe overlapping of the semi-circular arcs of adjacent forms, causing visual confusion and obscuring the distinctness of each form. To accommodate this tight arrangement and eliminate the overlapping, Mai condensed each form vertically by changing the arcs from semi-circles to quarter-circles. This kind of proportional adjustment is not uncommon given the challenge facing the artist as he proceeds from considerations of the forms as independent objects to considerations of how the forms should be related collectively, while also retaining their individual characteristics, in the final composition.

One additional characteristic of the forms required visual clarification. The forms themselves offer no distinction between their endpoints, so an observer has no way to know which node corresponds to 1 and which to 4 in any permutation of {1, 2, 3, 4}. Mai solved this problem by using a fixed set of colors to encode the permutation in the nodes. Each form begins at an orange node, which is connected by an upward arc to a red node, which in turn is connected via a downward arc to a blue node, and is completed by an upward arc to a green node. While the different configuration of arcs makes each form visually distinct, the colored nodes provide the link to permutations of {1, 2, 3, 4}. For the purposes of reproduction in this paper, the color-coding has been replaced with a shape-coding for the nodes where the progression: orange, red, blue, green, has been replaced with: hollow circle, hollow diamond, solid diamond, solid circle.

3. Possibilities for Arrangements of Forms

After establishing the eight forms to be displayed, their sizes and vertical distribution, and the distribution and color-coding of the nodes, the set of forms initially offered no obvious order for their vertical arrangement. Compositionally, this ordering was the final formal question to be addressed. Mai's initial observation that there were four symmetrical and four asymmetrical forms presented an opportunity for arranging the eight forms into two groups. It seemed clear that the first form in the last row of Figure 1 should be at the top of the vertical composition, since it was the simplest permutation of {1, 2, 3, 4} and the most basic of the eight forms. Since it is one of the symmetrical forms, this suggested that the remaining three symmetrical forms should be grouped immediately beneath, followed by the four asymmetrical forms beneath those. This initial composition is shown in Figure 2a; recall that the color-coding of nodes has been replaced with a shape-coding in this paper for purposes of clearer reproduction.

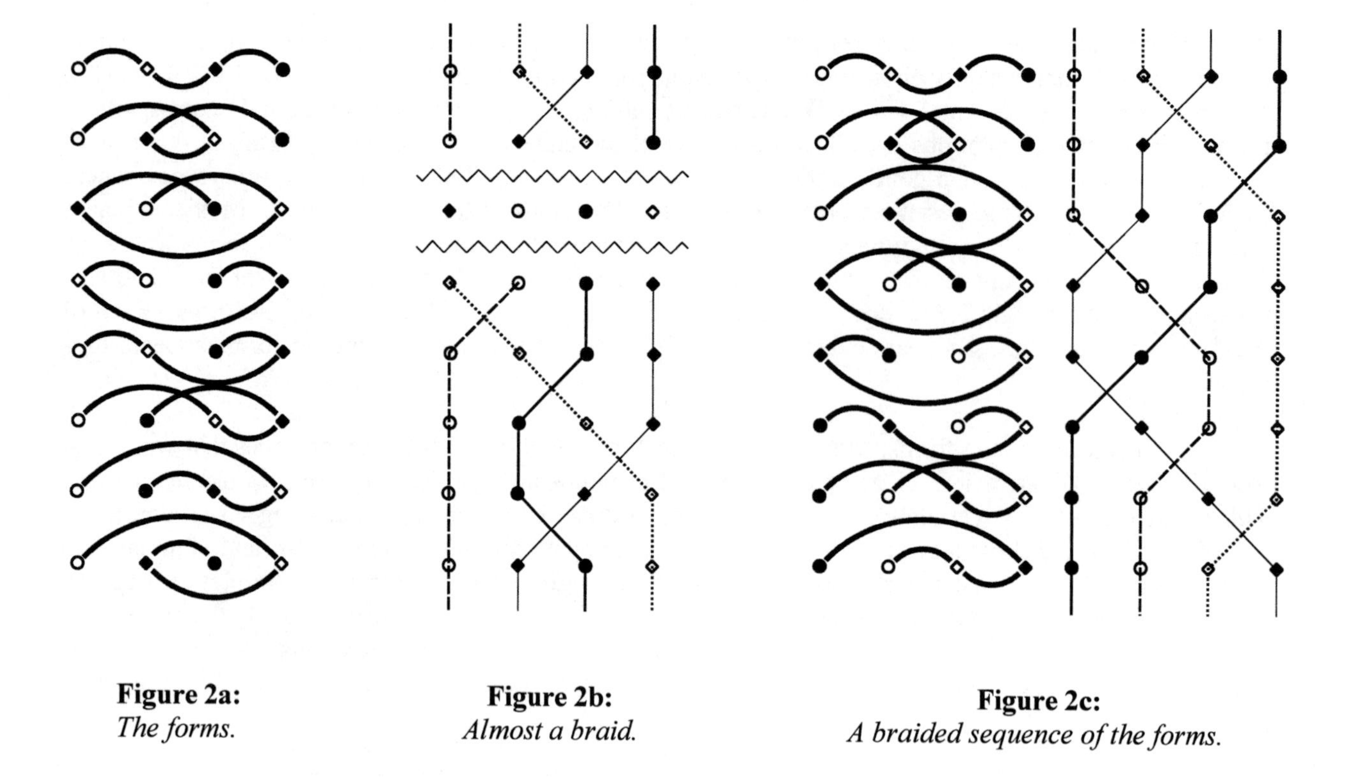

Figure 2a:
The forms.

Figure 2b:
Almost a braid.

Figure 2c:
A braided sequence of the forms.

At this intermediate stage, when Zielinski saw the composition represented in Figure 2a, she immediately noticed that connecting the corresponding nodes from form to form almost produced a braid. The "almost" carries two distinct qualifications in this case. First, the third form cannot be created via a single braid move from the second form as indicated by the first break shown in Figure 2b, because a mathematical braid is allowed to cross only one pair of adjacent strands at a time. In addition to this obstruction, the node arrangement in the fourth form cannot be produced by any single braid move from the third form. Secondly, a formal mathematical braid must distinguish between over and under crossings, but Mai's forms have no visual characteristics that convey such crossing information. With a bit of trial and error, Zielinski found an ordering, seen in Figure 2c, that did create a braid void of crossing information. As we shall see later, this sequence is one of twelve braided orderings that begin with Mai's preferred form. For the remainder of this paper, we shall use the word braid in this looser context in which crossing information is disregarded.

Despite the lack of crossing information, the artist was immediately interested in this new ordering schema and posed a new question: Can the braided sequence of forms be made cyclical, whereby the bottom form can also be transformed via one braid move to the top form? This is primarily an aesthetic question for Mai, who frequently works with circuity and closed visual forms as expressions of wholeness and transcendence. Mai pursued the question by using a tree structure to diagram possible sequences of the eight forms. Mai's graph showed him that cyclical braiding was not possible when one begins with his preferred form. Furthermore, it suggested that there were only two braided sequences of the forms; as we shall see, this is correct. Further analysis by Zielinski established that cyclical braiding is indeed impossible and that, although there are twelve braided sequences that begin with Mai's preferred form, there are only two sequences of the forms that can be generated as braids on the nodes. To establish these and other facts, we need additional representations of the forms as graphs.

First, we will pull back to the full set of 24 configurations of the eight archetypal forms. The graph in Figure 3 shows exactly which configuration can be produced from any other via a braid move that swaps the left-most pair of nodes, the middle pair, or the right-most pair. Hence, the edges in this graph are labeled with an *L*, *R*, or *M* according to which move was used. As long as one is following a path of edges in the graph, the sequence of forms given by those vertices is a braided sequence. Thus, this graph displays for the artist all possible arrangements of these forms that can be produced through a sequence of braid moves. Consequently, any cycle in the graph gives a braided cyclic ordering of the forms. Hence, a fuller answer to Mai's question comes from investigating the eight-cycles of the graph in Figure 3. Though it is not difficult to find eight-cycles in this graph, it is tedious to confirm that not one passes through each of the eight archetypal forms.

Just as with Mai's tree, this analysis takes quite a bit of time and concentration. Before we redirect our analysis to a simplified version of the graph below, we need to take note of a particularly interesting, and mathematically relevant, fact. A handful of the forms produce the same type of new form regardless of whether the left-most or right-most nodes are swapped. An example of this can be seen by examining the relationship between what the authors refer to as the kink-form and the spiral-form. A kink-form can be found to the left of the top-most 'M' in the graph below, immediately to the left of Mai's preferred form, and one of the four spiral-forms can be found immediately to the left of that kink-form. It is not hard to confirm that each kink-form is connected to two distinct spiral-forms, even though each spiral-form is connected to only one kink-form.

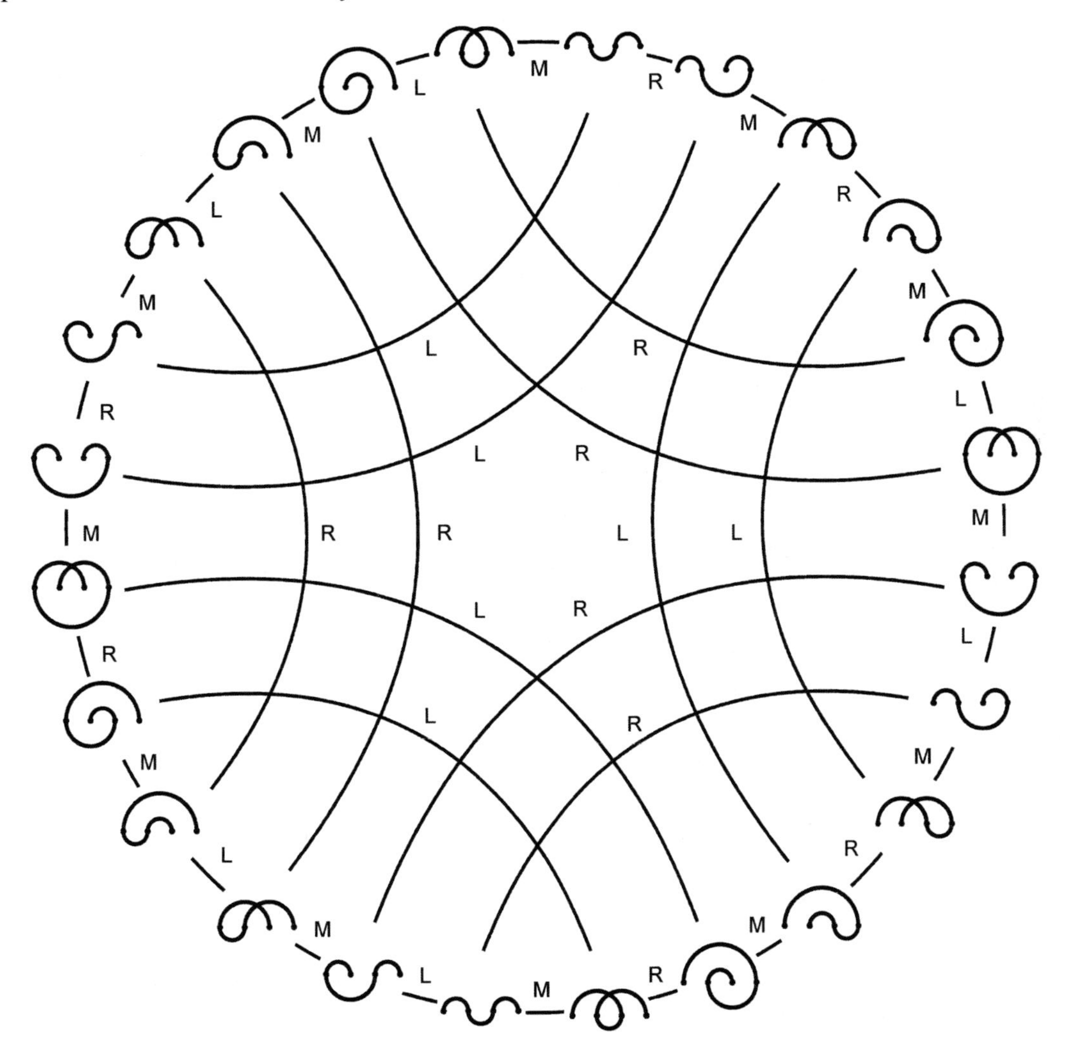

Figure 3: *All 24 configurations of the eight forms with related braid moves.*

A simpler graph arises when one disregards which braid move was used and which configuration of a particular form resulted. We do want to retain, however, the number of distinct paths that exists between each pair of archetypal forms. Thus, we will create a multi-digraph that summarizes the information in the more complex graph above. Our new graph has the prefix *multi-* since we will connect certain forms to others with multiple edges because of the phenomenon mentioned above. The prefix *di-* indicates that we have added directions to our edges. Each edge in Figure 3 is bi-directional, but we need some uni-directional edges in Figure 4a to indicate when we have choices and when we do not. The graph in Figure 4a shows each of the archetypes with edges connecting any two forms that have configurations that can be transformed to each other via a single braid move.

The question of looking for braided sequences of the eight forms boils down to looking for paths of edges in this graph that visit each vertex/form exactly once. Such a path is called Hamiltonian. A cyclic braided sequence of the eight forms would be represented by a Hamiltonian cycle in this graph, but not too much effort is required to see that no such cycle exists. One can see, however, that there are several Hamiltonian paths. Figure 4b shows all possible Hamiltonian paths that begin with Mai's preferred form. There are eight distinct paths shown in upper portion of Figure 4b and four more in the lower portion giving the total of twelve possible braided sequences of the forms mentioned earlier. Interestingly, some of these sequences differ only in their vertex colorings and not in either the order or the forms or even the configurations of the individual forms. Several other braided sequences of forms can be created beginning with forms different from Mai's preferred form. In fact, there are multiple Hamiltonian paths beginning from each of the forms that are connected to only two others in Figure 4a, but there are no such paths that begin at either of the two, left- and right-most, forms in Figure 4a, which are connected to three other forms. In total, there are 64 braided sequences of all eight forms.

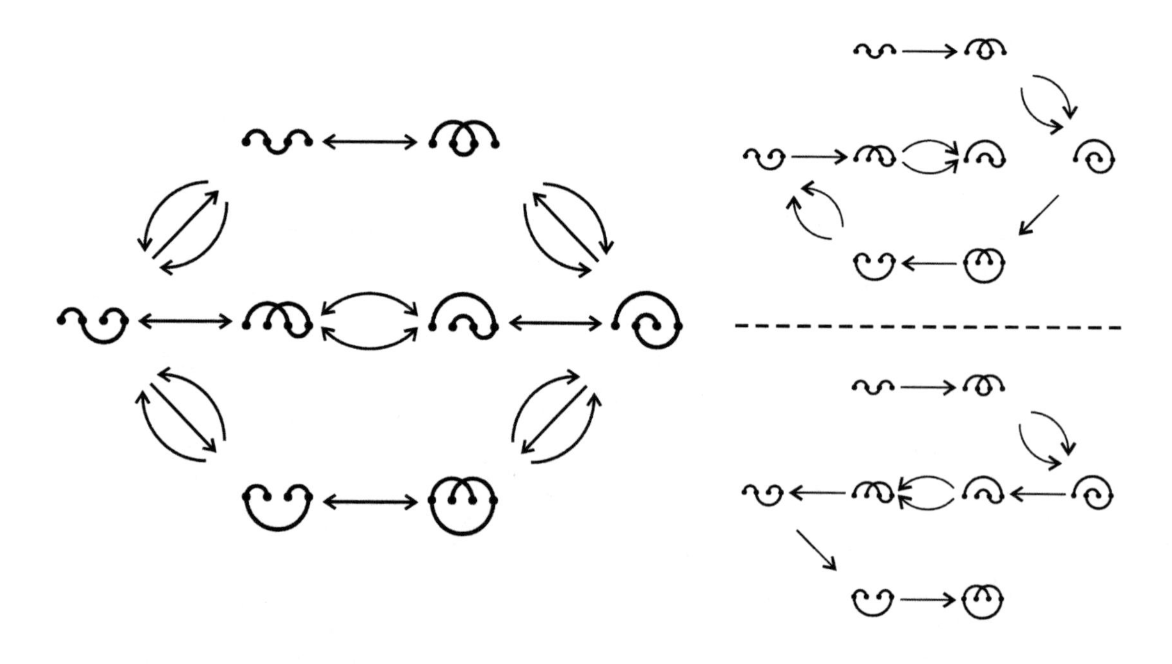

Figure 4a:
Multi-digraph of the eight archetypes and the number of braid moves that relate them.

Figure 4b:
All Hamiltonian paths beginning at Mai's preferred form

Figure 5: Epicycles, *digital print, 40 x 18"*

4. Particularizing the Abstract

Mai sees the task of art as not only to define new visual forms but also to suggest metaphoric associations between those forms and other events or experiences. Mai uses minimal figurative additions to his final compositions so as to stimulate imaginative allusion rather than to present straightforward illustration of a subject matter. The subject and title of this artwork refer to the efforts by early astronomers to understand the complex retrograde motions of planets, which at times trace out unusual looping and winding paths against the background of fixed stars. Working within the Ptolemaic model of the universe, early Greek astronomers posited epicycles, which are smaller circular orbits in combination with the larger orbit about the Earth, to explain a planet's back-and-forth motions across the sky [2]. Mai saw a similarity between his forms' nodes with their curvilinear connections on the one hand and the planets with their curvilinear retrograde motions on the other.

The permutational forms discussed in this paper, although clearly emerging from purely abstract, formal procedures, were lent this particular, figurative identity by means of just a few visual cues: a vertical color gradation in the background that matches the fading light of the sky at dusk, a slight blurring of the nodes to suggest the glow of planets as they appear to the naked eye, the combination of points and curves to suggest astronomical diagrams, and, of course, the title, *Epicycles*. The challenges faced by early astronomers to read accurately the order underlying the changing positions of the planets seems not altogether unrelated to the search by artists and mathematicians to discover new orders in the universe of form and number.

References

[1] J. Mai and D. Zielinski. Permuting Heaven and Earth: Painted Expressions of Burnside's Theorem. In *Conference Proceedings of Bridges: Mathematical Connections in Art, Music and Science, 2004*. Eds. Sarhangi, Reza and Séquin, Carlo.

[2] G. J. Toomer. Ptolemy and His Predecessors. In *Astronomy Before the Telescope*. Ed. C. Walker, pp. 74-76. New York: St. Martins Press, 1996.

On a Family of Symmetric, Connected and High Genus Sculptures

Ergun Akleman and Cem Yuksel*
Visualization Sciences Program,
Department of Architecture, College of Architecture
emails: ergun@viz.tamu.edu, cem@viz.tamu.edu

Abstract

This paper introduces a design guideline to construct a family of symmetric, connected sculptures with high number of holes and handles. Our guideline provides users a creative flexibility. Using this design guideline, sculptors can easily create a wide variety of sculptures with a similar conceptual form.

1 Introduction

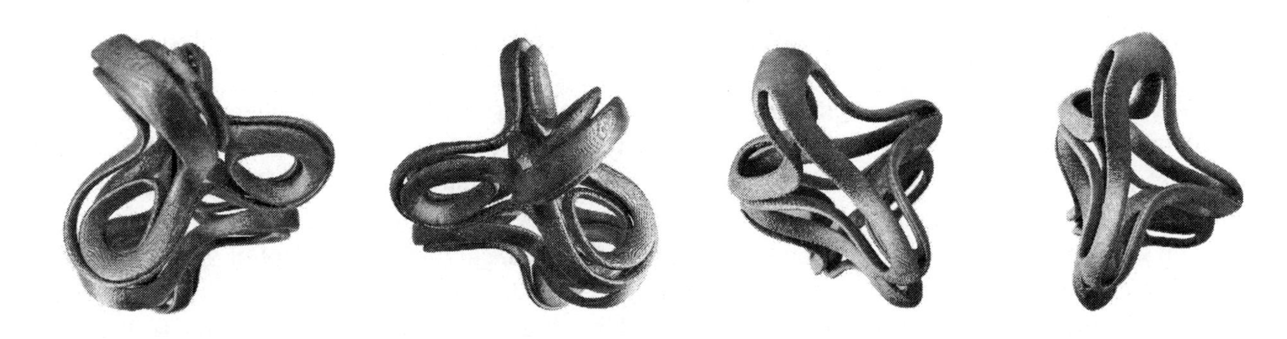

Figure 1: Photographs of 3D prints of two of the first symmetric high genus shapes (shapes with high number of handles and holes) designed by Ergun Akleman. For each sculpture, we took two photographs from slightly different point of views. The sculptures are photographed on a mirror. Background is eliminated. These shapes are made from ABS plastic and printed using a Fused Deposition Machine (FDM). They are later painted using an acrylic paint.

In this paper, we present a design guideline to create a new sculptural family with interactive topological modeling. Using this design guideline a large set of sculptures that have a similar conceptual form can easily be created (see Figures 1 and 2). We have tested the design guideline in a computer aided sculpting course [2]. We observe that, using the design guideline, students can rapidly create a wide variety of shapes. Although these shapes are completely different; they can be perceived as belonging to the same family and having a similar conceptual form. Figure **??** shows some examples of shapes that were created by some of the students using our design guideline, as one of the biweekly assignments of the computer aided sculpting course.

2 Motivation

Development of new methods to create new aesthetic forms are essential for artistic applications. In computer graphics and mathematics, it is very common to find methods to create aesthetically pleasing visual results

*Thanks to Ozan Ozener for rendering images and printing the sculptures.

which has never been created before. Although, most researchers usually view their results as proof-of-concepts, those results can be aesthetically pleasing and considered to be artworks. The methods developed by researchers are also useful for artists since they are reproducible and allow a wide variety of results to be produced. For instance, the fractal methods to create Julia sets are widely used by non-mathematicians to create artworks [1].

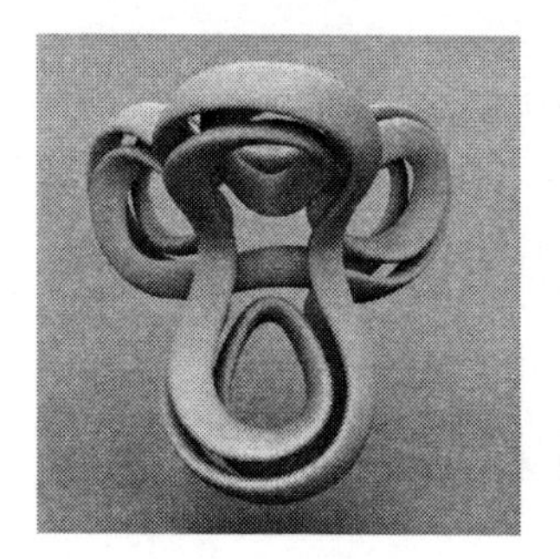
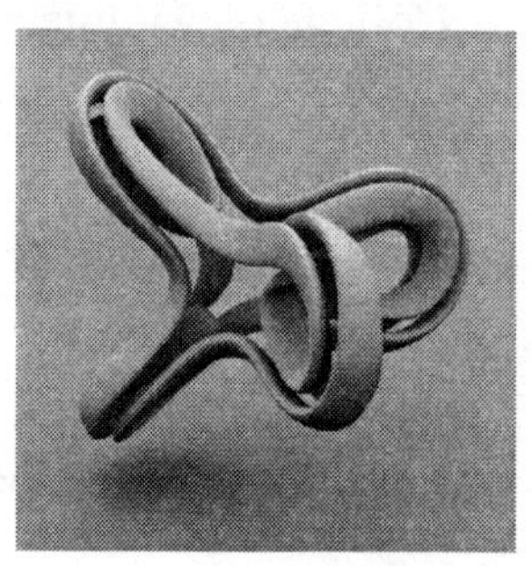
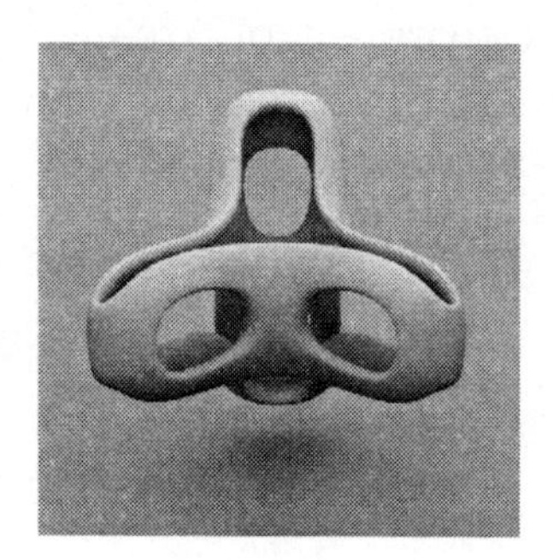
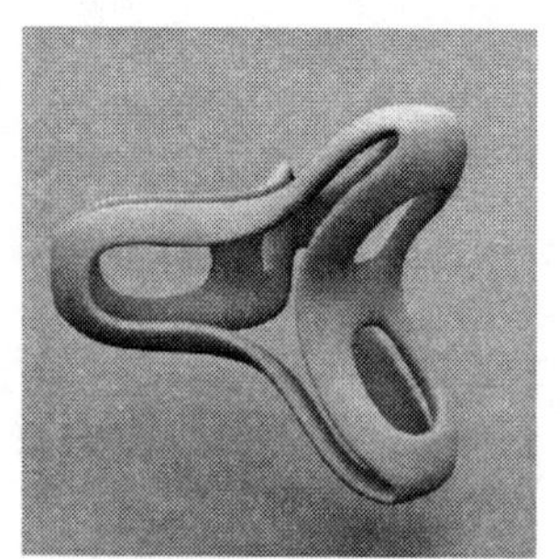

Figure 2: Computer rendered versions of the models from which sculptures in Figure 1 are printed. Images are rendered by Ozan Ozener. The images show two views of the same shape. The ribbon shape on the left is really a genus-1 surface, but, during the construction of this surface, genus changed several times.

With the advance of computer graphics, many artists have also begun to use mathematics as a tool to create revolutionary forms of sculptures. There are many contemporary sculptors such as George Hart [14], Helaman Ferguson [11, 10], Bathseba Grossman [12], Brent Collins, and Carlo Séquin [16] who successfully combine art and mathematics to create unusual and high-genus sculptures. These mathematical sculptors develop their own methods to model, prototype and fabricate an extraordinary variety of shapes. Most of these contemporary sculptors who successfully combine art and mathematics are, in fact, mathematicians or engineers who have developed their own methods. Identification of methods to create new aesthetically pleasing sculptural forms can eventually be a major direction for solid and shape modeling research.

One of the most exciting aspects of sculpting has always been the development of new methods to design and construct unusual, interesting, and aesthetically pleasing shapes. One of the most interesting sculptural shapes are high-genus surfaces [19, 20, 10, 17, 5]. Our goal in this work is to develop a reproducible guideline to design high-genus sculptures. We also want to provide creative flexibility to users for constructing a wide variety of sculptures that can still be perceived as belonging to the same family. We also want our resulting shapes to be physically realizable. In other words, we should be able to 3D print the resulting shapes. In this paper, we present a design guideline that satisfy all the goals to create new aesthetically pleasing sculptural forms.

- Using the design guideline it is possible to create a wide variety of shapes that do not specifically resemble any existing sculptural or architectural forms.
- On the other hand, despite the variety all the shapes that are created using the same design guideline must still be perceived as belonging to the same family.
- The resulting shapes can be physically realizable with a minimal effort from the user as shown in Figure 1.

3 Design Guideline

The design guideline is based on a set of topological mesh modeling operators. These operators guarantee that the resulting shape is topologically 2-manifold. If the user avoid self-intersection, which is easy to achieve, the resulting models are 3D-printer-ready, i.e. they can be printed by using a rapid prototyping machine as shown above. Our design guideline consists of six stages.

- **Stage 0: Initial Shape.** Start with a symmetric convex polyhedral shape [7, 13, 21]. Any platonic solid such as tetrahedron, cube or dodecahedron [18, 22] are perfect candidates for starting shape. In the example shown in Figure 3, the initial shape is a cube.
- **Stage 1: Extrusions.** Apply the same extrusion operation to all faces of the starting shape. Adding a twist or rotation to the extrusions can create an additional effect as shown in Figure 3. Moreover, the recently introduced local mesh operators can also be used in this stage. These local operators can extrude generalized platonic solids and are called tetrahedral, cubical, octahedral, dodecahedral and icosahedral extrusions [15].

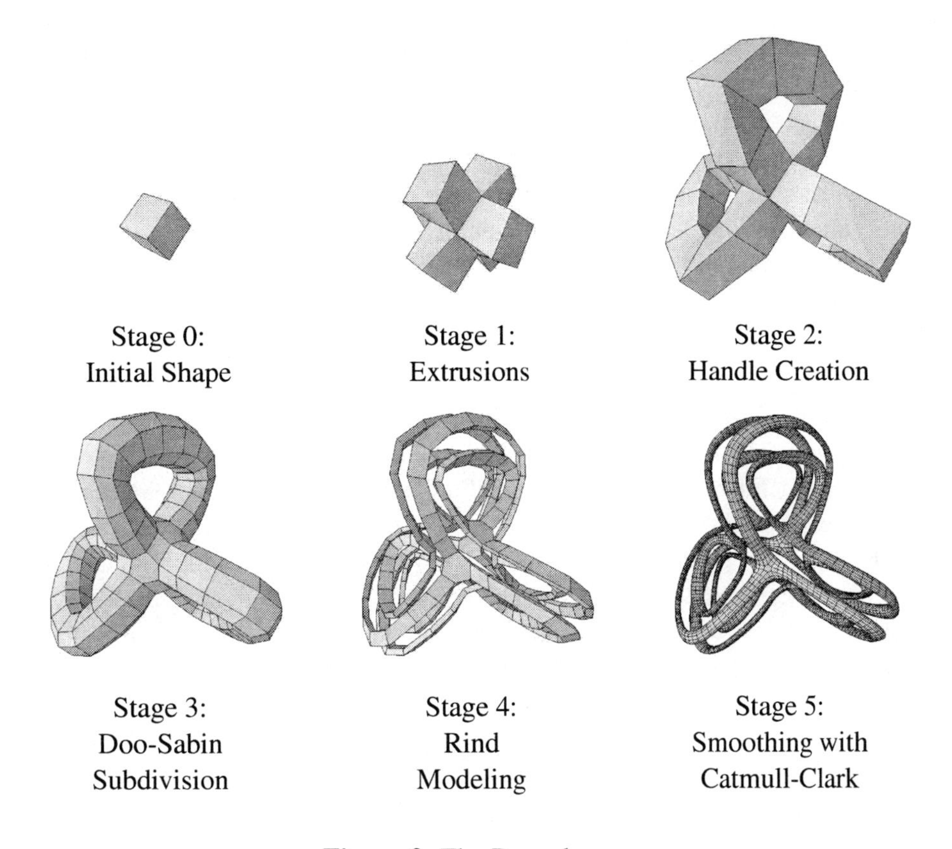

Figure 3: The Procedure.

- **Stage 2: Creating Handles.** Connect symmetrically related faces using multi-segment curved handles [17]. This operation, along with cubical extrusions, populate the mesh with quadrilateral faces with 4-valent vertices as shown in Figure 3. After the operation, there will be only a handful of extraordinary points (i.e. vertices with valence other than 4 and faces with sides other than 4).
- **Stage 3: Doo-Sabin Smoothing.** Apply Doo-Sabin subdivision once [8]. This operation will create visibly connected clusters of faces. These clusters of faces are formed as a result of extraordinary points. The faces in each cluster defines a route that connects one extraordinary point into another. These clusters can easily be seen in Figure 3.
- **Stage 4: Rind Modeling.** Rind modeling creates a rind structure by creating a smaller replica of initial shape [5]. In rind modeling, users can punch holes in this ring shape by clicking the faces. There exist two strategies: (1) Deleting all non-cluster faces by making the clusters clearly visible, (2) Deleting all

cluster faces by making the non-cluster structures visible. In the example shown in Figure 3 we made clusters visible.

- **Stage 5: Final Smoothing.** Final smoothing is useful to create a simple and smooth surface. Although, in final smoothing any subdivision scheme can be used, we use Doo-Sabin [8] or Catmull-Clark [6] schemes for smoothing.

Using both stages 2 and 3 is not really required. It is even possible to skip either one of them to create different results. As it can be seen in attached examples it is possible to create a wide variety of shapes using this design guideline.

4 Implementation

The operations used in all six stages have already been implemented and included in our existing 2-manifold mesh modeling system, called TopMod [3, 9, 5, 17]. Our system is implemented in C++ and OpenGL. All the examples in this paper were created using this system. You can download TopMod from http://www-viz.tamu.edu/faculty/ergun/research/topology/ for free and use the software for non-commercial applications. TopMod provides only Open-GL based interactive rendering. For high quality rendering, shapes can be exported in obj format and rendered in some 3D modeling and animation system. Our students use Maya and 3D Studio Max for rendering the final models.

5 Results

Figures 4 show a student (Cem Yuksel) work that illustrate creative flexibility that is provided by our guideline. Each figure show the steps that is used by Cem and final virtual sculptures. By adding handles after completing the basic guideline Cem created two conflicting forms that exists in the same sculpture. Lauren Simpson, to create the shape shown in Figure **??**, added tetrahedral extrusion, which is stellation operation, before adding handles. This particular step allowed her to create handles with triangular cross-sections. Finally, rind modeling on handles with triangular cross-sections provided a simple and elegant connections. Audrey Wells' sculpture shown in **??** is based on octahedral extrusions that also allow her to create handles with both quadrilateral (thicker) and triangular (thinner) cross-sections.

6 Conclusions and Future Work

Sculpting and Architecture can provide major research directions for solid and shape modeling. In both, the precision, which is very important for engineering shape design, is not a major concern. In this work, we have introduced a new sculptural approach with a motivation coming from strong aesthetics concerns. Similar to results from computer graphics research, our results are reproducible and allow even novice users to create interesting sculptures with a creative flexibility. Because of their strong symmetry, these shapes can possibly be built using a few building blocks. We are currently investigating physical construction of large versions (medium size shapes such as ones larger than $1m^3$) of such complicated shapes using low-cost materials such as concrete.

We have used the terms such as family and conceptual form to introduce the idea. Unfortunately, such terms that are related to aesthetic or perception are not mathematically well-defined. An interesting research direction is to provide rigorous definitions of such aesthetic related terms using quantitative studies.

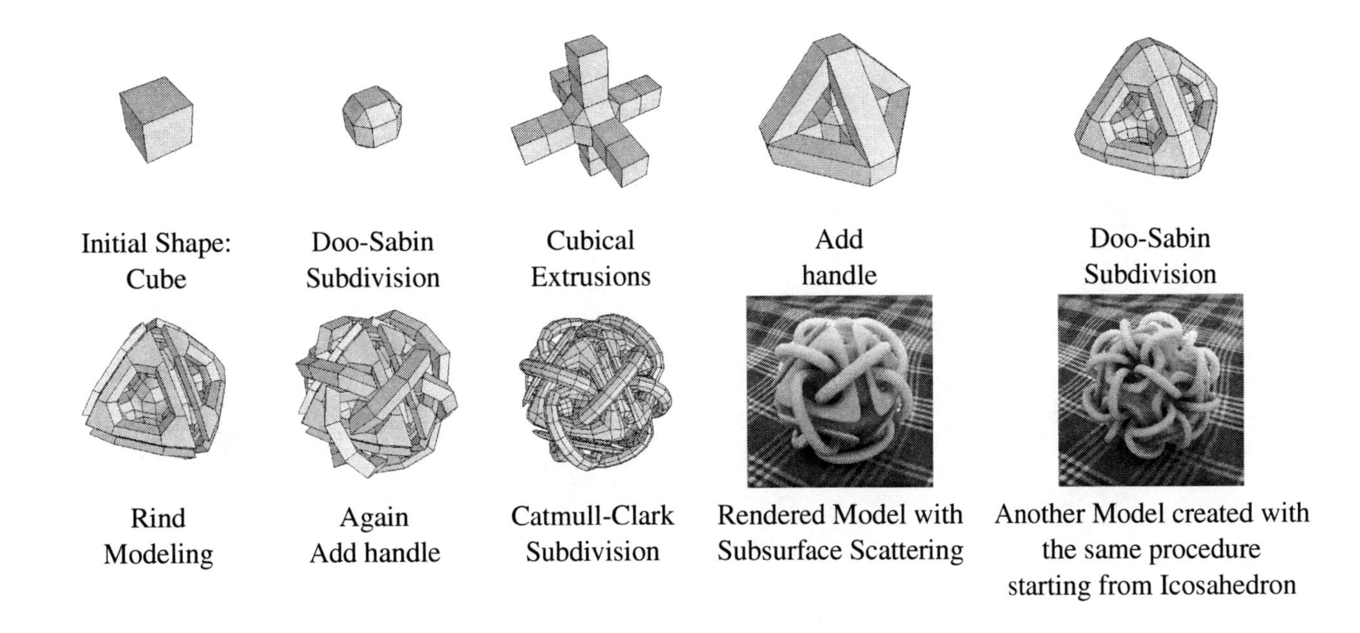

Figure 4: The Procedure used by Cem Yuksel and final virtual sculptures. Both rendered using subsurface scattering to give an illusion of ABS plastic.

References

[1] E. Akleman. Viza 656 - image synthesis; see volume rendering homeworks spring 2004-2005,. http://www-viz.tamu.edu/courses/viza656/index.html, 2005.

[2] E. Akleman. Viza 657 - computer aided sculpting,. http://www-viz.tamu.edu/courses/viza657/index.html, 2005.

[3] E. Akleman and J. Chen. Guaranteeing the 2-manifold property for meshes with doubly linked face list. *International Journal of Shape Modeling*, 5(2):149–177, 1999.

[4] E. Akleman and V. Srinivasan. Honeycomb subdivision. In *Proceedings of ISCIS'02, 17th International Symposium on Computer and Information Sciences*, volume 17, pages 137–141, November 2002.

[5] E. Akleman, V. Srinivasan, and J. Chen. Interactive rind modeling. In *Proceedings of the International Conference on Shape Modeling and Applications*, pages 23–31, 2003.

[6] E. Catmull and J. Clark. Recursively generated b-spline surfaces on arbitrary topological meshes. *Computer Aided Design*, (10):350–355, 1978.

[7] P. Cromwell. *Polyhedra*. Cambridge University Press, 1997.

[8] D. Doo and M. Sabin. Behavior of recursive subdivision surfaces near extraordinary points. *Computer Aided Design*, (10):356–360, 1978.

[9] J. Chen E. Akleman and V. Srinivasan. A minimal and complete set of operators for the development of robust manifold mesh modelers. *Graphical Models Journal, Special issue on International Conference on Shape Modeling and Applications 2002*, 65(2):286–304, 2003.

[10] H. Ferguson, A. Rockwood, and J. Cox. Topological design of sculptured surfaces. In *Proceedings of SIGGRAPH 1992*, Computer Graphics Proceedings, Annual Conference Series, pages 149–156. ACM, ACM Press / ACM SIGGRAPH, 1992.

[11] Heleman Ferguson. Sculptures,. http://www.helasculpt.com, 2005.

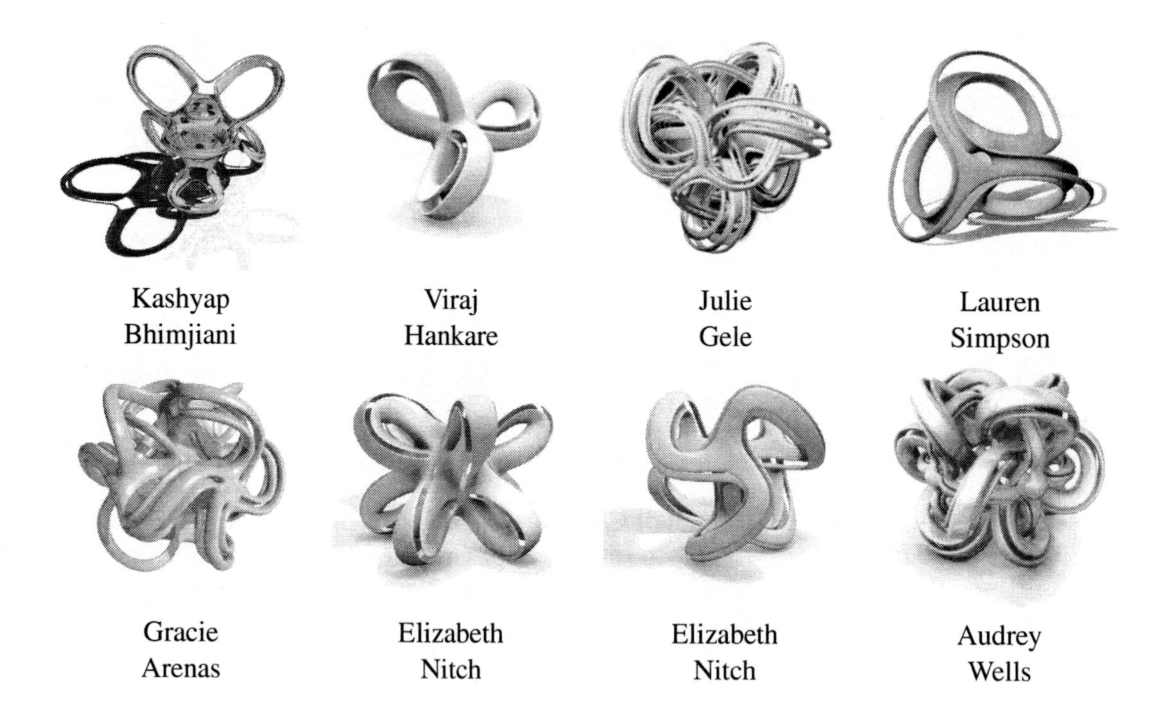

Kashyap Bhimjiani · Viraj Hankare · Julie Gele · Lauren Simpson

Gracie Arenas · Elizabeth Nitch · Elizabeth Nitch · Audrey Wells

[12] Batsheba Grossman. Sculptures,. http://www.bathsheba.com/, 2005.

[13] B. Grunbaum and G. Shephard. *Tilings and Patterns*. W. H. Freeman and Co, NY, 1987.

[14] George Hart. Sculptures,. http://www.georgehart.com, 2005.

[15] E. Landreneau, E. Akleman, and V. Srinivasan. Local mesh operations, extrusions revisited. In *Proceedings of the International Conference on Shape Modeling and Applications*, pages 351 – 356, 2005.

[16] Carlo Sequin. Sculptures,. http://www.cs.berkeley.edu/ sequin/, 2005.

[17] V. Srinivasan, E. Akleman, , and J. Chen. Interactive construction of multi-segment curved handles. In *Proceedings of 10th Pacific Conference on Computer Graphics and Applications*, pages 429–435, 2002.

[18] I. Stewart. *Game, Set and Math: Enigmas and Conundrums*. Penguin Books, London, 1991.

[19] S. Takahashi, Y. Shinagawa, and T. L. Kunii. A feature-based approach for smooth surfaces. In *Proceedings of Fourth Symposium on Solid Modeling*, pages 97–110, 1997.

[20] W. Welch and A. Witkin. Free-form shape design using triangulated surfaces. In *Proceedings of SIGGRAPH 1994*, Computer Graphics Proceedings, Annual Conference Series, pages 247–256. ACM, ACM Press / ACM SIGGRAPH, 1994.

[21] D. Wells. *The Penguin Dictionary of Curious and Interesting Geometry*. London: Penguin, 1991.

[22] R. Williams. *The Geometrical Foundation of Natural Structures*. Dover Publications, Inc., 1972.

Affine Regular Pentagon Sculptures

Douglas G. Burkholder
School of Computing Science & Mathematics
Lenoir-Rhyne College
625 7th Avenue NE
Hickory, North Carolina, 28603
E-mail: burkholderd@lrc.edu

Abstract

In this paper we shall describe how to apply symmetric linear constructions to a random non-planar pentagon to construct mathematically and artistically interesting sculptures, such as in Figure A. This process will always produce a nested set of affine regular stellar pentagons. This generalizes a procedure created by Jesse Douglas.

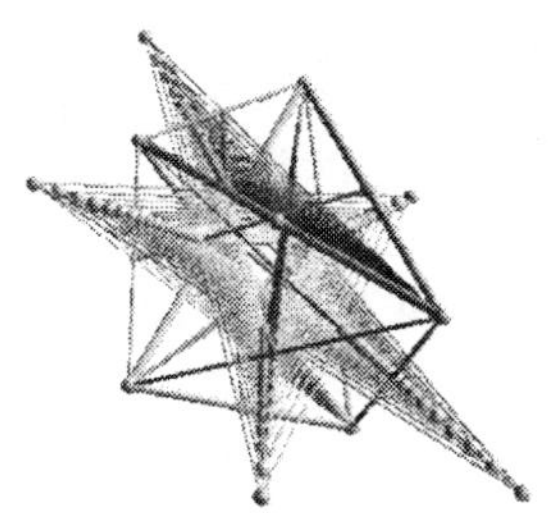

Figure A

Starting from a random non-planar pentagon, we shall describe a sequence of steps consisting of constructing line segments between pairs of known points followed by locating a new point on this line at a specified ratio of this length. At the completion of these steps, label the last point constructed v_1. We repeat the same process from the vantage point of each successive vertex of the original pentagon and label the last point of each sequence v_i respectively. By connecting the v_i in order we form a new pentagon. Although the final sculpture is not symmetric, since the process is symmetrical, this is called a *symmetric linear construction.* If this process is carefully selected, then the resulting pentagon will be planar and affine regular. By including variations on the theme, we can construct a collection of nested affine regular pentagons within the original pentagon. Affine regular pentagons, if oriented just right, will cast a shadow which is regular.

Given a random pentagon with vertices $p_1, p_2, \ldots, p_5$, any finite sequence of steps of the type listed above can be written as a weighted average $w_1p_1+w_2p_2+\ldots+w_5p_5$ of these 5 points. By symmetry, all of the vertices of the final pentagon constructed can be written as $v_i=w_1p_i+w_2p_{i+1}+\ldots+w_5p_{i+4}$, where subscripts are modulo 5. If these weights w_i measure the distances from successive vertices of a regular stellar pentagon to an arbitrary line, with distances below the line measured as negative distances, then the pentagon constructed from these weighted averages will be affine regular stellar [1]. For our sculpture we shall place the line parallel to and a distance of a units above the base, as shown in Figure 1. By letting a vary, we shall be able to construct a nested set of affine regular stellar pentagons within our sculpture. If the distance between the base p_4p_2 and the horizontal line p_1p_5 is 1 unit then the height of the pentagon is the golden ratio $\varphi=(1+\sqrt{5})/2$. Thus, the desired weights, in order, are $(1-a)$, $-a$, $(\varphi-a)$, $-a$, $(1-a)$. While these weights will always generate affine regular pentagons, we must normalize these by dividing by $(2+\varphi-5a)$ before we can convert them into a sequence of linear steps. This gives us the following formula for calculating the location of the new vertices:

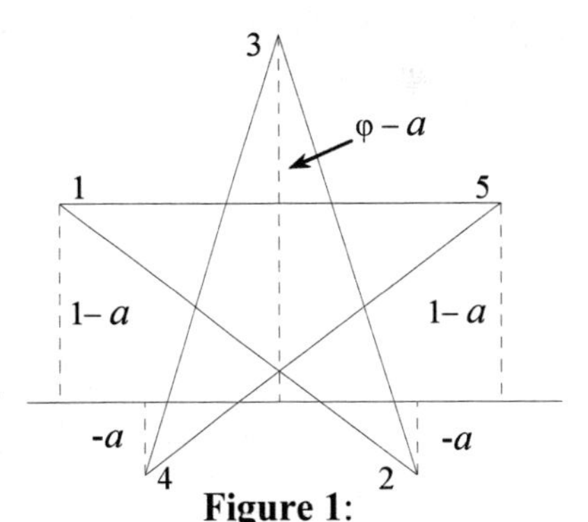

Figure 1:
Calculating the weights

$$v_1=\frac{1-a}{2+\phi-5a}p_1+\frac{-a}{2+\phi-5a}p_2+\frac{\phi}{2+\phi-5a}p_3+\frac{-a}{2+\phi-5a}p_4+\frac{1-a}{2+\phi-5a}p_5.$$

To convert these weights into a sequence of steps consisting of locating points on line segments we first rewrite the above in groupings of pairs of the form $(1 - c)x + cy$. The above is equivalent to:

$$v_1 = \left(\frac{2-4a}{2+\phi-5a}\right)\left[\left(\frac{1-a}{1-2a}\right)\frac{p_1+p_5}{2}+\left(\frac{-a}{1-2a}\right)\frac{p_2+p_4}{2}\right]+\left(\frac{\phi-a}{2+\phi-5a}\right)p_3 .$$

For any pair of points x and y, the point $(1- c)x + cy$ is located on the line xy at ratio c of the distance between x and y as measured from x towards y. For example, if c=½ then we find the midpoint; if c=1.5 then we go from x to y and then go an extra half the distance of xy past y; and if c= -½ then we travel half the distance of xy from x away from y. We start from any non-planar pentagon, such as the one shown in Figure 2. We can interpret the above formula, reading from the inside out, as: 1) Locate the midpoint $m_{15} = (p_1+p_5)/2$ as shown in Figure 3 below. 2) Construct the line segment p_2p_4 and locate the midpoint $m_{24} = (p_2+p_4)/2$. 3) Construct the line $m_{15}m_{24}$ between these midpoints and locate the point which is $-a/(1-2a)$ of the distance from the m_{15} towards m_{24}. Call this point q. 4) Construct the line from q to p_3 and locate the point which is $(\varphi-a)/(2+\varphi-5a)$ of the distance from q towards p_3. This last point is the desired point v_1. Repeating this in a symmetric manner for the other four points will construct points $v_2, \ldots v_5$. By connecting these in order, we obtain an affine regular stellar pentagon. Note that by letting a=0 we get the stellar construction first created by Jesse Douglas in [2]. To create the following sculptures, we repeat the above process for different values of a. Notice that steps 1), 2) and half of 3) do not need to be repeated for different values of a. Figures 4 and 5 consist of computer generated sculptures starting from the pentagon in Figure 2. Figure 6 consist of both the sculpture in Figure 4 and the sculpture in Figure 5.

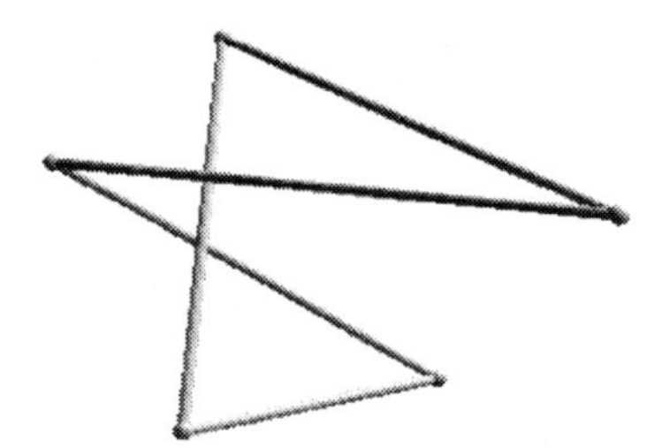

Figure 2: Non-planar pentagon

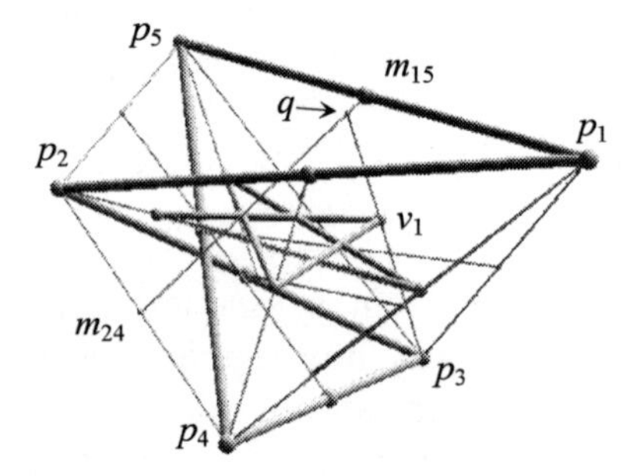

Figure 3: Construction $a = -0.1$

We finish with several observations. As the line in Figure 1 moves closer to the middle, $(2+\varphi-5a)$ gets smaller causing the resulting pentagons to get larger. This explains why values between 0.3 and 1.0 were ignored. As can be seen, the affine regular pentagons all lie upon the same plane, however, none of the lines constructed, other than the pentagons, lie on this plane. One wonders, what makes this plane special?

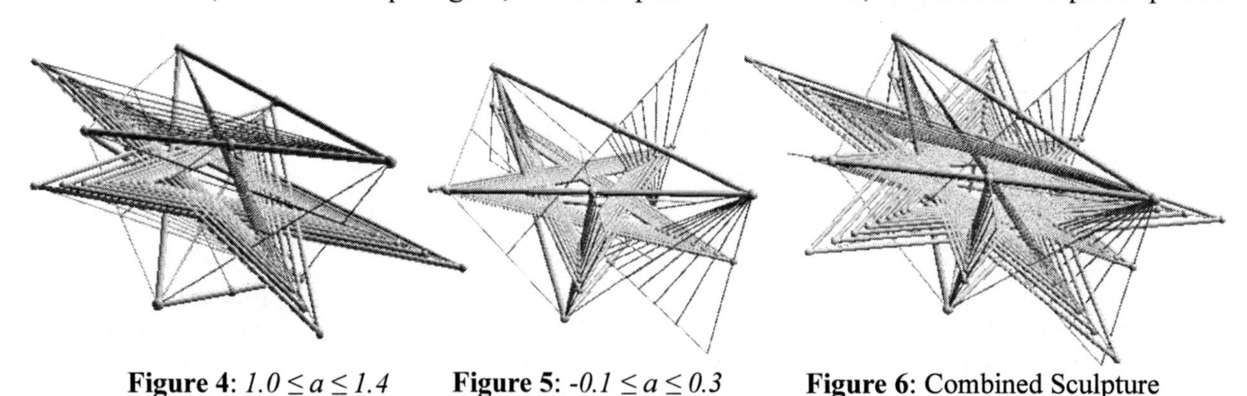

Figure 4: $1.0 \leq a \leq 1.4$ **Figure 5**: $-0.1 \leq a \leq 0.3$ **Figure 6**: Combined Sculpture

References

[1] Douglas Burkholder, "Parallelogons and Weighted Averages," Revista de Física y Matemática (FISMAT) de la Facultad de Ciencias, Escuela Poltécnica Nacional (Quito-Ecuador), XII (2004).

[2] Jesse Douglas, *A Theorem on Skew Pentagons,* Scripta Mathematica, Vol. 25, 1960.

The Effect of Music-Enriched Instruction on the Mathematics Scores of Pre-School Children

Maureen Harris
Children's House Montessori
13145 Riverside Drive East
Tecumseh,Ontario, Canada, N8N 2M7
Email: mharris@montessorimozarts.com

Abstract

While a growing body of research reveals the beneficial effects of music on education performance the value of music in educating the young child is not being recognized, particularly in the area of Montessori education. This study was an experimental design using a two-group post-test comparison. A sample of 200 Montessori students aged 3 to 5-years-old were selected and randomly placed in one of two groups. The experimental treatment was an "in-house" music-enriched Montessori program and children participated in 3 half-hour sessions weekly, for 6 months. This program was designed from appropriate early childhood educational pedagogies and was sequenced in order to teach concepts of pitch, dynamics, duration, timbre, and form. The instrument used to measure mathematical achievement was the Test of Early Mathematics Ability-3 to determine if the independent variable, music instruction had any effect on students' mathematics test scores, the dependent variable. The results showed that subjects who received music-enriched Montessori instruction had significantly higher mathematics scores. When compared by age group, 3 year-old students had higher scores than either the 4 or 5 year-old children.

The value of music in educating the young child is not being recognized, particularly in the area of mathematics. Despite the amount of literature available regarding the effect of music instruction on academic achievement, little has been written on different Montessori music pedagogies and the effect on students' math scores. If research of students in the school system indicates that learning through the arts can benefit the 'whole' child [21]; that math achievement scores are significantly higher for those students studying music [17]; and if Montessori education produces a more academically accomplished child [3]; then what is the potential for the child when Montessori includes an enriched music curriculum?

Education Today and Tomorrow

At the outset of the 21st Century many educators and parents are considering the kind of education young people need to become responsible and productive members of a global society. Major changes globally are making it increasingly more difficult to prepare the student to be the responsible citizen of the future with the life skills to live and work in a global world. Recognizing that schooling should enhance the development of creative and responsible citizens, we need to consider how such development takes place, and provide rich opportunities for learning for all students

Music and Brain Development

The role music plays in the education of the child is the focus of much discussion in education today. The baby at play is sculpting a brain that will be used for the rest of his or her life [16]. Research results would seem to indicate that the learning and remembering of a melody can occur not only before birth but actually before or at the beginning of the third trimester [6] The first 3 years in a child's life is a time when music can be used to stimulate the development of nerve connections between brain cells necessary for optimal cognitive development [12]

Music and Math

Researchers have found time and again an apparent link between the arts (music most commonly) having a positive impact on – reading, math, writing, self-esteem, and brain development. Rauscher & Shaw (1997), while studying higher brain function, found a connection to the brain linking music lessons to improved spatial-temporal reasoning abilities of four to six year olds. While music is viewed as a separate intelligence, there is a high correlation between mathematics and music and it is more than a coincidence that math and music are noted for their crossover talents. Music involves ratios, regularity, and patterns, all of which parallel mathematical concepts [14]. For example, the musical scale is similar to a neat logarithmic progression of frequencies. There are also similar connections between patterns of notes and patterns of numbers [4]. Students who developed the rhythms for the songs, began to think in multiples of four. They realized that if they had sixteen beats of music, they then had four sets of four beats. Students also grasped the concept of odd and even as the groups were subdivided into smaller units for particular steps or musical rounds [4].

Music enables students to learn multiplication tables and math formulas more easily; rhythm students learn the concept of fractions more easily; students who were taught using rhythm notation scored 100% higher on fractions tests, and a child may use the ability for logical thinking that was developed in the music class to solve problems quite unrelated to music [13]. The core question being: "is the ability to learn 'anything' enhanced when music, rhythm and movement are added and the child is engaged"

.

Engagement and Learning through the Arts

Engagement means that children are wholly involved, physically, emotionally, intellectually, and socially. Work in the arts requires that children learn how to pay attention to relationships and so many of the decisions that are made in life are decisions that cannot be made by appealing to formula, recipes, or algorithms. The arts promote that kind of perception and engender that sort of thinking. The tools the workforce of tomorrow will need are creative thinking, problem solving, risk-taking, teamwork and communication, and are precisely the skills the arts teach [15]. If we do not encourage students to master these skills through quality arts instruction today, how can we ever expect them to succeed in their highly competitive business careers tomorrow?

Life without Music

A study of five hundred thousand students in forty-five countries has shown that the United States is below average in mathematics [8]. A study titled "Musical training improves a child's ability in spatial-temporal reasoning, which is important in mathematics and science education" suggests music education be present in schools, preferably starting in preschool, to develop "hardware" for spatial temporal reasoning in the child's brain. The absolute crucial role of spatial temporal reasoning in learning difficult math and science concepts must be explored and exploited. Dr. Jean Houston of the Foundation for Mind Research says that children without access to an arts program are actually damaging their brain. They are not being engaged to non-verbal modalities that help them to learn skills like reading, writing, and math [18].

Montessori Music Research

The decision to support music cannot be made without knowing music's effect on academic achievement and its contribution to a student's education. The goal is to meet and exceed the challenge of giving young children the best possible preparation for the future [5]. Assuming that young children's involvement in music programs provides a conceptual foundation for subjects such as mathematics, a study examining the difference in math achievement scores between Montessori students who received traditional Montessori instruction and students who received music-enriched Montessori instruction predicts positive results [9]. A sample of 250 Casa students (ages 3-5) within the jurisdiction of a Montessori School board located in Southwestern Ontario was selected for the study. The researcher, an experienced Montessori teacher and music specialist, used the Test of Early Mathematics Ability 3 (TEMA-3) assessment for this study [7.a]. This instrument measures mathematical achievement 1) concepts of relative magnitude, 2) counting skills, 3) calculation skills, 4) knowledge of conventions and 5) number facts (reviewed by American Educational research Association, American Psychological Association, and National Council on Measurement in Education 1999). All schools were established Montessori programs that met recognized affiliation standards. The children in the study, aged 3-5 years, were divided into two groups, experimental and control. The experimental treatment was a 6-month 'in-house' music-enriched Montessori program designed from appropriate early childhood educational perspectives and based on Kodaly techniques. The program was sequenced to teach concepts of pitch, dynamics, duration, timbre and form as well as skills in moving, playing, listening, singing and organizing sound. Children participated in 3 half-hour sessions weekly. The comparison group received traditional Montessori instruction during this period. Children in both groups were post-tested on the TEMA – 3.

Significance of the Study

Based on these findings it appears that students who received music-enriched Montessori instruction had higher levels of mathematics achievement than students who received traditional Montessori instruction. When compared by age group, 3-year old students had higher scores than either the 4-year old, or 5-year old children. Suggested follow-up research - a longitudinal 3-year study following the progress of the 3-year-old students and testing them again at 4 years and 5 years to see what is the consistent positive effect of enriched music instruction on these students' math ability scores. Further significant findings indicated that the Montessori students performed in the high percentile range for mathematics based on the expected norms of the TEMA-3 testing tool [7]. Of the Montessori students in the experimental group (those receiving music-enriched Montessori instruction) 100% fell in the 90th -99th percentile range. These scores far exceed the expected norms of the TEMA-3.

These findings are significant because a grasp of proportional mathematics and fractions is a prerequisite to mathematics at higher levels, and children who do not master these areas of mathematics cannot understand more advanced mathematics critical to high-tech fields [22]. This study offers quantitative results that could help Montessori and early childhood educators recognize the value of music-enriched instruction for the young child, and implementing the instructional designs used in this study could lead to higher levels of student achievement [10]. As the quantity, quality, and availability of empirical studies increases, Montessori and Early Childhood educators will have the knowledge to make a stronger connection between their design decisions and the evidence of 'what works'. This is the time to explore how research and practice reflects the wider world of early childhood education.

The quality of early childhood education can have long-term effects on a child's attitude toward further education and educational achievement [1]. Evidence indicated that once children's achievement patterns were established, there was a high degree of continuity from that point forward, and early attainment set boundaries on later attainment[2] The goal is to meet and exceed the challenge of giving young children the best possible preparation for the future and to do this a basic part of their learning experiences must be involvement with the arts [19].

As we embrace the 21st century and face the challenges of the future with and opportunity and responsibility to change, research suggests that providing a quality arts education today will ensure optimal opportunity for our children to succeed in the highly competitive world of tomorrow

References

[1] Andersson, B. E. (1989). The importance of public day care for preschool children's later development. *Child Development, 69*, 857-866.

[2] Belsky, J., & MacKinnon, C. (1994). Transition to school: Developmental trajectories and school experiences. *Early Education and Development, 5*, 106-119.

[3] Clifford, A. J., & Takacs, C. (1991). *Marotta Montessori Schools of Cleveland follow-up study of urban center pupils.* Unpublished manuscript, Cleveland State University.

[4] Dean, J. (1992). Teaching basic skills through art and music. *Phi Delta Kappan, 73*, 613-618.

[5] Fiske, E. (1999). *Champions of change: the impact of the arts on learning.* Washington, DC: The Arts Education Partnership and the President's Committee on the Arts and Humanities.

[6] Gardiner, M., Fox, A., Knowles, F., & Jeffrey, D. (1996). Learning improved by arts training. *Nature.*

[7] Ginsburg, & Baroody. (1990). Test of early math ability 3rd ed. 2003. PRO-ED, Texas.

[8] Grandin, T., Peterson, M. & Shaw, G. L. (1998). Spatial-temporal versus language-analytic reasoning. The role of music training. Arts Education Policy Review 99, 6 (July/August), 11-14.

[9] Harris, M. A. (2005). Montessori Mozart Programme. *Montessori International Journal*, (75) 17.

[10] Harris, M. A. (2005) *Differences in mathematics scores between students who receive traditional Montessori instruction and students who receive music-enriched Montessori instruction.* Unpublished manuscript, University of Windsor, ON.

[11] Hepper, P. G. (1991). An examination of fetal learning before and after birth. *Irish Journal of Psychology, 12*, 95-107.Hodges, D. (2000). *Why are we musical? Support for an evolutionary theory on human musicality.* Paper presented at the 6th International Conference on Music Perception and Cognition, Keele, England.

[12] Hodges, D. (2000). *Why are we musical? Support for an evolutionary theory on human musicality.* Paper presented at the 6th International Conference on Music Perception and Cognition, England.

[13] Kelstrom, J. M. (1998). The untapped power of music: Its role in the curriculum and its effect on academic achievement. *NASSP Bulletin, 82*, 34-43.

[14] Marsh, A. (1999). Can you hum your way to math genius? *Forbes, 16*, 176-180.

[15] Milley, J., Buchen, I., Oderlund, A., & Mortatotti, J. (1983). *The arts: An essential ingredient in education.* CA: California Council of Fine Arts Deans.

[16] Olsho, L. (1984). Infant frequency discrimination. *Infant Behavior and Development, 7*, 27-35. Paper presented at the The Learning and the Brain Conference, Boston, MA.

[17] Rauscher, F., & Shaw, G. (1998). Key components of the Mozart effect. *Perceptual and Motor Skills, 86*, 835-841.

[18] Roehmann, F. L., & Wilson, F. R. (1988). *The biology of music making: Proceedings of the 1984 Denver conference.* St. Louis, MO: MMB Music Inc.

[19] Sylva, K. (1994). School influences on children's development. *Journal of Child Psychology and Psychiatry, 35*(1), 135-170.

[21] Upitis, R., Smithrim, K., Patteson, A., & Meban, M. (2001). The effects of an enriched elementary arts education program on teacher development, artist practices, and student achievement *International Journal of Education and the Arts, 2*(8).

[22] Vaughn, K. (2000). Music and mathematics: Modest support for the oft-claimed relationship. *The Journal of Aesthetic Education, 34*(3-4), 149-166.

Teaching Arabesque

Jean-Marc Castera
6, rue Alphand
75013 Paris, France
E-mail:jm@castera.net

Abstract

Presentation Presentation of 3 teaching experiences to introduce geometrical arabesque to people of different ages and cultural backgrounds. The first experience consists of imitating traditional examples, the second is based on specific method of drawing, the last uses a set of zellij-like tiles cut with a laser.

1. Imitation of traditional examples

1.1. Beginnings. This activity took place during an experimental course that I gave at the Paris-8 University. The course led to a practical application, the decoration of a hall at the university in the Arabic style. The initial idea was only to paint on a wall an adaptation of a motif from the Moroccan traditional style as an advertisement for the course. Afterwards, we created a fountain as well.

1.2. The wall. The motif was a 64-pointed star without interlaces. The wall was cleaned, then painted in white before drawing the motif with stencils. Coloring was done with the spontaneous help of people passing by. "Columns" were painted on each side of the wall at the entrances to the Departments of Mathematics and Philosophy (Fig. 1). Some of these drawing were reminiscent of the work of Escher (Fig. 3). Arches with muqarnas structures were placed atop each entranceway. That was our first approach of muqarnas: we used wood and plaster in a wild technique. Nowadays we would rather use the traditional technique of assembling units.

1.3. The fountain. After the wall was completed several people suggested that a fountain be constructed and so we did that. The objective of this project was to make a wild dream come true, then to learn the technique of "gebs" (plaster engraving) and finally to improve our knowledge of muqarna structures. But also the common space where people congregated, with nice daylight, was threatened by the administration of the university who had begun to build new rooms in the area which had the effect of reducing the common space and masking the light. We began to construct the fountain against the last built wall which prevented the spread of more rooms. We started to work without permission. Later on, because we wanted to have water in the fountain, we required permission from the university, and it was granted. The basin of the fountain was made with concrete, covered inside with colored glass mosaic, and outside painted in the zellij style (Fig. 5). The façade had engraved plaster, painted decorations and muqarnas corbels. For this second approach of muqarnas we experimented a better technique than the one used for the arches of the wall. We first made a negative model in plaster, and used it as a mold to make the two corbels that were fixed on the fountain and engraved. There was also built-in lighting, drinking water, and a palm tree. As a tribute to the city of Fez (Morocco) that fountain was dedicated to the famous "Nejarine."

Figure 1: *The wall, University of Paris 8.*

Figure 2: *The wall, and detail of a column.*

Figure 3: *The motives on the columns, a tribute to Escher.*

Figure 4: *Finally, the painting of the center.*

Figure 5: *The fountain, University of Paris 8.*

1.4. Evaluation. For the most part, the first construction (the wall) was a two-dimensional work. Some parts were done collectively while others were done by individuals (e.g., preparation of the drawings, etc.) The work was done rapidly and in a festive atmosphere. The second activity, the fountain, was much more complex, involving problems with materials, experimental techniques, and so forth. It took much time and energy. Fortunately, the results were worth the effort and do not reflect the pain of carrying out the task. In retrospect, this project may have been too demanding for an introduction to the art of arabesque.

2. Drawing

2.1. The method. A specific drawing technique is used. Its simplicity is in great contrast to the complexity of its application. The first Western analysis of zellij patterns used classical methods of geometry, compass and straightedge drawings. Although they are attractive, these techniques, using exact geometry, are somewhat limited. How does one improvise, imagine and create while constrained by such a rigorous application of geometry ? Our approach is very informal: no rules, no compass (except occasionally for constructing complex rosettes), rather we use freehand drawing on square sheets of paper giving us a high degree of freedom to improvise and engage in the creative process (**Fig. 6**). We limit ourselves to only the octagonal family. The price of this freedom is a diminishing of exactness: we use approximation. However, this approximation is very efficient, and it is then easy to convert our constructions to exact proportions at a later time. The materials used are pencil, eraser, square paper, concentration, patience and any other tools needed for coloring.

Figure 6: *Freehand drawing technique. On the left is a detail of a zellij panel using the same kind of approximation.*

2.2. Evaluation. I have been surprised by the good results we have achieved, even with young children. The simplicity and power of the method fits well with the minimalist character of traditional crafts: few tools simply employed. Only an elementary knowledge of geometry is required. In this activity, the rigor of geometry and freedom of imagination meet together. Though, a period of training is necessary before being able to work creatively, and this will not work for people who have difficulties in concentrating.

3. Return to hands-on

3.1. The method. I gave the participants a set of zellij-like tiles cut with a laser (Castera, J-M.: *Zellij Multipuzzle*, ISAMA-BRIDGE Proceedings, 2006). In the first workshops, I gave the participants some models to copy, but I soon realized that many people preferred to make their own models. Now I no longer give out models. I explain with few words the simple rules and let people experience the pleasure of (re)discovering patterns. Sometimes you need a shape that is not available in this game, but you can often construct these from combining smaller shapes much as in the game of tangrams.

3.2. Evaluation. This is the most efficient way to introduce the art of arabesque : direct immersion in the galaxy of zellij. The laser makes for sharp cutting, which reproduces the effect of actual zellij patterns which use cutting after firing. The only disadvantage is the need to clean up the tiles after their use. Also a guide is needed. The manufacturing is still in an experimental stage, and at the moment it is very expensive. Finally, I have found that it is more difficult to work with adults (even if they are highly trained academics) than with children.

Figure 1: i*mages of the first activity. Arabic World Institute in Paris, March 2005.*

Slide-Together Structures

Rinus Roelofs
Sculptor
Lansinkweg 28
7553AL Hengelo
the Netherlands
E-mail: rinusroelofs@hetnet.nl
www.rinusroelofs.nl

Abstract

About ten years ago I discovered an interesting way to construct a tetrahedral shape by sliding together four rectangular planes in a certain way. By using halfway cuts in the planes it was possible to slide them together, all at once, to become the enclosed tetrahedron. This way of constructing objects and structures, finite and infinite, has been one of my interests from then on. In this paper I will give an insight into some of the results of my research in this field. Besides halfway cuts I examined some other ways of slide-together structures.

1. Introduction

1.1. "Slide-together". At Bridges 2004 George Hart presented his "slide-togethers": polyhedral constructions built with simple flat paper elements, which were slid together [1]. Each "slide-together" was made from identical copies of a single type of regular polygon with slits cut at the proper locations. The pieces in these constructions had to be bent during the assembly. In my search I tried to focus on structures that can be built with rigid elements. So in most cases sliding the pieces together to form the final structure is possible without bending the pieces.

Figure 1 a,b,c: *Basic principle - halfway cut.*

1.2. Halfway cuts. With the use of halfway cuts as in Figure 1 we can combine several pieces to construct complex structures by just sliding them together. The simplest and most direct way to do this is by adding the pieces one by one. However, when more than one slide direction is used this is not always possible. And in some cases it is even necessary to put the pieces together all at the same time. In Figure 2 we see a construction built from six identical square pieces with four halfway cuts each. When making the assembly we could start with one piece and then add the other pieces one by one. Doing so you will notice that you will have a problem when you want to add the last piece. It is better to make groups of three pieces first and then slide together the two groups. In the completed structure you can distinguish three different slide directions. We can use any of these directions to split the structure into two parts. And this is also the case in the more complex structure of Figure 3, which is an extension of Figure 2.

Figure 2a: *Ring - 6 basic pieces*

Figure 2b: *Basic and double piece*

Figure 3: *3D - 48 pieces*

1.3. Four slide directions. In Figure 4 we see an assembly of four rectangular shapes with two halfway cuts each. The enclosed shape is a tetrahedron. The only way to put the pieces together is sliding them together all at the same time. The discovery of this structure was the start for my research. Many questions arose. For instance: are there more, similar, structures like this one? How can we make infinite slide-together structures? Is it possible to use even more slide directions? And which systems can be assembled only by sliding together all the pieces at the same time?
Besides the extension of the basic tetrahedral slide construction (Figure 5) I also found another way to build an 'infinite' tetrahedral structure as can be seen in Figure 6. This sculpture consists out of 104 flat triangles, slid together all at the same time.

Figure 4: *Tetrahedral structure*

Figure 5: *Wood, 36 pieces*

Figure 6: *Steel, 104 pieces*

2. 'Folded' Elements

2.1. Double pieces. The two infinite structures in Figure 3 and Figure 5 are both generated by doubling the basic pieces. A first idea in trying to find new shapes and structures was by introducing a 'folding' line into the pieces. Not as a hinge between the two 'half' pieces but as a rigid connection between the parts. In the first example with the basic structure of the ring of Figure 2 (three slide directions) the use of the 'folded' pieces leads to an object that looks like 2 connected cubes. And also this object can only be taken apart in two groups of three pieces before you can separate all the pieces from each other.

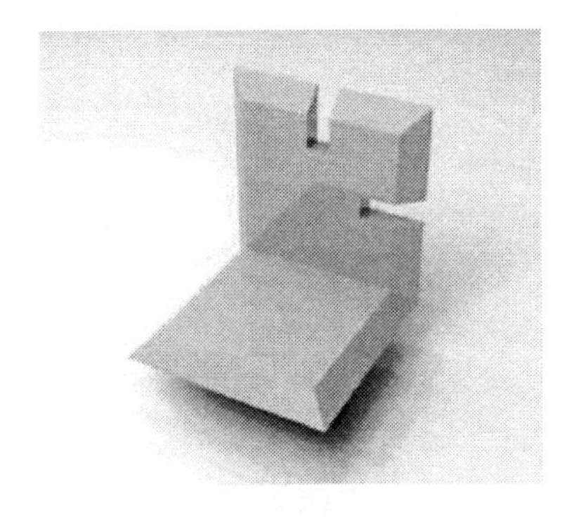

Figure 9a: *Folded piece.* **Figure 9b, c**: *Two connected cubes.*

2.2. Slide together cylinders. When we double the pieces of the tetrahedral structure, we can create a variety of different cylinders by changing the angle between the 'halves'. All these cylinders can only be taken apart by sliding away all the pieces at the same time. The cylinders can be seen as a stack of antiprisms. The first cylinder in the set is a stack of tetrahedra, which also can be seen as the first member of antiprisms. The next member, based on the three-sided antiprism, has the same structure as the ring of Figure 2, and therefore you might expect that it can be split into two parts by sliding along one of the sliding lines, but because of the doubling of the pieces the number of sliding lines has also been doubled, which changes the sliding system.

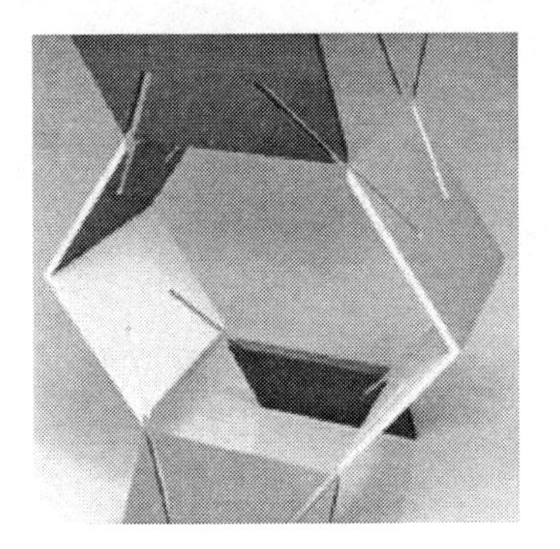

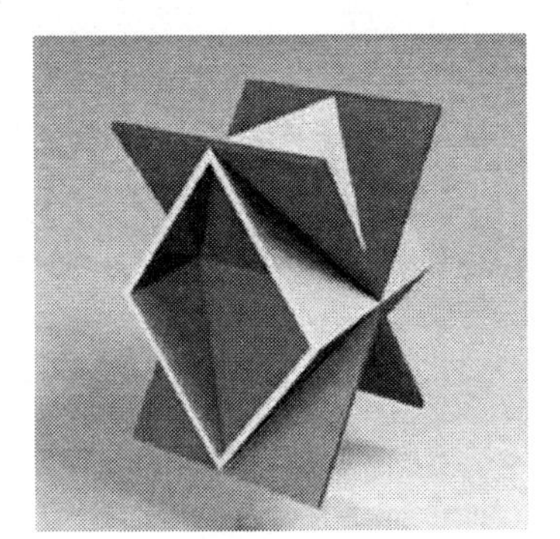

Figure 10, 11: *Sliding together.* **Figure 12**: *Cylinder.*

3. Flat Infinite Structures with Folded Elements

3.1. 2D infinite. The cylindrical structure is infinite in one direction. A cylinder can be cut and unrolled to become a 2D flat surface. In Figures 13, 14, 15 you can see how this step leads towards a 2D infinite slide together structure. From the two types of elements, which were used to build the cylinder, only one is left in the flat structure. Again you see the enclosed tetrahedra, which means that also in this structure four sliding directions are used. So we can conclude that these elements will have to slide together all at the same time too.

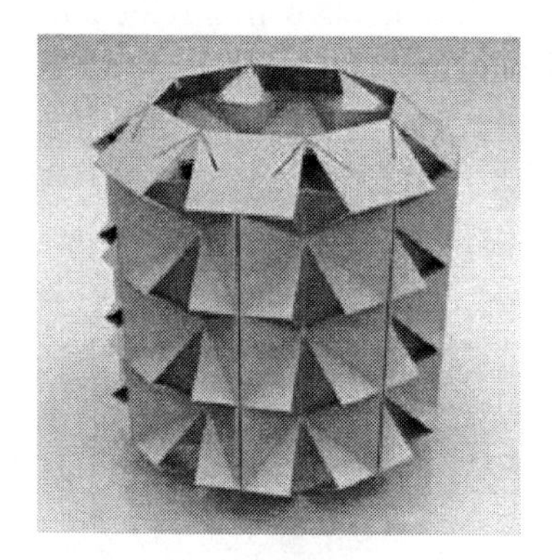

Figure 13, 14, 15: *Unrolling the cylinder: cylinder - dome - 'flat' structure.*

3.2. Variations. There are a few interesting variations of this type of slide-together structures. When we make the pieces a little bit more complicated by adding extra parts as in Figure 16 we will get nice and strong double layered structures (Figure 17 and 18).

Figure 16, 17, 18: *Double layer structures.*

In the case of Figure 19, a double connection, a parallel pair of sliding lines, between pairs of pieces is used. Special about these two examples is that you first will have to make rows, and then slide the structure together row by row.

Figure 19a, b, c: *Double connection.*

4. "Tile Rotation"

4.1. Bending the pieces. There is one special group 2D slide-together structures that can only be assembled when we allow bending the pieces during the assembly. I decided to add this group because of a very interesting property. The pieces of this group consist of two connected triangles which both have two halfway cuts (Figure 20) and in the final structure the double connection is used (Figure 21). Because now the pair of slide lines is not a parallel pair, bending of the pieces is needed. When we look at the final structure we can recognize a tiling pattern lying on the surface: a pattern with small and big square tiles (Figure 22). And when we look through the structure we can recognize the same tiling but now with the opposite orientation: left turning instead of right turning.

Figure 20, 21, 22: *Tile rotation 4,4.*

When we project the upper tiling onto the lower tiling we see that the center of each tile is in the same position and each tile seems to have been rotated around its own center. I found a few other tilings to which I could apply this operation, which I called 'tile rotation'. I wanted to examine whether it was possible to construct a slide-together structure from tilings with this property. In the tiling of Figure 23 we can change the orientation by rotating the hexagons and the triangles, each around their own center. And so from this tiling I was able to design the accompanying slide-together structure (Figure 24 and 25).

Figure 23, 24: *Tile rotation 6,3.*

Figure 25: *Tile rotation 3,6.*

5. Mortise-and-tenon Joints

5.1. Introduction. Another way of making slide-together structures is the use of mortise-and-tenon joints. The structure in Figure 26 consists of six equal elements and has some similarities with the ring structure of Figure 2. Here too we have got two possibilities. You either take it apart into two groups first or you take it apart by moving all the elements away from the center at the same time. And to become an infinite structure we can again double the elements. See Figure 27.

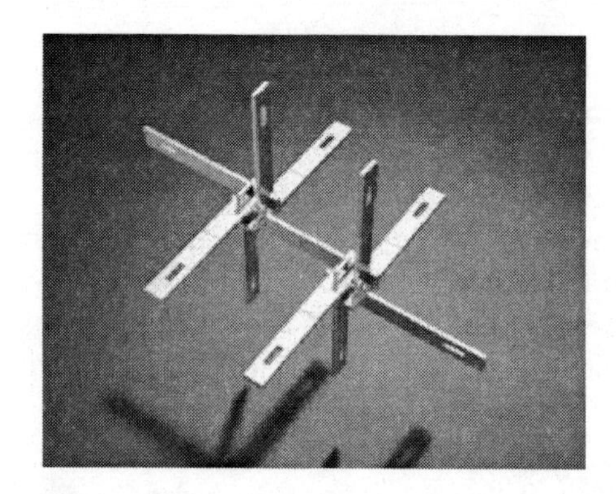

Figure 26, 27: *Mortise-and-tenon joints.*

5.2. One mortise two tenons. The idea of using two tenons instead of one to fill one mortise opened up a completely new field of slide together structures. In many 3d structures you can find intersecting planes. The idea was, when you have a set of intersecting planes, to split up one of the planes in such a way that you get two tenons crossing from opposite directions the intersection line with the other plane. An example of such a tiling of a plane can be seen in Figure 28a. Here the plane is divided into three equal pieces. The division line has a Z-shape to create the tenons. The Z-cut is made in such a way that the pieces can slide apart when moving from the center. Now each piece has two sides with a tenon and in the middle is a mortise of such a size that two such tenons, coming from different directions, fit in precisely. Sliding together 12 pieces creates an intersection of four planes (Figure 28b, c). The shape of the completed construction can be recognized as the cuboctahedron. Sliding together the 12 pieces is fairly easy to do. For me it was a surprise that taking the cuboctahedron apart is practically impossible.

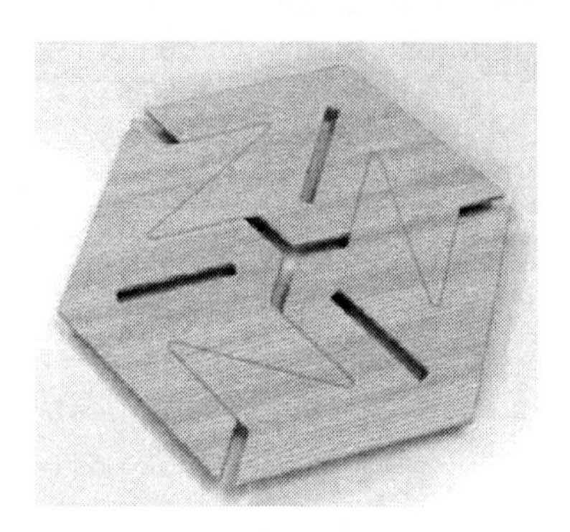

Figure 28a, b, c: *Cuboctahedron - twelve pieces.*

5.3. Polyhedra. The cuboctahedron is one of the Archimedean polyhedra in which you can recognize a set of intersecting planes. The four planes have one intersection point in the center of the polyhedron; each of these planes is divided into three pieces. And now all the pieces can be moved towards this center point and the tenons will slide smoothly into the mortises.

Similar situations can be found in a few other polyhedra: the octahedron with three intersecting planes and the icosidodecahedron with six intersecting planes. In Figure 29a you can see a subdivision of one of the six planes of the icosidodecahedron into five equal pieces. Thirty such pieces can be slid together to create the icosidodecahedron of Figure 29c.

Figure 29: *Icosidodecahedron - thirty pieces.*

5.4. General intersections. So far all the slide together structures with mortise-and-tenon joints were regular, all the pieces in the construction were equal. But the Z-cut method can be applied also in other cases. As an example a thirty piece cylindrical structure is shown in Figure 30. Five different types of pieces were used to make this object.

Figure 30a: *Pieces.*

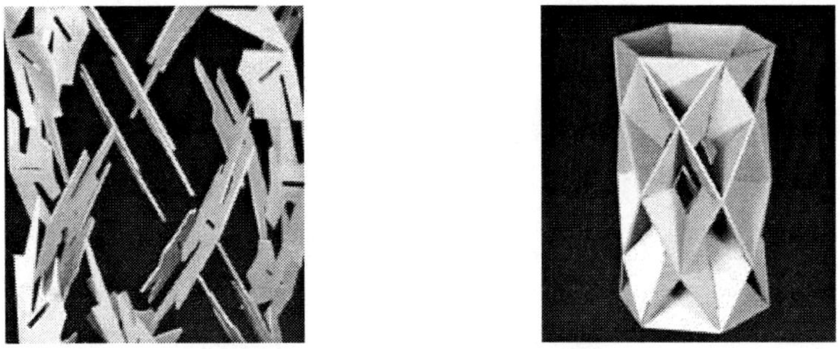

Figure 30b, 30c: *Cylinder - thirty pieces.*

6. Further Generalization

6.1. Parallel Faces. In section 5 in each structure any two planes had an intersection line. There were no parallel faces. The number of pieces then can be calculated by multiplying the number of faces n by (n-1). It is however no condition that each plane should intersect every other. Also in situations where we have parallel faces we can use the Z-cut method to construct a set of pieces for a slide-together structure. The cubical construction in Figure 31 is made of twenty four equal pieces. Each of the six faces is divided into four parts using the Z-cut, now leaving a square hole in the middle.

 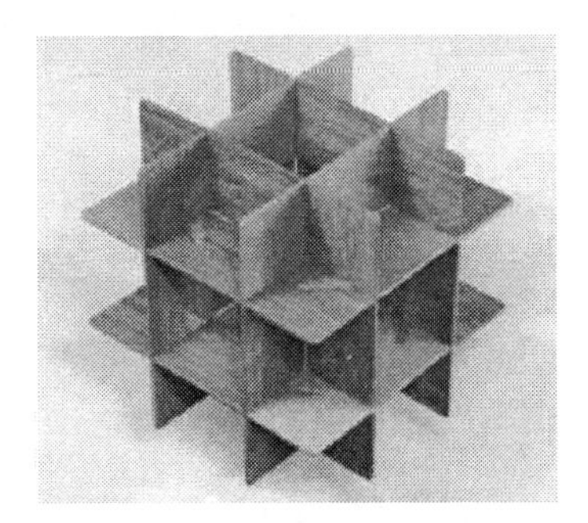

Figure 31a, b: *Cube - twenty four pieces.*

6.2. Outside the center. Looking at the construction of Figure 31 we can say that the connection between the pieces takes place at the vertices of a cube and that pairs of pieces follow the edges of the cube. Describing the structure in this way gives us a method of using the same steps starting with other polyhedra, for instance the tetrahedron or the dodecahedron. Also in these polyhedra three faces meet at every vertex, so the same translation can be used. In case of the dodecahedron (Figure 33) we end up with a 60-piece slide-together structure.

Figure 32a, b, c: *Tetrahedron - twelve pieces.*

Figure 33a, b: *Dodecahedron - sixty pieces.*

6.3. Parallel tenons. Instead of opposite tenons we can also use parallel tenons. In this way it is possible to create box like slide together structures as you can see in the example of Figure 34.

Figure 34a, b: *Cube - twelve pieces.*

7. Coxeter Polyhedra

7.1. Infinite structures. The question arose how to create infinite structures using the kind of pieces developed in the previous section. Staying in the field of polyhedra we can then take the infinite regular polyhedra, found by H.S.M. Coxeter [2] as the basic structures. Both methods, tenons in opposite directions, and parallel tenons can be applied to make slide together structures. It gives the idea of an implosion seeing all the pieces coming together to form the slide together structure.

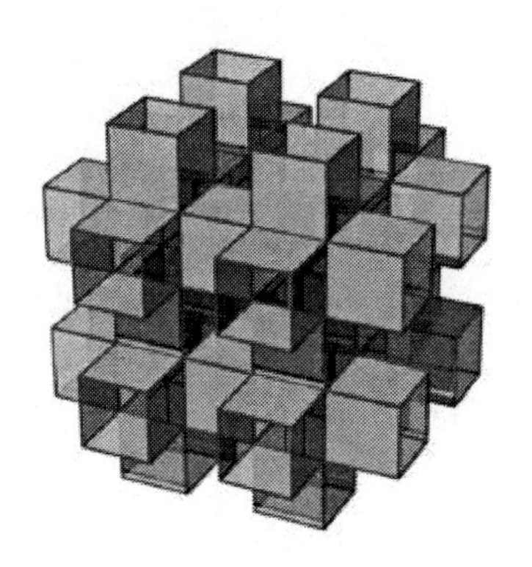 

Figure 35, 36: *A Coxeter glide together structure.*

8. Further steps

8.1. Alternate shapes. To open up the field for further research I experimented with some other shapes. The first group is a special set of star shaped polyhedra. The objects are constructed by applying a certain transformation to the octahedron, the cube and the dodecahedron. All three shapes can be assembled with a kind of pyramid shape. In each object the separate pieces will slide towards the center at the same time.

Figure 37a, b, c: *A octahedron - six pieces.*

Figure 38a, b: *A dodecahedron - twenty pieces.*

8.2. Escher's star - Rhombic star. Besides other shapes of the pieces also other types of movements can be studied to find new possibilities. The 12-pointed star that M.C. Escher uses in his print "Gravitation" [3] can be made from 12 flat pentagonal shapes. The question is whether it is possible to slide the pieces together to become the complete star. I decided to study the rhombic version of the 12-pointed star and found out that it was even possible to create an object with smaller holes then M.C. Escher used. The Rhombic star can be made as a slide together structure. But now we have to slide the pieces together not only by translation but we also have to rotate the pieces during the translation. It is a nice spiraling movement, which brings all the pieces smoothly together to a 12-pointed rhombic star. And in a way this brings us back to the beginning where as apart from the moving aspect there is a clear connection between this object and some of George Hart's sculptures [4].

Figure 39: *M.C.Escher's 12-pointed star.*

Figure 40a,b: *The 12-pointed rhombic star.*

References

[1] George W. Hart, *"Slide-Together" Geometric Paper Constructions,* Workshop at Bridges 2004, http://www.georgehart.com

[2] H.S.M. Coxeter, *Twelve geometric Essays,* Southern Illinois University Press 1968

[3] M.C. Escher, *Grafiek en Tekeningen,* van Tijl 1972

[4] George W. Hart, *Sculpture from Symmetrically Arranged Planar Components,* ISAMA-Bridges 2003

Special thanks to Tom Longtin for helping me with the realization of some of the models.

Repeated Figures

Susan McBurney
211 Rugeley Road
Western Springs, IL 60558 USA
E-mail: smcburney@prodigy.net

Abstract

This paper illustrates the development of two types of design from the beginning concept through execution onto enhancement for final presentation. Emphasis is on a structured, modular process suitable for instruction in either an art or beginning programming curriculum.

1. Initial Concept

The power and flexibility of modern computers and software have opened new avenues of exploration for artists and mathematicians alike. Sometimes even a very simple concept can lead to unexpected yet stunning results. This paper will present two types of designs developed by the author from basic building blocks manipulated with simple algorithms. Additionally it will discuss the enhancement of one of these designs leading to a finished piece of artwork.

The overall concept can be denoted as "Repeated Figures" since all the examples are built from just one figure repeated and manipulated to yield a more complex design.

An example of repeated figures can be seen in Figure 1. These designs were presented and explained at the 2006 Joint Mathematics Meeting in San Antonio, Texas, USA [1].

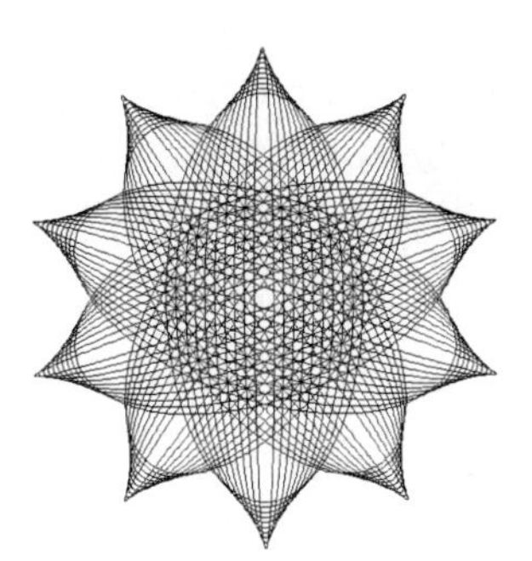

Figure 1

Each of these is made of "just ellipses", nothing else, grouped and rotated in a pattern. The surprise comes from the intricacy of the design that develops. No one could predict the rings of circles that appear in the finished design, yet all of the complexity derives from the interactions of very basic elements—ellipses.

2. The Sunflower Series

Now we will develop two additional types of figures step by step. First is the "sunflower" series. Once again the name comes after the development. That is, one doesn't start to draw a sunflower, but instead chooses a design element and then waits to see what happens. Only afterward does the finished product suggest a name.

In this case the underlying construct is a semi-circle and the language used to draw it is Logo, an easy-to-learn language freely available on the web. Logo has two basic semi-circle commands, arc and arc2, as illustrated below.

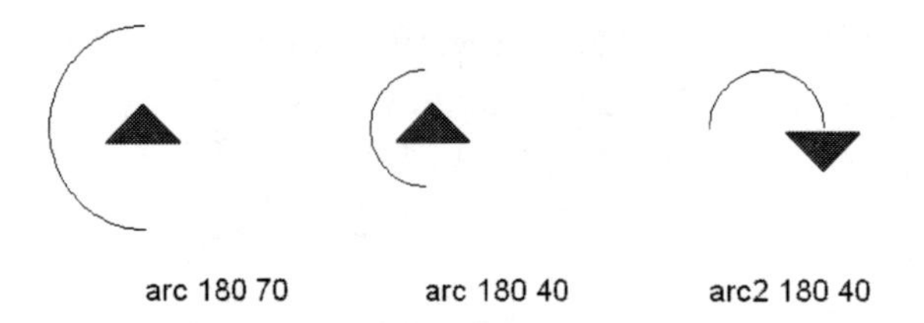

arc 180 70
arc 180 40
arc2 180 40

Figure 2: *Drawing semi-circles*

Drawing begins at the rear of the pointer. The first number specifies the number of degrees spanned by the arc and the second number determines the radius of the arc from the pointer. With the "arc2" command drawing begins at the base of the pointer which is then left at the termination of the arc.

Using the last command, arc2 180 40, consider what would happen if it were repeated five times, each time turning 108 degrees to the right before drawing the next arc. This is the result.

Figure 3: *repeat 5 [arc2 180 40 right 108]*

Some exploration and a little thought will allow students to discover a method to determine the amount of turn necessary at the end of each arc to complete a full figure after n turns. Thinking in terms of a regular polygon, draw one side, calculate the vertex angle of the polygon, and realize that you must turn 180° minus that number to continue. The vertex angle (VA) of any regular polygon is 180(n-2)/n so the turning angle (TA) = 180-VA degrees which reduces to 360/n degrees where n is the number of sides of the regular polygon. For the arc2 command, adjustments must be made to accommodate the position of the pointer at both the beginning and end of the arc. The desired TA is 180-360/n degrees, or 108° if n=5.

Now what would happen if this figure were repeatedly rotated about the beginning point two times, six, or twenty times?

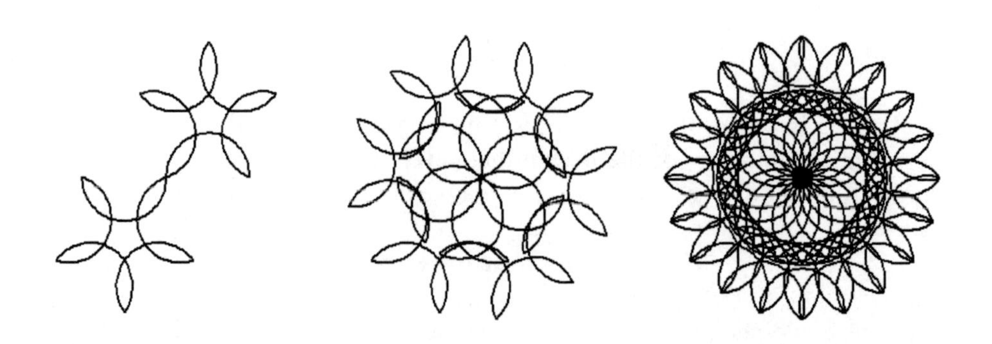

Figure 4:
repeat :n [repeat 5[arc2 180 40 right 108] right 360/:n]
n = 2, 6, 20

In three steps we have gone from drawing a simple arc to creating a design of interest. In particular the last design has four distinct elements that can be emphasized in different ways. They are the outside rim, the interim band with holes, the inner overlapping arcs, and the small interior circle at the center of the design.

Here are three "sunflower" designs of increasing complexity.

Figure 5: *Sunflowers, n=20, 30, 40.*

Continuing with this series, the design in Figure 6 is created by the command:
repeat :n [repeat 5[arc2 180 40 right 108] right 360/:n] with n = 36. Notice the change in shape of the outermost petals as well as the heavier interior band.

Once the design is obtained, it can be enhanced in other programs. This last design was imported into PhotoShop LE and manipulated in layers. The background is a solid dark green. On top of this there are two layers, each with the basic design transformed in some way. On the first layer the figure was resized, the background deleted, and the transparency set to 44%. The second layer contains a clipped, transparent copy of the figure moved toward the bottom right hand corner. Then a square was selected, filled with white, and the Texture, Grain filter applied. The final design is shown below in grayscale.

Figure 6: *n = 36*

Figure 7: *PhotoShop Enhancement*

3. Square Ball Series

A second set of designs resembles offset squares with raised hemi-spherical interiors. In this case the basic structure is the ellipse. A simple command in Logo, ellipse :a :b, will draw an ellipse with width a and height b about the center point. Drawing ten ellipses on top of each other with decreasing widths and increasing heights produces Figure 8. The following subroutine does the drawing automatically, with an initial width and height specified by the user.

```
to ells :a :b
repeat 10 [ellipse :a :b  make "a :a-10  make "b :b+5]
end
```

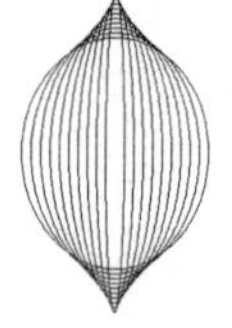

Figure 8

Repeating the figure four times and rotating it about the center leads to the following "square ball" designs.

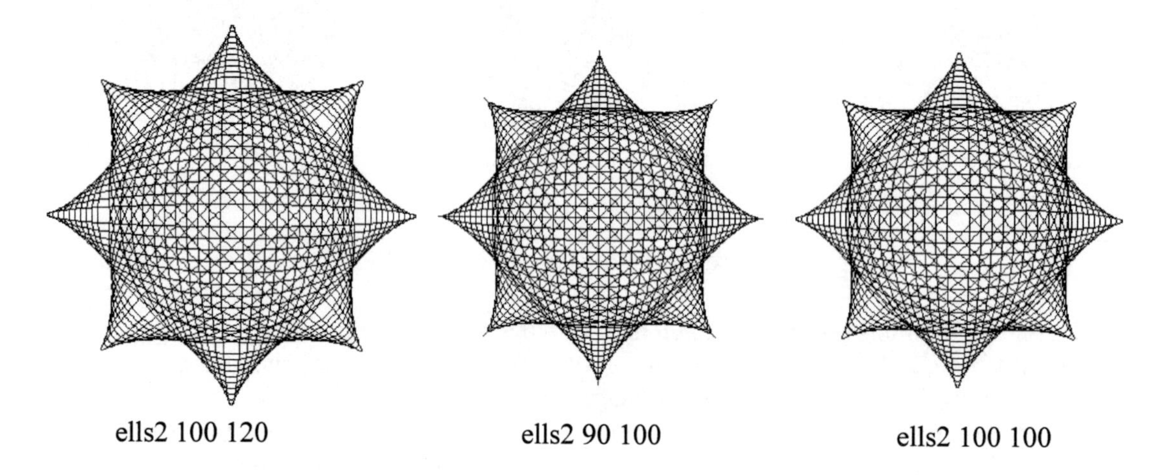

ells2 100 120　　　ells2 90 100　　　ells2 100 100

Figure 9: *Square Ball designs*

The subroutine for this combination is

```
to ells2 :a :b
repeat 4 [ells :a :b  right 45]
seth 0  ;sets the pointer to face up, toward the top
end
```

It is interesting to note the subtle differences among the designs. In the first one there is some open space around the interior and the center is open. The middle design is filled throughout and the points come exactly to a line at the peak. The center is nicely filled in as well. The right design combines these qualities to include some openness within a solid "ball". These designs of course can then be enhanced in PhotoShop or other program of choice such as Paint to add color, backgrounds, borders, etc.

Shown here are just two of the limitless possibilities for design development. It is hoped that these examples will encourage exploration into the wonderful worlds of math appreciation and computer programming.

References

[1] Susan McBurney, *A Collection of Curves*, Joint Mathematics Meeting 2006, San Antonio, TX, USA
[2] Robert Williams, *The Geometrical Foundation of Natural Structure*, Dover, 1979
[3] Robert J. Krawczyk, *Workshop*, ISAMA-CIT conference, Chicago, IL, USA, 2004

Seville's *Real Alcázar:* Are All 17 Planar Crystallographic Groups Represented Here?

B. Lynn Bodner
Mathematics Department
Cedar Avenue
Monmouth University
West Long Branch, New Jersey, 07764, USA
E-mail: bodner@monmouth.edu

Abstract

Contemporary with the Alhambra, the *Real Alcázar* of Seville, Spain was rebuilt in 1364 as a palace for Dom Pedro, Christian king of Castile (1334 - 1369) in the *Mudejar* style. (Muslims who chose to live under Christian rule were known as *mudéjares*). Although there have been alterations and additions over the centuries, this remarkably well-preserved palace was originally built by a Christian ruler in the Islamic style of Andalusia and retains its Islamic character, containing some of the most beautiful examples of *Mudejar alicatado* (Spanish, for cut tiles, derived from the Arab verb *qata'a,* "to cut") from this time period. Since all 17 planar crystallographic groups are now believed to be represented in the tilings of the Alhambra, one wonders if the same may be said of the ornament found in the *Alcázar.* This paper will briefly discuss the history of the *Alcázar,* illustrate and classify some of the planar designs as to the isometries they permit and then attempt to answer the salient question broached in the title of this paper.

The *Real Alcázar*

As early as 712, palaces were erected by the invading Islamic armies near the shores of Seville's *Rio Quadalquivir* and at the current site of the *Real Alcázar* (Spanish for "royal palace or fortress" from the Arabic word, *al-qasr*). 'Abd al-Raman III, ruler of the Córdoban caliphate, was also known to have ordered the building of a fortified palace here in 913. Although few remains of these Moorish structures exist, portions of the high encircling walls, built in the 11th century, may still be seen today (see Figure 0. below).

Figure 0. *View of the Lion's Gate in the wall surrounding the Real Alcázar*

On November 2, 1248, after a siege that lasted 16 months, Seville was reconquered by the Christians under King Ferdinand III of Castile and León. However, for several hundred years after the fall of the caliphate, Seville remained under the influence of Islamic artistic models, due in part to the continuing existence of the Nasrid dynasty in Granada, the last sultanate on the Iberian Peninsula. During the second reign of Muhammad V (1362 – 1391), considered the Golden age of Nasrid rule, many buildings of the Alhambra were erected and lavishly decorated with geometric mosaic tiles, vegetal patterns, inscriptions and stucco, giving that palace complex its present appearance. This Nasrid style continued to evolve in the *Mudéjar* style after the Reconquest. (Muslims who chose to live under Christian political authority were known as *mudéjares*, from the Arabic word, *mudajjan*, meaning "settled" or "tamed" [1]. The *mudéjares* "continued to wear distinctive garb, adhere to a religiously prescribed diet, follow social mores different from those of the Christians, among whom they lived, speak privately in Arabic, and heed their own religion." [2])

Dom Pedro, the Catholic king of Castile (1334 – 1369) valued the Moorish heritage, and so, in 1364, when he made significant additions to and remodeled the *Alcázar* as his palace, he used the best available craftsmen from Seville, Toledo, and Granada. Muhammad V, said to be a friend of King Pedro, sent Islamic artisans and materials for the construction and decoration of the *Alcázar*. Hence, the *Real Alcázar* is contemporary with and evokes the Alhambra. In 1492, when the Nasrids were driven out of Granada, it was the *mudéjares* tilers of Seville who made the repairs to the damaged dados of the Alhambra and who were also called upon to produce the *alicatado* for their own *Alcázar*. (*Alicatado* is Spanish for cut tiles, derived from the Arab verb *qata'a,* "to cut").

The *Alcázar* has been used as a residence by the Spanish royal family ever since 1248 and as such, is the oldest royal residence in Europe still in use. It is a remarkably well – preserved palace, having undergone many alterations and additions over the centuries, including renovations in the 15th century by Ferdinand and Isabella. The *Real Alcázar* also retains its Islamic character, containing some of the most beautiful examples of *Mudejar alicatado* from this time period.

Classifying Two-Dimensional Patterns

To analyze and classify patterns in the plane, we assume that the motif repeats infinitely in two directions. Using group theory, we find that there are only 17 possible categories, based on the isometries (distance-preserving transformations) permitted by the pattern. The isometries include translations, rotations, reflections across a vertical, horizontal or diagonal axis, and glide reflections, which involve translation and reflection. The classification notation used here was devised by Carl Hermann and Charles-Victor Mauguin, where each category begins with either a ***p*** for a primitive cell, or a ***c*** for a face-centered cell. The ***p*** or ***c*** is then followed by a digit ***n***, which indicates the highest order of rotational symmetry the pattern exhibits. For example, patterns with no rotational symmetry have ***n = 1***; patterns with ***n = 2*** have 2-fold rotational symmetry; and so on, for ***n = 3, 4,*** or ***6***. The third and fourth symbols indicate the existence of mirror reflections, glide reflections, or no reflections, and use the symbols ***m, g*** or ***1***, respectively.

In 1944, E. Muller found patterns at the Alhambra which represented 11 of the 17 planar crystallographic groups. In a 1986 paper, B. Grunbaum and Z. Grunbaum of the University of Washington (in the USA) & G. C. Shepard of the University of East Anglia (in the UK) reported to have found 13 of the patterns there. The four groups they did not find were *p2, pg, pgg* and *p3m*1. Subsequently, R. Perez-Gomez of the University of Granada & J. Montesinos of the University of Madrid have reported to have found the missing ones. Hence, one cannot help but wonder if all 17 of the planar crystallographic groups may also be found at the *Real Alcázar*. This paper will attempt to answer that question by first illustrating some of the recognizably Islamic planar patterns, and then classifying them as to the symmetry elements they permit. Note that colorings and interlacings may not be considered.

The first pattern (if color is not considered) may be classified as ***p*111** (or ***p*1** for short) to indicate that the pattern may be contained within a primitive cell, and there are no rotations, mirrors or glide reflections present. Notice the trapezoidal polygonal tiles at the top of the tiling, which seem to "point" to the left. This *alicatado* (see Figure 1. below) may be found as part of a dado (a mosaic found on the lower portion of a wall) in a hallway leading to the *Patio de las Doncellas*.

Figure 1. *A* ***p*1** *pattern*

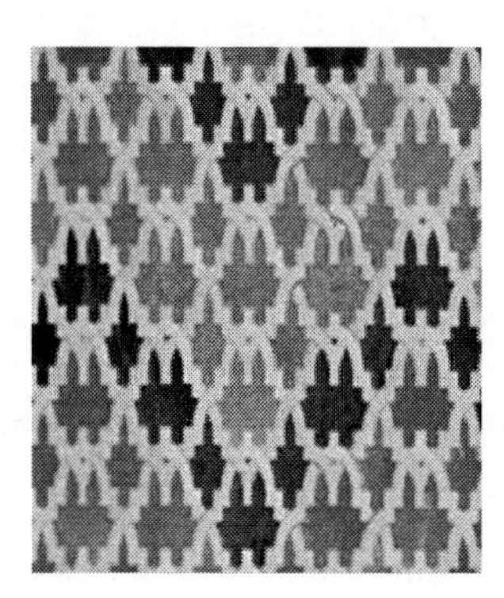

Figure 2. *A* ***pm*** *pattern*

The second pattern (if color and interlacing are not considered) may be classified as ***p*1*m*1** (or ***pm*** for short) to indicate that the pattern may be contained within a primitive cell, there are mirror reflections but no rotations, nor glide reflections off the mirror reflection axis. This *alicatado* may be found as part of a dado in the *Patio de las Doncellas* (see Figure 2. above).

The third pattern (see Figure 3. below), a floor tiling showing a zigzag motif, may be classified as ***p*1*g*1** (or ***pg*** for short) to indicate that the pattern may be contained within a primitive cell, there are no rotations nor mirror reflections, but there are glide reflections.

Figure 3. *A* ***pg*** *pattern*

Figure 4. *A* ***cm*** *pattern*

The fourth pattern (see Figure 4. above), part of an archway in the *Patio de las Doncellas* (and if idealized), may be classified as ***c*1*m*1** (or ***cm*** for short) to indicate that the pattern may be contained within a centered cell, there are no rotations, but there are mirror reflections and a glide reflection off the mirror reflection axis.

The fifth pattern (if one discounts the heraldic shields within the octagons) may be classified as ***p*211** (or ***p*2** for short) to indicate that the pattern may be contained within a primitive cell, there is a two-fold rotation but there are no mirror reflections nor glide reflections. Note that the interlacing (see Figure 5. on the next page) precludes the mirror reflections.

Figure 5. *A* ***p2*** *pattern*

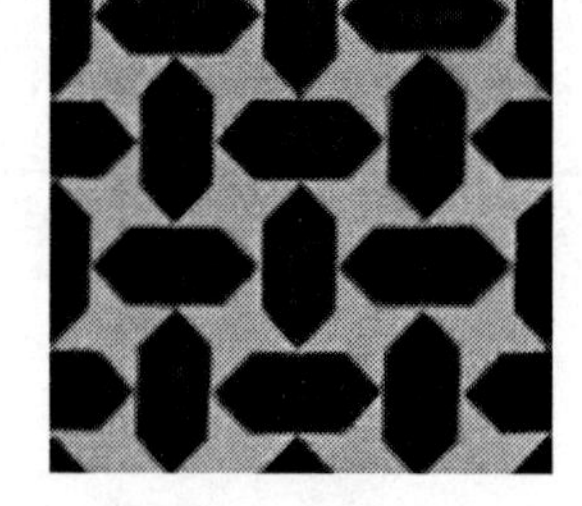

Figure 6. *A* ***pmm*** *pattern*

The sixth pattern (see Figure 6. above), part of a dado, may be classified as ***p2mm*** (or ***pmm*** for short) to indicate that the pattern may be contained within a primitive cell, there is a two-fold rotation and mirror reflections, with all the rotation centers falling on reflection mirrors.

The seventh pattern (if one disregards the coloring of the pattern) may be classified as ***p2gg*** (or ***pgg*** for short) to indicate that the pattern may be contained within a primitive cell, there is a two-fold rotation and a glide reflection but no mirror reflections. This (see Figure 7. below) is part of a dado.

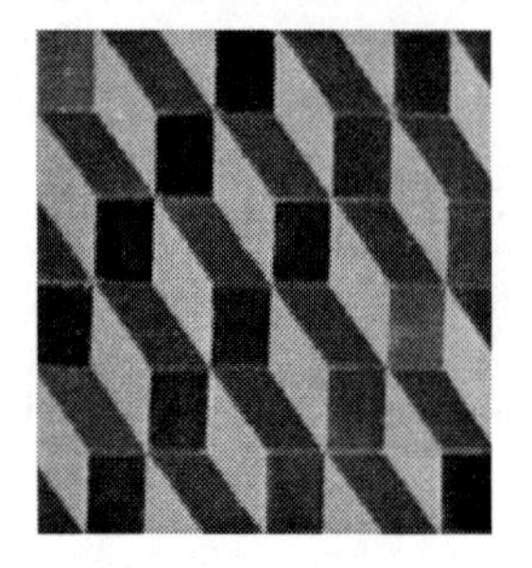

Figure 7. *A* ***pgg*** *pattern*

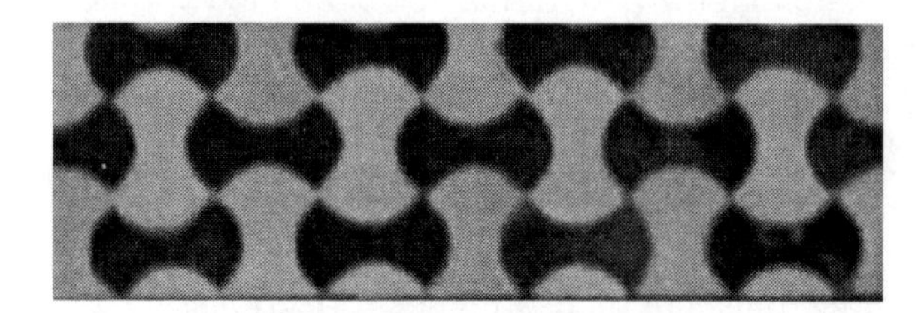

Figure 8. *A* ***cmm*** *pattern*

The eighth pattern, part of a dado (see Figure 8. above), may be classified as ***c2mm*** (or ***cmm*** for short) to indicate that the pattern may be contained within a centered cell, there is a two-fold rotation and mirror reflection axes both perpendicular and parallel to the main axis, with not all rotation centers on the mirrors.

The ninth pattern, part of a dado, may be classified as ***p311*** (or ***p3*** for short) to indicate that the pattern may be contained within a primitive cell, there is a three-fold rotation and no mirror reflection nor glide reflections. Note that the motif (see Figure 9. below) consists of one white and one adjacent black shape.

Figure 9. *A* ***p3*** *pattern*

Figure 10. *A* ***p31m*** *pattern*

The tenth pattern (see Figure 10. on the preceding page), preserved in the *Alcázar*'s museum, may be classified as ***p*31*m*** if one disregards the following: the six-sided polygonal shapes have different colorings and the light colored polygons are slightly larger than the dark-colored polygons. If one considers an idealized tiling with all the polygons congruent and similarly colored, the motif for the tenth pattern may be contained within a primitive cell, with a three-fold rotation and a mirror reflection, not all rotation centers on the mirrors. Note that in the idealized pattern, there is a point about which one may rotate 120 degrees and this point is not on a mirror of reflection.

The eleventh pattern, part of a dado (see Figure 11. below), may be classified as ***p*411** (or ***p*4** for short) to indicate that the pattern may be contained within a primitive cell, there is a four-fold rotation but there are no mirror reflections nor glide reflections.

Figure 11. *A* ***p*4** *pattern*

Figure 12. *A* ***p*4*m*** *pattern*

The twelfth pattern (see Figure 12. above), part of a dado, (if interlacing is not considered) may be classified as ***p*4*mm*** (or ***p*4*m*** for short) to indicate that the pattern may be contained within a primitive cell, there is a four-fold rotation and mirror reflections 45 degrees to one another.

The thirteenth pattern (see Figure 13 below), preserved in the *Alcázar*'s museum, (if interlacing is not considered) may be classified as ***p*4*gm*** (or ***p*4*g*** for short) to indicate that the pattern may be contained within a primitive cell, there is a four-fold rotation, mirror reflections 45 degrees to one another, and a glide reflection perpendicular to the main mirror reflection axis.

Figure 13. *A* ***p*4*g*** *pattern*

Figure 14. *A* ***p*6** *pattern*

The fourteenth pattern, part of a dado (see Figure 14. above), may be classified as ***p*611** (or ***p*6** for short) to indicate that the pattern may be contained within a primitive cell, there is a six-fold rotation, but no mirror reflections nor glide reflections.

The fifteenth pattern (see Figure 15. below), part of a dado, (if interlacing is not considered) may be classified as ***p6mm*** (or ***p6m*** for short) to indicate that the pattern may be contained within a primitive cell, there is a six-fold rotation and mirror reflections.

Figure 15. *A **p6m** pattern*

Discussion

As mentioned earlier, Grunbaum, Grunbaum, and Shepard reported to have found patterns represented by 13 of 17 the planar crystallographic groups at the Alhambra. The four they didn't find were *p*2, *pg*, *pgg* and *p*3*m*1, which Perez-Gomez and Montesinos reportedly did find. This author was unable to locate examples of patterns representing two of the 17 groups: *p2mg* (or *pmg* for short) which displays 2-fold rotation, has reflections, but no perpendicular reflections, and *p*3*m*1 which displays 3-fold rotation, where all rotation centers are on mirrors. It is interesting to note that one of these, *p*3*m*1 was also missed by Grunbaum, Grunbaum, and Shepard.

In *Symmetries of Islamic Geometric Patterns* [3], Abas and Salman provide a chart showing the distribution of the various symmetry groups of the Islamic patterns they have compiled. The three rarest categories of patterns found in mosaic tilings in Andalusia and the *Maghreb* involve glide reflections, indicating that the least preferred tilings in Islamic cultures may be classified as *pg*, *p*1 and *pmg*. Thus, it is no surprise that the author was unable to find an example representing the *pmg* class at the *Real Alcázar*. This is consistent also with the findings of the author in a previous presentation [4], where the only unrepresented frieze pattern at *Seville's Real Alcazar* was *p*1*g*1, which also employs a glide reflection.

References

[1] J. Reid. *The Moors in Spain and Portugal.* Faber and Faber, 1974.

[2] F. C. Lister and R. H. Lister. *Andalusian Ceramics in Spain and the New World.* The University of Arizona Press, 1987.

[3] S. J. Abas and A. S. Salman. *Symmetries of Islamic Geometrical Patterns.* World Scientific, 1998.

[4] B. L. Bodner. *Classifying the Frieze Patterns of Seville's Real Alcazar,* presented at the Joint Mathematics Meeting in January, 2006.

Math must be Beautiful

Carla Farsi
Department of Mathematics
University of Colorado
395 UCB
Boulder, CO 80309-0395, USA
E-mail: farsi@euclid.colorado.edu

Abstract

I present here a video installation inspired by the famous performance of Marina Abramovic "Art must be Beautiful., Artist must be Beautiful" It addresses the theme of teaching as a performance art.

1.1. The Art Work. My video installation "Math must be Beautiful, Mathematician must be Beautiful" will be shown at the Core New Art Space Gallery in Denver from October 5 to October 21, 2006. This installation was inspired by Marina Abramovic's 1975 famous performance "Art must be Beautiful, Artist must be Beautiful."

1.2. Marina Abramovic's Work. Marina Abramovic, born in Belgrade, Yugoslavia, in 1946, is one of the most influential contemporary artists of our time. She called herself, "Grandmother of Performance Art." Often, her legendary performances continue "at infinity" until they reach a conclusive critical point, often involving audience's participation, [2]. In mathematics, we could say that they continue until they reach a singular point of catastrophic nature.

1.3. Teaching and Performing. As all of us teachers know, teaching is in many senses a performing art. We not only need to try to make our students learn, but we also have to try to catch their attention, engage them in interactive learning processes, and groom them. Many times gendered behaviors are also a big component of classroom dynamics. We need to establish a rapport and dialog with our audience, and to present our teaching materials in an efficient, and perhaps entertaining way.. As BBC News reports, [1], the results of a recent Times Higher Education Supplement survey does nothing else than valuate our claim, "The responses from 648 students found many thought academics were "snooty" and had "objectionable facial hair." A comment from one of the students reads, "They try to be funny - I'm not at clown college."

1.4. The Video. This video is a parody of a mathematics lesson. At the same time, it is also a comment on how mathematicians communicate (or fail to communicate) among each other and with others. In this video, shot in my office, of which I am the lonely "star," I continually repeat the phrase, "Math must be Beautiful, Mathematician must be Beautiful." The first part of this sentence is paraphrasing some well-known quotations of the famous physicist Paul Dirac. For example, Dirac said, "If one is working from the point of view of getting beauty into one's equation, one is on a sure line of progress," [4], [5].

The formulas I am writing on the blackboard are either made up on the spot, or are famous mathematical equations such as the one that expresses energy in terms of the speed of light, which

I never use in my own work. Some other real-math formulas that can be seen in the background are left over from a real mathematical conversation I had with a colleague. In a sense, my work is a mathematically-slanted rehearsed and performed fantasy. "Math must be Beautiful, Mathematician must be Beautiful," is my mantra, a formula-like phrase that becomes a prayer. Besides my voice, we can hear the clicking of my colleagues' computers key-strokes. There is a feeling of distance introduced by the solo performance; it is perhaps reflected in real life by the isolation we mostly do mathematics in, or perhaps the introduction of the computer medium. Toward the end of the video, I add a feminine touch by putting on some lipstick in a hurried way, to hint at gender-related themes that so far, I believe, have only been addresses superficially. My performance simply ends as my voice becomes less and less audible, as it trails off gradually into the infinite (mathematical) land of repetitions.

1.6. Technical Remarks. I used Adobe Premiere 1.5 to edit my original video.

1.7. Final Remarks. I hope that this provocative video will help fuel the on-going discussion on our roles as professional educators.

Images

Figure 1: *An image from the first part of the clip*
Figure 2: *An image from the second part of the clip*

References

[1] BBC News, *Students Bemoan 'Unhip' Lecturers,* http://news.bbc.co.uk/1/hi/education/4647766.stm
[2] Eyestorm Article, *Marina Abramovic,* http://www.eyestorm.com/feature/ED2n_article.asp?article_id=38
[3] Faculty Teaching Excellence Program Web Pages University of Colorado at Boulder, http://www.colorado.edu/ftep/
[4] *Quotations by Paul Dirac*, http://www-history.mcs.st-andrews.ac.uk/Quotations/Dirac.html
[5] P. Dirac, *The evolution of the Physicist's Picture of Nature*, Scientific American **208** (5) (1963)

The Integrated Scale Desirability Function:

A Musical Scale Consonance Measure Based on Perception Data

Richard J. Krantz
Department of Physics
Metropolitan State College of Denver
Denver, CO 80217
krantzr@mscd.edu

Jack Douthett
5044 Rockcress Drive NW
Albuquerque, NM 87120
douthett@comcast.net

1 Introduction

Based on previous work by Krantz and Douthett [1, 2, 3] and Sethares [7, 8], briefly reviewed in Section 2, a universal measure of the equal-tempered musical scale consonance is developed in Section 3. Preliminary results applying this, so-called, Integrated Scale Desirability Function for scales made up of complex tones show that higher frequency components of complex tones are necessary to explain the emergence of our usual 12-tone equal-tempered musical scale. These results are discussed in Section 4.

The formalism is also applied to the well-known Bohlen-Pierce scale. Here, too, it appears that higher frequency components are necessary for the emergence of the historically important 13-note to the "tritave" Bohlen-Pierce scale. These results are also discussed in Section 4.

Subsequently, in Subsection 4.4, we show that complex interval spectra can be "sculpted" so that non-standard equal-tempered musical scales turn out to be consonant as measured by the Integrated Scale Desirability Function.

2 Background

2.1 The Generalized Desirability Function: Previously, Krantz and Douthett [1, 2] articulated an approach for comparing the reasonableness of c-tone equal-tempered musical scales that simultaneously approximate multiple just musical intervals. This so-called Desirability Function is:

$$D(c,N) = 10 - 20\sum_{i=1}^{N}\left|\left\{c\,\log_2(R_i) + 0.5\right\} - 0.5\right|, \tag{1}$$

where c is the chromatic cardinality of the equal-tempered scale (the number of notes to the octave), N is the number of target intervals to be approximated, the R_i*'s* are the frequency ratios of the target intervals, $\{x\}$ is the fractional part of x, and $|x|$ is the absolute value of x.

This expression was used to determine which c-tone equal-tempered scales with octave closure best approximated the musically important intervals of the pure fifth, major third, and minor third. Subsequently, the Desirability Function was generalized to measure which c-tone equal-tempered scales with non-octave closure best approximated appropriately weighted multiple-intervals [3]. This Generalized Desirability Function is:

$$D_b(c,N) = 10 - 20\sum_{i=1}^{N} p_i\left|\left\{c\,\log_b(R_i) + 0.5\right\} - 0.5\right|, \tag{2}$$

where b, the base of the logarithm represents the interval of closure and the p_i are the respective normalized weights of the R_i*'s*.

This formalism was then used to describe and compare the desirability of non-traditional scales such as the Bohlen-Pierce scale [4, 5] and Balzano's 20-fold scale [6].

2.2 Dissonance Function, Perception, and Normalized Weights: In 1993 W. A. Sethares [7] developed a Dissonance Function based on the data of Plomp and Levelt [9]. Sethares fit the consonance/dissonance data of Plomp and Levelt to develop a standard consonance/dissonance curve, the so-called "roughness curve" which described the perceived tonal dissonance of two pure tones played simultaneously. Figure 1 shows a typical roughness curve.

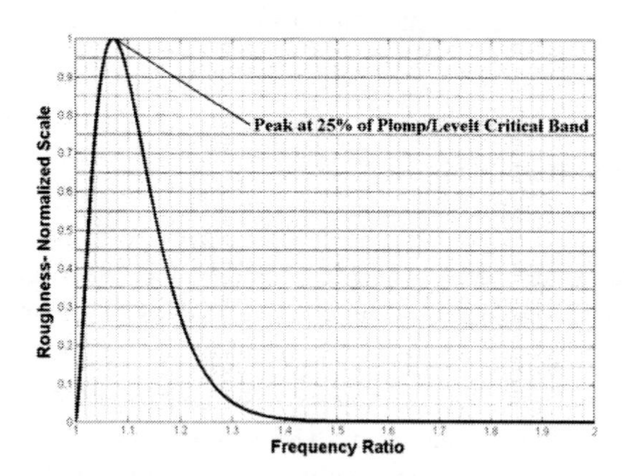

Figure 1- Typical Roughness Curve

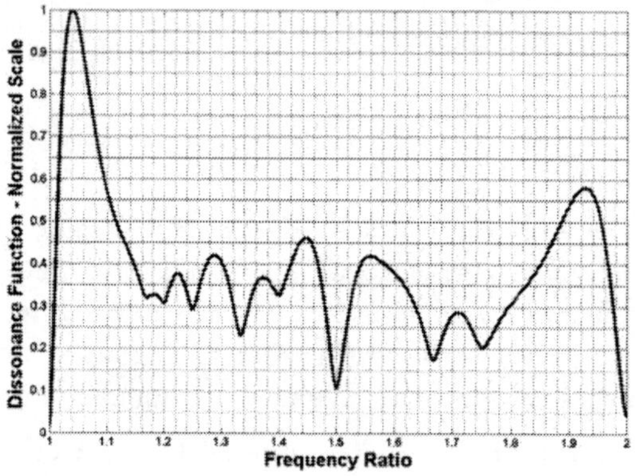

Figure 2 – Typical Dissonance Function

The peak of the curve occurs at 25% of the critical bandwidth, the frequency bandwidth at which two pure tones are perceived as distinct (see Zwicker et al. [10]). Below the peak beats are heard. Above the peak the tones are perceived more and more as two distinct tones. This dependence of perception based on beat frequency and the distinction between two pure tones, critical band theory, was first formulated by Helmholtz [11].

To generate a dissonance measure for intervals involving complex tones, those with harmonics, Sethares added the roughness curves for all pairs of pure tones weighted by the spectral amplitude of each pure tone. The result was the so-called Dissonance Function. Figure 2 shows a typical Dissonance Function. Peaks represent the most dissonant complex intervals and the sharp valleys represent the most consonant complex intervals.

Consider a Sethares-like Dissonance Function, $D_F(R)$, where R represents the frequency ratio of the fundamentals of the two complex tones of the interval and F represents the timbre of the complex tones. If,

$$M_F = \max\left\{D_F(x) \mid x \geq 1\right\}. \tag{3}$$

Then,

$$p_{F,b}\left(k,n\right) = \frac{M_F - D_F\left(b^{k/n}\right)}{n\left(M_F - C_{F,b}\left(n\right)\right)}, \tag{4}$$

represents the probability of the kth partial out of n being "consonant" where $C_{F,b}(n) = \sum_{k=1}^{n} D_F(b^{k/n}) \Big/ n$

3 The Integrated Scale Desirability Function

3.1 Discrete Case: We may now use the probabilities defined in equation (4) as the probabilities in equation (2) to define an "n^{th} partial desirability sum":

$$S_{F,b}(n,c) = 10 - 20\sum_{k=1}^{n} p_{F,b}(k,n)\left|\left\{\frac{ck}{n}+0.5\right\}-0.5\right| \tag{5}$$

where c is the chromatic cardinality of the equal tempered scale. Using equation (4) and simplifying yields:

$$S_{F,b}(n,c) = 10 - \frac{20}{n}\sum_{k=1}^{n}\frac{(1-D'_F(b^{k/n}))}{(1-C'_{F,b}(n))}\left|\left\{\frac{ck}{n}+0.5\right\}-0.5\right| \tag{6}$$

where $D'_F(b^{k/n}) = \dfrac{D_F(b^{k/n})}{M_F}$, the normalized Dissonance Function, and $C'_{F,b}(n) = \dfrac{C_{F,b}(n)}{M_F} = \dfrac{1}{n}\sum_{j=1}^{n} D'_F(b^{k/n})$.

3.2 The Continuous Case: We may define an Integrated Scale Desirability Percentage (ISDP) for a fixed base, b, timbre, F, and chromatic cardinality, c, as:

$$P_{F,b}(c) = 10\lim_{n\to\infty} S_{F,b}(n,c). \tag{7}$$

Substituting equation (6), with $k = n\,x_k$ *and* $\Delta x = 1/n$, in to equation (7) and taking the limit yields:

$$P_{F,b}(c) = 100 - \frac{200}{M_F - C''_{F,b}}\int_0^1 (M_F - D'_F(b^x))\left|\{cx+0.5\}-0.5\right|dx \tag{8}$$

where $C''_{F,b} = \lim_{n\to\infty} C'_{F,b}(n) = \int_0^1 D'_F(b^x)dx$. With $A_{F,b} = 50(M_F - 2C''_{F,b})$; $B_{F,b} = 200/(M_F - C''_{F,b})$; and a simple change of variables the Desirability Percentage, equation (8), may be re-written as:

$$P_{F,b}(c) = A_{F,b} - \frac{B_{F,b}}{\ln(b)}\int_1^b \frac{D'_F(y)}{y}\left|\{c\,\log_b(y)+0.5\}-0.5\right|dy. \tag{9}$$

In this form it is apparent that the ISDP is a weighted average of the normalized Dissonance Function from the interval of the unison, $y_{min} = 1$, to the interval of closure, $y_{max} = b$. The weighting function: $\left|\{c\,\log_b(y)+0.5\}-0.5\right|/y$ is a generalization of the original Desirability Function of Krantz and Douthett [1]. The Normalized Dissonance Function is based on the original, perception-based, Dissonance Function of Sethares [7].

3.3 An Important Property of the ISDP: In order to define a useful measure for the Integrated Scale Desirability we consider how, if at all, the timbre and base affect the "expected" desirability percentage of $P_{F,b}(c)$. We define the expected desirability percentage (EDP) as:

$$\left\langle P_{F,b}\right\rangle = \lim_{c\to\infty}\frac{1}{c}\sum_{j=1}^{c} P_{F,b}(j). \tag{10}$$

Remarkably, the expected desirability percentage is 50%; independent of the timbre, as reflected by the dissonance function used, or the closure interval, represented by the base b. This makes comparison to the EDP a universal measure for all scales.

3.4 The Integrated Scale Desirability Function: Given the universal property of the EDP; we, therefore, redefine the constants in equation (9) so that a value of zero represents the EDP measure, a value of 5 represents the most consonant scale possible, and a value of -5 represents the most dissonant scale possible. We define this function the Integrated Scale Desirability Function (ISDF), $K_{F,b}(c)$:

$$K_{F,b}(c) = A'_{F,b} - B'_{F,b} \int_1^b \frac{D'_F(y)}{y\ln(b)} \left| \left\{ c \log_b(y) + 0.5 \right\} - 0.5 \right| dy \,. \tag{11}$$

Shown in Figure 3 is a particular example of dissonance function, $D'_F(y)$ (solid), for intervals consisting of complex tones each having 7 harmonics of equal amplitude and the weighting function for a 12-tone equal-tempered chromatic scale (dashed) with octave closure (b=2). The product of these two functions is the integrand in equation (11).

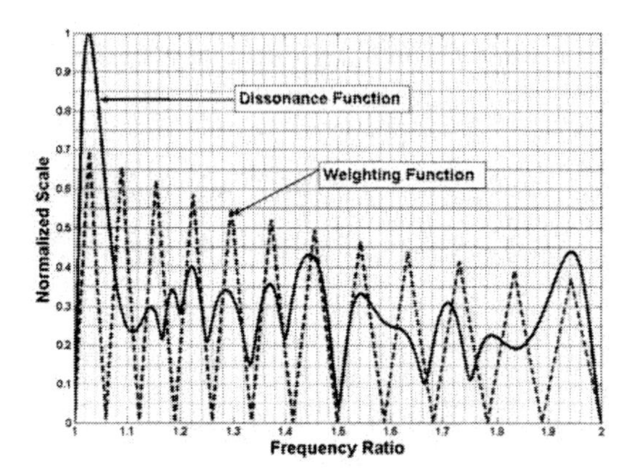

Figure 3 – 12-Tone Dissonance/Weighting Function

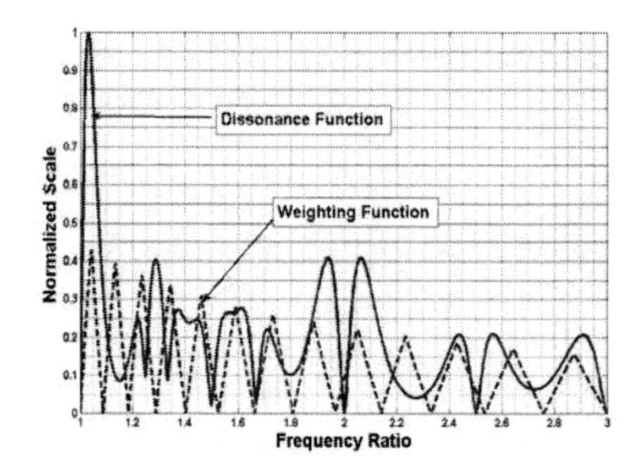

Figure 4 – 13-Tone Dissonance/Weighting Function

For comparison, Figure 4 shows the dissonance function (solid), for intervals consisting of complex tones each having 5 harmonics of equal amplitude and the weighting function for a 13-tone equal-tempered chromatic scale (dashed) with tritave, 3:1 (octave plus a fifth), closure. This represents the weighting function for the, well-known, Bohlen-Pierce scale [4, 5].

Comparison of the details of Figures 3 and 4 shows that the ISDF will depend, crucially, on the overlap of the dissonance function and the particular weighting function.

4 Preliminary Integrated Scale Desirability Function (ISDF) Analysis

In this section we use the ISDF to analyze the effects of: 1) the fundamental frequency, 2) the number of partials in the complex tones, and 3) the relative amplitudes of the partials making up the complex tones; on the overall scale consonance of equal-tempered scales of varying chromatic cardinality.

In each figure the fundamental of each complex tone is listed. For convenience, except in subsection 4.4, the fundamental of each tone is assumed to be the same. The closure interval for the equal-tempered scale is listed. For example, a closure interval of 2 represents equal-tempered scales with octave closure. A closure interval of 3 represents equal-tempered scales closed at an octave plus a fifth, the closure interval for the Bohlen-Pierce scale.

Also listed are the partials included in the complex fundamental tone and the partials included in the complex interval tone. For example, in Figure 5 below, the complex fundamental tone, as well as, the complex interval tone is each made up of 5 harmonics. These are listed after each respective spectrum. The relative amplitudes, of each harmonic in the complex spectrum, are also given. In Figure 5, each partial has the same amplitude.

4.1 The Effect of the Fundamental: In this subsection we show the effect of changing the fundamental frequency on scale consonance.

4.1.1 Octave Closure: Keeping spectral components and the amplitudes constant we changed the starting frequency of the complex tones making up the intervals in the dissonance function. We see, in Figures 5 and 6, that for octave closure the familiar 12 tone equal-tempered scale does not emerge as a consonant scale until higher frequency components are present. This indicates that our perception of consonance and dissonance depends on the presence of higher frequency components as observed by Sethares [7, 8].

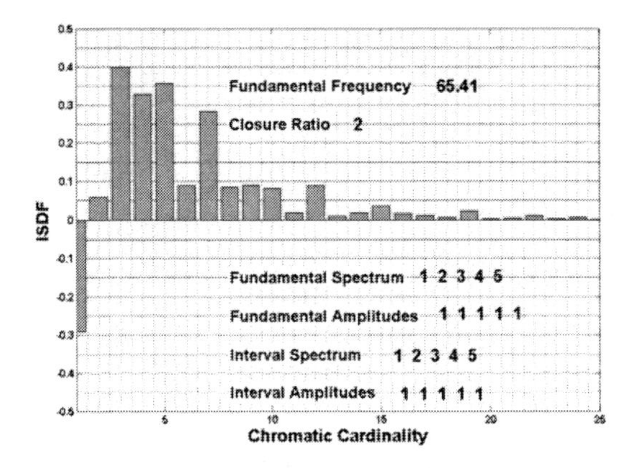

Figure 5 – ISDF Octave Closure (65.41 Hz)

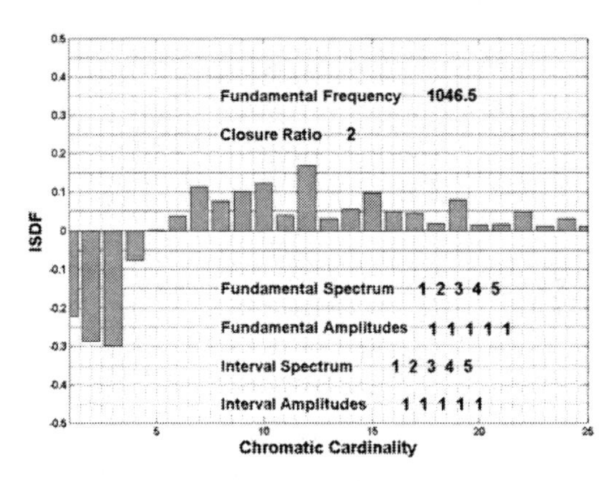

Figure 6 – ISDF Octave Closure (1046.5 Hz)

4.1.2 Tritave Closure: Shown in Figure 7 and 8 is a comparison of the effect of changing the starting tones on scales with 3:1 closure ratios. This closure interval is the basis for the Bohlen-Pierce scale. In keeping with the Bohlen-Pierce scale, we include only odd harmonics in the complex tones. Again, the amplitudes of the partials included are equal.

Although not as dramatic as the case for octave closure, we note again that the consonance of scales with larger chromatic cardinalities do not start to emerge until higher frequency components are included. Even in non-standard scales, it appears that the presence of higher frequency components is necessary for the perception of consonance and dissonance. Further analysis indicates that mixing in partials of all equal amplitudes, a clearly artificial case, obscures the effects of more complicated spectra. More on this in subsection 4.3.

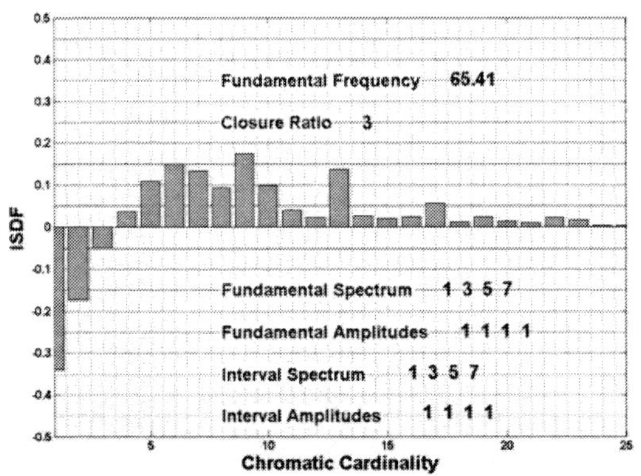

Figure 7 – ISDF Tritave (Odd Harmonics: 65.41 Hz)

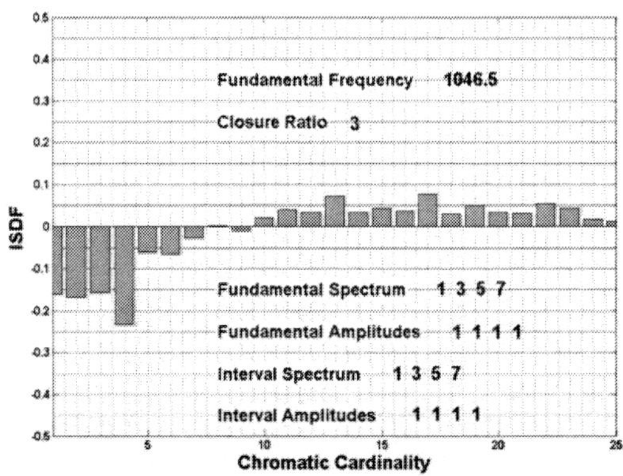

Figure 8 - ISDF Tritave (Odd Harm: 1046.5 Hz)

4.2 The Effect of the Number of Partials in the Complex Tones: In this subsection we show the effect of changing the number of partials on scale consonance.

4.2.1 Octave Closure: As shown in Figures 9 and 10, the effect of adding partials, of equal amplitudes, is similar to the effect of adding higher frequency components, as shown in Figures 5 and 6 above. Again, this indicates that adding higher frequency components adds to the perception of consonance of equal-tempered scales of higher chromatic cardinality, suppressing the consonance of scales with small cardinalities.

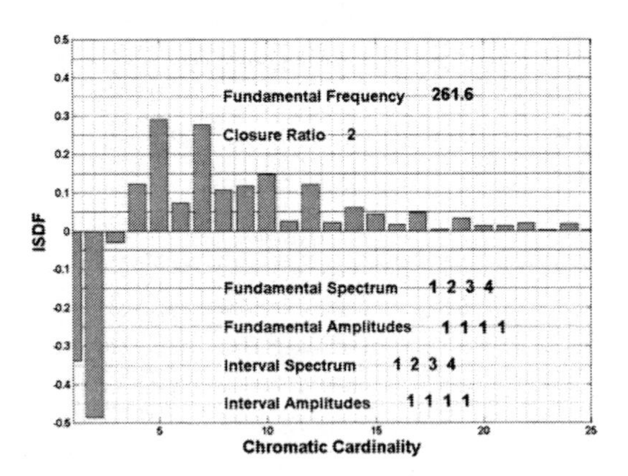

Figure 9 – ISDF Octave Closure (4 Harmonics)

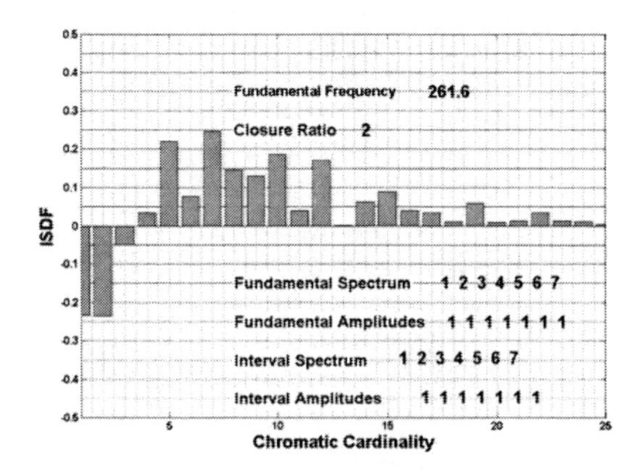

Figure 10 – ISDF Octave Closure (7 Harmonics)

4.2.2 Tritave Closure: The effect of the shifting of consonance to higher cardinalities with the inclusion of more partials of equal amplitude is also born out for the non-standard tritave scales as shown in Figures 11 and 12. As more partials are added, smaller chromatic cardinalities become more dissonant while sifting the more consonant cardinalities to larger numbers of equal-tempered notes per scale. Again, it appears that mixing in of higher partials with equal amplitudes obscures the effects due to including partials with varying amplitudes.

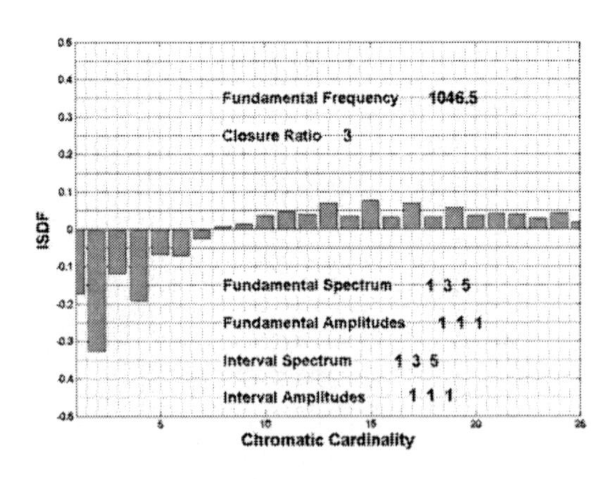

Figure 11 – ISDF Tritave (3 Odd Harmonics)

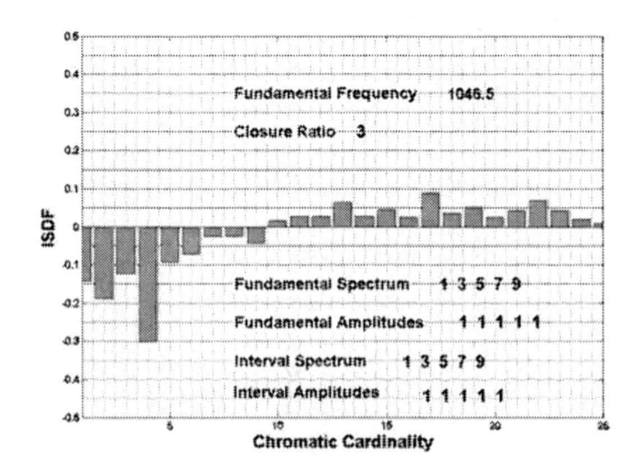

Figure 12 – ISDF Tritave (5 Odd Harmonics)

4.3 The Effect of the Amplitudes of the Partials in the Complex Tone: In this subsection we investigate the effect, on scale consonance, of including more partials with decreasing amplitudes.

4.3.1 Octave Closure: As shown in Figures 13 and 14 including more partials, even with smaller relative amplitudes, shifts the consonant scales to larger cardinalities. As more partials are included, the 12-tone equal-tempered scale emerges as the most consonant cardinality with octave closure. It appears that more complicated spectra are needed for our familiar 12-tone scale to be perceived as most consonant compared to other possibilities.

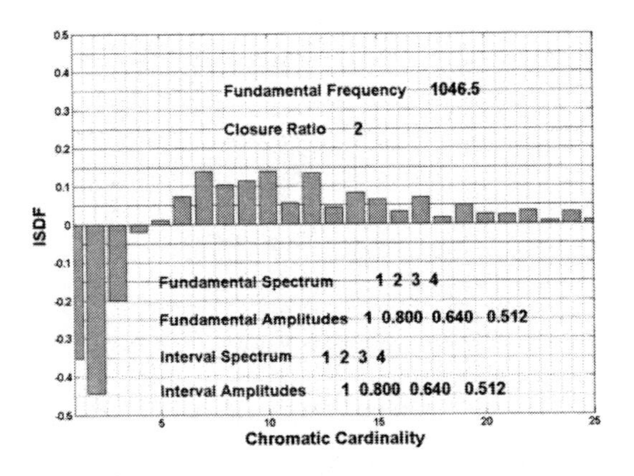

Figure 13 – ISDF Octave (4 Decreasing Harmonics)

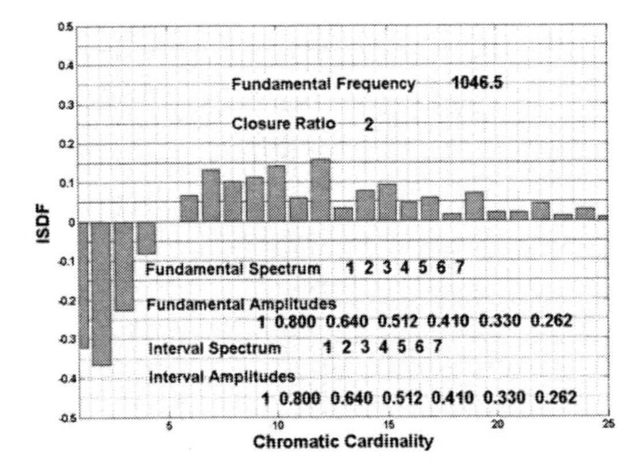

Figure 14 – ISDF Octave (7 Decreasing Harmonics)

4.3.2 Tritave Closure: A similar conclusion, that more complicated spectra move the consonant cardinalities to more notes per scale, is born out for the non-standard tritave scales shown is Figures 15 and 16 as well.

As we include more partials, 13 equal-tempered notes to the tritave emerges, even more strongly, as the most consonant of scales. It was 13 notes to the tritave that was used in the original Bohlen-Pierce scale [4, 5].

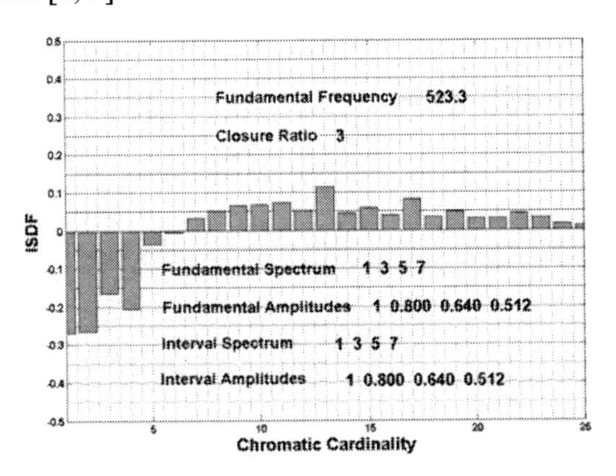

Figure 15 – ISDF Tritave (4 Decreasing Odd Harm.)

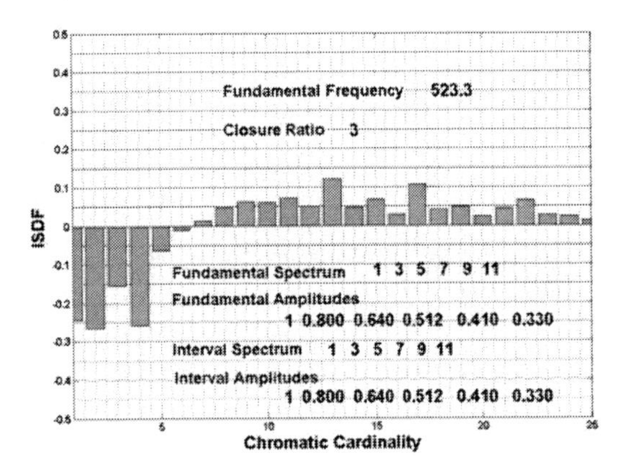

Figure 16 – ISDF Tritave (6 Decreasing Odd Harm.)

4.4 Sculpting Spectra: Shown in Figure 17 is the ISDF for a relatively complex fundamental and interval spectra. As long as we assume octave closure and spectra based on the harmonic series (each subsequent partial is a multiple of the starting tone) and the interval start tone is a pure fifth above the fundamental start tone, 12-tone equal-temperament emerges as the most consonant scale.

Even so, it appears that 19-equal-tempered notes to the octave could be important. This is born out in Figure 18. Even though each spectra is based on the harmonic series, 19-notes to the octave emerges as a consonant scale if the interval spectra starts on a tone a minor third above the fundamental start tone. Clearly, if spectra are "sculptured" one can get non-standard scales to be consonant. It should be pointed out that, historically, 19-tones to the octave has been used in compositions based upon the minor third.

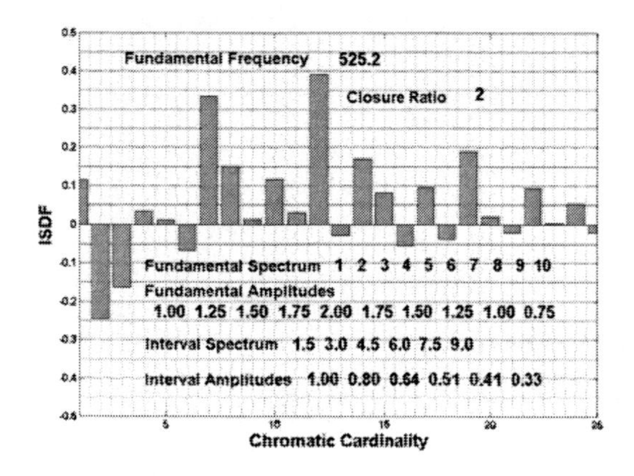

Figure 17 – Sculpted Harmonics Example 1

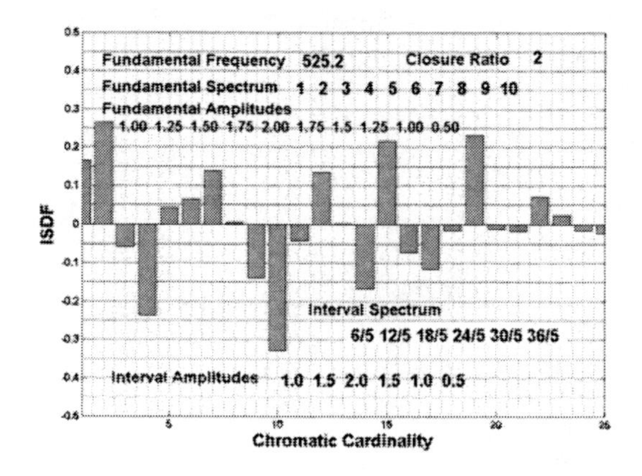

Figure 18 – Sculpted Harmonics Example 2

5 Summary

Based on previous work [1, 2, 3], we have developed a universal measure of the equal-tempered musical scale consonance, the so-called Integrated Scale Desirability Function which is based on the perception-based [9, 10] Dissonance Function of Sethares [7, 8]. Preliminary analysis shows that for scales made up of complex tones higher frequency components are necessary to explain the emergence of our usual 12-tone equal-tempered musical scale.

Preliminary application of the formalism to the historically important Bohlen-Pierce scale [4, 5] also shows that higher frequency components are necessary for the emergence of this well-known 13-note to the "tritave" scale.

We also show that complex interval spectra can be "sculpted" so that non-standard equal-tempered musical scales turn out to be perceived as consonant by our measure.

6 References

[1] R. J. Krantz and J. Douthett, "A measure of the reasonableness of equal-tempered musical scales," J. Acoust. Soc. Am. Vol. 95, No. 6, 3642-3650, June 1994.

[2] J. Douthett, "The Theory of Maximally and Minimally Even Sets, The One-Dimensional Antiferromagnetic Ising Model, and the Continued Fraction Compromise of Musical Scales," Ph.D. dissertation (University of New Mexico, Albuquerque, NM, May 1999).

[3] R. J. Krantz and J. Douthett, "Construction and interpretation of equal-tempered scales using frequency ratios, maximally even sets, and P-cycles," J. Acoust. Soc. Am. Vol. 107, No. 5, Pt. 1, 2725-2734, May 2000.

[4] M. V. Mathews, J. R. Pierce, A. Reeves, and L. A. Roberts, "Theoretical and Experimental Explorations of the Bohlen-Pierce Scale," J. Acoust. Soc. Am. Vol 84, 1214-1222, 1988.

[5] H. Bohlen, "13 Tonstufen in der Doudezeme," Acustica Vol. 39, 76-86, 1978.

[6] G. Balzano, "The Group Theoretic Description of 12-Fold Pitch Systems," Comput. Music J. Vol. 4, 66-84, 1980.

[7] W. A. Sethares, "Local consonance and the relationship between timbre and scale," J. Acoust. Soc. Am. Vol. 94, No. 3, 1218-1228, Sept. 1993.

[8] W. A. Sethares, Tuning, Timbre, Spectrum, Scale, Ch. 5, (Springer, 1998).

[9] R. Plomp and J. M. Levelt, "Tonal Consonance and Critical Bandwidth," J. Acoust. Soc. Am., Vol. 38, 548-560, 1965.

[10] E.Zwicker, G. Flottorp, and S. S. Stevens, "Critical bandwidth in loudness summation," J. Acout. Soc. Am. Vol. 29, 548, 1957.

[11] H. Helmholtz, On the Sensations of Tone, (1877). Trans A. J. Ellis, Dover, New York (1954).

Tiled Artworks Based on the Goldbach Conjecture

Sharol Nau
Studio at 212 East Fourth Street
Northfield, MN 55057, USA
www/sharolnau.snakedance.org
whitecrow@snakedance.org

Abstract

A simply, stated though still unproved, mathematical conjecture by Christian Goldbach is utilized to make two-dimensional artworks. Tile patterns with even numbers of tiles are divided into two sets. Each set consists of a prime number of tiles that reflects Goldbach's conjecture that any even number greater than two has at least one pair of primes that sum to that number.

The Artworks

I assembled the following groups of modular shapes for basic, sometimes tedious, designs to create, as it turn out, an extensive series of artworks. These arrangements are cartoons for the event of construction of aesthetically interesting artworks. The division of the picture plane in the shape of a rectangle or hexagon is an easy, if time consuming procedure. The method is much like a writer's use of language. Words are formed by the arrangement of the same 26 letters; however emotion and experience and intent are stated on a personal level. My contribution to contemporary art is to make the most of these images that carry the minimal usage of triangles and squares into a state of visual interest, something that compels the viewer to examine further or merely kick back and enjoy. My methods challenge the simplistic nature of these designs. By themselves the divisions that are chained together by connecting the sets of triangles that represent a prime number might be of little interest to many viewers. The important connection of shape, pattern, rhythm, texture, repetition, variation and color that results in the production of a compelling image is of interest. These artworks are handmade using a variety of materials that contribute surface qualities ranging from areas that are smooth to areas that are abrasive to the touch. How the designs were initiated is secondary. The artworks become complicated forms through the manipulation of traditional and nontraditional media.

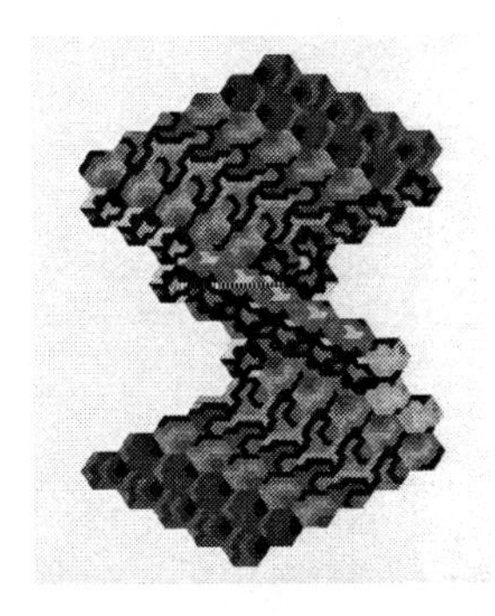

Figure 1. Hexagon Design
Inkjet print, 10x8 inches

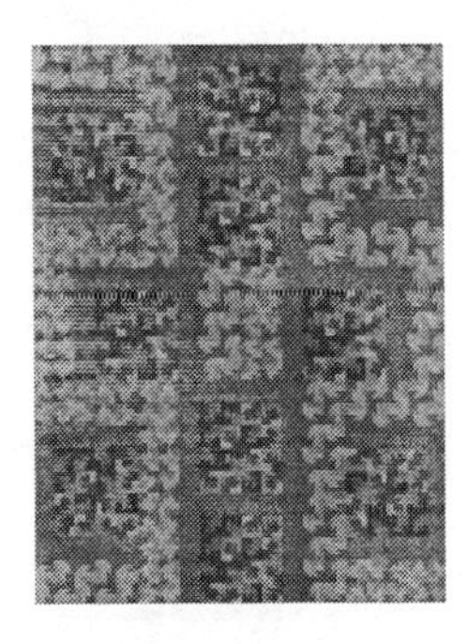

Figure 2. Goldbach Tiling
Ink jet print, 10x8 inches

Figure 3. Goldbach Tiling, lino-cut print, and collage on canvas, 48x48 inches

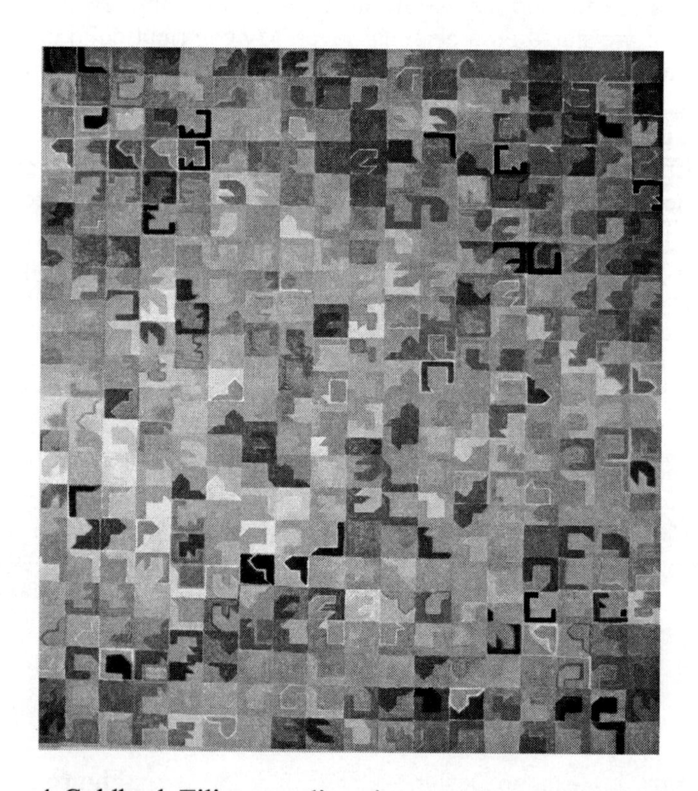

Figure 4. Goldbach Tiling, acrylic paint on canvas, 64x54 inches

The Tilings

The structured base developed by utilizing Christian Goldbach's (1690-1764) conjecture that any even number greater than two has at least one pair of primes that sum to that number brings forth the potential for hundreds of designs.

My extensive series of drawings, paintings, and assemblages based on the partitioning of a rectangle into an even number of triangles, a rectangle partitioned into an even number of squares and a hexagon partitioned into an even number of equilateral triangles is the subject of this writing. Upon completing the initial divisions, the triangles or squares corresponding to a pair of prime numbers are rearranged until a potentially satisfying aesthetic adjustment is produced. Color, texture and choice of material heighten the visual effect.

The even number thirty-two is convenient when working with a rectangle broken up using triangles. The near relationship of 19 and 13 to the Golden Ratio offers a pleasing design element. These designs have fallen into two categories.

The scattered arrangement allows for the distribution of the triangles corresponding to a pair of primes to be separated into several groups which offer several smaller shapes similar in color or texture (figure 5). This scatter arrangement when used as a repeated motif results in an interactive all-over pattern (figures 6 and 7). A rhythm is created through visual groupings and the viewer may easily pick out recurring images such as triangles, squares, trapezoids and undefined shapes created by the merging and separation of the basic triangles. The eye is allowed to search in and out and around the entire composition. Balance teases somewhere among the formal, the informal and radial. The interactive complexity provides a compositional unity with stability as well a movement that is fairly independent of enhancements such as color and texture.

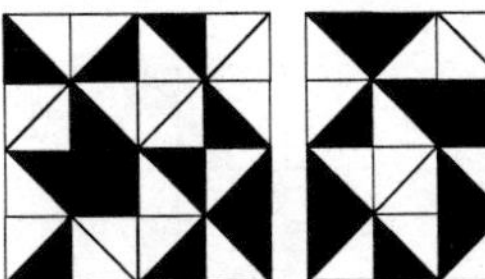

Figure 5. Scattered motif

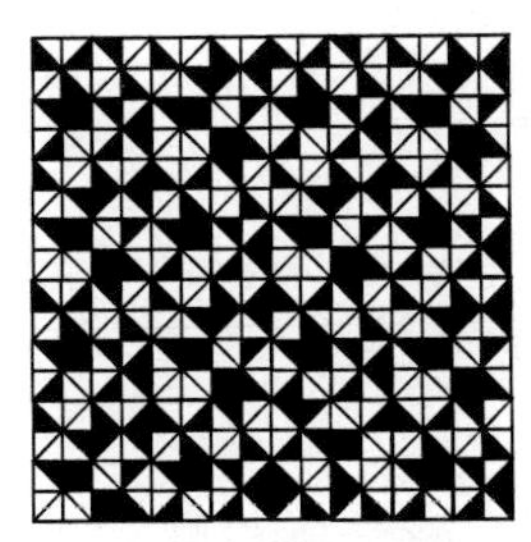

Figure 6. Scattered arrangement, repeated motif

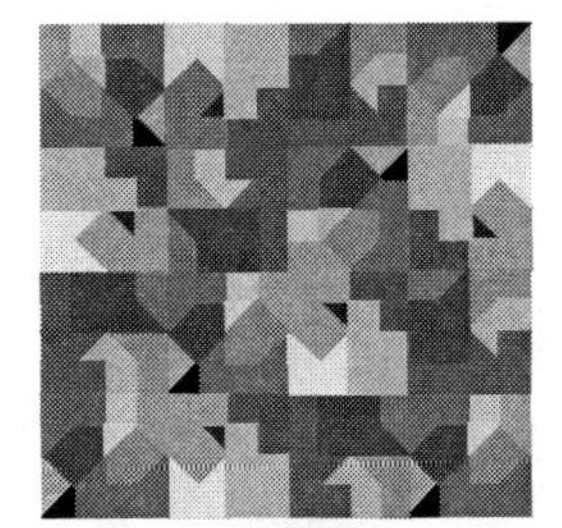

Figure 7. Scattered arrangement, repeated motif

Using the same three numbers, viz. 19+13=32, and a **tight arrangement** restricts each prime to one shape, two per rectangle (figure 8). The result is a grid that visually separates each grouping when multiples are utilized (figures 9 and 10). The rhythm is more along a staccato theme. Repetition is held together by the abutting and sharing of edges rather than the involuntary interlocking that becomes evident in the scattered arrangement.

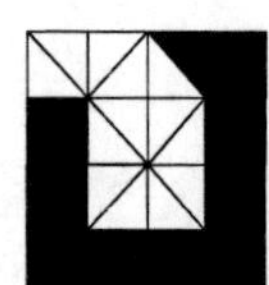
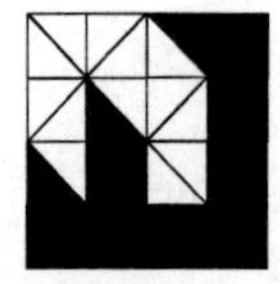
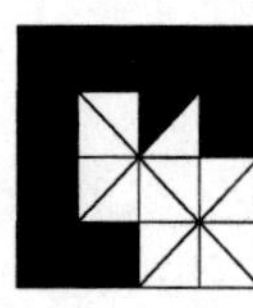
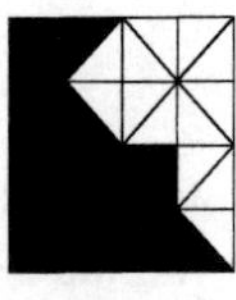

Figure 8. Five tight arrangements

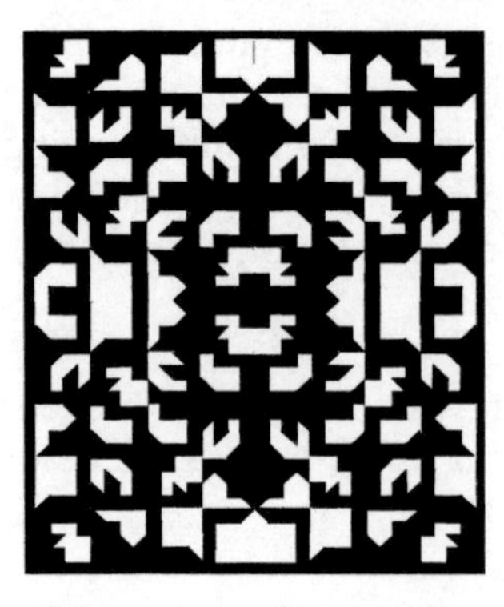

Figure 9. Tight arrangement, repeated motif

Figure 10. Tight arrangement, repeated motif

The even number 42 is used for its aesthetic qualities, which are Spartan, when combined with theme and variation in the construction of paintings, drawings or assemblages using **the square motif** and the Goldbach conjecture. Pairs of primes include 23 and 19 (figure 11) or 31 and 11 (figure 12). These groupings require an intense approach with materials to make them interesting. Mirrors and the build-up of the surface give artworks the sparkle that's required.

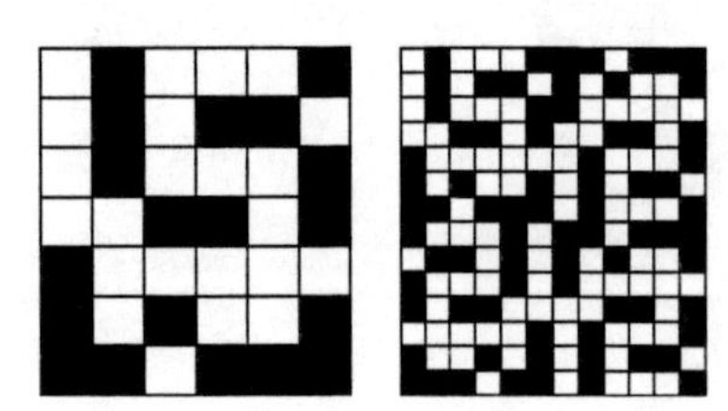

Figure 11. Arrangement of squares, 23+19=42

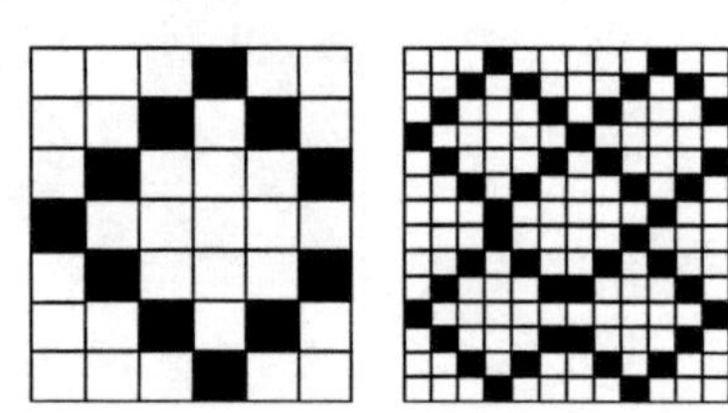

Figure 12. Arrangement of squares 11+31=42

The hexagonal motif uses fifty-four equilateral triangles to offer further possibilities for design. Experimental pairs of primes include 17 and 37 as well as 23 and 31.

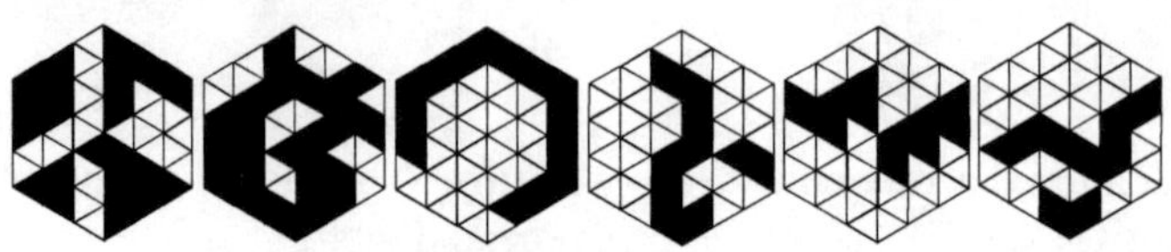

Figure 13.

Other even numbers have been used for creating patterns. Numbers in the hundreds and at times even in the thousands are more of challenge for creating an artwork. Many of these illustrations in their basic black-and-white form are depleted of the intellectual and emotional expression necessary to bring them to life. A closer consideration of my intent can be viewed by visiting my web site or better yet by visiting and seeing in person the completed works.

Sculpture Puzzles

George W. Hart
Dept. Computer Science
Stony Brook University
Stony Brook, NY, 11794, USA
http://www.georgehart.com
E-mail: george@georgehart.com

Abstract

A series of novel sculpture-puzzles is illustrated, with mathematical explanation. Each consists of a set of identical parts that snap together into a symmetric form. The parts are flat, so they can be cut out or stamped from sheet materials such as wood, metal, plastic, or cardboard. High accuracy is required for the parts to mate properly, so computer-controlled fabrication technologies are useful. The examples shown were made by laser-cutting, by solid freeform fabrication techniques, or by scissors and paper. Their intricate geometric forms make for challenging assembly puzzles and attractive artworks. A template and instructions show how to make one from paper.

1. Introduction

Twelve copies of the shape outlined in Figure 1a can be cut out and assembled to form the construction shown as a virtual model in Figure 1b. If made of a thin, slightly flexible sheet material, the parts would snap together and directly hold each other in position without any glue or connectors. Although far from regular, the flat shape is directly related to the underlying regular pentagon and the assemblage is directly related to an implicit underlying dodecahedron. The parts meet each other in threes at the dodecahedron vertices and in twos at the midpoints of the dodecahedron edges.

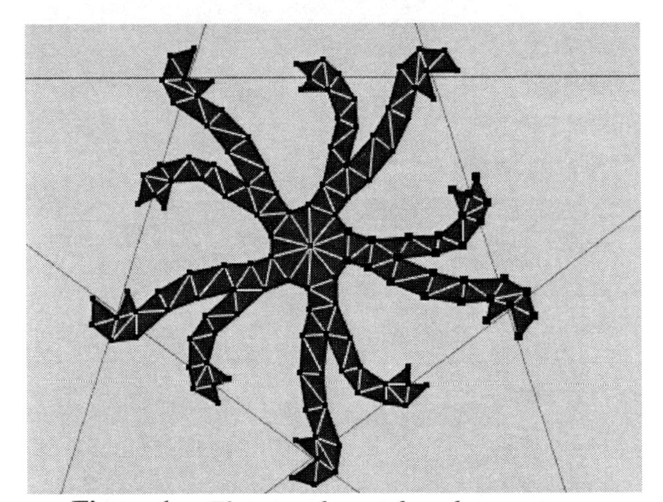

Figure 1a: *Flat piece layout, based on pentagon.*

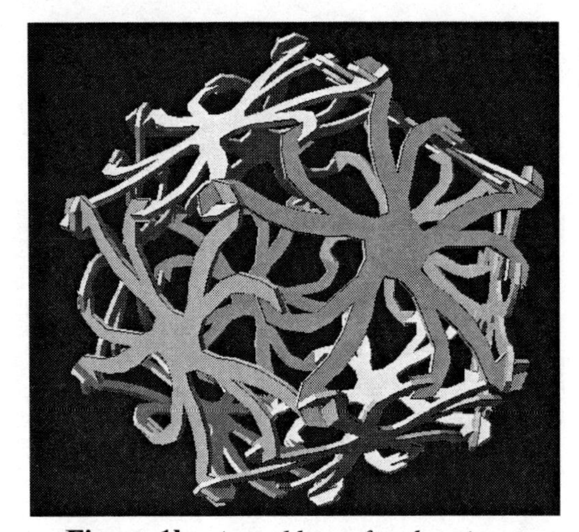

Figure 1b: *Assemblage of twelve pieces.*

To understand how the parts are rigidly locked together, it is important to notice that the tips of the arms in Figure 1a point alternately clockwise and counterclockwise. If all arms faced clockwise or if all faced counterclockwise, then the parts could still be positioned together but any part could "unscrew" from the assemblage of Figure 1b, so the structure would not hold together. With half the mating surfaces facing each direction, they are locked. But there must be some flex in the material to insert the pieces.

The designer has freedom in choosing how the arms are shaped in Figure 1a, as long as the tips of each arm end up at the appropriate location facing the chosen direction. With a symmetric result in mind, the twelve parts were chosen to be identical, each with 5-fold symmetry about their centers. This is not essential to the mathematical connectivity; it simply leads to what I consider an attractive sculptural form. There are only two types of arms and they exactly repeat, which gives a visual coherence to the final form. The use of non-identical parts with different types of arms could add a permutational aspect to future puzzle designs.

The construction of Figure 1 is pedagogical, to show the key ideas of the puzzles illustrated below. It is not intended as a puzzle itself, because it would be overly simple to solve, assuming one can visualize a dodecahedron. But variations lead to more interesting puzzle designs. To generalize the idea, notice that the tips of the arms meet at 3-fold or 2-fold rotational axes of the dodecahedron. In general, this paper considers flat shapes that lie in the face planes of a symmetric polyhedron with arms that extend to two different types of points where symmetry axes pass through the planes. One type of arm has a mating surface facing clockwise and the other type faces counterclockwise. This is a highly restrictive set of design rules, but has enough freedom to allow the exploration of a wide range of interesting sculpture puzzles. Some of these I have physically made and assembled, as illustrated herein.

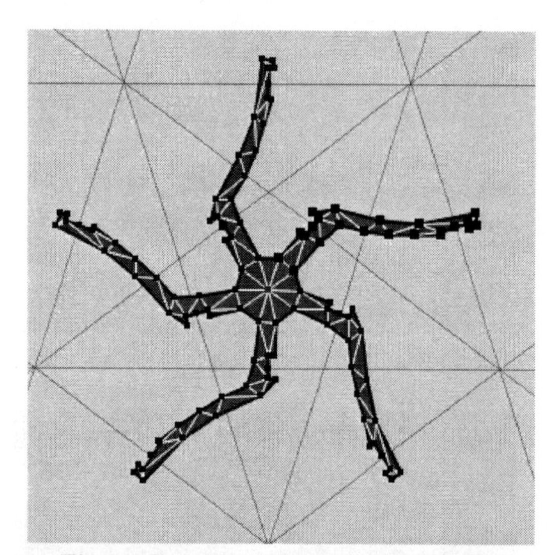

Figure 2a: *Part based on dodecahedron.*

Figure 2b: *Construction in transparent acrylic.*

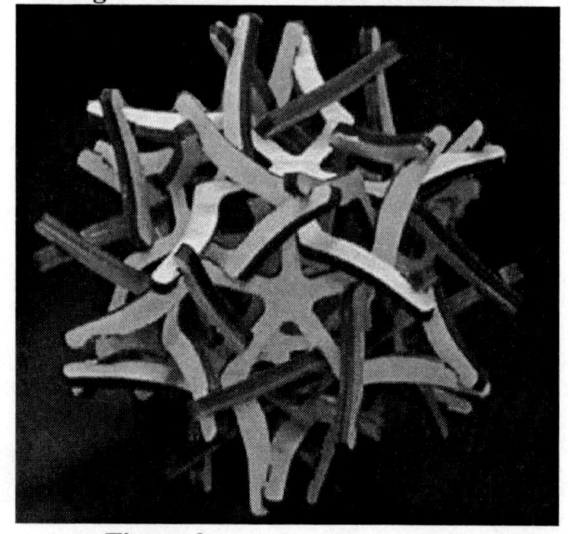

Figure 2c: *Construction in wood.*

Figure 2d: *Construction in acrylic.*

2. Sculpture Puzzle Examples

For several years, I have been designing puzzles that are variations on the ideas behind Figure 1. These puzzles are closely related to my sculptures made from symmetrically arranged planar components [2], but are designed to snap together and be disassemblable. I have been experimenting with various materials to explore their appearance and usability. Figure 2 shows three realizations of one design, laser cut from three different materials, each between 3 and 7 inches in diameter. The acrylic versions are very attractive, but the user must be careful not to stress the parts while manipulating them, because acrylic plastic is brittle. The wood version is made of a multi-layer Baltic birch plywood, so is much more robust. The ideal material might be a springy stainless steel, but I have not yet produced such a version. In any material, it is quite a challenge to assemble the parts, as the arms are very likely to become improperly tangled and many hands seem to be needed to hold the parts in proper relative orientation.

Figures 1a and 2a are screenshots from a software tool described in [2] in which I can draw a rough polygonal outline that I later manually tweak and smooth out. In particular, the width of the notches must be adapted to the thickness of the material and the angle at which the planes meet. In the design of Figure 2a, both types of notches are at 2-fold axis points in the plane. Instead of two types of arm (as in Figure 1a) Figure 2a has a single type of dual-purpose arm, with a counterclockwise-mating "elbow" partway out and a clockwise-mating surface at the extremity.

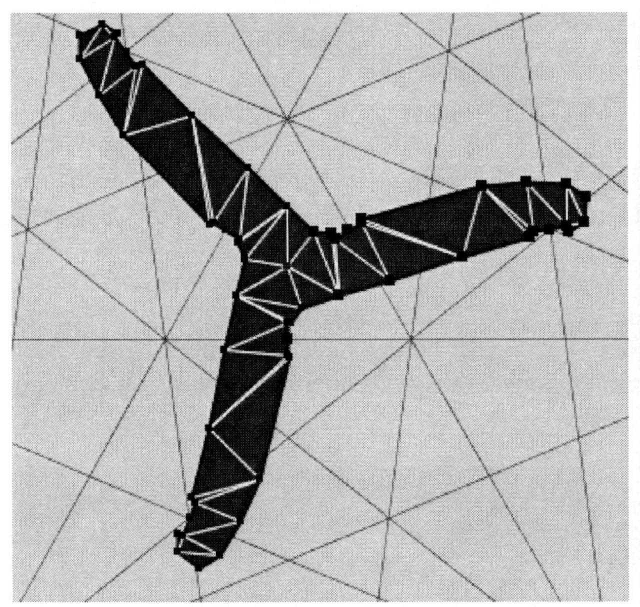

Figure 3a: *Part based on icosahedron.*

Figure 3b: *Assemblage of twenty pieces.*

Figure 3 illustrates another puzzle based on the same idea, but with twenty three-armed pieces arranged like an icosahedron. The background of lines in Figure 3a is the stellation diagram of the icosahedron [1]. The parts mate with each other in pairs near the center and at their tips. Both these locations are 2-fold axis points in the stellation diagram. The example in Figure 3b is laser-cut from ABS plastic, about 6 inches in diameter. It can be understood as five subsets of four pieces each, which make five tetrahedral sub-arrangements. So the design can be seen as based on the uniform compound of five tetrahedra. I can assemble it alone, but it is challenging fun to assemble this as a two-or-more person activity.

The 12-piece design in Figure 4 is based on the dodecahedron, but now the mating points are at an inner 3-fold axis where the parts spiral around each other and an outer 2-fold axis where pairs snap together. The pieces for the 3-inch model in Figure 4b are made of nylon, using a selective sintering process in which a computer-controlled laser melts nylon powder to form the parts. Although this solid freeform fabrication process can directly make 3D forms—it could even make the entire model in its assembled state—here it was used to make a stack of 2D parts for me to assemble. Nylon is strong and flexible, so it is very pleasurable to snap them together and discover what a very stable model they make. A ten-inch laser-cut aluminum version of this design is illustrated below in Figure 12.

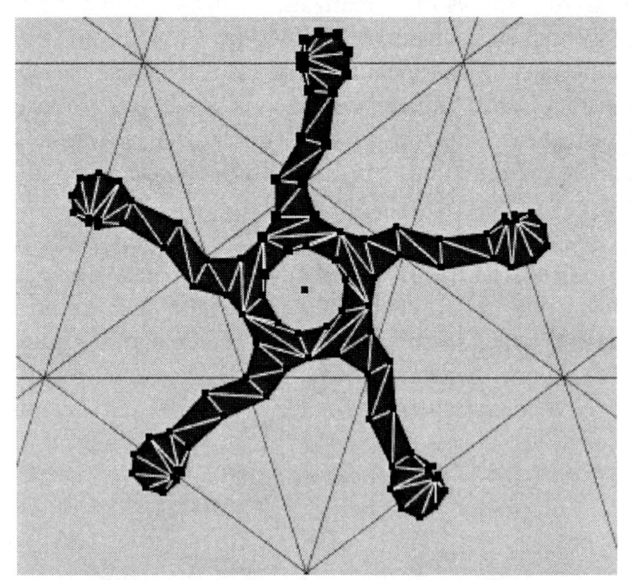

Figure 4a: *Part based on dodecahedron.*

Figure 4b: *Assemblage of twelve pieces.*

The design in Figure 5a is based on the stellation diagram of the rhombic triacontahedron. The parts meet in orthogonal groups of three (like the corners of a cube) at "mouths" on the 3-fold axis points. The inner mating points are 2-fold axis points where they meet in pairs. Also made of nylon by selective laser sintering, thirty parts assemble to make the 5-inch model shown in Figure 5b. This puzzle is very tricky to assemble because everything tends to fall apart until everything is all together. But the result is quite stable. To visualize it during assembly, it is helpful to know that the parts' inner segments (from 2-fold point to 2-fold point) form pentagonal rings that are interlocked in the manner of Holden's six-pentagon "orderly tangle" [3].

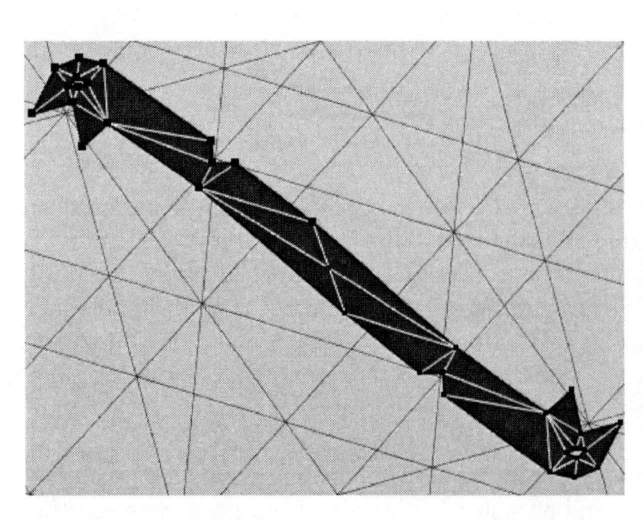

Figure 5a: *Part based on rhombic triacontahedron.*

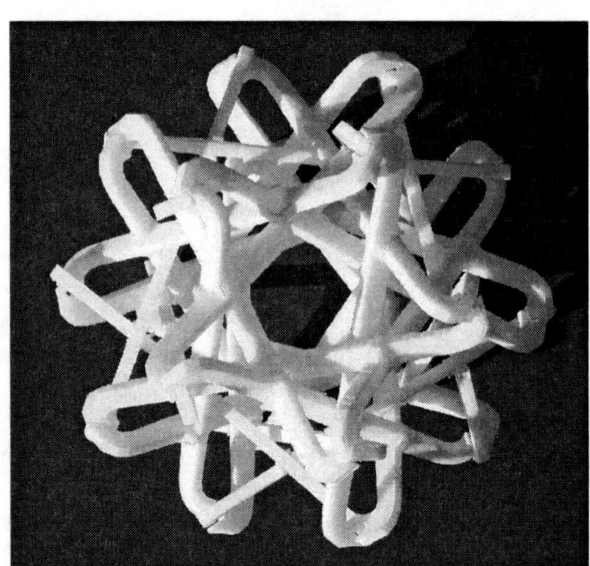

Figure 5b: *Assemblage of thirty pieces.*

From the octahedron comes the eight-piece design shown in Figure 6. The inner connection point is a 2-fold axis point at the midpoint of the octahedron edge. The outer connection point is a 3-fold axis point above the center of a face. The model in figure 6b is made of eight parts cut with scissors from card stock. It is easy to visualize the form during assembly, because the corners point in the directions of a cube's vertices, but it is amazingly tricky to assemble alone. They tend to fall apart until the final one is positioned. But after assembly, it is amazingly stable. One can throw it high in the air and catch it without any fear of spontaneous deconstruction. Because it has just eight smooth parts, it is the easiest to make by hand of the puzzles in this paper. A template for it is included as an appendix. Copy it to cardstock, carefully cut the parts, and enjoy the frustration! The result is about four inches across from tip to tip.

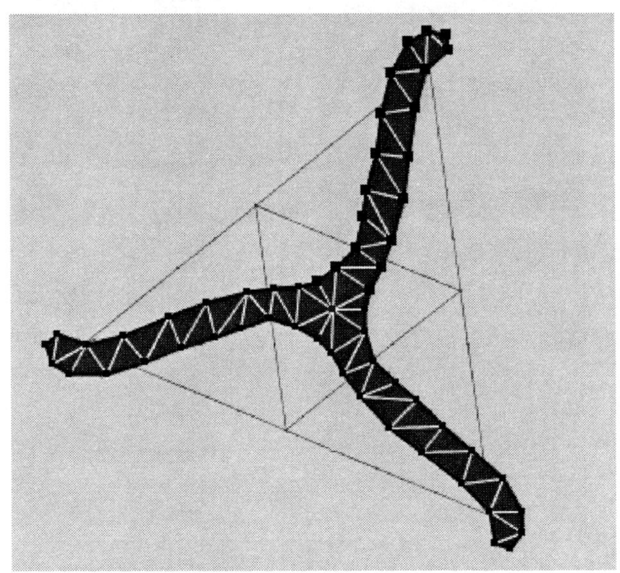

Figure 6a: *Part based on regular octahedron.*

Figure 6b: *Assemblage of eight pieces.*

Figure 7 shows another octahedral puzzle. As shown here, this one turned out not to be very stable. The 4-inch card stock model of Figure 7b easily falls apart, so I put little drops of glue on the outer tips to secure it. I expect that extending the tips of the arms will fix the problem.

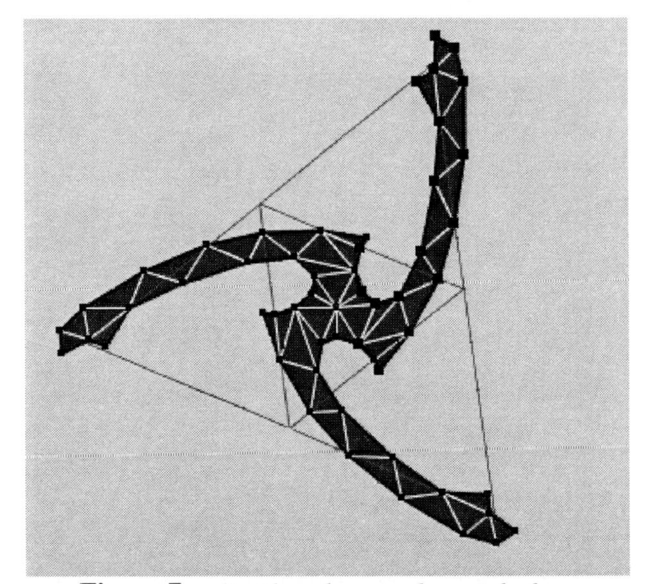

Figure 7a: *Part based on regular octahedron.*

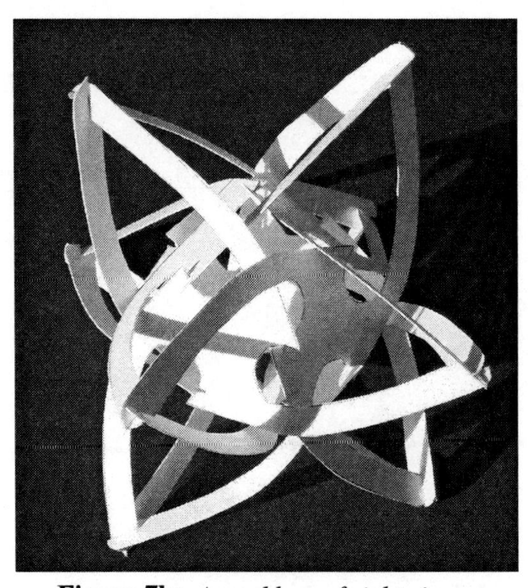

Figure 7b: *Assemblage of eight pieces.*

Another thirty-part design based on the rhombic triacontahedron is shown in Figure 8. It is similar to the design of Figure 5 in that the "mouths" are at 3-fold axis points where the parts meet in orthogonal planes. But the inner mating points here are at 5-fold points, which make for a very challenging assembly. Again, this is highly stable when complete, but very unstable until the last few parts are positioned. The 12-inch model in Figure 8b is laser-cut from plastic foam. Fifteen parts of two colors are randomly intermixed, with no pattern to the color arrangement. A visualization aid for assembly is that the parts' inner segments (from 5-fold point to 5-fold point) form the edges of an icosahedron.

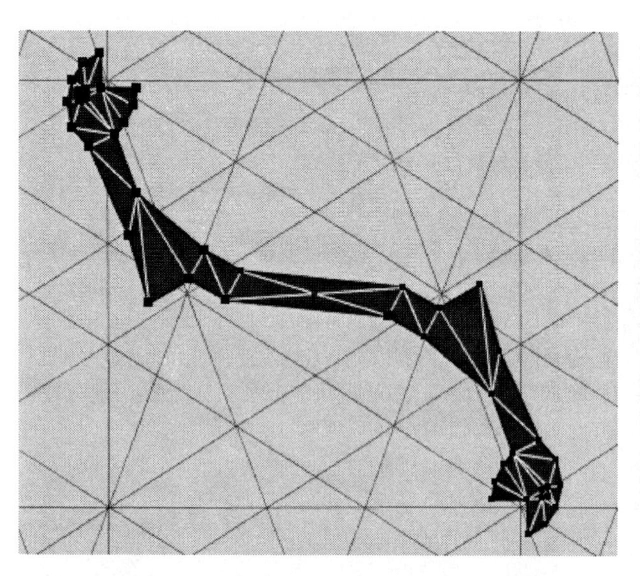

Figure 8a: *Part based on rhombic triacontahedron.*

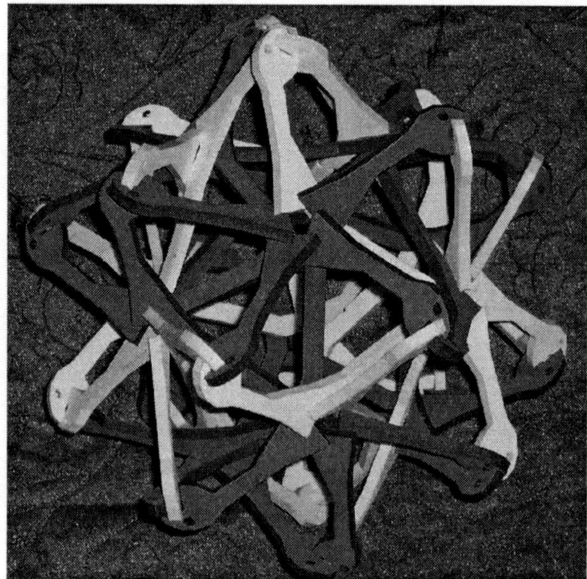

Figure 8b: *Assemblage of thirty pieces.*

Figure 9 illustrates a twenty-part design based on the icosahedron. The components in Figure 9b are made from ABS plastic on a fused deposition modeling machine. Figures 10 and 11 show virtual images of more complex designs I have not yet made. Each has potential pitfalls to be carefully worked out concerning stability, thickness, springiness, or assemblability. I expect to explore more ambitious designs after gaining experience with simpler puzzles. Structures such as Figures 10 and 11 require considerable insight to judge if rigid parts can be maneuvered past each other for assembly.

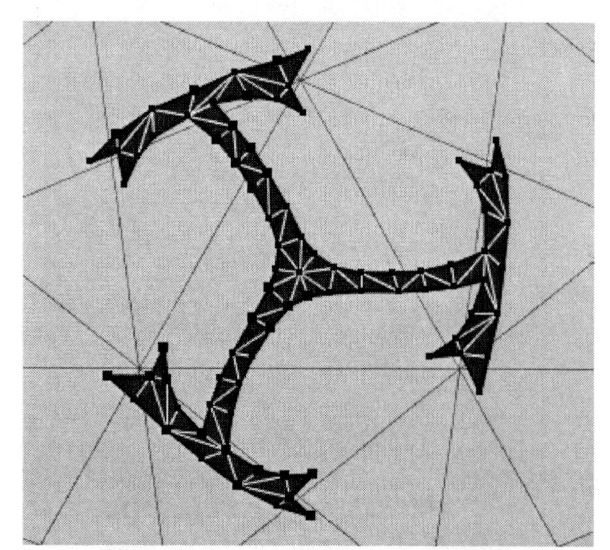

Figure 9a: *Part based on icosahedron.*

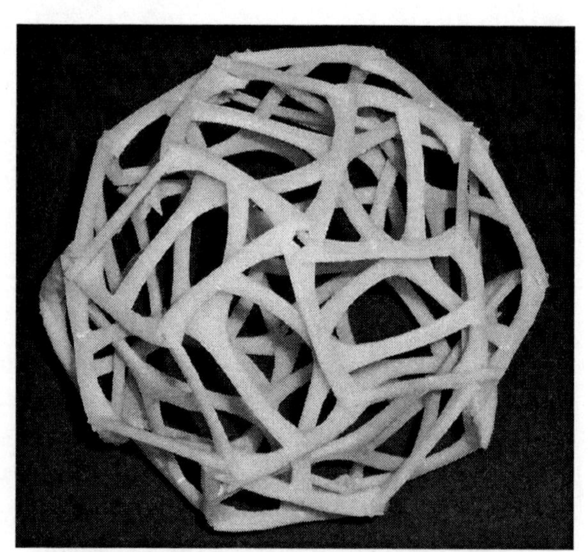

Figure 9b: *Assemblage of twenty pieces.*

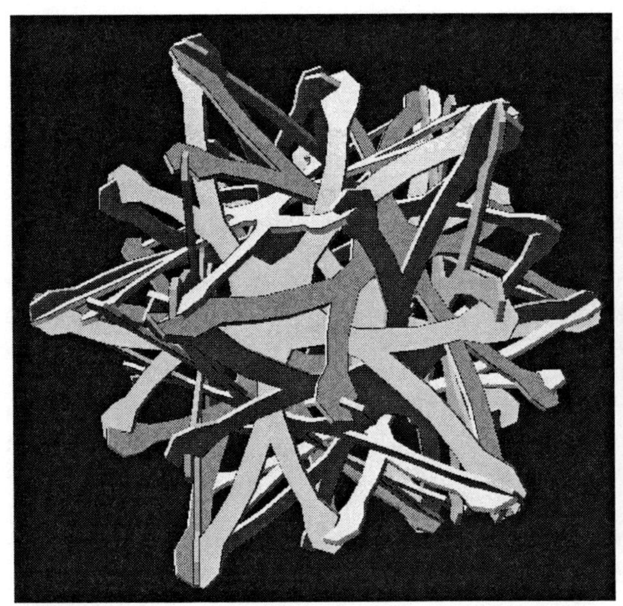

Figure 10: *12-Part design based on dodecahedron.*

Figure 11: *12-Part design based on dodecahedron.*

3. Conclusions

Puzzle expert Jerry Slocum gives the following definition: *A mechanical puzzle is a self-contained object, composed of one or more parts, which involves a problem for one person to solve by manipulation using logic, reasoning, insight, luck, and/or dexterity* [6]. The sculpture puzzles presented here are intended to be good puzzles by that definition (and maybe good multi-person puzzles too). They are also intended to be attractive as sculptures when assembled, with a symmetric intricacy that appeals to people with an aesthetic taste for mathematical structure.

In my constructive sculpture, connectors are an important issue that I carefully address [4]. I put much effort into the design of connections between parts, e.g., bevels and glues, or brackets and screws. From the perspective of simplicity, the ultimate solution is to have no connector, so the parts directly mate together. Of course, a consequence of such a design is that by reversing the process the parts can be directly disassembled. This would not be desirable for a public sculpture where vandalism is a concern. But for a puzzle sculpture, it can be an elegant design. Many puzzles and sculptures are of a snap-together nature. Works of Rinus Roelofs stand out as particularly clever in this regard [5].

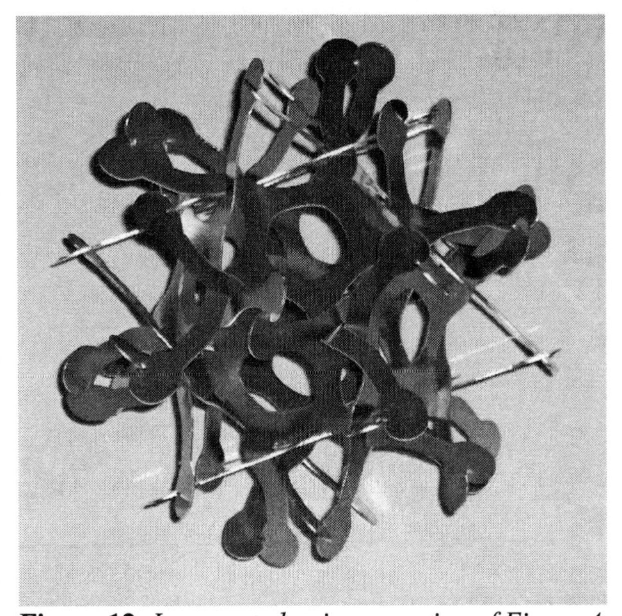

Figure 12. *Laser-cut aluminum version of Figure 4.*

While expensive to produce in the prototyping stages illustrated here, such as Figure 12, I hope to develop some designs to the point where they can be produced inexpensively. Because they are made of flat parts, they should be stampable in plastic or metal. But thickness, stiffness, burrs, surface finish, etc, are all to be addressed.

References

[1] H.S.M. Coxeter, et al., *The Fifty-Nine Icosahedra,* U. Toronto Pr., 1938.
[2] G.W. Hart, "Sculpture from Symmetrically Arranged Planar Components", in *Meeting Alhambra, (Proceedings of ISAMA-Bridges 2003,* Granada, Spain), Javier Barrallo et al editors, Univ. of Granada, 2003, pp. 315-322.
[3] G.W. Hart, "Orderly Tangles Revisited", in *Renaissance Banff, (Proceedings of Bridges 2005: Mathematical Connections in Art, Music, and Science,* Banff, AB), Reza Sarhangi and Robert Moody editors, 2005, pp. 449-465.
[4] G.W. Hart, "The Geometric Aesthetic," in *The Visual Mind II,* Michele Emmer (ed.), MIT Press, 2005.
[5] Rinus Roelofs, "Three Dimensional and Dynamic Constructions Based on Leonardo Grids", in *Renaissance Banff, (Proceedings of Bridges 2005,* Banff, Alberta), Reza Sarhangi and Robert Moody editors, 2005, pp. 161-168.
[6] Jerry Slocum, et al., *Puzzles Old and New,* Univ. Wash. Pr., 1986, p. 4.

Acknowledgments: For fabrication assistance with the models illustrated here, I thank Erik Demaine and Martin Demaine (Figure 2c), George Miller of www.puzzlepalace.com (Figures 2d and 8b), Chris Palmer (Figure 2b), Jim Quinn (Figures 4b and 5b), and Abhi Shelat (Figure 3b).

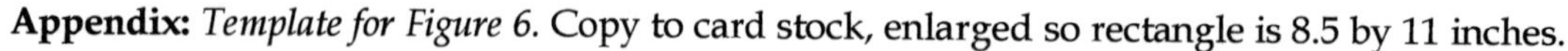

Appendix: *Template for Figure 6.* Copy to card stock, enlarged so rectangle is 8.5 by 11 inches.

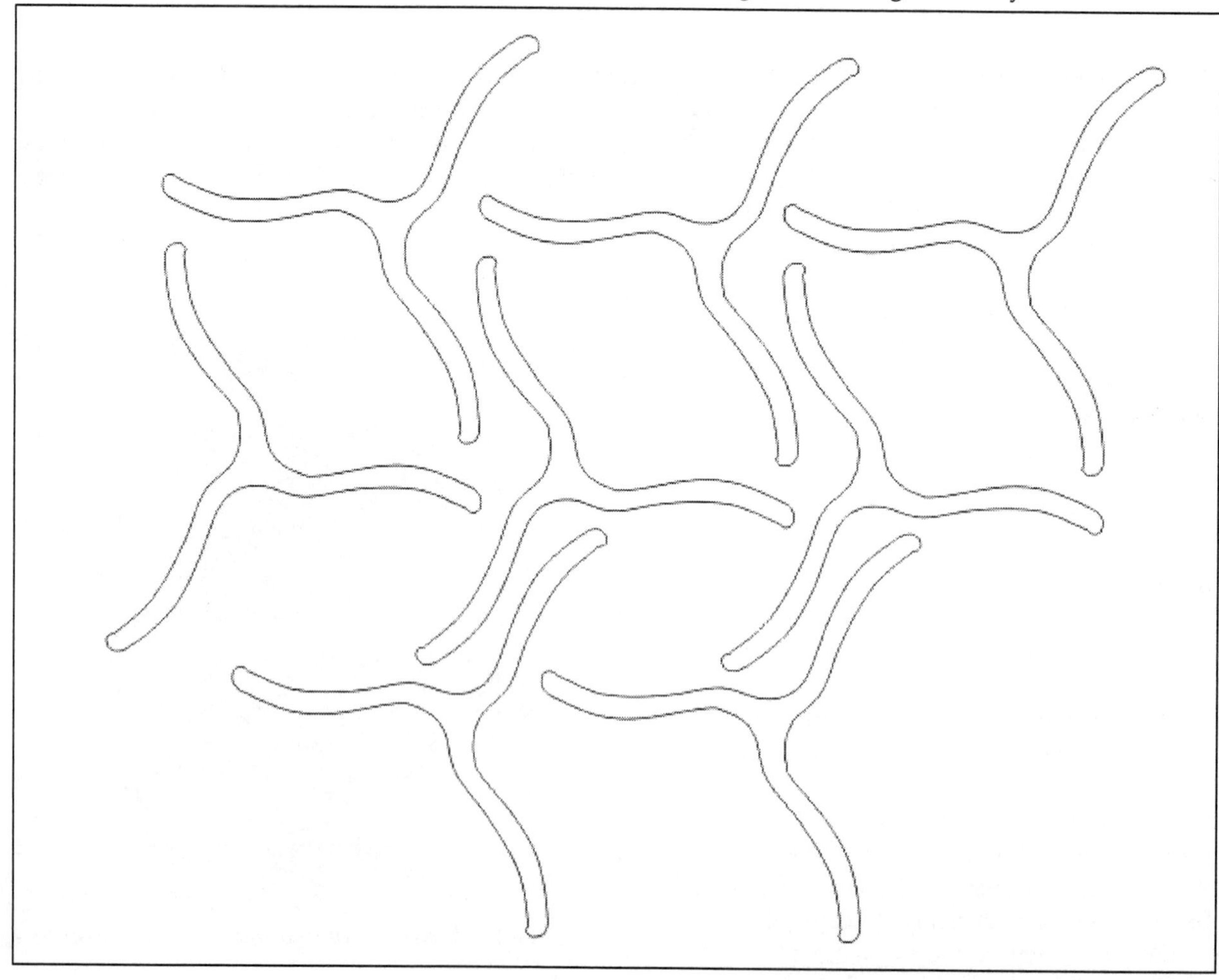

The Mechanical Drawing of Cycloids, The Geometric Chuck

Robert Craig
8633 E. 96th Terrace
Kansas City, MO 64134
E-mail rtcraig@yahoo.com

Abstract

This paper discusses cycloids and their construction using the 19th century mechanical drawing instrument known as the Geometric Chuck. The first part of the paper is a brief history and description of the Geometric Chuck. The last part of the paper is devoted to a discussion the definition of cycloids and examples showing the results that various settings of the Geometric Chuck have on the cycloid patterns produced. This paper is an attempt, in part, to respond to the comment in the Savory book "As this book does not aim at giving a scientific account of the principles on which it works. It might be an exceedingly interesting subject for the scientific person, ...the scientific knowledge required to understand a three-part chuck would be so great that I doubt if there is the person existing who could describe the course of a line that would be produced...."[6]

1. History

1.1 Definition. "The Geometric Chuck is an arrangement of mechanism for producing two or more circular movements in parallel planes. The combination of these movements with different velocity ratios and different radii results in the formation of a great variety of highly interesting curves and geometric figures"[1]. The Geometric Chuck is just one of many devices that were designed to be used as accessories to the Rose Engine Lathe. This lathe, similar to the modern lathe in name only, was designed to produce ornately decorated items. The craft of producing such items on the Rose Engine Lathe is known as Ornamental Turning. A foot treadle was used to power the lathe but some larger workshops used steam power.

There are two main differences between the Rose Engine Lathe and the modern lathe. First, the rate of rotation of the Rose Engine Lathe was very slow. With a Geometric Chuck attached the rotation would not be much more than 100 revolutions of the drive spindle per minute. This slow rate of rotation was required by the fact that the mass of the Geometric Chuck is not centered on the drive chuck and a fast rotation would destroy the out of balance Geometric Chuck. Second, in the modern lathe the work piece is turned and the tool is held relatively stationary in a tool rest and applied to the spinning material. The work piece in the Rose Engine Lathe may or may not be rotating when the tool is applied and often the tool is mounted in a device that moves the tool in a prescribed manner to produce a design. These devices are driven by another set of belts and pulleys off of the foot treadle.

1.2 Development. The Geometric Chuck, as it is known today, is the result of the designs of a number of inventors. They are generally known by the inventors' names, Plant's Chuck (Fig. 1), Ibbetson's Geometric Chuck, and Holtzapffel Geometric Chuck (Fig. 3,4). All of these devices are very similar having various improvements added to make the setup and execution of patterns easier and more consistent. According to Northcott, "Ibbetson himself says he derived the idea and the name from the geometric pen (Fig. 2) of Suardi, who published an account of it in 1752[1]. The *Manuel du tourneu* shows a device, "La Machine Epicycloie", that has many of the basic parts of the Geometric Chuck[2].

The Holtzapffel Chuck is arguably the most advanced in features and flexibility. There are a number of problems in maintaining consistent starting and stopping locations when scaling patterns and changing gear ratios that the Holtzapffel Chuck solved by various adjustment devices that were incorporated into the design.

Ibbetson attempted to keep the details of his Geometric Chuck a secret, believing the use of it for the engraving of spirals on bank notes, to prevent forgery, of utmost importance[1][3][4]. The Geometric Chuck was and still is used in this application, but Ibbetson's attempt at secrecy was a failure.

Figure 1 Plant's Compound Geometric Chuck[6]

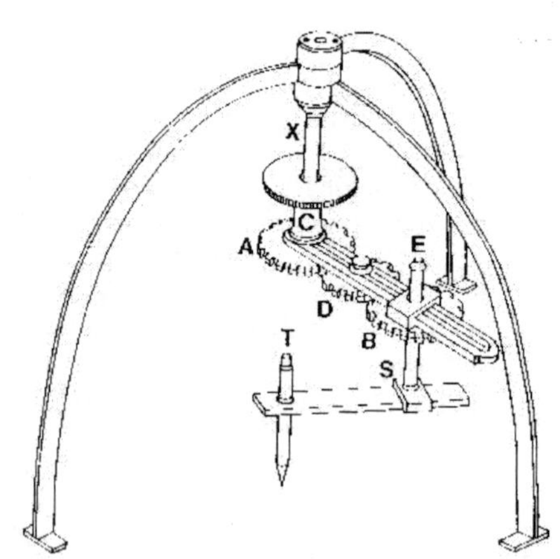

Figure 2 Suardi's Geometric Pen[11]

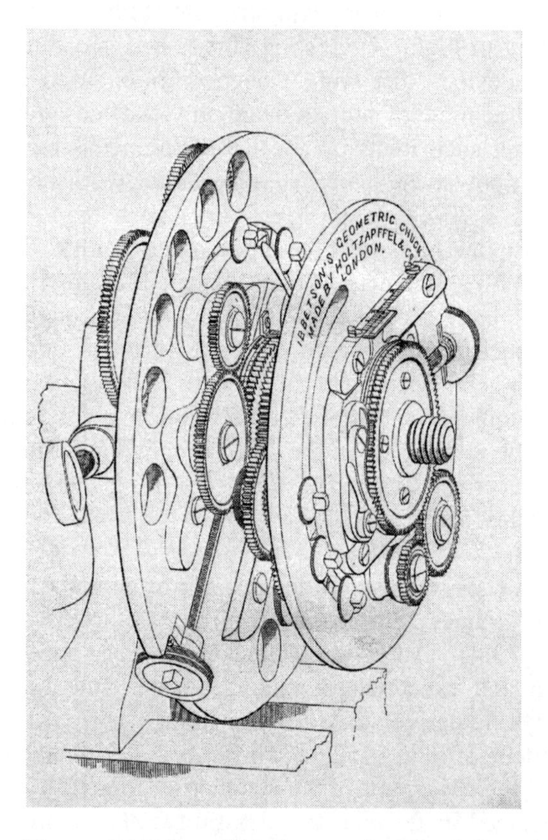

Figure 3 Holtzapffel Geometric Chuck 2 stages[5]

Figure 4 Holtzapffel Geometric Chuck 7 stages[12]

The number of Geometric Chucks produced was low and it was quite expensive due to the precision and handwork required to build such a complex mechanism. Its popularity was also limited because of the complexity of the device and amount of time and experimentation that was required to produce pleasing patterns. Thomas Bazley in *Index to the Geometric Chuck* states, "The impression of this work consists of 150 copies only, probably about equivalent to the number of persons who take an interest in this peculiar branch of amateur mechanism"[5]. This book contains detailed directions on the workings and settings of the chuck including 3500 patterns and the parameters used to produce them. The book *Geometric Turning* by Savory is another text that contains many examples and instructions[6]. This book is less scientific and more artistic in its approach than the Bazley. The author states, "If rules are meant to make a subject difficult and to frighten, they are very successful rules….." This comment is in response to the 39 rules by Mr. Perigal in the Northcott book[1]. Mr. Perigal makes these rules even more difficult to understand by using terminology that was unfamiliar to the average person and by the absence of any drawings, figures, or examples.

2. Description of the Geometric Chuck

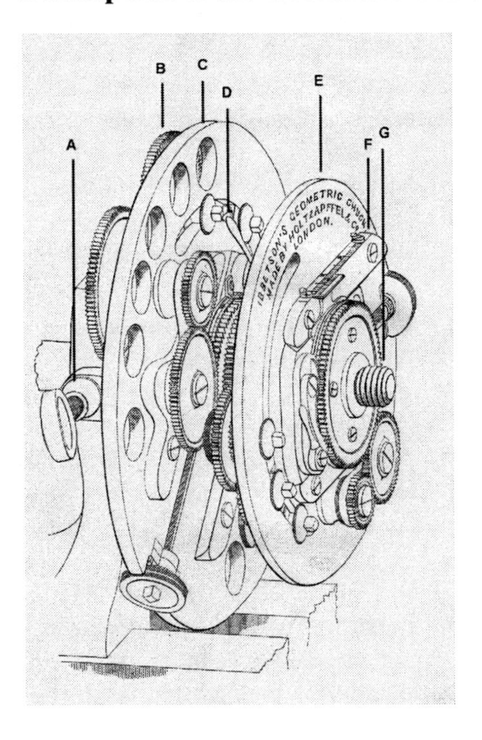

Figure 5 [5]

The figure above is a drawing of a two part Geometric Chuck. It consists of the following primary parts. A. Attachment to the head of the lathe. B. Train of wheels on arbors. C. Foundation plate of first stage. D. Spring detent. E. Foundation plate of second stage. F. Eccentric slide. G. Front terminal wheel and mandrel of stage two.

A brief description of the purpose of these parts follows:

A. The foundation plate of the first stage is screwed onto the mandrel of the lathe. The first motion wheel is clamped to the headstock of the lathe with the large thumbscrew. Without this attachment the whole mechanism would be free to rotate and no drive motion would be transmitted to the train of wheels.

B. The train of wheels consists of various sizes of wheels on arbors that are attached to the curved slot in the foundation plate. The slot allows space for the different sizes and numbers of wheels used in making the patterns. The motion of these wheels is transmitted through the foundation plate just below "D". The motion then is transmitted to the front terminal wheel by the two wheels on the front face of the foundation plate. They are mounted on a movable link whose purpose is to keep the wheels in contact with the front terminal wheel as it is moved on the eccentric slide (F, second stage). The train of wheels is what gives rotation to the first motion wheel of the next stage and determines the number, direction and spacing of cusps generated by this stage.

C. The foundation plate has a number of functions. The arbors of the wheels are attached to it, the eccentric slide is attached to the front face and the first motion and front terminal wheels are attached to it. The spring detent (D) has its base attached to the foundation plate.

D. The spring detent. There are two types of these. The one whose lever is just visible behind the foundation plate of the second stage is used to clamp the front terminal wheel of stage one to first motion wheel of stage two. By doing this the chain of wheels from stage one are connected to stage two. This detent also allows the train of wheels to be thrown "out of gear" which effectively stops the transmission of motion from the train of wheels from this stage on. The second type of detent is visible near the front terminal wheel of stage two. This detent allows the front terminal wheel to be locked so that it will not spin freely. This would be used when facing the surface of a plate for engraving.

E. The foundation plate of the second stage, similar in function to the foundation plate of the first stage.

F. Eccentric slide. There is an eccentric slide on each foundation plate. This slide is used to give the eccentric motion to each stage. It is adjusted by the large knob, shown at the back of stage two and at the bottom of stage one. The eccentric slide has a graduated index engraved on it, usually in 100ths of an inch. The position of the eccentric slide determines the size of the loops and cusps generated by stage.

G. Front terminal wheel and mandrel of stage two. This is where the next stage is attached or the piece of material to be engraved is mounted. Often a flat plate with paper is mounted to this and a pen or pencil held in the slide rest of the lathe is used to draw the pattern. This was done as a way to test the setup and function of the Geometric Chuck before the actual engraving was done.

The figures show the Geometric Chuck mounted horizontally, as it would be in a lathe. Compound Geometric Chucks such as the one shown in Fig 4 are often used in a vertical position. The reason for this is that the weight of the numerous stages becomes more than the head of the lathe can safely support. Another reason for this using this orientation is the fact that when the numerous stages are set at various eccentricities, the whole mechanism is thrown out of balance and rotation in the horizontal is unstable. Rotation of the Geometric Chuck in the vertical orientation is accomplished by attaching the first stage to a base, which usually contains some type of hand wheel mechanism. Since the Geometric Chuck is always driven at low revolutions per minute this is quite adequate.

3. Types of Cycloids

The common definitions of the epicycloid and hypocycloid are:
Epicycloid: The path traced out by a point P on the edge of a circle of radius b rolling on the outside of a circle of radius a [7]. (Fig. 6)
Hypocycloid: The curve produced by fixed point P on the circumference of a small circle of radius b rolling around the inside of a large circle of radius a [8]. (Fig 7)

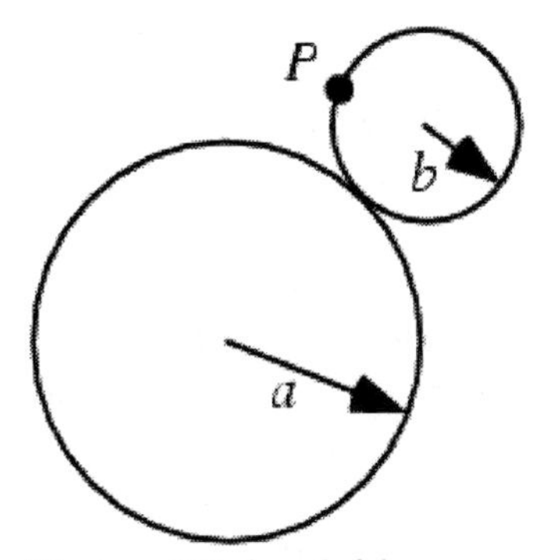

Figure 6 Epicycloid

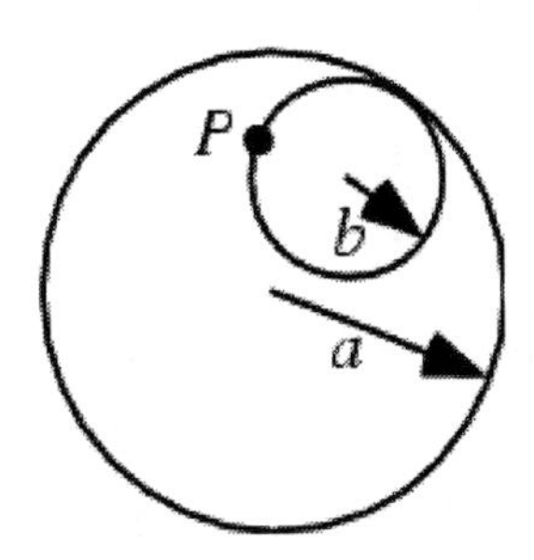

Figure 7 Hypocycloid

Two important points should be noted here.

First, *circumference* of the circle with radius b rolls without slipping inside or outside the *circumference* of circle of radius a. Thus, the rotational velocity of the circle with radius b is fixed by the relationship of radius b to radius a.

The specific relationship of rotational velocity to radius, while possible, is seldom maintained when creating patterns with the Geometric Chuck. The variety of patterns is limited if this relationship must be maintained. The actual physical size of the patterns seldom exceeded 4-5 inches and the smallest patterns were determined by the minimum line thickness a tool could inscribe or draw and still maintain clarity.

Second, for an epicycloid, the direction of rotation of the point P on the circumference of the circle of radius b is the same as the direction of rotation of the *center* of that circle.

Figure 8 shows the relationship of the stages of the Geometric Chuck as it is most often used.

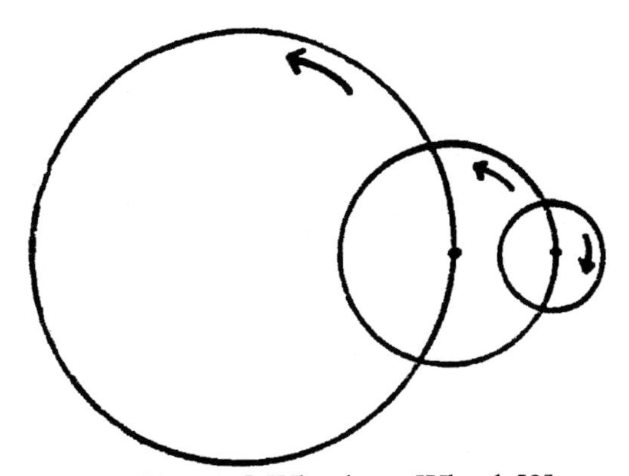

Figure 8 Wheels on Wheels[9]

Note that the *centers* of the wheels follow the *circumference* of the wheel to which they are attached. The direction of rotation of the wheel is *independent* of the direction that the center of that wheel is moving. Thus, the direction of rotation direction of a point on the circumference of a wheel may or may not be the same direction as the rotation of the center of that wheel.

These last two points are often confused at worst or unclearly stated at best. The cycloids generated the Geometric Chuck or other wheels on wheels devices are not governed by the same rules as those generated by the popular drawing toy the Spirograph® Hasbro Inc. [9][10]

The definition of the epicycloid as it relates to wheels on wheels and most settings of the Geometric Chuck is as follows:

Epicycloid: The path traced out by a point P on the circumference of a circle of radius b whose center is following the circumference of a circle of radius a.

On the Geometric Chuck point P is physically the center of the front terminal wheel and mandrel of a stage. The radius of the circle, which point P is on, is the relationship of the center of the stages first motion wheel and the center of the front terminal wheel and mandrel. The point P of the final stage in a compound Geometric Chuck is additionally determined by the relationship of the center of the front terminal wheel and mandrel of the last stage and the point of the cutting or drawing tool.

For the remainder of this paper the generation of cycloids using the common definitions located at the top of section 3 and with the restrictions that they imply will be referred to as "Type A" and cycloids generated by the definition above will be referred to as "Type B".

4. Effects of Changing Wheel Sizes

This section will show the effects of changing wheel sizes. For Type A cycloids both the number and position of cusps are changed. (Fig. 9) For Type B cycloids only the relationship of the cusps to one another is changed. (Fig. 10) The patterns in figures 9 and 10 were generated using a two-stage system for the sake of clarity. A multistage (compound) system would have the same characteristics. Part of the "art" of creating pleasing patterns using the Geometric Chuck is the selection of wheel sizes. The smaller the wheel size of a stage, relative to the other stages, the less impact on the pattern that stage will have. For complicated patterns using several stages the pattern can become very cluttered if too many stages have similar wheel sizes.

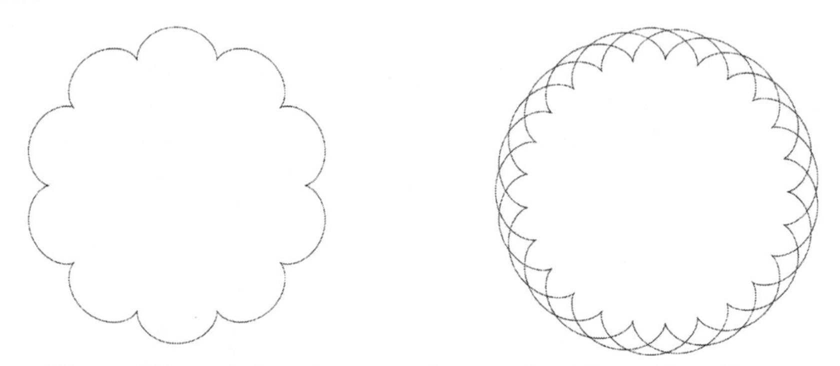

Figure 9 Type A changing second stage wheel from 10 to 12

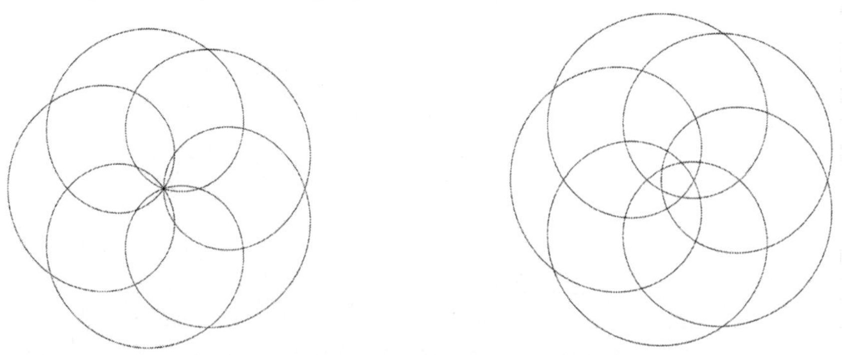

Figure 10 Type B changing second stage wheel from 10 to 12

5. Effects of Changing Wheel Rotation Direction

The changing of the direction of rotation of a stage causes inward-facing loops to face outward and vice versa. Figure 11 is an example of a Type B cycloid. The patterns in figure 11 were generated using a three-stage system, in this case not only did the orientation of the cusps change but also the symmetry of the loops is modified by the direction change. When using the Geometric Chuck, changing the wheel rotation direction is accomplished by adding an idler wheel in the chain of wheels of the stage that needs to be reversed. The idler wheel, being of the same size as the wheel that precedes or follows it in the chain, has no effect on the rotation rate of the stage. This aspect of the setup and usage of the Geometric Chuck often has a profound impact on the visual appeal of the patterns produced as seen in figure 11.

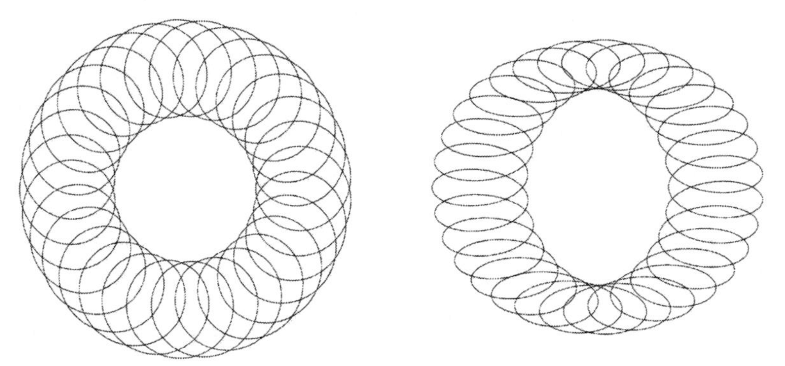

Figure 11 The rotation direction of stage three is reversed

6. Ratios, Cusps Loops and Symmetry

The subject of the relationships of rotational rates, wheel sizes and symmetry is well documented so it will not be covered in great detail here. References [1][5][9][10][11] all contain sections related to this. Rotation direction and the relationship of the wheel sizes of the stages determine Type A cycloid symmetry. The wheel size determines the rotation rate of the stage, which in turn controls the number of cusps or loops and how densely they are drawn. With Type B cycloids the number of cusps, loops and symmetry are determined by the relationship of the rates of rotation and direction of rotation of the different stages of the Geometric Chuck. Type B cycloid symmetry of compartments (the empty space between lines) can be controlled by the size of wheels that are chosen. As stated earlier if several stages have wheels of similar sizes the pattern often becomes very cluttered.

7. Asymmetrical Patterns

The majority of the patterns that were created in the 18th century were based on symmetric patterns. This was probably due to the tastes of the day. Considering the amount of time required to setup and generate patterns using the Geometric Chuck experimentation was not that easy to do. When more stages are added to the Geometric Chuck asymmetric patterns seem to be easier to find. Using computer software, experimentation and adding stages is quite simple. Most 18th century Geometric Chucks only had two or three stages.

Figures 12 and 13 are two patterns that are examples of asymmetry. Figure 12 is a Type A cycloid generated using the digits of the dates of Bridges 2006 for the wheel sizes, (84826, excluding the zeros). This is a five-stage pattern and all wheels are rotating in the same direction with the exception of the first one. Figure 13 is a five-stage Type B pattern. The wheel sizes are 100,11,35,48,70. The rotation rates are 22,66,42,22 and the fourth and fifth stages are rotating in the opposite direction of the others.

Figure 12

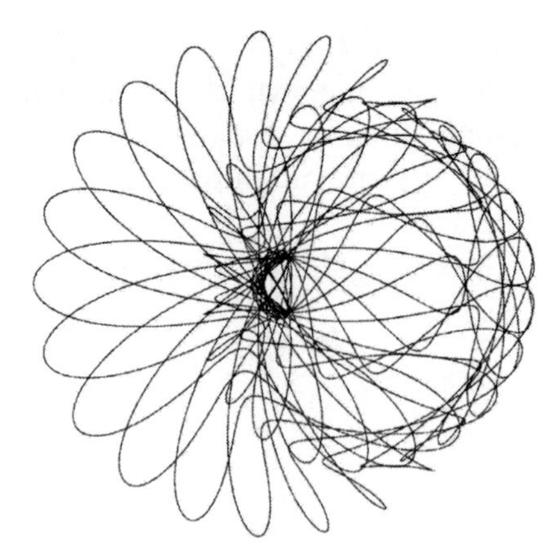

Figure 13

8. The Art of Using the Geometric Chuck

The art of the using Geometric Chuck has two components. The first is knowing how to set up the various variables of the device, wheel chains and wheel size (eccentric slides) to obtain the desired patterns. The second is the combining and scaling of patterns together on one engraving or drawing.

9. Conclusions

The Geometric Chuck and the cycloid patterns that it is capable of generating are an interesting combination of mechanics, mathematics and art. The relationship of direction of wheel rotation and rotational velocity of multistage Geometric Chucks, especially those of 4 or more stages is an area for more study. I would like to thank Craig Kaplan for math, programming and graphics advice and John Sharp for his assistance in the locating of reference materials.

References

[1] Northcott. W. Henry, *A Treatise on Lathes & Turning,* Longmans Green & Co., 1868, pp. 270-288

[2] Bergeron, L. -E. ,*Manuel du tourneu 2nd ed. rev., cor., et considérable augm.,* Paris, 1816, Plate XXXIX

[3] "Making Money," *Harper's Monthly Magazine* No. CXLI, Vol. XXVI February, 1862, pp. 310-313

[4] http://www.citiesofscience.co.uk/go/London/ContentPlace_2854.html

[5] Bazley, Thomas S., *Index to the Geometric Chuck,* Waterlow and Sons, 1875, pp. v

[6] Savory, H.S.,*Geometric Turning:Comprising a Description of the New Geometric Chuck,* Longmans, Green & Co, 1873, pp. 4-5

[7] EricW.Weisstein."Epicycloid."From MathWorld--A Wolfram Web Resource http://mathworld.wolfram.com/Epicycloid.html

[8] EricW.Weisstein."Hypocycloid."From MathWorld--A Wolfram Web Resource http://mathworld.wolfram.com/Hypocycloid.html

[9] Farris, Frank A., "Wheels on Wheels on Wheels-Suprising Symmetry," *Mathematics Magazine* Vol.69, No. 3, (June 1996) pp. 185-189

[10] Giblin, Peter and Trout, Matthew, "Cusps on Wheels on Wheels on Wheels," *Mathematics Magazine* Vol. 71, No. 4, (October 1998) pp. 309-313

[11] Edwards, Ross, *Microcomputer Art,* Prentice-Hall of Australia Pty Ltd, pp. 33,65,67

[12] Reed, AV and Tweddle,Norman, "Great Grandpa Did Contour Turning," *American Machinist,* November 10,1952, pp. 119-121

Sashiko: the Stitched Geometry of Rural Japan

Barbara Setsu Pickett
Department of Art
5232 University of Oregon
Eugene, OR 97403-5232, USA
email: bpickett@uoregon.edu

Abstract

Shashiko comes, not from the imperial courts, but from the humble origins of rural Japan. This textile tradition requires only needle, thread and countless hours of patient stitching. No fancy machinery or clever devices are used. It is just cloth, single or layered, held together by running stitches. The results are beautiful: geometric patterns interlock with precision and grace, stunning tessellations emerge. Some of the traditional patterns are easy to decipher but others are less obvious. This paper will examine how these patterns are drawn on the cloth and what design principles the stitcher uses to guide the needle.

1. Background

1.1 Origins of the technique. Peasant farmers and fisherman knew how to be frugal. Cloth and thread were precious in pre-modern Japan and garments were carefully patched and reinforced to extend their useful life. At first repairs were done with the same fiber dyed the same color. Efforts were made to camouflage the patch and make it blend in. Often the damaged cloth was of indigenous hemp or other local bast fiber dyed deep blue with indigo. Winters in Japan can be severe, especially in the North. To better fend off the chill and piercing wind, peasants donned layers of clothing and quilted clothing.

1.2 White, cotton thread. Cotton arrived in Japan in the 1300s and 1400s: it was an exotic, expensive foreign good, and only the aristocracy had sufficient wealth to make an ostentatious display of it. In the 1700s it became more prevalent but still was considered a status thread. By the 1800s white cotton thread became more commonplace and gained great popularity in the countryside. Instead of trying to hide the stitch, sewers in the 18th and 19th century drew attention to them. They made nets of fine white stitches arranged in regular overall geometric pattern floating on seas of midnight blue. The contrast and the aesthetic of white on blue or blue on white became a folk tradition. Many of the traditional geometric patterns refer to and are stylizations of the natural world. They have names like pine bark, mist, blowing pampas grass and tortoise shell. Like characters in Kanji, they are echoes of abstracted pictographs, reduced to the bare essentials. The steady, predictable rhythms make a quiet beauty and serenity.

2. Patterns Made with Overlaying Sets of Parallel Lines

1.1 Types of grids. A few of the patterns are made from simply parallel lines. The lines bend or turn sharp corners but never intersect. Their spacing adds to the rhythms. There is a snap-to-the-grid adherence. The grid can be squares, or equilateral triangles of isometric graph paper. The line of running stitches follows and matches the grid. Quilters fondly refer to this as 'stitch in the ditch'. Patterns like mist and mountain are formed from these parallel lines.

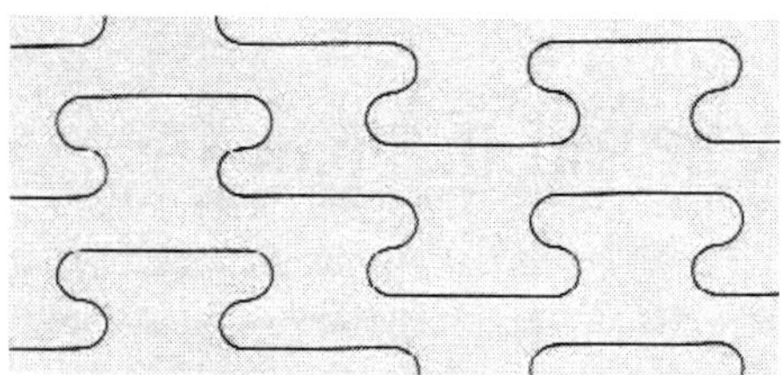
Figure 1: mist (kasumi tsunagi)

Figure 2: mountains (yama sashi)

1.2 Intersecting sets of parallel lines. Patterns like Straight Feather, Basketweave, Fish Scales, Pine Bark, and Woven Bamboo overlays grids of parallel lines. Again the layout grid is obvious. The stitcher marks the fabric with guidelines. She then sews sets of parallel lines. The pattern emerges when the eye sees the intersections. Quilters fondly refer to this as 'stitch in the ditch'.

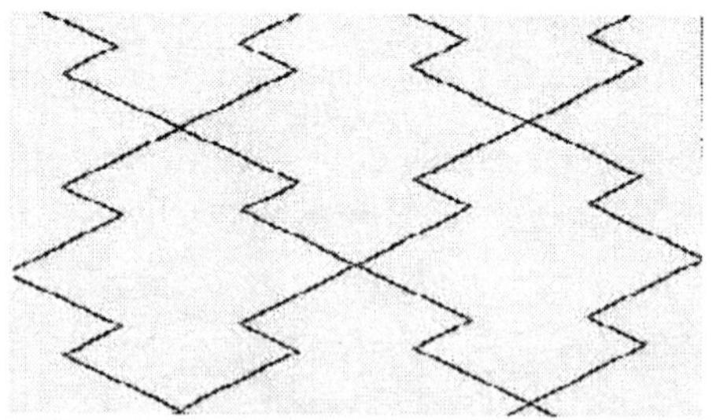
Figure 3: pine bark (matsukawa bishi)

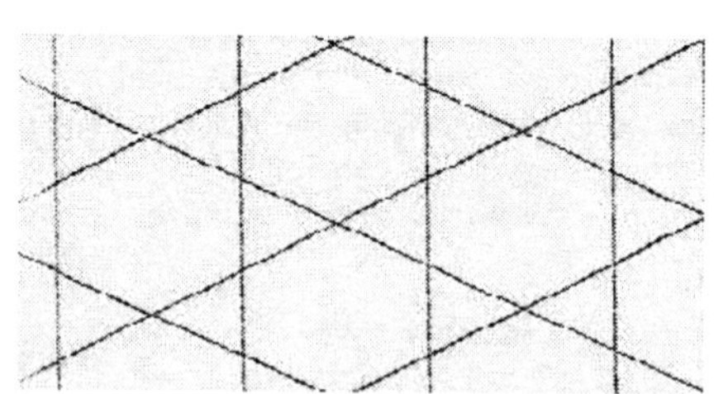
Figure 4: woven bamboo (kagome)

1.3 Motifs that conform to either grid. Sometimes a motif can be drawn in two ways, locked to the square or isometric grid. The convoluting lines of running stitches still follow to the ditch, but the grid itself becomes a series of pivot points. The overall results may look like interlocking motifs, but it is achieved by stitching maze-like lines that jig and jog but eventually reach the opposite side. The Silk Weave and the Cypress Fence are based on both grids.

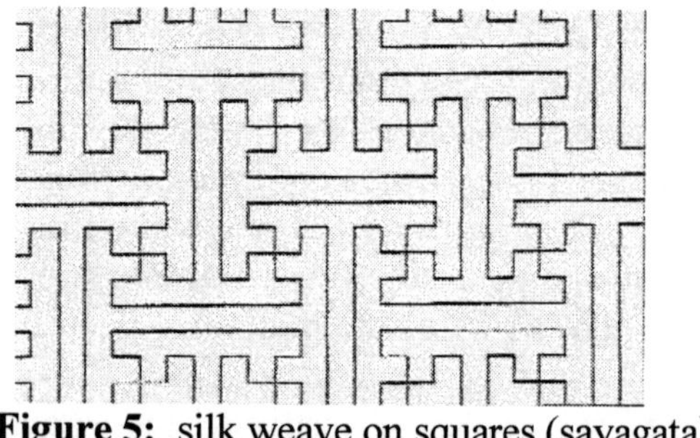
Figure 5: silk weave on squares (sayagata)

Figure 6: silk weave on triangular grid

3. Patterns Made by Grid Lines, Connecting Lines, and Pivot

3.1 Hydrangea motif. The hydrangea motif is based on the square grid. A network of carefully drawn crosses, evenly spaced with line segments of equal length is the foundation. These lines are the grid lines. From the center of the crosses, short, connecting lines are sewn to make the diamond shapes.

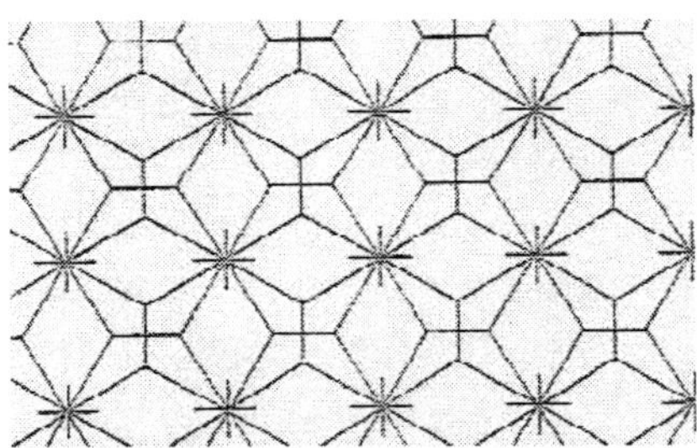
Figure 7: hydrangea (ajisai)

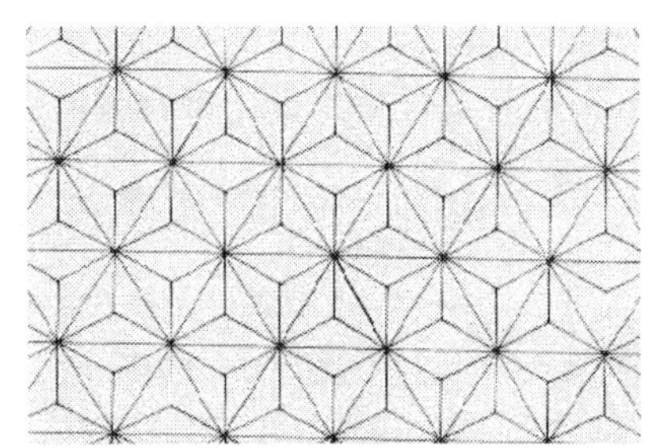
Figure 8: hemp leaf (asa no ha)

3.2 Hemp leaf motif. The hemp leaf motif may look similar to the hydrangea but there is a fundamental difference. It stems from the equilateral triangle. Here the triangular grid rules. Rows and rows of isosceles triangles make the foundation. Intermediary connecting lines go from the tips of the angles to its center. In sewing this pattern one stitches series upon series of straight parallel lines.

4. Patterns that Use Hidden Elements

4.1 Salient features. This group of patterns I find especially intriguing and satisfying. They have the grid lines, the short connecting line segments, the pivot points, but in addition they employ stealth. The lines disappear and reappear. Like a diving whale, it is difficult to discern the course. However, if one examines the base grid carefully and see where the lines fall, the underlying relationships and design strategies become apparent.

4.2 Counterweights. For this design motif one draws rows of circles tangent to one another on a square grid. Each circle would have a four-square diameter and a two-square radius. Next a second set of circles is drawn as an overlay with the same diameter/radius. The centers of these circles are in the exact middle of the space left void in the first set. If all the lines are sewn the design motif called Seven Treasures (shippo tsunagi) emerges. But for the counterweight design some arcs are eliminated. In the first set of circles only the NE and SW quadrants are stitched and in the second set, only the NW and SE quadrant get thread. When stitching the two sets are combined and the needle makes an undulating line diagonally across the cloth.

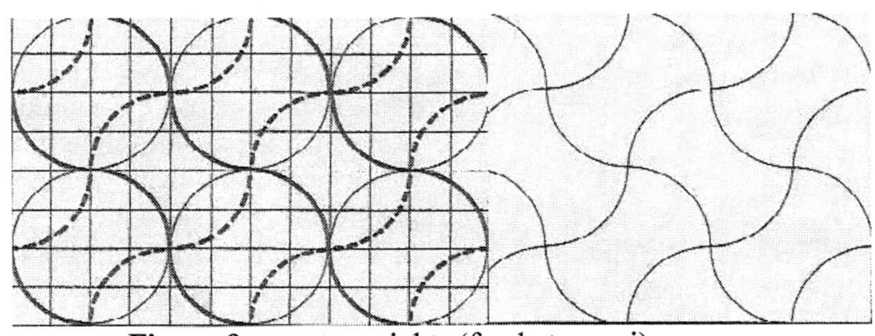
Figure 9: counterweights (fundo tsunagi)

Figure 10: 7 treasures (shippo)

4.3 Plovers The plover or flock of 1000 birds depends on similar design strategies. Again the base is the square grid. The individual motif is six squares wide by six squares tall. First one establishes a baseline and draws a series of semicircles above and below that line using a radius of 1 square. A sashay of arcs emerges that are 1 square high by 2 squares wide. For the second baseline, one leaves three rows of squares free below the first baseline and uses the fourth line for the pivot point of the second row of arcs. The second undulating line should be parallel to the first one. This process is repeated until the entire design plane is complete. Next the cloth is turned 90 degrees and the second set of wavy lines with the same proportions are precisely overlaid upon the first. The tips match up perfectly. In stitching the needle follows the wavy lines across the fabric, selvage to selvage or down the bolt of material. Each individual motif has a four-block square free in the middle.

Figure 11: plovers (chidori tsunagi)

Figure 12: connecting circles

4.4 Connecting semicircles. This design is drawn with a base set of semicircles that is overlaid by a second set of smaller arcs. A series of semicircles are drawn above and below a baseline using a radius of 2 squares. The second baseline is 8 lines below the first. The second wavy line is parallel to the first. One fills up the design plane. Going back to the initial position one drawn a second set of wavy lines to mirror the first set. If calculations were accurate, this second set fits snuggly into the space left by the first set. One then turns the cloth 90 degrees and does the same operations. Now with the compass set with a .5 square radius, one draws smaller arcs subdividing the main arc. The resulting motif looks like a stylized bat to me.

5. Conclusion

There are many more patterns to discover and explore. In the Japanese textile tradition of sashiko stitching simple geometry elevates the plying of needle and thread to a celebrated, widely admired folk art. The simple task of mending took on a new meaning and importance. The cloth's utility was extended and those white on blue fabrics beautified domestic life.

References

[1] Matsunaga, Karen Kim. Japanese Country Quilting. Tokyo: Kodansha International, 1990
[2] Parker, Mary. Sashiko. New York: Larks Books,1999.

Literatronic: Use of Hamiltonian Cycles to Produce Adaptivity in Literary Hypertext

Juan B. Gutierrez
Department of Mathematics
Florida State University
208 Love Building Tallahassee
Tallahassee, FL, 32308, USA
E-mail: jgutierr@math.fsu.edu

Abstract

Literatronic is an adaptive hypermedia system for hypertext fiction. Its adaptive features are based on an algorithm that simulates a Hamiltonian cycle on a weighted graph. The algorithm maximizes narrative continuity and minimizes the probability of loosing a reader's attention. The metric for this optimization is defined as the minimization of hypertextual friction and hypertextual attraction. We consider the challenges involved with modeling such hypertext, and we offer specific examples of this type of adaptivity.

1. Introduction

Literary hypertext is a genre of electronic literature. Its fundamental attributes have been mostly faithful to the origin of electronic text: a set of linked episodes that contain hypermedia elements. The reader chooses links to move from one page to the next, building in this fashion a story from a large set of possible stories.

Whether some features of literary hypertext could be reproduced in printed media or not has been subject of debate by opponents and proponents of digital narratives. However, as the electronic media evolves, some traits truly unique to literary hypertext have appeared. Significant effort has been invested in creating hypertexts responsive to reader's actions by using conditional links; additionally there have been efforts to create systems capable of producing fiction, with varying degrees of success. Both of these approaches have in common that they grant greater autonomy to the computer, thus making of it an active part of the literary exchange. These changes have not only redefined the concepts of reader and author, but also they have created a new element in the reading and authoring processes: the "digital author", which is a media element that acts on the message. The "digital author" positions itself between the reader and the author to perform a task that did not exist in printed media: the adaptive optimization of the text according to rules set by the author and the interaction of the reader.

In order to analyze a work of fiction, specially when it is a literary hypertext, it is convenient to differentiate between the story that is told as a linear sequence of temporally episode-based moments and the way that story is told, not necessarily as a linear sequence; i.e. *fabula* and *plot*. Sometimes the distinction is difficult because of phenomena such as the explanation of events that have happened in the past but have not yet been identified (*analepsis*) and the anticipation of something still to occur (*prolepsis*) [9, 5]. Given a set of pages in a hyperfiction, the plot corresponds to the sequence of pages the reader selects among many options. The fabula remains unchanged; however, the perception of it might vary depending upon the selected plot.

In this paper, we explore a mathematical model that transforms the computer in some sort of "digital author" who actively participates in the construction of the plot. The assumptions at the base of this work are: (i) the interface should focus on rendering the fabula in a way that it makes the most sense to readers according to some logic defined by authors; we can think of it as narrative continuity, regardless of the temporal or spatial continuity of events, i.e. a linear plot with variation in analepsis and prolepsis at author's discretion, (ii) readers want to explore the entire literary space, and (iii) the computer should assume the burden of assembling the pieces rather than giving them to the reader, i.e. it should become a digital author.

This paper is organized as follows: Section 2 gives references and establishes the background for this work. Section 3 describes the mathematical model and how it was implemented. Section 4 describes some results obtained with this model. Section 5 discusses some ideas presented in this paper and describes future directions.

2. Background

Literary hypertext, primarily an aesthetic form, differs from other forms of hypertext such as scholarly communication, learning systems and other types of hypertext systems in many aspects. For instance, it is not necessary to include an abstract in a work of fiction, require the existence of hyperlinks in keywords, or expect the user to complete a successful training cycle. It has been proposed and widely accepted that: (i) reader's choice, intervention and empowerment are the key elements of literary hypertext[1] [15], (ii) hypertext reading fosters both passive and active reading, where links provide decision points [16] and (iii) the suggestive power of literary hypertext lies in the lyric quality of links; lyric because of its particular intensity when searching for meaning, similar to the way we read poetry [17].

Since literary hypertext seems to be qualitatively different to other forms of hypertext, the natural question we must ask is: What are the characteristics of literary hypertext? There are two aspects of this question:

First, speaking of hypertext in general, there are several well known problems with interfaces [10]: (i) Users often do not know how to get to desired pages or how to return to previously-visited pages, (ii) they can become frustrated when they keep "rediscovering" the same page over and over, (iii) at some point the number of links may overwhelm the reader, and (iv) an author's reasons for including specific links may not always be clear to readers. In addition to the intrinsic difficulty of literary hypertext, readers are expected to be [5]: (i) tolerant enough with the up to this point experimental nature of literary hypertext, (ii) skilled enough to cope with different reading devices, and (iii) open enough to accept the non-linear result.

Secondly, there is experimental evidence reported by Gee [8] that refutes the premises described above regarding literary hypertext. According to the experiment: (i) multilinearity causes disorientation and results in readers skimming rather than reading; (ii) readers want a single starting point, (iii) they prefer narrative structures more or less linear with moderate branching, and (iv) they do not seem to be clamoring to be co-authors or to be empowered. The work by Gee has special significance. It seems to be the only factual study about readers of literary hypertext. Since most of the systems used to compose literary hypertext work with client-side technology, and those that work with server-side technology does not seem to track user interaction, there is no way to consolidate user's behavior beyond anecdotal and subjective information. However, these results might be confounded with the fact that only one tool was used, and generalization to other systems of reading might not be valid.

---

[1] That is an early 90s utopian assumption that has been challenged many times. It basically points out that links can give readers more control than printed text does.

Significant work has been done by proponents of adaptive literary hypertext. Mostly, it can be divided into two types of systems [6]: (i) Conditionality-based systems: certain rules are triggered when specific conditions are met. Examples of conditional systems are Storyspace [1] and Connection Muse [14]; some variations around this theme are what Bernstein [2] calls calligraphic vs. sculptural hypertext, that is, hypertext built by addition of links between episodes, or by the removal of them. (ii) Adaptivity -based systems: the choice of what conditions to meet depends on reader's interaction. An example of an adaptive system, though not literary-oriented, is AHA! [7].

Another way to achieve a dynamical response of text, though not necessarily adaptive, is algorithmic literature. Algorithmic generation of text driven by rules and adaptive generation has been explored by Bootz in poetry [4]. A similar goal is pursued by StoryEngine, a client constructed on top of Auld Linky structural link server [2]. The most classic attempts to use mathematics as a tool to model narrative with mathematics is Queneaus's *Cent Mille Milliards de Poèmes.*

Little, however, has been said about the computer as an author. We must make a careful distinction about exactly what part the computer authors. It is possible, in principle, to build a device that creates narratives automatically (dialogs, actions, etc.) in such a way that responds to user interaction [3]. However, in practice, this type of algorithmic literature has been more focused in interactive fiction[2]. One of the most recent examples is Façade[3].

What we call the digital author is the role of the computer when it builds links between episodes in such a way that corresponds to user interaction. While the concept of adaptivity is not new, the novelty in the approach described in this paper is that adaptivity is defined for the literary case, instead of using existing frameworks for adaptivity that were not created for the literary problem. As we will show in this paper, we develop concepts such as hypertextual friction, hypertextual attraction, and we use the well-known concept of Hamiltonian cycles, which do not have a counterpart in existing adaptive hypermedia systems.

3. Mathematical Model

In this section we deal with the formal representation of literary hypertexts. A *graph* consists of two finite sets: a set V of points, called *vertices*, and a set E of connecting lines, called *edges*, such that each edge connects two vertices, called the endpoints of the edge. In a graph $G = (V,E)$ we can walk from a vertex v_l along some edges to some other vertex v_k. The vertices belonging to an edge are called the ends, endpoints, or end vertices of the edge. An edge $e = (v_i, v_j)$ is considered to be directed from v_i to v_j ; v_j is called the head and v_i is called the tail of the edge. A graph with directed edges is called a *directed graph*; otherwise, it is called undirected. Any sequence of adjacent edges is known as a *walk.* A *trail* is a walk in which a vertex may occur at most once. A *path* is a specific kind of trail that does not visit any vertex more than once, except that its first and last vertices may be the same, in which case we call this a *cycle.* If we choose a path that passes exactly once through each vertex of an undirected graph G, we call this a *Hamiltonian path.* If the Hamiltonian path finishes in the initial vertex of an undirected graph G, thus allowing repetition of only one vertex, we call this a *Hamiltonian cycle.*

The graph representation of a hypertext places pages on the vertices. The edges are the hyperlinks. The goal is to find a reading path that visits every page exactly once, i.e. a Hamiltonian path. Tradition in literary hypertext does not require visiting every page that belongs to the hyperfiction; on the contrary,

[2] Interactive fiction refers to software containing simulated environments in which players use text commands to control characters.

[3] Façade, a one-act interactive drama. http://www.interactivestory.net/

dead-ends and conditionally-accessible pages became the norm in earlier hypertexts. However, this was influenced by the tools used to produce hypertext rather than by a conscious effort to produce an aesthetic effect. My position as a fiction author is that every word should be carefully crafted and deserves a place within the text. If a sentence is not needed, it should be eliminated. Likewise, if a sentence is missing, a vital part of the text is missing.

After reading all available pages, the last page of the hypertext could display the laconic sentence "The End" and dismiss the reader. But the reading of a hypertext does not necessarily ends in the last page, as printed text does. As we will show below, it is possible to create edges dynamically that simulate a Hamiltonian path, effectively creating many possible readings. Therefore, after the Hamiltonian path has been completed, the reader can be taken back to the initial page, completing in this way a Hamiltonian cycle. This serves two purposes: (i) it becomes an indication to the reader that the entire text was read, and (ii) it emphasizes the fact that there are many readings of the same text. Closing the path is not necessary, only convenient.

The reading along a path makes sense if and only if the author creates directed edges between pages that have narrative continuity. It is obvious that while the sequence of text v_j, v_k defined by the author makes sense, the sequence v_k, v_j does not necessarily have to make sense. Thus, we can enounce more precisely the problem we are trying to solve: we want to find a directed Hamiltonian cycle in the hypertext. We must emphasize the fact that "narrative continuity" is a subjective measure set by the author; it means that if an edge exists between the vertices v_j and v_k it should be possible to aggregate sequentially the text and have a seamless reading according to the author's plan.

There are two technical complications with the model we have described: (i) given a page v_i, and some other pages to which it is connected, it will occur most of the times that some connections make more sense than others. We need some type of measure that describes the narrative continuity between pages, and (ii) a Hamiltonian cycle does not always exist for a given graph. Even if the Hamiltonian cycle existed, and the problem could be solved, nothing would guarantee that the mathematically optimal solution would make sense from the point of view of the narrative. A solution to each one of these problems will be explained in the following two subsections.

3.1. A Measure of Narrative Distance. Let us assume for the moment that page repetition is allowed. We will refer from now on to vertices as pages. The problem regarding narrative continuity is easily solved if we introduce an additional element in the graph: the narrative cost between pages. We will call it *hypertextual cost.* It is a positive real number proportional to how disruptive the transition from page v_i to page v_j is. It is necessary to define only the hypertextual cost between pages that are directly connected; the hypertextual cost between pages that are not connected is assumed to be infinite. The graph $G = (V,E)$ along with its associated costs is called a weighted graph. Since small costs maximize narrative continuity, the shortest paths between every pair of pages v_i and v_j in the graph will render the "best" plot for the fabula that starts in v_i and finishes in v_j. It is possible, in principle, to calculate this distance for each pair of pages. This is the well-known problem of finding multiterminal network flows.

Minimum paths do not necessarily solve the Hamiltonian cycle problem. In fact, left unchecked, reading a hypertext following minimum paths would result in repeated pages most of the cases. Some authors of classic hypertext used precisely repetition as narrative devices in their works. In fact, many critics and practitioners see in it an inescapable consequence of the essence of literary hypertext. However, arguably repetition has a negative impact in readers' attention thus encouraging readers' desertion.

We will call hypertextual attractors to those pages that disrupt the narrative continuity by repetition. We will call hypertextual attraction to the metric used to define a hypertextual attractor. Attractors pose a serious problem to authors because normally they wear down a reader's interest. Let c_{ij} represent the hypertextual cost between the origin page v_i and the destination page v_j. Let e_i represent the number of edges connected to page v_i Let V represent the total number of pages. Let A_i be the hypertextual attraction of page v_i. Then,

$$A_i = \frac{1}{\sum_{m=1}^{V} \sum_{n=1}^{V} c_{mn}} \sum_{j=1}^{e_i} \frac{c_{ij}}{2}. \tag{1}$$

The numerator in equation 1 is the sum of costs of all connections to the page i, while the denominator is the total sum of hypertextual costs. Small values of hypertextual attraction mean that the cost of including this page in a reading path is low, therefore the page v_i is very likely to appear in many reading paths. The measure of attraction is problematic if, for example, one of the costs c_{ij} is much higher than the others; in this case, the total measure of attraction could seem appropriate, but in fact the page could be very attractive by the action of many small costs. Therefore, attraction must be used in conjunction with page stiffness, which is the ratio of the highest distance to the smallest one among those edges connected to a single page.

We will call hypertextual friction between two pages to the probability of loosing readers' attention during a navigation event that goes from the origin page v_i to the destination page v_j; we assume it is directly proportional to the hypertextual cost between the two pages, directly proportional to the hypertextual attraction of page v_j, and inversely proportional to the total number of pages read R. The latter means that narrative costs are compensated by reader's increased knowledge. High measures of friction indicate a high probability of reader's desertion. Let D_{ij} represent the hypertextual friction. Let c_{ij} represent the hypertextual cost between the origin page v_i and the destination page v_j. Let A_i be the hypertextual attraction of page v_i as defined in equation 1. Then,

$$D_{ij} = \frac{c_{ij} A_j}{R}. \tag{2}$$

For authors, the problem of maintaining a manageable volume of narrative, and at the same time minimizing the effect of hypertextual attractors and hypertextual friction as defined in equations 1 and 2, poses an enormous practical challenge.

3.2. The Simulation of a Hamiltonian Cycle. Finding a Hamiltonian cycle in an undirected graph is a problem NP-complete. Finding it in a directed graph is even more difficult if not impossible in most cases. Therefore, we will use the special characteristics of the problem at hand to produce an approximate solution. The most important feature of literary hypertext is that if there is no direct connection between an origin page v_i and a destination page v_j, but the reader has read the intermediate pages needed to traverse the path from v_i to v_j, then a virtual edge between v_i and v_j exists. This is possible because the reader already has the intermediate information needed to fill the narrative gap between those two pages.

Here is where the computer comes into play as a digital author. Let us assume that the reader observes in each page a number of hyperlinks that allows visiting other pages. For simplicity, let us limit the number of links to three. Since it is possible to find the solution of the multiterminal network flow, the narrative cost and the shortest path between each pair of pages is known. We can assume without loss of generality that links shown to the reader are sorted by narrative cost: the first option corresponds to the page connected by the lowest narrative distance; the second option corresponds to the next bigger cost, and so on. After reading a page, the reader will know that: (i) reading is possible through several paths, and (ii) some paths offer greater narrative continuity than others. When the user selects a destination page, there are shown in the new page three links pointing to the closest pages, which do not contain pages already visited. It is in this sense that the system exhibits adaptation. The effect is that the hypertext is read through a Hamiltonian cycle, and the reading is sorted according to shortest narratives distances. A reading process that follows the previous algorithm is guaranteed to be optimum from the point of view of the author. Note that even though in practice the reader reads a hypertext that corresponds to a Hamiltonian cycle, such mathematical solution has not been found.

4. Methods and Results

The system Literatronic was implemented on the web following the algorithms described above. It is available at http://www.literatronic.com. It was developed with server-side technology, meaning that all readers' interactions are processed and stored in a central database. It has been possible for authors to mine user interaction data in order to find readers' patterns, points of desertion, attractive pages, etc.

The costs c_{ij} between the origin page v_i and the destination page v_j were arbitrarily chosen. The books in Literatronic were initially designed with five values of hypertextual cost: 5 for immediate connection, 10 for near connections, 15 for medium connections, 20 for far connections and 30 for improbable connections. Statistical analysis of hypertextual friction, hypertextual attractors and user interaction of approximately 4000 anonymous users and approximately 200 registered users, showed that a distance of 30 has little effect in the reading, and a distance of 5 behaves almost exactly as a distance of 10. Currently only three hypertextual costs are being used: $c_{ij} = \{10, 15, 20\}$.

The multiterminal network flow was solved using the Gomory-Hu algorithm [11] because of its simplicity at development time, and because it not only finds the minimum costs between every pair of pages, but also the minimal paths. It is worth noticing that calculations performed by the Gomory-Hu algorithm have to be done when the author decides to test the hypertext. It is analogous to "compiling" the hypertext before reading it. Gomory and Hu created an elegant and simple N^3 algorithm; calculation time can be noticeable for medium-size and large hypertexts (more than 100 pages).

5. Discussion

Following the algorithms described in this paper, it was possible to build an information system for literary hypertext that positions itself between the reader and the author to perform a task that did not exist in printed media: the adaptive optimization of the text according to rules set by the author and the interaction of the reader.

A high value of hypertextual attraction it indicates that a single page could potentially be linked from many other pages. The question authors must ask themselves in this case is: Will the reading make sense in all linked cases? A high value of hypertextual friction could indicate that the lyrical quality of links is being stretched. The question authors must ask themselves in this case is: Does this connection make sense? The system takes the burden of performing these checks during the writing process and advises the author accordingly. As mentioned earlier, the author can use precisely hypertextual friction and hypertextual attractors as narrative devices. The system would just make sure the author is aware of it.

Adaptivity in this context has a precise meaning: the system adapts the connections in the graph in order to produce a Hamiltonian cycle in which hypertextual friction and hypertextual attraction have been minimized. As we have seen, the challenge for the author is to set proper values to the hypertextual cost of every link. Even though it is possible to use an automated optimization method, author's discretion is always the main criteria to assign costs between pages.

The Hamiltonian cycle is not actually found but simulated by creating edges on the fly in the graph. From the system's perspective, the Hamiltonian cycle is not even requested; from the user perspective, the Hamiltonian cycle is exactly what is offered. Pages already visited are removed from the list of possibilities to continue reading. However, if too many pages are removed from the path, the effective cost between the current page and the destination page could be too big, leading to an increase of hypertextual friction, or loss of narrative continuity.

One of the main arguments opponents of hypertext have shown in the past against it is the fragmented story that is offered to the reader. In this case, the reader receives a plot that is optimized, from the narrative perspective. That is to say, the reader receives a linear text most of the times. This raises a question: Are multilinearity and fragmentation the goal of hyperfiction, or are they the product of the state of the art when the first literary hypertexts were produced? Is fragmentation a paradigm that we want to preserve? Our intention is not to answer this complex question here. However, we want to indicate that we have the ability to produce a text that exploits the essence of digital media, and that at the same time preserves the essence of narrative in a classical sense: immersion.

The experiment reported by Gee [8] must be regarded carefully. An approach is to consider that the subjects of the experiment were still too close to the paper universe, and therefore could not appreciate the full extent of hypertextual possibilities. Another approach would be to consider that in fact readers enjoy linear texts more, regardless of the media that supports them. Although it is not possible to generalize from one single experiment, these results are consistent with the author's own experience with literary hypertexts. We consider these results a strong signal of what readers expect, and what can be expected from them.

The future direction of Literatronic will be focused on extracting information from the growing database of user interaction. As of May, 2006, there are 185 members. Between November 2003 and January 2005, Literatronic.com has received about 4,000 unique visitors. This user interaction data will be analyzed anonymously to calibrate processes of supervised learning (though neural networks) and non-supervised learning (through clustering techniques). The goal will be to increase the probability of occurrence of paths favored by readers. At that point, there will be a fuzzy line between a human author and a digital author.

An adaptive literary hypertext system is a fascinating possibility. The idea is very suggestive and promises advances in this area. Even if it did not offer any answers, it would help us to improve our perspective about hyperfiction. Quoting Stephen Jacobson: "Before you do it right, you have to do it all" (Wearable Robots. Technology Review, July/August 2004. Cambridge, MA).

6. Acknowledgements

The platform Literatronic was built as part of Juan B. Gutierrez's graduate research project at the Department of Mathematics, Florida State University, between August 2002 and August 2005. The hyperfiction used in this project, Extreme Conditions was financed by the Ministry of Culture of Colombia (grant COLCULTURA/SECAB 014/1996) and the Repository of Artistic Projects of the Institute of Culture of Bogota (grant 514/1997, grant 410/1998). The third version of Extreme Conditions, the version used in this experiment, was finished in November, 2003.

References

[1] M. Bernstein. Patterns of hypertext. In Proceedings of the Nineth ACM Conference on Hypertext and Hypermedia, pages 21–29. ACM Press, 1998.

[2] M. Bernstein, D. E. Millard, and M. J. Weal. On writing sculptural hypertext. In HYPERTEXT '02: Proceedings of the thirteenth ACM conference on Hypertext and hypermedia, pages 65–66, New York, NY, USA, 2002. ACM Press.

[3] J. D. Bolter and M. Joyce. Hypertext and creative writing. In HYPERTEXT '87: Proceeding of the ACM conference on Hypertext, pages 41–50, New York, NY, USA, 1987. ACM Press.
[4] P. Bootz. Vers un multim´edia contraint et a-m´edia. http://motsvoir.free.fr/, 176(1):101– 108, June 2002.
[5] L. Calvi. 'lector in rebus': The role of the reader and the characteristics of hyperreading. In Proceedings of the Tenth ACM Conference on Hypertext and Hypermedia, pages 101–109. ACM Press, 1999.
[6] L. Calvi. Adaptivity in hyperfiction. In Proceedings of the Fifteenth ACM Conference on Hypertext and Hypermedia, pages 101–109. ACM Press, 2004.
[7] P. de Bra, A. Aerts, B. Berden, B. de Lange, B. Rousseau, T. Santic, D. Smits, and N. Stash. Aha! the adaptive hypermedia architecture. In Proceedings of the fourteenth ACM conference on Hypertext and hypermedia, pages 81–84. ACM Press, 2003.
[8] K. Gee. The ergonomics of hypertext narrative: Usability testing as a tool for evaluating and redesign. ACM Journal of Computer Documentation (JCD), 25(1):3–16, 2001.
[9] G. Genette. Paratexts Thresholds of Interpretations. Cambridge University Press, New York, NY, 1997.
[10] G. Golovchinsky. Queries? links? is there a difference? In Proceedings of the SIGCHI Conference on Human Factors in Computing Systems, pages 407–414. ACM Press, 1997.
[11] R. E. Gomory and T. C. Hu. Multiterminal network flows. Journal of the Society of Industrial and Applied Mathematics, 9:551–570, 1961.
[12] J. B. Gutiérrez. Hipertexto en contexto. Signo y Pensamiento, XIX(36):36–40, 2000.
[13] J. B. Gutiérrez. Literatrónica: Hipertexto literario adaptativo. In 2o Congreso ONLINE del Observatorio para la CiberSociedad, Barcelona, Spain, 2002. Observatorio Para La Cibersociedad. URL: http://www.cibersociedad.net/congres2004/grups/fitxacom publica2.php?grup=2&id=242&idioma=es.
[14] R. Kendall. Toward an organic hypertext. In Proceedings of the Eleventh ACM Conference on Hypertext and Hypermedia, pages 161–170. ACM Press, 2000.
[15] G. P. Landow. Hypertext 2.0. Johns Hopkins University Press, Baltimore, MD, 1997.
[16] I. Snyder. Hypertext: The Electronic Labyrinth. New York University Press, New York, NY, 1999.
[17] S. P. Tosca. A pragmatic of links. In Proceedings of the Eleventh ACM Conference on Hypertext and Hypermedia, pages 77–84. ACM Press, 2000.

Responsive Visualization for Musical Performance

Robyn Taylor
University of Alberta
Edmonton, Alberta, Canada
E-mail: robyn@cs.ualberta.ca

Pierre Boulanger
University of Alberta
Edmonton, Alberta, Canada
E-mail: pierreb@cs.ualberta.ca

Daniel Torres
University of Alberta
Edmonton, Alberta, Canada
E-mail: dtorres@cs.ualberta.ca

Abstract

We present a framework that facilitates the visualization of live musical performance using virtual and augmented reality technologies. In order to create a framework suitable for developing technologically augmented artistic applications, we have defined our system in a way that is modular and incorporates intuitive development processes when possible. In this paper we present a method of musical feature extraction and provide three examples of music visualization applications that we have developed using our system. Our visualizations illustrate features in live singing and keyboard playing using responsive virtual characters, responsive video imagery, and responsive virtual spaces.

1. Introduction

Each time a song is interpreted, the notes and words may be the same, but the experience is unique. Interaction between a vocalist and her fellow musicians, the musicians and the audience, the energy of the room, and the adrenalin rush of performing all contribute to the serendipitous nature of live music, making each performance preciously ephemeral and distinct.

When an interpretive artist performs a piece, she knows *what* she will be singing, but gives herself freedom to decide in the moment *how* it will be sung. Concert or theatrical productions that accompany live musical performance with pre-recorded visuals lack the flexibility to let the artist manipulate imagery as well as sound. In order to provide artists with a truly responsive and spontaneous audio-visual environment, it is important for the visualization medium to be flexible enough to convey the subtle nuances of the live performance. Each repetition of the visualization experience will be unique, since no two live performances will ever be the same.

While traditionally used for scientific purposes, responsive visualization technologies offer artists an exciting new way to create audio-visual art pieces. Advancements in graphics rendering and computer processing power, the accessibility and ease-of-use provided by visually programmed development environments, and the relative affordability of sophisticated visualization hardware have made responsive technologies available to artists wishing to create audio-visual entertainment.

Examples of existing music visualization artworks created by linking scientific visualization technologies with musical control systems include Jack Ox's *Color Organ* [8] which visualizes harmonic structures and instrument timbres within an immersive CAVE environment, Levin and Lieberman's *Messa di Voce* [6] which augments the vocalizations of live performers in a theatrical setting, and *The Singing Tree* [7] created by Oliver et al. as part of MIT MediaLab's Brain Opera which allows participants to experience audio-visual feedback that responds to the sound of their singing.

Our goals, when creating our system, were to devise a schema for extracting musical feature information from live performance, and to facilitate the mapping of audio data onto visual parameters so as to enable artists to create dynamic audio-visual performances that are controlled in real-time through live musical input.

2. Overview

To augment a live musical performance through responsive visualization, musical feature data within a stream of real-time input must first be parsed into discrete and measurable units. Subsequently, this musical feature data can be mapped to responsive imagery.

Our system separates the musical feature extraction tasks from the visualization mechanisms. The musical feature extraction module runs on a Macintosh G5 system, while visualization engines may run on Windows, Linux, Macintosh or Irix machines. Visualization engines receive information about the musical performance by connecting to the musical extraction module via a network connection implemented using the Virtual Reality Perception Network (VRPN) [12]. This allows musical and visual content to be easily re-mapped in different ways. This modular approach encourages code re-use.

Since the system is intended to be used by teams of artists and developers who are creating technically augmented performable works, this system should be as accessible as possible for users who may have little or no formal training in computer programming. We used visual programming environments whenever possible to create our system modules. These environments allow programs to be designed by drawing connections between system modules in order to visualize the way data flows within the application. These environments are often very intuitive for non-technical users.

In this paper we describe the musical feature extraction process and present three examples of mappings between music and responsive imagery that were generated using our system. In each implementation a different mapping is used to illustrate properties of interpreted music:

- Visualization of the emotional content of a musical piece through the behaviour of a virtual character
- Visualization of a singer's vocal timbre using responsive video
- Visualization of vocal dynamics inside a reactive immersive environment

3. Musical Feature Data Extraction

In order to visualize live music, it is essential that the stream of live music be parsed into discrete parameters. Cycling `74's visual programming development environment Max/MSP [3] is specially designed for sound processing, allowing a programmer to easily manipulate audio and MIDI data. We use Max/MSP to create our Musical Perception Filter Layer, which is illustrated in Figure 1.

Our Musical Perception Filter Layer (see Figure 1) extracts the following parameters from live musical performance:

- Pitch: Vocal pitches are identified both in terms of their raw pitch values and in terms of their scale degrees relative to the tonic of the key signature of the sung melody.
- Loudness: Vocal amplitude is transmitted in dB.
- Timbre: A descriptor of the user's timbre is devised. Open vowels (like /a:/) are differentiated from closed vowels (like /i:/).
- Chord: The chords the user plays on the digital piano are identified.

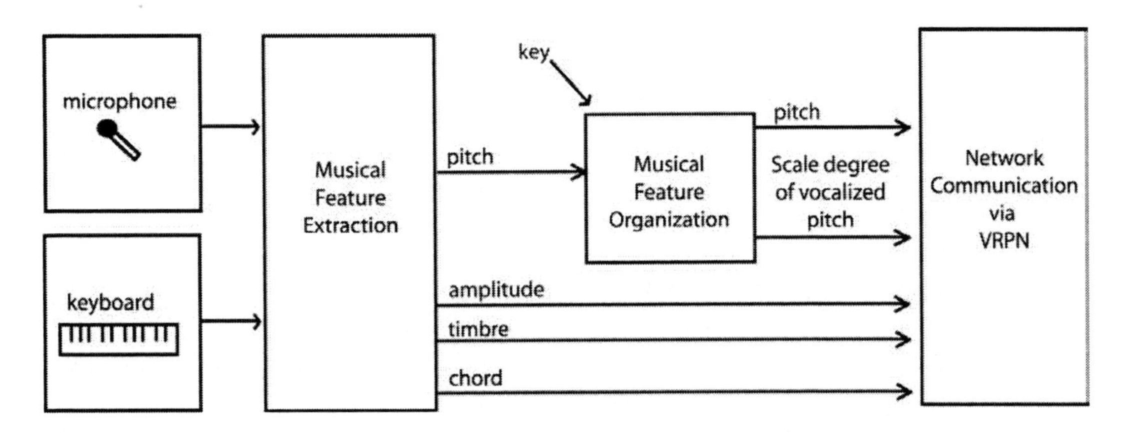

Figure 1: *The Musical Perception Filter Layer*

3.1 Vocal Pitch and Loudness Extraction. Pitch and amplitude information is extracted from sung vocal input using the Max/MSP `fiddle~` object created by Puckette *et al.* [9]. The `fiddle~` object analyzes the incoming sound signal to determine the pitch of the singer's vocalization, using Fourier transformation to convert the signal's complex waveform into a harmonic spectrum. This spectrum contains data describing the frequency and amplitude of each sinusoidal component contained in the incoming sound. The fundamental frequency of the signal is reported as the pitch of the incoming sound, and the signal's amplitude is reported to be the loudness.

3.2 Tonal Encoding of Pitch Data. We use Western tonal music as input melodies to our music visualization systems, so once our raw pitch data is extracted, we then organize it in a way consistent with tonal music theory. Our system organizes vocal input into a tonal context using strategies devised by Deutsch and Feroe [5]. We encode each pitch in terms of its intervallic relationship with the tonic note of the key signature of the melody. This facilitates any music-theoretical analysis we may later wish to do upon the sung input.

3.3 Timbral Descriptors. Since the `fiddle~` object outputs the frequency and amplitude data describing each of the partials forming the harmonic spectrum of the user's singing voice, we can analyze this harmonic spectrum to obtain information about the singer's timbre. Upon examining the harmonic spectrum, we assess the distribution of energy amongst the partials in the sound. If we compare the harmonic spectrum that describes the singer's vocalization to known information about the harmonic spectra characterizing different vowel sounds, we can describe an aspect of the singer's vocal timbre by providing a rough estimation of the singer's vowel choice. The vocalist can manipulate this timbral descriptor by modifying the vowel choice he or she employs while singing.

3.4 Chord Identification. Keyboard input monitoring is done in a separate sub-patch. Our Max/MSP sub-patch monitors MIDI events in order to determine what chords are being played on the keyboard. To do this, we determine which pitches are being played on the keyboard and identify them in terms of pitch class. We consider the note C to have a pitch class of 0, Db to have a pitch class of 1, and so on. We examine the pitch classes that are played on the keyboard and compare them with a list of the pitch classes found in a collection of `known' major and minor chords. This method permits chords to be played in any inversion. More chord types (diminished, augmented, major and minor sevenths, etc.) could easily be classified and added to the list of known chords, making expansion of this module trivial.

4. Visualizing Music through Virtual Character Behaviour

Our first example of a music visualization application visualizes sung melodies through the responsive behaviours of an artificially intelligent virtual character. In this implementation a character created using Torres and Boulanger's ANIMUS Framework [13] [14] is used to visualize the emotive content of music by expressing simulated emotion through animated behaviour [10][11].

4.1 The ANIMUS Framework. The ANIMUS Framework is used to create believable virtual characters who can respond to events in their environment. Responsive character behaviour is generated using a three-layer process which simulates the perceptual and cognitive processes used to process live musical input and formulate and express an emotional response in real-time:

- **Perception Layer**: The musical features extracted in the Musical Perception Filter Layer are communicated to the virtual character's perception layer via the VRPN connection. In this way, the virtual character is able to perceive salient musical features within the stream of live musical input.

- **Cognition Layer**: In the cognition layer, the musical information obtained in the perception layer is used to influence the virtual character's emotional state. Subtleties of musical phrase and vocal intonation are used to interact with the creature and modify its simulated mood. In the cognition layer, the artistic concept for the link between the nuances of vocal performance and the character's emotive response is defined.

- **Expression Layer**: In the expression layer, the 3D creature's cognitive state is visualized to the audience using dynamically generated animation. The animations are generated by interpolating between keyframe poses that are associated with emotional states.

The ANIMUS Framework is designed with the intention of facilitating artist/scientist collaboration. The system allows artistic designers to create program skeletons outlining how each of the three ANIMUS layers should function. Technical team members then implement the specific functionality in order to create synthetic characters that are capable of communicating believably with an audience of viewers.

4.2 The *Alebrije* Character. Alebrije is a lizard-like virtual character (see Figure 2) that was created during the development of the ANIMUS Framework. We extended the Alebrije character so that his responses could be used to visualize the emotive content of sung music. The animated imagery is life-sized and can be displayed upon a stereoscopic display, making it suitable for use in augmented theatrical productions where the virtual actors must be consistent in scale with the live performers.

Figure 2: *A singer transitions Alebrije from a neutral to a sad position*

In order to simulate Alebrije's awareness of the emotional signifiers in interpreted music, the research of Deryck Cooke [1] correlating melody and composers' emotional intentions is used to formulate his musical cognition system. Aspects of Cooke's research serve as the basis for Alebrije's cognition processes in this implementation. Cooke associates emotive meaning to the tonal structure of Western melodies, associating each tone in the musical scale to an emotional context. Alebrije's interpretation of the emotional meaning of a melody is consistent with Cooke's metric.

Our Alebrije character is capable of distinguishing 'sad' melodies from 'happy' ones, and displaying responsive behaviours that communicate his simulated emotional state. We are currently exploring the possibility of using musically responsive virtual characters like Alebrije in live performance settings.

5. Visualizing Music through Responsive Video

Our second music visualization example uses responsive video to create a performable multimedia piece that is manipulated by a performer's singing and keyboard playing. Cycling 74's Jitter [2] is a video processing package that can be integrated into the Max/MSP environment. The Jitter package allows users to create responsive video applications by describing (using visual programming) how data flows between Max/MSP and Jitter objects. Jitter can be used for a variety of visualization tasks such as manipulation of video playback, generation of basic 3D animation, or the modification of still images. Using the Jitter environment to colour-edit and layer a selection of video clips, we have created a visualization which illustrates the vocal timbre of a live performer and the harmonic relationships between chords played on a digital keyboard. This visualization system has been used in a live concert setting to perform an interactive piece called *Deep Surrender*.

5.1 Visualizing Musical Parameters. In *Deep Surrender*, the Musical Perception Filter Layer is used to extract feature data describing the vocal timbre of a singer, and the chords played on a digital piano. In this visualization, chords are related to one another with regards to their positions on the music theoretical device, The Circle of Fifths, as was previously explored in Ox's *Color Organ* music visualization environment [8]. The chords that are played affect the colour balance of the video display. To define the relationship between chords and colours, the Circle of Fifths was mapped to the standard colour wheel, making chords that are similar to one another on the Circle similar in colour.

To visualize vocal timbre, the amplitude of the fundamental frequency and second and third partials extracted from a singer's vocalization are mapped to a Red-Green-Blue colour selector. The harmonic spectrum of the user's singing defines the hue that is output by the colour selector. Varying the vowel sound varies the harmonic spectrum of the singing, and therefore outputs different colours. The colours produced by the performer's singing affect the layered videoclips that form the visualization.

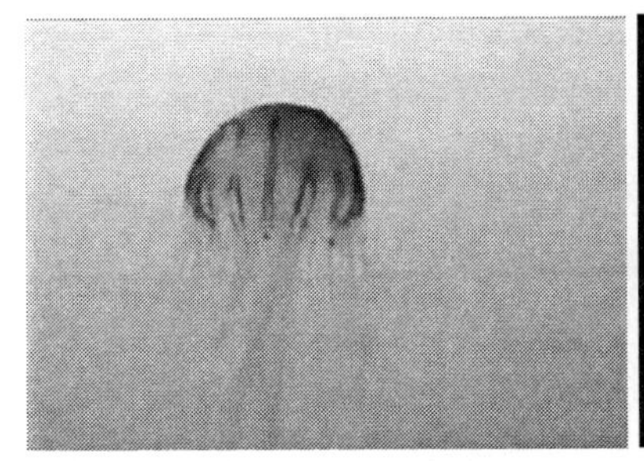
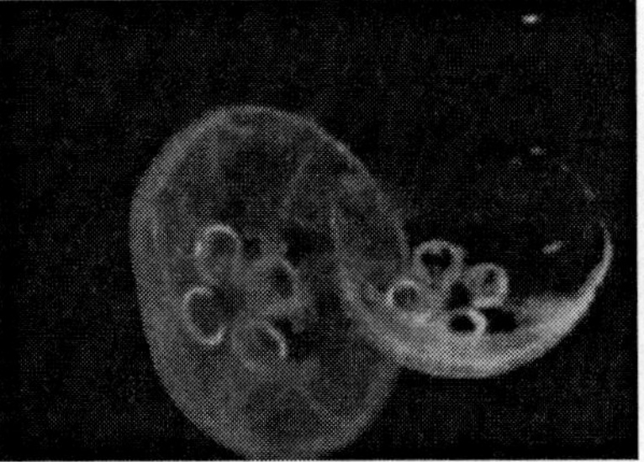
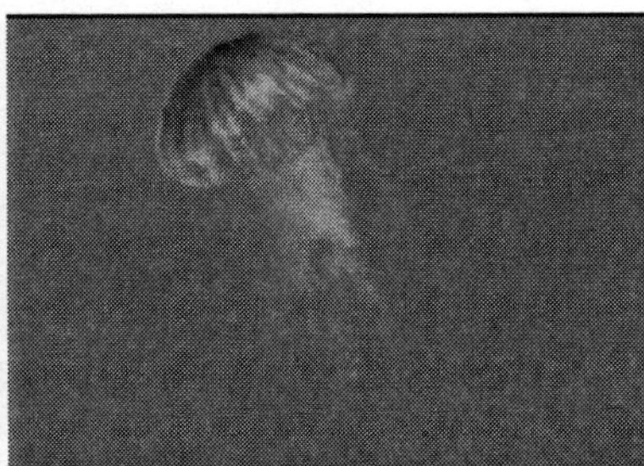

Figure 3: *Three phases of the Deep Surrender performance*

5.2 Deep Surrender. *Deep Surrender* is a multimedia piece written for soprano, synthesizer, and responsive video. The intention of the piece is to illustrate the way an artist can harness anxiety and adrenalin to produce a beautiful performance. This illustration is done through the visual metaphor of a jellyfish – a creature both beautiful and terrifying. The artist's musical performance affects the jellyfish representation, in order to visualize how the artist interacts with and overcomes her anxiety.

In addition to having been performed in concert at the University of Alberta, *Deep Surrender* is routinely performed in the laboratory in order to show visitors how visualization can be used for artistic purposes. It was also performed during several media interviews, including a live performance on CBC Radio.

6. Visualizing Music in Immersive Spaces

Our third music visualization example illustrates vocal dynamic inside a responsive virtual space. The Virtools environment [4] is a visual programming environment which allows designers of virtual reality applications to create immersive visualizations. A musical control system for Virtools applications has been created, connecting extracted musical parameters to behaviours inside the Virtools environment.

6.1 Virtools. Virtools' intuitive authoring environment (see Figure 4) allows different visualization metaphors to be easily defined, tested, and modified. Connecting our musical feature extraction system to the Virtools environment allows us to rapidly develop music visualization applications. This illustrates one of the benefits of our proposed architecture: the connection between Max/MSP's music processing environment and Virtools' virtual reality simulator allows both the musical and visual aspects of music visualization projects to be implemented using visual programming techniques.

6.3 Visualization in Immersive Spaces. The Virtools rendering system is capable of visualizing a virtual environment inside an immersive space consisting of three large screens which display stereoscopic imagery to the users who stand within the enclosure. See Figure 5 for an example of an immersive visualization room. Immersive environments enhance the realism of the virtual experience, as the audience members experience depth perception inside the life-sized space.

6.2 Particle Manipulation through Vocal Dynamics. We have created an example implementation of music visualization within a virtual environment created in Virtools which allows a vocalist to use his or her voice to formulate particle clouds (see Figure 6) generated by Virtools' particle generation and interaction routines. The size and colour of the particle clouds varies in response to the pitch and loudness of the vocalist's singing.

The particle system is one of the built in behaviours that Virtools includes as part of its extensive behaviour library. Virtools authors may not need to manually code any part of the application if the library contains all the behaviours they need. If they do need to custom-create Building Blocks, Virtools makes this possible via a simple SDK that allows developers to code new Behaviors in C++.

Virtools' ease-of-use and extensive Behavior libraries make it a valuable tool in the creation of artistic visualizations. The fact that a visual metaphor can be created and customized in a rapid and intuitive fashion makes it easy to experiment with numerous effects and parameterizations when creating an application. Being able to display the visualizations in an immersive environment increases their expressive and communicative potential.

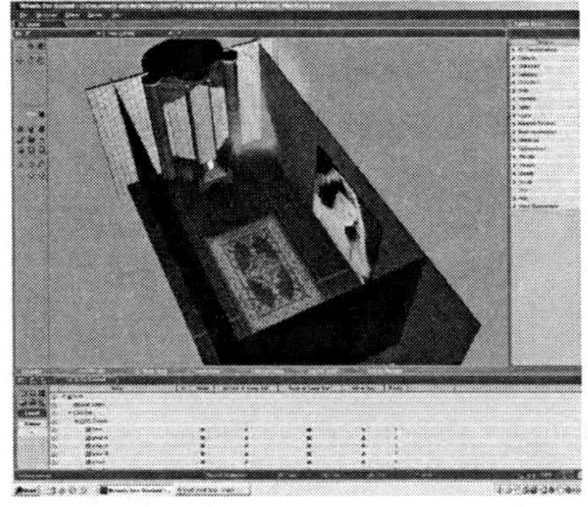
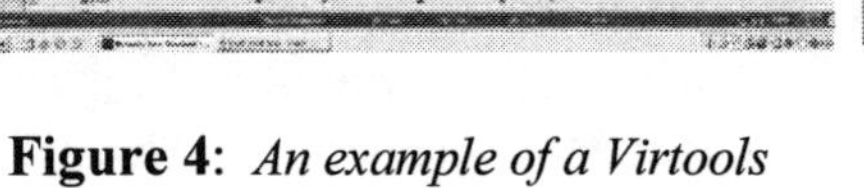

Figure 4: *An example of a Virtools Composition*

Figure 5: *A participant inside theUniversity of Alberta's immersive VizRoom*

Figure 6: *A particle cloud triggered by vocalization*

7. Discussion

This paper has presented a music visualization system which operates in a distributed fashion, facilitating easy re-use of the musical analysis module contained in the Musical Perception Filter Layer.

The Musical Perception Filter Layer was implemented in the musical development environment Max/MSP [3]. Pitch, amplitude, timbral information and chord data were used to interact in real-time with aspects of a virtual environment.

Three experimental examples were implemented in order to show how the features extracted by the musical analysis module could be mapped to different visualization metaphors:

- Interactive ANIMUS virtual characters [13][14] were used to illustrate the emotive capacities of music [10][11] through visible behaviours that illustrated emotional responses to music. The cognitive system used to trigger these behaviours was consistent with Deryck Cooke's research correlating melody and emotion [1].
- Cycling `74's Jitter environment [2] was used to create responsive video streams which responded to vocal and keyboard input. This visualization was used to create a multimedia performance piece, *Deep Surrender*.
- The Virtools [4] visual programming environment was used to create a visualization which can be performed in an immersive virtual space. This visualization used particle dynamics to illustrate vocal performance.

This framework simplifies the process of generating multiple visualization metaphors to express extracted musical feature data. The virtual character and immersive environment visualizations are currently being used to develop artistic pieces, and the responsive video production has already been used in live performance.

Care has been taken to develop this system in ways that enable rapid development of creative visualizations and collaborative work between artists and programmers. Max/MSP, Jitter, and the Virtools environment can be visually programmed, making possible the rapid prototyping of visual metaphors. The fact that the system allows visualization and musical feature extraction to be conducted in visually programmed environments makes it more accessible to artists who may have no formal training in computer programming. Although the ANIMUS environment does not yet support visual programming,

ANIMUS is designed with the idea of task delegation in mind, allowing artists and programmers to work alongside one another to develop creative works.

We look forward to the continued use of this system for the purpose of creating artistic works that combine live music and responsive visualization.

Acknowledgements

The source video footage for the *Deep Surrender* video production was filmed by Melanie Gall.

The textures on the models used in the Virtools simulation are from http://www.ktn3d.com/.

The use of the VRPN library was made possible by the NIH National Research Resource in Molecular Graphics and Microscopy at the University of North Carolina at Chapel Hill, supported by the NIH National Center for Research Resources and the NIH National Institute of Biomedical Imaging and Bioengineering.

References

[1] Deryck Cooke. *The Language of Music*. New York: Oxford University Press, 1959.
[2] Cycling '74. Jitter, 2004.
[3] Cycling '74. Max/MSP, 2004.
[4] Dassault Systémes. Virtools, 2005.
[5] Diana Deutsch and J. Feroe. The internal representation of pitch sequences in tonal music. *Psychological Review*, 88:503-522, 1981.
[6] Golan Levin and Zachary Lieberman. In-situ speech visualization in real-time interactive installation and performance. In *Proceedings of The 3rd International Symposium on Non-Photorealistic Animation and Rendering*, pages 7-14. ACM Press, 2004.
[7] William Oliver, John Yu, and Eric Metois. The Singing Tree: design of an interactive musical interface. In *DIS '97: Proceedings of the conference on Designing interactive systems: processes, practices, methods, and techniques*, pages 261-264. ACM Press, 1997.
[8] Jack Ox. 2 performances in the 21st Century Virtual Color Organ. In *Proceedings of the fourth conference on Creativity & Cognition*, pages 20-24. ACM Press, 2002.
[9] M. Puckette, T. Apel, and D. Zicarelli. Real-time audio analysis tools for Pd and MSP. In *Proceedings of the International Computer Music Conference*, pages 109-112. International Computer Music Association, 1998.
[10] Robyn Taylor, Pierre Boulanger, and Daniel Torres. Visualizing emotion in musical performance using a virtual character. In *Proceedings of the Fifth International Symposium On Smart Graphics*, pages 13-24. Springer LNCS, 2005.
[11] Robyn Taylor, Daniel Torres, and Pierre Boulanger. Using music to interact with a virtual character. In *Proceedings of the International Conference on New Interfaces for Musical Expression*, pages 220-223, 2005.
[12] Russell M. Taylor II, Thomas C. Hudson, Adam Seeger, Hans Weber, Jeffrey Juliano, and Aron T. Helser. VRPN: A device-independent, networktransparent VR peripheral system. In *Proceedings of the ACM symposium on Virtual reality software and technology*, pages 55-61. ACM Press, 2001.
[13] Daniel Torres and Pierre Boulanger. The ANIMUS Project: a framework for the creation of interactive creatures in immersed environments. In *Proceedingsof the ACM symposium on Virtual reality software and technology*, pages 91-99. ACM Press, 2003.
[14] Daniel Torres and Pierre Boulanger. A perception and selective attention system for synthetic creatures. In *Proceedings of the Third International Symposium On Smart Graphics*, pages 141-150, 2003.

The Necessity of Time in the Perception of Three Dimensions: A Preliminary Inquiry

Michael Mahan
PO Box 913
Valley Center, CA 92082
E-mail: mike@mmahan.com

Abstract

In working with 3-D computer models I came to realize that there would not be much advantage to presenting them as a three dimensional representation rather than on a flat screen. In either case, they would have to be manipulated, over time, in some way to offer much information. This paper is a non-rigorous exploration of why that is true. It begins by presenting some of the mechanisms by which we orient ourselves in space and how we perceive it. The most important of these are visual, but they do not yield much information in a static situation, since they are vulnerable to misinterpretation and illusion. The paper then goes on to examine the importance of a changing point of view in the perception of space, how points of view have been depicted in art, and how time affects point of view. The example of motion pictures provides foundation for the idea that certain perceptions are essentially free of time, while others occur over time. It goes on to discuss time and how it becomes essential to the perception of space. Finally, it offers some insight into the perception of time.

1. Introduction

1.1 How This Paper Came to Be. I was working creating 3-D computer models of four dimensional objects, when I came to realize that, even though these models existed in the computer and in my mind's eye, the only way they could be presented was on a monitor, which was only 2-D. As I started thinking, hypothetically, about how they could be presented in true 3-D, holograms for example, I came to the remarkable conclusion that this would make little difference in the perception of the object. Whether the representation was a hologram, or simply an image on a computer screen, it would still have to be manipulated to get any useful information about its appearance. It is only through movement in space that three dimensions can be perceived, and movement is a function of time as well as space. When I suggested to a colleague that I believed that time was essential to 3-D perception, I was met with mild dispute. Although that discussion went no further, I conceived of this paper as a defense of my position.

1.2 A Note on Methodology. Although I have a degree in mathematics, I am not a mathematician, nor am I a scientist. I have no background in the psychology or physiology of perception. I am an artist and a theorist, whose focus has been on relationships of language, image, and reality. This paper is not intended as a scientific inquiry, nor is it a rigorous logical exercise. But perception is a funny thing, easily influenced by both subjective and objective forces. So what I present here is simply an epistemological essay, offered to stimulate the imaginations of artists, mathematicians, and scientists.

2. Perception and Space

2.1 Spatial Orientation. I will first examine static ways humans perceive space. We exist in a world of three spatial dimensions. We sense the world as spatial with the senses of sight, hearing, touch, and to a lesser extent smell. We learn to orient ourselves visually in relationship to objects in that world using a number of mechanisms. Most obvious is stereo vision, but this is not necessarily the most important. The eyes have a slightly different location in space giving each one a slightly different view. The mind combines these to create a 3-D picture of the world. Holding a pencil in each hand and bringing them together is more difficult with one eye closed, but not impossible because there are other factors involved in depth perception. Because our angle of vision is increased using both eyes, we sense space as surrounding us.

Depth of field is the range of distances that a lens holds effectively in focus. The closer the point of focus is to the lens, the more limited is the depth of field. We can see this in a photograph when the background and/or the foreground is blurred. The lens of the eye changes focus so quickly that we are not likely to see objects as blurred, yet the mind interprets changes of focus as differences in distance.

Optical perspective may be the most powerful indicator of depth. Parallel lines seem to converge in the distance. At the beginning of the Renaissance artists discovered mathematical rules for creating perspective drawings that transformed the nature of art and perception of space. In the middle ages representative art was dominated by the human figure. Multiple figures were situated on a common ground and sized in proportion to their relative importance. Mathematical perspective allowed a new degree of realism and the introduction of architectural and landscape settings. Texture becomes less distinct as distance increases, whereas visual density increases. At a close distance we may not "be able to see the forest for the trees," while at a distance smaller trees in greater density make the forest appear.

2.2 Illusions. All the factors that allow us to judge depth can be manipulated to create illusions of depth as well. Paintings, drawings, and photographs become realistic representations of space by playing on the mechanisms of our spatial perception. A *trompe l'oeil*, representation is a deliberate deception. (Figure 1.) Various 3-D illusions can be created by introducing different images to the left and right eyes. Stereo photography produces images with two lenses spaced similarly to the eyes. As soon as time and motion are introduced into perception most illusions fail.

Figure 1: *Trompe l'oeil mural. Narbonne, France.*

2.3 Representations of Higher Dimensions. I started thinking about the relationship of time to perception when I was creating computer representations of hyperobjects. A two dimensional representation of a cube is a square within a square. By analogy a three dimensional representation of a 4-D cube is a cube within a cube. This is a simple analogy and a very powerful metaphor. The only trouble is that it represents a straight on point of view in single point perspective. In addition it is dependent on the concept of a wire frame cube. In straight on single point perspective, an opaque cube is simply a square, indistinguishable from a 2-D square. From an analogous point of view an opaque 4-D cube in 3-D is simply a cube, indistinguishable from any other. (Figure 2.)

3. Points Of View

3.1 Of Cubes and Cubism. From other points of view the cube becomes various combinations of one, two, or three quadrilaterals. But in no case can more than three sides be seen at any one time. The invention of photography freed painting from pictorial realism and brought on the advent of modernism. The impressionists explored the nature of light, but artists, beginning with Paul Cézanne, began to experiment with nonlinear perspectives. (Figure 3.) "For Cézanne the important point was the discovery that the

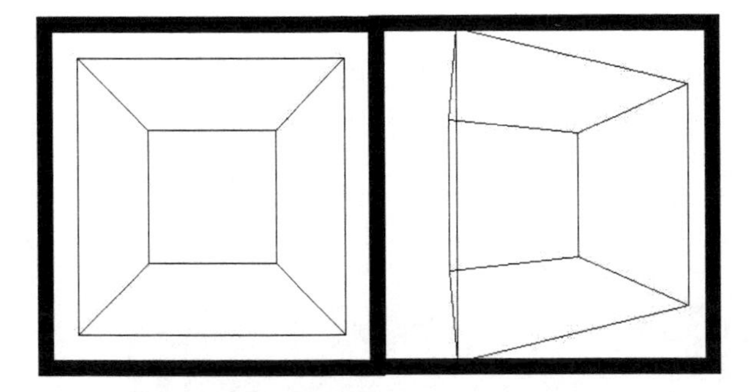

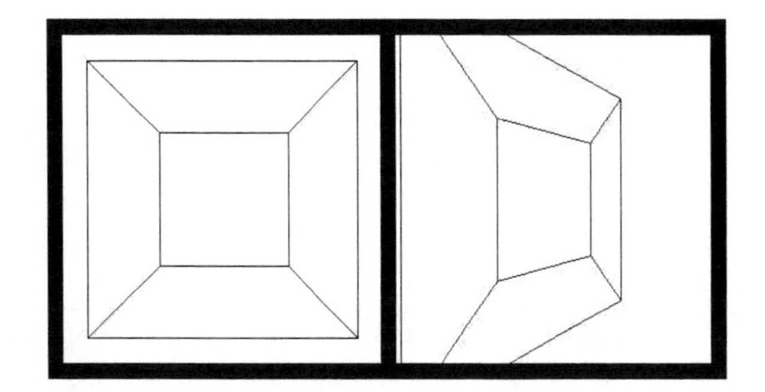

Figure 2: (*From left to right.) Front view of a computer model of a cube. The same model rotated slightly. A computer drawing of a cube in front view. That drawing rotated slightly. Comment: Until the front views are moved, they cannot be distinguished from each other. It is not until they are rotated that one can be recognized as 3-D and the other as 2-D.*

eye took in a scene both consecutively and simultaneously, with profound implications for the construction of the painting."[1] "Consecutive" and "simultaneous" are both words that describe time. Analytic Cubism continued these experiments by combining various views of an object into one image. (Figures 4 and 5.)

Figure 3: *Paul Cezanne.* ***Gardanne.*** *1885-86*

3.2 Time and Points of View. If "point of view" is taken to mean a subject's position relative to an object, then changing the point of view does not necessarily mean movement of the subject, but simply a change in

Figure 4: *Pablo Picasso.* ***Reservoir at Horta.*** *1909*

Figure 5: *Georges Braque.* ***Houses at L'Estaque.*** *1908*

relationship that can come about from movement of the object. The obvious implication of differing view points is different locations in space, which of course implies movement, and consequently time. In the 1870s and 1880s, photographer, Eadweard Muybridge did a number of studies of figures and animals in motion. (Figure 6) Although intended to explore the nature of motion itself, these works feature a point of view that changes with time. In his paintings, Marcel Duchamp uses time as a tool of exploration the way the cubists used space. (Duchamp has apparently denied the connection to Muybridge's motion studies.) (Figure 7)

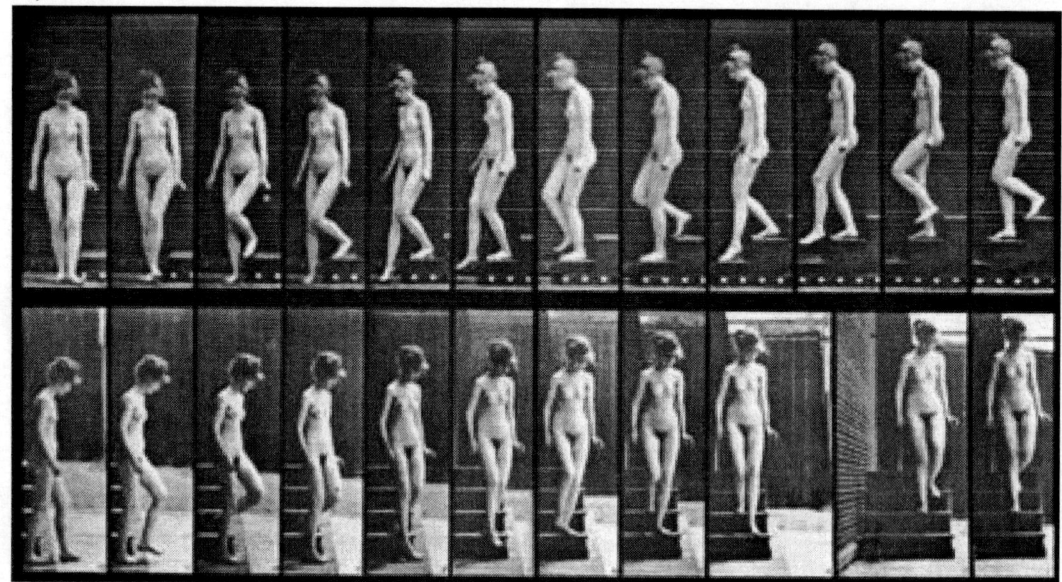

Figure 6: *Eadweard Muybridge.* ***Woman Walking Downstairs*** *from* ***The Human Figure in Motion.*** *(Note that the subject matter could be described as "nude discending a staircase.)*

3.3 Motion Pictures. Muybridge is seen as a seminal figure in film history, and his works are often regarded as proto-movies. Indeed the individual images have been combined into animated versions. A movie consists of a number of discrete movements each represented by a frame. The mind holds each image and merges it into the next 24 time per second giving the illusion of continuous motion. For this reason, it is meaningful to distinguish between instantaneous perception and perception over time. If the time period is so short that there is no subjective awareness of spatial movement, then there is no subjective awareness of the passage of time, and the perception can be regarded as instantaneous.

In the 1950s a few 3-D films were made using optical stereo. But this gimmick never really caught on, because time already gives *movies* a third dimension.

Figure 7: *Marcel Duchamp.*
Nude Descending a Staircase.
1912.

4. Time

4.1 Time and Dimension. Time is often characterized as a fourth dimension, yet it seems different than the other three. We seem to be able to move in any direction in space but only forward in time. Since there are no fixed points in space, the motion of an object can only be seen in relation to another object which is also moving. Our seeming ability to move freely in space only holds in relation to our immediate surroundings. We cannot control our movements on a cosmic level. We can pace back and forth across the room, or we can fly around the world, but the room and the world, with us in it, are all moving around the sun, which in turn is hurling through the expanding universe. We can perceive only movement on the smallest of scales. Our position in time does not change relative to our immediate environment. But according to Einstein's theories, our movement in time is relative to the speed of our movement in space.

Another distinction between time and the other dimensions is that our perception of time seems to lack a spatial component. While we can see or feel an object's position in space we cannot in time. Two objects can seemingly exist at the same point in time but not space. Without motion the idea of time is meaningless. Motion is a change in space over a period of time. Without change time does not exist.

4.2 A Metaphysical Aside. Nonscientific and nonrigorous discussions of time and dimension will invariably, at this point, lead to questions about the metaphysics of space and time. For example, "What if there is only one point in space in which everything exists at different times?" Questions like this may be interesting but their primary focus is objective in nature. One could also call into question the reality of two dimensions. Since concrete reality is dependent on the existence of matter, a very plausible argument could be made that our world consists of no less than three dimensions. A two dimensional object, if it could exist at all, would be invisible, since it could not reflect light. But since, in essence, all that is visible of a 3-D object is its surface, the idea of 2-D perception is conceptually valid. A similar argument could be made against perception outside of time, but again, I would suggest that this argument uses a technicality to trivialize essential concepts.

4.3 Time and Perception. Our perception of two dimensional objects is essentially visual; all the information, including the boundaries of the object, is on the surface of a plane. We can see both the object and its boundaries. Our visual perception of 3-D objects is only of the surface, i.e., the boundary, and we cannot see all of that at once. Stereo vision is the only significant visual mechanism we have to distinguish between two and three dimensional objects, and this does not provide a significant increase in the amount of information about that object. Information is dramatically increased by movement of either the object or the subject in space, and movement is, of course, dependent on time. Immediate tactile examination of an object, while providing a sense of space, does not provide a meaningful amount of information. It is not until perception continues over a period of time that information increases. When this occurs, experience and memory become perceptual mechanisms that condense perceptual time for familiar objects and environments.

The importance of memory in spatial orientation cannot be underestimated. We can easily navigate familiar environments in the dark, which is what the blind must do all the time. It has been shown that mice navigate mostly by memory. If a barrier is placed in a mouse's normal path it will learn to jump over it. When the barrier is removed the mouse will continue to jump at the same location. We often examine unfamiliar objects by turning them with our hands, but we can turn familiar objects with our minds. We become so familiar with the 3-D world that we can easily recognize space in a 2-D representation such as a painting or photograph. Memory is obviously linked to time.

5. The Epistemology of Perception.

5.1 Perception of Space. Although the concept of a two dimensional object is mathematically valid, and easily grasped in the mind, it cannot exist in the physical world. In order to perceive two dimensions, three are necessary. A 2-D entity can move in 2-space, or it can be viewed from the perspective of three dimensions. Similarly, I have shown that time is necessary to the perception of three dimensions. If we consider time to be a fourth dimension, then these propositions become analogous. Extending this analogy could lead to the conclusion that perception of n dimensions requires $n + 1$ dimensions. The +1 being either time or another spatial dimension. This could be seen to suggest an incompleteness conjecture for perception, but I think that, given the methodology used here, this would be stretching my analogy too far. First of all, perception is not an axiomatic system. Secondly, current thinking in theoretical physics does not allow an infinite number of dimensions. Finally, perception is too closely linked to a very subjective view of the physical world. But it would certainly offer a subject for further investigation.

5.2 Perception of Time. While the necessity of time as another dimension in 3-D perception does not lead to any conclusions that can be extrapolated to further spatial dimensions, it does lead to some very interesting answers to questions about perception of time itself. While the other dimensions seem to have no preferred direction, time seems to be asymmetrical. "…what else might lead nature to prefer retarded waves (forward in time) over advanced waves (backward in time), given that both varieties apparently comply with her laws of electromagnetism?"[2] I answer this question with one of my own: If time is necessary for the complete perception of three dimensions, then why shouldn't some further dimension be necessary for complete perception of time?

Notes

[1] H. H. Arnason, *History of Modern Art*, 1986, Harry N Abrams, Inc., New York
[2] Paul Davies, *About Time: Einstein's Unfinished Revolution*, 1995, Simon & Schuster, New York.

References

Herschel B Chipp, *Theories of Modern Art: A Source Book by Artists and Critics*, 1968, University of California Press, Berkeley.
George Gamow, *One two three...infinity: Facts & Speculations of Science*, 1961, The Viking Press, New York.
Oliver Grau, *Virtual Art: From Illusion to Immersion*, 2003, The MIT Press, Cambridge, Massachusetts.
Charles Harrison and Paul Wood, editors, *Art in Theory 1900-1990: An Anthology of Changing Ideas,* 1992, Blackwell, Oxford.
Leonard Shalain, *Art and Physics: Parallel Visions in Space, Time and Light,* 1991, William Morrow, New York.

An Interactive/Collaborative Su Doku Quilt

Eva Knoll
Mount Saint Vincent University
Halifax, Nova Scotia
eva.knoll@msvu.ca

Mary Crowley
Mount Saint Vincent University
Halifax, Nova Scotia
mary.crowley@msvu.ca

Abstract

After introducing Su Doku, a popular number place puzzle, the authors describe a transformation of the puzzle where each number is replaced with a distinct colour. The authors investigate the nature of the experience of solving this transposed version. This, in turn, inspires a design process leading to the creation of an interactive quilt. This process, involving issues of choice of medium, level of interactivity, colour theory and aesthetics, is described. The resulting artefact is a textile diptych accompanied by a collection of coloured buttons, constituting a solvable puzzle and its solution.

1. Introduction

Sometimes called the "Rubik's cube of the 21st century" [1], the mathematical puzzle Su Doku has been sweeping the world. The rules of the puzzle are simple:

> The puzzle is most frequently a 9×9 grid, made up of 3×3 subgrids called "regions" [...]. Some cells already contain numbers, known as "givens" or "clues". The goal is to fill in the empty cells, one number in each, so that each column, row, and region contains the numbers 1-9 exactly once. Each number in the solution therefore occurs only once in each of three "directions". [2]

For the rare reader who has not heard of this puzzle, the online encyclopaedia entry at wikipedia [2] can answer many questions both historical and analytical. In this paper, we describe a method of solution that uses colours instead of numbers and show that the solution results in a pleasing template for a quilt. Using this idea, we design a quilted diptych that depicts a specific puzzle [3] in a way to make it interactive. The numerical puzzle on which we based our work, shown in Figure 1, has been classified as "easy".

1			8	3				2
5	7				1			
			5		9		6	4
7		4			8	5	9	
		3		1		4		
	5	1	4			3		6
3	6		7		4			
			6				7	9
8				5	2			3

Figure 1: *The starting point of the puzzle*

2. The strategy versus the technique

Under normal circumstances, solving a Su Doku puzzle involves scanning, marking up and analysing [2] digits in the grid. Digits need to be manipulated mentally and placed, in various combinations in the appropriate cells of the grid, based on logical reasoning. Different people use different strategies to find the solution to a given puzzle, and, as is often the case in problem solving, these strategies appear to be more different than they really are, since the logic that leads different people to the solution is based on the same starting point and the same rules.

We distinguish two components to the solving of a Su Doku puzzle: strategy and technique. The strategy refers to the conceptual processes used to determine the content of each cell, such as, for example, elimination. In contrast, the technique refers to the physical manipulation of instruments, which, in the traditional solving process, are pencil and paper. Our main interest is in the latter: we investigate a technique that emphasises sensory-motor skills over logico-mathematical skills. In the technique we are describing, each number in the puzzle is replaced by a colour.

3. The colour technique

At first glance, it would seem that simply replacing the digits with colours does not change the solving method. In an experiential sense, however, it can. For instance, instead of holding a single pen or pencil in her writing hand, the solver would hold 9 different coloured ones in her other hand. For each blank cell, she puts down the pens corresponding to invalid colours. The pens remaining in her hand represent the possible solutions for that cell. If she is holding a single pen, that colour is the solution and the cell can therefore be coloured in. If she is holding several pens, she needs to record each of these possibilities. In a later stage, the solver determines the final solution for each cell by elimination. The consequence of the change of medium is that when it comes time to eliminate possibilities, this takes place through *separating off* pens (physical, motor), rather than *selecting* numbers (mental).

To illustrate these steps in the colour technique, we describe the solving of the puzzle for the central "region" of our example (number 5 in Figure 1, if counting sequentially left-to-right and top-to-bottom). Borrowing from the quilter's vocabulary, we refer such a 3×3 region as a block. Within the block, as the order in which the cells are investigated is open, we choose ones that highlight possible scenarios.

Figure 2: *The starting point in colour*

In Figure 2, the numbers have been replaced by colours (The colour version of these pictures can be found in the CD, and we recommend that you colour the photographs on this and the next two pages yourself to follow the discussion more easily). The chosen colours are: 1-*yellow*, 2-*orange*, 3-*red*, 4-*pink*, 5-*violet*, 6-*dark-blue*, 7-*light-blue*, 8-*dark-green*, 9-*light-green*.

Figure 3: *Step 1, a "sure"*

In Figure 3, to determine the solution for the cell that is indicated with the arrow (block 5, cell 9), the solver is able to eliminate *dark-green, yellow* and *pink* because they occur in the block itself. Additionally, investigating the column eliminates *light-green*, and *orange*. Finally, *violet*, *red* and *dark-blue* are also eliminated because they appear in the same row as cell 5-9. This leaves only *light-blue* in the hand, making it the solution for this cell. Holding a single pen this early in the solution process is not a frequent occurrence, and we call this a "sure". The colour is noted down in the cell (see Figure 4).

Figure 4: *Step 2, two possibilities*

For cell 5-1, starting with the nine colours again, the block rule eliminates *dark-green*, *yellow*, *pink*, and *light-blue* (because of the "sure" in the previous move), leaving *orange*, *red*, *violet*, *dark-blue* and *light-green.* The row rule allows the solver to remove *violet* and *light-green* as well, leaving *orange*, *red* and *dark-blue*. The column already contains *dark-blue*, and therefore, after all three rounds of elimination, the solver is left with *orange* and *red*. In this case, the final colour for that
cell is not yet determined. The solver records these possibilities as little marks in the cell (see Figure 5).

Figure 5: *Examining cell 5-2*

In cell 5-2, because of the block rule, she can again eliminate *dark-green*, *yellow, pink* and *light-blue* leaving *orange*, *red*, *violet*, *dark-blue* and *light-green* in the hand. The row rule allows the solver to remove the *violet* and *light-green* pens, and the column rule eliminates *red*. This leaves the solver with *dark-blue* and *orange* in hand, which the solver records in the cell (see Figure 6).

Figure 6: Examining cell 5-8

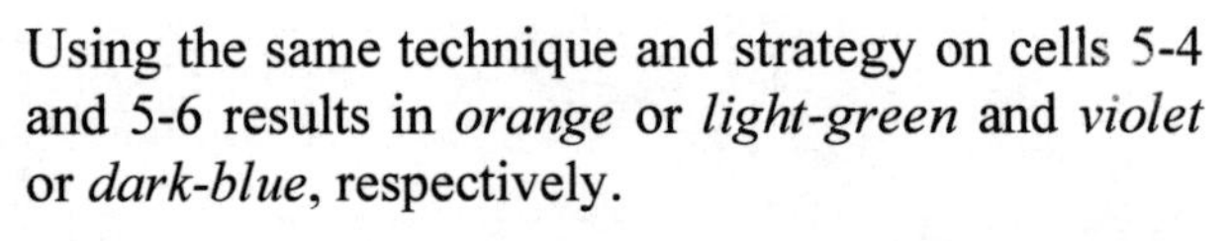

Using the same technique and strategy on cells 5-4 and 5-6 results in *orange* or *light-green* and *violet* or *dark-blue*, respectively.

In the last unexplored cell, 5-8, the solver is again left with *orange*, *red*, *violet*, *dark-blue* and *light-green* after applying the block rule. Applying the row and column rules then leaves the solver with *orange* and *light-green* in hand. The marks are noted in the cell.

The solver has now evaluated all the cells in the block, by considering only the givens and "sures". She can now determine some of the cells further through deductive reasoning and comparing.

Figure 7: *Solving by elimination*

For example, cell 5-6 is the only one in the block where *violet* is possible. This sets up a domino effect: by elimination, since cell 5-6 is not *dark-blue*, cell 5-2 is the only possible place for it. Further examination reveals that cell 5-1 is the only possible place for *red* (see choices in Figure 6).

Cells 5-4 and 5-8 cannot be determined at this stage: their resolution depends on the elimination of colours in their respective rows and columns outside of block 5.

The above sequence of moves shows the kinds of reasoning and movement that can occur while solving a Su Doku using coloured pens, instead of a single pen and symbols (be they numbers or otherwise). This demonstration is only the beginning of the design journey, however.

4. Harmonising the colours

As a consequence of the rules of the puzzle, each of the colours is distributed, though randomly, evenly across the grid. For this reason, a finished puzzle with all its colours possesses an overall balance, or harmony, depending on the choice of the colours. It is this choosing of colours and the accompanying design issues that we discuss next.

Traditionally, the colour wheel [4] is subdivided into the three primary colours (red, yellow and blue), the three secondary colours, which combine two primary colours (red-yellow for orange, yellow-blue for green and blue-red for violet), and the six tertiary (combining a primary with an adjacent secondary, for example blue and violet giving indigo) as shown in Figure 8, below. This total of twelve, however, does not suit the needs of colouring the Su Doku, if we want to maintain the puzzle's inherent balance. It is possible, however, to shift the colours so they are still regularly distributed along the wheel, but total nine. We have shown this in the outermost ring in Figure 8.

Because we tend to think of colours nominally, rather than by their position in the spectrum, we have chosen nine names that will help remember the order: Yellow, Orange, Red, Magenta, Violet, Indigo, Blue, Teal and Green. Conceptually, they are not equally spaced on the circle, but when the specific shades are chosen, they can be shifted to accommodate for this condition.

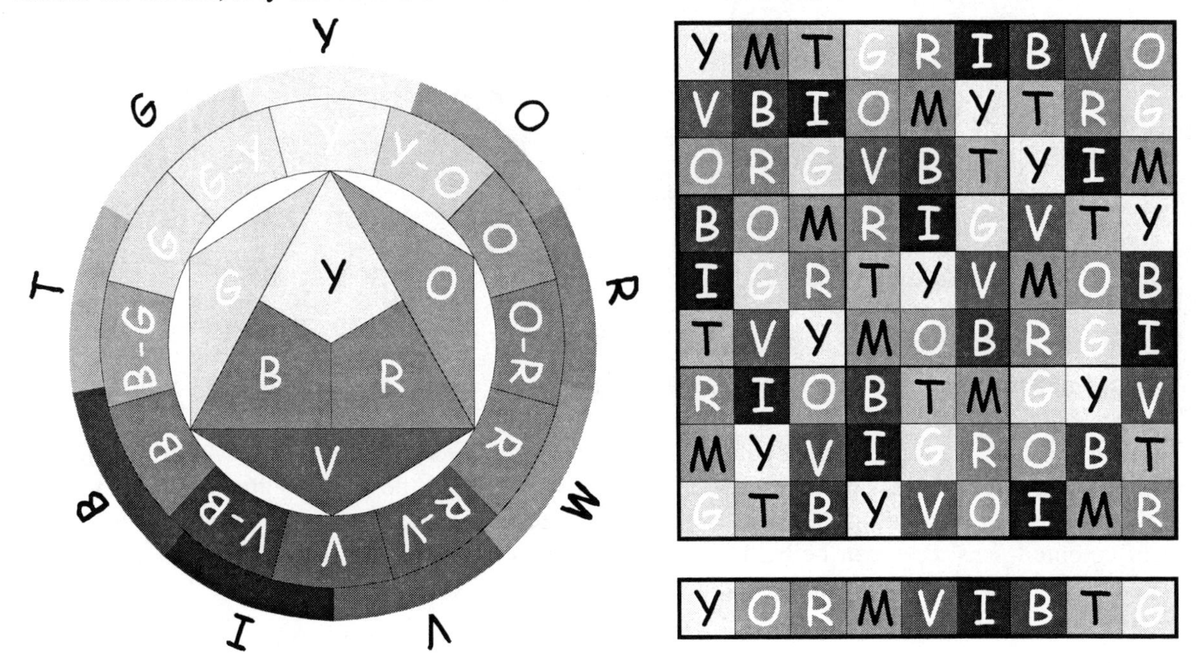

Figure 8: *The traditional colour wheel, and the subdivision into 9 [3]*

Figure 9: *The solution, using the 9 colours*

Figure 9 shows the solution to the puzzle with the nine colours chosen from the outer ring of the colour wheel of Figure 8. In the photographs depicting the process in section 3, the choice of colours was of course limited by the availability of the pens, and by the quality of the camera.

5. Developing the quilt – some design issues

All the laying out of colours and setting up of 3×3 blocks explained above is immediately recognisable to any quilter that has used this simple block design. Working out how to preserve the challenge of the Su Doku in a quilt, however, drew out our ingenuity.

We wanted the finished product to have a panel depicting the starting point of the puzzle and a panel depicting the finished solution. The puzzle could then be solved by using some additional moveable parts that would replace the written numbers in the conventional technique and the pens in the colour technique. This desire for interactivity is the engine that drove many of the decisions and added another dimension that a quilter normally does not need to work into the design.

The discussion extended from using beads corresponding to the same nine colours as the cells (which proved problematic due to the available materials), to creating patches that would, on one side, have the final colour for each corresponding cell, and on the other, the three, four or sometimes five possible colours for that cell, as illustrated in Figure 10.

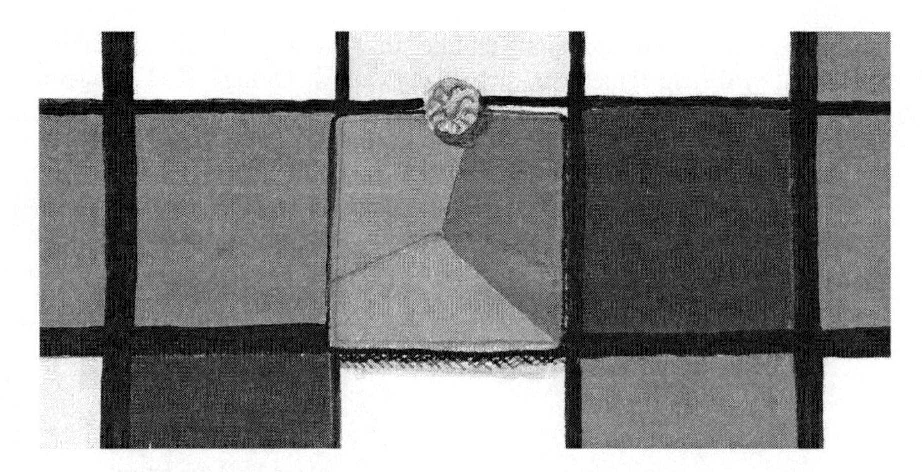

Figure 10: *Prototypes of the patches for one possible design*

We pursued this last idea for a while, but later abandoned it because it reduced both the challenge posed by the puzzle and its interactivity. The pre-made patches already depicted predetermined combinations for the cells. In addition, the solver cannot eliminate choices progressively, one at a time, within a single cell. Ultimately, we settled on a design that provided interactivity through the use of buttons covered with the same material as that composing the quilt itself, as illustrated in Figure 11. These buttons can be used in place of coloured pens: they can be held, discarded and positioned to record the possibilities for each. This medium provides no hints to the solver.

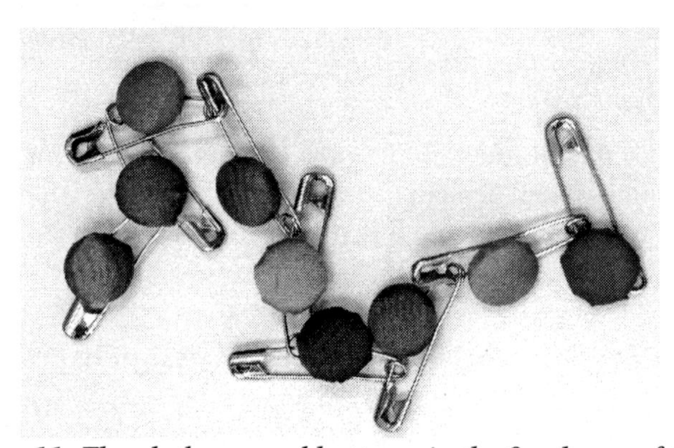

Figure 11: *The cloth covered buttons, in the 9 colours of the quilt*

We determined the total number of buttons per colour by calculating how many empty cells could contain each colour, if considering only the givens. This allows a solver to work in any order desired and leaves the choice of strategy open. Table 1 summarises the results of this calculation. Column 1 lists the Su Doku numbers. Column 2 indicates their frequency in the selected puzzle and column 3 the maximum possible number of buttons needed. This tabulation reveals a pattern: number 2, with the least givens (2), requires the most buttons (23). In contrast, numbers 3, 4 and 5, with the most givens (5) require the fewest buttons (4, 7 and 7 respectively).

Another important component of the design process involved choosing the colours. We still had to decide what colour was what number. In the case of the pens, we had simply assigned the colours to follow the spectrum (clockwise in Figure 8) starting from 1-yellow.

Table 1, column 5 shows the colours of the textiles, in order. We chose this sequence for two reasons. Firstly, the number of buttons required for colour 2 stands out: if all the possible colours per cell were displayed simultaneously, colour 2 would dominate. We decided that magenta, the most vibrant of the nine colours, would therefore provide the most pleasing energy to the starter quilt if it displayed all the buttons simultaneously.

Secondly, having chosen to follow the spectrum, we decided on the direction to follow around the colour wheel. Of the nine colours, more than half of the materials are in the violet/blue/green section of the spectrum. Consequently, we elected to counter-balance this bias by emphasising, through the initial number of givens, the magenta/red/orange/yellow section. This section then totalled 17 givens, against 18 for violet/blue/green, and meant that we were following the wheel counter-clockwise.

#	Givens	Maximum # of Buttons	Pen colours	Initial colour (first quilt)	Used colour (first quilt)
1	4	12	Yellow	Violet	Violet
2	2	**23**	Orange	Magenta	Magenta
3	5	**4**	Red	Red	Red
4	5	7	Magenta	**Orange**	**Blue**
5	5	7	Violet	Yellow	Yellow
6	4	8	Dark-blue	Green	Green
7	4	10	Light-blue	Teal	Teal
8	3	17	Dark-green	**Blue**	**Orange**
9	3	15	Light-green	Indigo	Indigo
Total	35	103			

Table 1: *Frequency of distribution and colour choices*

Using the colour associations of column 5 results in 62 buttons in the violet/blue/green section against 41 in the magenta/red/yellow/orange. This would produce a preponderance of dark buttons. To remedy this, blue was substituted for orange and vice versa, changing the ratio from 62:41 to 52:51 and giving the distribution of column 5. In this scenario, the sequence of frequencies along the spectrum becomes: Violet (12), Magenta (23), Red (4), Orange (17), Yellow (7), Green (8), Teal (10), Blue (7) and Indigo (15). The impact of this last design decision on the number of givens was minor: from 17:18 to 19:16.

We made two more colour decision. To mimic the original printed puzzle, we selected an off-white material, and for the frame, the colour of number 9, indigo because it was taken from the cells and contrasted well with the blanks.

6. Additional design decisions

There were many other decisions to be made along the way. Scale, for example, was important. The puzzle had to be manageable. The size of the buttons was determined by available supplies and affected the size of the cells. Ultimately, we chose a 2" x 2" square for the single cell, and this determined the overall puzzle dimensions.

At an early stage, the discussion even involved the making of a triptych, with the starting point on one side, an intermediate stage in the middle, and the final solution on the other side. In the end, we settled on a diptych, and the discussion moved on to whether the panels should be connected. Should they be sown back-to-back, allowing them to be hung in space and admired from two sides, or should the solver be able to set them side by side? In the end, we made two separate panels (see Figure 13), which allowed for flexibility both in hanging and playing.

Figure 13: *The finished quilts*

7. Conclusion

When we embarked on this journey from puzzle to quilt, we did not know where it would lead. The initial transition from numbers to colours connected with our mutual love of colour. It also showed results that were reminiscent of works of Constructivist and Swiss Concrete art of the previous century, such as Joseph Albers, Max Bill, Karl Gerstner, Paul Klee and Richard Paul Lohse [5]. In addition, the geometry, as well as the colour distribution underlying the puzzle, invoked thoughts of quilting prompting us to re-interpret the puzzle as a coloured quilt. Using buttons instead of pens preserved the sensory-motor flavour of the puzzle-solving experience.

The design process itself was satisfying and multi-faceted. The variety of the avenues we explored, the complexity of the debates that the issues provoked and the range of media we considered provided us with continuing inspiration. Translating the puzzle into a quilt created a piece of mathematical art.

References

[1] http://www.conceptispuzzles.com/articles/Sudoku/

[2] http://en.wikipedia.org/wiki/Sudoku

[3] Gould, Wayne (Su Doku Grand Master). New York Post Su Doku: The Official Utterly Addictive Number-Placing Puzzle Book 1. New York: Harper Collins.

[4] Itten, Johannes (1985). Der Farbstern. Otto Maier Verlag Ravensburg.

[5] Guderian Dietmar (n.d.). Mathematik in der Kunst der Letzten Dreiβig Jahre. Ebringen i. Br.: Bannstein-Verlag.

Patterns on the Genus-3 Klein Quartic

Carlo H. Séquin
Computer Science Division, EECS Department
University of California, Berkeley, CA 94720
E-mail: sequin@cs.berkeley.edu

Abstract

Projections of Klein's quartic surface of genus 3 into 3D space are used as canvases on which we present regular tessellations, Escher tilings, knot- and graph-embedding problems, Hamiltonian cycles, Petrie polygons and equatorial weaves derived from them. Many of the solutions found have also been realized as small physical models made on rapid-prototyping machines.

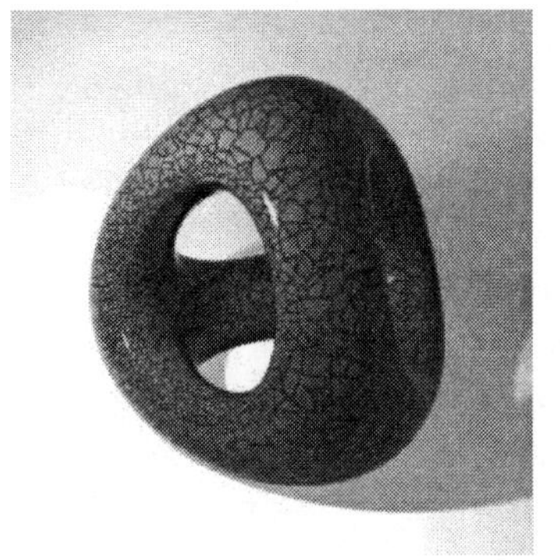

Figure 1: *Quilt made by Eveline Séquin showing regular tiling with 24 heptagons (a); virtual tetrus shape with cracked ceramic glazing programmed by Hayley Iben (b); and dual tiling with 56 triangles (c).*

1. Introduction

The Klein quartic, discovered in 1878 [5] has been called one of the most important mathematical structures [6]. It emerges from the equation $x^3y + y^3 + x = 0$, if the variables are given complex values and the result is interpreted in 4-dimensional space. This structure has 168 automorphisms, where, with suitable variable substitution, the structure maps back onto itself. To make this more visible, we can cover the 4D surface with 24 heptagons. Every automorphism then maps a particular heptagon onto one of the 24 instances, in any one of 7 rotational positions. This means that all 24 heptagons, all 56 vertices, and all 84 edges are equivalent to each other. In 4D this is a completely regular structure in the same sense that the Platonic solids are completely regular polyhedral meshes.

If we try to embed this construct in 3D so that we can make a physical model of it, we lose most of its metric symmetries, the regular heptagons get distorted, and only the symmetries of a regular tetrahedron are maintained. However, the topological symmetries are fully preserved. Helaman Ferguson's sculpture *The Eight-Fold Way* at MSRI in Berkeley celebrates this shape. A template for one of the 24 heptagons obtained from Bill Thurston allowed my daughter Eveline to stitch together the quilt shown in Figure 1a. The underlying shape is a rounded tetrahedral frame; we will call such a smooth, symmetrical genus-3 torus with full tetrahedral symmetry a *tetrus* (Fig.1b). Polyhedral approximations of this shape, such as the structure exhibiting the dual tiling of Figure 1a, are called *tetrads* (Fig. 1c).

In this paper, we use the tetrus shape as a canvas to study knot-, link-, and graph-embedding problems, as well as patterns derived from the underlying regular heptagonal tiling, and explore what artistically satisfying structures may result. In particular, I will subject the Klein surface to some of the same analyses and embellishments that I have applied to the 3 and 4 dimensional regular solids in the past. We will look for symmetrical colorings, Escher-like tilings, Hamiltonian cycles on the edges of that tiling, and interwoven ribbons forming orderly tangles [3] on that surface. Some of these patterns may lead to rather attractive abstract geometrical sculptures.

2. Regular Two-Manifold Meshes

Geometrically, Klein's highly symmetrical quartic can bee seen as a hyperbolic "Platonic" solid of genus 3. It is a completely regular 2-manifold composed of 24 heptagons, 84 edges, and 56 valence-3 vertices. Embedded in 4-dimensional space it exhibits 168 automorphisms and 168 anti-automorphisms (mirrored mappings). This is the maximal number of symmetries for a compact Riemann surface of genus 3. (The next genus, for which this maximal limit of 84*(g–1) automorphisms can be reached, is genus 7, leading to a group with 504 automorphisms [4].)

Helaman Ferguson's marble sculpture carries its name *The Eight-fold Way* because of an intriguing property: You can start at any vertex and move along consecutive edges, while alternately taking the left or the right branch at each subsequent vertex, and you will then end up where you started after exactly eight moves. These particular paths, called *Petrie polygons*, which always hug any face for exactly two consecutive edges before moving away from it, can also be drawn on the Platonic solids. On the tetrahedron the Petrie polygon has only four segments; on the cube and octahedron we obtain a "6-fold way" (Fig.2a), and on the dodecahedron and icosahedron we obtain zig-zag loops with 10 edges (Fig.2b). On the Klein quartic, the six Petrie polygons wrapping individually around each arm are easy to follow (yellow in Fig.2c), the other ones are trickier to trace without making a mistake. Looking for these Petrie polygons is a good way to check connectivity, when making a polyhedral model of the Klein quartic (Fig.1a); because it is very easy to join the four tripodal hubs with the wrong amount of twist.

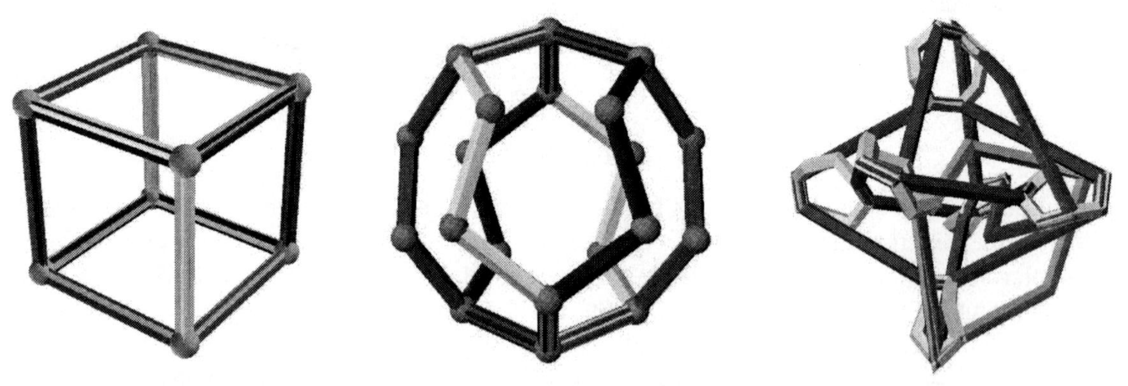

Figure 2: *Petrie polygons on the cube (a), on the dodecahedron (b), and on the Klein quartic (c).*

3. Equatorial Weaves

These Petrie polygons serve as a starting point for other intriguing constructions. If we connect subsequent midpoints of the edges in each Petrie polygon, we obtain 21 geodesically smoother cycles on the Klein quartic. This network can now be turned into a "woven" structure by having each strand alternatingly passing over and under subsequent strands that it crosses. Applying this technique to the Petrie cycles of the dodecahedron, and realizing each strand with a tube of finite thickness, one obtains a very pleasing and intricate structure, that is being sold in many museum stores (Fig.3a). Notice that all six star-shaped equatorial orbits are perfectly planar.

The question now arises, whether we can construct a similar assembly for the Klein quartic. Each loop comprises 8 vertices. If we restrict the equatorial loops around the 6 arms of the tetroid frame to be regular octagons (yellow in Fig.3b), then the overall figure is defined in its gross shape by two main parameters; the diameter of these arm loops and their distances from each other. Among the 21 Petrie cycles, there are: the six yellow "arm loops", three "major loops" in a Borromean configuration (white in Fig.3b,c), and 12 "shoulder loops," each running over one of the three shoulders of the four tripodal hubs, and then diving inside the structure and winding around exactly one arm of the tetrus. Our first goal is too keep all of these loops planar. The arm loops are planar by construction. The major loops share 4 points with 4 arm loops that all lie in the same coordinate plane; all we thus need to do, is to keep the 4 intermediate points also in this plane. The shoulder loops cross only three arm loops. These three points are just sufficient to define the planes for these shoulder loops. The intersection points between the shoulder loops and the major loops can be calculated in closed form (with some non-trivial expressions) from the intersection of the two loop planes involved and from some symmetry constraints. The intersections between pairs of shoulder loops leave 2 free parameters, which determine the height of the inner and outer tetrahedral hubs. Figure 3b shows the resulting network of 21 planar octagons.

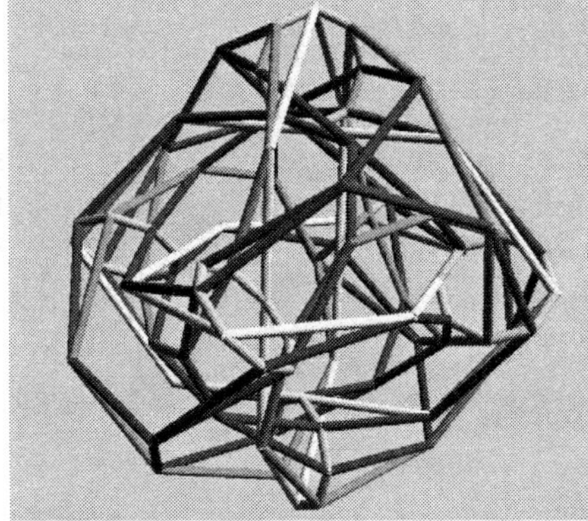

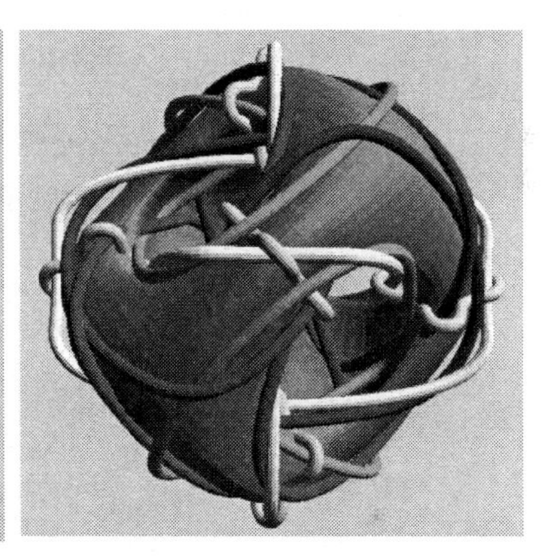

Figure 3: *Equatorial weave on a dodecahedron (a) and on the Klein quartic; network of Petrie cycles (b) and corresponding {over-under-}weaving loops on a tetrus surface (c).*

Starting from this basis, we now apply suitable in-plane distortions to these Petrie cycles so as to achieve the desired {over-under-} weave for all paths. We turn the paths into smooth B-splines, and sweep circles along them to form tubes of a finite cross section. The result of this equatorial weave, superposed onto a smooth tetrus body, can be seen in Figure 3c.

4. Knot and Link Embedding

Figure 4a singles out the three white Borromean loops found in the equatorial weave of Figure 3c. When drawn in the plane, this would be a 6-crossing configuration. In the typical knot tables [1], this link is identified as $6^3{}_2$. On the tetrus (or on any other genus-3 surface) it can be embedded crossing free.

Among all the knots, only the unknot (0_1) can be drawn without crossings in the plane. On the torus, all the torus knots and links can be drawn crossing-free. Among those the simplest knot is the trefoil knot (3_1), and the simplest link is $2^2{}_1$, composed of two interlocking unknots. The latter partitions the torus surface into two regions that can be colored differently.

It is thus natural to ask: Which is the simplest knot (lowest index) in the Knot Book [1] that can be drawn crossing-free on the Klein surface, but which cannot be embedded in a surface of lower genus. Progressing through the table of knots, we find that we can embed all knots up to 6_2 on either a 1- or 2-hole torus. Knot 6_3 (Fig.4b) is the first one that needs a genus-3 surface to be embedded crossing-free (Fig.4c).

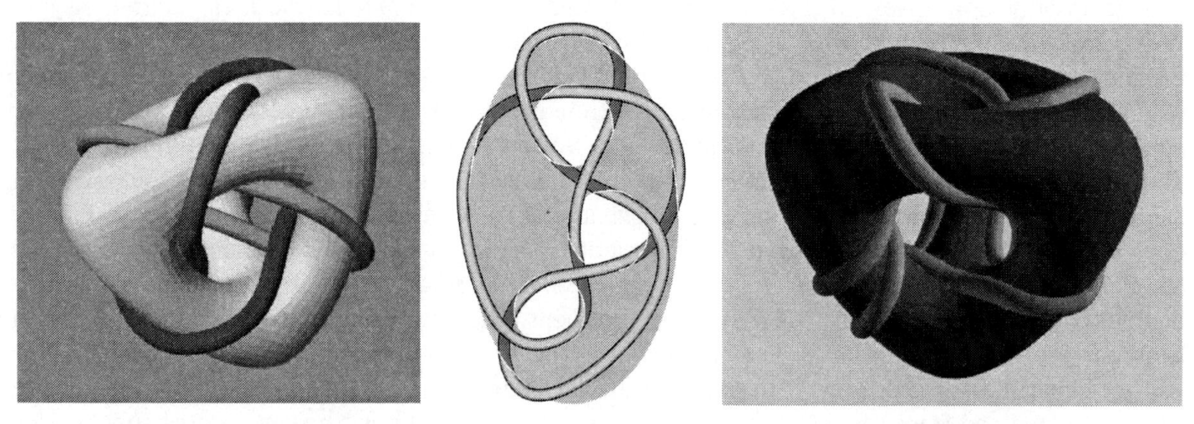

Figure 4: *Borromean link (a), and knot 6_3 on a 3-hole disk (b) and on the Klein quartic (c).*

These findings were established by looking at the graphs of the various knots and trying to see them embedded in the surface of a disk with holes. In an alternating {over-under-} knot, every single segment of the graph between two subsequent crossings makes a transition from the front of the disk to its back, or vice versa. Thus all these knot segments need to go around an outer border of the disk or through one of its holes to make such a transition. We thus draw the necessary border-loops into the knot diagrams as follows: Draw one outer border-loop around the whole knot figure, touching each of the outermost knot segments exactly once. Also draw similar border-loops into each region of the knot diagram that has an even winding number; expanding those loops so that they touch exactly once each of the knot segments forming that region. The number of these inner border loops then tells us the genus of the surface that is sufficient for a crossing-free embedding of that knot. (But this may not be the lowest possible genus!) Figure 5 shows representative examples of this construction for some genus-1 and genus-2 cases.

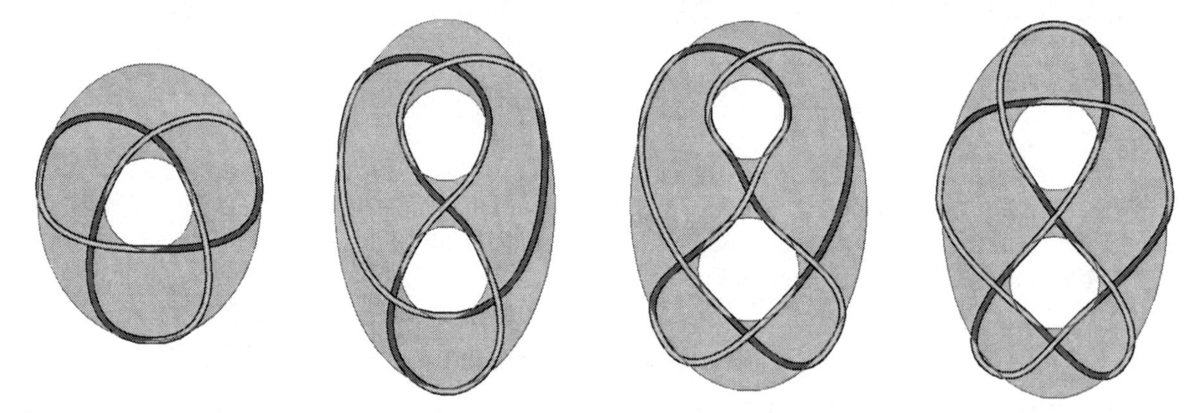

Figure 5: *Embedding analysis for knots 3_1 (trefoil), 4_1 (figure-8 knot), knot 5_2, and knot 6_2.*

The knot 6_3 does not partition the Klein surface into separate regions (Fig.4c). The simplest knot/link that can achieve this is the link 6^2_2. Its embedding diagram on a disk is shown in Figure 6a, and a smooth geodesic embedding on the Klein surface appears in Figure 6b. This figure exhibits C_2 symmetry; the intersections of the symmetry axis with the two links are marked with small bright circles. Figure 6c shows the resulting partitioning of the Klein surface into two differently colored domains.

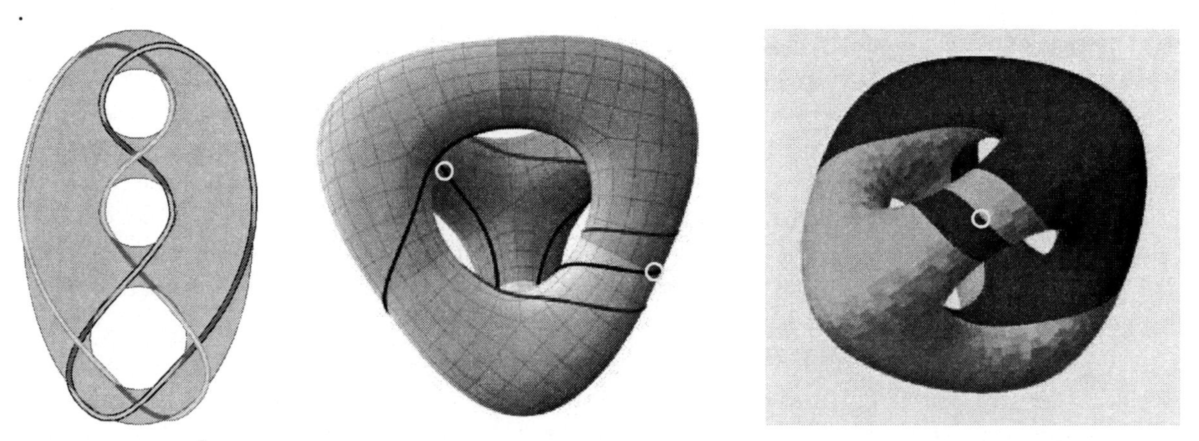

Figure 6: *Link $6^2{}_2$; embedding analysis (a), geodesic embedding on the Klein surface (b), and the resulting partitioning of the tetrus surface (c). Circles indicate C_2 symmetry points.*

5. Hamiltonian Cycles

Given the edge graph of the Klein quartic with 24 heptagons, it is natural to ask whether this graph admits a Hamiltonian cycle. We expected the answer to be positive, and thus, in addition, we aim to find as symmetrical a path as possible. On all the Platonic solids we can find such a cycle as the perimeter of a strip of simply connected faces, using exactly half of them; and for all 5 solids such a cycle exists that dissects the surface into two disjoint congruent regions. On the tetrus tiled with heptagons this is not possible. I conjecture that on a surface of odd genus no simple knot, that requires that genus for crossing-free embedding, can partition the surface into disjoint regions. I started a manual search on the 4-fold symmetrical parallel projection of the 84-member edge graph (Fig.7a) and boldly searched for a path with bilateral symmetry across both the x- and the y- axis. On the fifth try I found one with full D_2 symmetry (Fig.7a,b,c).

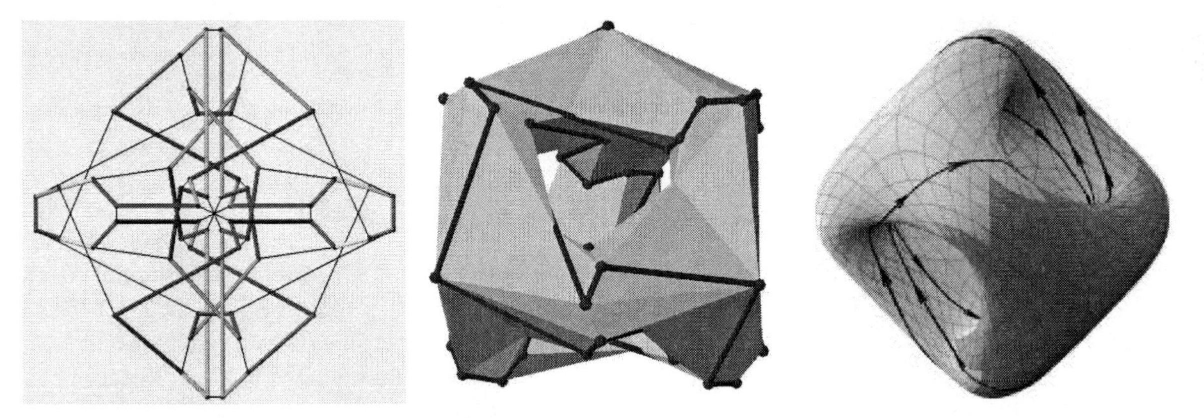

Figure 7: *Hamiltonian cycle on the Klein quartic with 24 heptagons: Path found on the projected edge graph (a), path shown on tetroid (b), and geodesically minimized path on a smooth tetrus shape (c).*

It is now interesting to explore what kind of knot this cycle forms. It turns out to be simply the unknot, but with a non-reducible embedding on the tetrus surface. And – no – it does not partition the surface!

6. "Viae Tetrus" – Circumnavigating the Klein Quartic

Tracing along the above Hamiltonian cycle with a tube or ribbon does not yield a good representation of the underlying shape in the same way that in Escher's "Bond of Union" (1956) a single ribbon defines the shape of the two heads that it is wrapped around. Of course, in Escher's drawing it helps that the ribbon also carries some of the salient features of the two faces, i.e., eyes and lips, as decorations. More to the point, most of my "Viae Globi" sculptures of a few years ago (Fig.8a) [7] make the underlying sphere readily apparent. We now look for ways to describe the tetrus shape in an artistically pleasing way by winding a ribbon over its surface. Ideally, I would like to find a solution that preserves the tetrahedral symmetry of the tetrus shape. However, I have come to the conclusion that this is not possible; deep down it has something to do with the fact that there exists no Eulerian cycle on the tetrahedron edges. Figure 8b shows a "multi-Eulerian" cycle on the tetrahedron, which visits every arm exactly twice. Such cycles are not difficult to find, if we use an even number of visits on each edge. However, this cycle has only 3-fold symmetry and does not preserve the full tetrahedral symmetry.

To obtain a good representation of the tetrus shape it is preferable to use **four** passes through each arm. We thus start with four intertwined helical paths. This leads to a good surface definition, since every cross section through an arm encounters four ribbon branches. The one parameter we can adjust most easily to obtain the desired connectivity, is the amount of twist we give to each of these cork-screws. It turns out that we can achieve the desired connectivity, that links everything into a single cycle, by using only two different twist numbers: On three of the arms, the helical paths twist through an angle of 180° and on the other three arms through 270° degrees. The result is shown in Figure 8c.

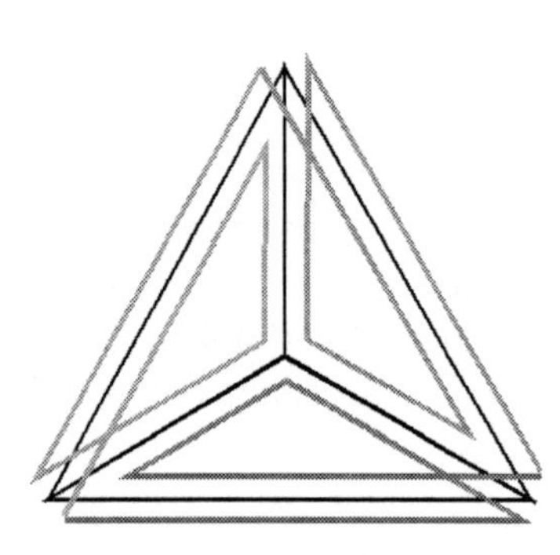

Figure 8: *'Altamont' from the Viae Globi series (a). Multi-Eulerian cycle on the tetrahedron (b). A single ribbon wound around a tetrus (c).*

7. Symmetrical Graph Embedding

The tetrus shape is a convenient canvas for the embedding of more complex graphs, in particular for those that have an inherent tetrahedral symmetry. One such graph is the tripartite graph $K_{4,4,4}$, the dual of Dyck's graph [8]. Each of its twelve nodes is connected to eight other ones in a 4-fold symmetrical manner (Fig.9a). It can be embedded in an orientable genus-3 surface forming a regular map, and the corresponding triangulated 2-manifold has 48 edges and 32 three-sided facets. The main challenge is to find good locations on the tetrus for the twelve nodes of this graph. A key issue is how to place the sets of nodes that are **not** connected to one another. For every node there are three others to which there are no direct connections. After some study, it becomes quite plausible to place each set of four such nodes onto one of the three D_2 symmetry axes of the surface. We believe that this results in an optimal overall embedding of the graph. Thus each tetrus arm carries two vertices, one on the inside and one on the outside. A physical model of the genus-3 Klein surface has been built on a rapid prototyping machine. Nodes, edges, and facet colorings have then been painted by hand onto this model (Fig.9b).

A virtual rendering of that same structure, using translucent surface panels and four internal light bulbs, thus simulating a genus-3 Tiffany lamp [8], is shown in Figure 9c.

Figure 9: *Graph embedding: The graph $K_{4,4,4}$ (a), painted onto a tetrus (b), and a virtual rendering of a corresponding "Tiffany lamp" (c).*

8. Regular Tilings and Colorings

After studying 1-manifolds embedded on this highly regular genus-3 canvas, we now look at various tilings of this surface, in particular, the tiling patterns implied by the symmetry group associated with Klein's quartic. The tilings that makes this surface such an important mathematical object are the topologically completely regular tessellation into 24 heptagons (Fig.10b), and its dual consisting of 56 triangles joined in 24 valence-7 vertices (Fig.1c). To maintain as much symmetry as possible, the points of 3-fold symmetry in these tiling patterns should be aligned with the four tripodal poles on a genus-3 surface with tetrahedral symmetry. Figure 10b shows the adjacencies of these heptagons on the tetrus. The two underlying dual tiling patterns, {7,3} and {3,7}, can also be embedded in the Poincaré disk, where infinitely many tiles can be accommodated within the circle limit; but only 24 heptagons are needed to cover the Klein quartic, after that, the same pattern (tiles #1– #24) repeats itself (Fig.10c).

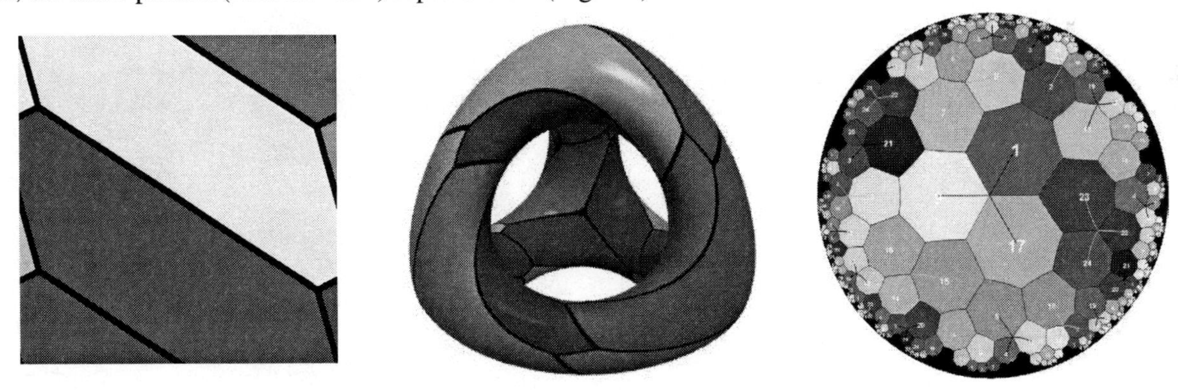

Figure 10: The basic tile texture (a) to produce Klein's heptagonal tiling on a Catmull-Clark subdivision tetrus with 48 quadrilaterals (b), and the {7,3} tiling of the Poincaré disk (c).

What is the smallest number of colors needed, so that no two tiles of the same color are adjacent? How many different color patterns are there that respect the symmetry of the tetrus? Clearly, that number must be an integer divisor of the number of tiles, and the occurrences of tiles of the same color should be distributed over the tetrus structure as evenly as possible. For the group with 24 heptagons a most appropriate solution is to use 8 colors. Each heptagon can then be surrounded by all the other seven colors. Along the two sides of any Petrie polygon we also find all 8 colors. At the 56 vertices we will see all possible color triplets; and every possible color pair shows up exactly 3 times at the 84 edges. Figure 11a shows this coloring scheme spread out in a {7,3} hyperbolic Poincaré disk. There are three groups of 8 colors; each group is marked with a different central dot. Heptagons with the same color combinations map onto the same face, when this hyperbolic tiling is wrapped around the Klein quartic. Taking the special tetrahedral embedding in 3D space into consideration, this color mapping places 4 colors entirely on the "outside" of the tetrus structure and the other 4 colors on its "inside" (Fig.11b). If we don't like this color separation, we also find a highly symmetrical arrangement with only 6 colors. In this case, each color appears on two outer and two inner heptagons, and each color appears in every one of the four tripodal hubs, either on an outer heptagon or on an inner one (Fig.11c).

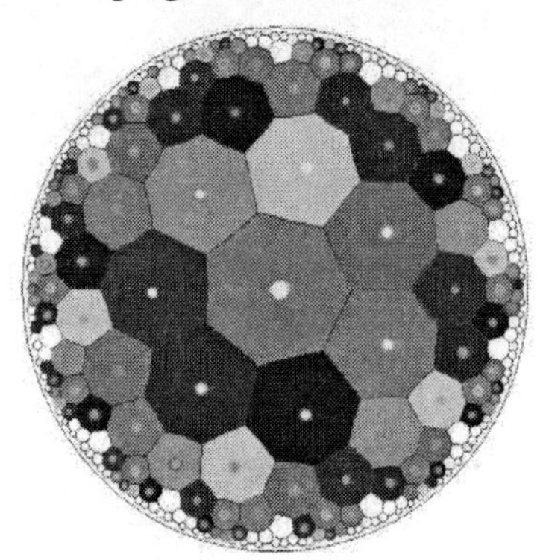

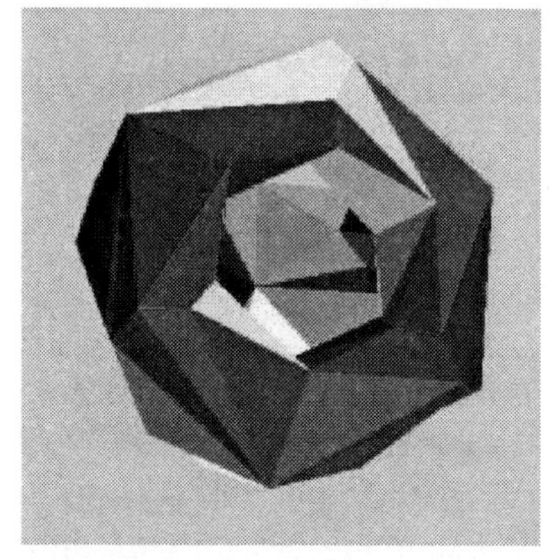

Figure 11: 8-color pattern on the Poincaré disk (a) and on a painted tetroid (b); tetroid coloring with tiles of 6 different colors (c).

Finally, even though 4 colors are not sufficient to color all heptagons so that no two adjacent tiles have the same color, this choice still offers an intriguing possibility. Adjacent pairs of equal colors (Fig.12a) can be chosen in such a way that we always combine an inner tile with an adjacent outer tile to form twelve identical double tiles. This can be done in four different ways. An extreme case where an inner and an outer tile are just joined by a short edge between them is shown in Figure 12b. A partial assembly of a tetroid built from such tiles is shown in Figure 12c.

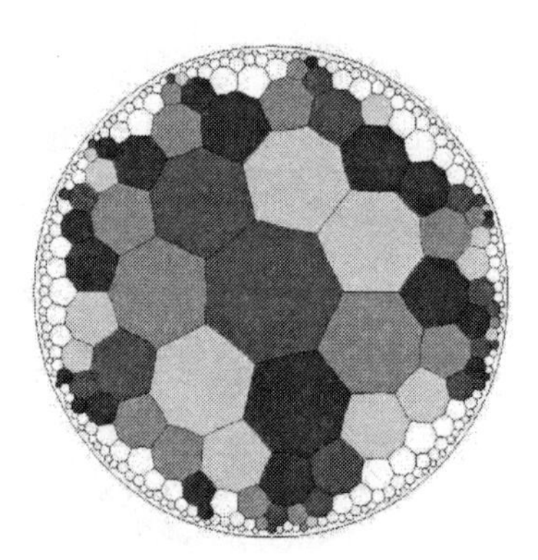

Figure 12: 4-color pattern on Poincaré disk (a); corresponding double tile on Klein's quartic (b), and a partially assembled tetroid from such double tiles (c).

Similarly, there are several options for pleasing regular or semi-regular color patterns for the dual tiling with 56 triangles. The possible contenders are, 4, 7, and 8 colors (Fig.1c). Four colors allow a coloring so that no two adjacent triangles are of the same color. Seven colors allow a mapping in which all 7 colors show up at each vertex. The possible tilings for the assembly of 84 quadrilaterals are being investigated.

9. Escher-like Tilings

Inspired by Douglas Dunham's pattern of *168 on a Polyhedron of Genus 3* [2], we now explore the ways in which the above regular tilings can be deformed to form Escher-like patterns on the tetrus shape. Ideally, we would like to have a tool such as the Escher-sphere editor created by Jane Yen [10]. The principle of generating an Escher tile is the same as in the plane: We distort the edges of a fundamental region (a heptagon) on the Klein quartic in such a way as to maintain all its symmetries, i.e., C_2 symmetry around the edge midpoint, C_7 symmetry around the tile center, and C_3 symmetry around the vertices. The so distorted tiles will then fit together again and seamlessly cover the whole surface. We just need to find a good way to map these tiles onto the tetrus surface. A difficulty arises from the fact that not all tiles have the same shape, even though they are topologically equivalent; the mapping onto the tetrus shape distorts the geometry of each heptagon in different ways, depending on where it lies. Fortunately, this distortion is not totally random; among the 24 heptagons there are only two different types of tiles: *outer* and *inner* heptagons. In the dual structure with 56 triangles we find 6 different tile shapes: 2 at the inner and outer poles, and 4 types around each arm. For the tessellation with 84 quadrilaterals resulting from the diamonds straddling the original heptagon edges, we find 8 different tile shapes, straddling respectively: the outer and inner pole edges, the long seams between inner and outer heptagons, and five different edges around each tetrus arm. These differently shaped tiles represent additional challenges when designing a decorative motif: We have to make sure that it works well for all occurring distortions of the tile. Fortunately, Escher's "creature tiles" are very deformable.

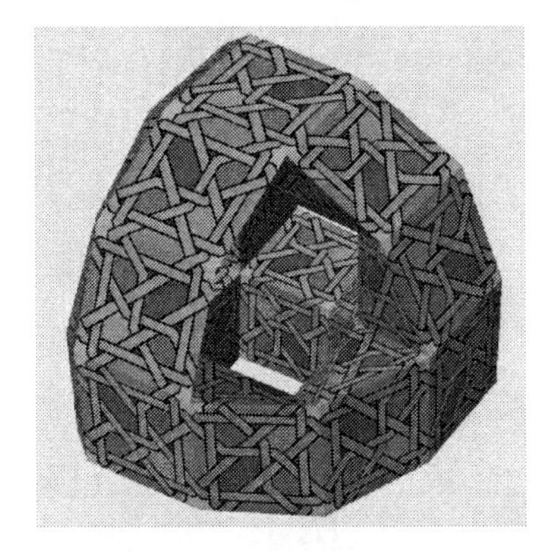

Figure 13: Texturing the tetrus. A polyhedral tetroid suitable for Catmull Clark subdivision (a); a quadrilateral tile with two "newts" (b), and the result of texture-mapping it onto a tetrus (c).

There are also some implementation and rendering difficulties. To obtain a smooth subdivision surface one needs nice, simple tiles, – not some concave animal-shape. Even just using the basic heptagonal regions shown in Figure 11b as a control polygon, leads to a badly wrinkled subdivision limit surface. For a Catmull Clark subdivision surface, we prefer to start with simple quadrilateral tiles lined up with the symmetry planes of the tetrus and/or with its principal directions (Fig.13a). But in that case the basic heptagon, and any Escher tile derived from it, will overlap several of the quadrilateral facets in odd ways (Fig. 10b), and designing suitable textures becomes more challenging.

Figure 13b shows a quadrilateral tile inspired by a combination of Escher's notebook patterns #25 and #35. It has been designed to fit directly onto a description of the tetrus structure using 48 quadrilaterals, covering a fundamental domain composed of four of them. For this case, texture mapping was easy, but the resulting pattern (Fig.13c) does not match the topological structure of Klein's symmetry group.

To capture the full topological symmetry with 168 automorphisms, we create a heptagonal tile with seven replicas of the fundamental tile (Fig.14a) – a fantasy fish inspired by Escher's pattern #55. The assembly of this tile in the {7,3} tessellation of the Poincaré disk (Fig.14b) has the same symmetry as Dunham's pattern of butterflies [2]. Since this display was created by texture mapping and I wanted to use only a single tile, I had to select a very special coloring pattern and carefully assign the orientation for each tile, in order to have all colors properly match up at the tile borders. Figure 14c shows the result of mapping 24 of these tiles around the tetrahedral genus-3 Klein quartic. Note that all yellow fish point towards the inner and outer tripodal poles of the tetroid. Other tiles and mappings can be found on my web site [9].

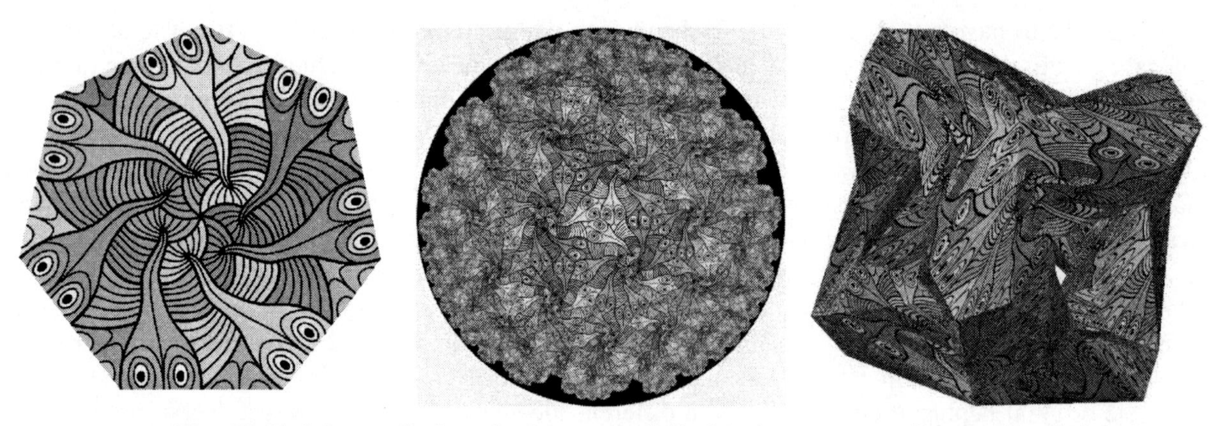

Figure 14: The {7,3} fish tessellation: the heptagonal tile (a), the corresponding Poincaré disk (b), and a symmetrical mapping of this pattern onto the tetrus (c).

10. Work in Progress and Conclusions

Klein's quartic and the genus-3 tetrus structure provide a rich domain for experimentation with geometrical operators and artistic effects. Beyond the examples presented in this paper, there are many more ways in which the Klein quartic can be celebrated. Work in progress includes a tangle of 24 knots, where each knot corresponds to one of the heptagonal tiles and links with all adjacent ones. A special knot has been designed to gracefully match the distorted shapes of these tiles on the tetrus surface. Other work concerns the development of a dissection puzzle for the tetrus shape into interlocking snap-together parts (Fig.12c).

Acknowledgements

This work was supported by the CITRIS Institute, one of the California Institutes for Science and Innovation (CISI) I also would like to thank the following students for their contributions: Hayley Iben for Figure 1b, Pushkar Joshi and Allen Lee for their help with mapping Escher tiles onto the tetrus and the Poincaré disk, and Vasily Volkov for the geodesic refinement of several knots and links. Thanks also to Douglas Dunham for a careful review of the manuscript.

References

[1] C. Adams, *The Knot Book*. W. H. Freeman and Co., New York, 1994.

[2] D. Dunham, *168 Butterflies on a Polyhedron of Genus 3*. Bridges 2002, Baltimore, Conf. Proc., pp 197–204.

[3] A. Holden, *Orderly Tangles*. Columbia University Press, New York, 1983.

[4] A. Hurwitz, *Ueber algebraische Gebilde mit eindeutingen Transformationen unter sich.* Math. Annalen 41 (1893) pp 403-442.

[5] F. Klein, *Ueber die Transformationen siebenter Ordnung der elliptischen Funktionen.* Math Ann. Vol 14, 1879. (Translation into English by S. Levy [4]).

[6] S. Levy, *The Eightfold Way: The Beauty of Klein's Quartic Curve*. Cambridge University Press, 1999.

[7] C. H. Séquin, *Viae Globi - Pathways on a Sphere*. Proc. Mathematics and Design Conference, pp 366–374, Geelong, Australia, July 3-5, 2001.

[8] C. H. Séquin, and L. Xiao, *K12 and the Genus-6 Tiffany Lamp*. Proc. ISAMA CTI 2004, pp 49–52, Chicago, June. 17-19, 2004.

[9] C. H. Séquin, *Tilings on Klein's quartic and on the Poincaré disk.* -- http://www/~sequin/GEOM/TILES/

[10] J. Yen and C. H. Séquin, *Escher Sphere Construction Kits*. Proc. Interactive 3D Graphics Symposium, pp 95-98, Research Triangle Park, NC, March 19-21, 2001.

The Lorenz Manifold: Crochet and Curvature

Hinke M Osinga and Bernd Krauskopf
Bristol Centre for Applied Nonlinear Mathematics
Department of Engineering Mathematics
University of Bristol, Bristol BS8 1TR, UK
E-mail: H.M.Osinga@bristol.ac.uk, B.Krauskopf@bristol.ac.uk

Abstract

We present a crocheted model of an intriguing two-dimensional surface — known as the Lorenz manifold — which illustrates chaotic dynamics in the well-known Lorenz system. The crochet instructions are the result of specialized computer software developed by us to compute so-called stable and unstable manifolds. The implicitly defined Lorenz manifold is not only key to understanding chaotic dynamics, but also emerges as an inherently artistic object.

1 Introduction

Many people know a version of the saying that a butterfly flapping its wings in Brazil can cause a tornado in Texas. This is also referred to as the *butterfly effect*, and it was first introduced by Edward Lorenz in 1963 [8] to illustrate extreme sensitivity of the weather. If a small effect such as the wing flap of a butterfly can be responsible for creating a large-scale phenomenon like a tornado, then it is impossible to predict the behaviour of a complex system. Of course, there are many butterflies that may or may not flap their wings...

Our work concentrates on the fact that one can still extract information from such a complex system and make predicitions at a more qualitative level via the computation of so-called *invariant manifolds*. The unpredictability of a chaotic system can be translated into the complexity of the geometry of these manifolds, which are two-dimensional surfaces in three-dimensional space in many cases. These surfaces emerge as inherently artistic shapes that are already implicitly contained in the mathematical model. We have developed computational methods to find and visualize such manifolds. Even better, our method can be interpreted as crochet instructions. This allows us to visualise the chaotic dynamics with a real-life crocheted model.

2 The Lorenz system

The classic model studied by Lorenz when he discovered the butterfly effect was a simplified set of seven equations describing the rising and cooling of hot air (thermal convection) in the atmosphere. He later managed to create the same dynamical effect in an even simpler model that is now known as the Lorenz system:

$$\left\{ \begin{array}{rcl} \dot{x} & = & \sigma(y-x), \\ \dot{y} & = & \varrho x - y - xz, \\ \dot{z} & = & xy - \beta z. \end{array} \right. \tag{1}$$

The classic values of the parameters, as introduced by Lorenz, are $\sigma = 10$, $\varrho = 28$, and $\beta = 2\frac{2}{3}$. The system is given in the form of a set of three ordinary differential equations where the vector

Figure 1: *The Lorenz manifold as computed by our algorithm.*

$(\dot{x}, \dot{y}, \dot{z})$ represents the (instantaneous) velocity of a particle at position (x, y, z). The Lorenz system is deterministic, which means that for each particle position Eqs. (1) uniquely describe the future and past of the dynamics. Remarkably, this is not enough to make predicitions for even relatively short time scales; we refer to [2] for a popular account and to [12] for more details on deterministic systems.

An important feature of system (1) is its symmetry: the behaviour of a particle starting at (x, y, z) is essentially the same as that of a particle starting at $(-x, -y, z)$. That is, any solution path in space can be transformed into another solution path by rotating it by 180° about the z-axis. The z-axis itself is invariant under the dynamics, which means that a particle on the z-axis will stay on it.

3 The Lorenz manifold

The origin of the Lorenz system (1) is an equilibrium, that is, the velocity vector at $(x, y, z) = (0, 0, 0)$ is the zero vector. It is a saddle point with a two-dimensional stable manifold, also called the Lorenz manifold. This manifold consists of all points that approach the origin in forward time. The chaotic dynamics of the system is essentially organised by how particles that pass close to

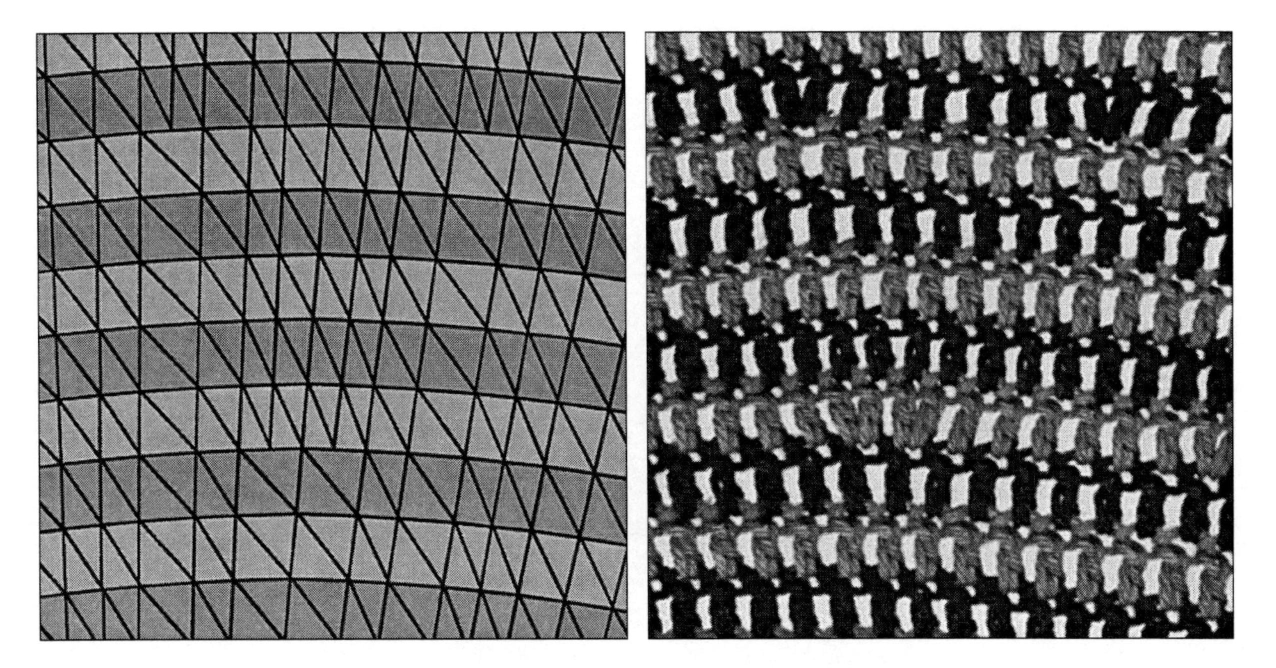

Figure 2: *The computed and corresponding crocheted mesh of part of the Lorenz manifold.*

the origin get pushed away again. Namely, the Lorenz manifold is the two-dimensional boundary surface that separates points that locally get pushed towards negative x-values from those that are locally pushed in the positive x-direction.

It has long been an open problem to find and visualize the Lorenz manifold, because there is no explicit formula for it. The first sketches of the Lorenz manifold appeared in 1982 in [1], but computational methods have been developed only quite recently; see [7] for an overview. The method we developed grows the surface as a set of concentric rings, starting from a small circle about the origin, in such a way that the surface grows with equal speed in all radial directions until a prescribed (geodesic) distance is reached. Details can be found in [4, 5], where we discuss the accuracy of the computations, while [9, 6] provide a less technical explanation. Figure 1 shows the Lorenz manifold computed with our method up to geodesic distance 110.75.

Our method represents the manifold as a triangulation between computed mesh points. At each step a new ring of mesh points and triangles is added. The images in this paper highlight consecutive rings by alternating light and dark blue. As the manifold grows, mesh points are added (or removed) to ensure an even distribution of mesh points and, hence, a faithful representation of the mathematical object.

This systematic way of building up the mesh can be interpreted directly as crochet instructions, which we published in [10]. Our crocheted model is built up round by round of crochet stitches of the appropriate length, where stitches are added or removed as dictated by the algorithm. Figure 2 shows a direct comparison between a part of the computed and the corresponding crocheted mesh. Notice in particular where mesh points and, consequently, crochet stitches have been added.

The result of crocheting the entire Lorenz manifold (up to geodesic distance 110.75) is shown in Figure 3. It is a floppy object that is impossible to lay down flat on a table. While the lower part is actually almost flat, the upper part (positive z-values) ripples considerably. This is due to the negative curvature of this part of the surface (compare with Figure 1), which is locally encoded simply by the way stitches are added during the crochet process. Negative curvature is generated during a growth process by a faster growth in the lateral direction. This can be found in nature, for example, in curly-leaf lettuce. The principle can also be used to crochet examples of hyperbolic

Figure 3: *The crocheted Lorenz manifold before mounting.*

planes [3], which are characterized by constant negative curvature. The Lorenz manifold, on the other hand, is not a (hyperbolic) plane, rather its curvature varies from point to point on the surface.

The model of the Lorenz manifold is obtained from the floppy crocheted object by mounting it with a stiff z-axis, a rim wire, and two additional supporting wires; precise mounting instructions can be found in [10]. Figure 4 shows the result seen from approximately the same viewpoint as in Figure 1. The mesh structure is brought out by the white background. Figure 5 shows a close-up view of the crocheted Lorenz manifold near the z-axis, where a black background is used to emphasize the surface geometry. In this region the local curvature is maximal, which creates the helical structure after mounting.

4 Mathematics or art?

Our motivation for creating the crocheted model of the Lorenz manifold was to have a three-dimensional hands-on model of this intriguing surface. However, apart from simply appealing to the specialists, the crocheted object is an excellent tool to communicate the beauty of complicated mathematical objects and ideas. This found resonance with the general public who perceived our creation as a piece of mathematical art. What is more, many people followed our published crochet and mounting instructions to produce their own Lorenz manifold. More information and photographs can be found on our dedicated website [11].

Stable and unstable manifolds of chaotic systems have complex and beautiful geometry. However, they are 'hidden' in the governing mathematical equations and must be brought to light with specialised algorithms. We hope that the example of the Lorenz manifold may serve as an inspiration to artists.

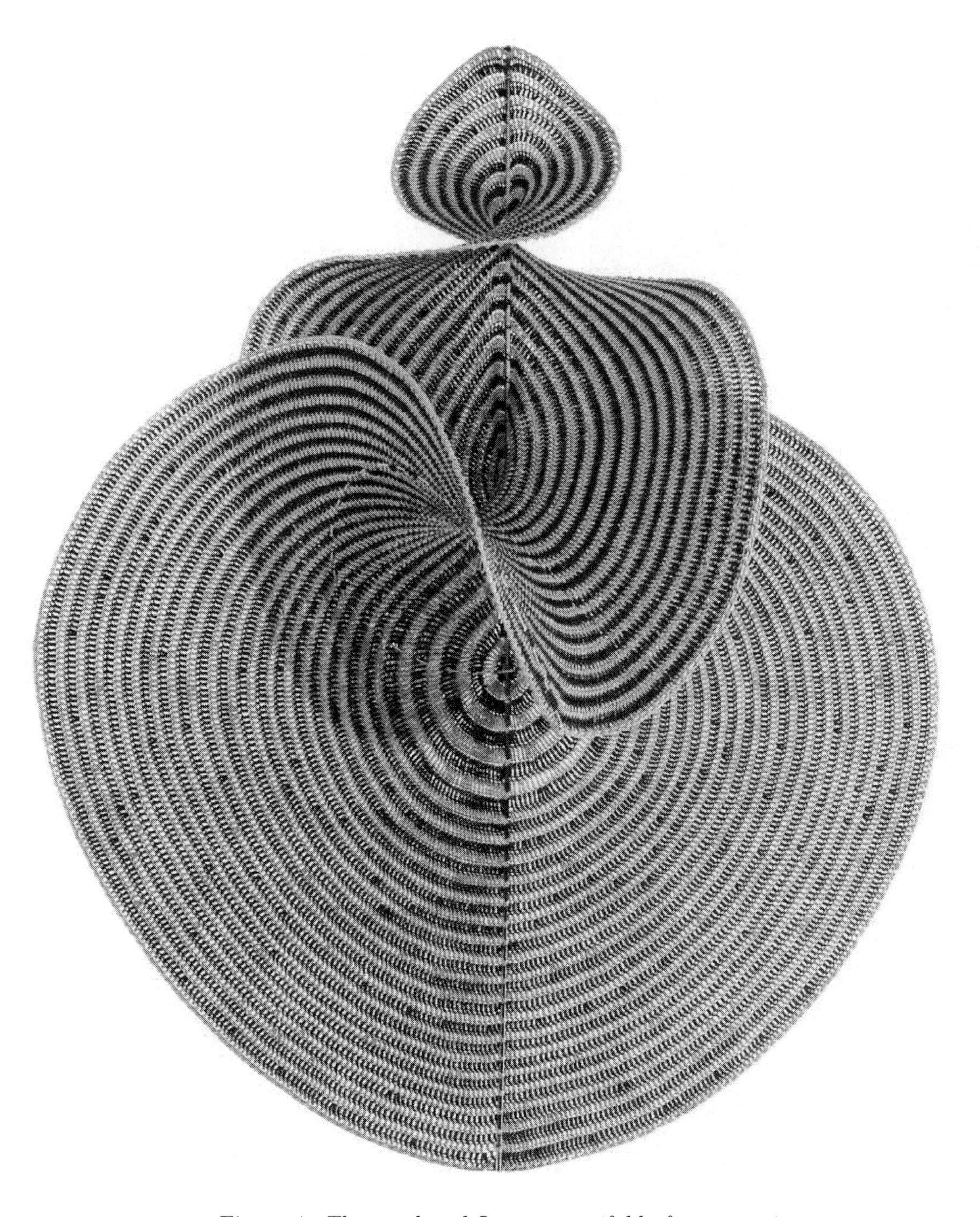

Figure 4: *The crocheted Lorenz manifold after mounting.*

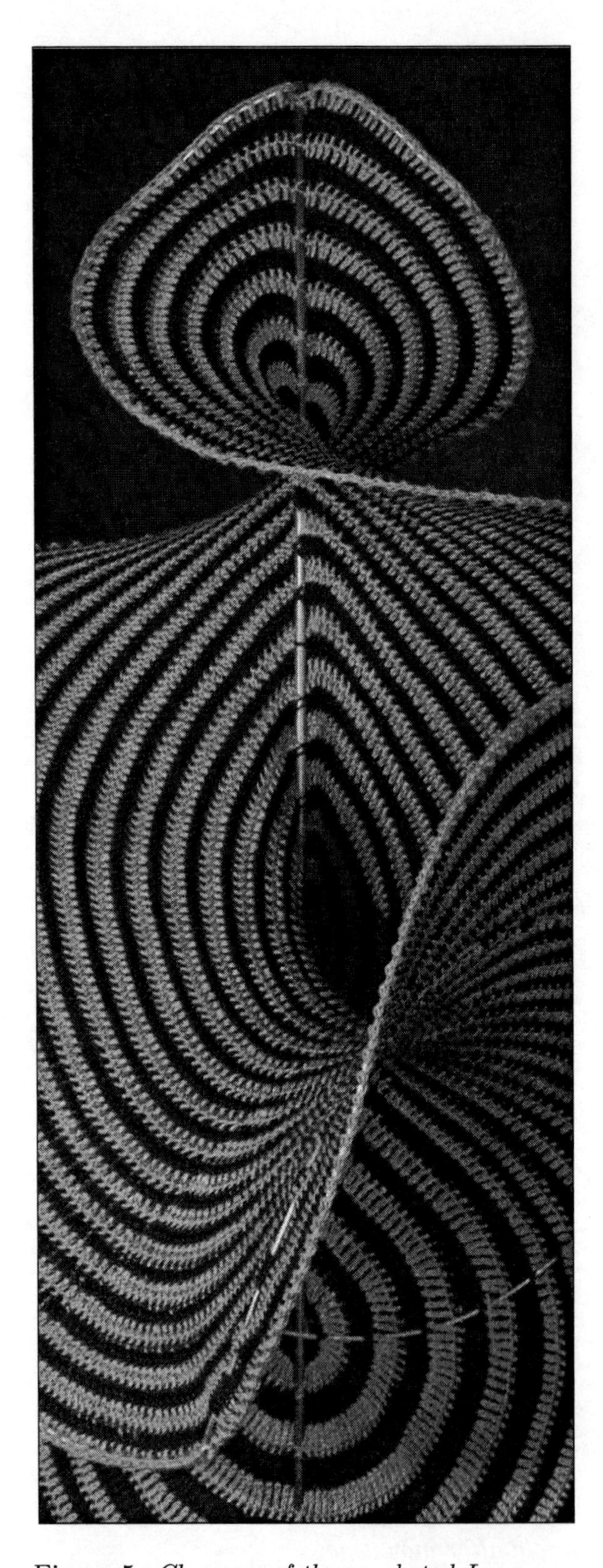

Figure 5: *Close-up of the crocheted Lorenz manifold.*

References

[1] R. H. Abraham and C. D. Shaw. *Dynamics — The Geometry of Behavior*, Part Three: Global Behavior, Aerial Press, Santa Cruz California, 1982-1985.

[2] J. Gleick. *Chaos, the Making of a New Science*, William Heinemann, London, 1988.

[3] D. W. Henderson and D. Taimina. Crocheting the hyperbolic plane. *The Mathematical Intelligencer*, **23**(2): 17–28, 2001.

[4] B. Krauskopf and H. M. Osinga. Two-dimensional global manifolds of vector fields. *CHAOS*, **9**(3): 768–774, 1999.

[5] B. Krauskopf and H. M. Osinga. Computing geodesic level sets on global (un)stable manifolds of vector fields. *SIAM J. Appl. Dyn. Sys.*, **2**(4):546–569, 2003.

[6] B. Krauskopf and H. M. Osinga. The Lorenz manifold as a collection of geodesic level sets *Nonlinearity*, **17**(1): C1–C6, 2004.

[7] B. Krauskopf, H. M. Osinga, E. J. Doedel, M. E. Henderson, J. Guckenheimer, A. Vladimirsky, M. Dellnitz and O. Junge. A survey of methods for computing (un)stable manifolds of vector fields. *Int. J. Bifurcation and Chaos* **15**(3): 763-791, 2005.

[8] E. N. Lorenz. Deterministic nonperiodic flows. *J. Atmos. Sci.*, **20**: 130–141, 1963.

[9] H. M. Osinga and B. Krauskopf. Visualizing the structure of chaos in the Lorenz system. *Computers and Graphics*, **26**(5): 815–823, 2002.

[10] H. M. Osinga and B. Krauskopf. Crocheting the Lorenz manifold. *The Mathematical Intelligencer* **26**(4): 25–37, 2004.

[11] H. M. Osinga and B. Krauskopf. Website: Crocheting the Lorenz manifold. `http://www.enm.bris.ac.uk/staff/hinke/crochet/`, 2005.

[12] S. H. Strogatz. *Nonlinear Dynamics and Chaos*. Addison Wesley, 1994.

Playing Musical Tiles

Rachel W. Hall
Department of Mathematics and Computer Science
Saint Joseph's University
5600 City Avenue
Philadelphia, PA 19131, USA
E-mail: rhall@sju.edu

Abstract

In this survey paper, I describe three applications of tilings to music theory: the representation of tuning systems and chord relationships by lattices, modeling voice leading by tilings of n-dimensional space, and the classification of rhythmic tiling canons, which are essentially one-dimensional tilings.

1. Introduction

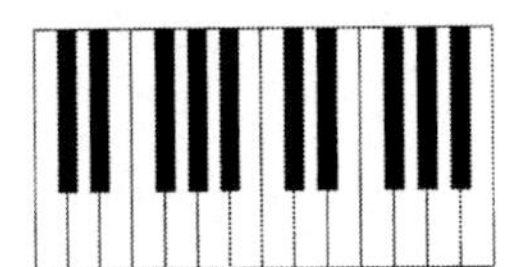

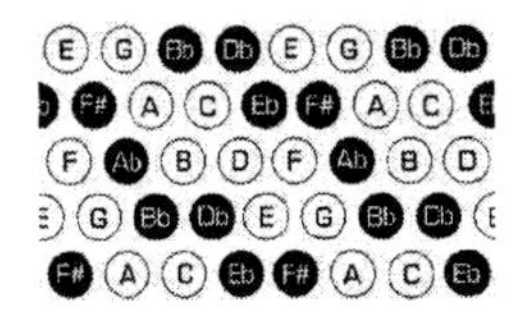


```
4/4        |:...*...*...*...|...*...*...*...:|
doumbek     B   O mmB O   m B   O mmB O   m
variation   Bm mBm mBm mBm mBm mBm mBm mBm m
reqq        O   O mmm O   O O   O mmm O   O
variation   O mmO mmO mmO mmO mmO mmO mmO mm
```

Figure 1: *Musical tilings.*

Figure 1 shows three "musical tilings": from left, a piano keyboard, a chromatic accordion keyboard, and a notation of the Egyptian *raqs sa'idi* rhythm [18]. Although, at least on the surface, it may not be clear that the tilings are related to each other, or indeed have any musical significance, in fact each example reveals deeper symmetries present in music, both in the domains of pitch and rhythm. These symmetries and their relationship to tilings are the subject of this article.

A *tiling* is a partition[1] of some space into congruent pieces, called tiles. There are many ways, both periodic and aperiodic, to tile the plane and higher-dimensional space. One-dimensional tilings, though less well known, are partitions of the real line into congruent collections of intervals. Tilings have long been of interest to visual artists and mathematicians alike. In addition, music theorists and mathematicians (going back to Euler) have discovered connections between tilings and musical structures. In this survey paper, I describe three applications of tilings to music theory: the representation of tuning systems and chord relationships by lattices, modeling voice leading by tilings of n-dimensional space, and the classification of rhythmic tiling canons, which are essentially one-dimensional tilings.

2. Euler and the Tonnetz

There are many ways to construct a scale. Solving the one-dimensional wave equation, which describes the behavior of string and wind instruments, produces a sequence of sinusoidal functions whose frequencies are the positive integer multiples of some constant. The scale is based on rationally related frequencies (octaves (2:1), fifths (3:2), and so on) or approximations to these frequencies. One approach builds upon a selection of the first few ratios among terms in the sequence. Fifths and octaves generate the Pythagorean scale; we

[1]Tiles are allowed to intersect in a set of measure zero—for example, tiles in two-dimensional space are allowed to share an edge, but not a two-dimensional area.

multiply the starting frequency by integer powers of two and three (including negative integer powers) to produce elements of the scale.[2] An alternate technique, an example of *just intonation*, is to generate a scale with fifths and major thirds (5:4). Although he did not originate just intonation, Euler [9] was the first to represent it as an infinite lattice, a portion of which appears in Figure 2 (left). Read left to right, rows are sequences of perfect fifths, and columns are sequences of perfect major thirds, read top to bottom—therefore, the lower left to upper right diagonals of the squares are perfect minor thirds (6:5). Since sequences of either fifths or thirds are geometric, we see that this lattice is drawn on a logarithmic scale. A similar lattice appeared in the late nineteenth century works of Oettingen [24] and Riemann[3] [25]. Riemann's lattice, called the *tonnetz*, depicts the major and minor third relationship more explicitly (Figure 2, center).[4] Since each triangle in the tonnetz represents a major or minor triad, vertices in its dual hexagonal lattice correspond to triads, and edges connect triads that have two notes in common. Figure 2 (right) labels the triads ("A" indicates A major and "am" indicates A minor), with major triads shaded.

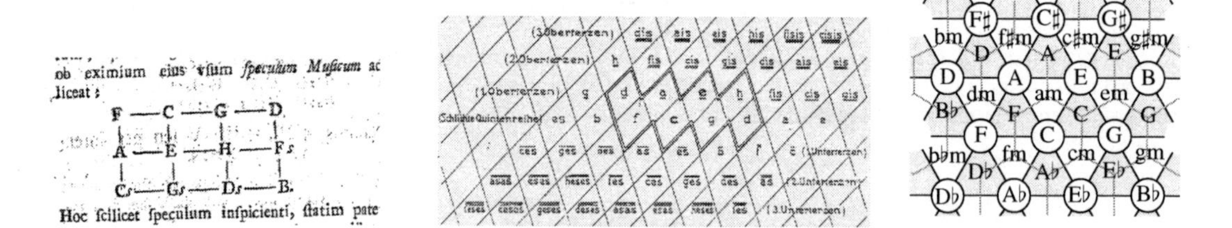

Figure 2: *Euler's Speculum Musicum [9, p. 350], Hugo Riemann's Tonnetz [25], and the tonnetz lattice with its dual.*

It is tempting to classify Figure 2 (right) as the tiling p3m1,[5] but one should be careful about what is actually represented. The symmetries of this lattice are transpositions by fifths, major thirds, or minor thirds (geometrically, translations along the lines of the lattice), so the tiling is p1, as is Euler's lattice. The fundamental region is a small rhombus. Inversion in the fifth, which exchanges major and minor thirds, introduces horizontal reflections, giving the tiling pm. The dual lattice, consisting of hexagons whose vertices are labeled by the alternating major and minor triads in Figure 2 (right), is p3m1 if we ignore everything except chord quality. Although this was not the original intention, the tonnetz can also represent notes in equal temperament. In this case, we have additional symmetries: enharmonic equivalence (for example, E♭ and D♯ share the same frequency), and, because twelve fifths equal seven octaves, octave equivalence. Using these two symmetries instead of transpositions by fifths and thirds gives a different p1 tiling whose fundamental region contains exactly one copy of each note in the twelve-tone scale.

3. Voice Leading and Continuous Transformations

We can describe relationships between chords in many ways—the circle of fifths is just the best-known example. If a sequence of chords is played by several voices, each sounding a single note, we can track the motion of individual voices in the progression from one chord to the next. This association is called *voice leading*. Although the conventions of voice leading have changed through the ages, some common principles persist. When leading between two chords, it is desirable that each voice move as short a distance (in pitch[6])

[2]Note that the octave will not "close up" in the familiar circle of fifths. That is, if we start with 440 Hz, no matter how many multiples we generate, this process never returns 440 Hz again. What is produced by repeating this method an infinite number of times is not the circle of fifths, which occurs only in equal temperament, but a dense subset of all frequencies.

[3]Hugo Riemann, not to be confused with the mathematician Bernhard Riemann.

[4]Interestingly, the tonnetz array forms the keyboard layout of a concertina patented in 1844 by the English physicist Wheatstone [17]. His instrument appears to be designed for equal temperament—though not all concertinas were—and he may have been motivated by the chord possibilities in the dual lattice. However, I have found no evidence that Wheatstone built this instrument, or that he had a role in the development of the tonnetz on the Continent.

[5]For an introduction to plane tilings, see `http://en.wikipedia.org/wiki/Wallpaper_group`.

[6]Pitch is determined by the logarithm of frequency—precisely, if we arbitrarily decide that middle C is 0, then 440 Hz (the A above middle C) corresponds to note 9, and pitch $= 9 + 12\log_2(\text{frequency}/440)$. In this system, integers correspond to notes in the chromatic scale of twelve-tone equal temperament.

as possible. In order to achieve this, voice-crossing—occurring when two voices change positions in the ordering of voices from low to high—is avoided. If we restrict ourselves to twelve-tone equal temperament, the closest distinct chords are those that differ by a semitone in one voice only (note that the chords in the dual tonnetz differ by either one or two semitones in one voice). This notion of closeness gives a structure to the space of chords of n voices. In fact, if we consider a "chord" to be an *ordered* multiset[7] of integer pitches, with each coordinate representing the pitch in one of the voices, we can map n-voice chords to the lattice $\mathbb{Z}^n$. The closest distinct chords are those that differ by one semitone in exactly one voice.

Our perception (and musical practice) gives this lattice of n-voice chords many symmetries: if two voices exchange pitches, if one voice shifts by an octave, or if all voices shift by the same amount, the respective resulting chords will sound quite similar to the original. How can we model voice leading in a way that respects these relationships? Quite a few music theorists have described voice leading using lattices or graphs: see Roeder [26], Douthett and Steinbach [8], Straus [27], Cohn [5], and Tymoczko [30]. The innovation that Callender [3], Quinn, and Tymoczko [29, 28] (henceforth, CQT) introduced is to embed the lattice of n-voice chords into *continuous* n-dimensional space (since pitch is continuous, not discrete) and study the effects of musically relevant symmetries. If we identify points in $\mathbb{R}^n$ that represent "similar" chords, what shape is the resulting space? Of course, the answer depends on which similarities we consider. CQT describe families of *chord spaces*, all of which are quotients of $\mathbb{R}^n$ under various isometries or combinations of isometries. Many discrete models of voice-leadings relationships embed nicely into these spaces. We will consider an example that Callender develops in detail in [3], and then touch on its relationship to CQT's general construction of chord spaces.

3.1. Representation in $\mathbb{R}^n$. In the discussion that follows, an "n-voice chord" means a vector in $\mathbb{R}^n$. We now represent operations on chords as rigid transformations of $\mathbb{R}^n$: transposition moves each voice by k pitches; permutation exchanges the pitches in two voices; octave shift moves one voice by some integer number of octaves; and inversion sends each voice to its additive inverse. Each of these operations describes a musical similarity of some sort. For example, all major triads in root position are equivalent under transposition. We call the set of vectors equivalent to $\mathbf{v}$ under all combinations of the four operations the *multiset class* of $\mathbf{v}$.

Using $\mathbf{e}_i$ to represent ith standard basis vector of $\mathbb{R}^n$, $\mathbf{1}$ to represent $\langle 1, 1, \ldots, 1\rangle$, and P_{ij} to represent the exchange of pitches in voices i and j given by $P_{ij} : \langle \ldots, v_i, \ldots, v_j, \ldots\rangle \to \langle \ldots, v_j, \ldots, v_i, \ldots\rangle$, we can write the operations as below. The CQT notation for these operations is $\mathbf{T}$, $\mathbf{P}$, $\mathbf{O}$, and $\mathbf{I}$.

Transposition	Permutation	Octave Shift	Inversion
$\mathbf{T} : \mathbf{v} \to \mathbf{v} + k\mathbf{1}, k \in \mathbb{R}$	$\mathbf{P} : \mathbf{v} \to P_{ij}(\mathbf{v})$	$\mathbf{O} : \mathbf{v} \to \mathbf{v} + 12n\mathbf{e}_i, n \in \mathbb{Z}$	$\mathbf{I} : \mathbf{v} \to -\mathbf{v}$

3.2. Continuous Transformations and Callender's T-class Space. Although Callender's construction of $\mathbf{T}$-class space in [3] does not explicitly discuss voice leading, it is consistent with the CQT model.[8] He begins with the composer Kaija Saariaho's *Vers le blanc* (Figure 3). This piece abandons the idea of pitch as discrete altogether; it consists of a *continuous* transformation from the chord C-A-B to the chord D-E-F over the course of fifteen minutes. Lines on the score indicates the position of the voices—note that the bottom two voices are briefly in unison towards the end of the piece.

[7] I'm glossing over some important issues here—for one, chords are usually considered to be sets, not multisets. See [28] for a full explanation.

[8] In the general literature, voice leadings are represented by associations between sets, rather than multisets, of pitch classes (pitches modulo 12). Callender's construction actually gives us equivalence classes of multiset voice leadings modulo transposition.

Figure 3: *Saariaho's Vers le blanc.*

Callender's model of continuous transformations is as follows.[9] As above, an n-voice chord is a vector of real numbers $\langle v_1, v_2, \ldots, v_n\rangle$, where v_i represents the pitch of the ith voice. For example, Saariaho's composition is a continuous interpolation from $\langle -12, -3, -1\rangle$ to $\langle -8, -10, -7\rangle$; it can be written as $\langle -12, -3, -1\rangle + (t/15)\langle 4, -7, -6\rangle$, where t is time in minutes and $0 \le t \le 15$. Callender begins by mapping the n-dimensional space of chords onto $(n-1)$-dimensional "$\mathbf{T}$-class space." The space of three-voice chords is a convenient example. Mapping each chord $\mathbf{v}$ to its transposition equivalence class ($\mathbf{T}$-equivalence class, or $\mathbf{T}$-class for short), defined to be $\{\mathbf{v} + k\mathbf{1} | k \in \mathbb{R}\}$, can be visualized as orthogonal projection onto the plane $\{\langle v_1, v_2, v_3\rangle | v_1 + v_2 + v_3 = 0\}$. For example, $\mathbf{e}_1$ maps to $\langle 2/3, -1/3, -1/3\rangle$. Note that the images of $\mathbf{e}_1$ and $\mathbf{e}_2$ form a basis for $\mathbf{T}$-class space; we will call them $\mathbf{a} = \langle 2/3, -1/3, -1/3\rangle$ and $\mathbf{b} = \langle -1/3, 2/3, -1/3\rangle$. The image of $\mathbf{e}_3$ is $\mathbf{c} = -\mathbf{a} - \mathbf{b}$. Thus, the projection of Saariaho's composition onto the plane is $-12\mathbf{a} - 3\mathbf{b} - \mathbf{c} + (t/15)(4\mathbf{a} - 7\mathbf{b} - 6\mathbf{c}) = -11\mathbf{a} - 2\mathbf{b} + (t/15)(10\mathbf{a} - \mathbf{b})$.

Now let's consider the effect of permutation of voices. Exchanging two voices corresponds to reflection in the planes $v_i = v_j$; in the plane $v_1 + v_2 + v_3 = 0$ this becomes reflection in one of the lines containing $\mathbf{a}$, $\mathbf{b}$, or $\mathbf{c}$ (that is, the projections of the coordinate axes onto $\mathbf{T}$-class space). These lines intersect at the origin at 60°, as shown in Figure 4. Each equivalence class under permutation and transposition ($\mathbf{PT}$-class) has a unique representative in the shaded sector, which is the projection of the vectors $\mathbf{v}$ where $v_1 \le v_2 \le v_3$. The symbols $\bullet$, $\square$, $\circ$, and $\triangle$ indicate, respectively, the $\mathbf{T}$-classes of the major triad $\langle 0, 4, 7\rangle$, minor triad $\langle 0, 3, 7\rangle$, diminished triad $\langle 0, 3, 6\rangle$, and augmented triad $\langle 0, 4, 8\rangle$ and their equivalents under permutation. In addition, the long arrow indicates the projection of *Vers le blanc*. Note that this projection crosses the line $r\mathbf{c}$ at the moment the bottom two parts are in unison.

We now consider the effect of octave shift. One generally perceives a C major chord played with the C in the highest voice as similar to one with the C in the lowest voice (root position). So, we identify all chords that are equivalent under octave shift; that is, $\mathbf{v} \equiv \mathbf{w}$ if and only if $\mathbf{v} - \mathbf{w} \equiv \mathbf{0} \pmod{12}$. The projection of the planes $v_i = 12n$ ($n \in \mathbb{Z}$) are shown in Figure 5 (left); octave equivalence introduces glide reflections in $\mathbf{T}$-class space. At this point, we have the tiling known as p31m, with the shaded kite-shaped fundamental region. We draw "mirror compositions" that lie in the same $\mathbf{OPT}$-equivalence class as *Vers le blanc*. It is evident that the composition begins and ends in the same $\mathbf{OPT}$-class!

Inversion is the last transformation to consider. The map $\mathbf{v} \to -\mathbf{v}$ exchanges minor triads and major triads; in $\mathbf{T}$-class space, reflection in the line $\mathbf{a} = \mathbf{b}$ is an inversion. Figure 5 (right) shows the tiling (p6m) of $\mathbf{T}$-class space for three-voice chords produced by permutation, octave shift, and inversion—note that there is now one symbol for minor and major triads. The fundamental region, originally depicted in Callender [3, Fig. 10, p. 12], contains exactly one representative of each multiset class (that is, $\mathbf{OPTI}$-equivalence class). The lattice points in the fundamental region correspond to multiset classes in twelve-tone equal temperament.

We now desire a *metric* on $\mathbf{T}$-class space—that is, a natural notion of the distance between two transposition classes. Callender defines the distance between $\mathbf{T}$-classes to be the distance between their projections into $\mathbf{T}$-class space, and sets $|\mathbf{a}| = |\mathbf{b}| = |\mathbf{c}| = 1$. There are many possible metrics with this property—Callender's preferred candidate is the Euclidean distance in $\mathbf{T}$-class space. Multiset classes now inherit a metric from $\mathbf{T}$-class space: the distance between multiset classes is the distance between their unique representatives in the triangular fundamental region for $\mathbf{OPTI}$-classes. Since the distance between any two points in $\mathbf{T}$-class space is never less than the distance between the respective members of their multiset

[9] I have changed his notation somewhat, but the essentials remain the same.

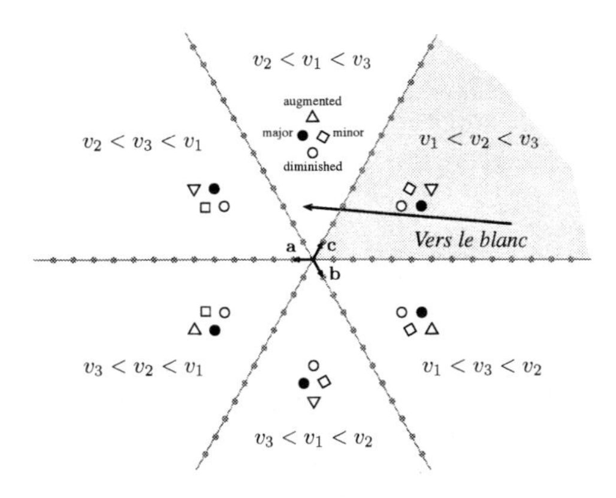

Figure 4: $\mathbf{T}$-*class space for three-voice chords, with permutation symmetries.*

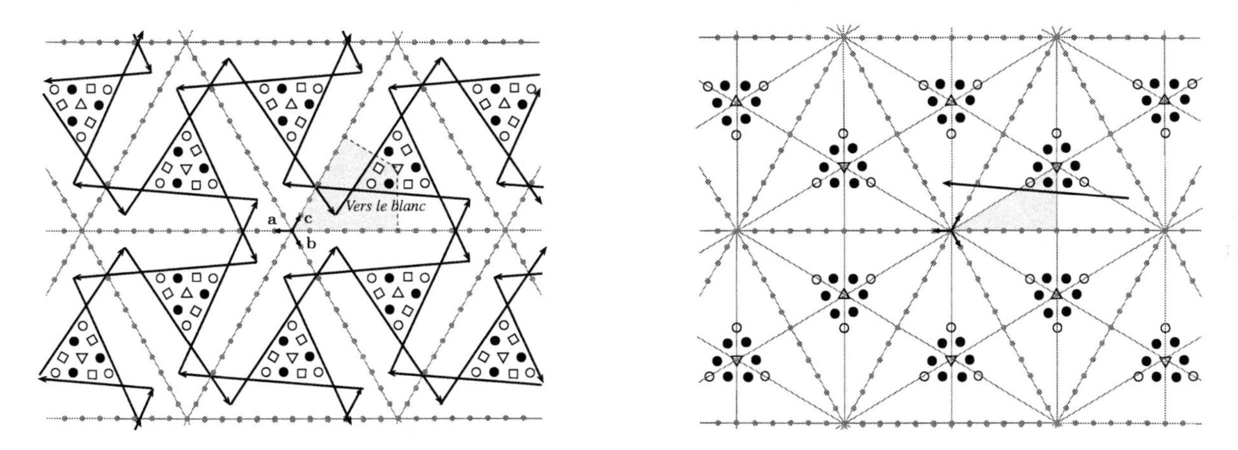

Figure 5: $\mathbf{T}$-*class space, with permutation and octave shift (left), and the addition of inversion (right).*

class in this fundamental region, his metric on multiset classes is well-defined. Callender uses this metric to investigate the shifts in our perception of chords in *Vers le blanc* as the voices draw near, but do not intersect, a number of lattice points in the fundamental region—see his article for more details [3].

Although not explored in [3], an important issue arises when we try to define the distance between $\mathbf{OPT}$-equivalence classes. In this case, the $\mathbf{OPT}$-classes on the dotted boundary of the kite-shaped fundamental region in Figure 5 (left) have *two* representatives on the boundary. In order to get rid of the doubles, we must identify the fundamental region's dotted edges and compute the minimal distance in the resulting cone. This cone is the quotient space of $\mathbf{T}$-class space modulo permutation and octave equivalence; in other words, it is the quotient of $\mathbb{R}^3$ under transposition, permutation, and octave equivalence. As such, it is an example of the more general quotient space construction developed by CQT and explored in the next section.

3.3. Generalized Chord Spaces and Orbifolds. Tymoczko [28] recognized that chord spaces are properly *orbifolds*, meaning quotients of $\mathbb{R}^n$ under the action of a finite group of isometries. For example, $\mathbf{T}$-class space is the orbifold $\mathbb{R}^n/\mathbf{T} \simeq \mathbb{R}^{n-1}$. The fundamental region of any tiling of $\mathbb{R}^n$, with the edge identification dictated by the tiling's symmetry group, is an orbifold, and that orbifold completely determines the tiling. Callender's kite-shaped fundamental region for $\mathbf{OPT}$-classes, with edge identifications, is the orbifold $\mathbb{R}^3/\mathbf{OPT}$ (**3*3** in Conway's orbifold notation); the fundamental region for $\mathbf{OPTI}$-classes is $\mathbb{R}^3/\mathbf{OPTI}$ (the orbifold ***632**). Other nice examples (from [28]) are the spaces of two-note chords modulo octave equivalence ($\mathbb{R}^2/\mathbf{O}$, a torus corresponding to the tiling p1 and orbifold **o**) and octave equivalence plus permutation ($\mathbb{R}^2/\mathbf{OP}$, a Möbius strip corresponding to the tiling cm and orbifold ***x**). Equivalence classes of voice leadings correspond to lines in these orbifolds. In their forthcoming paper [4], CQT further develop universal models for voice-leading—that is, a set of "parent spaces" in which many of the lattice models proposed by other music theorists embed. These parent spaces are orbifolds that are quotients of $\mathbb{R}^n$ under the action of **O**, **P**, **T**, and **I**, and combinations of these. Tymoczko's interactive program Chord-Geometries, available at `http://www.princeton.edu/~dmitri`, will help the reader visualize and explore some of the possibilities.

3.4. Open Problems. The geometric representation and exploration of chord relationships and voice leadings is an active area of research. There is much interest in measures of voice-leading size and in developing efficient algorithms to find minimal voice leadings (one such algorithm is described in [28]). CQT have thoroughly described the orbifolds $\mathbb{R}^n$ modulo various combinations of $\mathbf{O}$, $\mathbf{P}$, $\mathbf{T}$, and $\mathbf{I}$ for $n \leq 4$, and in higher dimensions in some cases [4]. However, there is more work to be done. In addition, Tymoczko posed the question of finding coordinate systems for these orbifolds that have some natural musical interpretation. He proposes using products of so-called "deep scales," but there are other possibilities [32]. There are also issues dealing with the conflict between the CQT representation of chords as multisets and their more common representation as sets (no duplication allowed) that have yet to be resolved geometrically.

4. Tiling Canons

Vuza showed that, upon mapping beats to integers, a rhythm forms a tiling canon if and only if its inner rhythm and outer rhythm correspond to sets A and B forming a tiling of the integers [33]. I will summarize the literature on tiling canons and integer tilings and state some open problems.

4.1. Rhythmic Tiling Canons. A *canon* is a musical figure produced when two or more voices play the same melody, with each voice starting at a different time. Canons appear in the works of J.S. Bach and others. *Rhythmic canons* are canons in which rhythms, and not necessarily melodies, are duplicated by each voice. The composer Olivier Messiaen (1908–1992), who coined the term "rhythmic canon," used rhythmic canons in his work (*Harawi*, "Adieu," and others). He describes the sound of a rhythmic canon as a sort of "organized chaos" [21, p. 46]. Using the symbols `x` to represent a note onset and `.` a rest, the canon in *Harawi* looks like this:

```
Voice 1:   x..x....x.......x....x..x...x..x......x..x...x.x.x..x....x..
Voice 2:     x..x....x.......x....x..x...x..x......x..x...x.x.x..x....x..
Voice 3:       x..x....x.......x....x..x...x..x......x..x...x.x.x..x....x..
                                       *                *.*.*.*
```

In this notation, simultaneous events are in vertical alignment. A rhythmic canon is *complementary* if, on each beat, no more than one voice has a note onset. For the most part, Messiaen's canon is complementary; asterisks mark deviations from this rule. A *tiling canon* is a complementary canon of periodic rhythms that has exactly one note onset (in some voice) per unit beat. Each voice plays a rhythm pattern, called the *inner rhythm*, and the voices are offset by amounts determined by a second pattern called the *outer rhythm*. For example, the inner rhythm `|: x.x..... :|` and outer rhythm `|: xx..xx.. :|` form a tiling canon. Rhythmic canons and tiling canons were first recognized as integer tilings and studied mathematically by Vuza [33] (see [2, 10] for further background); all these articles consider canons of periodic rhythms.

4.2. Integer Tilings. A *tiling of the integers* consists of a finite set A of integers (the *tile*) together with an infinite set of integer translations B such that every integer may be written in a unique way as an element of A plus an element of B. If the pair (A, B) forms a tiling of the integers, we write $A \oplus B = \mathbb{Z}$, where $\mathbb{Z}$ denotes the integers. The example

$$A = \{0, 2\} \text{ and } B = \{\ldots -4, -3, 0, 1, 4, 5, \ldots\} = \{0, 1\} \oplus 4\mathbb{Z}$$

corresponds to the tiling canon in the previous section. Tilings of the integers were first studied in 1950 by Hajós [14] and de Bruijn [7] in connection with factorizations of abelian groups. Newman (1977) and others showed that all integer tilings are periodic [23]—which is not the case in higher dimensions. One-dimensional aperiodic tilings are possible only if we allow reflections (a *monohedral* tiling). Restricting one's attention to the integers, rather than the real numbers, may appear to be an oversimplification of the one-dimensional tiling problem. However, Lagarias and Wang [20] showed in 1996 that all tilings of the real number line by finite sets of intervals may be reduced to tilings of the integers.

Although many have studied this problem, the complete classification of such tilings is an open question; indeed, for a given finite set of integers A, it is not known whether A is a tile in a tiling (that is, whether there exists a B such that (A, B) defines a tiling), although there are results in some special cases. If the number of elements of A is a prime power, there is a simple criterion for determining whether A tiles (see Newman [23]). In 1999, Coven and Meyerowitz answered the question for sets A whose cardinality has at most two prime factors, and, in their 2005 article, Granville, Laba, and Wang [13] solved the problem for sets A whose cardinality has three prime factors. Klingsberg and I approached the problem from a different angle. Instead of specifying the number of elements in the tile set A, we started with the period N of the tiling, and counted the number of tilings of $\mathbb{Z}_N$, the integers mod N, by equally-spaced tiles. These are the tilings of the form $A \oplus n\mathbb{Z}$, where n divides N. We proved that a periodic rhythmic canon of ℓ voices, each spaced by n notes from the previous, is complementary if and only if its inner rhythm is ℓ-asymmetric, as defined in our articles [16, 15]; if, in addition, the inner rhythm has n notes, then it is a tiling canon. Our formulas in [15] give the number of such tilings for each N; we have since extended our results to enumerate tilings by symmetric tiles (that is, $A = -A$). The ℓ-asymmetry condition was originally defined to classify certain African rhythms.

4.3. Open Problems. As mentioned before, the complete classification of integer tilings is an open question—in particular, if the cardinality of A is divisible by more than three primes, it is not known whether A tiles, except in special cases. Laba [19] proved that solving the one-dimensional tiling problem is equivalent to proving (or disproving) Fuglede's Conjecture [12] in dimension one—a question posed in 1974 that is still unsolved. Another area of study concerns enumerating all tilings of a given period. The requirement in our articles [16, 15] that the tiles be equally spaced greatly simplifies the problem of enumerating them. Fripertinger [11] has enumerated all tilings up to period forty. A special class of tilings occurs when both the inner and outer rhythms of a tiling canon are primitive,[10] producing a *tiling canon of maximal category*. Vuza proved that no nontrivial tiling canons of maximal category exist for period of less than seventy-two [33, Theorem 2.2, part one, p. 33]; this result was proved independently by Hajós [14]. There is no known formula for the number of tiling canons of maximal category of a given period.

The *inversion* of a rhythm pattern is that pattern played backwards. Beethoven used a modified tiling canon, in which the rhythm patterns are inversions of each other, in his string quartet Op. 59, no. 2. This type of tiling canon corresponds to a monohedral tiling of the integers. The problem of finding tiling canons using one rhythm and its inversion is equivalent to a mathematical problem considered by Meyerowitz [22]. He proved that any set of three integers forms a monohedral tiling. The general question of which sets can form monohedral tilings remains open. Incidentally, monohedral tilings can be aperiodic, creating interesting possibilities for composers. In the pitch domain, certain *tone rows*—orderings of the twelve-tone scale—called *derived rows* are based on monohedral tilings of period twelve. Such tilings appear in the work of Schoenberg, Babbitt, and others. The idea is to start with a generator of n pitches, where n divides 12, and tile $\mathbb{Z}_{12}$ with the generator and transpositions of its retrograde (mirror image in the time domain), inversion (mirror image modulo 12 in the pitch domain), and retrograde inversion. Given an arbitrary period N, how many derived rows are possible, and how does the tiling determine the symmetries of the derived row?

5. Conclusion

Tilings are a locus of cross-fertilization of mathematics and the visual arts. Regular tilings of the plane were known to artists long before they were classified by mathematicians. Aperiodic tilings, first discov-

[10] A canon of period N is primitive if N is its smallest period.

ered by mathematicians, have now been used in art and architecture. I hope that investigation of tilings in music theory will inspire composers, and interest in the musical applications of tilings will lead to further investigation by mathematicians.

Acknowledgments. I am grateful for the help I received from Paul Klingsberg, Cliff Callender, and, especially, Dmitri Tymoczko, who made substantial, patient, and helpful comments on this paper, and drew my attention to many examples of which I was unaware.

References

[1] E. Amiot, Why rhythmic canons are interesting, *Perspectives in Mathematical and Computational Music Theory*, G. Mazzola, T. Noll, and E. Lluis-Puebla, eds., epOs-Music, Osnabrück, 2004.

[2] M. Andreatta, *Méthodes algébriques en musique et musicologie du XXe siècle : aspects théoriques, analytiques et compositionnels*, Ph.D. thesis, École des Hautes Etudes en Sciences Sociales, Paris, 2003.

[3] C. Callender, Continuous transformations, *Music Theory Online* **10** (3) (2004).

[4] C. Callender, I. Quinn, and D. Tymoczko, Generalized chord spaces, preprint, 2005.

[5] R. Cohn, A tetrahedral graph of tetrachordal voice-leading space, *Music Theory Online* **9** (4) (2003).

[6] E. M. Coven and A. Meyerowitz, Tiling the integers with translates of one finite set, *J. Algebra* **212** (1999) 161–174.

[7] N. G. de Bruijn, On bases for the set of integers, *Publ. Math. Debrecen* **1** (1950) 232–242.

[8] J. Douthett and P. Steinbach, Parsimonious graphs: a study in parsimony, contextual transformations, and modes of limited transposition, *Journal of Music Theory* **42** (2) (1998) 241–263.

[9] L. Euler, De harmoniae veris principiis per speculum musicum repraesentatis, in *Novi commentarii academiae scientiarum Petropolitanae*, St. Petersburg, 1774, p. 330-353.

[10] H. Fripertinger, Enumeration of non-isomorphic canons, *Tatra Mt. Math. Publ.* **23** (2001) 47–57.

[11] —, Tiling problems in music theory (preprint, 2004).

[12] B. Fuglede, Commuting self-adjoint partial differential operators and a group theoretic problem, *J. Functional Analysis*, **16** (1974) 101–121.

[13] A. Granville, I. Laba, and Y. Wang, A characterization of finite sets that tile the integers, preprint, 2005.

[14] G. Hajós, Sur les factorisation des groups abéliens, *Časopis Pest. Mat. Fys.* **74** (1950) 157–162.

[15] R. W. Hall and P. Klingsberg, Asymmetric rhythms and tiling canons, to appear in the *American Mathematical Monthly*.

[16] —, Asymmetric rhythms, tiling canons, and Burnside's lemma, in *Bridges: Mathematical Connections in Art, Music, and Science*, R. Sarhangi and C. Séquin, eds., Winfield, KS, 2004, pp. 189–194.

[17] B. Hayden, Fingering systems, *Concertina Magazine* **16** (1986).

[18] L. Morris, The rhythm catalog, 1997, available at `http://www.sunnykeach.com/drum/index.html`.

[19] I. Laba, The spectral set conjecture and multiplicative properties of roots of polynomials, *J. London Math. Soc. (2)* **65** (2002) no. 3, 661–671.

[20] J. C. Lagarias and Y. Wang, Tiling the line with translates of one tile, *Inventiones math.* **124** (1996) 341–365.

[21] O. Messiaen, *Traité de rythme, de couleur, et d'ornithologie*, Editions musicales Alphonse Leduc, Paris, 1992.

[22] A. Meyerowitz, Tiling the line with triples, in *Discrete Models: Combinatorics, Computation, and Geometry (Paris, 2001)*, Discrete Math. Theor. Comput. Sci. Proc., AA, Maison Inform. Math. Discrèt. (MIMD), Paris, 2001, pp. 257–274 (electronic).

[23] D. J. Newman, Tesselation of the integers, *J. Number Theory* **9** (1977) 107–111.

[24] A. v. Oettingen, *Harmoniesystem in dualer Entwicklung*, W. Gläser, Leipzig, 1866.

[25] H. Riemann, Ideen zu einer 'Lehre von den Tonvorstellungen,' *Jahrbuch der Bibliothek Peters*, **21–22**, 1914–1915.

[26] J. Roeder, A geometric representation of pitch-class series, *Perspectives of New Music* **25** (2) (1987) 362–410.

[27] J. Straus, Uniformity, balance, and smoothness in atonal voice leading, *Music Theory Spectrum* **25** (2003) 305–352.

[28] D. Tymoczko, The geometry of musical chords, preprint, 2006.

[29] —, Scale theory, serial theory, and voice leading, preprint, 2006.

[30] —, Scale networks in Debussy, to appear in the *Journal of Music Theory*, 2006.

[31] —, Voice leadings as generalized key signatures, *Music Theory Online* **11** (4) (2005).

[32] —, personal correspondence, April 2006.

[33] D. T. Vuza, Supplementary sets and regular complementary unending canons (parts one through four), *Perspectives of New Music* **29** (1991) 22–49, **30** (1992) 184–207, **30** (1992) 102–125, **31** (1993) 270–305.

Mathematics and the Architecture: The Problem and the Theory in Pre-Modern Cultures

Zafer Sagdic
Architecture Department
Yildiz Technical University
80630 Besiktas
Istanbul, Turkiye
zafersagdic@hotmail.com

Abstract:

There is always a mystery on pre-modern architecture practice on the relation between dimensions and ratios. The reasons of using certain proportions used on the design of religious buildings/ spaces are the result of the application of numerical symbolism and Pythagorean triangle. Thus, the paper will be focused on the unity of theory in pre-modern architecture practice via giving some special examples of pre-modern architecture through the human history, such as Antique Egyptian and Antique Greek temples, Roman churches, Gothic cathedrals, and so on.

1.The Problem and The Theory

In modern architectural practice, it is known that, the formation of spaces are the result of diverse needs and requirements, such as; being suitable for certain activities, offering enough space to accommodate needed number of users, giving answer to technical regulations and parallel to reality resisting the influence of climate, earthquakes, storms, etc. Also, in modern thought, scientific thinkers interpreted the nature has an inherent order beyond that which man brings to his observations. In pre-modern cultures, theologians transformed nature to mean god, and through this they were able to give architecture a formalised higher purpose. This way of thinking allowed them to interpret nature through the use of power of algorithms- mathematics. The artist- form makers- usually found himself somewhere between the two positions, making the best use of each for the purposes determined by his culture. At certain times in history the artist relied heavily on religious scholasticism for interpretations of natural phenomena. It is seen on the pre-modern cultures that some certain numerical values are used to serve a religious purpose, especially on the design of religious buildings, such as mosques and churches. It should not be forgotten that the mathematics is a kind of language used through the mystery of existence. According to Galileo, "the big book of nature only can be read by the one who knows the language of it; the language of mathematics". If we are using mathematics to the formation of result of "real" needs and requirements, it should be zoomed on the nucleus of designing purpose of the religious buildings in the pre-modern world. Parts of the main religious building may be grouped to symbolise religious beliefs, and some symbolic numbers have been utilised for centuries in these spaces.

"The form of a building is its internal physical structure, as described under some appropriate conceptualisation". This definition is in the spirit of the general usage of the term aesthetics. According to Clive Bell (1914), all of the relations and combinations of line, spaces and even the colours are build up the "significant form". So, what architects did in pre-modern space formulation, is that, they always use the organisations and completion of their formal experiences are mixing with the mathematical and geometrical rules of architecture, the rules such as proportion, balance, line recession and so on. Architects are the coordinators and organisers of these experiences and finally the ones who definite form in a building. According to Arpat [1], certainly after so many years of research and hundreds of analytical calculations performed on plans of religious architecture, an answer to this enigma had to start to take shape: "Logical Thinking". It is a possible source and driving force, powerful enough to generate a secret worldwide influence in many fields of human activities, including architecture, regardless of differences between cultures and religions, over several centuries. In 1958, Monroe Beardley's "Aesthetics" named text suggests that, "the form of an aesthetical object is the total web of relations among its parts". Numerical symbolism, which is the nucleus of pre-modern aesthetic, beauty understanding and architectural practice, is nearly as old as history itself. And definitely, it existed nearly as far back as Babylonian and Antique Egyptian designs. The rules, that nature has in, were symbols of "beauty". If we get a view of historical development of human being, we can understand that architects designing artificial-environment and living-spaces by using these rules; from the Antique Egyptian, Antique Greek periods, till medieval architecture and Renaissance and maybe till today.

Here, the definition of design should be given. Design is the process of generating form fort he purpose of enriching human existence. In modern life, while design process designers are facing with some needs and requirements as the basic point of their designs, which is an effective rationale way. Obviously, pre-modern design creating based on geometric models of natural phenomena form designates inanimate physical entities; pattern will designate religion.

The Greek Philosopher, Pythagoras (560-480 BC) who travelled to Babylonian and Antique Egypt acquired some secrets of numerical symbolism and mathematical based design formulation in these countries and founded a religious philosophy. The numerical relations of the movements of heavenly bodies and of the relations inherent in the 3-4-5 triangle. Thus, the Pythagorean theory was discovered afterwards. Pythagoras also found that there are two other important way of creating numerical symbolism in pre-modern cultures: first is a relationship between sound and number, and the second is gematria. Gematria is a mystical art that has played a key role in the history of architecture, says Arpat [1]. Gematria involves assigning numerical values to the letters of the alphabet, therefore numbers are deriving from words, names and passages of main script. It is used in all three major religions to calculate the corresponding symbolic numbers of the holiest name, especially the name of God.

The pre-modernite accepted and developed rather than studied and restored the heritage of the past, unlike today. Even they stuck with using of the same religious based designs through the universal language mathematics century by century, they used new geometrical patterns and new architectural orders on their creations. The creative experience of a work of art depends not only the natural sensitivity and the visual training of the spectator, but also on his cultural equipment. This cultural perfection shows how the artist-form maker- can be used his geometry knowledge and/or gematria knowledge on his art pieces. In this point of view, patterns, iconography, hieroglyph, miniatures, making pictures of especially religious scenes in art and etc. concern themselves with the subject matter or meaning of works of art, as opposed to their form. Thus, all of these branches of art are methods of interpretation which arise from synthesis rather than analysis in modern cultures.

2.Antique Egyptian Architecture

Complex religion of Egyptians was the reason of making huge temples and monumental sized tombs dedicated to after-life mentality and half animal gods. They gave importance to buildings symbolised after-life. So that, there are two different important tomb types in Egypt: **1.**monumental sized tombs named **Pyramids** of Pharaohs, **2.mastabas** for rich and important people, monks and merchants.

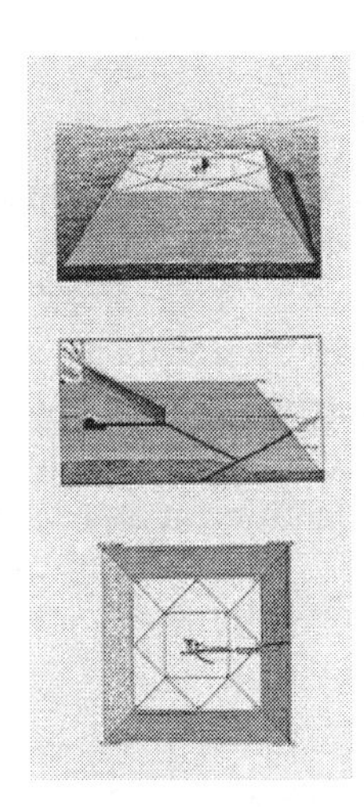

Figure 1: *Pyramids of Giza and Sphinx, Egypt.* **Figure 2.**: *Details of Egyptian mastabas*

In Egyptian culture nearly every image is symbolic, but that did not change the truth of mathematical and geometrical rules applied on designs came aside. It is known that Egyptians used very complex mathematical and geometrical knowledge on their royal tombs (Pyramids), which were temples for the name of worship of the deceased pharaoh. There are many labyrinths, mathematically well organised on 3rd dimension, inside pyramids and also the location of each pyramid is very specifically chosen according to religion based geometrical rules. The second important building type in Egypt is the temple. Each of Egyptian temples planned according to symbolical zoning: praying zone for ordinary people, praying zone for young monks, and zone of Pharaoh. All of Egyptian temples also designed according to symbolically based geometrical rules. The Khonsu Temple of Karnak (1198 BC.) and the Amman Temple, which is quite similar to Khonsu Temple, but six times in size, were stood inside a large walled enclosure which had to be also containing service buildings and sacred lake. Both have the similar mathematical rules used in site plans with a large door between two tall pylons, an open court and a colonnaded room, or hyposytle hall. Maybe the Egyptian style was seen too specific and too conservative for the needs of the other civilisations such as more poorer ones like Minoans and Mycennaeans in the Aegean sea till Antique Greek style appeared, no huge and some meanings, menthalitical and religious reasons loaded buildings made.

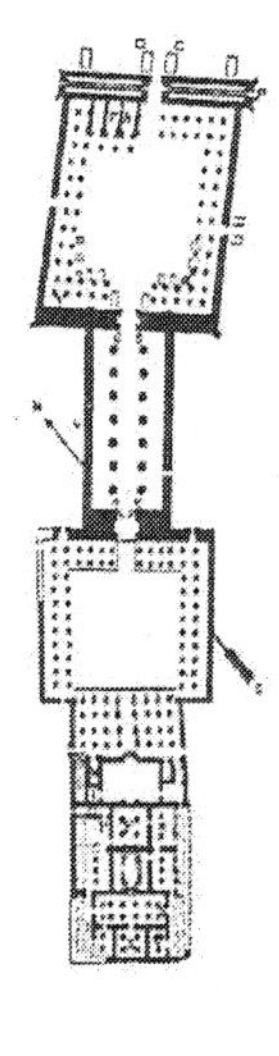

Figure 3-4: View from and plan of *Luksor Temple, Egypt.*

3.Antique Greek Architecture

The Ancient Greeks were fascinated by the concept of form and in their consideration of the forms of things they drew a fundamental distinction between chaos and cosmos. In their architectural products, the attempt at discovering a cosmology and disciplines in compositions is seen. All these definitions about Antique-Greek architecture insist on influential tradition of formalist criticism by the creation and usage of round column, graceful symmetry, repetition of rhythms, well-used proportions, harmonies and etc. It is not possible to add a further comment about "roundness" for the technical reason that round is a predicate symbol, not the identifier of an object in the universe of discourse.

This can be remedied by treating "round columns" as the value of the function. While the symmetry was used in Antique-Greek buildings such as temples, rationality was used for impressing evaluating functions. In other words, the symmetry was impressed by the gracefully of roundness; so what they had "the graceful symmetry".

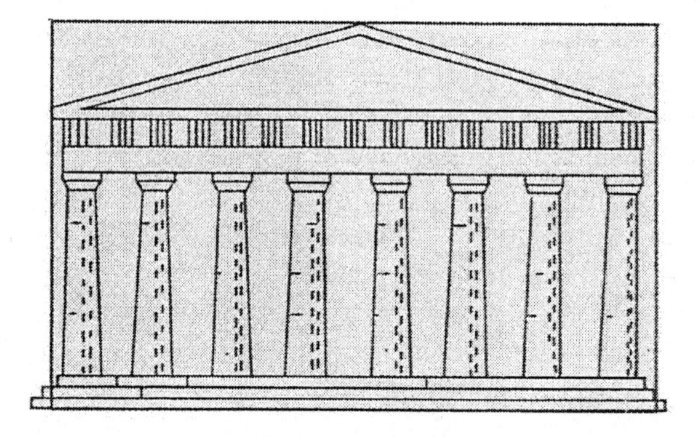

Figure 5-6: *View of façade of Parthenon Temple, Greece.*

By the repetitions of the columns, they also caught the rhythm; and by the dimensional compositions they caught the ratios, well-used proportions and finally harmony.

4.Roman Architecture

Roman Empire brought a new administration-system to Mediterranean world, its own city state democracy. Within this democracy, they built lots of monumental sized buildings, especially basilicas for law, amphitheatres, Roman baths and so on. These huge and monumental seem buildings were built by Emperors for inheriting the democratic obligation to maintain the buildings, their own popularity and Empire's power. Perhaps the most famous building of them, is the Coliseum in the capital, Rome; began in AD. 70 in the Emperor Vespesian period. This monumental sized structure has 50.000 spectator places for the brutal entertainment of the amphitheatre. Doric, Ionic, Corinthian and Composite columns orders are set in an ascending sequence. Although this arrangement was not universal, but adoption of this vertical sequence of the orders is seen in Coliseum.

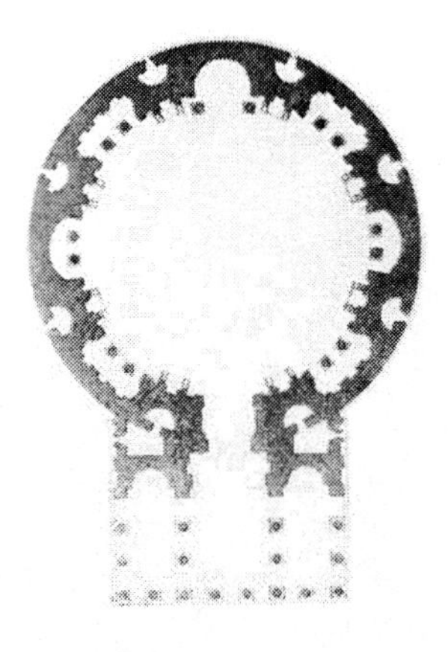

Figure 7-8: *Façade-section and plan of Panthenon Temple, Rome.*

With the systematic adaptation of columns orders, high capacity carried structure and monumental sized façades, Coliseum is still a memorial of mathematical perfection. Pantheon was one of the biggest and mathematically well-proportioned temple of Roman architecture, was built in AD.120 in Emperor Hadrian's period. This temple gives us the most complete impression of great buildings of Emperors. The huge brick clay concrete drum covered with 40 meters wide dome creating a dramatic top-lit space decorated with rows of coloured marble columns and arches.
So, it can be seen that, lots of high capacity carried, monumental sized façades having and mathematically well-proportioned buildings in the Roman architecture.

5.Gothic Architecture

Maybe products of Gothic were more menthality, more objective than all the other architectural styles' products; designs of Gothic Cathedrals were based on mathematical rules and complex than this, geometrical rules. Gothic architecture was based on logical solutions: using of one unique element, such as using of the rose window at the entrance façade or the symbol exists at the first stone step of the cathedral. This formation has a logical systematic, which is not only on the structure of the cathedral, but also on the layout and elevations with even the smallest details of ornaments. While designing a Gothic Cathedral, to make a good design, there was not any certain formula. However, there were some certain ways within classical architecture of organising the layout and detailing cathedrals. The plan is usually symmetrical on either sides or axis, starting from the main entrance and through the cathedral. It was believed that the symmetry reflected the balance of nature and the human form. The axis is more than the geometric structure of the plan and described the way, that, someone moves through, views and understands the cathedral. It can be seen in Greek temples focused on the image of god with a single axis, while Gothic cathedrals have a series of secondary axes. If Roman type of buildings are checked, it is seen that a rigid axial layout could be more informal; have individual groups of spaces at various angles by a series of secondary axes (=AD120). The design of an enclosed area of Gothic cathedrals, can make the space comprehensible by defining its geometry. Space elements; walls, floors and ceilings can be decorated to emphasise the geometry of the space. The construction of the cathedral introduces the geometric framework and all interior volumes of the cathedral are the parts of external design. The modification of detailed façade gives emphasis to any part of cathedral, not only for indicating its functional importance but also for improving the overall composition.

Figure 9-10: *Notre-Dame Cathedral and pattern of north rose window.*

The Renaissance designs establish a tradition of separating volumes in space organisation while Gothic cathedrals have much more logically organised volumes because of the logical systematic that was told at upwards. Superstitions and mystic forms of Christianity were reasons of cryptic world of symbols and myth having figures used in Gothic. The architects of Gothic used more complex systematic of dimensions and numbers of architectural features of ancient buildings.

The bishops of northern France asked to architects to open up the walls of their churches for large areas of stained glass to cast a mystic light on elegant internal rows of classical columns. To achieved these idea, architects reduced the weight of stone vaults by crossing arches diagonally between columns. These arches or ribs carried outwards to half arches and flying-buttresses leaning against the outside walls.

Figure 11-12: *Chartres Cathedral and pattern of North rose window.*

7.Renaissance Architecture

In the fourteenth century within the Italian city-life, an interest to the Rome led to the rescue of forgotten manuscripts from monastic libraries and a new philosophy, humanism, reconciling Christian and pagan thought.

Renaissance, the great artistic revolution blossomed from this fertile ground in the early fifteenth century. Architects studied the forgotten ruins of Rome to create buildings that served the needs of a society only gradually emerging from the feudalism of the Middle ages. The palazzo of Medici family, added classical order to the traditional scheme, in 1440. The whole composition is crowned by huge classical overhanging eaves and the cornice of traditional battlements. Alberti designed St. Andrea church in l470, he proposed on clear classical image to follow and drew his inspiration from large Roman vaulted interiors such as Constantine Basilica. He followed the established long church plan while changing side-aisles into chapels to support the heavy stone roof. And finally in the façades he used the design of a Roman triumphal arch, changing its role from the glorification of an empire to Christian God. St. Maria della Consolazione church, was designed by an architect Cola da Caprarola in 1508, has a series of circles and squares and one central domed central space plan. With this great artistic revolution, a fresh consciousness of history was infused with the vigour of originality to create a new classical architecture of great beauty. Proportion and symmetry are two basic concepts within the relationship of mathematics and design. Pascal, who is known as the genius of mathematicians, apparently thought of proportion and symmetry as "synonymous". The first written architectural source we have today, from Roman Empire period, is the "Ten Books of Architecture" which was written by Marcus Vitrivius Polyo. He wrote this book to take constructions activities under a specific control, in Julius Cesars' period. Vitrivius put 3 basic rules that could be maybe the theory of him. **1) Firmitas:** firmness of statics and construction, **2) Utilitas:** appropriateness to aims; functionality; suitablity, **3) Venustas:** aesthetical necessity and proportions. Also venustas insists some other sub-rules such as symmetry, decor, etc, in 1673, when Claude Perrault published his French translation of "The Ten Books of Architecture", he rendered "symmetria" as "proportion". But in his deep-notices, he commented that symmetry referenced to "the relationship which parts on the left side have with those on the right, those high up with those low down, those in back with those in front".

8.Conclusion

Actually, modern mathematics has further generalised and formalised the concept of geometric symmetry, grounding it upon the idea of a group of geometric transformations. It can be said that, an architectural composition is symmetrical to the extent, which has symmetry operations as isometric transformations such as translations, rotations, reflections and compositions. This symmetry is defined as a property of a set of transformations. It could be exemplified as a bilaterally symmetrical plan is transformed into itself by reflection across its axis or a pinwheel is transformed into itself by rotation and etc. Thus, not only symmetry, but also axes passing through a point, repeating linear patterns or two or more dimensional patterns could be given as examples. The rules that nature has in, were symbols of "Beauty". If we get a view of historical development of human being, we can understand that architects design artificial-environment and living-spaces by using these rules. From the ancient times, the "Beautiness" applied to compositions that have classical formal qualities of rhythm, proportion and symmetry; as Vitrivius mentioned in his book. He put "eurythmia" word for it which could be translated as "grace". He defined it as "...beauty and fitness in the adjustments of the members. This is found when the members of a work are of a height suited to their breadth, of a breadth suited to their length and in a word when they all correspond symmetrically". Such an idealist view can be also seen in St. Augustine's aesthetic theory; "...If I ask to a workman why, after constructing one arch, he builds another like it over against it, he will reply, I dare say that in a building like parts must correspond to like. If I go further and ask why he thinks so, he will say that it is fitting, or "Beautiful" or that it gives pleasure to those who behold it." Also in

Alberti's "Ten Books", it is seen that such a similar definition was used under drawing directly upon Aristotelian ideas: "... just as all the individual members harmonise in an animal organism, so all the separate parts of a building should harmonise...Each part of a building must correspond to all the others; so, as to contribute to the success and beauty of the whole. The building can not be beautiful in only one of its parts while the others neglected; all must harmonise in order to appear as a single, well-articulated body, not a jumble of unrelated fragments". In 1725, one of the British philosopher Francis Hutcheson wrote the "Beauty" to show how it depends on formal principles. In his doctrine, richly varied compositions are organised in accordance with some underlying unifying principle are "Beautiful".

In pre-modern thought, it is seen that artists- form makers- always used the organisations and completion of their formal experiences mixing w,th the mathematical and geometrical rules, the rules such as proportion, balance, line recession, symmetry, numerical symbolism, gematria and so on. Even sometimes the method has been changed on designs, never the menthalitic point of view on the theory was changed. Thus, it can be said that, there is a unity of design theory in pre-modern architecture practice.

An architectural design was based to visual symbolism on the multi-God focused religions in pagan period. Geometrical formulation were important in that point of view. However, numbers were not always coupled with geometry. Some cases they were applied to dimensions just by simple multiplication or divisions. Religious shaped numerical symbolism is shaped in centuries, where unity of God is believed. The planning of religious spaces in medieval era seems to have originated in different concepts, such as in the images corresponding in the human body proportions, in the musical harmony of cosmos, in regular polygons, in gematria, and etc.

So, all these and more are the proof of mathematical theorems, apply to extensive sets of apparently diverse figures or curves; the forms of plants and animal; that there is always a relationship, such a very precise relationship between "Mathematics and Design"; more than this, the relationship through "Mathematics, Nature and Design".

References

[1] A.Arpat, Secrets of Architecture, Evans Communications, Canada, 2004.
[2] C.Yildirim, Matematiksel Dusunme, Remzi Kitapevi, İstanbul, 1996.
[3] D.Wells, You are a Mathematician, Penguin Books, 1995.
[4] J. P. King, *Matematik Sanati*, tr. Nermin Arik, Istanbul, 1992.
[5] R.Adam, *Classical Architecture*, A Times Mirror Company, 1991.
[6] L.Mumford, *The City in History*, A Pelican Book, NewYork, 1979.
[7] M.Safdie, Form and Purpose, Houghton Mifflin Company, Boston, 1982.
[8] PG.Rowe,Design Thinking, The MIT Pres, Cambridge, 1987.
[9] T.Pappas, Yasayan Matematik, tr.Yildiz Silier, Sarmal Yay., İstanbul,1993.
[10] Vitrivius, Mimarlik Uzerine On kitap (The Ten Books of Architecture), İstanbul,1990.
[11] W.J.Mitchell, *The Logic of Architecture*, The MIT Press, Cambridge, 1994.
[12] Z.Sagdic, *Mathematics and Design Relationship in Architecture Within Chronological Period*, pp.383-390,Spain,1998.
[13] A.T.Mann, The Sacred Architecture, Element Pub., USA,1993.
[14] E.Panofsky, Meaning in the Visual Arts, Doubleday & Company Inc., Garden City, 1955.

*All of the images are from A.T.Mann, The Sacred Architecture, Element Pub., USA,1993.

Towards Pedagogability of Mathematical Music Theory: Algebraic Models and Tiling Problems in computer-aided composition

Moreno Andreatta*, Carlos Agon Amado*, Thomas Noll+, Emmanuel Amiot°

*Ircam/CNRS UMR 9912
1, place I. Stravinsky – 75004 Paris, FRANCE
E-mail: {andreatta, agon}@ircam.fr

+Technische Universität, Berlin, GERMANY
E-mail: noll@cs.tu-berlin.de

°CPGE Perpignan, FRANCE
E-mail: manu.amiot@free.fr

Abstract

The paper aims at clarifying the pedagogical relevance of an algebraic-oriented perspective in the foundation of a structural and formalized approach in contemporary computational musicology. After briefly discussing the historical emergence of the concept of algebraic structure in systematic musicology, we present some pedagogical aspects of our *MathTools* environment within *OpenMusic* graphical programming language. This environment makes use of some standard elementary algebraic structures and it enables the music theorist to visualize musical properties in a geometric way by also expressing their underlying combinatorial character. This could have a strong implication in the way at teaching mathematical music theory as we will suggest by discussing some tiling problems in computer-aided composition.

1. Introduction

In an interesting article on the notion of music theory's new pedagogability and the state of music theory as a research field [12], Richard Cohn suggests the opportunity to transgress the boundaries between teaching and research by showing that this opportunity is linked to the ways that music theorists have been recently crossing boundaries, particularly between fields of knowledge. One example of this necessity of transgressing the boundaries between disciplines is given by the difficulty to establish a strict separation between music theory, analysis and composition, in particular from the perspective of an algebraic-oriented approach towards a structural formalized computational musicology [6].

Historically, the emergence of algebraic methods in music has been a long process that has occurred, surprisingly, independently from stylistic considerations and geographical contexts. According to Iannis Xenakis, the structure of mathematical *group* enables an "universal formulation for what concerns pitch perception" [25]. In other words, any division of the octave in a given number n of equal parts can be represented as a group, the cyclic group of integers modulo n, with respect to the addition modulo n.

In a large study dedicated to the emergence and development of algebraic models in 20th century music and musicology [3], one of the authors of this paper proposed to try to transgress the boundaries between the American serial and set-theoretical tradition (Krenek, Babbitt, Perle, Forte, Lewin,...) and an European structural tradition (Xenakis, Vieru, Riotte, Mesnage, Mazzola,...). In fact, in the *set-theoretical* (Forte) and *transformational* (Lewin) approaches Milton Babbitt's initial formalization of the twelve-tone system has some very deep intersections with the music-theoretical constructions proposed around the 60s by some European theorists/composers, in particular Anatol Vieru and Iannis Xenakis. The system was conceived as a "collection of elements, relations between them and operations upon them"[8].

In this paper, we focus on a relatively elementary theoretical approach which is essentially based on the structure of mathematical *group*. In particular we will concentrate on three groups that are the basis of our structural approach to the enumeration, classification and computer-aided implementation of musical tiling structures: cyclic, dihedral and affine groups. Nevertheless, more abstract approaches that use the ring structure of polynomials [2] and category/topos theory [20] clearly show that the tiling constructions that we present in the following sections might be described by means of a more powerful theoretical and computational framework.

2. Some preliminary definitions

This section introduces the three basic families of groups that constitute the theoretical framework of our computational model of tiling musical structures. Let $(G, \cdot)$ be a group with inner binary operation "$\cdot$".

2.1. Definition and music-theoretical interpretation of cyclic groups

A cyclic group of n elements (i.e., of order n) is a group $(G, \cdot)$ in which there exists (at least) one element g such that each element of G is equal to $g \cdot g \cdot \ldots \cdot g$, where the group law is applied a finite number of times. In other words, G is generated by g. A cyclic group of order n can be represented by the set $\{0,1,\ldots,n\text{-}1\}$ of integers (modulo n), and it is usually indicated as $\mathbf{Z/nZ}$. In general a cyclic group of order n is generated by all integers d which are relatively primes to n and it can be represented by a circle where the integers 0, ..., n-1 are distributed uniformly. In the usual twelve-tone clock, one may go from an integer to another simply by rotating the circle around his centre by an angle equal to a multiple of $\pi/6$. Musically speaking, rotations are equivalent to *transpositions*. Let T_k be the transposition of k minimal divisions of the octave (i.e., halftones in the case n=12). For any integer k relatively prime to n the transposition T_k generates the whole cyclic group.[1] As originally shown by Halsey and Hewitt in one of the first application of combinatorial algebra to music theory [16], musical transpositions define mathematical *actions* so that classifying transposition classes of chords is finally equivalent to study orbits under the action of the group $\mathbf{Z/nZ}$ on itself. Polya enumeration theory provides therefore the underlying framework that enables to approach tiling problems in computer-aided composition in all generality. We will not enter in the mathematical aspects of this theory that has also been generalized to the computations of orbits of more complex musical structures (melodic patterns, twelve-tone rows, mosaics, ...). See [14] for a detailed discussion of this approach.

2.2. Definition and music-theoretical interpretation of dihedral groups

A dihedral group $(\mathbf{D_n}, \cdot)$ of order $2n$ is a group generated by two elements a, b with relations:

1. $a^n = b \cdot b = e$ where a^n means $a \cdot a \cdot \ldots \cdot a$ (n times)
2. $a^n \cdot b = b \cdot a^{n\text{-}1}$

In other words, the dihedral group $\mathbf{D_n}$ consists of all $2n$ products $a^i \cdot b^j$ for i=1,...,n, and j=1,2. Geometrically, the dihedral group corresponds to the group of symmetries in the plane of a regular

[1] This is a well-known music-theoretical statement that we find already in Babbitt's dissertation [8].

polygon of n sides. These symmetries are basically of two types: rotations and reflections (with respect to an axis). Musically speaking, reflections are inversions with respect to a given note that is taken as a fixed pole. As in the case of the cyclic group, the dihedral group can be considered as generated by transpositions and inversions. Orbits under the action of the dihedral group on the equal-tempered system are also called *pitch-class sets* in the American tradition [13].

2.3. Definition and music-theoretical interpretation of affine groups

The affine group $\mathbf{Aff_n}$ consists of the collection of affine transformations, i.e., the collection of function f from $\mathbf{Z/nZ}$ into itself which transforms a pitch-integer x into $ax+b$ (modulo n) where a is an integer relatively prime with n and b belongs to $\mathbf{Z/nZ}$. In the special case of $n=12$, the multiplicative factor a belongs to the set $U=\{1,5,7,11\}$ of units of $\mathbf{Z/12Z}$. Note that an affine transformation reduces to a simple transposition when $a=1$. On the other side, inversions are affine transformations with $a=11$. All musical structures which are equivalent up to transposition will also be in the same orbit under the action of the affine group on $\mathbf{Z/nZ}$. This means that affine orbits are a natural generalization of pitch-class sets, a music-theoretical statement which is shared by the American [21] and European traditions [19].

3. Tessellations and Tiling in music

Tiling problems in music theory, analysis and composition have a relatively old history in music theory. From a pitch-perspective, the study of some tiling problems in music is historically related to Hugo Riemann original representation of the tone space by means of a translation of deformed squares generated by oblique major and minor thirds axis (see Figure 1).

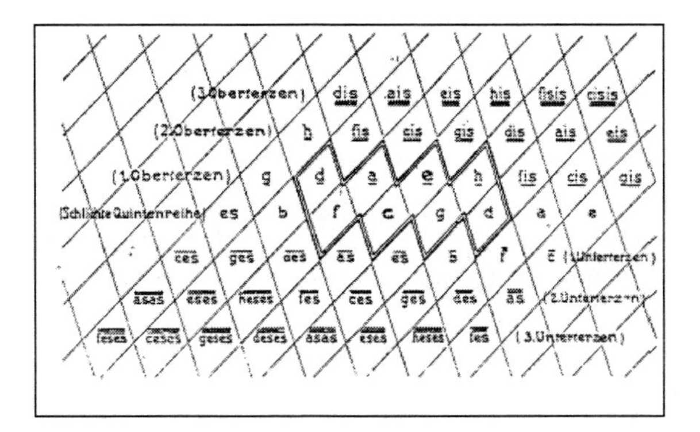

Figure 1: *Hugo Riemann's* Tonnetz [22].

Many attempts have been made in order to provide some alternative models of the tiling process of the tone space, particularly by the so-called Neo-Riemannian American and European tradition.[2]

Surprisingly, despite the well-known canonical equivalence (isomorphism) between a equal-tempered division of the octave and the cyclic character of any periodic rhythm [24] the study of some tiling properties of the timeline by means of translations of a given rhythmic tile (or some usual transformations of it) is a relatively new research area inside mathematical music theory. We have already sketched the history of the emergence of tiling rhythmic structures in music composition starting from Olivier Messiaen original attempt at defining musical canons independently of any considerations about pitch values [3]. In this paper we focus on different tessellations' canonic structures as they have been implemented in the "MathTools" environment in *OpenMusic*.

[2] See [11] for a survey and a historical perspective on Neo-Riemannian theory. See [18] for one of the first examples of analytical computational models based on the tiling properties of the tone space.

4. Tiling problems in *OpenMusic* new "MathTools" Environment

Algebraic models for computer-aided music theory, analysis and composition have been implemented as a package of mathematical tools in the last version of *OpenMusic*, a visual programming language developed at Ircam.[3] There are six main families of tools, which are: circle, sieves, groups, sequences, polynomials, canons. Recently [6] we presented some aspects of these mathematical tool strictly connected with the problem of paradigmatic classification of musical structures (circular representation, groups and polynomials). After briefly summarizing the potentialities of the circular representation of musical structures (in the pitch as well as in the time domain) we now focus on two families of tiling musical canons whose mutual intersection is at the present an active field of research for composers involved in computer-aided composition.[4] We will take, as a point of departure, a recent paper by R.W. Hall and P. Klingsberg [15] that makes the bridge between the combinatorial theory of asymmetric periodic rhythmic patterns and various models of tiling canons that have been proposed in the music-theoretical literature. This shows some theoretical potentialities and some pedagogical benefits of the *MathTools* environment.

4.1. Circular representation of tessellation structures and pitch-time isomorphisms

By using the geometrical representation of the equal-tempered system as a circle divided in 12 parts, a musical chord of *m* distinct notes can be represented as a *m*-polygon inscribed in a circle. To each chord one can associate a sequence of *m* integer numbers counting the successive intervals in the chord (*interval structure*). Such a structure is an invariant enabling us to identify, in a unique way, a given chord up to transpositions. Figure 2 shows two rhythmic patterns which are particularly interesting from a tiling perspective. They are called "3-asymmetric" in Hall and Klingsberg's terminology since they generalized the oddity property [9]. From a group-theoretical point of view these two rhythmic patterns are what mathematical music theorists usually call "self-inverse partitioning" musical structures. In fact they are "inversionally" related. This means that they are basically the same orbit under the action of the dihedral group on the twelve-tone system. Moreover, the twelve-tone system can be partitioned into a disjoint union of transposition classes of the two structures. We show their circular representation, the intervallic structure (1 2 3 6) and one possible rhythmic interpretation (with the free choice of a period and of a metric unitary step, in this case respectively the eight and the sixteenth note)

[3] See [1] for several analyses of recent musical pieces composed by means of *OpenMusic*.

[4] As pointed out by the French composer Georges Bloch in a recent article devoted to the compositional aspects of the tiling canon process, "this is a intuitively evident structure that has interested numerous composers, among them Messiaen. But the construction of such a canon was beyond the means of composers working without computers, and research on the characteristics of such objects requires a musical representation tool such as *OpenMusic*" [1].

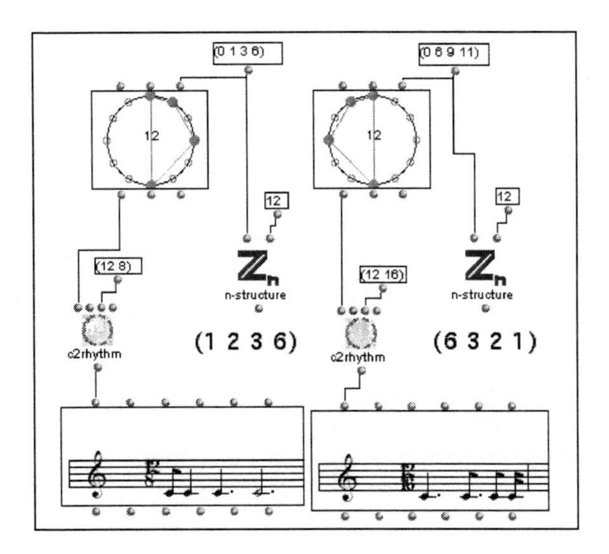

Figure 2: *Circular representation, intervallic structure and rhythmic interpretations of two inversionally related 3-asymmetric rhythmic patterns.*

By interpreting each rhythmic structure in the pitch domain, we obtain the tiling of the entire equal-tempered space with transposition classes of the same chord structure (Figure 3).

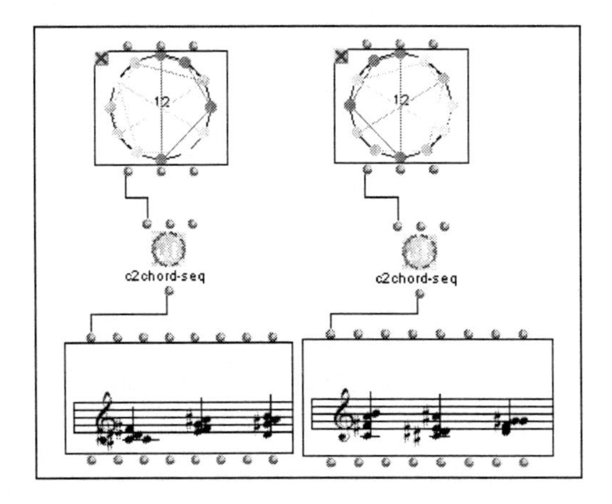

Figure 3: *Pitch-interpretation of the two previous 3-asymmetric rhythmic patterns*

4.2. Some canonical representation of tessellation structures

Instead of tiling the chromatic space with transposition classes of the same chord, as discussed previously, we can interpret the tiling process in the time domain. This corresponds to the construction of rhythmic canons in which each voice is the translation of a given rhythmic pattern, with the property that once the last voice appears, each instant of time is affected to one (and only one) rhythmic beat (*tiling rhythmic canons*). Figure 4 shows this first case.[5]

[5] The tiling property is made evident by revealing the underlying metric grid.

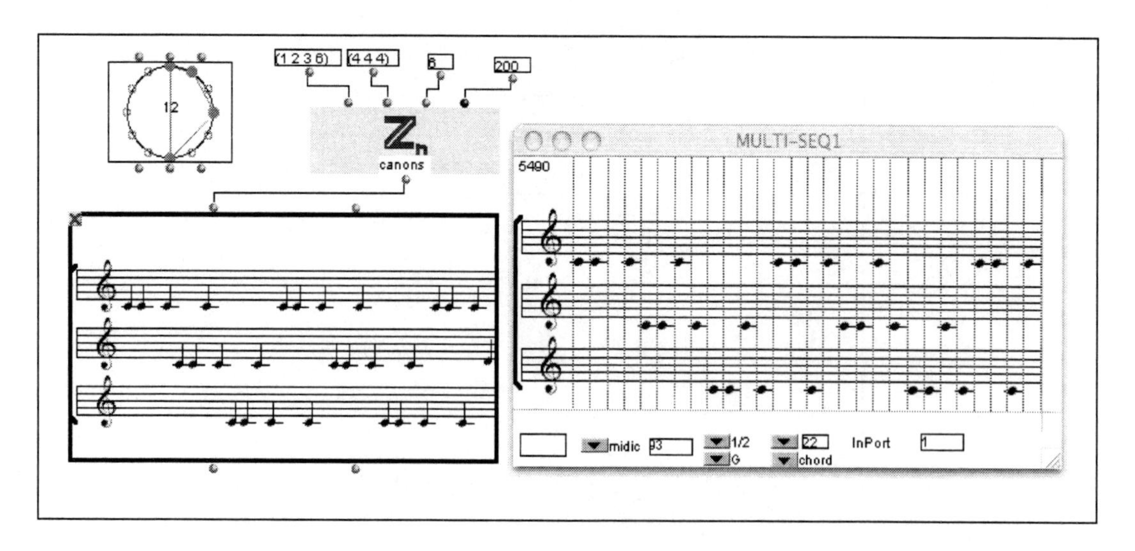

Figure 4: *A tiling rhythmic canon in the* "Math Tools" *environment*

By adding the pitch content to this global structure, it is possible to build melodic-rhythmic tiling canons, as shown in Figure 5. In this case the pitch dimension is isomorphic to the rhythmic content, as the circular representation clearly shows.

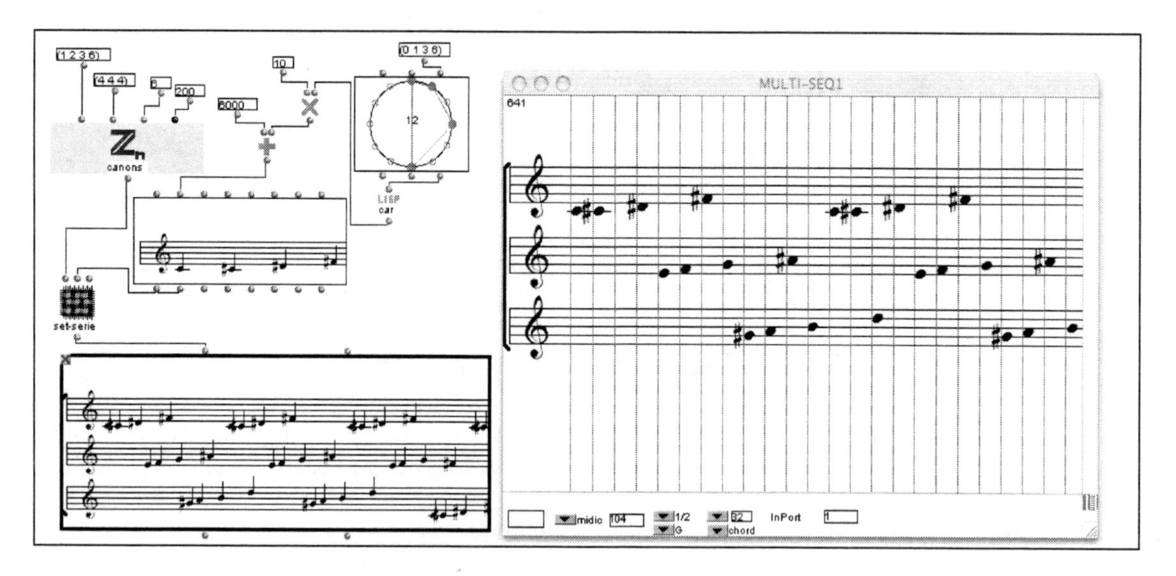

Figure 5: *A melodic-rhythmic tiling canon*

This model of canonic constructions makes use implicitly of the action of the cyclic group on the space of rhythmic structures. By changing the "paradigm" and by using for example the action of the affine group it is possible to build canonic structures in which the voices are augmentations (by a given factor) of an initial rhythmic pattern. This provides the basis of a second class of tiling rhythmic canons called "Augmented Canons" [4]. It is particularly interesting to apply this second model of tiling canons to the first of the two 3-asymmetric rhythmic patterns that we presented before. There are in fact many multiplicative factors that can be applied to this type of patterns in order to canonically tile the space. More precisely, we can easily see that there are three choices of multiplicative factors (Figure 6).

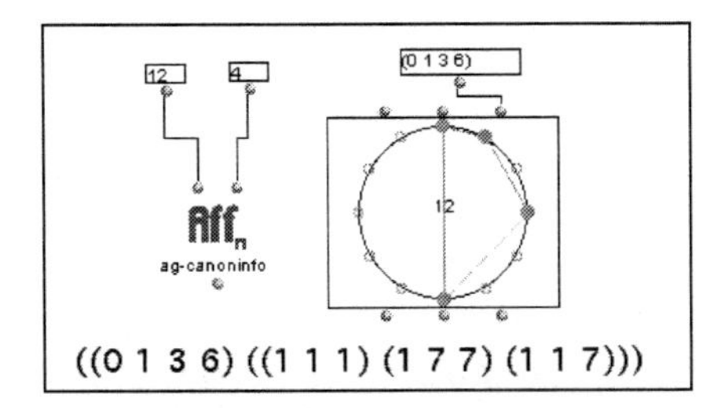

Figure 6: *Potential augmentation factors for a given rhythmic pattern*

Apart from the first solution, in which the multiplicative factors are equal to the identity (which means that the corresponding augmented canon simply reduces to the previous regular tiling canon of figure 7), there is one non trivial factor by which one could augment the initial rhythmic pattern in order to tile the time line. By taking the third solution, (1 1 7), we may construct an augmented canon in 9 voices: the first two voices are the identical patterns and the remaining 7 voices are 7 repetitions of the initial voice but augmented by a factor equal to 7.

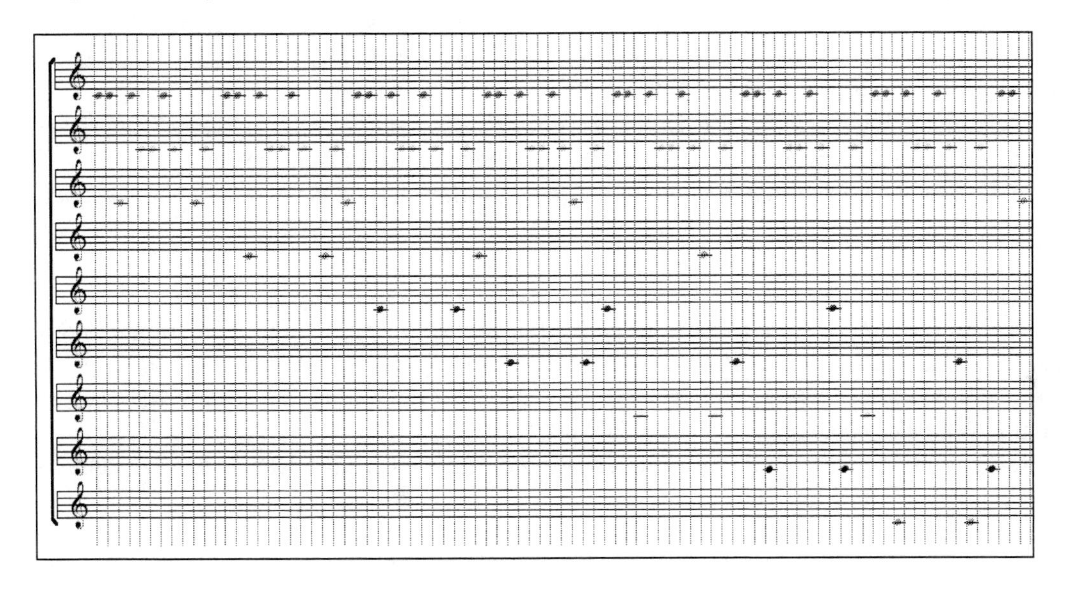

Figure 7: *An augmented tiling canon based on a 3-asymmetric rhythmic pattern*

4. Conclusions

The *OpenMusic* implementation of several families of algebraic-oriented mathematical tools suggests some interesting aspects of the notion of pedagogability in mathematical music theory. We discussed some examples concerning tiling problems in computer-aided composition. Two classes of tiling rhythmic canons are at the moment an active field of research: regular canons (where the voices are a temporal translation of a given rhythmic pattern) and augmented canons (where the voices are augmentations of the first voice). This implementational model is presently taught in many computer-aided composition courses, as well in conservatories (CNSMDP in Paris, Conservatoire of Adria in Italy), schools of music (Escola Superior de Musica de Catalunya,...), and Master programs (ATIAM at Ircam, “Art/Science/Technology” in Grenoble...). It provides a good example of pedagogability of a field of study - Mathematical Music Theory - which naturally aims to transgress the boundaries between teaching and research activities.

References

[1] C. Agon, G. Assayag, J. Bresson (ed.): *The OM Composer's* Book, vol. 1, Series *Musique/Sciences*, Ircam/Delatour France, Paris, 2006.

[2] E. Amiot, M. Andreatta, C. Agon, *Tiling the (musical) line with polynomials: some theoretical and implementational aspects*, Proceedings of the ICMC, Barcellona, 2005.

[3] M. Andreatta, *Méthodes algébriques dans la musique et musicologie du XXe siècle: aspects théoriques, analytiques et compositionnels*, PhD Thesis, Ehess/Ircam, 2003.

[4] M. Andreatta, T. Noll, C. Agon, G. Assayag, *The Geometrical Groove: rhythmic canons between Theory, Implementation and Musical Experiments*, Proceedings of the JIM, Bourges, 2001

[5] M. Andreatta, C. Agon, and E. Amiot, *Tiling problems in music composition: Theory and Implementation*, Proceedings of the ICMC, Göteborg, 2002, pp. 156-163, 2002.

[6] M. Andreatta, C. Agon, *Algebraic Models in Music Theory, Analysis and Composition: Towards a Formalized Computational Musicology*, Understanding and Creating Music, Caserta, 2005

[7] G. Assayag, C. Rueda, M. Laurson, A. Agon, O. Delerue, *Computer Assisted Composition at Ircam : PatchWork & OpenMusic*, Computer Music Journal, 23(3), 1999.

[8] M. Babbitt, *The function of Set Structure in the Twelve-Tone System*, PhD, Princeton University, 1946.

[9] M. Chemillier, C. Truchet, "Computation of words satisfying the rhythmic oddity property", *Information Processing Letters*, n° 86, 2003

[10] J.-M. Chouvel, *Représentation harmonique hexagonale toroïde*, Musimédiane (Revue audiovisuelle et multimédia d'analyse musicale), n°1, Dec 2005 (http://www.musimediane.com)

[11] R. Cohn, *Introduction to Neo-Riemannian Theory: A Survey and a Historical Perspective*, Journal of Music Theory, Vol. 42, No. 2, pp. 167-80, 1998.

[12] R. Cohn, *Music Theory New's Pedagogability*, Music Theory Online, 4(2), 1998.

[13] A. Forte, *The Structure of Atonal Music*, New Haven: Yale University Press, 1973.

[14] H. Fripertinger, *Enumeration and construction in music theory*. In H.G. Feichtinger and M. Dörfler, *Computational and Mathematical Methods in Music*, Vienna, Austria, December 2-4, 1999, pp. 179-204

[15] R.W. Hall & P. Klingsberg, *Asymmetric Rhythms, Tiling Canons, and Burnside's Lemma*, Bridges Proceedings, Winfield, Kansas, pp. 189-194, 2004.

[16] D. Halsey & E. Hewitt, *Eine gruppentheoretische Methode in der Musik-theorie*, Jahresber. der Dt. Math.-Vereinigung 80, pp. 150-207, 1978.

[17] D. Lewin, *Generalized Musical Intervals and Transformations*, New Heaven: YUP, 1987.

[18] H.C. Longuet-Higgins, *Two letters to a musical friend*, Music Review, pp. 244-248; 271-280, 1962.

[19] G. Mazzola, *Gruppen und Kategorien in der Musik*, Heldermann, Berlin, 1985.

[20] G. Mazzola, *The Topos of Music*, Birkhäuser Verlag, 2002.

[21] R. Morris, *Composition with Pitch-Classes: A Theory of Compositional Design*, YUP, 1987.

[22] H. Riemann, *Ideas for a Study On the Imagination of Tone*, (translated by Robert Wason and Elizabeth West Marvin, *Journal of Music Theory* 36/1, pp. 81-117, 1992).

[23] A. Vieru, *The book of modes*. Bucharest: Editura Muzicala, 1993 (Orig. ed. 1980).

[24] D. T. Vuza, *Some mathematical aspects of David Lewin's Book Generalized Musical Intervals and Transformations*, Perspectives of New Music, 26(1), pp. 258-287, 1988.

[25] I. Xenakis, *La voie de la recherche et de la question*, Preuves, 177, 1965.

Streptohedrons (Twisted polygons)

David Springett
8, Strath Close
Rugby, Warks, England, CV21 4GA
E-mail: david@cdspringett.fsnet.co.uk
website: davidspringett.fws1.com

Abstract

Imagine a simple form, a cone with a symmetrical cross-section. Now split that cone from apex to base, twist the two halves and re-join. Before your eyes a new, complex form is produced. Imagine more intricate geometric solids which are split, twisted and re-joined, magically producing shapes which coil and twirl - shapes not seen before, unexplored shapes. Remove the inner form of some of these twisted shapes and a path or ribbon remains. These shapes, these ribbons, this idea, will excite the Mathematician, the Sculptor and artist alike.

In June 2001 I was shown two intriguing shapes. The first was in the form of two cones, fixed base to base, which had been split from apex to apex. The two halves had then been separated, one half rotated 90° before being glued to create a form with two edges and one side. (See figures 1, 2 and 3). I later discovered that this double, twisted, cone was developed by Colin Roberts in 1967 and was called a Sphericon. [1]

Figures 1, 2, and 3, creating a Sphericon

The second shape was a cone, its cross section being an equilateral triangle, which had been split vertically from apex to the centre of the base. The two halves had then been rotated so that the edge of one half lay against the baseline all the other half. The two halves had been glued together produce a form which had one side and one edge. (See figure 4).

This twisted double cone rolls drunkenly down a slope, the twisted single cone also rolls in a bizarre manner but stops on one of its half bases. For some reason I also remembered a third similar shape, similar in that it was also a solid of revolution which had been spit along axis twisted and rejoined. This was in the form of a cylinder with a wide disc at its centre. (Figures 5 and 6). It was then that I looked more carefully at the cross-section of these three solids of revolution. The first, the Sphericon, I had mistakenly viewed as a diamond shape. (Third form the right, top line of figure. 7 and photo. 1 second row, extreme right). I should have seen it as a square with the centre of rotation through its points. The second has an equilateral triangle as its cross-section. The third has a cross-shaped cross-section. (Figure 8). I put aside the third shape, for the moment, for it did not yet fit in with the pattern I had just seen.

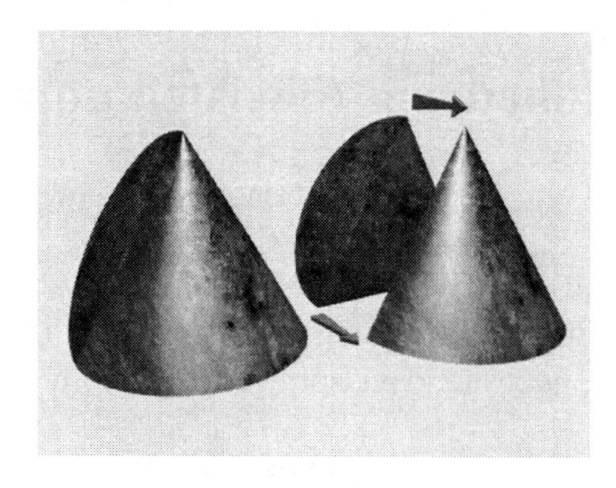

Figure 4, rotated cone

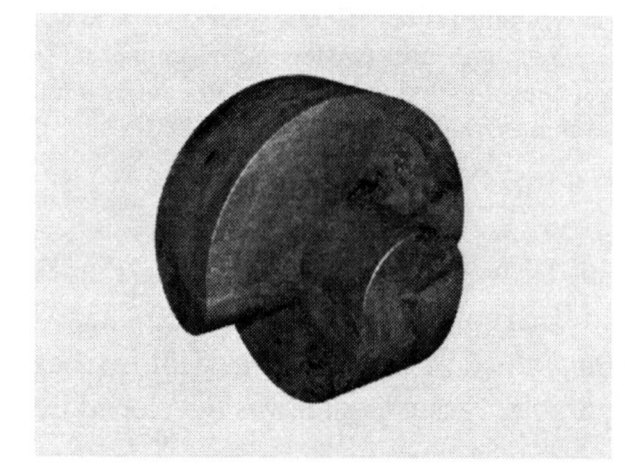

Figures 5 and 6

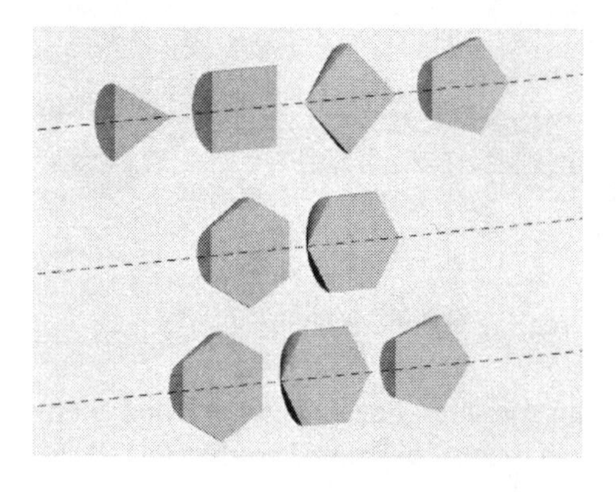

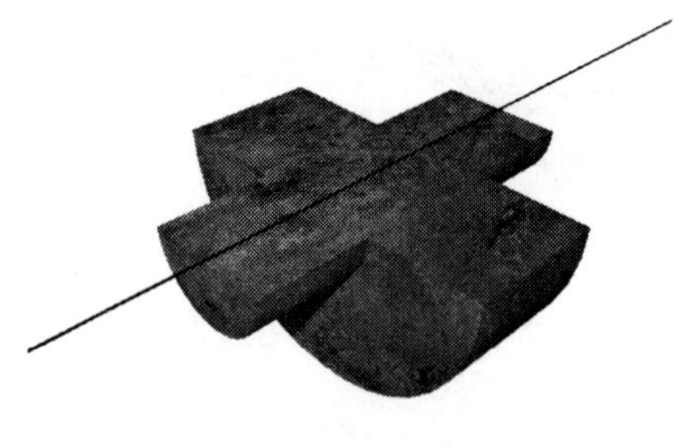

Figures 7 and 8

A triangular cross-section, a square cross-section, obviously the next shape to work with would be a Pentagon. A regular Pentagon can be used as the base for a solid of revolution, the axis will be through the apex then crossing the midpoint of the base. (Top row, extreme right, figure 7).

When I split, twisted and rejoined this piece I discovered two very interesting details. First it produced two forms. Move the apex of one half to join the rim of the of the other half for shape number one. Move the apex of one half to join the base of the other half for shape number two. A second detail I discovered was that the shape could have a right-hand or left-hand twist. (Photo. 2 bottom row, second from left).

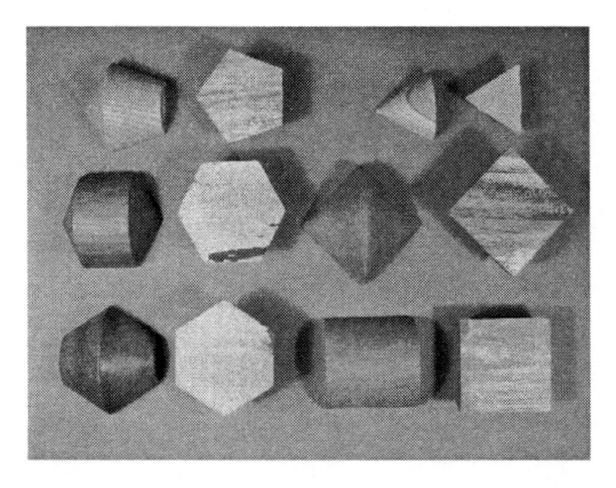

Photos 1 and 2

A triangle, a square, a Pentagon, it was natural to move on to produce a solid of revolution based on a regular hexagon. So with the axis lined up through the flat of the base, through its centre and the flat of the top I produced an hexagonal form. Only a single form could be made by splitting, twisting and rejoining, but again it can have a left or a right-handed twist. (Figure 9 and photo.2 bottom line, first left).

Figure 9 hexagonal shape

Photo 3 right and left handed forms

Looking again I realized that a hexagon has two different centres of rotation, one through its flats as I had produced above and a second through its points. I produced this second hexagonal shape, with the centre of rotation through its points, and discovered a second hexagonal twisted form. Again this could have a left or right-handed twist. (Fig. 9, photo.2 second left).

I was surprised when I looked at the two hexagonal forms which were lying on the table in in their split halves, each had the same size hexagonal base shape. I took one of the halves which had been rotated through the points and joined it to one of the halves which had been rotated through the flats to produce a hybrid. This hybrid could be right or left-handed. (Photo. 3). Having realized that polygons with equal number of sides can have two centres of revolution I returned to the Sphericon. It has a square as its base form and its centre of revolution is through the points but if the square is rotated through its flats then a simple cylinder is produced.

A rather unexciting form is produced when the two halves of the cylinder are twisted and refitted (Photo. 4). Its outer edge follows the curve similar to the seam line on a tennis ball. A more interesting hybrid is formed using one half of the Sphericon and one half of the cylinder.

Figure 10 star pentagon

Photo 4 twisted cylinder

After a few days of thought it suddenly occurred to me that a regular star form could be used as the basis for further twisted solids of revolution. I first produced a regular pentagonal star drawn by joining points 1, 3, 5, 2 and 4 on a regular Pentagon. (Figure 10).

Remembering that the pentagonal solid, when split, twisted & re-joined, produced two forms, I twisted the top point of the star to meet with the edge. This produced form number one. (Photo. 5). Next I moved the top point of the star to meet with the base. This produced form number two. (Photo. 5). Of course there were left and right hand versions of both twisted forms.

Photo 5 twisted start pentagon

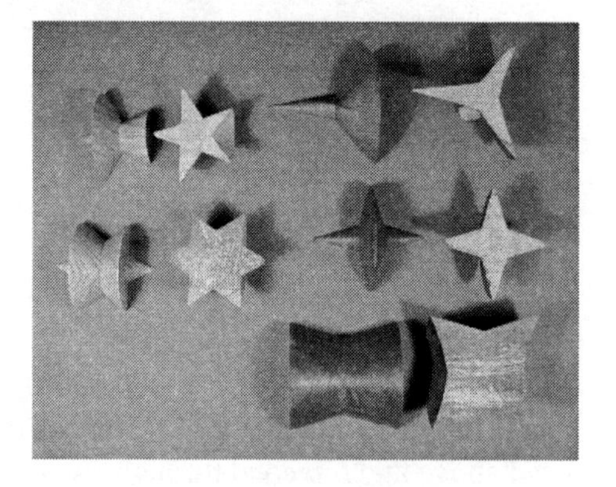

Photo 6 other stars

A three pointed star followed, looking much like the Mercedes symbol. From this I was able to produce one form only but it can have left or right hand versions. (See figure 11 and photo 6 plus photo 7 bottom right). The four pointed star looks quite ordinary. It has two forms, one with the axis of revolution through its points, the other with the axis of revolution through the valleys. (Fig. 11 and photo 6). And, of course, the right and left-handed versions of each of these, added to which there is the hybrid version.

A six pointed star was more exciting producing a more "twisted" form. Again there are two axes of revolution, the first through the points and the second through the valleys and there are right and left hand version of each of these plus the hybrid. (Figure 11 and photo. 7 top right).

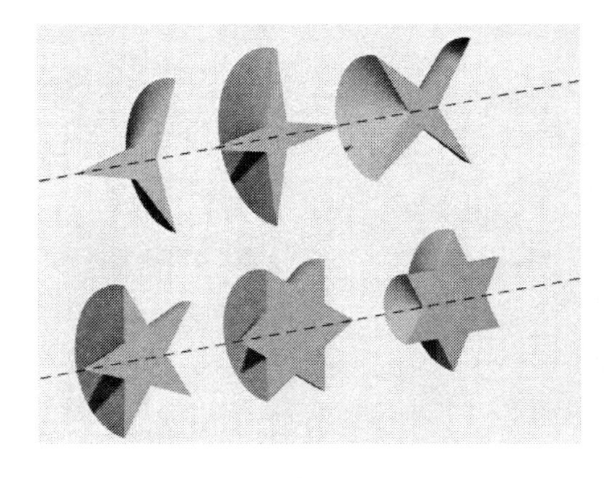

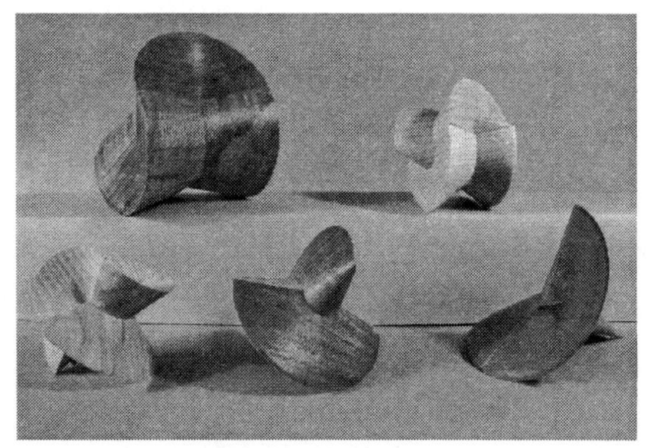

Figure 11 axes **Photo 7** models

At this point I returned to the cross shaped, cross section pieces (figure 6) trying to work out how this fitted in with the pieces I had produced so far. It became apparent that this form was similar to the star shapes but the arms would have parallel sides and ends rather than a point.

I produced a three armed version (photo. 8 top right & photo. 9 bottom right) which when split & twisted gave a very satisfying twisted form, either right or left-handed. The five arm piece was particularly interesting (two forms plus left and right handed) with a snaking valley twisting around the form. (Photo. 8 top left and photo. 9 top right). The six arm shape was produced with the axis running through the arms (left and right handed versions) or the axis running through the valleys (left and right handed versions) and of course the hybrid, all stunningly sinuous. (Photo 8 bottom left and photo. 9 top left).

Photo 8 various arm splits **Photo 9** models

I returned to the simple cross and turned it so that the axis of revolution was set through the valleys. (Photo. 8 bottom right). The resulting shape was unexpected and produced one form only (Photo.9, bottom left) but combined with half the original form (figure.8) it produced an interesting hybrid. (Photo. 10).

I have run an undercut groove into the corners of a hexagonally based shape allowing ball bearings to run in that groove. When the shape had been split the ball bearings were inserted into the groove, the halves twisted and re-glued producing an amusing piece with the ball bearings running around the twisted track. (Photo.11 middle top).

Photo 10

Photo 11

Properties

Before I explain the areas I wish to explore I will detail some of the properties of these pieces.

1. These Streptohedrons will fit inside a sphere (the size of that sphere relates to the base size of the polygon) and their extreme edges or points will touch the inside of that sphere whether they are in their left or right handed form, whether they are twisted or aligned, or whether they are in the hybrid form.
2. These forms are based on regular polygonal figures which have rotational symmetry.
3. These streptohedron can be produced in left or right-handed forms.
4. Those forms based on equal sided regular polygons (star and arm forms included) will produce hybrid forms.
5. Those forms based on odd sided regular polygons (star and arm forms included) will NOT have a hybrid form.
6. All these streptohedron forms roll in an unusual "drunken" manner similar to the "two disc rollers" or "wobblers".

Photo 12

Figure 12

Areas to explore

I have begun to introduce curves into these shapes as can be seen in photo. 11 middle left. This piece is based on the cross form seen in figure 8 but it has concave curves cut into the faces. Figure 11 shows that same cross base with convex (almost spherical) ends. Also in photo. 11, middle bottom, it will be seen that curves have been worked into three of the six faces of an hexagonal form.

On the right, in photo. 12 is a figure which is based on a Reuleaux triangle which has constant diameter but not a fixed centre. On the left of photo. 12 can be seen a stepped form, based on an equilateral triangle, which looks like an Art Deco Ziggurat. I have not developed any of these final pieces using their different centres of revolution, produced hybrids or extended the curves more deeply.

At present I am producing some of these forms so that the centre is removed producing a more ribbon-like form. (Photos 13 and 14). The negative shapes within these "ribbons" are worth exploring. Photos 15 and 16 show a ribbon form based on an equilateral triangle. The negative centre is a twisted hexagon, precisely the same as shown in photo 2 bottom left.

Shapes and their relationships continue to present themselves. Returning to the beginning ... the double cone (figure 1 and photo 17). This is a solid of revolution based on a square. Now take the same size square and, using on corner as the axis rotate it to produce a ring as shown in split form in photo 18. Next fit double cones into the hollow ends of that ring. (Cross-section shown in photo 19). This produces a larger double cone (photo. 20). If this double cone is split, twisted and rejoined it makes a sphericon (photo 21). Pull this sphericon apart and two symmetrical halves are revealed (photo 22). I can only show the next stage in half-section at present. Apply another ring, (same size as in photo 18), to the end of the double cone filling the end hollow with a third double cone (photo 23). If a further, larger, ring (shown as white cross-section and wire in photo24) is applied a double cone of a greater size is made. This much larger double cone can be split, twisted and rejoined to make a sphericon or it may be pulled apart revealing symmetrical parts.

And finally, for the moment, that central shape looks familiar. It is the cross, cross-section figure seen in photo 10. The variety of the shapes I have uncovered seem endless and linked just like the pathways on some of these forms.

Photo 13

Photo 14

References

1. Ian Stewart, Scientific American October 1999
2. David Springett, “Streptohedrons (spheriforms)”, *Society of Ornamental Turners Bulletin* No 111 Sept. 2004
3. David Springett “Streptohedrons” a two part article *Woodturning Magazine* issue 135 Spring 2004 & issue 136 May 2004 (GMC publication)
4. David Springett, *Woodturning Wizardry*, 1993, reprinted and revised 2005, Guild of Master Craftsmen Publications Ltd

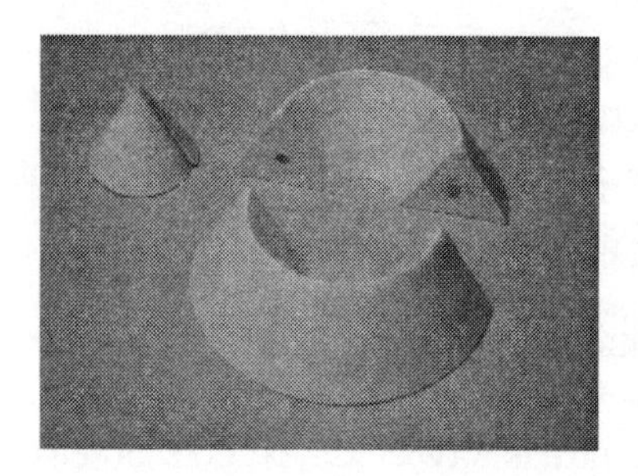

Photos 15, 16, and 17

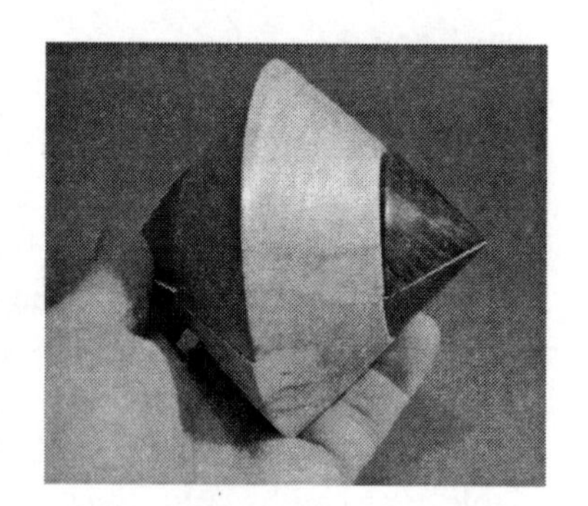

Photos 18, 19 and 20

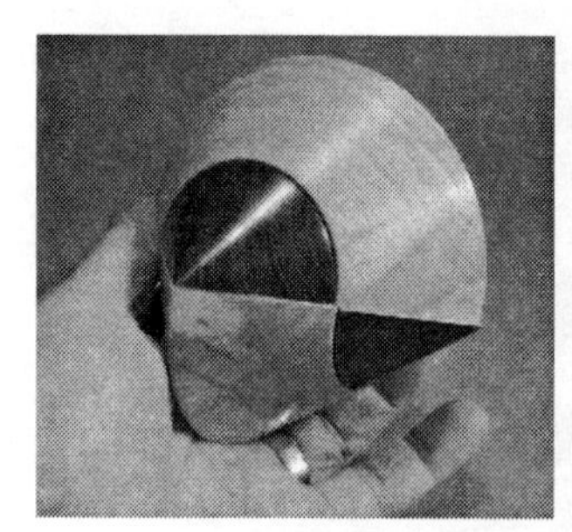

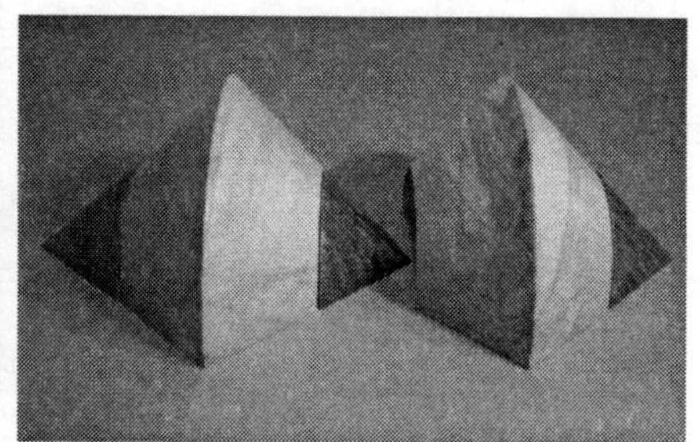

Photos 21, 22 and 23

Photos 24 and 25

Acknowledgements

The figures are by Robin Springett and first appeared in *Woodturning Magazine* issues 135 and 136 in spring 2004. Reproduced by permission of GMC Publications, Lewes, UK.

Fractal Tilings Based on Dissections of Polyominoes

Robert W. Fathauer
Tessellations Company
3913 E. Bronco Trail
Phoenix, AZ 85044, USA
E-mail: tessellations@cox.net

Abstract

Polyominoes, shapes made up of squares connected edge-to-edge, provide a rich source of prototiles for edge-to-edge fractal tilings. We give examples of fractal tilings with 2-fold and 4-fold rotational symmetry based on prototiles derived by dissecting polyominoes with 2-fold and 4-fold rotational symmetry, respectively. A systematic analysis is made of candidate prototiles based on lower-order polyominoes. In some of these fractal tilings, polyomino-shaped holes occur repeatedly with each new generation. We also give an example of a fractal knot created by marking such tiles with Celtic-knot-like graphics.

1. Introduction

Fractals and tilings can be combined to form a variety of esthetically appealing constructs that possess fractal character and at the same time obey many of the properties of tilings. Previously, we described families of fractal tilings based on kite- and dart-shaped quadrilateral prototiles [1], v-shaped prototiles [2], prototiles that are segments of regular polygons [3], and prototiles derived by dissecting polyhexes [4]. Many of these constructs may be viewed online [5]. These papers appear to be the first attempts at a systematic treatment of this topic, though isolated examples were earlier demonstrated by M.C. Escher [6] and Peter Raedschelders [7].

In Grünbaum and Shephard's book *Tilings and Patterns* [8], a tiling is defined as a countable family of closed sets (tiles) that cover the plane without gaps or overlaps. The constructs described in this paper do not for the most part cover the entire Euclidean plane; however, they do obey the restrictions on gaps and overlaps. To avoid confusion with the standard definition of a tiling, these constructs will be referred to as "*f*-tilings", for fractal tilings.

The tiles used here are "well behaved" by the criterion of Grünbaum and Sheppard; namely, each tile is a (closed) topological disk. Most of the *f*-tilings explored in References 1-5 are edge-to-edge; i.e., the corners and edges of the tiles coincide with the vertices and edges of the tilings. However, they are not "well behaved" by the criteria of normal tilings; namely, they contain singular points, defined as follows. Every circular disk, however small, centered at a singular point meets an infinite number of tiles. Since any *f*-tiling of the general sort described here will contain singular points, we will not consider singular points as a property that prevents an *f*-tiling from being described as "well behaved". These *f*-tilings provide a rich source of unique fractal images and also possess considerable recreational mathematics content.

The prototiles considered in this paper are derived by dissecting polyominoes. A polyomino is a shape made by connecting squares in edge-to-edge fashion. Following common usage, the first nine polyominoes will be called monomino, domino, tromino, tetromino, pentomino, hexomino, heptomino, octomino, and nonomino, respectively. Polyominoes made from a number of squares $n > 9$ will be called n-ominoes. For a discussion of different types of polyominoes and conventional tilings using them, see References 8 and 9. Most polyominoes have adjacent straight-line segments longer than the edges of the constituent squares, as shown in Figure 1. For this reason, there are relatively few true edge-to-edge f-tilings using prototiles created by dissecting polyominoes. In order to provide a richer variety of examples, pseudo-edge-to-edge f-tilings will be considered as well. In these cases, the edges of the constituent squares are considered to define pseudo-edges of the polyominoes, as shown in Figure 1.

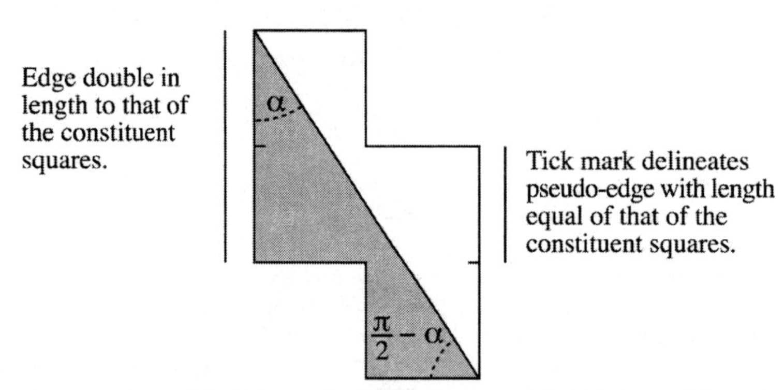

Figure 1: *A tetromino with tick marks indicating pseudo-edges and irrational angle α.*

Each prototile has one or two long edges and two or more short edges. The angles between the long edges and adjacent short edges are for the most part irrational, but sum to a multiple of $\pi/2$, as shown in Figure 1. The f-tilings are constructed by first matching the long edges of identical prototiles, then cloning the prototile, reducing its size by the ratio of the length of a short edge to that of a long edge, fitting multiple copies of these smaller tiles around the first generation of tiles according to a matching rule, and finally iterating this process an infinite number of times. For a given f-tiling, every tile is similar to the single prototile. The group of first-generation tiles forms the dissected polyomino. In practice, the appearance of the overall tiling changes little after 4-7 generations, since the tiles become extremely small relative to the first-generation tiles. The figures shown here are constructed by iterating until the individual tiles become very small on the scale of the page size. Fractal tilings may also be constructed that possess more than one prototile, but these will not be considered here.

2. Candidate Prototiles

Three requirements simplify the search for polyomino-based prototiles that allow f–tilings.
1. The generating polyomino must have 2-fold or 4-fold rotational symmetry. While polyominoes without rotational symmetry may be used, they generate no new prototiles, and therefore they do not need to be considered. While not proven here, this is readily apparent by examining a table showing all polyominoes possible for a given number of squares. In addition, mirrored variants of polyominoes are not considered to generate distinct prototiles, as they would result in f-tilings that are an overall mirror of the f-tilings constructed from non-mirrored variants.
2. For a prototile generated by bisecting a polyomino, each bisecting line, which will form the long edge of the prototile, must originate and terminate at corners of the polyomino and pass through the centroid of the polyomino. If the endpoints aren't at corners or pseudo-corners, the short edges will not all be of the same length. The long edges of the prototile must be longer than the short edges or pseudo-edges of the prototile.
3. For a prototile generated by dissecting a polyomino into four equal parts, each dissecting line, which will form one long edge of the prototile, must run from the centroid of the polyomino to a corner. The four dissecting lines are related to each other by rotations of 90° about the centroid. Again, the long

edges of the prototile must be longer than the short edges or pseudo-edges of the prototile. For reasons shown below, the number of short edges or pseudo-edges must be even. This rules out any polyomino made up of $4n + 1$ squares [10]. The only other polyominoes with 4-fold symmetry are made up of $4n$ squares, so only these need be considered.

Figure 2 shows candidate polyominoes made up of 1 to 5 squares and prototiles that meet these criteria. Prototiles colored green and red allow *f*-tilings, while those colored black do not. Only green prototiles allow true edge-to-edge *f*-tilings.

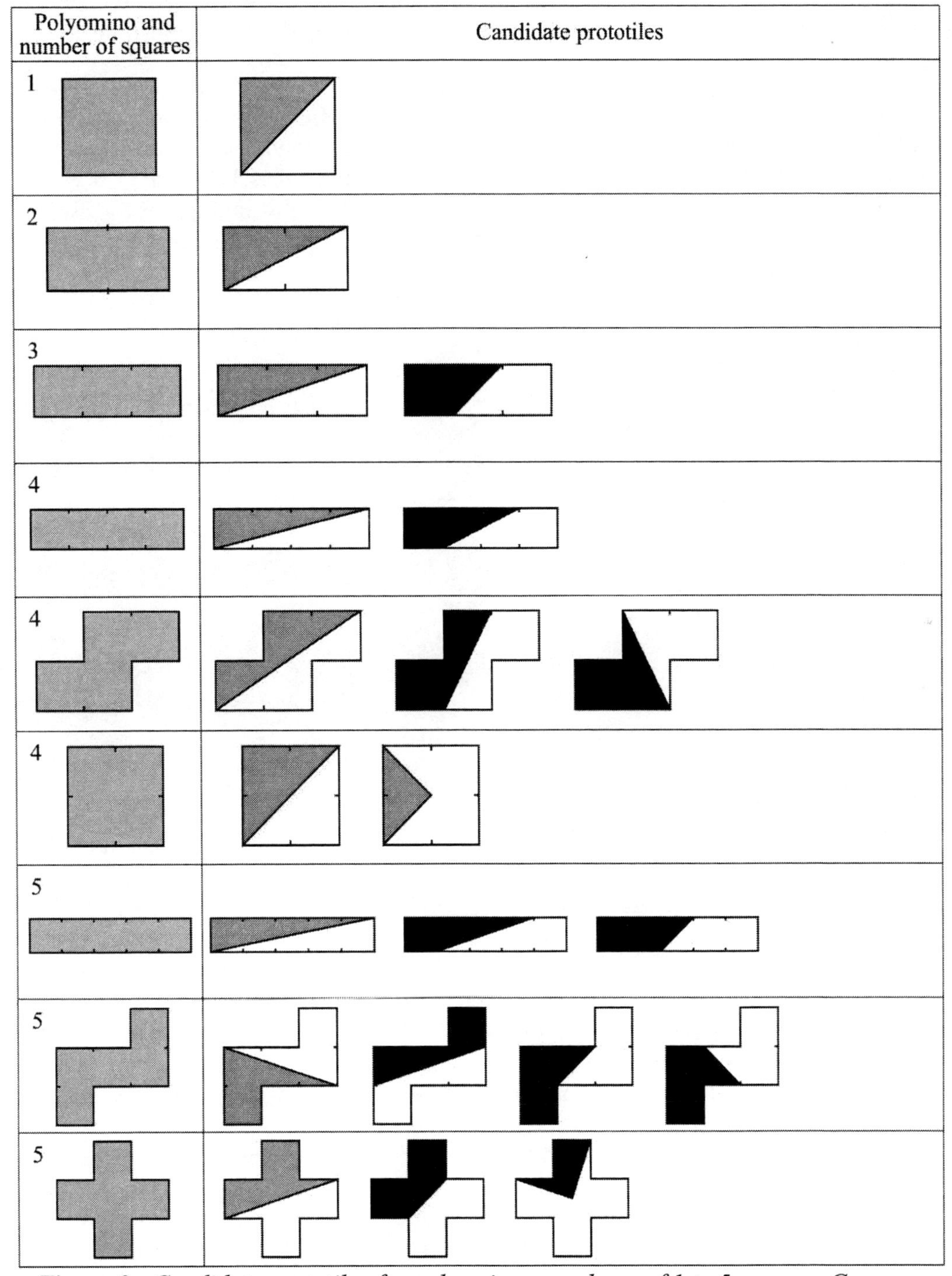

Figure 2: *Candidate prototiles for polyominoes made up of 1 to 5 squares. Green ones tile edge-to-edge and red ones pseudo-edge-to-edge, while the black ones do not tile.*

3. *f*–tilings Based on Dissected Polyominoes

3.1. Bounded *f*-tilings. In this section, we give a number of examples of *f*–tilings constructed from prototiles derived by dissecting polyominoes. Our first examples have overall 2-fold rotational symmetry. The starting point for an *f*-tiling of this sort is a pair of tiles that form the generating polyomino. These are surrounded by smaller tiles in (pseudo-) edge-to-edge fashion, with the construction process proceeding iteratively. The only option is whether or not the tiles are mirrored between successive generations, but no mirrored variants are considered here due to space limitations. The process by which *f*-tilings are constructed is described in greater detail in Reference 1.

In Figure 3, we show two examples of *f*-tilings of this sort. The left *f*-tiling is edge-to-edge, with a reduction factor of $1/\pi 10$ between tiles of successive generations. This number is easily obtained by noting that the diagonal is that of three squares in a row and then applying the Pythagorean theorem. (If the squares have edges of length 1, the long edge of the prototile has length $\pi(1^2 + 3^2)$.) Between successive generations the tiles are rotated by $\arctan(1/3)\ \pi\ 18.43°$ (plus multiples of $\pi/2$ as required for fitting a particular edge). The right *f*-tiling in Figure 3 is pseudo-edge-to-edge, with a reduction factor of $1/\pi 13$ (which can be obtained in similar fashion).

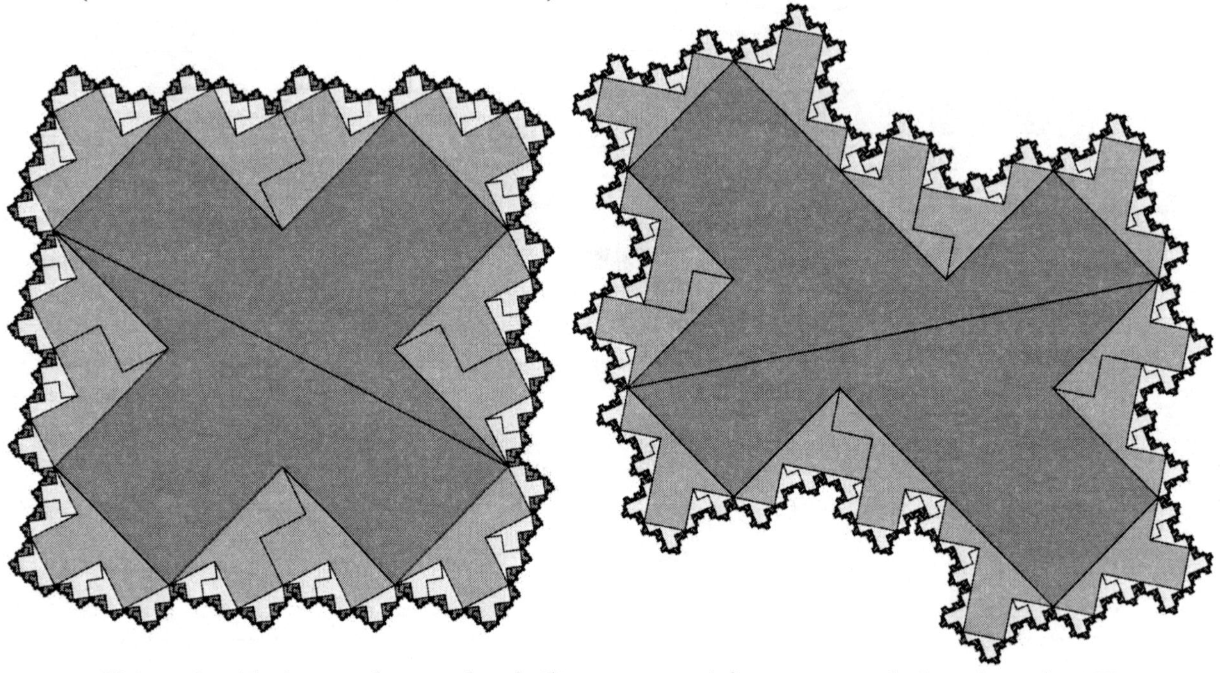

Figure 3: *(a) A true edge-to-edge f–tiling generated from a prototile based on the "X" pentomino.* (b) *A pseudo-edge-to-edge f–tiling generated from a prototile based on a hexomino.*

In general, the boundaries of *f*-tilings are fractal curves, though there are cases in which the boundaries are non-fractal polygons. The boundaries are similar to Koch islands and related constructs, in which a line segment is distorted into multiple smaller line segments, which are in turn distorted according to the same rule, etc. [11].

In Figure 4, we show a 2-fold rotational *f*-tiling in which the boundary is an octagon with two different edge lengths. In order to make the figure more interesting visually, the tiles, which are isosceles right triangles, are decorated with Celtic-knot-like markings. The markings on this *f*-tiling thus form a fractal knots (or fractal links, to use knot terminology more properly) design.

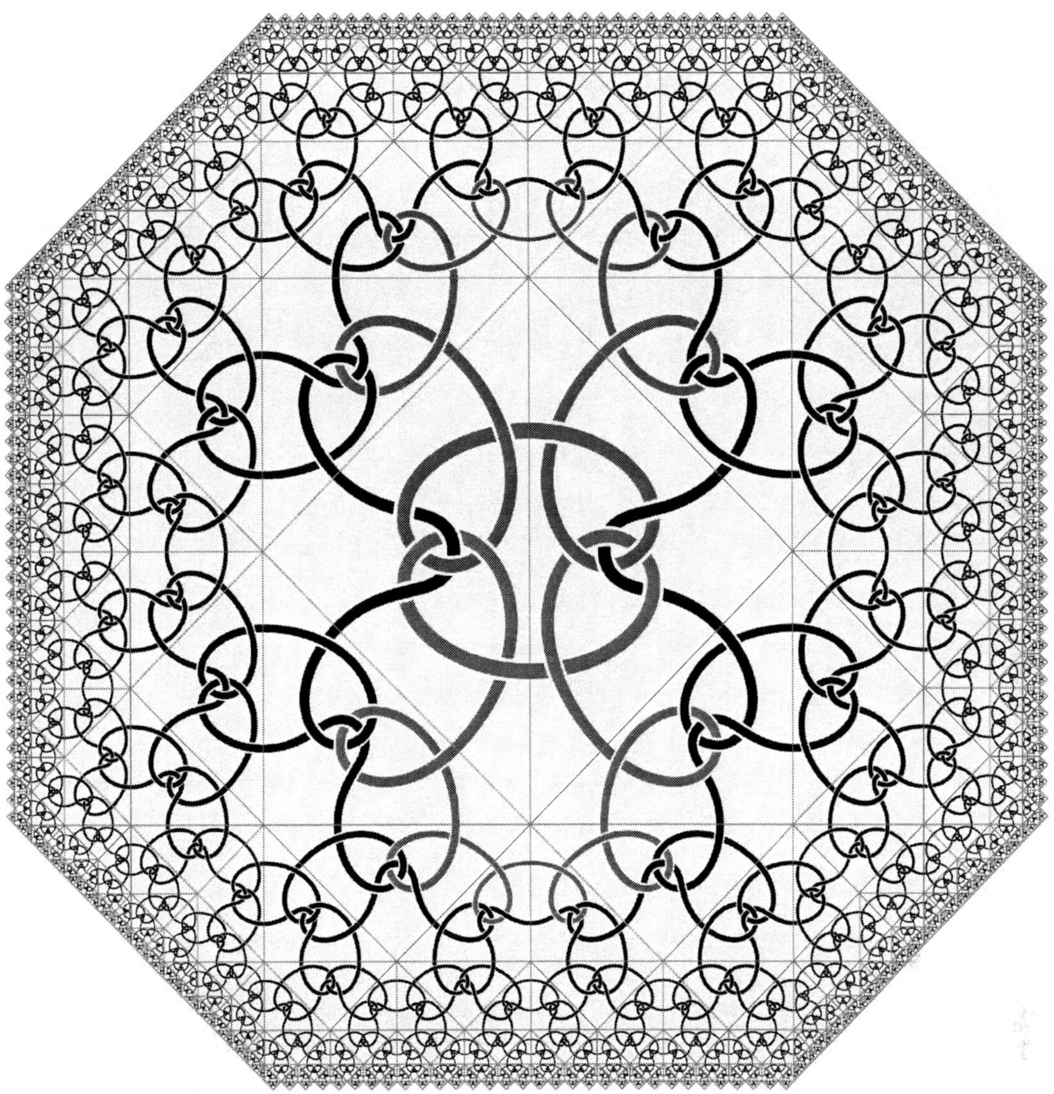

Figure 4: *A true edge-to-edge f–tiling generated from the monomino. In order to make a more esthetically-pleasing figure, the tiles are decorated with markings that create a fractal knots design.*

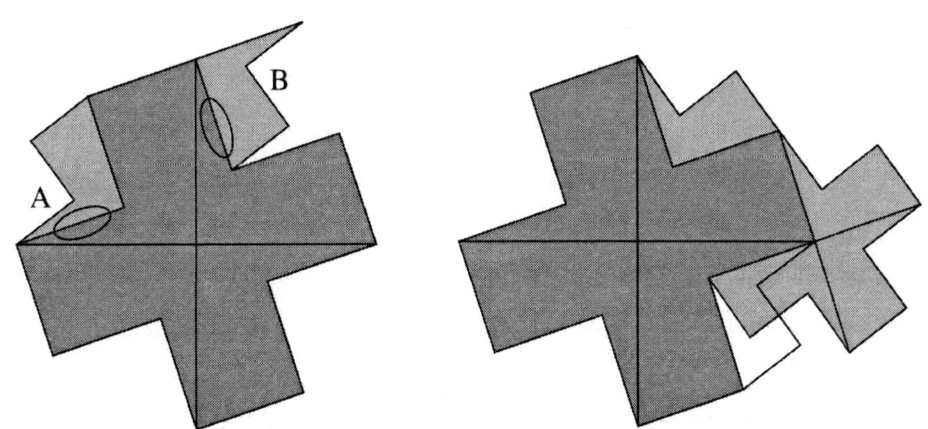

Figure 5: *For prototiles generated from 4-fold polyominoes, there are two choices, A and B, for arranging smaller tiles around large tiles. In order to avoid overlaps using a consistent matching rule, 4-fold prototiles must have an zeven number of short edges.*

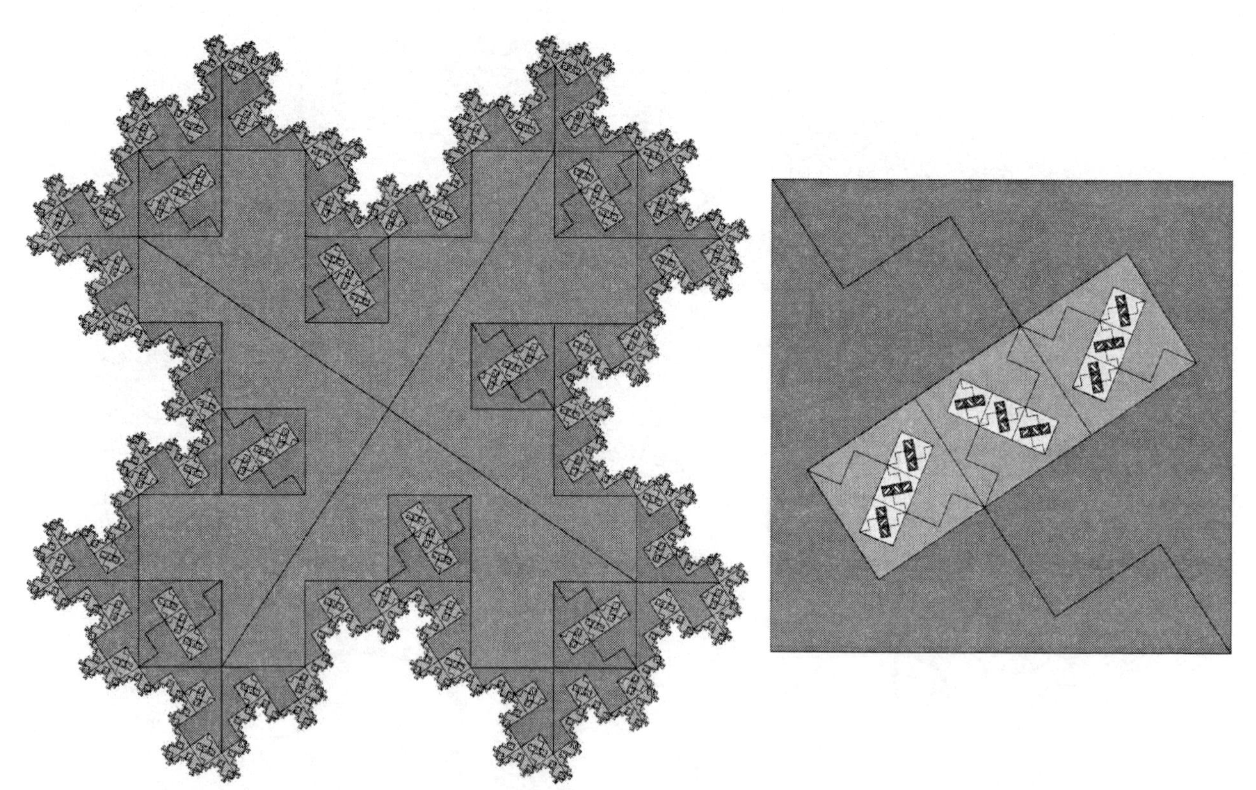

Figure 6: *An f–tiling generated from a prototile based on a 20-omino. Note the formation of "I" tromino-shaped holes. The zoomed-in portion of the f-tiling at right highlights this feature. The scaling and rotation between successive generations of tiles is* $1/\sqrt{13}$ *and* $\sqrt{}$ *33.7°.*

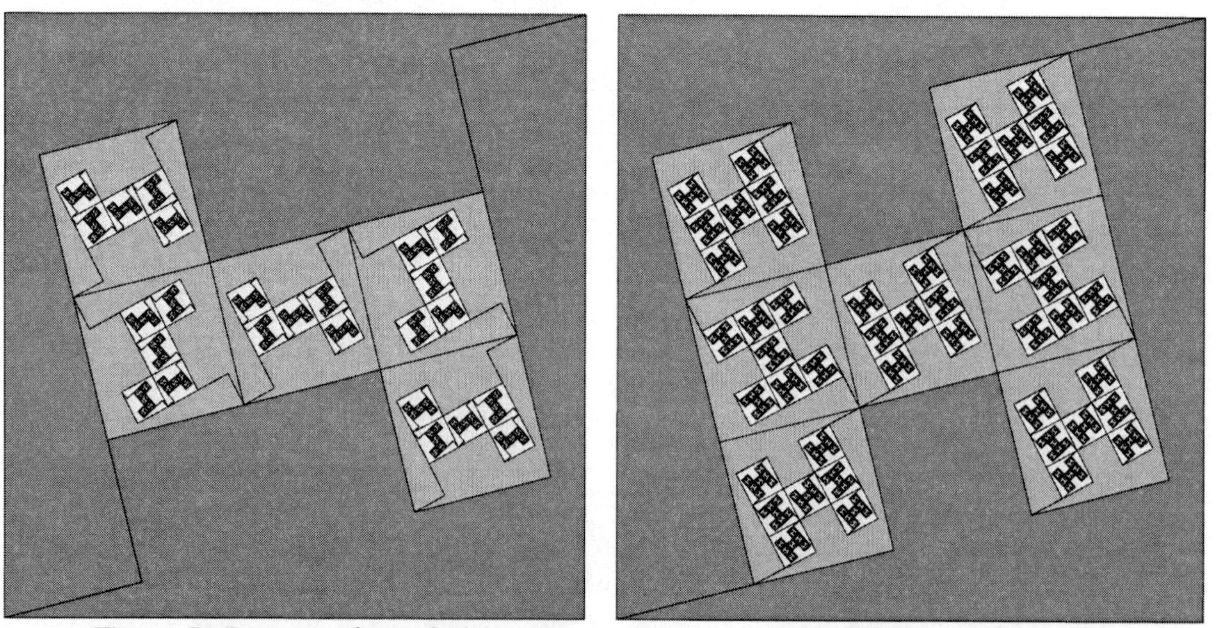

Figure 7: *Portions of two f–tilings generated from prototiles based on 24- and 20-ominoes, respectively. Note the repeated formation of "Z" pentomino- and "H" heptomino-shaped holes. The scaling between successive generations of tiles is* $1/\sqrt{17}$ *for both f-tilings.*

Next we show some examples with overall 4-fold rotational symmetry. For these prototiles, there are two choices of matching rules for arranging tiles of a given generation around tiles of the next larger generation. This is illustrated in the left side of Figure 5, where the two choices are labeled A and B. The right side of Figure 5 shows why the number of short edges cannot be odd if a single matching rule is used. The fact that there are two long edges for each smaller tile requires an even number of short edges for each larger tile if overlaps are to be avoided.

An *f*-tiling of this sort is shown in Figure 6, where the prototile is derived from a 20-omino. The left side of Figure 6 shows the full *f*-tiling, while the right shows a magnification of a hole formed by two tiles fit together within a square region. The hole is shaped like the "I" tromino, and further iterations create three smaller holes of the same sort within each larger hole. This tripling of holes with the same shape will clearly continue *ad infinitum* as additional generations of tiles are added, with the hole just being filled in the infinite limit. Figure 7 shows similar square regions for two additional 4-fold *f*-tilings. The left figure shows "Z" pentomino-shaped holes using a prototile derived from a 24-omino, while the right figure shows "H" heptomino-shaped holes using a prototile derived from a 20-omino.

In some cases, an entire *f*-tiling constitutes a supertile that tiles the plane; the *f*-tilings in Figure 8 possesses this property. Plane filling may also be achieved by using square building blocks such as those shown in Figure 7.

4. Conclusions

We have presented several examples of fractal tilings (*f*-tilings) based on prototiles derived by dissecting polyominoes with 2-fold and 4-fold rotational symmetry. There are an infinite number of polyominoes and an infinite number of *f*-tilings of this sort. However, in general they become increasingly less interesting for higher-order polyominoes due to the fact that the scaling factor between successive generations becomes more extreme. Decoration of tiles with markings to create fractal knots is one means of increasing the visual interest of these constructs. Fractal tilings that result in recurring polyomino-shaped holes have also been demonstrated.

References

[1] Robert W. Fathauer, *Fractal tilings based on kite- and dart-shaped prototiles*, Computers & Graphics, Vol. 25, pp. 323-331, 2001.
[2] Robert W. Fathauer, *Fractal tilings based on v-shaped prototiles*, Computers & Graphics, Vol. 26, pp. 635-643, 2002.
[3] Robert W. Fathauer, *Self-similar Tilings Based on Prototiles Constructed from Segments of Regular Polygons*, in Proceedings of the 2000 Bridges Conference, edited by Reza Sarhangi, pp. 285-292, 2000.
[4] Robert W. Fathauer, *Fractal Tilings Based on Dissections of Polyhexes*, in Renaissance Banff, Mathematics, Music, Art, Culture Conference Proceedings, 2005, edited by Reza Sarhangi and Robert V. Moody, pp. 427-434, 2005.
[5] Robert W. Fathauer, http://members.cox.net/fractalenc/encyclopedia.html.
[6] Bruno Ernst, *The Magic Mirror of M.C. Escher*, Ballantine Books, New York, 1976.
[7] Peter Raedschelders, "Tilings and Other Unusual Escher-Related Prints," in M.C. Escher's Legacy, edited by Doris Schattschneider and Michele Emmer, Springer-Verlag, Berlin, 2003.
[8] Branko Grünbaum and G.C. Shephard, *Tilings and Patterns*, W.H. Freeman, New York, 1987.
[9] Solomon W. Golomb, *Polyominoes*, Princeton University Press, Princeton, New Jersey, 1994.

[10] The only pentomino with 4-fold rotational symmetry yields prototiles with 3 short edges. Adding 4 squares to this pentomino yields prototiles with either 3 or 5 short pseudo-edges. It can easily be seen that adding four squares to any 4-fold polyomino will either add 2 short pseudo-edges, leave the number of short pseudo-edges unchanged, or subtract 2 short pseudo-edges. The number of short pseudo-edges for prototiles is therefore always odd.
[11] H.-O. Peitgen, H. Jürgens, and D. Saupe, *Fractals for the Classroom – Part One*, Springer-Verlag, New York, 1992.

Figure 8: *A plane-filling f–tiling generated from a prototile based on an octomino. The f-tiling was carried through six generations, with each generation a different color.*

Vortex Maze Construction

Jie Xu Craig S. Kaplan

David R. Cheriton School of Computer Science
University of Waterloo
{jiexu,csk}@cgl.uwaterloo.ca

Abstract

Labyrinths and mazes have existed in our world for thousands of years. Spirals and vortices are important elements in maze generation. In this paper, we describe an algorithm for constructing spiral and vortex mazes using concentric offset curves. We join vortices into networks, leading to mazes that are difficult to solve. We also show some results generated with our techniques.

1. Introduction

Mazes have been a part of human society for thousands of years [5]. They can form the locus of a powerful legend, as in the journey of Theseus into the labyrinth of Crete. They can serve as a focus for meditation and prayer, most famously on the floor of the cathedral at Chartres. They are a compelling and occasionally frustrating source of entertainment for both children and adults. Countless books of mazes of all kinds are available (see, for example, the masterpiece by Christopher Berg [2]), and designers such as Adrian Fisher [4] are revitalizing the construction of corn and hedge mazes. A good maze is simultaneously a work of art and a complex puzzle.

Figure 1: *Simple circle spiral and vortex.*

We may approach the problem of maze design in many different ways, and indeed enthusiasts such as Walter Pullen have catalogued many possible construction algorithms and visual styles [7]. During the design process, one factor we would like to understand (or, ideally, control) is the difficulty of the maze. Although some work has been done on characterizing the difficulty of mazes [6], we feel that the problem is very subtle and far from solved.

As part of our larger goal of studying the difficulty of mazes, in this paper we give a construction technique that we feel produces mazes with a high degree of difficulty. Our construction makes use of two specific visual devices: the spiral and the vortex. These are identified by artist Christopher Berg as being crucial to the design of effective mazes [1]. A spiral is a single passageway that winds inward to a dead end. A vortex has a similar spiral structure, but features multiple passageways meeting at a central junction. A vortex is a powerful obfuscation device in maze construction, as it obscures the relationships between the paths that meet there. Moreover, the regularity of the concentric paths in spirals and vortices make them difficult to solve. Figure 1 shows an example of spiral and vortex.

The rest of the paper is organized as follows. Section 2 introduces our approach to create spiral and vortex mazes. Technique for constructing network of vortices is given in Section 3. Section 4 presents some visual effects. Some results appear in Section 5. Finally, we provide some future work in section 6.

2. Vortex Construction

In this section, we give an algorithm for constructing vortices within regions of the plane. The input to the algorithm is a region (for example, a convex polygon) and a set of labels on the boundary of the region. The output is a vortex inside the region that connects all labels at a central junction. We do not provide an explicit algorithm for constructing spirals. We believe that the dead end of a spiral path is relatively easy to detect by visual inspection, and can therefore be avoided while solving. And in any case, a spiral can simply be considered as a vortex with a single label on its boundary.

Given a region boundary, we construct a sequence of *offset curves*, closed paths each of which is offset by a constant parallel distance from its outer neighbor. We sample at regular intervals along these curves to obtain candidate vertices for the piecewise linear paths that will make up the walls of the vortex. In order to avoid "ridges" in the result (as in the top row of Figure 3), we select the closest corner as the start point instead of using the labeled location directly. Then, we find the closest sample point on the outermost offset curve and draw a wall connecting the two points. We then proceed around the offset curve clockwise or counter-clockwise (at the designer's discretion), building a wall as we go. When the wall cannot be extended any further we drop into the next deeper offset curve and continue. The vortex is complete when no disconnected sample points remain. Figure 2 illustrates this process.

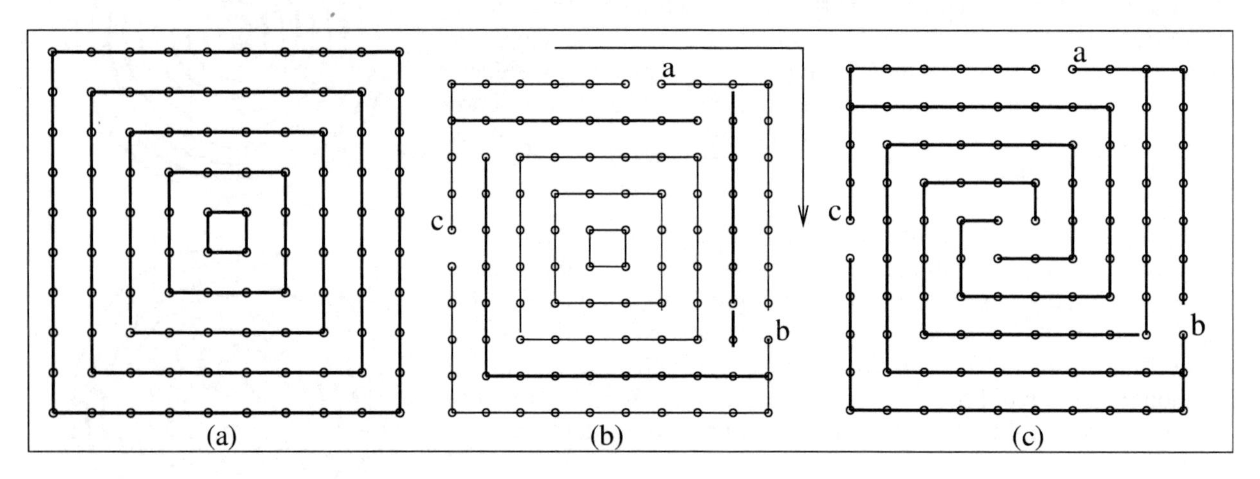

Figure 2: *Process of creating vortex. The region and sample points shown in (a), labeled exits are in (b) and the final result in (c).*

At this point, the vortex has a central junction that connects all labels on the boundary. We may, however, wish to construct a vortex where the labels are partitioned into subsets, and there is a path from one label to another precisely if they are in the same subset. We can implement this more sophisticated routing by selectively closing passageways connecting labels from different subsets.

We can now render the piecewise linear paths directly to get a polygonal maze. Alternatively, the sample points can be used as control points for a cubic spline, giving a smoother, more fluid appearance. Figure 3 shows some basic vortex mazes.

Figure 3: *Sample circular vortices with three, four, and six arms are shown in the top row. Because there are no corners in circles, our algorithm produces ridges. The bottom row shows examples of vortices constructed in differently-shaped polygonal regions.*

3. Networks of Vortices

Armed with the vortex as a primitive, we can now create very complex mazes by assembling networks of interconnected vortices. Berg mentions this technique as an especially fiendish maze style [1].

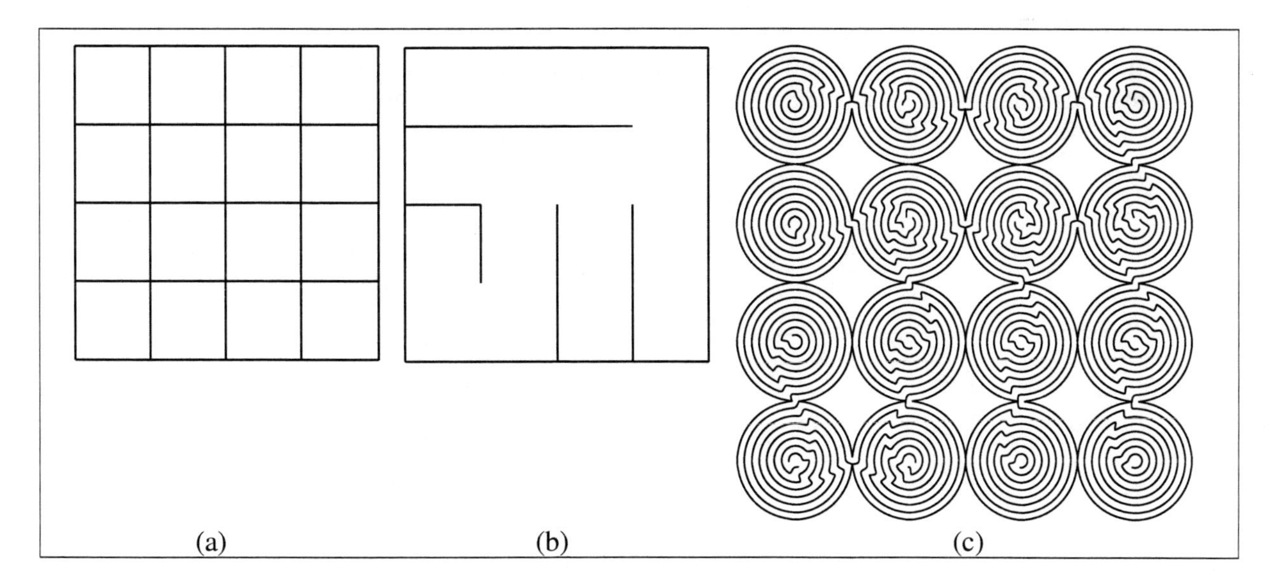

Figure 4: *The construction of a networks-of-vortices maze from a grid. The original 4x4 grid is shown in (a). the template maze in (b), and the final vortex maze in (c).*

Two adjacent regions that have a label on their shared boundary can become interconnected by breaking the boundary at the label's position. We can then construct a network-of-vortices maze in a simple two-level process. First, we use a simple maze construction algorithm such as that of Shivers [8] to build a small maze on a square grid, which we call the *template maze*. We then render each cell of the grid as a vortex, where there is a label at each compass heading precisely if the cell's wall in that direction has been erased. Figure 4 gives an example of this construction technique on a small square grid. Because of the obfuscating effect of the vortices, the final maze is much more difficult to solve than its template.

We can add to the complexity and appeal of this approach by further subdividing the grid cells in the template maze. There is no reason not to allow a diagonal line to cut a cell in half. The grid-based maze construction algorithm adapts naturally to this new grid. In fact, this division can create the interesting situation where a solver has to consume one path through a vortex, only to return later and consume another path on the way to the end. Figure 5 shows an example of this modified approach.

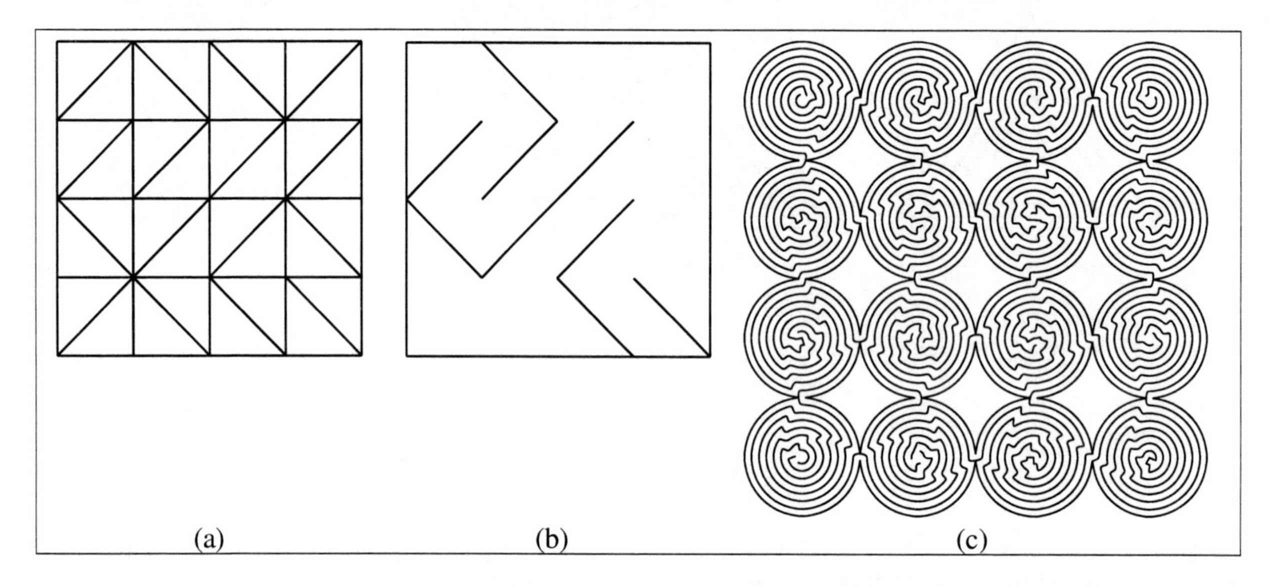

Figure 5: *An example of a more complex vortex maze that can be constructed by splitting grid cells.*

Rather than letting a random algorithm construct an arbitrary template maze, we can give the user some control over the connectivity of the template. We allow the user to sketch a *solution path* the topology of which will ultimately be reflected in the interconnections between vortices (see Figure 6).

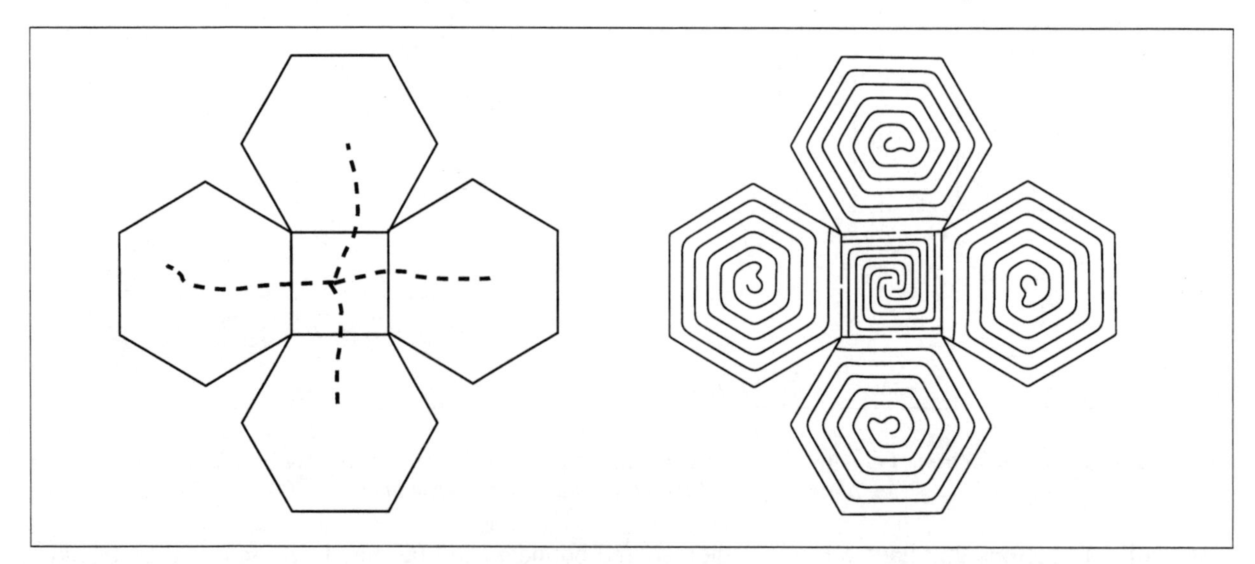

Figure 6: *The user-supplied solution paths and the corresponding maze. There is a winding path from the centre of the square to the centre of each hexagon.*

4. Visual Effects

We have considered a number of other modifications to the algorithm that can make the resulting mazes more attractive. We mention two of them here. We can control the clockwise or counter-clockwise rotation of the vortex, and by choosing orientations carefully we can add some life and contrast to a maze with multiple vortices. We can also control the distance between adjacent offset curves. A smaller distance creates a denser, and hence darker vortex. By choosing offset distances appropriately, we can produce some variation in shading. Figure 7 gives an example of varying both vortex orientation and density.

Figure 7: *An example of varying the aesthetic parameters of the vortices in a maze. We alternate the orientations and densities of adjacent vortices.*

5. Results

Our algorithm is implemented as a C++ program that produces Postscript output. We use the CGAL library [3] to store and manipulate the geometry of the maze. We have used the program to produce the finished results in this paper. Some examples are given below. In some cases, they are marked with suggested start and end positions.

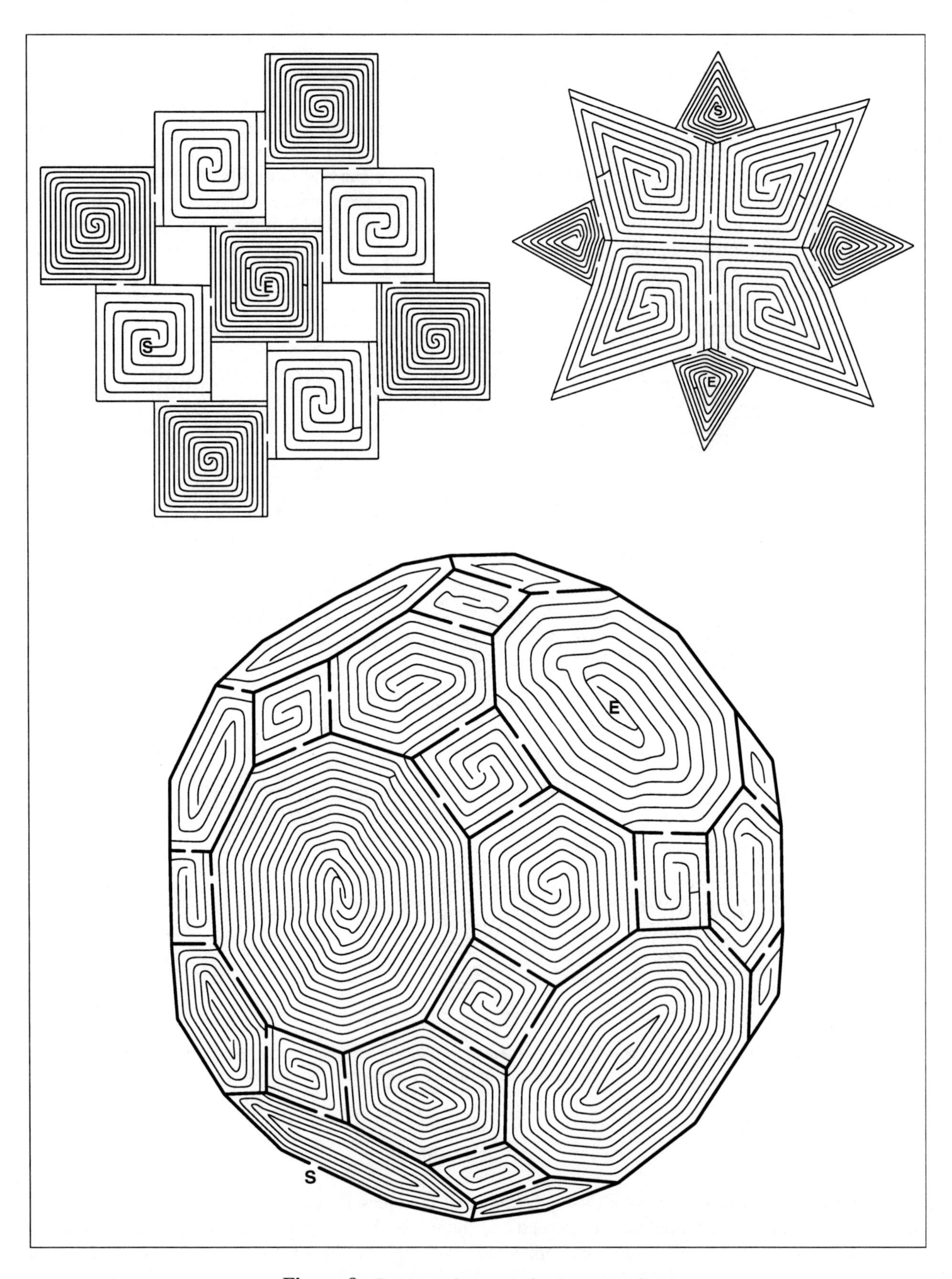

Figure 8: *Some results created using our system.*

Figure 9: *More results created using our system.*

6. Future Work

In this paper, we describe a system that can generate attractive spiral and vortex mazes quickly. Our system greatly decreases the burden of maze designers by providing high-level design control. And our system can be easily incorporated into other maze generation algorithms.

There are still some artifacts in our mazes. Some turns are too steep, and some passages are too narrow. We believe that some extra work may be needed to adapt our vortex construction algorithm to non-convex regions. We hope to add machinery to our algorithm to overcome these defects.

We are also interested in defining the complexity of mazes more precisely. Spirals and vortices might provide some clues on the way to that definition, because we believe they should be considered very difficult. In general, the problem is a very subtle and complex one, as it relies both on formal mathematical measures of the maze's geometry and on ineffable properties of human perception.

It might also be interesting to apply spiral and vortex construction to other areas. For example, bronzework of the Shang dynasty in China and many Chinese jades exhibit characteristic spiral patterns [9]. We believe it would be possible to extend our construction technique to create decorative patterns in the style of these artifacts.

References

[1] Christopher Berg, *Amazeing Art,* http://www.amazeingart.com, 2005.

[2] Christopher Berg, *Amazeing Art: Wonders of the Ancient World,* Harper Collins, 2001.

[3] CGAL: *Computational Geometry Algorithms Library*, http://www.cgal.org, 2005.

[4] Adrian Fisher, *Adrian Fisher's Maize Maze Website,* http://www.maizemaze.com, 2005.

[5] Hermann Kern, *Through the Labyrinth: designs and meanings over 5000 years,* Prestel, 2000.

[6] Michael Scott McClendon, *The Complexity and Difficulty of a Maze,* Bridges: Mathematical Connections in Art, Music, and Science, July 27-29, 2001.

[7] Walter D. Pullen, *Think Labyrinth,* http://www.astrolog.org/labyrnth/maze.htm, 2005.

[8] Olin Shivers, *Maze Generation,* http://www.cc.gatech.edu/~shivers/mazes.html, 2005.

[9] Jessica Rawson, John Williams, David Gowers, *Chinese Jade from the Neolithic to the Qing,* Published for the Trustees of the British Museum by British Museum Press, 1995.

Models of cubic surfaces in polyester

Sergio Hernández
Hcsoft Programación S.L.
C/ Párroco José Mª Belando 8, 2ºA
30007 Murcia, Spain
E-mail: sergio@hcsoft.net

Carmen Perea
Etadística, Matemáticas e Informática
Universidad Miguel Hernández
EPSO, 03312 Orihuela, Spain
E-mail: perea@umh.es

Irene Polo Blanco
Institute of Mathematics and
Computer Science,
University of Groningen, P.O. Box 800
9700 AV Groningen (The Netherlands)
E-mail: irene@iwinet.rug.nl

Cayetano Ramírez "Tano"
Sculptor
C/ Rafael Alberti, 4
30640 Abanilla-Murcia, Spain
E-mail: tano@hcsoft.net

Abstract

Historically, there are many examples of model building of mathematical surfaces. In particular, models of a very special cubic surface called the Clebsch diagonal have been built in plaster and clay since the 19th century. The sculptor Cayetano Ramírez has succeeded in building this surface using polyester. With this material, the resulting sculpture shows all the mathematical properties of the surface. We first give a short mathematical introduction and an overview of the models that have been built in the past to respresent it. Next, we proceed to describe the work of Cayetano, explaining the techniques used by him in the whole procedure.

1. Introduction. Real Cubic Surfaces

In the second half of the nineteenth century, the interest of mathematicians in algebraic geometry grew enormously. This interest began after the fascinating discovery of Salmon and Cayley in 1849:

Any smooth cubic surface contains precisely 27 lines.

By a smooth cubic surface it is meant the set of roots of a polynomial of degree 3 in projective space and containing no singularities. The cubic surface given by the equations

$$x_0^3 + x_1^3 + x_2^3 + x_3^3 + x_4^3 = 0$$
$$x_0 + x_1 + x_2 + x_3 + x_4 = 0$$

is known as the *Clebsch diagonal surface*. It was defined by Clebsch in 1871 in [2], and it is one of the most famous cubic surfaces. The reason is its special property that all the complex 27 lines are real. It is also the only cubic surface with 10 Eckard points, i.e., points on the surface where three of the 27 lines intersect.

In 1863 (see [12]), Schläfli had classified the projective real cubic surfaces with respect to their number of real lines. Moreover, Cayley [1] and Segre [13] worked on cubic surfaces with singularities.

More recently [8], Knörrer and Miller classified the real cubic surfaces in 45 types, according to their topology.

In the last years, thanks to the development of visualization software that allows us to produce high quality raytraced images, it has been possible to easily visualize surfaces. See for example the webpage [14] that contains some movies and images. Surfex uses S. Endrab's Surf (see [5]) to produce high quality raytraced images of the surfaces. In the article [10] we can find how to obtain a nice affine equation to visualize the 45 topological types of real cubic surfaces of [8].

2. Artistic Models of Real Cubic Surfaces Through History

Due to the great development of algebraic geometry at the end of the 19th century, mathematical models were built in order to illustrate geometrical properties of some surfaces. This interest in model building was specially popular in Germany, Felix Klein being one of its main developers. According to him [7], in Whitsun 1868, Christian Wiener for the first time built a model of a surface of degree 3 with 27 real lines. However, this model was still very asymmetric. In the summer of 1872, and under the guidance of Alfred Clebsch, the mathematician and sculptor Adolf Weiler built a model of the Clebsch diagonal surface (see [3]). In the Proceedings of the Royal Society of Sciences in Göttingen [7], Clebsch and Klein reported that Clebsch submitted two models built by Weiler: one model of the diagonal surface, and one model representing its 27 lines.

In 1879, Carl Rodenberg (see [11]), a student of Felix Klein, published his thesis that characterized the cubic surfaces with singularities. Together with it, he presented a series of 27 plaster models of cubic surfaces, among which the diagonal surface was present.

Mathematicians such as Felix Klein or Eduard Kummer starting building models of some surfaces. These models were mostly made of plaster and string. Many of the models built were reproduced and sold by the German companies L. Brill (and later M. Schilling) from 1888 to universities and museums around the world. By the turn of the century there were a large number of models of surfaces available

In 1986, Gerd Fischer [4], published a two volume book with a compilation of these German models, together with mathematical explanations and photographs. In particular, The University of Groningen contains a numerous collection of German models. The webpage [15] shows pictures and mathematical description of this collection.

We find the next evidence of modeling the Clebsch diagonal surface in 1999, when the sculptors Claudia Carola Weber and Ulrich Forster built the surface in clay (see [6]). This big sculpture (1,40 m wide and 2,50 m high) can be found at the Cafeteria of the University Heinrich Heine, in Düsseldorf. Lately, 3D printers have also been used in order to rebuild Rodenberg series, again in plaster.
During the spring of 2005, the sculptor Cayetano Ramírez built again the diagonal surface in plaster with the 27 lines on it. The model can be found in the Department of Mathematics at the University of Groningen (see figure 1).

Figure 1: *Plaster model of the Clebsch surface, by Cayetano Ramírez, May 2005*

3. New Artistic Models of Polyester of Real Cubic Surfaces

The building of mathematical models can have educational or research purposes: the visualization of the surface helps much in the understanding of its geometrical properties, as in the case for the German mathematical models of the end of the 19th century. Seeing is of basic importance also in art: mathematical surfaces have been used as ideas to generate art forms, like in the case of the clay model in Düsseldorf.

Computer representation often enable us to visualize the exact mathematical surface (see figure 2).

For example, the Clebsch diagonal surface has the following properties:
1. It contains 27 lines.
2. It contains 10 Eckard points (i.e., points where three of the lines meet)
3. It contains 7 passages (or "holes").

This surface can be represented in the computer showing these 3 properties. When it comes to plaster or clay models, the interior of the surface has to be filled in order to make the result solid (see figure 1). In this way, some of the properties can be lost, as it is the case for the seven passages of the Clebsch diagonal surface: only three of them are really visible in the models (compare figure 1 and figure 2).

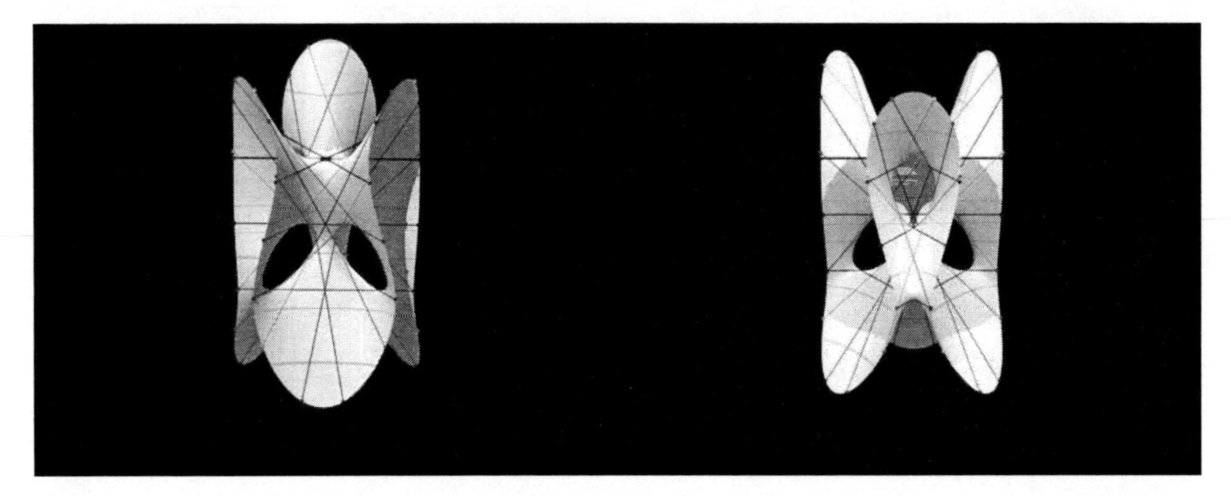

Figure 2: Clebsch diagonal cubic surface with the 27 lines. (Representation realized by the authors in POV-Ray 3.1)

Our present project consists of building the Clebsch diagonal surface in polyester. With this new construction, the model of polyester provides us with both the computer and the model advantages: it shows the precise mathematical surface with the 27 lines, the 10 Eckard points, and the seven passages. Moreover, we obtain a new artistic form made of a thin and transparent surface, that enable us to see the inside of the figure.

Before the final sculpture was built, several test figures where constructed as preliminary tests, obtaining imperfect Clebsch that showed the desired properties (see figure 3).

Figure 3: Sketch of the Clebsch diagonal cubic surface in polyester

For the final result, special care was take to assure an almost-perfect mathematical surface. The process starts by creating a perfect plaster model which would later be covered by a thin layer of liquid resin, and polyester fibre. However, in order to obtain a perfect surface, instead of constructing the whole model in plaster, it is more practical to build it first in polythene cork, and to cover it later with a thin layer of plaster.

For this purpose, we used computer images of different sections of the model (see figure 4), as well as, the outline of the Clebsch model intersected with a cylinder (see figure 5).

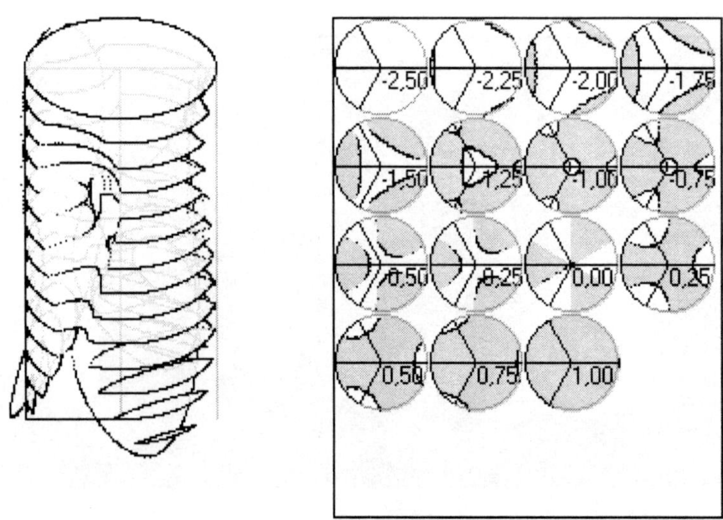

Figure 4: Transversal sections of the Clebsch diagonal cubic surface

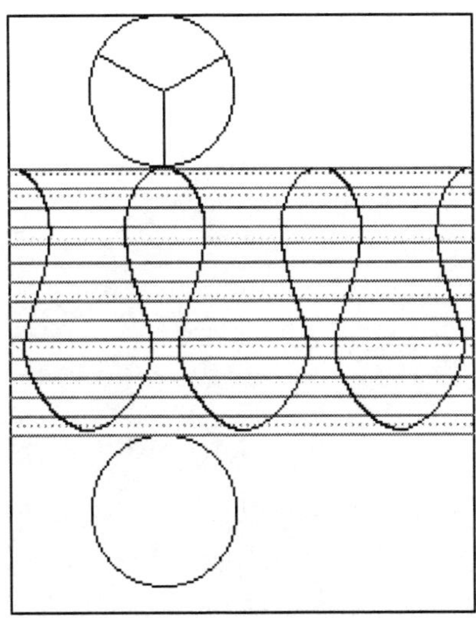

Figure 5: Outline of the Clebsch model intersected with a cylinder

Using the plans of figures 3,4, and 5, we construct a wood pattern to make sure that the model is correct (see figure 6 and 7)

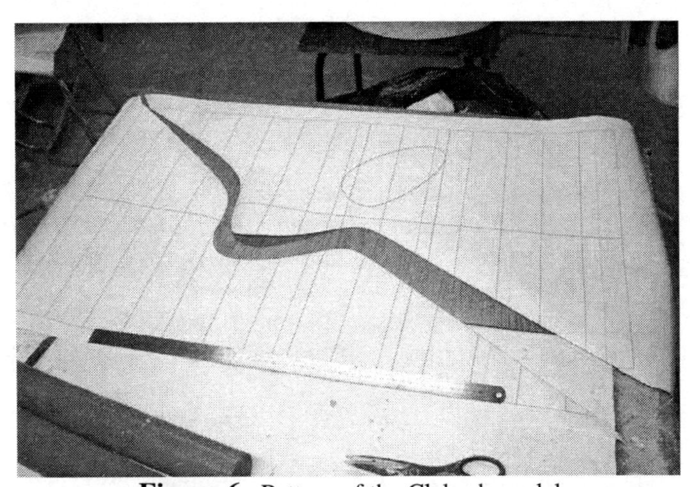

Figure 6: Pattern of the Clebsch model

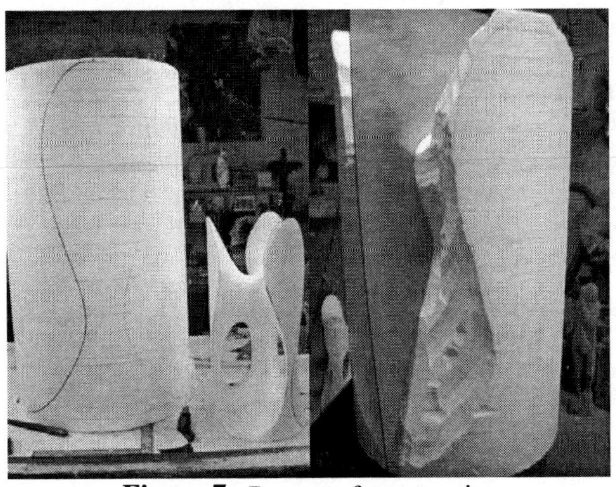

Figure 7: Process of construction

Using a similar procedure as the one described for the model of the Clebsch diagonal surface, we construct other cubic surfaces like the ones in figure 8.

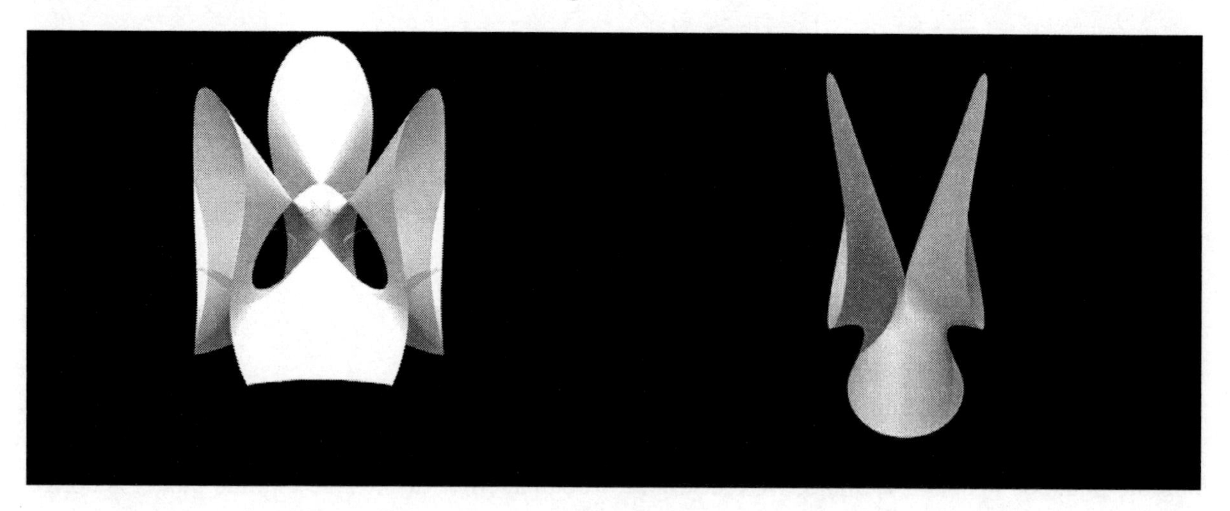

Figure 8: Real cubic surfaces represented by the authors with Pov-Ray 3.1

References

[1] A. Cayley, *A Memoir on Cubic Surfaces,* Phil. Trans. Royal Soc., CLIX, pp. 231-326, 1869.
[2] A. Clebsch, *Ueber die Anwendung der quadratischen Substitution auf die Gleichungen 5ten Grades und die geometrische Theorie des ebenen Fünfseits,* Math. Ann., Vol. 4, pp. 284-345. 1871.
[3] A. Clebsch and F. Klein, *Über Modelle von Flächen drotter Ordnung,* Königl. Gesellschaft der Wissenschaften un der G. A. Univers. zu Göttingen, 20. pp. 402-404, 1872.
[4] G. Fischer. *Mathematische Modelle / Mathematical Model.* Vieweg, 1986. Bildband, Kommentarband.
[5] S. Holze, O. Labs, and R. Morris. *SURFEX – Visualization of Real Algebraic Surfaces.* www.surfex.AlgebraicSurface.net, 2005.
[6] R. Kaenders, *Die Diagonalfläche aus Keramik.* DMV-Mitteilungen, 4/99, 1999.
[7] F. Klein, *Gezammelte Mathematische Abhandlungen,* Band II. Verlag von Julius Springer, Berlin, 1922.
[8] H. Knörrer and T. Miller, *Topologische Typen reeller kubischer Flächen.* Mathematische Zeitschrift, 195, 1987.
[9] O. Labs and D. van Straten. *The cubic Surface Homepage.* www.CubicSurface.net, 2000.
[10] O. Labs. *Illustrating the Classification of Real Cubic Surfaces,* Preprint. University of Mainz, 2005.
[11] C. Rodenberg. *Zur Classification der Flächen dritter Ordnung,* Math. Ann. XIV 1879.
[12] L. Schläfli, *On the Distribution of Surfaces of the Third Order into Species, in reference to the presence or absence of Singular points and the reality of their Lines.* Philos. Trans. royal Soc., CLIII, pp. 193-241, 1863.
[13] B. Segre, *The non-Singular Cubic Surfaces.* Clarendon, Oxford, 1942.
[14] www.CubicSurface.net
[15]www.math.rug.nl/models
[16] http://surf.sourceforge.net

"Geometry" in Early Geometrical Disciplines: Representations and Demonstrations

Elaheh Kheirandish
Department of the Classics
Harvard University
Boylston Hall 204
Cambridge, MA, 02138, USA
E-mail: kheirand@fas.harvard.edu

Abstract

This paper discusses various manifestations of geometry in early geometrical disciplines with reference to specific cases from the Islamic "Middle Ages", a period of intense scientific activity falling intermediately between the initial reception of Greek scientific material in the early Islamic period (8th-9th centuries AD), and their subsequent diffusion within both Islamic and to European lands (12-13th centuries AD). The paper begins with the classification of mathematical sciences in ancient Greek and early Arabic sources, and proceeds with the identification and distinction of aspects of geometry such as geometrical "representation" and "demonstration" through a case study of specific geometrical disciplines. The case study covers sample problems from four early geometrical disciplines: optics, mechanics, surveying and algebra: optics and mechanics are subdivisions of plane and solid geometry in Aristotelian classifications, surveying and algebra are the respective subdivisions of each in early Arabic Classifications. The samples include geometrical representations (definitions, figures, models) and geometrical demonstrations (illustrations, constructions, proof), as representatives of a range of Arabic and Persian scientific sources from the Islamic Middle Ages.

Introduction

This paper treats some historical aspects of geometry, a subject with an outstanding presence in both the sciences and the arts, and it is divided into three sections. The first section introduces early disciplinary classifications and divisions through the well-known "*Catalogue*" of Al-Fārābī (d. ca. 950 AD)[1]. The focus is on geometrical disciplines such as optics and mechanics whose long entries under "mathematical sciences", along with sub-entries such as surveying and algebra [2,3], reflect an extension of the classical "quadrivium" beyond the foursome of "arithmetic, geometry, astronomy and music". The second section highlights distinctions between geometrical "representation" and "demonstration" through the discipline of geometry itself: examples of "geometrical representations" are presented as definitions and figures involving line, circle, and square and rectangle with reference to the *Elements* of Euclid (ca. 300 BC), one of the earliest and most circulated works in both Arabic and Persian [4]; examples of "geometrical demonstrations" are presented with reference to the formal divisions of a Euclidean proposition, itemized in Euclid's first book as: enunciation, illustration, definition or specification, geometrical construction or set up, proof or demonstration and conclusion [5]. The distinction between "geometrical representation" and "geometrical demonstration" is further extended to "geometric" optics through specific geometrical "models" based on the *Optics* of Euclid: as geometrical representations such as angles and cones represent various ancient Greek and medieval Arabic visual models (e.g. Euclid and Ptolemy: 4^{th}c. BC - 2nd c. AD; al-Kindī and Ibn al-Haytham: 9th-11th centuries)[6], demonstrations themselves represent a shift from the geometry-based illustrations of Greek antiquity to the physical set ups of the Islamic Middle Ages. The paper's third and final section concludes with a presentation of four sample problems from optics, surveying, mechanics and algebra, and a discussion of their respective representations and demonstrations (Samples and Figures 1-4)[7-10].

1. Disciplinary Classifications

Synopsis: The Mathematical Sciences in the important and influential *Catalogue* of Al-Fārābī (d. ca. 950 AD) include, in addition to the four disciplines forming the basis of the classical "quadrivium" (arithmetic, geometry, astronomy, and music), three others: the sciences of optics, mechanics and weights. While optics and mechanics are classified as the respective subdivisions of plane and solid geometry since Aristotle, surveying and algebra appearing under the respective entries on optics and mechanics in the same *Catalogue*, may be considered "geometrical" from yet other standpoints.

2. Representation vs. Demonstration

Synopsis: An explicit distinction between geometrical representation and demonstration is not found in early historical sources; but the distinction is clear in texts as early as Euclid's *Elements of Geometry* and *Optics* (ca. 300 BC), and its multiple variations: here, representations appear in the form of definitions and figures involving a point, line, circle, square and other geometrical figures, and demonstrations and their components appear under the formal divisions of a Euclidean proposition: enunciation, illustration, definition or specification, geometrical construction or set up, proof or demonstration and conclusion.

3. Case Studies

Synopsis: The following four samples provide case studies of the different manifestations and functions of geometry in disciples such as optics, mechanics, surveying and algebra. These include geometrical representations (definitions, figures, models), and demonstrations (illustrations, constructions, proofs), the latter with components such as geometrically based calculations (proportionality of triangles), and superimpositions (application of areas).

Sample 1: Optics (Euclid's *Optics*, Arabic Prop. 10, tr. early 9thc.: Kheirandish, 1999, v.1: 30-32)
Right-angled Figures Viewed from a Far Distance

The first sample is an optical problem about the circular appearance of far rectangular objects, which going back to many classical sources, appears as the first of a few the problems included in Al- Fārābī's "optics" entry in his widely circulated *Catalogue*. The geometrical representations are visual rays and angles, and the geometrical demonstration is a Euclidean style proof based on other proofs and assumptions (Sample 1, Figure 1).

Sample 2: Surveying (Euclid's *Optics*, Arabic Prop. 20, tr. early 9thc.: Kheirandish, 1999, v. 1: 58-60)
Determination of Heights Through Reflecting Visual Rays

The second sample is a surveying problem about the determination of the height of an object by means of a plane mirror. Problems of surveying such as determinations of height, width and depth are also included in Al-Fārābī's entry, but here the use of a plane mirror makes this a problem of indirect, rather than direct vision. The geometrical representations are, again, visual rays and angles, and the geometrical demonstration, a Euclidean style proof, this time based on the proportionality of two similar triangles for the calculation of unknown quantities based on known ones (Sample 2, Figure 2).

Sample 3: Mechanics (Heron's *Mechanics*, Arabic tr.9thc.: Kheirandish, work in progress)

Lesser Weights Lifting Greater Weights

The third sample is a mechanical problem as early as the Pseudo-Aristotelian *Mechanical Problems* (847a24-847b: ca. 4th c. BC, Anonymous Arabic translation: 9th c.?), which is set in terms of the "wondrous" effect of small weights lifting great weights through mechanical devices (mechané). This is also a central problem in the *Mechanics* of Heron of Alexandria (Book II: 2nd c. AD, Arabic translation: Qusṭā ibn Lūqā, d. ca. 912-13), there expressed in terms of "powers" that include, in addition to the "wheel and axle", "pulley", "wedge" and "screw", the "lever", one whose calculation is described in terms of the ratio of the load to be moved and the force meant to move it. The geometrical representation of a mechanical device such as the lever is through geometrical shapes and relations of lines, angles, and triangles, and the geometrical demonstration of the function of the mechanical principle itself is based on the proportionality of similar triangles (Sample 3, Figure 3).

Sample 4: Algebra (Euclid's *Elements*, Book II. Prop. 4, tr. early 9th c.: Kheirandish, 1987: 18-19)
Algebraic Identity Corresponding to a Quadratic Equation

The fourth sample is a problem that appears as the fourth proposition of the second book of Euclid's *Elements*, and is treated as a more general problem as early as in a ninth-century Arabic text by Thābit ibn Qurra (d. ca. 901). While the identity in Euclid's Book I (I. 47) corresponds to the problem known as the "Pythagorean Theorem" where the square of the hypotenuse is equal to the sum of the square of the two sides, the problem in Book II (II.4) deals with an identity corresponding to a second degree algebraic equation where the square of the sum of two segments on the sides of a square is equal to the square of each, plus twice their products. The geometrical representations here are geometrical shapes such as lines, squares, and rectangles, and the geometrical demonstration is based on the application of the areas where the area of the larger square is equal to the area of the two smaller squares, plus the two side rectangles (Sample 4, Figure 4).

References

[1] Abū Naṣr Al-Fārābī, *Enumeration of the Sciences (Iḥṣā' al-'ulūm)*, ed. 'Uthmān Amin [= Osman Amine], Cairo: Librairie Anglo-Égyptienne, pp. 93-110. 1968 (also, 1931, 1949).
[2] Al-Fārābī, "Science of Optics": tr. A. I. Sabra's *The Optics of Ibn al-Haytham*: Books I-III On Direct Vision, Translated with Introduction and Commentary, 2 vols. (London: The Warburg Institute, University of London), edited by J. B. Trapp, 2 vols., Vol. 2: lvi-lvii. 1989.
[3] Al-Fārābī, "Sciences of Mechanics" tr. George Saliba, "The Function of Mechanical Devices in Medieval Islamic Society", *Science and Technology in Medieval Society*, ed. Pamela O. Lang, *Annals of the New York Academy*, pp. 144-145. 1985; "Sciences of Weights" tr. Marshall Clagett, "Some General Aspects of Physics in the Middle Ages," *Isis*, 39, pp. 29-44: 32. 1948.
[4] John E. Murdoch, "Euclid," *Dictionary of Scientific Biography (DSB)*, 4, esp. pp. 438–443. 1971.
[5] Thomas Heath, *The Thirteen Books of Euclid's Elements,* 3 volumes, Vol. I, pp. 129-131. 1926.
[6] Elaheh Kheirandish, "The Arabic 'Version' of Euclidean Optics: Transformations as Linquistic Problems in Transmission," *Tradition, Transmission, Transformation*, eds. F. Jamil Ragep and Sally P. Ragep with Steven Livesey, (Leiden: Brill), pp. 227–247: 247. 1996.
[7] Elaheh Kheirandish, *The Arabic Version of Euclid's Optics*, 2 volumes, Springer-Verlag: Sources in the History of Mathematics and Physical Sciences, no. 16, Vol. 1, pp. 30-32. 1999.
[8] Elaheh Kheirandish, *The Arabic Version of Euclid's Optics*, 2 volumes, Vol. 1, pp. 58-60. 1999.
[9] Elaheh Kheiranish, "The Arabic and Persian Traditions of the *Mechanics* of Heron and Pappus of Alexandria " (Work in Progress).
[10] Elaheh Kheirandish: *Problems Corresponding to Quadratic Equations in Early Mathematical Texts: A Chapter in the History of Algebra* (unpublished Master's thesis), 1987, pp. 18-19.

Sample 1: Optics (Euclid's *Optics*, Arabic Proposition10) **Right-angled Figures Viewed from a Far Distance** From: Elaheh Kheirandish, *The Arabic Version of Euclid's Optics,* 2 vols., Springer-Verlag, 1999, vol. 1, pp. 30-32: The enunciation is that right-angled figures when seen from a distance are seen as circular. The illustration is that of a right-angled figure (ABGD)(Figure 1), and the eye, that when at a distance, its visual rays do not pause at a single point but shift, such that what is seen is figure EZHTKLMN but not what is in between, namely corners. The conclusion is that because what is hidden from sight, namely the object's angles, is the part of the figure with least magnitude, and that any figure with no angles is circular, then all figures are seen from a far distance as circular, and so is the right-angled figure, ABCD.	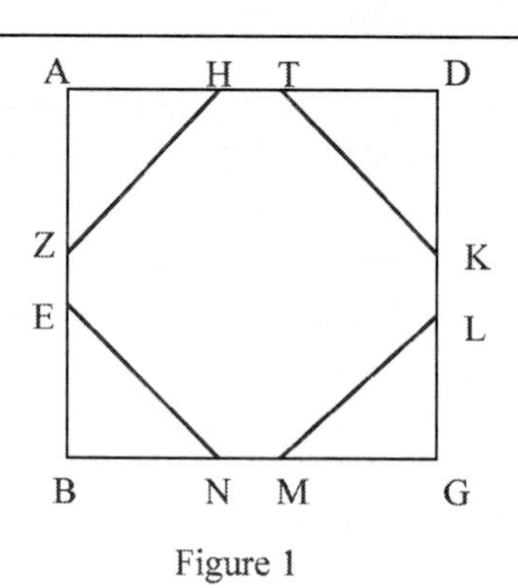 Figure 1
Sample 2: Surveying (Euclid's *Optics*, Proposition 20) **Determination of Heights Through Reflecting Visual-rays** From: Elaheh Kheirandish, *The Arabic Version of Euclid's Optics*, Springer-Verlag, 1999, vol. 1, pp. 58-60. The problem is to determine the height of an object by means other than the sun. The illustration is that of a point and two lines, representing the eye (G), an upright object (AB), and a horizontal plane mirror (ED)(Figure 2). The geometrical construction is that of line EB, of a visual ray GH, deflecting at point H to fall on the farthest point of AB (A) such that a line from H to T meets a perpendicular from the eye (G), and the deflection of ray GH at H at equal angles results in two similar triangles, AHB and GHT. The conclusion is that because the ratio of HB to AB is known, from the measurable values of GT and TH, the magnitude of AB can be determined.	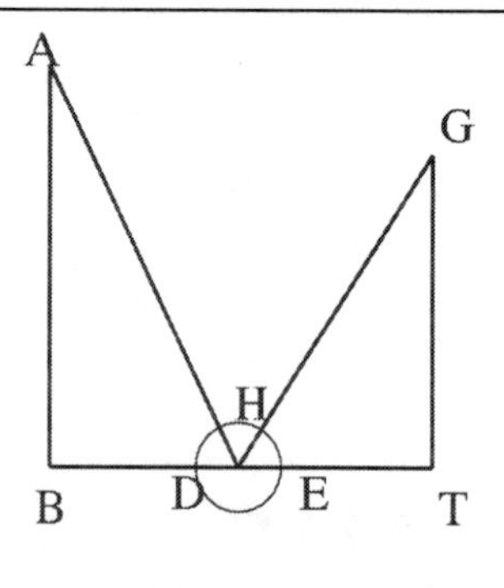 Figure 2
Sample 3: Mechanics (Aristotelian Mechanical Problems) **Lesser Weights Lifting Greater Weights (Heron Book II)** From: Elaheh Kheirandish, "The Arabic and Persian Traditions of the Mechanics of Heron" (Work in Progress). The five powers or simple machines, described by Heron, are the wheel and axle, pulley, lever, wedge and screw. In the case of the lever a heavy load (A) is moved by a little force (C), acting on it through the fulcrum (B) (Figure 3). According to Heron, the size of such a machine is set up according to the size of the load to be moved with it, and its calculation take place according to the ratio of the load one wants to move to the force that is meant to move it.	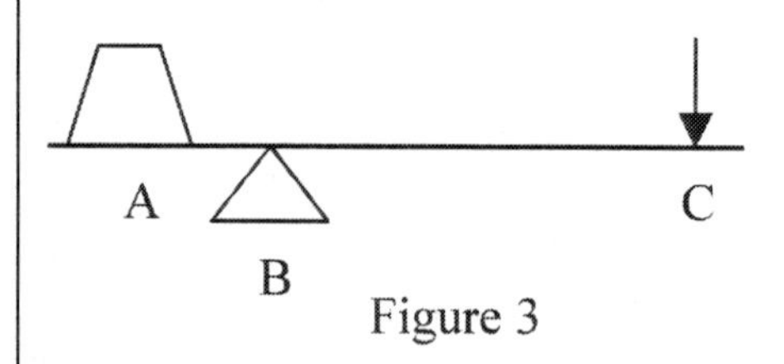 Figure 3
Sample 4: Geometric Algebra (Euclid's *Elements*, II. 4) **Algebraic Identity Corresponding to Quadratic Equation** From: Elaheh Kheirandish: *Problems Corresponding to Quadratic Equations in Early Mathematical Texts* (unpublished), 1987, pp. 18-19. The problem in Book II (II.4) deals with an identity that corresponds to a second degree algebraic equation, where the square of the sum of two segments on the sides of a square is equal to the square of each, plus twice their products (Figure 4). Similar to the problem, where the square drawn on the hypotenuse of a right-angled triangle is equal to the sum of those drawn on the two sides, the demonstration in this proposition is through the equality of areas involving an outer square and its inner areas such that the outer square area $(a+b)^2$ is equal to the sum of the inner areas a^2+b^2+2ab (C cuts AB at random: AC=a; CB=b).	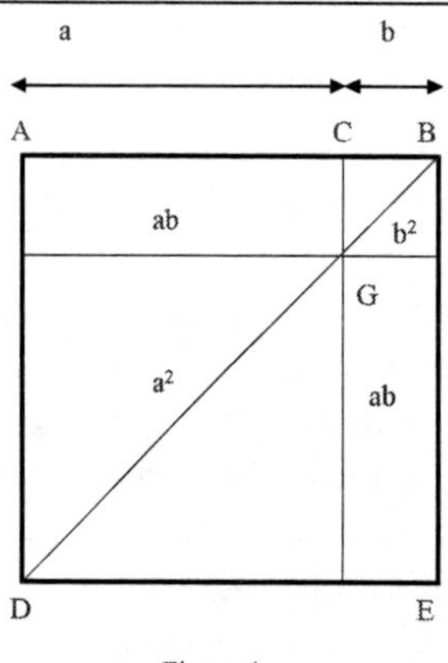 Figure 4

Ant Paintings using a Multiple Pheromone Model

Gary R. Greenfield
Department of Mathematics & Computer Science
University of Richmond
Richmond, VA 23173, U.S.A.
ggreenfi@richmond.edu

Abstract

Ant paintings are visualizations of the paths made by a simulated group of ants on a toroidal grid. Ant movements and interactions are determined by a simple but formal mathematical model that often includes some stochastic features. Previous ant paintings used the color trails deposited by the ants to represent the pheromone, but more recently color trails and pheromones have been considered separately so that pheromone evaporation can be modelled. Here, furthering an idea of Urbano, we consider simulated groups of ants whose movements and behaviors are influenced by both an external environmentally generated pheromone and an internal ant generated pheromone. Our computational art works are of interest because they use a formal model of a biological system with simple rules to generate abstract images with a high level of visual complexity. We strive to show how designing ways to make ant paintings becomes an artistic pursuit.

1. Introduction

Ant paintings trace their origins to the ant colony simulation experiments of Ramos [8, 9, 10]. He investigated the use of ant colony simulations for image processing purposes. A group led by Monmarché [1] appears to be the first to actually use the term "ant painting" to describe the abstract images made by virtual ants that roam over a toroidal canvas. In their model, ants deposit scent by laying down one color while searching for the scent (i.e. color) deposited by other ants. Monmarché et al. used a small number of virtual ants, typically 4-6. Ant behavior was controlled by a *genome* that determined what color ants should deposit, what color ants should seek, and their exploration tendencies. Ant paintings were evolved using an *interactive* genetic algorithm. An unusual feature of their simulation was that the sensory mechanism of the virtual ants was modelled in such a way that ants were responsive only to the luminance values of the colors representing scent instead of their tristimulus values. In [3] we evolved ant paintings using a model where ants were responsive to tristimulus color values. We also introduced a *non-interactive* genetic algorithm by designing fitness functions to evolve ant behavior based on arithmetic expressions that required us to measure the exploration and exploitation capabilities of the ants. Three examples of the ant paintings made by ants that we evolved using this model are shown in Figure 1.

Figure 1: *Three ant paintings made using a model where color deposited by the ants is interpreted as scent.*

The "scent" in ant paintings evolved using such methods had limited diffusion properties and no evaporation properties. Urbano [14] addressed the latter shortcoming by considering a model where each individual cell in the *environment* exuded scent — the attractant — until it was visited by an ant. By diffusing and evaporating this exuded scent; by using two competing species of ants; and by marking each cell according to which species of ant reached it first, Urbano's technique yielded ant paintings that were "finished" once there were no more unvisited cells left to exude scent. The visual characteristics of these ant paintings are influenced both by the number of ants and their initial placement. Figure 2 shows three ant paintings using this model that we made by setting the parameter values for the model so that each non-visited cell produced five units of scent every time step, seven percent of the scent in each cell was evaporated every time step, and two percent of the scent in each cell was diffused over its Moore neighborhood of radius one — the eight neighboring cells that surround the 3×3 square with the cell at its center — after every time step.

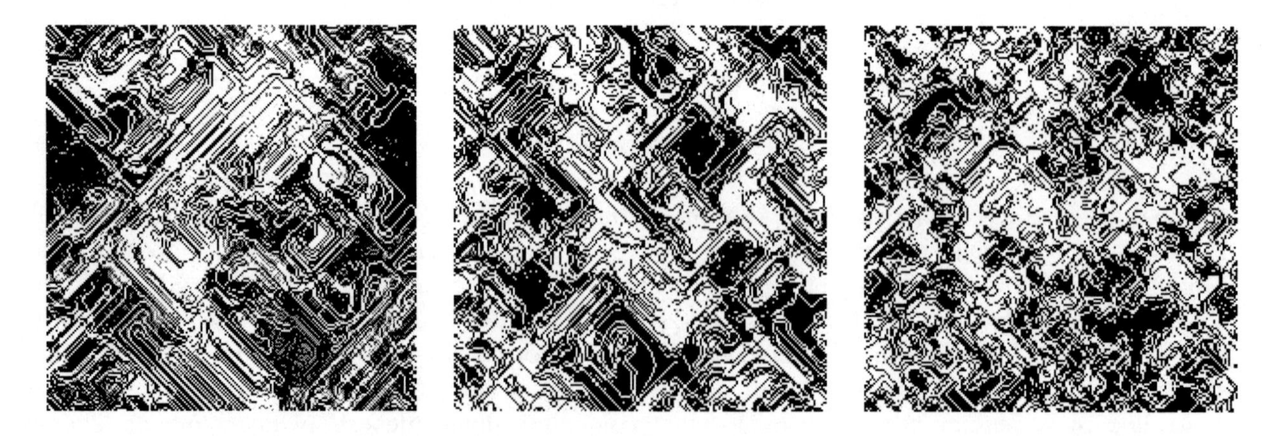

Figure 2: *Ant paintings in the style of Urbano using 50, 100, and 500 randomly placed ants, respectively, to mark each cell of a 200×200 toroidal grid on the basis of which one of two species of ants reaches it first. Cells in the grid attract ants by exuding scent.*

Of related interest are the efforts of Moura et al. [5, 6] to build autonomous robots that execute "swarm paintings", because both Urbano and Moura appeal to the concept of *stigmergy* to help explain why their virtual and physical software controlled entities are able to exhibit creative or artistic tendencies. Stigmergy [13] refers to the situation where the behavior of agents in swarms is controlled wholly by external, environmental factors. Recent work of Semet et al. [12] used virtual ants that were responsive to environmental cues that were provided by *users* in addition to interactions with other ants in order to develop image processing techniques for producing non-photorealistic visual effects. This brief, but by no means complete, exposition of previous work brings us full circle, since as early as 1993 Tolson began experimenting with swarms of agents that were externally controlled by having a user introduce visual cues into the environment in order to produce visual special effects for video stills and animations [11].

In this paper, we will consider a swarm of virtual ants that is responsive to two scents — one produced by the environment and one produced by the ants themselves. Further, by allowing ants to make two types of marks, we will describe how we are able to make ant paintings where ants simultaneously couple the method of Urbano to create a background image together with more traditional ant colony simulation evaporation-diffusion foraging methods [2] to conduct further processing of this image. Our goal is to explore and develop new forms of ant painting. This paper is organized as follows. In Section 2 we introduce our multiple pheromone model, in Section 3 we explain how we make an ant painting, in Section 4 we discuss the methods we use to control the behavior of our ants, in Section 5 we present examples of the ant paintings that were produced using our model, and in Section 6 we offer our conclusions. Before proceeding with the technical details we wish to point out that it is still an open problem to decide how to evaluate the creativity of a swarm of agents [4], and it is still a highly contentious issue to decide precisely what it should mean for a computational work of art to possess aesthetic merit [7, 15].

2. The Multiple Pheromone Model

We treat pheromones as units of virtual substances that are found within the cells of an $N \times N$ toroidal grid. These pheromones can be detected and measured by the M virtual ants that roam on the grid. In this paper the linear dimension of the grid will always be $N = 200$ and the number of ants will always be $M = 100$ with the understanding that ants may be further distinguished as belonging to different castes or species. Ants maintain a compass heading that is one of the following eight compass directions: N, NE, E, SE, S, SW, W, or NW. Before advancing from one cell to the next ants sense the amount of each type of pheromone in the three cells that are at the three compass headings that are currently directly in front of them. Based on this data they must choose to advance to one of these three cells. To simulate "noise" within the system, with probability $1/200 = 0.005$ at every step one of the three cells to advance to is randomly selected. The reason we limit ant's turning ability to forty-five degrees per time step is to try and discourage ants from going in circles, a common problem in many ant colony simulations. After every time step, for each cell, E percent of each pheromone is evaporated and D percent of each pheromone is diffused (in equal parts) to the eight neighboring cells comprising the cell's Moore neighborhood of radius one. For all the results shown here we set $E = 7$ and $D = 3$.

In our model, individual cells exude a pheromone — the attractant — until they are "harvested" by being visited by an ant, while ants deposit pheromone in every cell they visit — the repellant — throughout the entire course of the simulation run. The ant repellent pheromone is the same for both species. In this regard, our ants are not biologically plausible. We let P_c denote the number of units of cell pheromone each cell generates per time step and P_a denote the number of units of ant pheromone each ant generates per time step. In our model cells can be simultaneously occupied by more than one ant.

3. Making an Ant Painting

For visualization purposes (i.e. to develop the aesthetic of an ant painting) we divide the ants into two groups of roughly equal size. We think of these two groups as representing different castes or species. For convenience we denote these two groups as A and B. All cells of the grid are initially gray. If an ant of type A (respectively B) is *first* to visit a gray cell it is re-colored white (respectively black). However, if a cell has already been re-colored *and* the amount of cell pheromone has dropped below the cell threshold T_c, then the cell is re-colored (i.e. post-processed) to a shade of blue when the visiting ant is of type A, and to a shade of red when the visiting ant is of type B. Thus in our model the ants themselves first create a black and white visual substrate, called the "underpainting," which is then used for further image processing as the ants create what we call the "overpainting." An ant painting is pronounced finished when either all cells of the grid have been visited or 2000 time steps have occurred.

Initial positioning of the ants influences the look of the substrate the ants will subsequently overpaint. Figure 2 shows a substrate where the ants are randomly placed in the environment. Figure 3 shows the stylistic differences that occur when ants are segregated by type and initially clustered around two distinct points. Our clustering results do not duplicate those of Urbano because there is an ant avoidance mechanism in effect (see next section). The reason that diagonal movement is favored by the ants is due to the fact that for ease of computation we are using grid coordinates as ant positions. This means ants travelling diagonally can establish better separation between themselves and other ants and become more successful at finding unvisited cells or picking up cell pheromone gradients.

Figure 3: *Ant paintings in the style of Urbano using 100 ants on a* 200×200 *grid where the two species of ants are segregated and initially clustered around separate points. The left and right images use cluster points located one-fourth and three-fourths away from the left edge of the horizontal bisector. The image in the middle uses the centers of the first and third quadrants for cluster points.*

Another stylistic difference in substrate we experimented with was to allow ants of type B to advance three cells during each time step. Figure 4 shows substrate backgrounds of this type. There are many other possibilities one could consider. However, because we were more interested in ant interactions and the effect of using two scents for substrate post-processing, we did not continue with this line of inquiry.

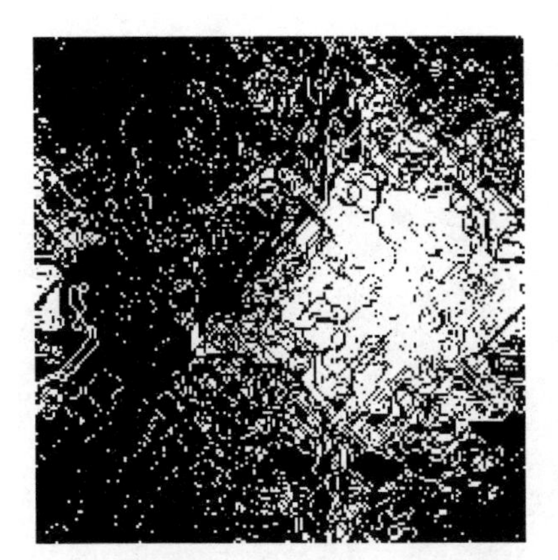 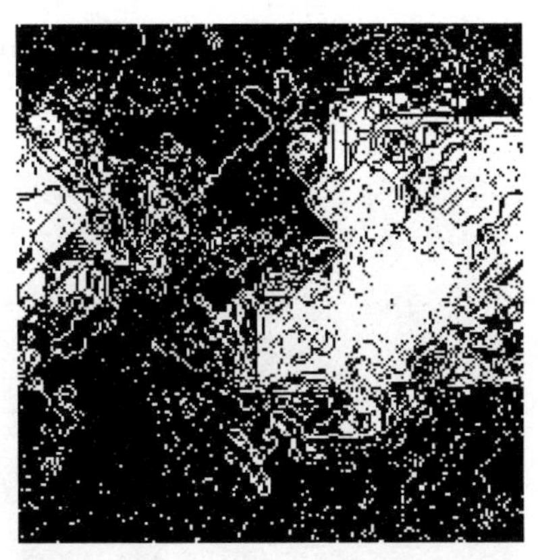 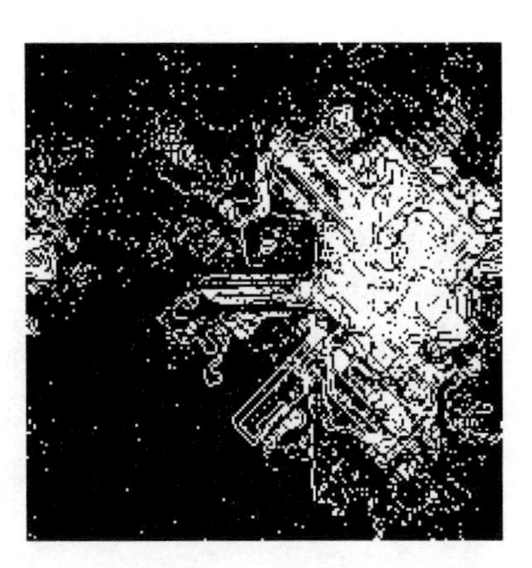

Figure 4: *Ant paintings for use as substrates made by 100 ants on a* 200×200 *grid where the two species of ants are segregated and initially clustered around separate points but ants of one species move three times as fast as the other.*

4. Virtual Ant Behavior

The purpose of generating, diffusing, and evaporating scent is to create scent *gradients* for ants to follow. Thus in order to define ant behavior we must define how they will react in the presence of such gradient fields. In a failed experiment, we first tried allowing the ants to advance to the cell within their three-cell field of vision that had the largest *combined* scent. This had no discernable effect on substrate creation, but when the threshold for over-painting came in to play, the result was that the ants conglomerated and became aligned along a few closed contours thereby creating faint difficult to detect trails in a sea of black, white, and gray. Ant paintings failed to reach completion because the small numbers of cells still generating attractant were not sufficient to draw ants away from their blind devotion to self-reinforcing trail following. The flaw in this model is clear. If cell-produced pheromone is supposed to act as an attractant, then ant-produced pheromone should act as a repellent so that the search for cells that have not yet been visited can be encouraged. Ant pheromone is used to deflect ants away from paths of already visited cells. This prevents either stagnation setting in due to ant pheromone trail following or image corruption setting in due to excessive post processing of non-gray cells prior to all cells having been visited. If we let S_c be the *maximal cell* pheromone value in the ant's current three cell field of vision, and s_a be the *minimal ant* pheromone value within the ant's current three cell field of vision, then the simple behavioral rule we define for our virtual ants is to advance in the direction where S_c was detected whenever $S_c > T_c$, but to advance in the direction where s_a was detected otherwise. Using this rule, up to the initial locations and headings of the ant population, ant paintings will be uniquely determined by the amount P_a of pheromone that ants are able to release into the environment during each time step, the amount P_c of pheromone that never visited cells are able to release into the environment during each time step, and the cell scent threshold value T_c that controls when overpainting of already visited cells occurs. Because of the non-linear interactions between these three parameters in our model, it can be difficult to anticipate what the resulting ant paintings will look like.

5. Some Examples of Multiple Pheromone Ant Paintings

To reveal the overpainting process that the ants are now capable of performing on the black and white substrate they have created, in Figure 5 we show examples of ant paintings where the threshold parameter T_c was set to a trace value and ant overpainting was halted by "timing out." As is to be expected, even though color is not diffused, such paintings reveal ant trails similar to the ones found in the ant paintings shown in Figure 1.

Figure 5: *Ant paintings demonstrating the image post processing to red and blue that occurs after the substrate image has been formed by coloring gray cells either black or white.*

We did not systematically explore the parameter space available to us. However, we did fix the cell threshold T_c at 40, the ant pheromone production parameter P_a at 1, and vary the cell pheromone production parameter P_c from 5 to 45 in non-uniform increments in order to determine how to best control the overpainting of the substrate in such a way as to achieve ant paintings that we felt held the best visual interest. Figure 6 shows some of the test results that led us to settle upon our preferred value of 45 for P_c. The ant paintings in Figure 6 were obtained by initially separating and clustering the two species of ants around points on the horizontal bisector.

Figure 6: *Ant paintings where the cell pheromone generation parameter P_c is varied, left to right, from 20 to 30 to 45 units to show how we arrived at a value that gave the desired balance between revealing the details of both the underpainting and overpainting.*

Finally, in Figure 7 we show a trio of ant paintings that were made using the parameter values above that we finally settled upon, but invoked different initial configurations and/or cell advancement rates for the two species of ants.

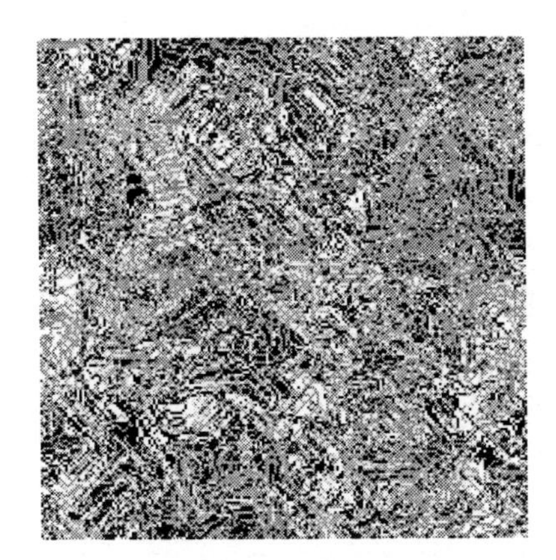
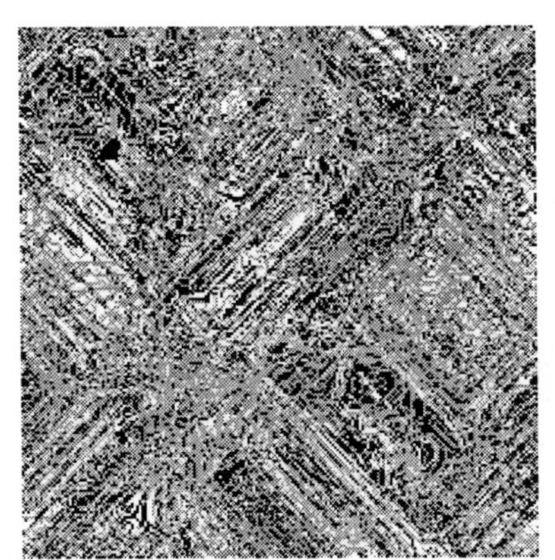
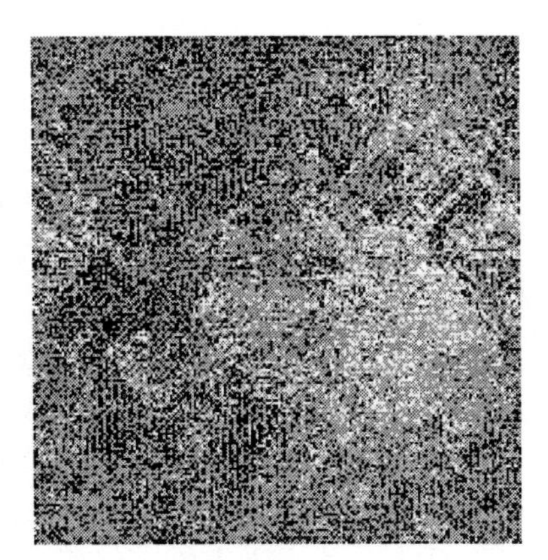

Figure 7: *A trio of ant paintings with the parameters P_c, P_a, and T_c fixed. The one on the left used random initial placement of the ants; the one in the center clustered* all *the ants initially about a point in the lower left of the painting; and the one on the right separated and clustered the ants on the horizontal bisector, but also advanced one species of ant three cells per time step while only overpainting the last of the three cells visited.*

6. Conclusions

We further developed a model of Urbano for generating ant paintings where cells exude ant attractant by incorporating mechanisms for additional ant communication though the use of pheromones that are deposited by the ants themselves and by allowing ants to make two types of marks. The result is that the ants first create a substrate image, or underpainting, and then post process that image by overpainting. We have described how we chose the parameters for our model and we have presented examples of ant paintings that were made according to a variety of initial conditions. We used only one simple rule to govern the behavior of our ants. It is to be expected that either more complex ant behavior rules, or additional ant behaviors triggered by the presence of additional pheromone substances would yield imagery of even greater artistic interest or potential. **Because grayscale reproduction of our ant paintings introduces contrast artifacts, the reader may wish to consult http://www.mathcs.richmond.edu/∼ ggreenfi/ to see the images in full color.**

References

[1] Aupetit, S., Bordeau, V., Monmarché, N., Slimane, M., Venturini, G.: Interactive evolution of ant paintings, *2003 Congress on Evolutionary Computation Proceedings* (eds. B. McKay et al.), IEEE Press, **2**, 2003, 1376–1383.

[2] Dorigo, M., Stutzle, T., *Ant Colony Optimization*, MIT Press, Cambridge MA, 2004.

[3] Greenfield, G., Evolutionary methods for ant colony paintings, *Applications of Evolutionary Computing, EvoWorkshops 2005 Proceedings*, Springer-Verlag Lecture Notes in Computer Science, LNCS 3449 (eds. F. Rotlauf et al.), 2005, 478–487.

[4] McCormack, J., Open problems in evolutionary music and art, *Applications of Evolutionary Computing, EvoWorkshops 2005 Proceedings*, Springer-Verlag Lecture Notes in Computer Science, LNCS 3449 (eds. F. Rotlauf et al.), 2005, 428–436.

[5] Moura, L., Pereira, H., *Man + Robots: Symbiotic Art*, Institut d'Art Contemporain, Lyon/Villeurbanne, France, 2004.

[6] Moura, L., Ramos, V., Swarm paintings — nonhuman art, *Architopia: Book, Art, Architecture, and Science* (eds. J. Maubant et al.), Institut d'Art Contemporain, Lyon/Villeurbanne, France, 2002, 5–24.

[7] Ramachandran, V., Hirstein, W., The science of art: a neurological theory of aesthetic experience, *Journal of Consciousness Studies*, **6**, 1999, 15–52.

[8] Ramos, V., Almeida, F., Artificial ant colonies in digital image habitats - a mass behaviour effect study on pattern recognition, *Proceedings of ANTS'2000 - 2nd International Workshop on Ant Algorithms: From Ant Colonies to Artificial Ants* (eds. M. Dorigo et al.), Brussels, Belgium, 2000, 113–116.

[9] Ramos, V., Merelo, J., Self-organized stigmergic document maps: environment as a mechanism for context learning, *AEB2002, First Spanish Conference on Evolutionary and Bio-Inspired Algorithms* (eds. E. Alba et al.), Mérida, Spain, 2002, 284–293.

[10] Ramos, V., Self-organizing the abstract: canvas as a swarm habitat for collective memory, perception and cooperative distributed creativity, *First Art & Science Symposium, Models to Know Reality* (eds. J. Rekalde et al.), Bilbao, Spain, 2003, 59–59.

[11] Robertson, B., Computer artist Michael Tolson, *Computer Artist* **2**, 1993, 20–23.

[12] Semet, Y., O'Reilly, U., Durand, F., An interactive artificial ant approach to non-photorealistic rendering, *Genetic and Evolutionary COmputation - GECCO 2004* (eds. K. Deb et al.), Springer-Verlag Lecture Notes in Computer Science, **3102**, 2004, 188–200.

[13] Theraulaz, G., Bonabeau, E., A brief history of stigmergy, *Artifical Life* **5:2**, 1999, 97–116.

[14] Urbano, P., Playing in the pheromone playground: experiences in swarm painting, *Applications of Evolutionary Computing, EvoWorkshops 2005 Proceedings*, Springer-Verlag Lecture Notes in Computer Science, LNCS 3449 (eds. F. Rotlauf et al.), 2005, 478–487.

[15] Zeki, S., *Inner Vision, An Exploration of Art and the Brain*, Oxford University Press, New York NY, 1999.

Verbogeometry
The Confluence Of Words And Analytic Geometry

Kaz Maslanka
Mathematical Poet
San Diego, CA, 91914, USA
E-mail: kaz@kazmaslanka.com

Abstract

Verbogeometry is a form of art which is interested in creating an aesthetic experience with poetic structures of mathematical / verbal metaphors. I am introducing Verbogeometry as a subset of a small movement of mathematical poetry occurring globally but mostly in America and Finland. This particular mathematical poetry movement has some connections to the visual poetry movement in the English speaking world. This paper on Verbogeometry is a primer and also an ongoing investigation.

The Mechanics of Verbogeometry:

1. Word-Axes and Word-Planes. One of the tenets of Verbogeometry and Mathematical poetry is that one performs mathematical operations on values of quality as if they were quantity. Of course this seems to be nonsense but then one must realizes that *paradox is the mathematical structure of metaphor*. Metaphor is the area of our interest and when quality and quantity are synonymous then math equations automatically transcend the normal duty of denotation and enter into the realms of connotation. This concept also bears some relationship to different concepts of infinity pioneered by the Russian-German Mathematician Georg Cantor, namely the infinities in gradation relative to the infinities in Counting.

Another tenet of verbogeometry is that it recognizes antonyms in only a few varieties. A simple antonym is a word whose antonym is a direct negation. (Example: just / unjust, probable / improbable or fertile / infertile etc.) A complex antonym is a word whose antonym is not a direct negation. (Example: just / unfair, probable / doubtful or fertile / barren etc.) There are also gradable (gradient) antonyms which are pairs that express relationships in a continuum, such as up and down. Complementary antonyms are pairs that express an either/or relationship, such as dead or alive. Verbogeometry uses all antonyms as if they were gradable. It is easy to find examples where poets use complementary antonyms as if they were gradable to create certain metaphors. Example: "Bob showed up half dead to work today."

Within the boundaries of verbogeometry it is important to understand that we view words as objects floating in space. When we focus on single words, with no context, they are alone inert relating to no other words. However, when we focus on words that have a synonymous partner we can easily imagine a line in space between the words. Probable and improbable are good examples of simple antonyms that we can view connected by a line. (Figure 1)

Life is full of dualities; it is hard to think about qualities without thinking about opposing ideas. We can view our 'probable / improbable' one-dimensional line as a number line but instead of values of numbers on the line, we think in terms about having different levels of meaning between the two words residing at each end of the line. Due to a number-line being a one dimensional axis it is easy to visualize a word-axis as an axis for a single spatial dimension. We call any pair of words connected by a line a word-axis. Two perpendicular number lines or word axes make a two-dimensional word-axis as well as defining a word-plane. We also have the ability to view the word-plane as an infinite number of coordinates delineated by the word-pairs much like the infinite number of coordinate pairs contained within a Cartesian coordinate system described in the realm of analytic geometry.

Figure 2 shows a visualization for the physics equation; distance = velocity multiplied by the time. Notice that the y-axis displays velocity and the x-axis displays time. When we multiply and blend the words in an infinitesimal weave, we arrive at the concept of distance in a tessellated product of the concepts of velocity and time. In another words by positioning the two axes perpendicular to each other, we view every value on one word-axis in relation to every value on the other word-axis. This method affords us a way to 'feel' the entire word plane or axis system with all its different augmented values and gradations. When we multiply two word-axes together we conceptually tessellate a two dimensional plane with different semantic values of the two words blended and augmented. If we were to take a normal Cartesian coordinate system and multiply the x positive integers (1 through 12) times the y positive integers (1 through 12) we get a tessellated plane as in Figure 3. (Figure 3) (Notice the intensity of blue relative to the value of the numbers).

To help us further visualize this concept let us create a word axis using the words red and green. In this instance, we are going to use red and green as nouns instead of adjectives. (We will use colors as adjectives later.) Let us multiply a red-green axis times another red-green axis and view it visually. (Figure 4) Multiplication of colors is similar to color addition except the disparate intensities of the colors are greater and follow a similar pattern shown in figure 3. The value of the 'numbers' is subjective and not as important as the relationship between the 'numbers'.

Figure 4 helps us to visualize different word meanings spread across a word plane. Let us create another example using two word-axes. However, let us use two different colored word axes instead of our previous example where both axes were the same colors. Following that, we will superimpose a set of different word-axes upon our color axes to compare how the system works. To facilitate visualizing two different word axes lets look at the example with the word-axes red-green and blue-orange multiplied by each other and mapped on a Cartesian coordinate system. (Figure 5) (Note these diagrams are visual aids not scientific data.)

What is nice about using colors for our examples is that words used for colors function as both a noun and an adjective depending on our intent. When we map a word plane with word-axes that comprise colors and we use them as adjective synonyms, then this word-plane serves as a paradigm or a pedagogical tool serving as a general model for understanding all two axes synonym word-planes. Example: Let us create a word-plane using the two word-axes of noble/ignoble and just/unjust. (Figure 6)

The next step would be to superimpose the noble/ignoble; just/unjust word-plane onto our previous word-plane of blue/orange; red/green. In essence, we are pretending color blue to mean ignoble, orange to mean noble, red to mean just and green to mean unjust. Now we can see the meanings blend into each other in the different areas of our word-plane. (Figure 7)

We can see the color purple as blend of ignoble and just, red-orange as a blend of just and noble, yellow-green as a blend of noble and unjust and blue-green as a blend of ignoble and unjust. For the record, I certainly am not trying to say there is a relationship of ignobility and injustice with the color blue-green! This example is just a tool to help us with our own concept of visualizing a word-plane. However, we could create a different but, in my opinion, limited set of color metaphors for noble/ignoble and just/unjust. Or we could look at our color example as adjectives on their own merit. This method would automatically help us see them as metaphors. Example: She was red hot. He had a blue day. He was so green he did not know what was happening. Multiplying adjective word-axes together instantly create metaphors.

2. Word-coordinate Pairs: We have witnessed a word-axis with different values of an antonym pair along a particular axis 'x' or 'y' in one dimension. (Figure 1) We also have seen a word-plane with values of two antonym pairs along two axes 'x' and 'y' in two dimensions. (Figure 6) Furthermore, we can have word-cubes along the x, y, and z-axes in the third and word-hypercubes in the fourth dimension or we can have antonymic pairs in innumerable dimensions. There is no limit to the dimensional palette for our expressions. Each antonym word-pair adds a new spatial dimension to our expressive construction.

Let us talk about the spatial accuracy in defining the location of words in space. Once again, let us look at figure 6 and notice the antonym word pairs, just/unjust and noble/ignoble. However, let us focus our attention to the word-axis just/unjust. We know that we have defined a one-dimensional word-axis with different values of just and unjust but we do not know exactly where each of the words is located along the axis. We have no quantitative value for just or unjust. However, we do have a qualitative value and we know that the word exists somewhere on the axis. What is most important to us in verbogeometry is not the value as such, but the spatial relationship of the values to each other in space. Because the value or the meaning of a word is relative to the context in which it is used, each viewer individually creates his or her own context for meaning. Therefore, exact quantification of the word or its location in space is not possible. However, in some cases, it may be possible to restrict the context to a level where repeatable correlations exist, but those studies are more akin to denotation for the purpose of science. Scientific experimentation "proves" the equation to be mathematically correct and workable within a range of acceptability. In other words, experimental data defines viability of the relationships between the concepts in a scientific equation. On a side note: (When scientific equations are in the intuitive stages of development, there may be an argument to claim that they are in the realm of art, I personally might accept this view if it were not for the fact that their intention is not to make art.) In verbogeometry, we construct equations based on relationships between the qualities of our experiences to evoke meaningful aesthetic expressions of which most are connotative but some may be denotative.

Let us get back to the Cartesian coordinate system for a moment and reiterate the idea of coordinate pairs. A point on a two dimensional coordinate system would have values for x and y and would be expressed as such: (x,y) (Figure 1.) A point on a three-dimensional axis system would have values for x, y and z and would be expressed as (x,y,z). A four dimensional point would be expressed as (x,y,z,w) in the fifth dimension as (x,y,z,w,v) etc.

Before we get into multidimensional word-axes let just look at a simple two-dimensional word-plane with two word-axes. (Figure 8.)

The vertical axis is a synonym word-pair of praise and punishment and the horizontal axis is a synonym word-pair of love and hate. It is very important to realize that not only does the words love and hate define the identity of the horizontal or x-axis they also hold conceptual points in space along the axis and the same for praise and punishment with respect to the y-axis. The words which are conceptual points in space define a metaphoric value along its respective axis and can be notated as a coordinate pair similar to (x,y) So you may ask what would a coordinate word-pair look like? Let us look at the two points identified as point 1 or P1(love,praise) and point 2 P2(hate,punishment) (Figure 8.)

3. Midpoint Formula in Verbogeometry *Any analytic geometry equation can use coordinate word-pairs instead of numbers to express poetic forms*. Let us use the midpoint formula to express the exact point between the two points P1(praise,love) and P2(punishment,hate) from figure 8. Before we look at coordinate word-pairs let us refresh the use of the midpoint formula in analytic geometry. To find the midpoint between two points on a Cartesian coordinate system we add the x coordinates together and divide by 2 to find the x value for the midpoint and we also add the y coordinates together and divide by 2 to find the y value for the midpoint or
$(x1 + x2)/2 = x0$ and $(y1 + y2)/2 = y0$.

Let us take a different approach and replace the numeric variables in the midpoint equation with the words/concepts of love, hate, praise and punishment. We will use the form of coordinate word-pairs

P2(love, praise) and P1(hate, punishment). The midpoint formula now shows us that P0(x0, y0) will be formed by the substitution of (x1 + x2)/2 = x0 with (love + hate) = x0 and (y1 + y2)/2 = y0 with (hate + punishment) = y0. Now we have expressed the exact point between love, praise and hate, punishment. (Figure 9.)

4. Verbogeometry with Trigonometry One method in trigonometry to solve the angle theta (Figure 10) is to find the inverse tangent of y/x. Now let us look at a similar expression within verbogeometry and looking for the value of angle theta. This time we define the y distance as the difference between the concepts of barren and infertile (think number-line again) and let us define the x distance as the difference between the concepts of infertile and fertile. (Figure 10)

If we want to know the angle of theta we have to take the inverse tangent of y/x or the inverse tangent of (barren - infertile)/(fertile - infertile) (Figure 10)

Notice that 'barren and infertile' is a complex antonym and 'fertile and infertile' is a simple antonym as previously defined. Simple antonyms and synonyms can be seen existing in orthogonal spaces. *This is interesting because we can see that there exists in verbogeometry a geometric construction where a line expressed as a simple antonym is normal (90 degrees) to a plane containing all of the complex antonyms related to the line which is expressing the simple antonym.* To illustrate this idea lets look again at the relationship between the simple antonyms 'fertile' and 'infertile' and the synonyms 'barren', 'fruitless', 'unproductive', 'sterile', 'impotent' which reside on the plane that is normal (90 degrees) to the line created by the simple antonyms. Furthermore you can draw lines from all of the synonyms back to the complex antonym 'fertile'. (Figure 11)

This idea also lends itself to prismatic structures where we have a group of parallel simple antonyms whose endpoints construct polygonal faces on two parallel synonym-planes. (Figure 11) Example: Let us define one synonym plane containing the words 'pleased', 'content', 'affected', 'satisfied', 'enchanted' and 'sympathetic'. The other plane contains the following simple antonyms for the previous group of synonyms: 'displeased', 'discontent', 'disaffected', 'dissatisfied', 'disenchanted' and 'unsympathetic'. Due to the synonyms of one plane have corresponding simple antonyms which create lines 90 degrees from the synonym-plane then the simple antonyms are synonyms of each other and reside on their own individual synonym plane and because the lines are 90 degrees to each other the planes must be parallel. The former verbiage is a lot easier to understand visually (Figure 12)

5. Distance Formula and Verbogeometry. As we have seen, to calculate the distance between two points, we need to describe our points by its coordinates using the nomenclature of the coordinate pair. Let me reiterate, describing a point in verbogeometry is no different from numerical coordinates except we use words. Lets look again at the example in figure 9 where we used the midpoint formula to find the exact point between the points: P1(love,praise) and P2(hate,punishment) but instead of putting them in the midpoint formula lets put them in the distance formula. (Figure 13)

Here we have an expression for the distance between the points P1(love,praise) and P2(hate,punishment) in two dimensions. But we can also use verbogeometry in any number of dimensions including hyper-dimensions. But before we look at hyper dimensional verbogeometry lets look at another example which we will express in the third dimension. The following example uses a three dimensional Cartesian coordinates system with 3 simple antonym word-axes. (Figure 15) The first axis is noble / ignoble the second axis is just / unjust and the third axis is loyal / disloyal.

Now lets look at the expression for the distance between the points P1(noble,just,loyal) and P2(ignoble,unjust,disloyal) (Figure 14)

Notice the green line in figure 15 is the visual representation for the mathematical expression above. However, it would be much easier to visualize if we were able to rotate the axis. Figure 15 is an isometric view, which I chose to use because it is best for viewing the axis but unfortunately at the expense of viewing the spatial orientation of the green line.

Now let us look at verbogeometry in a hyper-dimension. Let us look at the distance formula used in seven dimensions: Figure 16 shows the mathematical poem 1+1+1+1+1+1+1+1 =1 This is a metaphorical piece that creates a metaphoric path from the concept of confusion, to where seven deities meet. The piece uses the analytic geometry distance formula in a seven dimensional space where each dimension is a gradation from confusion to a point where a deity exists. Lets look at the coordinate pairs for these two points P1(confusion, confusion, confusion, confusion, confusion, confusion, confusion) and P2(Allah,Buddha,Jesus,Spider woman,Vishnu,Yahweh,Zeus) The detail of figure 16 looks like the following:

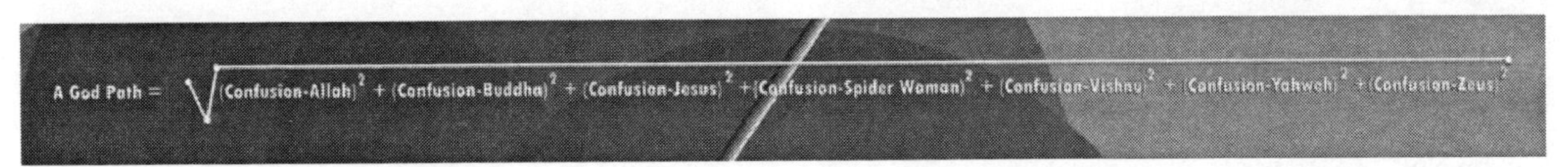

The latter expression can also be written as: A God Path = ((Confusion-Allah)^2 + (Confusion-Buddha)^2 +(Confusion-Jesus)^2 +(Confusion-Spider Woman)^2 +(Confusion-Vishnu)^2 +(Confusion-Yahweh)^2 +(Confusion-Zeus))^.5

In conclusion what I have shown is scratching the surface of the possibilities of verbogeometry. Verbogeometry can be taken in vast directions that I have not covered or will be able to cover. I hope, in the future, more people join in to explore the possibilities of verbogeometry.

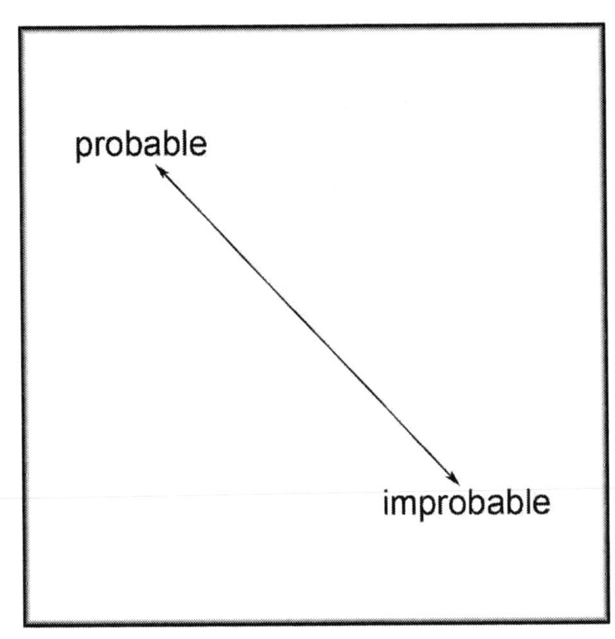

Figure 1. Word Axis

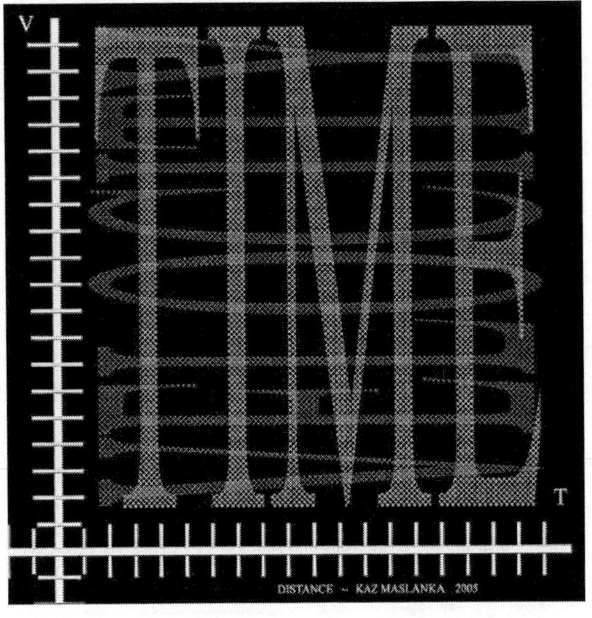

Figure 2. Distance = (velocity)(time)

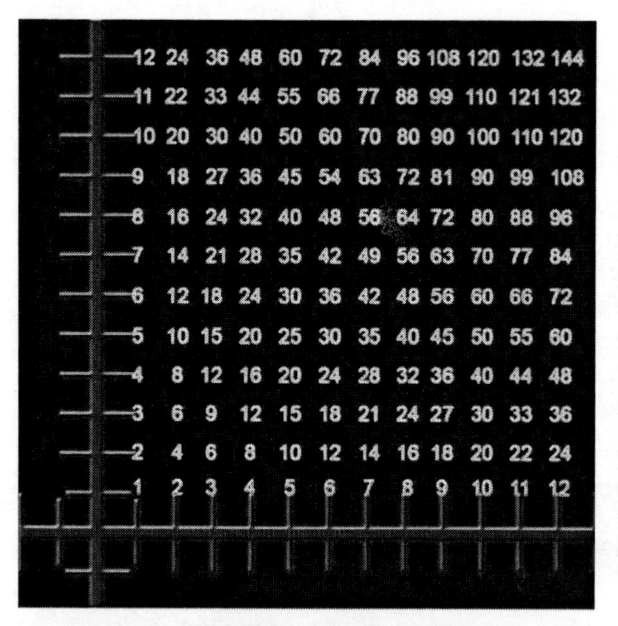

Figure 3. Multiplication table

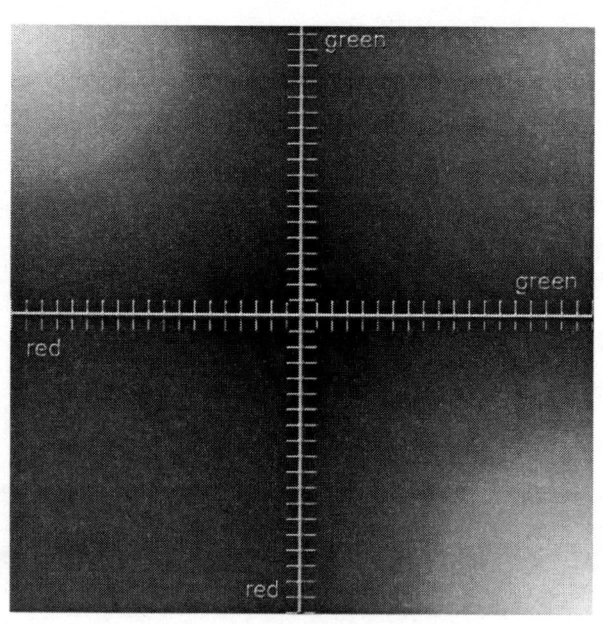

Figure 4. Two Red-Green Axes

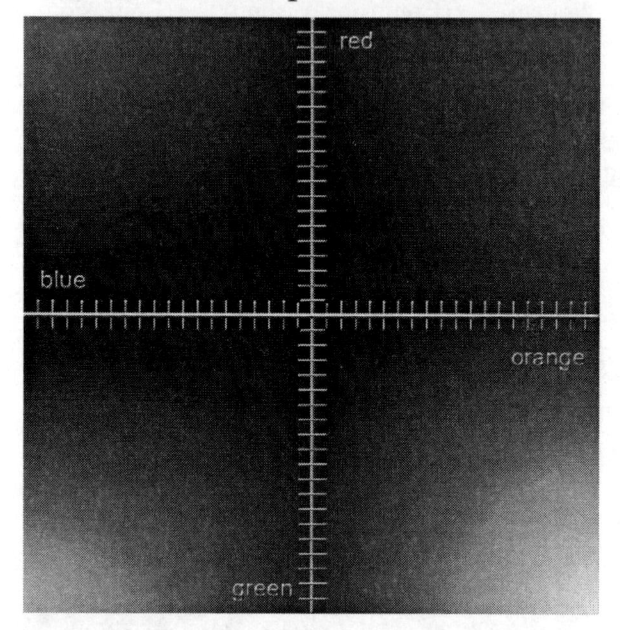

Figure 5. Red-Green Blue-Orange Axes

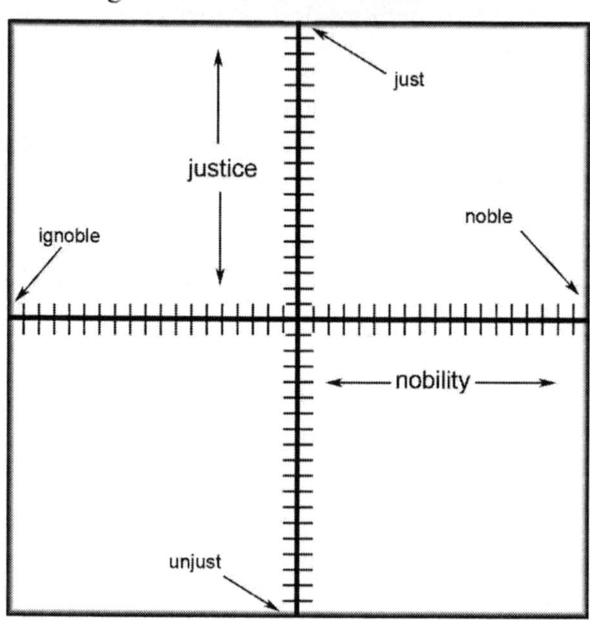

Figure 6 Just-Unjust Ignoble-Noble

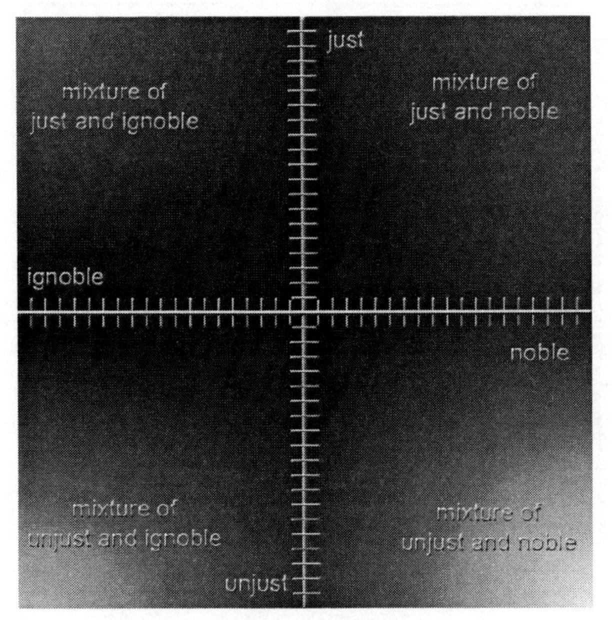

Figure 7. Two Axes Mixture

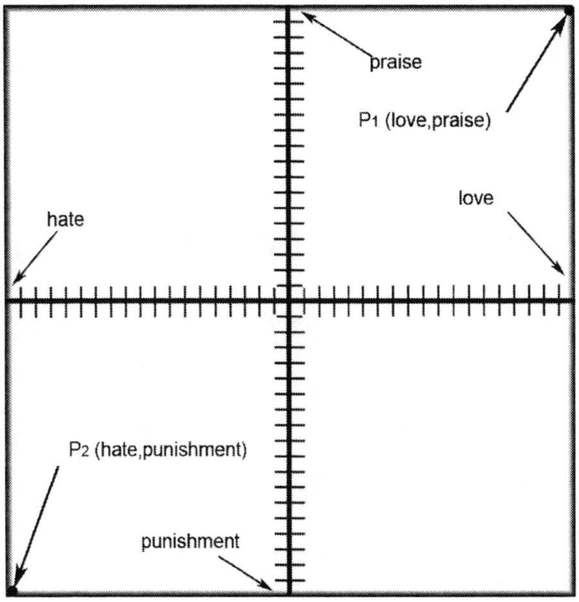

praise

P1 (love,praise)

love

hate

P2 (hate,punishment)

punishment

Figure 8. (love,praise hate,punishment)

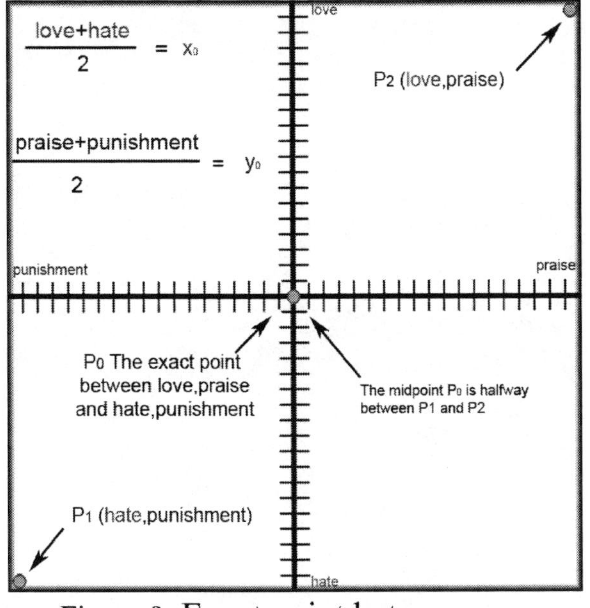

Figure 9. Exact point between (love,praise) and (hate,punishment)

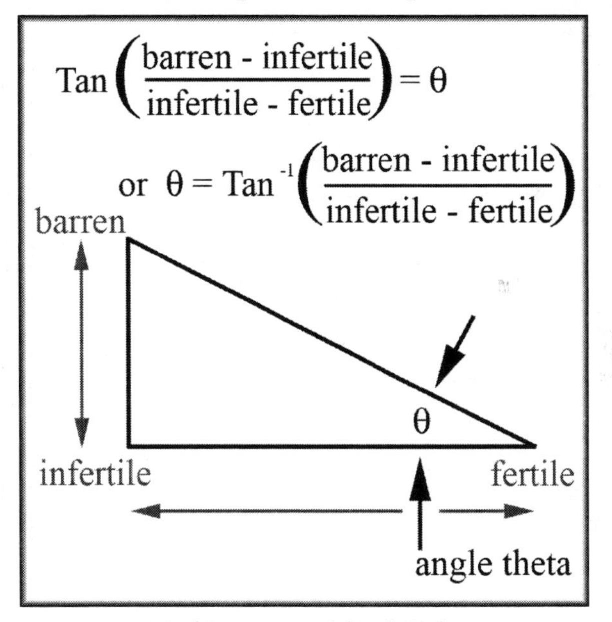

Figure 10. Verbogeometrical Trigonometry

distance between P1(love,praise) & P2(hate,punishment)

$$d = \sqrt{(\text{love} - \text{hate})^2 + (\text{praise} - \text{punishment})^2}$$

Figure 13 Distance between (Love,Praise) and (Hate,Punishment)

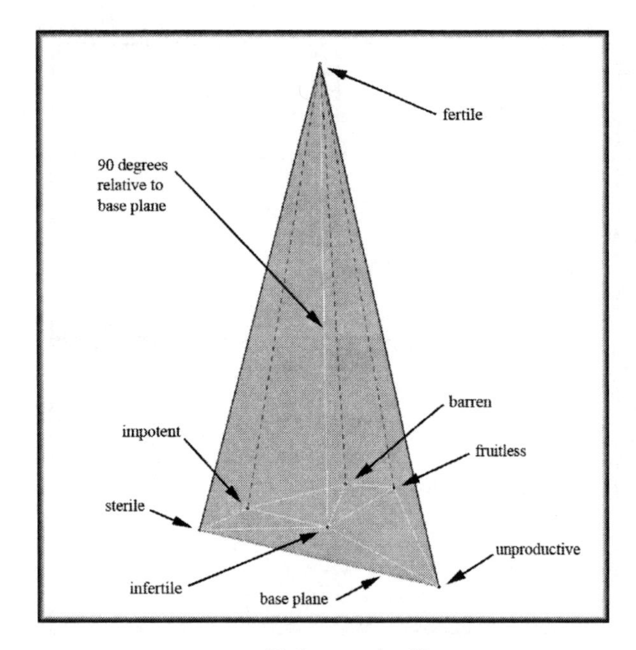

Figure 11. Prismatic Structures

Figure 12 Prismatic Structures

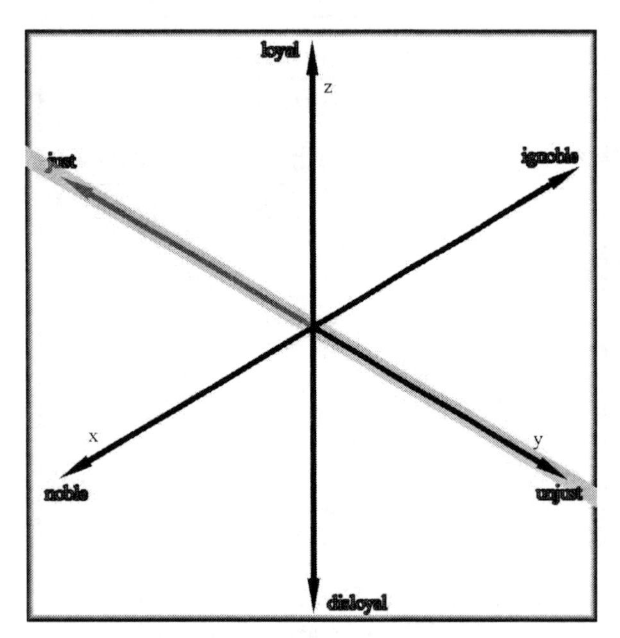

Figure 15. Three Dimensions

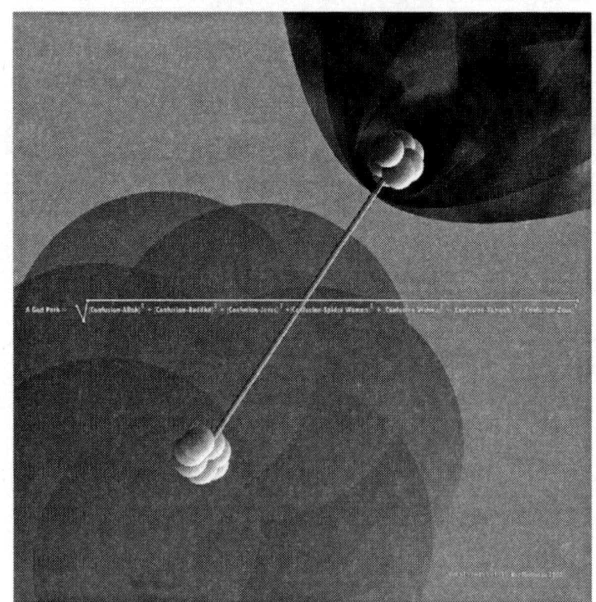

Figure 16. Seven Dimensions

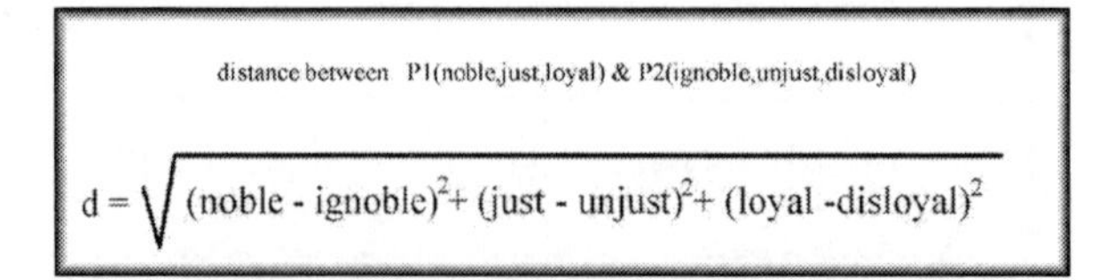

distance between P1(noble,just,loyal) & P2(ignoble,unjust,disloyal)

$$d = \sqrt{(\text{noble} - \text{ignoble})^2 + (\text{just} - \text{unjust})^2 + (\text{loyal} - \text{disloyal})^2}$$

Figure 14. Distance between (noble,just,loyal) and (ignoble,unjust,disloyal)

This paper also at the following URL:
http://www.kazmaslanka.com/verbogeometry/verbogeometry.html

Zome-inspired Sculpture

Paul Hildebrandt
Zometool Inc.
1400C 1st Avenue
Longmont, CO 80501 USA
Email: paulh@zometool.com

"There's something irritating about doing something right by accident" -- S. Rogers

Abstract

An invitation to build 1) permanent, Zome-inspired sculptures 2) designed and built as a collaborative effort under the name of fictitious artist(s), 3) as much about art as mathematics, 4) which could serve as the basis for large-scale architectural projects for the 21st century 5) to be installed at Bridges venues, as possible, on an ongoing basis.

I'll give a little background about Zome, survey some sculptures and artists, and discuss the guidelines above in more detail. There are no designs yet. This is an invitation to get started!

1. Introduction

Zome seems best described by analogy. I call it a language for understanding the structure of space; Zometool Inc. cofounder Marc Pelletier has called it "frozen music."

If Zome is music, it's currently "played" on two instruments: kits of plastic balls and struts (distributed as an educational toy, mathematical manipulative and research tool), and vZome software. But just as music isn't limited to two instruments, Zome cannot be restricted to its current software and hardware incarnations. There are many ways to express Zome's elegant mathematics, and sculpture is a good medium to begin exploring possibilities.

If Zome is a language, most of us are illiterate. A few mathematicians and scientists have jotted down "shopping lists," [1] while artists like Jean Baudoin and Steven Rogers have composed "Haiku" in this new language. No one has ever written a short story, let alone a book. This is an invitation to do so; to form a team of artists to explore the richness of Zome's geometry as an artistic medium, to build sculpture(s) that inspire a deeper appreciation of both mathematics and art, and plant a seed for bolder projects. We may not "write a book," but even a few coherent paragraphs would be a good start.

2. Background

2.1. What it is -- Without the crutch of analogy, describing Zome concisely is a challenge for me. In Zome Primer, his classic introduction to this discovery, Steve Baer describes Zome as a [structural] "system based on the 31 lines that pass through the center of an icosahedron and either a vertex, edge midpoint or face midpoint, [which] is new and unusual." [2] David Richter claims that Zome geometry can be derived from the E_8 lattice [3]. Marc Pelletier's assertion that Zome is a projection of a 61-dimensional hypercubic lattice into 3-space works best for me, but leaves a lot of other lay people scratching their heads.

In its current physical form, the 61-zone system, Zome balls and struts represent points and lines which define regular 2, 3 and 5-fold symmetries in space. Each point in the 61-zone system generates an array of 122 vectors determined by the edge midpoints (blue lines), vertices (yellow lines), face midpoints (red lines) of the regular dodecahedron and the edge midpoints (green lines) of the 5 cubes associated with same. Lines follow the 122 vectors with respective lengths in Divine Proportion (τ) powers of unity, cosine 30°, cosine 18°, and cosine 45°:

$A\tau^n$, where A = unity, cosine 30°, cosine 18°, or cosine 45°
$\tau = (5^{1/2}+1)/2$, and
$n = 0,1,2...$

The integers 2, 3, and 5 are the first three primes as well as Fibonacci numbers, and their incestuous spatial relationships have been explored by artists and mathematicians for centuries. Ratios of consecutive Fibonacci numbers are rational approximations of the Divine Proportion – which resonates throughout Zome geometry. Various claims and counterclaims have been made concerning Fibonacci numbers, the Divine Proportion and our inherent sense of beauty. Whether these relationships play into your sense of beauty is a subjective question, but they do for me.

Figure 1: *Clark Richert's "True Story of the Quasicrystal" (with Drop City in the background)*

2.2. History -- Zome was born of a confluence of art, architecture and mathematics. Steve Baer and associates discovered the 31-zone system while searching for a structural system based on the geometry of the icosahedron, and artists at Drop City adopted the geometry as an alternative to geodesic domes. The mathematics of Zome was derived from explorations in art and architecture, and its appeal seems to be related to the natural beauty and structural versatility of icosahedral symmetry and the Divine Proportion.

Artists Clark Richert and Gene & JoAnn Bernofsky conceived of Drop City in Spring of 1964. Their antecedent art events, "Droppings," were highly informed by the "Happenings" at Black Mountain College and the work of R. Buckminster Fuller. "We envisioned a whole city as a live-in work of Drop Art." [4] So they bought 7 acres of pasture near Trinidad, Colorado, and founded Drop City. Living in sculptures turned out to be somewhat impractical, so they built geodesic domes from car tops purchased at a nearby junkyard.

In his artistic explorations, Richert discovered quasiperiodic tilings in 2 and 3 dimensions at about the same time Sir Roger Penrose sought them in his mathematical research [5]. These mathematical and artistic explorations preceded by roughly two decades the physical reality of quasicrystals, which were discovered by Dan Shectman in 1984 and previously thought to be impossible.

While Drop City was growing, Steve Baer started Zomeworks Corporation to seek commercial applications of the 31-zone system, which he and his associates discovered in 1968. Most of the applications were architectural, although Zomeworks also built playground climbers, a monkey cage at the Albuquerque zoo, and they manufactured Zometoy, the predecessor to the Zome modeling system in its current form. Years later, when asked how he derived the Zome's exquisite mathematics from his structural research, Baer remarked, "that's where the pipes ran into each other [so we had to cut them in Divine Proportion ratios.][5]"

In a curious parallel to Richert's and Penrose's explorations, French artist and master carpenter Jean Baudoin independently discovered the 15-zone system (Zome's blue lines) about the same time Baer and associates discovered the 31-zone system (Zome's blue, yellow and red lines.) Baer introduced Richert et al to the concept of zomes [6] when he visited the Drop City in 1969, spawning a rich cross-pollination of ideas that ultimately led to Zome in its present forms [7].

3. Some Sculptures and Artists

There is a tendency to see Zome as a medium for modeling purely mathematical ideas. It is a uniquely elegant means of mathematical expression, but not to the exclusion of more fanciful artistic explorations. Following is a discursive survey of sculptures and artists related in some way to Zome, ranging from purely mathematical models to more artistic expressions. If you see your name in this section, you're invited to participate. If you don't, but think it should have been included, you're also invited. If you'd just like to join in, please do. I'd especially like to hear from Johannes Kepler, Paul Donchian and Jean Baudoin.

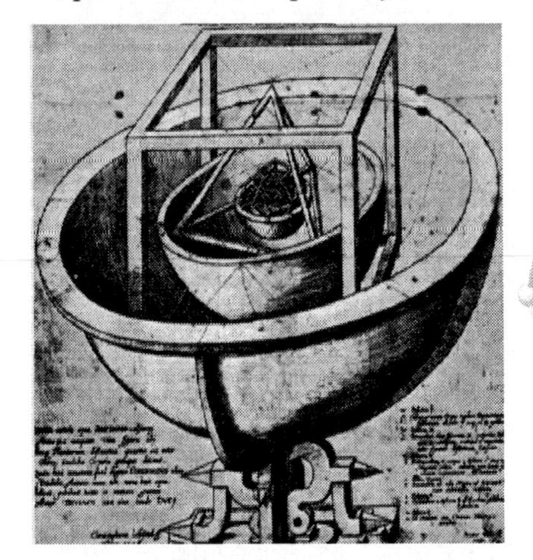

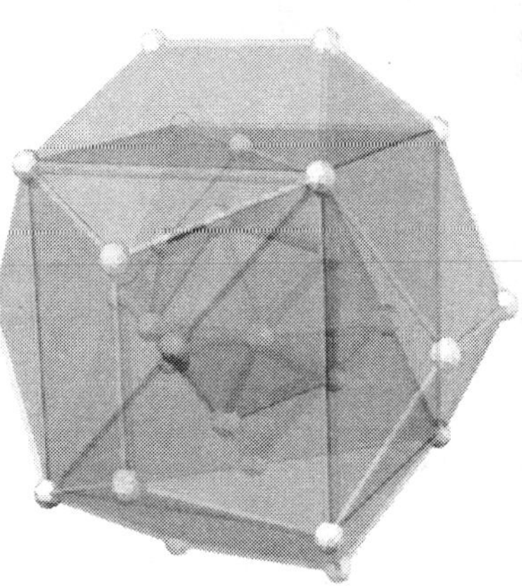

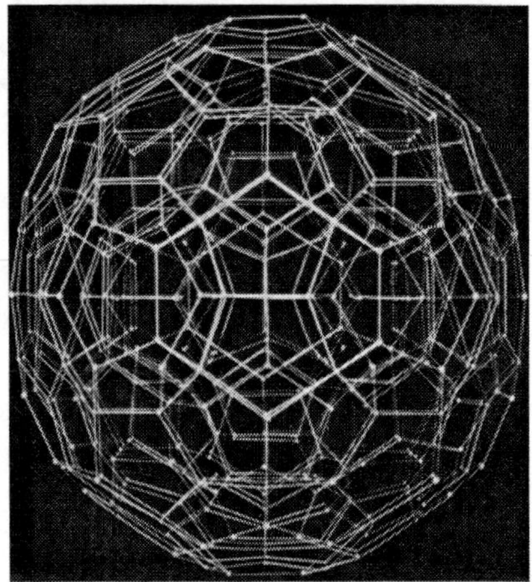

Figure 2: *Classic mathematical models: Kepler's "Weltgeheimnis" (left), Conway's "Cosmogram" (center) and Donchian's 120-cell (right)*

3.1 History -- Johannes Kepler's "Weltgeheimnis [8]," which nested the five Platonic solids in five concentric spheres, was a kind of 17th century theory of everything, providing a complete explanation for the number of known planets and the distances between them [9]. There must be scores of ways to model the relationships among the 5 Platonics in Zome. My favorite is John Conway's "cosmogram," a simple, 3-dimensional way to see how the Divine Proportion "falls out" the natural inter-dependence among 2-, 3- and 5-fold symmetries, which could be called the "Rosetta Stone" of Zome.

Paul S. Donchian [1895-1967] built dazzling models of projections of 4- and higher-dimensional figures. According to Donald Coxeter, after a "series of startling and challenging dreams of the previsionary type... he determined to make a thorough analysis of the geometry of hyperspace. His aim was to reduce the subject to the simplest terms, so that anyone like himself with only elementary mathematical training could follow every step. For this purpose he devoted many years to the task of making... exquisite models [10]." Some of these would qualify as sculptures, such as the projections of the 120-cell {5, 3, 3} and 600-cell {3, 3, 5} featured in Regular Polytopes [11] . He painstakingly soldered these models together from wire segments, taking up to 2 years to complete one. Many of Donchian's surviving models are suffering from benign neglect at the Franklin Institute in Philadelphia [12].

Figure 3: *Marc Pelletier with his 120-cell before installation at the Fields Institute for Coxeter's 95th birthday (left) and a stellated dodecahedron in an icosahedron by Pelletier and Chris Kling (right).*

3.2 Recent events -- Zome cofounder Marc Pelletier built several sculptures of the120-cell from welded stainless steel rods. One was anonymously donated to the Fields Institute in Toronto, to honor Coxeter on his 95th birthday on February 15th 2002. Pelletier also developed, with Chris Kling, an architectural-scale version of Zome's versatile blue lines, introduced at the Bridges Conference in Towson University, Towson, Maryland in July 2002. They built a model of a stellated dodecahedron inside an icosahedron, using 32 aluminum nodes of some 500 manufactured. A spectacular array of far more beautiful and complex sculptures would be possible using this giant Tinkertoy kit. Kling's company, Aurodyn Inc., is advancing architectural applications of Zome geometry. Pelletier has also collaborated with Chris Palmer, exploring decorations of zonohedral structures based on traditional Islamic patterns. Their work demonstrates Baer's assertion that all aspects of a piece, from gross structure to fine detail, can be derived from one simple, elegant system.

Figure 4: *Dan Duddy's Sierpinski tetrahedron (left) and George Hart's "Truncated Ambo 600-Cell," under construction (right).*

Professors David Richter and George Hart, and mathematics student Dan Duddy build large-scale "rZome"[13] mathematical models. Among his larger projects, Duddy built a 6' Sierpinski tetrahedron from 6,000 Zome components in about 30 hours. Hart and Richter have spent much time exploring the 15 rZome constructible "shadows" of 4-dimensional figures in the H4 group [14]. Richter and Duddy organized a team to build the cantellated 120- and 600-cells at Bridges 2005 in Banff; Hart has led numerous workshops to build these and other interesting mathematical models, sometimes involving scores of human-hours with part counts exceeding 10,000. An artist in his own right, Hart's permanent sculptures wax whimsical.

Jean Baudoin, Fabien Vienne, Sam Verbiese and Steve Rogers have mined Zome's richness as an artistic medium. Rogers has built dragons with a Chinese "flavor" based on projections of 6- and 10-dimensional cubes, and explored a world of nanotechnology during his "carbon phase." In 2004, Rogers also discovered that models such as the hyperbolic parabaloid could be generated as ruled surfaces in Zome, opening another world for artistic and mathematical explorations [15].

Figure 5: *Sculptures by Steven Rogers (left), Jean Baudoin (middle), and Sam Verbiese (right.)*

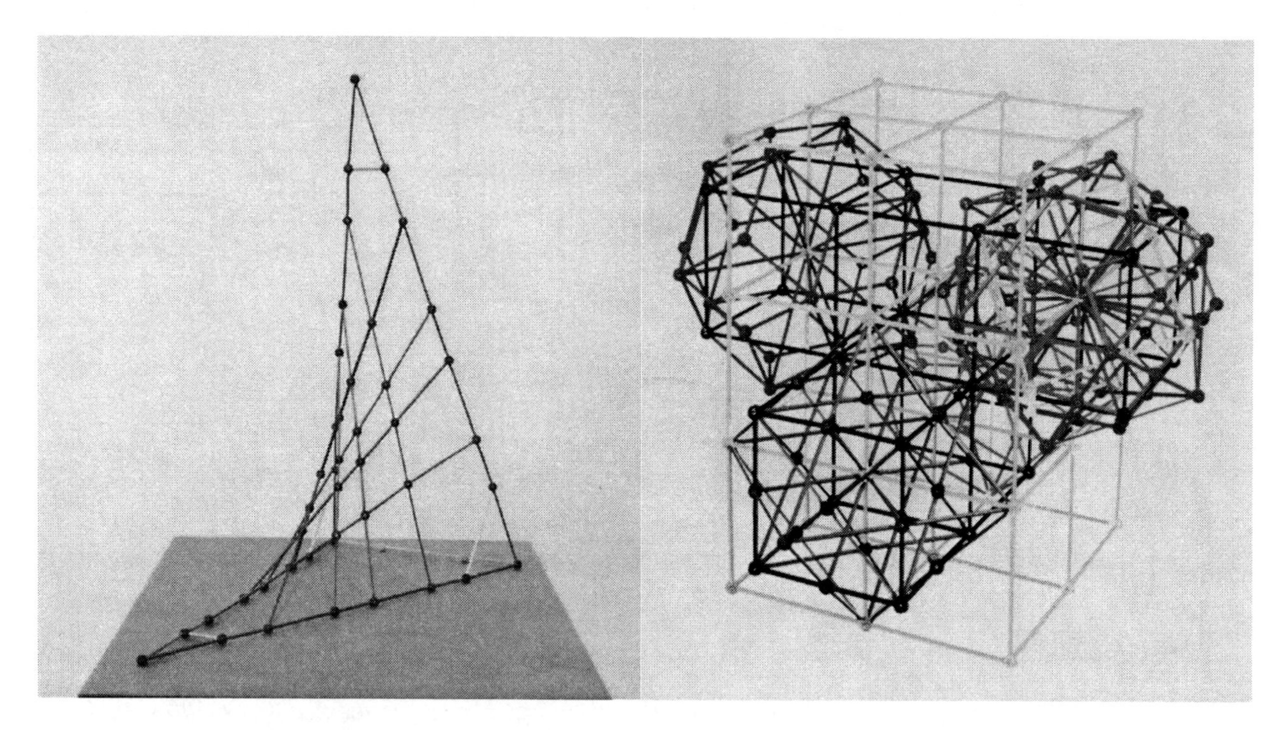

Figure 6: *Two Zome sculptures from the 2004 biennial "Formes Utiles" exhibit -- a ruled surface informed by Steve Rogers, and Fabien Vienne's CubeSpace.*

Samuel Verbiese is a versatile Belgian artist who has installed a number of Zome-inspired scultures, including a meta-zome [16] representation of the Atomium (a monument originally built in Brussels, Belgium, for the 1958 World Fair [17].) The artist generally calls his Zome models "expressions of art" but hesitates with the MetaZome Atomium, as it was an expansion of a known structure. However, it relates to the idea of using Zome-inspired sculpture to catalize a visionary architectural project in section 4.4.

As a co-discoverer of Zome geometry, Jean Baudoin outlined some of the most poetic manifestations of the system over the course of 30+ years. His design for la Géode in the Cité de la Science, Paris, was a powerful expression of Zome's architectural versatility. Unfortunately, the proposal placed 2nd in the design competition, after a more regressive utopian design [18]. It's a tragedy that Baudoin's work remains largely unknown [19]. The artist died in 2002.

Baudoin and Fabien Vienne co-designed the Zome's GreenLines. Vienne is also an accomplished artist and designer whose work with Scott Vorthmann, Marc Pelletier and others has helped to extend the understanding of Zome geometry from 61 to 181 zones. He was responsible for installing several Zome sculptures in the 2004 biennial "Formes Utiles" exhibit in the Musées Art Moderne in St. Etienne, France.

4. Guidelines for the Zome-inspired Sculpture Project

I hope many artists and mathematicians will participate in designs for Zome-inspired sculptures. Here's an explanation of guidelines outlined in the abstract:

4.1 Permanent, Zome-inspired sculpture – rZome components are an excellent medium for 3-dimensional sketches, but not ideal for permanent installations. The material is subject to UV degradation and the primary colors are corny (although a great shorthand for 2-, 3-, and 5-fold symmetries.) rZome components also impose artificial limits on design: for example, the fractal nature of Zome can be shown in a MetaZome structure, but as an infinitely iterative process, it's not practical -- after two generations, the models falter under their own weight. But the sculpture should be "Zome-inspired," because (1) the anonymous donor who is funding the project has a keen interest in popularizing Zome, and (2) keeping the sculpture within Zome's mathematical framework can give the project visual and spatial coherence.

4.2 A collaborative effort under the name of a fictitious artist – Like Coxeter, the Zome sculpture team should be "guided almost completely by a profound sense of what is beautiful." [20] Collaborative work suggests that no one person is responsible for the design or execution of the sculpture any more than any one person was solely responsible for the discovery and development of Zome – it's an ongoing process, just as I hope the sculpture will pave the way for more and better art and architecture. No fictitious name for the team has yet been suggested. Any ideas?

4.3 As much about art as mathematics – If Bridges Conferences are a call "to regain the lost mutual understanding" among artists and mathematicians, as well as "between mathematicians and the general public," this project hopes to tip the scales in favor of the artists and the general public. Bridges seems mathematics-heavy and art-light. Most artists and ordinary folks paging through the proceedings would find some images interesting, some obscure, and much of the text abstruse. I'd love to see more art for the sake of balance. After all, drawing a distinction between "mathematical" and "non-mathematical" art is silly, since everything that streams into our senses can be interpreted as mathematics.

4.4 To serve as the inspiration for a visionary architectural project for the 21st century – Where is the Eiffel Tower of the 21st century? Engineering feats like the Eiffel Tower, built for the 1889 Paris Universal Exposition, or the Ferris Wheel, constructed for the 1893 Chicago fair, inspired confidence in many fairgoers in the capacity of the world's fair organizers to engineer a better future. [21] While serving on Zometool Inc.'s board of directors in the 90s, Steve Baer suggested building a Zome-inspired architectural wonder at an apocryphal Y2K Expo to be located on the site of Drop City [22]. Fuller's geodesic sphere for the U.S. Pavillion at Expo '67 in Montreal was probably the last truly visionary structure paraded before the eyes of the world. It's time to change that.

4.5 To be installed at Bridges venues, as possible, on an ongoing basis -- Installations could be patterned after George Hart's "geometry barn raisings," as regularly scheduled conference events. [23]

5. Conclusion

Based on relationships among 2-, 3- and 5-fold symmetries and the Divine Proportion, Zome offers rich and scarcely explored possibilities as an artistic and architectural medium. Installing permanent, Zome-inspired sculptures at Bridges conference venues offers opportunities to more deeply mine those possibilities, and perhaps write a few coherent paragraphs in the "language" of Zome.

Notes and References

[1] including Roger Penrose, Richard Smalley, Dan Shechtman and Linus Pauling
[2] Steve Baer, *Zome Primer*, Zomeworks Corporation, 1970, page 2
[3] David Richter, "Two Results Concerning the Zome Model of the 600-Cell," *Renaissance Banff Conference Proceedings*, 2005, pages 419-426
[4] "the real concept was to create an artists' community whose purpose was to provide food, clothing, studio space and housing for artists who would live their art" – Clark Richert, from personal interview in February, 2006
[5] Penrose points out that the critical difference between his tilings and others based on 5-fold symmetries (including Richert's and Kepler's) is his matching rules, which force periodicity, a point missed by many writers for popular media (and me, until I asked him about it in March 2006)
[6] from conversation with Baer after Zometool Inc. board meeting in 2002
[7] The word "Zome" was originally coined by Steve Durkee in 1968 as a conflation of the words "zonohedral" and "dome." See the Dome Cookbook, Lama Publications, 1970 (revised), page 24
[7] Richert suggested Zome's green lines in the early 70s; in 1997 Baudoin and Fabien Vienne designed a version to be compatible with the original shape-coded Zome node. Marc Pelletier suggested extending Zome's geometry from 61 to 121 zones in the 80s and Vienne, Scott Vorthmann, Brian Hall, David Richter and others are pushing Zome "to infinity and beyond"
[8] Literally, "universe-secret"
[9] In proving his theory wrong, Kepler discovered his three laws of planetary motion and launched the modern study of celestial mechanics
[10] H.S.M. Coxeter, *Regular Polytopes*, 1947, Dover, pg. 260
[11] Ibid, pages 161 and 177
[12] Marc Pelletier has been archiving photos of Donchian's work, a presentation of which is available at http://www.fields.utoronto.ca/audio/01-02/sculpture/pelletier/index.html?8;onesize#slideloc.
[13] "rZome" means "real Zome" vs. "vZome" for "virtual Zome"
[14] Scott Vorthmann integrated H4 mathematics into his vZome software package, available at http://www.vorthmann.org/zome/
[15] Rogers also contributed a number of key innovations leading to Zome's current form, among them, a method for modeling the first truncated triacontahedral Zome node and the design for Zome's "clicks"
[16] I.e., you can use the balls and struts to build bigger balls and struts. The term "MetaZome" was coined in 2002 by Andrew Mihal, at the time an Engineering PhD student at UC Berkeley.
[17] see http://www.atomium.be
[18] The winning entry was a mirrored sphere informed by the work of 18th century visionary architect, Claude Nicolas Ledoux (1736 - 1806)
[19] Pelletier rescued and archived a great deal of Baudoin's work, although even more has been lost
[20] Siobhan Roberts, "Donald Coxeter: The Man Who Saved Geometry," *Toronto Life*, January 2003 (quoting Robert Moody)
[21] Rydell, Robert W. et al, *Fair America*, Washington, Smithsonian Institution Press, 2000, p135 With their spectacular technological and ethnological narratives, fairs engaged in a mission of "manufacturing consent," the alternative to which, in the eyes of the ruling elites who organized the fairs between 1876 and 1916, was social and political revolution. If promoters of world fairs resisted social revolution, they boosted technological revolution, witnessed by the reaction of the Parisian "protectors of art" to the Eiffel tower: "We, the writers, painters, sculptors, architects and lovers of the beauty of Paris, do protest with all our vigour and all our indignation, in the name of French taste and endangered French art and history, against the useless and monstrous Eiffel Tower." --Maupassant, Emile Zola, Charles Garnier (architect of the Opéra Garnier), and Dumas the Younger (among many others.)
[22] Ironically, Drop City was portrayed by the media as the antithesis of naïve faith in salvation through technology (although the Droppers, informed by R. Buckminster, considered themselves pro-technology.) Clark Richert recalls a myth that on July 20, 1969, as Commander Neil Armstrong stepped onto the moon, network TV news planned to cut to Drop City for a counterpoint state of affairs on planet earth, a move that the editors nixed in the last moment. Media-wary Droppers once worked out a deal with CBS reporter Terry Drinkwater, agreeing to grant interviews if he didn't use the words "drugs" or "hippies" in his story. Of course, the segment opened with "hippies taking LSD" at Drop City.
[23] Thanks to Phillip Kent, photos of a proposed space at the London Knowledge Lab can be viewed at http://www.lkl.ac.uk/bridges/LKL%2Dphotos/. We are working on a "test" installation for Bridges 2006.

"Civilization's greatest achievements all but wreck the societies in which they occur." -- A. N. Whitehead

Developable Sculptural Forms of Ilhan Koman

Tevfik Akgün
Faculty of Art and Design
Design Communication
Department
Yildiz Technical University
Istanbul, Turkey
akgunbt@yildiz.edu.tr

Ahmet Koman
Molecular Biology and
Genetics Department
Boğaziçi University
Istanbul, Turkey
akoman@boun.edu.tr

Ergun Akleman
Visualization Sciences Program,
Department of Architecture
Texas A&M University
College Station, Texas, USA
ergun@viz.tamu.edu

Abstract

Ilhan Koman was one of the innovative sculptors of the 20th century [9, 10]. He frequently used mathematical concepts in creating his sculptures and discovered a wide variety of sculptural forms that can be of interest for the art+math community. In this paper, we focus on developable sculptural forms he invented approximately 25 years ago, during a period that covers the late 1970's and early 1980's.

1 Introduction

$\pi+\pi+\pi+\pi+\pi+$

Hyperform

Figure 1: The photograph $\pi+\pi+\pi+\pi+\pi+$ series is from a Koman exhibition at Gronningen, Copenhagen (Denmark) in 1986. Hyperform is realized in stainless steel for the steel company Atlas Copco, (Entrance hall, Atlas Copco, Sweden, 1973) (Photos by Koman family)

In this paper, we want to present developable forms invented by sculptor Ilhan Koman during the 1970's. The first is the $\pi+\pi+\pi+\pi+\pi+$ a series of forms derived by the increase of the surface of a circle without changing its radius. A second one is called rolling lady and created using four identical cones. The third is the "Hyperform" that he produced by self-joining a single sheet of metal. Extension of the length of the rectangular sheet in a defined manner led to multiple derivatives. Figure 1 shows photographs of $\pi+\pi+\pi+\pi+\pi+$ and Hyperform developable sculptures.

2 Developable Surfaces

Developable surfaces are defined as the surfaces on which the Gaussian curvature is 0 everywhere [20]. The developable surfaces are useful since they can be made out of sheet metal or paper by rolling a flat sheet of material without stretching it [14]. Most large-scale objects such as airplanes or ships are constructed using un-stretched sheet metals, since sheet metals are easy to model and they have good stability and vibration properties. Moreover, sheet metals provide good fluid dynamic properties. In ship or airplane design, the problems usually stem from engineering concerns and in engineering design there has been a strong interest in developable surfaces. For instance, modeling packages such as Rhino provides developable surface analysis [14, 15].

Although, once invented, it is easy to physically construct developable surfaces using sheet metal or paper, it is not that easy to provide computational models to represent developable surfaces. Sun and Fiume developed a technique for constructing developable surfaces [19], but, their method is useful only to represent ribbons and it is hard to use to represent general developable surfaces. Haeberli has recently introduced a method to represent a shape with piecewise developable surfaces and developed a Lamina Design Software [7]. The current results seem to be limited but the Haeberli's approach Lamina has a great potential for developable surface design. Mitani and Suzuki introduced a method to approximate any given shape using developable surfaces to create paper models [11]. Because of the approximate nature of their models, there exists gaps between individual pieces and therefore, their method is not suitable for engineering application.

Sheet metal is not only excellent for stability, fluid dynamics and vibration, but also one can construct aesthetic buildings and sculptures using sheet metal or paper. Developable surfaces are frequently used by contemporary architects, allowing them to design new forms. However, the design and construction of large-scale shapes with developable surfaces requires extensive architectural and civil engineering expertise. Only a few architectural firms such as Gehry Associates can take advantage of the current graphics and modeling technology to construct such revolutionary new forms [12]. Some architectural structures can be easier to construct with developable surfaces. For instance, Fishback and Tuazon introduced Randome, a dome like structure that is constructed from developable surfaces [6].

Developable surfaces are particularly interesting for sculptural design. It is possible to find new forms by physically constructing developable surfaces. Recently, very interesting developable sculptures, called D-forms, were invented by the London designer Tony Wills and introduced by Sharp, Pottman and Wallner [16, 13]. D-Forms are created by joining the edges of a pair of sheet metal or paper with the same perimeter [16, 13]. Pottman and Wallner introduced two open questions involving D-forms [13, 5]. Sharp introduced anti-D-Forms that are created by joining the holes [17]. Ron Evans invented another related developable form called Plexagons [4]. Paul Bourke has recently constructed computer generated both D-Forms and plexons [2, 4] using Evolver developed by Ken Brakke [1].

3 Ilhan Koman Biography

Ilhan Koman was born in 1921, Edirne, Turkey, studied at the Art Academy in Istanbul, opened his first workshop and exhibition in Paris, 1948, moved to Sweden in 1958, where he taught at the Konstfack School of Applied Art in Stockholm until his death in 1986. Figure 2 shows two photographs of Ilhan Koman. At right Koman is photographed in front of his and architect Chet Kanra's sculpture, called *"From Leonardo To..."* in Stockholm. His works cover a wide spectrum of styles and materials, including 12 public monuments in Sweden and 4 in Turkey. He is represented in several museums including Moderna Museet, Stockholm, Museum of Modern Art, New York, Musée d'Art Moderne de la Ville de Paris. During the last twenty years of his life, he worked on inventing diverging geometrical forms which he developed as prototypes to be realized in large-scale projects. Among the geometrical shapes he developed, tetraflex was a flexible polyhedron which he registered at the Swedish Patent Office in 1971. At the time he had created and proposed these

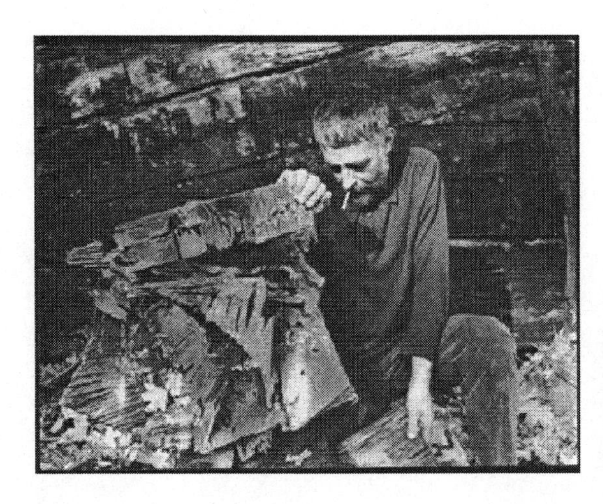 

Figure 2: Two photographs of Ilhan Koman (left photo by Christer Strmholm, 1970's; right photo by Tayfun Tunçelli, 1980's).

structures as modules for architecture, constructions in space and aviation fuel tanks [9, 10]

4 Vertex Angle Deflections and $\pi+\pi+\pi+\pi+\pi+$ Form

Ilhan Koman's developable sculptures in $\pi+\pi+\pi+\pi+\pi+$ form [8, 10] vividly illustrate vertex angle deflections which are less known but very useful mathematical devices to understand local behavior of surfaces [3]. Vertex deflection is a measure to show how much a given region is deflected from plane. In practice, local behavior can be changed by vertex deflections with nip and tuck. For instance, By simply take out some angle (nipping or pinching) we can create a convex or concave region with paper (See Figure 3). To create saddle regions we can add angles to a planar paper surface by tucking as shown in Figure 3.

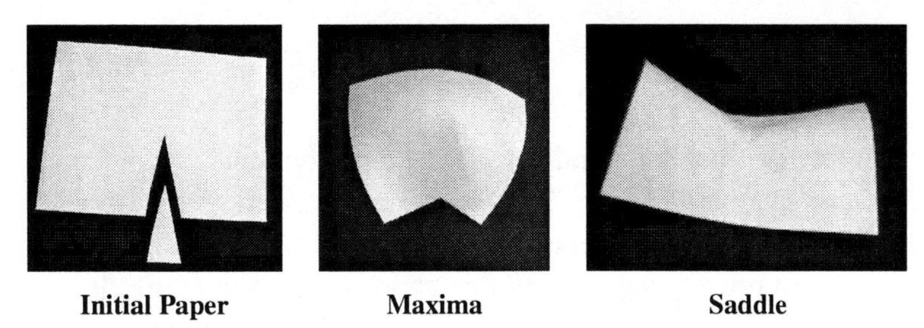

Initial Paper **Maxima** **Saddle**

Figure 3: Creation of maxima and saddle with subtracting and adding angles to a flat surface.

To construct $\pi+\pi+\pi+\pi+\pi+$ sculptures, Koman started from a circular sheet of plastic or metal and added angles (tucking) by constructing a series of wild saddle structures by forcing the angle deflection in a given vertex much smaller than -2π [8, 10]. As shown in Figure 4, when the angle decreases lower than -2π, it is possible to create a wide variety of saddle shapes. We observe that Koman's developable structures can give very useful visual intuition about the local behavior at a vertex for piecewise linear meshes.

Vertex angle deflections can formally defined around vertices of surfaces that are piecewise linear or piecewise developable. A mesh M is called *piecewise linear* if every edge of M is a straight line segment, and every face of M is planar. Piecewise linear meshes are useful since the surface of each face is well-defined [?]. Most common piecewise linear meshes are triangular meshes. However, piecewise linear meshes are a significant generalization of triangular meshes, which not only allow non-triangular faces, but also allow faces to have different face valence.

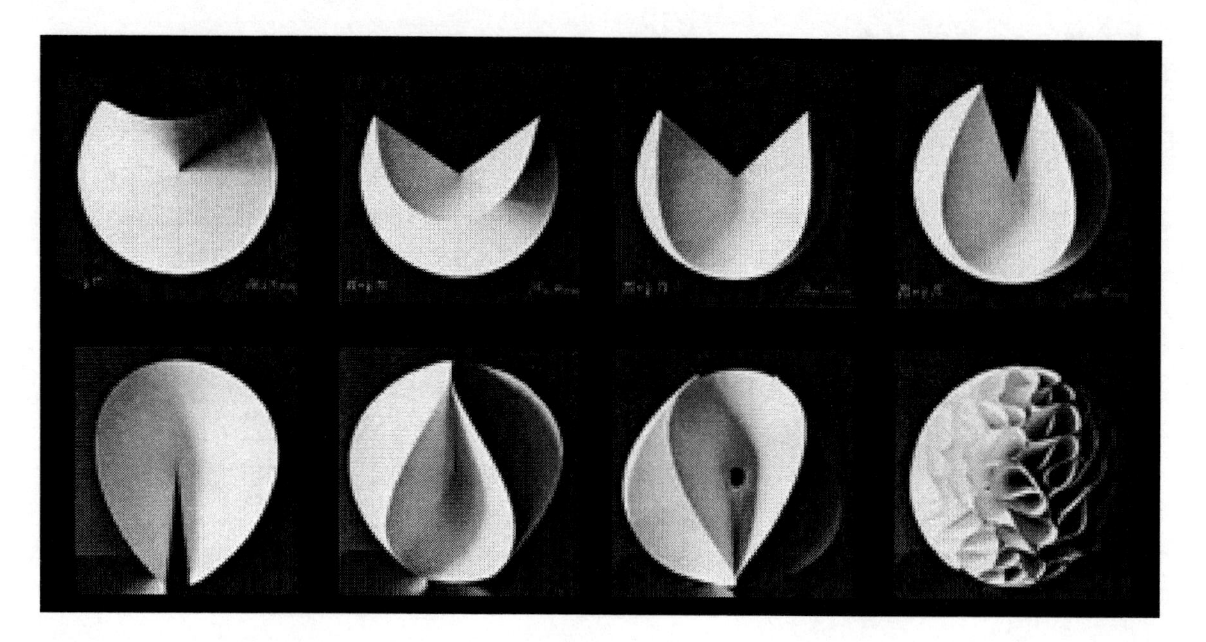

Figure 4: Ilhan Koman's Developable Sculptures. These were constructed with sheet metal in the 1980's. In these sculptures, the connections are almost invisible (photos by Tayfun Tunçelli).

Vertex angle deflections are defined based on the corner angles around a vertex (μ_i) as

$$\bar{\theta}(\mu_i) = 2\pi - \sum_{j=1}^{m_i} \theta_j$$

where θ_j is the internal angle of the j-th corner of vertex μ_i and m_i is the valence of the vertex μ_i. Our recent result show that sum of vertex angle deflections

$$\sum_{i=1}^{v} \bar{\theta}(\mu_i) = 2\pi(2-2g)$$

where g is the genus (number of holes and handles) and $2-2g$ is the Euler Characteristic of the surface. This result is in sync with Gauss-Bonnet theorem [21] which states that the integral of the Gaussian curvature over a closed smooth surface is equal to 2π times the Euler characteristic the surface.

Vertex angle deflection can provide a good intuition about surface curvature. It is interesting to point out that the value of $\bar{\theta}(\mu_i)$ is somewhat related to curvature and, in fact, Calladine used vertex angle deflections to estimate curvature in triangular meshes [3].

- $\bar{\theta}(\mu_i) > 0$: then vertex μ_i is either convex or concave.
- $\bar{\theta}(\mu_i) = 0$: then vertex μ_i is planar.
- $\bar{\theta}(\mu_i) < 0$: then vertex μ_i is a saddle point.

Ilhan Koman's sculptures are easy to make and they can be great tool to teach the effect of vertex angle deflections. Koman was planning to create an extremely large version of one of his wild $\pi+\pi+\pi+\pi+\pi+$ with a platform in the center such that people will go the center of the resulting spherical shape will only see space curve. Unfortunately, he could not complete this dream. It is interesting to note that, Sharp created a simple saddle shape from $\pi+\pi+\pi+\pi+\pi+$ series during his Bridges talk (A saddle with vertex deflection is around $-\pi$) [18]. It is highly possible that $\pi+\pi+\pi+\pi+\pi+$ form is passed from people to people and became an anonymous discovery. It is also possible that it exists several co-discoveries.

5 Rolling Lady

Koman discovered *Rolling Lady* form in early 1980's. This form consists of four cones that are pasted together to create two wheels. As the name suggest this sculpture can roll with a beautiful motion. These sculptures can be created with four identical cut cones as shown in Figure 5. During his Bridges talk, Sharp also mentioned a similar structure that can be constructed from two perpendicular circles [18]. It is also interesting in this case to find out whether or not there is a co-discovery. It is highly possible that Rolling Lady, because of its extreme simplicity and elegance, was quickly created by many people and became an anonymous discovery.

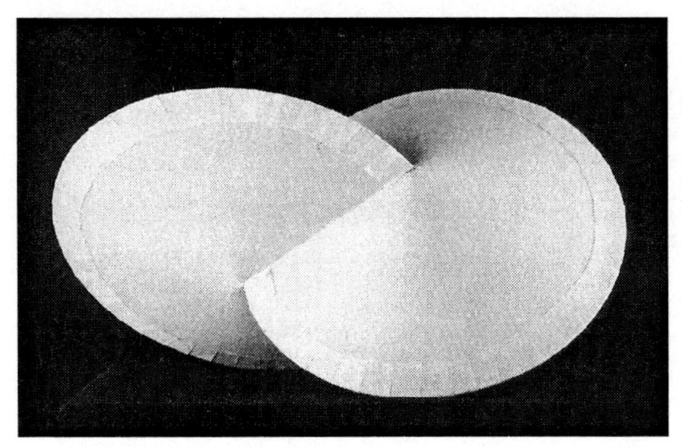

Cardboard (Stockholm, 1980's)

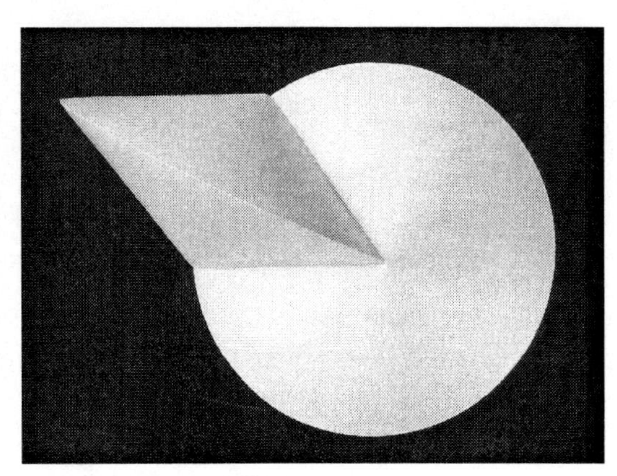

Metal Foil (1983)

Figure 5: Rolling Ladies (photos by Yildirim Arici)

6 Hyperforms

Hyperforms are one of the most elegant and difficult to understand forms Koman discovered. Simplest hyperforms are shown in Figure 6. These simplest hyperforms are created from rectangles formed by four equal squares, twisting the corners by 2π and joining them together. Figure 7 shows how to create these simplest hyperforms.

Figures 8 and 9 show more complicated hyperforms. Elegant beauty of these sculptures is even apparent in the photographs.

Unfortunately, we do not have much information how they were really created. Koman wrote that more complicated hyperforms can be constructed from rectangles that consist of more than four equal squares. He simply states that as the number of squares increases, the degree of twist at each level increases accordingly, $(2\pi, 4\pi, 6\pi \ldots)$, creating derivatives of the form. He also provided a number he called proportion for some of these sculptures. It is not clear how proportion is related to these final forms.

7 Conclusions and Future Work

Ilhan Koman's developable sculptures can easily be created in classroom and can be used to teach an appreciation of the complexity of 3D structures. It will also be interesting to develop mathematical models to create hyperforms using computer graphics.

Ilhan Koman's discoveries do not end with developable sculptures. He has also invented a variety of moveable rigid bodies in the same class as flexagons and kaleidocycles; self-carrying structures; forms

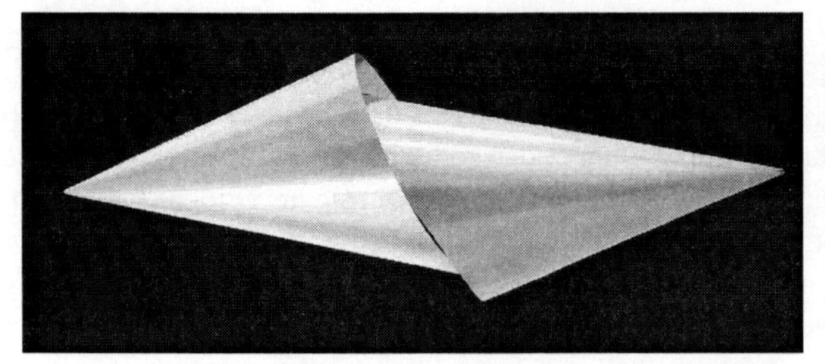

Hyperform (1978)
Stainless Steel
$60 \times 60 \times 145$ cm

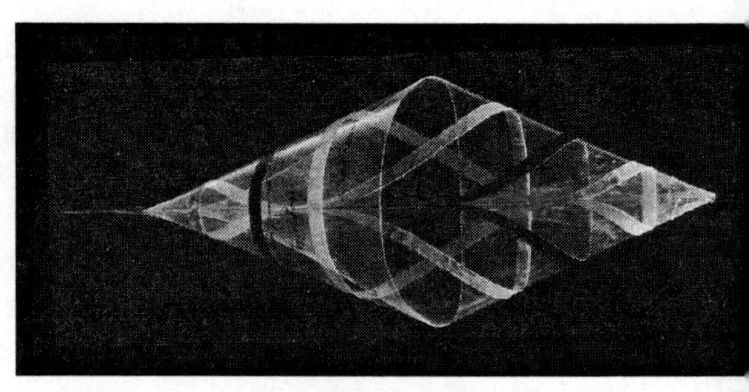

Hyperform (1983)
Transparent Plastic
$12 \times 12 \times 27$ cm

Figure 6: Simplest hyperforms that are created from $w = 4h$ rectangles where w is the width and h is the height of the rectangle (photos by Yildirim Arici). (These sculptures are hanged from ceiling. The images are rotated $\pi/2$ counter-clockwise to use less space in paper.)

created from self-similarity and walking sculptures. We hope that we will be able to continue to explore other mathematical sculptures of Ilhan Koman.

References

[1] Kenneth A. Brakke, Surface Evolver, http://www.susqu.edu/facstaff/b/brakke/evolver/evolver.html

[2] Paul Bourke, D-Forms, http://astronomy.swin.edu.au/ pbourke/surfaces/dform/

[3] C. R. Calladine, "Theory of Shell Structures", Cambridge University Press, Cambridge, 1983.

[4] Ron Evans, Plexons created by Paul Bourke, http://astronomy.swin.edu.au/ pbourke/geometry/plexagon/

[5] Erik D. Demaine and Joseph O'Rourke, "Open Problems from CCCG 2002," in Proceedings of the 15th Canadian Conference on Computational Geometry (CCCG 2003), pp. 178-181, Halifax, Nova Scotia, Canada, August 11-13, 2003.

[6] Richard Fishback and Oscar Tuazon, "An Introduction to the Randome", Bridges 2004, pp. 347-348, 2004.

[7] Paul Haeberli, http://laminadesign.com/index.html

[8] Ilhan Koman Foundation For Arts & Cultures, "Ilhan Koman - Retrospective", Yapi-Kredi Cultural Activities, Arts and Publishing, Istanbul, Turkey, 2005.

[9] Ilhan Koman and Franoise Ribeyrolles, "On My Approach to Making Nonfigurative Static and Kinetic Sculpture", Leonardo, Vol.12, No 1, pp. 1-4, Pergamon Press Ltd, New York, USA, 1979.

[10] Koman Foundation web-site; http://www.koman.org

[11] Jun Mitani and Hiromasa Suzuki, "Making Papercraft Toys From Meshes Using Strip-Based Approximate Unfolding. ACM Trans. Graph. 23(3). pp. 259-263, 2004.

[12] Frank Gehry, http://www.gehrytechnologies.com/

[13] Helmut Pottmann and Johannes Wallner, "Computational Line Geometry", Springer-Verlag, 4, p. 418, 2001.

[14] http://www.rhino3.de/design/modeling/developable/index.shtml

[15] http://www.rhino3.de/design/modeling/developable/marine_design.shtml

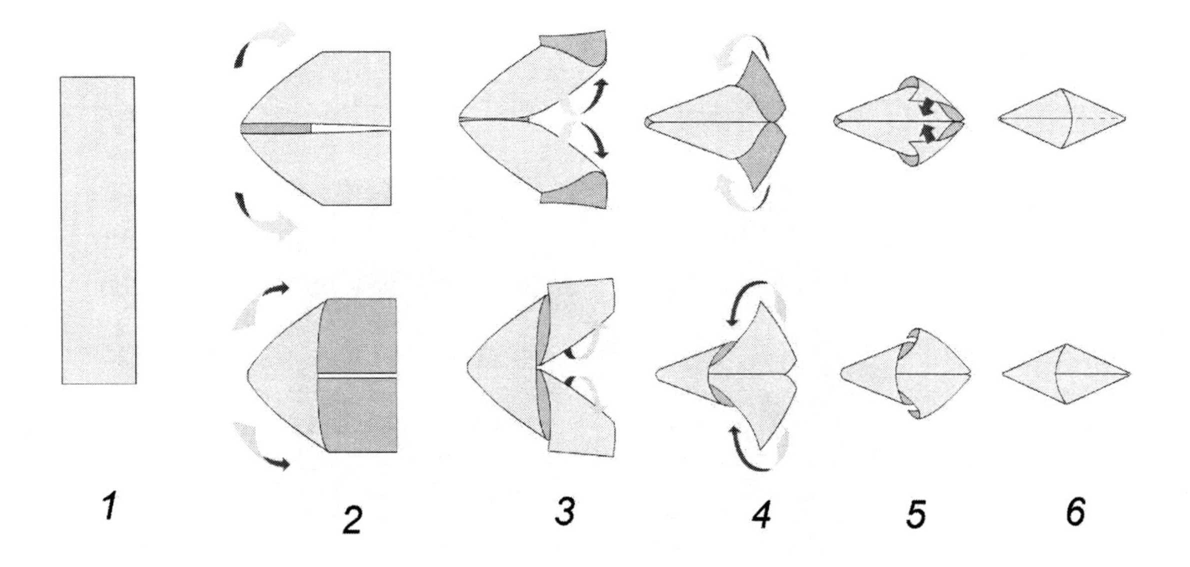

Figure 7: Construction of simplest hyperform from a rectangle formed by four equal squares. To illustrate the process we show both sides in each stage. The final form is obtained by joining edges in stage 5. (Illustration by Özlem Bakir)

[16] John Sharp, D-Forms and Developable Surfaces, Bridges 2005, pp. 121-128, 2005.

[17] John Sharp, D-Forms: Surprising new 3D forms from flat curved shapes,Tarquin 2005.

[18] John Sharp, Presentation of "D-Forms and Developable Surfaces", in Bridges 2005, Banff, Canada, 2005.

[19] Meng Sun and Eugene Fiume, A technique for constructing developable surfaces, Proceedings of the conference on Graphics interface '96, Toronto, Ontario, Canada, pp. 176 - 185, 1996.

[20] Eric W. Weisstein. "Developable Surface." From MathWorld–A Wolfram Web Resource. http://mathworld.wolfram.com/DevelopableSurface.html.

[21] "Eric W. Weisstein", "Gauss-Bonnet Formula", 2005, From MathWorld–A Wolfram Web Resource. http://mathworld.wolfram.com/Gauss-BonnetFormula.html

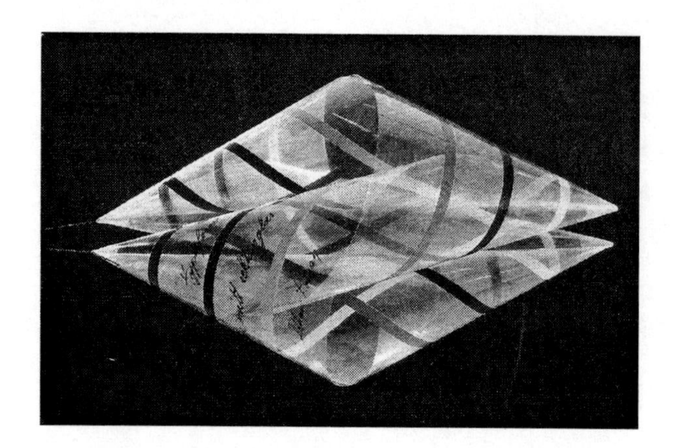

Figure 8: Twin Hyperform that is constructed from transparent plastics (1970-75). The actual dimensions are $24 \times 24 \times 50$ cm. (This sculpture is hanged from ceiling. The image is rotated $\pi/2$ counter-clockwise to use less space.)

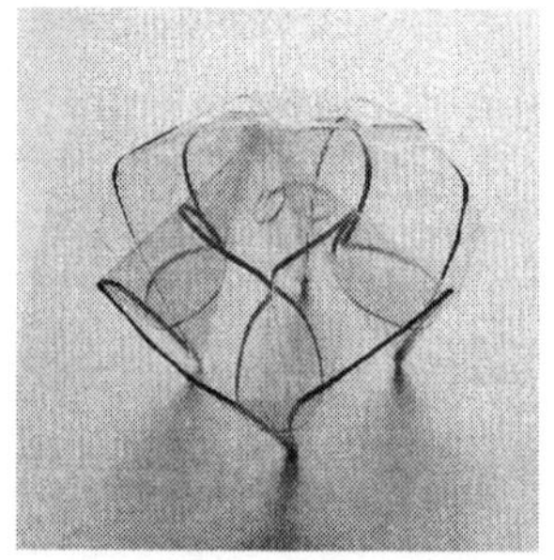	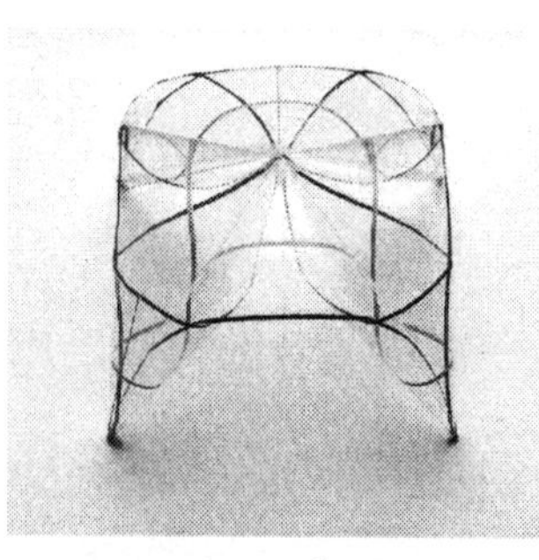	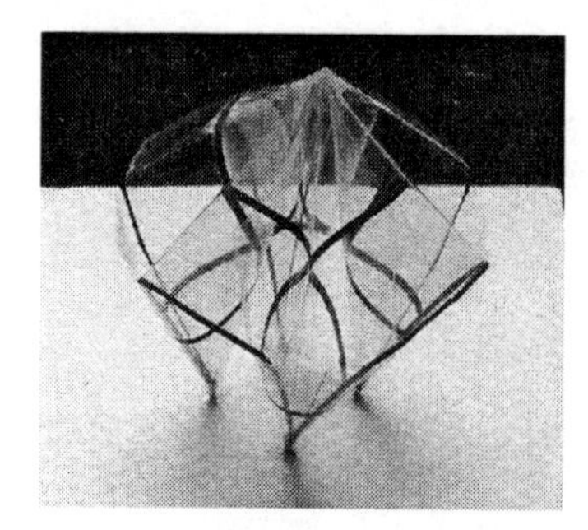	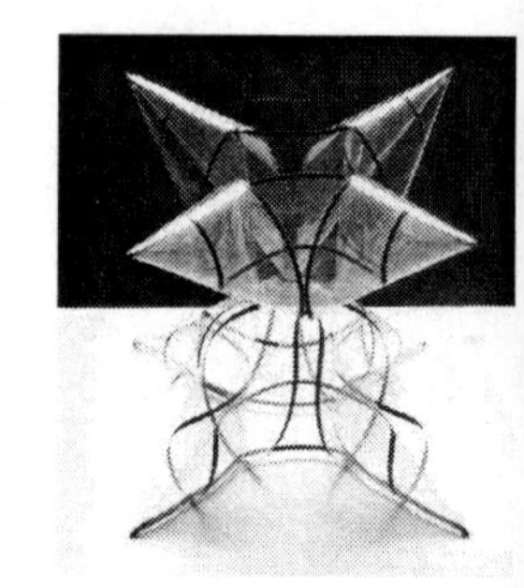
Proportion $\sqrt{2}h$ (1983) $60 \times 60 \times 15$ cm	Proportion $3/8$ (1983) $50 \times 50 \times 13$ cm	Proportion $3/2h$ (1983) $60 \times 60 \times 15$ cm	Quadruple Hyperform(19 $55 \times 55 \times 50$ cm

Figure 9: Complicated Hyperforms constructed with Transparent Plastic on stand that are constructed (photos by Yildirim Arici).

Constellations of Form: New Compositional Elements Related to Polyominoes

James Mai
School of Art
Campus Box 5620
Illinois State University
Normal, IL 61790-5620, USA
E-mail: jlmai@ilstu.edu

Daylene Zielinski
Mathematics Department
Bellarmine University
2001 Newburg Road
Louisville, KY 40205, USA
E-mail: dzielinski@bellarmine.edu

Abstract:

A predominant theme of artist James Mai's compositions is the development of finite sets of related objects derived from permutational processes. Each element is distinct, yet all of them share particular features. Thus, he develops families of objects that are at once diverse since each object is visually distinct and integral since the set of objects is exhaustive. These objects provide the elements for combination and composition in paintings and digital prints. Recent permutational investigations by Mai have yielded objects we call *point arrays* and *strutforms,* which are related to polyominoes via dual graphs. These new objects, however, have greater variety than polyominoes and offer some new opportunities for a different interpretation of tilings. The results of these investigations are visible in the digital print, *Heart of Sky*, which includes the complete sets of 3- and 4-strutforms in a "close-packed" or minimal area arrangement. Mai is currently working on compositions with the set of 5-strutforms.

1. The Evolution of Strutforms

A consistent theme in the work of artist James Mai is the drive to produce complete sets of visually distinct forms that share a particular visual and/or mathematical quality defined by the artist. The objects we are about to introduce were developed by just such an organic artistic and mathematical investigation. Mai's initial investigation into these objects began as permutations of arrays of points in a square grid prompted by the following question: How many visually distinct ways can four points be arrayed at adjacent intersections of a square grid? Requiring that each form be visually distinct omits duplication of forms that are merely rotated or reflected versions of others in the same family. Thus, each of the remaining forms is unique. With four points, the family of unique forms is small; there are only five such arrays, which are shown in Figure 1a.

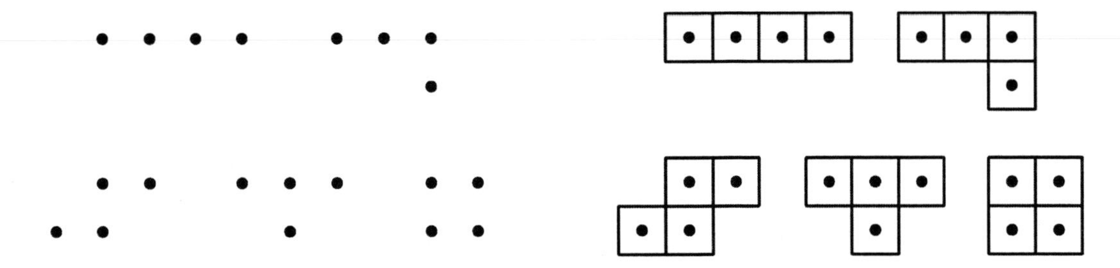

Figure 1a: *4-point arrays.* **Figure 1b:** *The five tetrominoes.*

Mathematically, a slight change in perspective reveals a deep connection to polyominoes, which are collections of n unit squares joined along edges to form a polygon [1]. If we place the points at the centers of adjacent squares in the grid rather than at intersections of lines in the grid, we can now see the tetromino underlying each 4-point array as shown in Figure 1b. In fact, this figure makes it obvious that Mai's 4-point arrays and the set of tetrominoes can be put into one-to-one correspondence. So, it should come as no surprise that there are exactly five 4-point arrays, since there are exactly five tetrominoes. For this same reason, there are exactly twelve 5-point arrays, which are displayed in Figure 2a. Positioning points in the center of each square in a polyomino is the first step to forming its dual graph, so Mai's investigation of point arrays coincided with enumerating the vertex structures of the dual graphs of polyominoes.

As Mai continued his exploration of these point arrays, he added edges between pairs of points that were connected along grid lines. This set of forms is displayed in Figure 2b. Although he originally did this as a visual aid so that he could quickly determine which point arrays were visually distinct, Mai quickly discovered that shifting his focus from points to the connecting edges produced a richer set of forms because the last form in Figure 2b is different from the others. It alone has five edges, although each of the other forms has only four edges. As Mai sought to bring this unique 5-edge form into accord with the rest of the 4-edge forms his investigation had yielded, he revised his original motivating question in the following manner: How many ways can four edges along grid lines be arranged so that they connect adjacent intersections points? A complete set of these forms is shown in Figure 3 below. We will refer to these objects as 4-strutforms, and it is at this juncture in Mai's investigation that his forms evolve beyond their polyomino ancestors.

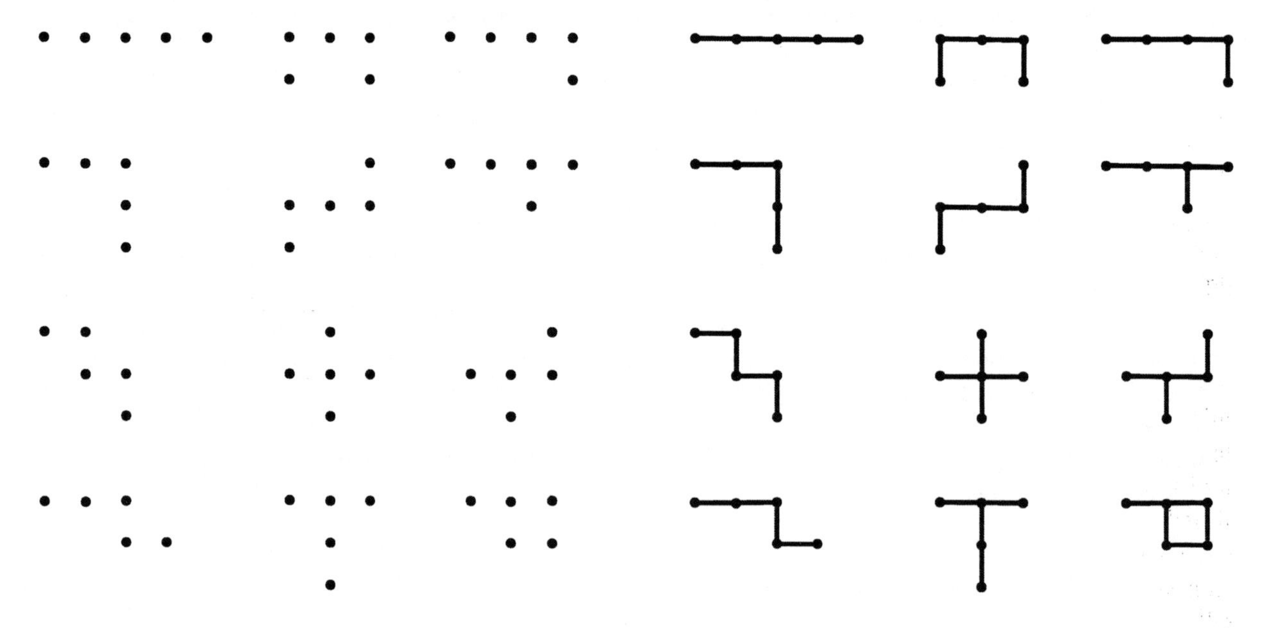

Figure 2a: *The twelve 5-point arrays.* **Figure 2b:** *The dual graphs of the pentominoes.*

At the point when Mai first introduced the edges to his point arrays, he had created the dual graphs of the polyominoes, since adding grid line edges to Mai's point arrays is tantamount to adding edges between points centered in adjacent squares in the underlying polyomino. However, Mai's decision to limit the number of edges to four forces the unique form with five edges in Figure 2b to fracture into four visually distinct 4-strutforms, which are shown in the first row of Figure 3. Furthermore, one of the tetrominoes has a dual graph with four edges, and thus this graph is also a 4-strutform and appears as the last form in Figure 3. Hence while there are only twelve pentominoes, there are sixteen 4-strutforms, only eleven of which are dual graphs to a pentomino.

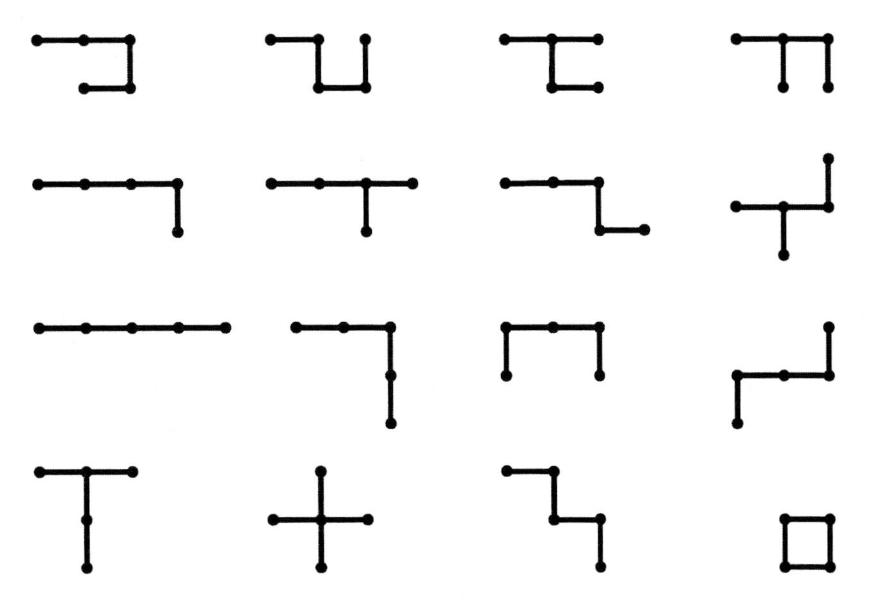

Figure 3: *Mai's sixteen 4-strutforms.*

2. Strutforms as Mathematical Objects

The preceding discussion of the development of n-strutforms makes it clear that they are closely related to the dual graphs of (n +1)-ominoes, but this relationship must be placed on firm mathematical ground. To preserve as much of the artist's original vision as possible, we will begin with a definition that is not as mathematically economical as it could be. We shall define an n-strutform as a connected graph with exactly n edges, whose vertices are centered in squares that form a grid; furthermore, edges can only occur between pairs of vertices that are centered in adjacent squares. Thus, our formal mathematical definition shall be as follows: an n-strutform is any connected subgraph with exactly n edges of the graph created by an n x n square grid.

The process for creating the set of n-strutforms is relatively straightforward. One begins with the dual graphs of the (n +1)-ominoes, which are formed by replacing each unit square with its center-point and adding edges between each pair of points that are centered in adjacent squares of the original polyomino. Next, we need to check each of these dual graphs for cycles, which are paths of edges in a graph that begin and end at the same vertex. If the dual graph of an (n +1)-omino has no cycles, then it will have exactly n edges and automatically be an n-strutform. If, however, the dual graph does have cycles, then it will have more than n edges, and at least one edge will need to be removed to create a n-strutform. Since strutforms must be connected, it is not valid to remove an edge or edges that cause the dual graph to become disconnected. The number of edges that must be removed depends on the cycle index of the graph, which is the minimal number of edges that must be removed from the graph so that it no longer has any cycles. For example, if we remove any one edge of the graph shown in Figure 4a, there will be at least one cycle still remaining; however, if we remove two of the vertical edges, the remaining subgraph will have no cycles, yet still be connected. Thus, the cycle index of the graph in Figure 4a is two. Luckily, the cycle index of a graph is simple to compute; the cycle index of any graph is the number of edges minus the number of vertices plus one. Note that the graph in Figure 4a has seven edges and six vertices, so its cycle index is two, not three, even though there are three distinct cycles in this graph. So, one must remove two edges from this graph to create 5-strutforms or one edge to create 6-

strutforms. This one graph, which is a 7-strutform on its own, generates six distinct 5-strutforms and three distinct 6-strutforms.

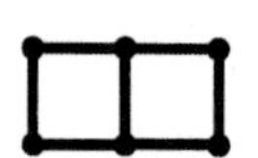

Figure 4a:
A graph with cycle index of two.

Figure 4b:
Graph with a cycle.

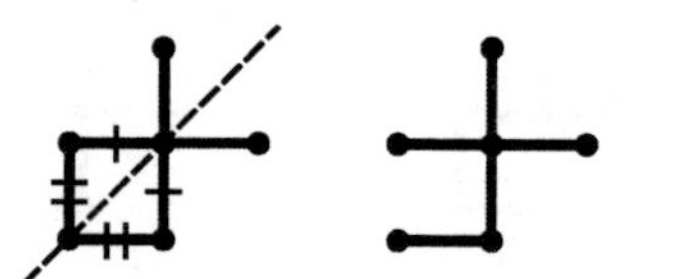

Figure 4c:
Dual graph and two 5-strutforms.

The mere presence of a cycle in an $(n+1)$-omino's dual graph does not insure that it will fracture into multiple n-strutforms because the symmetry of the graph also plays a role. For instance, the graph in Figure 4b, which is the dual graph of the square tetromino, has a cycle, but removing any one edge produces the same 3-strutform because the reflective symmetries alone of the original graph, which are shown as dashed lines, make each of its edges equivalent. An asymmetric dual graph of any $(n+1)$-omino with exactly one cycle, will fracture into as many n-strutforms as there are edges in the cycle. This is why the last graph in Figure 2b generates four distinct 4-strutforms. The graph in Figure 4c, with its one symmetry that makes two pairs of the edges in its single cycle equivalent, will produce only two distinct 5-strutforms. In fact, the question of how many distinct n-strutforms will be generated by any particular polyomino's dual graph is a question best answered by Burnside's Theorem. The process of applying this result to questions of this type is demonstrated in [2].

When removing edges from a graph, it is important to remember that the vertices remain. Because n-strutforms must be connected, we cannot remove all edges incident to any one vertex, thereby creating an isolated vertex. Hence, the 4-strutform in Figure 4b above cannot be generated from the last graph in Figure 2b; removing the single edge that is not part of the cycle would create an isolated vertex and, thus, a disconnected subgraph. So, we cannot rely on the dual graphs of the $(n+1)$-ominoes alone to create the full set of n-strutforms. In fact, whenever the dual graph of any k-omino where $k-1<n$ has cycle index equal to $n-(k-1)$, then that dual graph will be an n-strutform. This is exactly why the graph in Figure 4a, which is dual to a hexomino ($k=6$) and has a cycle index of two, is a 7-strutform ($n=7$).

3. Visual Encoding and Composing

Once any system of forms is complete, Mai develops a compositional organization that seeks to be as clear, complete, and efficient as possible in reference to the system. In the development of *Heart of Sky*, Figure 5, clarity was sought by encoding in color, scale, and distribution the many shared and distinct characteristics of the strutforms, including not only the different organizations of points and struts, but also the acknowledgment of symmetrical and asymmetrical forms. For Mai, compositional completeness demands the inclusion of every distinct form in the system, without the repetition of any form by symmetric translation, rotation, or reflection. Efficiency was sought by "close-packing" the set of strutforms in the minimum area.

When considering the close-packing possibilities, Mai examined the characteristics of both the individual strutforms and their families for appropriate possibilities. The sixteen unique 4-strutforms required 79 points in the lattice since fifteen of these strutforms require five points each and one requires only four points. This offered the possibility of a composition in a 9 x 9 square point lattice leaving two points unused. On further consideration, Mai opted to combine the 4-strutforms with the 3-strutforms bringing the total number of points needed for the strutforms to 99, which leaves one unused point in a 10 x 10 square point lattice. Mai visually encoded two additional types of information about the strutforms in the final work. First, he acknowledge the difference between the 3-strutform and the 4-strutform families by orientation within the grid. Hence, each group is set at a different angle within the point lattice and each group is a different pair of complementary colors—red and green for the 3-strutforms and blue and orange 4-strutforms. Secondly, Mai wanted to distinguish between asymmetrical strutforms and those having either rotational or reflective symmetries. This also was accomplished through color by reserving the two primary colors, red and blue, for those strutforms with at least one symmetry.

Figure 5: Heart of Sky, *digital print, 40 x 40 inches.*

4. Strutforms, Tilings, and an Open Question

Since strutforms are a richer set of objects than the related polyominoes, it makes sense to investigate what advantage strutforms may have over the traditional polyominoes. In particular, the topic of primary interest where polyominoes are concerned is tilings. Since strutforms do not have any area, traditional tilings do not make sense; however, we can take a cue from Mai's artwork, *Heart of Sky,* and investigate a different interpretation of tilings. What we see in *Heart of Sky* is the 3- and 4-strutforms arranged on a point lattice. Indeed, if we recall the origin of the strutforms, it seems quite appropriate to position them on the vertex structure of the dual to the square grid. Therefore, we will call any arrangement of strutforms that covers each point in a lattice with the vertices of the strutforms without any overlapping a *point lattice covering.*

It is a well-known result that if any 2 x 2 block is removed from the standard checkerboard, the remainder of the board can be tiled with the twelve pentominoes; for a proof of this see [1]. To translate this into the setting of point lattice covings using strutforms, we will replace the unit squares of the checkerboard with a point at the center of each square and remove the grid lines completely so that we are left with an 8 x 8 point lattice. In this manner, any tiling with pentominoes has a dual point lattice covering with 4-strutforms. Since the set of 4-strutforms includes the dual graph of the 2 x 2 square tetromino, shown in Figure 4b, this special strutform can be used to cover the 2 x 2 section of the dual point lattice that corresponds to the 2 x 2 hole left by the pentomino tiling of the standard checkerboard. Thus, we arrive at the following result.

Regardless of where the 4-strutform that is dual to tetromino is placed in an 8 x 8 point lattice, the rest of the lattice can be covered with other 4-strutforms.

Further inspection of Mai's *Heart of Sky* suggests yet another possibility for using strutforms to cover point lattices. Notice that Mai has rotated his point lattice through a 45-degree angle. Beyond this, he has visually distinguished between the 3- and 4-strutforms in *Heart of Sky* by using a different orientation for the two sets. This is tantamount to the artist allowing the unit measurement of the composition to vary. The unit length for the family of 4-strutforms is the distance between the adjacent grid squares arranged at a 45-degree angle; however, the 3-strut forms are oriented differently so that the length of one strut is now equal to the diagonal of one grid square rather than just the length of one side of a grid square. This longer unit strut length allows the 3-strutforms to interact in a distinctly different manner: the 3-strutforms in *Heart of Sky* can interlock with each other as seen in center just below the isolated point. Therefore, strutforms can interact in a point lattice covering in a manner that their dual polyominoes cannot in any grid tiling.

We can take advantage of this interlocking behavior without changing the unit strut length by using a double point lattice, created from the standard square grid and its dual by using both the points at grid line intersections and the points at the center of each grid square. In the standard checkerboard, this would create a double grid of 145 points—64 points from the dual grid and 81 points from grid line intersections including those on the outer edges of the board. This suggests an open question.

Can the double point lattice of the standard checkerboard be covered with the 4-strutforms having unit strut length equal to the side of one square in the checkerboard?

There are clearly some restrictions as to how the 4-strutforms can be used in such an endeavor. Most obvious is that the 4-strutform that played such a vital role in our first result above is now anathema. If the 4-strutform that comes from the dual graph of a tetromino is placed anywhere in this doubled grid, it will isolate a point, as shown in Figure 6, making a covering impossible. There are similar kinds of restrictions, such as how far many of the other 4-strutforms must be from the edge of the doubled point lattice to avoid creating other isolated points, but the answer to the question posed above has eluded the authors thus far.

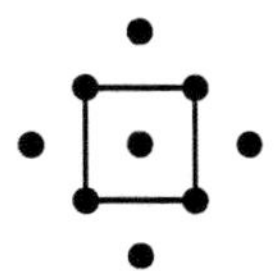

Figure 6: *Tetromino-based 4-strutform isolating a point in a double point lattice.*

5. Figurative Allusions

Mai's shift in focus from point arrays to strutforms did more than push his objects from a mathematically known set into the unknown, it also presented the artist with an overarching metaphor for the final composition. From the beginning, recognition of the distinct point arrays required the visual grouping of disparate parts into understandable wholes, a perceptual process that Mai found akin to recognizing constellations in the night sky.[1] Adding struts to his original point arrays, and thereby creating connected graphs, strengthened this identification with constellations and, from an early stage, helped guide the subject matter and metaphoric references that would eventually emerge in the finished composition.

With this astronomical reference in mind, the sole, unused point in the 10 x 10 lattice seemed equivalent to Polaris, the star around which the night sky revolves. This helped to guide the compositional decisions as Mai worked intensively to find an arrangement that would not only close-pack the strutforms but also locate that single, isolated point near the top and on the vertical axis of the lattice, as suggestive of the North Star. Only one solution was found that fulfills all of the above conditions, realized in the final large-scale digital print. The title, *Heart of Sky*, derives from the name given by ancient Maya astronomers to the north celestial axis about which the night sky appears to turn [3].

6. Further Investigations

This line of investigation continues with the development of the 5-strutform family. Mai's 35 unique 6-point arrays are coincident with the vertex structure of the dual graphs of the 35 hexominoes. Among these, there are 27 hexominoes whose dual graphs have no cycles; these translate directly to 5-strutforms, which are seen in Figure 7a. Eight of the hexominoes, however, have dual graphs that contain cycles, each of which creates several distinct 5-strutforms. These strutforms are shown in Figure 7b grouped in families that are all subgraphs of the same dual hexomino. Also shown at the bottom right of Figure 7b is the one 5-strutform that is dual to a pentomino. Since 54 of these require six vertices and the remaining one uses five vertices, any composition of these 5-strutforms will require 329 vertices in a point lattice. The system is complete; final compositional solutions are still in progress.

Mathematically, the task of enumerating n-strutforms for a given n is non-trivial; note that the process of computing the number of n-strutforms for a particular n is at least as difficult as the related, and still unsolved, problem of enumerating n-ominoes. Further note, that enumeration results for spanning trees will be helpful for small values of n, but for larger n, many of the n-strutforms will not be trees. There has been much work done on enumeration of lattice animals, of which polyominoes are a subset, because this problem relates to questions about percolation in physics. Readers interested in this line of investigation would do well to start with the works of A. R. Conway, M. Delest, and E. Jensen.

---

[1] This perceptual structuring, vital to artists, is elucidated by the Gestalt principles of grouping [4].

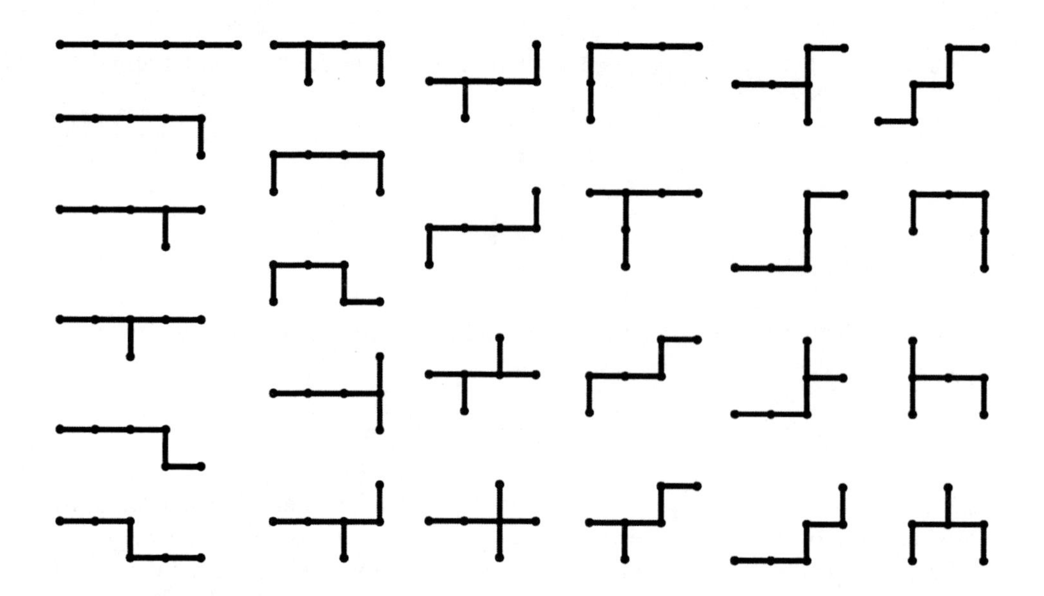

Figure 7a: *Mai's 5-strutforms that correspond directly to dual graphs of hexominoes.*

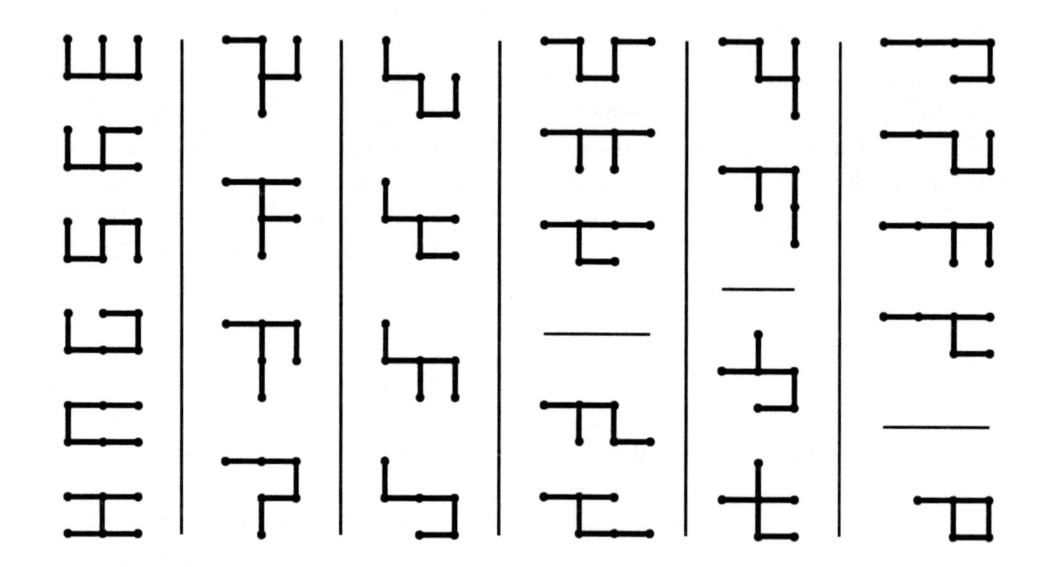

Figure 7b: *Mai's 5-strutforms that are subgraphs of dual graphs of hexominoes and the single 5-strutform that is dual to a pentomino.*

References

[1] S. W. Golomb. *Polyominoes: Puzzles, Patterns, Problems, and Packings.* Revised Edition. Princeton University Press, 1994.

[2] J. Mai and D. Zielinski. Permuting Heaven and Earth: Painted Expressions of Burnside's Theorem. In *Conference Proceedings of Bridges: Mathematical Connections in Art, Music and Science, 2004.* Eds. Sarhangi, Reza and Séquin, Carlo.

[3] Friedel, D., L. Schele, and J. Parker. *Maya Cosmos: Three Thousand Years On the Shaman's Path.* New York: William Morrow & Co., 1993.

[4] Palmer, S. E. *Vision Science: Photons to Phenomenology.* Cambridge: The MIT Press, 1999.

Transformations of Vertices, Edges and Faces to Derive Polyhedra

Robert McDermott
Center for High Performance Computing
University of Utah
Salt Lake City, Utah, 84112, USA
Email: mcdermott@chpc.utah.edu

Abstract

Three geometric transformations produced a large number of polyhedra, each originating from an initial polyhedron. In the first transformation, vertices were slid along edges and across faces producing ***nested*** polyhedra. A second transformation produced ***dual*** polyhedra, whereby edges of the initial polyhedron were rotated and scaled and the end points of these edges derived the dual polyhedra. In a third transformation, faces of an initial polyhedron were rotated and scaled producing ***snub*** polyhedra. The vertices of these rotated and scaled faces were used to derive other polyhedra.

This geometric approach which derives new vertices from previous vertices, edges and faces, produced precise results. A CD-ROM accompanying this paper contains three animations and data for all the derived polyhedra. This CD-ROM can be obtained by sending me email.

1. Introduction

While I was working at NYIT Computer Graphics Laboratory from 1980 to 1990, Haresh Lalvani approached me to produce precise data for an icosahedron and a dodecahedron. He was interested in constructing a quasi-crystal structure that contained more than 600 instances of these two polyhedra. Being an architect, he imagined exact representations. These polyhedra needed to be precise in their spatial coordinates when so many polyhedra were assembled. Any error would have a tendancy to accumulate significantly and prevent the structure from registering all the polyhedra into their respective locations.

I searched for such spatially accurate polyhedra in [1], and [2]. Unable to find precise vertex coordinates, I undertook to develop such polyhedra and subsequently developed other polyhedra: Platonic, Archimedean, Prisms, Anti-Prisms and their Duals. I began by positioning the vertices of a tetrahedron at the corners of a unit cube. These vertices were represented by zeros and ones, making them precise.

2. Vertices are Translated Along Edges & Across Faces to Derive *Nested* Polyhedra

From the vertices of the tetrahedron in **Figure 1a**, a ***nested*** set of polyhedra were derived. I could derive a truncated tetrahedron in **Figure 1b** by translating vertices 1/3 of the length along the edges of the initial tetrahedron. This translation is defined computationally by taking the two end points of the edge, adding the coordinates together, and dividing by the ratio of the proportion along the edge. This very simple computation produced very little error.

By continuing to translate vertices to the halfway point along the initial edges, I derived an octahedron, **Figure 1c**. Now I turned the direction of translation from along the edges to across the face to derive other polyhedra. When I translated the vertices 1/5 the distance from the mid-edge points to the opposite vertices, another truncated tetrahedron was produced **Figure 1d**. When the translated vertices reached the mid-faces or 1/3 the distance from the mid-edge points to the opposite vertices, a dual tetrahedron was derived, **Figure 1e**. In **Figure 2** these five polyhedra are displayed individually and then ***nested*** together.

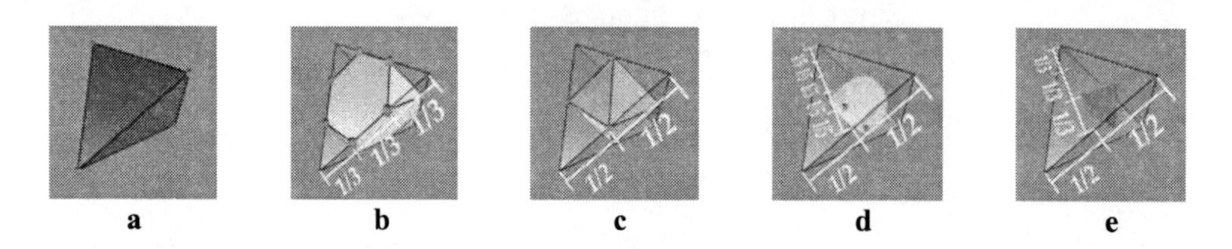

Figure 1: Tetrahedron (**a** & **e**), truncated tetrahedron (**b** & **d**), octahedron (**c**).

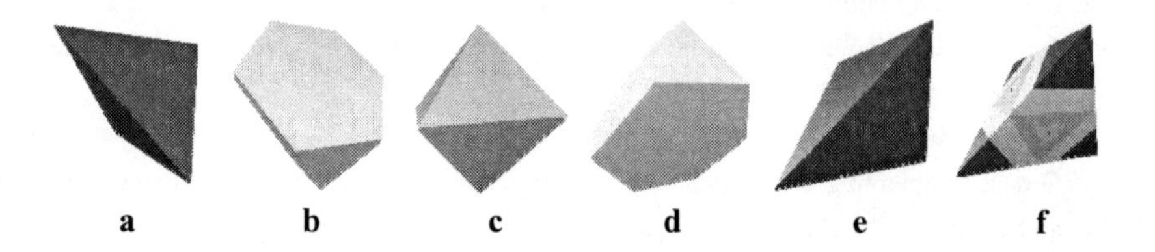

Figure 2: Tetrahedron (**a** & **e**), truncated tetrahedron (**b** & **d**), octahedron (**c**), & ***nested*** (**f**).

I then applied this ***nesting*** concept to an octahedron. I translated vertices to the 1/3 point of edges to derive the vertices of a truncated octahedron. When vertices were at the 1/2 point of the edge a cuboctahedron was derived. Again, I translated vertices across the faces. When vertices are at 1/5 that distance a truncated cube was derived. When the vertices were at the mid-faces a cube was derived. These five polyhedra can be viewed individually or ***nested*** in **Figure 3**.

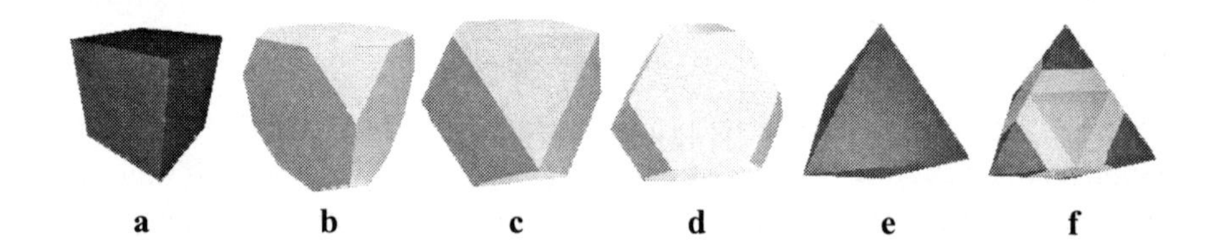

Figure 3: Cube (**a**), truncated cube (**b**), cuboctahedron (**c**), truncated octahedron (**d**), octahedron (**e**) & ***nested*** (**f**).

In addition, the vertices of an icosahedron were translated along its edges and faces to derive a truncated icosahedron, an icosidodecahedron, a truncated dodecahedron and a dodecahedron. When I initially performed this work, I utilized Coxeter's [1] formulation for the coordinate values of vertices to generate an icosahedron. Five polyhedra and a ***nesting*** of these polyhedra derived from an icosahedron are illustrated in **Figure 4**.

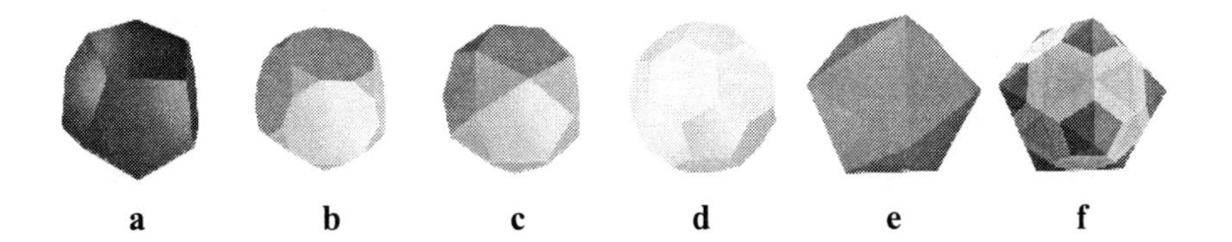

Figure 4: Dodecahedron (**a**), truncated dodecahedron (**b**), icosidodecahedron (**c**), truncated icosahedron (**d**), icosahedron (**e**) & ***nested*** (**f**).

3. Platonic & Some Archimedean Polyhedra

The Platonic and Archimedean polyhedra produced above were the only polyhedra able to be derived from such a simple translation of vertices along edges to their mid-point and across faces to their mid-point.

4. Translating Vertices Along Edges and Across Faces to Square Faces

A slightly new concept was used to translate vertices along edges, and then translate vertices across faces to form square faces. This concept was applied to a cuboctahedron and an icosidodecahedron to derive truncated and rhombic polyhedra. The vertices were translated 1/3 of the distance along edges to form a polyhedron that was topologically equivalent to a truncated cuboctahedron. However, the truncated faces formed were rectangles, not squares. **Figure 5(a-d)** are such derived polyhedra with rectangular faces. These rectangles were the result of each vertex having two triangles and two squares. The new edges derived from the triangles differ in length from the new edges derived from the squares; hence the rectangles.

To transform these rectangles into squares, I used the mid-point of each of these rectangles and translated vertices of rectangles toward these mid-points to form squares. **Figure 5(e-h)** are such polyhedra with square faces. Consequently, a truncated cuboctahedron **Figure 5e** and a rhombicuboctahedron **Figure 5f** were derived from a cuboctahedron, and a truncated icosidodecahedron **Figure 5g** and a rhombicosidodecahedron **Figure 5h** from an icosidodecahedron.

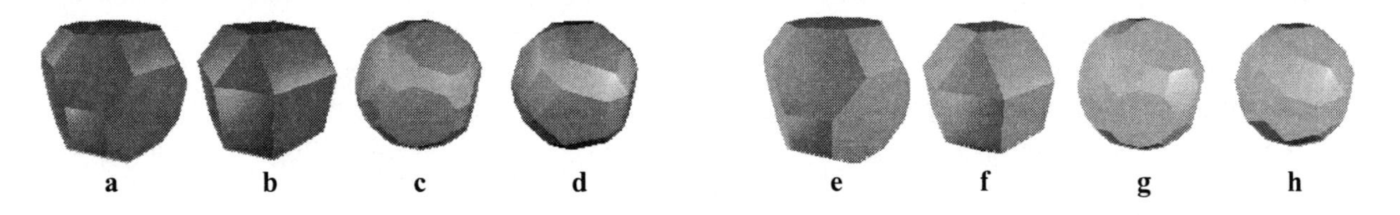

Figure 5: Rectangle faces (**a, b, c, d**), square faces (**e, f, g, h**).

5. Rotating and Scaling of Edges to Define *Dual* Polyhedra

I next investigated the transforming of edges to see what would result. The axes of rotation were defined from the center of a polyhedron orthogonal to points on edges. I rotated each edge about each of these axes. The angle of rotation was 90 degrees, so that the rotated edges were orthogonal to the initial edges. For example, I rotated the six edges of a tetrahedron in **Figure 6a**, by 90 degrees to form another tetrahedron in **Figure 6c**. This second tetrahedron was the dual of the initial tetrahedron. For the tetrahedron there is no scaling of the rotated edges. The end points of the rotated edges intersect exactly at the vertices of the dual tetrahedron.

A dual relationship for two polyhedra is an interchange between faces and vertices. That is, faces of an initial polyhedron become vertices in the dual polyhedron, and vertices in the initial polyhedron become faces of the dual polyhedron. When an initial polyhedron and dual are displayed simultaneously it is referred to as a polyhedron ***compound***.

Figure 6 (a-c) are rotating edges without scaling for tetrahedron. **Figure 6 (d-f)** are edges both rotated and scaled for a truncated tetrahedron and a triakis tetrahedron. **Figure 7 (a-f)** are two sets of an initial polyhedra **(a&d)**, polyhedra compounds **(b&e)** and polyhedra duals **(c&f)**.

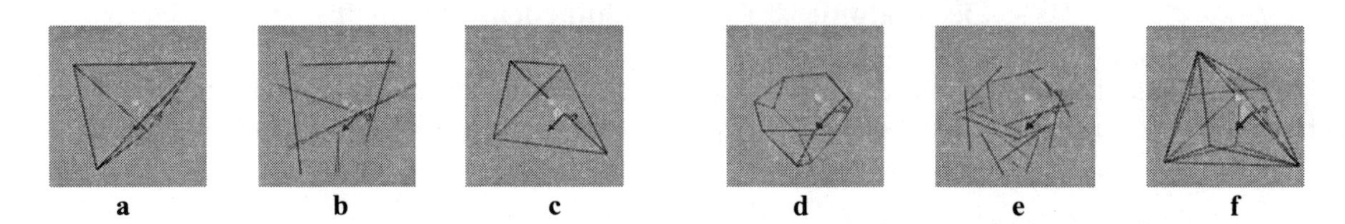

Figure 6: Edges of a tetrahedron (**a-c**), a truncated tetrahedron, & a triakis tetrahedron (**d-f**).

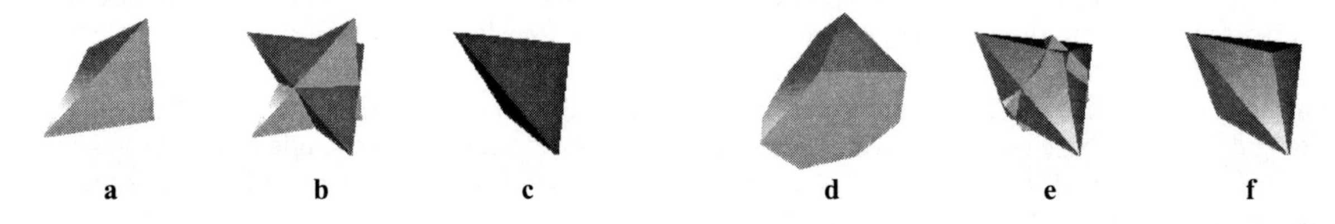

Figure 7: Tetrahedron (**a** & **c**), composite (**b**), truncated tetrahedron (**d**), composite (**e**) & triakis tetrahedron (**f**).

6. Dual Polyhedra for Platonic, Archimedean, Prisms and Anti-Prisms

The tetrahedron is a self dual. Therefore, scaling was not required to derive its dual when using the concept of rotating edges for ***duals***. The rotating and scaling edges for ***duals*** is a generalized concept for deriving dual polyhedra. The scaling of the edges is relative to the point where the axis of rotation for the edge and that edge intersect. In most cases there are two scale factors, one for each half edge. A precise scaling value was derived by computing the distance between the end points of edges that were rotated by 90 degrees about their axis of rotation. Each half edge with a like configuration of adjoining polygons had the same scale factor. Once the scale factors were applied for all half edges, the rotated edges intersected precisely in a single vertex of the dual polyhedron.

This concept of simultaneous rotation by 90 degrees and scaling of half edges was applied to Archimedean, Prism and Anti-Prism to define all of their dual polyhedra.

7. Rotating and Scaling of Faces to Derive *Snub* Polyhedra

After translating vertices and rotating and scaling of edges, simultaneously rotating and scaling faces was investigated. The axis of rotation for transforming faces was a line from the center of the polyhedron to a point on a face where the line was orthogonal to the face. Scaling of the faces was performed with respect to the point where the axis of rotation intersected the face. There was only one scale factor used for all faces that were transformed.

The four faces of a tetrahedron, as in **Figure 8a**, were rotated by 60 degrees and scaled by ½ to derive the six vertices of an octahedron, as in **Figure 8b**. This was a second method of deriving an octahedron from a tetrahedron. Note that the tetrahedron has vertices with three fold symmetry and an octahedron has vertices with four fold symmetry.

8. A Snub Tetrahedron is An Icosahedron

Starting with a tetrahedron as in **Figure 8a**, I used the four triangle faces, consisting of twelve vertices, rotated them by 60 degrees, and scaled them by ½. These vertices formed an octahedron, as in **Figure 8b**. When the vertices of these newly transformed triangles were rotated and scaled further, they appeared to move along the edge of the triangle of the octahedron, to a point, where the ratio of the two pieces of this edge were in the golden mean ratio to form the vertices of an icosahedron **Figure 8c**.

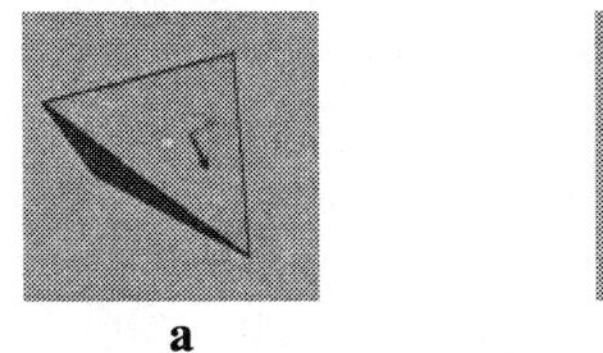
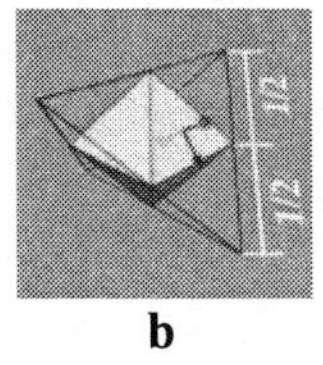

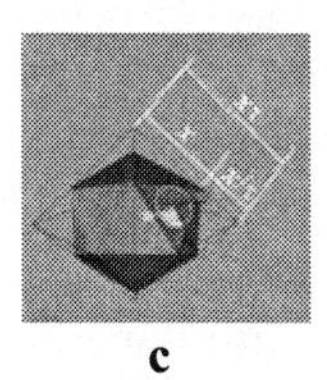

a b c

Figure 8: Tetrahedron (**a**), Octahedron (**b**) and Icosahedron (**c**).

Faces of a tetrahedron could be rotated either clockwise or counter-clockwise to produce an identical octahedron. However, when the faces of this octahedron were rotated in a clockwise direction an icosahedron **Figure 9c** was derived. When these vertices were rotated in a counter-clockwise direction a icosahedron with a different orientation **Figure 9d** was derived. When the **Figure 9c** and **Figure 9d** are displayed simultaneously with **Figure 9b**, the direction of the rotation can be more easily seen. When **Figure 9a** is added to **Figure 9e**, the results can be seen in **Figure 9f** showing a full set of polyhedra.

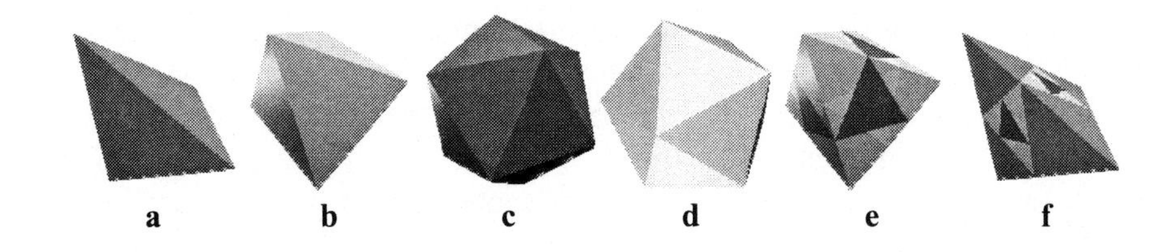

a b c d e f

Figure 9: Tetrahedron (**a**), Octahedron (**b**), Chiral Icosahedron (**c&d**), & ***Snubs*** (**e&f**).

9. Two Snub Polyhedra Complete the Set of Archimedean Polyhedra

Six faces of a cube were scaled by ½, and rotated by 45 degrees for vertices of a cuboctahedron . These six faces were further rotated and scaled for the twenty-four vertices of a snub cube. Similarly, the twelve faces of a dodecahedron were rotated by 36 degrees and scaled by ½ to form the vertices of an icosidodecahedron. These twelve faces were further rotated and scaled to form the sixty vertices of the snub dodecahedron. This ployhedron is also known as a snub icosahedron or a snub icosidodecahedron.

10. Statistical Study of Data Model Vertices

When I computed the data for Platonic, Archimedean, Prism, Anti-Prism and their Dual polyhedra, I printed 18 decimal digits tables for the double precision, floating point numbers. The tables consisted of the three coordinates of each vertex, a radius for each vertex, three coordinates of the mid-edges point, a radius for each mid-edge, three coordinates of the mid-faces point, and a radius for each mid-face point. I also computed and printed all edge lengths between vertices and all dihedral angles between faces.

11. Conclusions

When Haresh positioned the last polyhedron in his quasi-crystal structure, it registered with the previously positioned polyhedra without significant error. This result reinforced the precision of the polyhedra derived with this method of such simple computation. From the printed tables I could visually observe that the data was showing accuracy to 16 digits and occasionally 17 digits. Since I could observe their accuracy, these polyhedra exhibited no directional error and had the potential to be applied to many problems containing polyhedra. Throughout my derivations of all these polyhedra, I used consistent geometric relationships between polyhedra for each of the tetrahedral, octahedral and icosahedral families. All vertices simultaneously moved along edges by a constant value, all half edges were simultaneously rotated and scaled by constant values, and all faces were simultaneously rotated and scaled by constant values.

In my work with transformations the concept of *moving vertices along edges and across faces* for ***nesting*** polyhedra was illustrated using transparency and color for different polyhedra. When animated, these ***nested*** polyhedra were seen to have their vertices coincident with the edges and faces of an initial polyhedron. Also, the concept of *edge rotating and scaling* for ***dual*** polyhedra was also illustrated using transparency and color for different polyhedra. Edges of the initial polyhedron intersected with edges of a dual polyhedron. In addition, they were orthogonal to each other and this relationship could be seen very clearly.

Finally, I used transparency and color for the different polyhedra to illustrate *face rotating and scaling* for ***snub*** polyhedra. The tetrahedron, with an octahedron inside, and an icosahedron inside the octahedron, clearly shows a relationship between vertex symmetry of three, four and five in a single composite **Figure 9f**. Since the golden mean ratio along the edge of the octahedron existed there was a serendipitous nature to such a precise relationship. This relationship is perhaps more subtle than can be easily seen by most viewers.

Acknowledgements

I would like to thank Azam Kahn of Alias/Wavefront for Maya animation software. I would also like to thank Sean Curtis, a student staff at CHPC for producing the three animations cited in this paper. I would also like to thank my spouse, Deborah, for proof reading this paper.

References

[1] Coxeter, H.S.M., Regular Polytopes, Dover Publications, Inc., NY, 1973.
[2] Pearce, Peter & Susan Pearce, Polyhedra Primer, Van Nostrand Reinhold Co., NY, 1978.

Chromatic Fantasy: Music-inspired Weavings Lead to a Multitude of Mathematical Possibilities

Jennifer Moore
49 Cerrado Loop
Santa Fe, NM 87508, USA
doubleweaver@aol.com

Abstract

As part of my thesis work for my MFA in Fibers at the University of Oregon, I wove five panels that were inspired by Johann Sebastian Bach's 'Chromatic Fantasy'. The many possible combinations of these weavings led me to create a flipbook of their images, as well as a computer-animated video of the weavings dancing to the music from which they were inspired.

Introduction

As a college student in the 1970's, I studied art, music and science. I was interested in finding ways that these fields connected to each other and looked to Leonardo da Vinci as a role model. When I signed up for a weaving class and sat down at a floor loom for the first time, it felt just like playing the pipe organ, except that the harmonies appeared in color instead of in musical tones. I became intrigued with the idea of finding relationships between weaving and music and expressing these relationships in my artwork.

A dozen years later I returned to graduate school at the University of Oregon to pursue a Master of Fine Arts in Fibers. When it came time to choose a topic to focus on for my thesis project, I decided to explore the relationships between weaving and music through their underlying mathematical structures. In my studies, I learned about the golden proportion and its relationship to musical frequencies. I also studied the movements of symmetry and found them to have wonderful correlations in musical compositions. In my weaving I work with many of the same elements found in music - rhythm, proportion, texture, harmonics - and strive to bring the beauty and order of music into visual form.

J. S. Bach's "Chromatic Fantasy"

Because I had worked extensively with the music of J. S. Bach as an organ student and was fascinated with the mathematical ideas and symbolism in his music, I chose to study it in more depth. As I read about Bach and his life, I became interested in a piece that he wrote for the harpsichord called "Chromatic Fantasy".

The word "chromatic" has both a musical and a visual meaning. In music, a chromatic scale contains twelve tones within an octave, as opposed to the seven-tone diatonic scale that we are accustomed to hearing in western music. Because of the inclusion of the half-tones in a chromatic scale, the harmonies can sometimes sound dissonant to our ears, but they also have a great richness and depth to them.

Visually, the word "chromatic" refers to color, and the full range of the color spectrum. As I thought about these two meanings of the word "chromatic", I became intrigued by the analogies that I saw between them and decided that I wanted to weave a visual Chromatic Fantasy.

Making Music Visible

While I often use mathematical concepts such as harmonic proportions, symmetry patterns, tessellations and fractals in my weaving, my work is inspired by my interpretations of music, rather than attempting to be direct representations of specific pieces of music. I listened to a recording of Bach's "Chromatic Fantasy" over and over and let the images come to me as I listened. This piece of music is a progression of single notes without any chords, but they are played so rapidly that the tones blur together like broad paint strokes of color. I imagined the notes flowing through space, with the fundamental tones grounding the piece, and the overtones floating up above them.

I work in a technique called doubleweave pick-up, in which two layers of fabric are woven simultaneously on the loom, one above the other. Individual threads are counted and exchanged between the two layers, enabling the weaver to create crisp design areas and to harmonize the colors of one layer against the other.

One of several significant differences between art and music is that art is experienced in space, while music is experienced in time. To incorporate an element of temporality into my piece, I decided to weave multiple panels that would fit together to make a larger composition and create a visual flow from one panel to the next. I ended up weaving a total of five panels using a full, rich spectrum of color against a black background.

Figure 1: 'Chromatic Fantasy' 44" x 168"

A Multitude of Possibilities

After I finished weaving the five panels, I stretched each one over foam-core board and framed them individually. Each panel measured 44" x 33", so the entire composition ended up nearly four feet high and fourteen feet across. I laid them all out in the order that I had planned for the composition. Then I began to experiment with arranging the panels in different configurations. I discovered that because each of the panels had a series of small diagonal lines going through the center of each of its sides, the panels could be placed in any order and any of them could be rotated 180 degrees, and the design would still fit together.

I quickly realized that there were quite a number of ways to arrange the five panels and that it would be difficult to keep track of which ways I had tried. Because there are five different panels and each one has two possible orientations, there are ten possible images for the first position. This leaves eight possible images for the second position, six images for the third position, four images for the fourth position and two possible images remaining for the last position. This leads to the equation: 10 x 8 x 6 x 4 x 2 = 3,840 different ways of arranging the five panels in a line!

Clearly, if I wanted to explore working with all these possibilities, I would have to come up with a less cumbersome way than moving five large panels around.

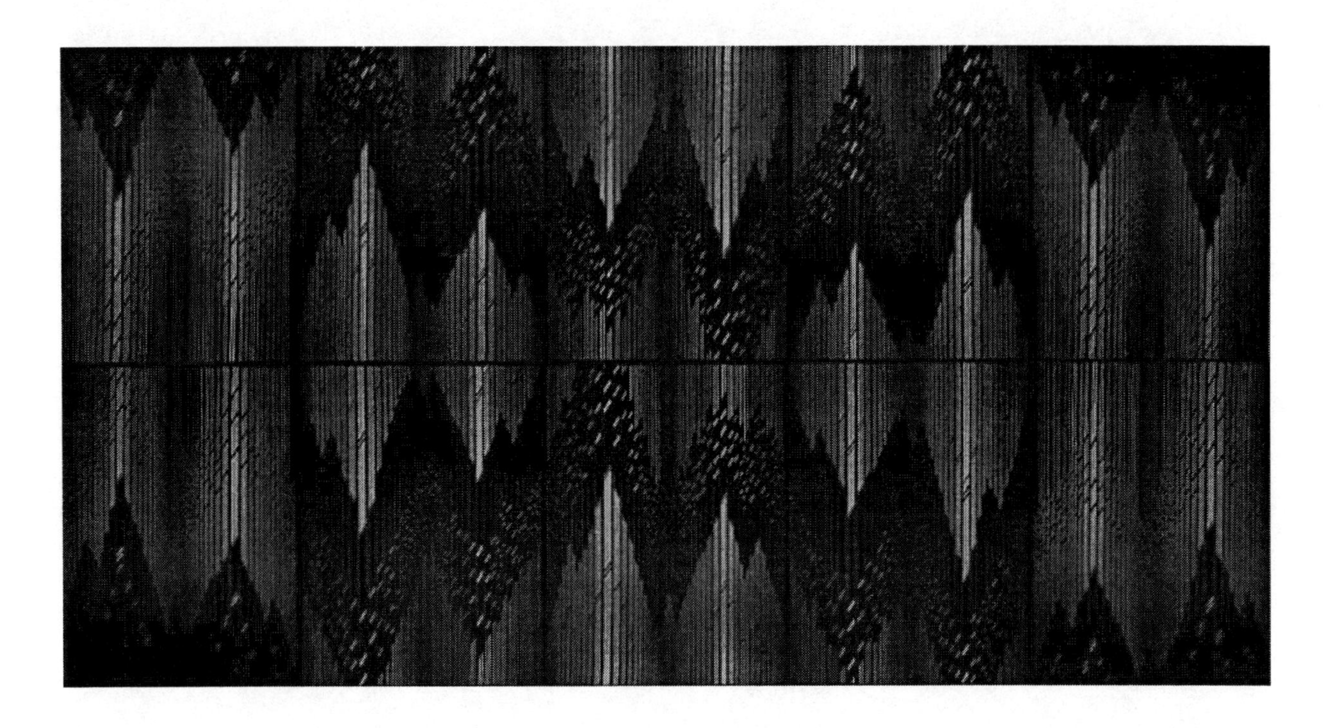

Figure 2: The five 'Chromatic Fantasy' panels in each of two orientations

Chromatic Fantasy Flipbook

I took a photograph of the woven panels and made a color photocopy of it. Then I cut apart the images of the individual panels, which allowed me to lay out the images and move them around freely. As I experimented with this, an idea came to me of a flipbook in which stacks of the different images could be hinged and flipped over to create new compositions.

I made numerous copies of the photograph of the weavings, cut apart the individual images, and mounted them on card stock to make them sturdier. Since there were a total of ten different images (the five original weavings each in two different orientations), I created two rows of five stacks of images.

Each of the stacks contained all ten of the different images. I punched holes along the outside edges of each of the stacks so that they could be hinged and flipped over. I attached all of this to a backing board that holds everything together and made a pair of covers for the top.

In this configuration of ten stacks with ten images in each stack, there are 10^{10}, or 10 billion possible combinations. I think of it as being something like a visual abacus.

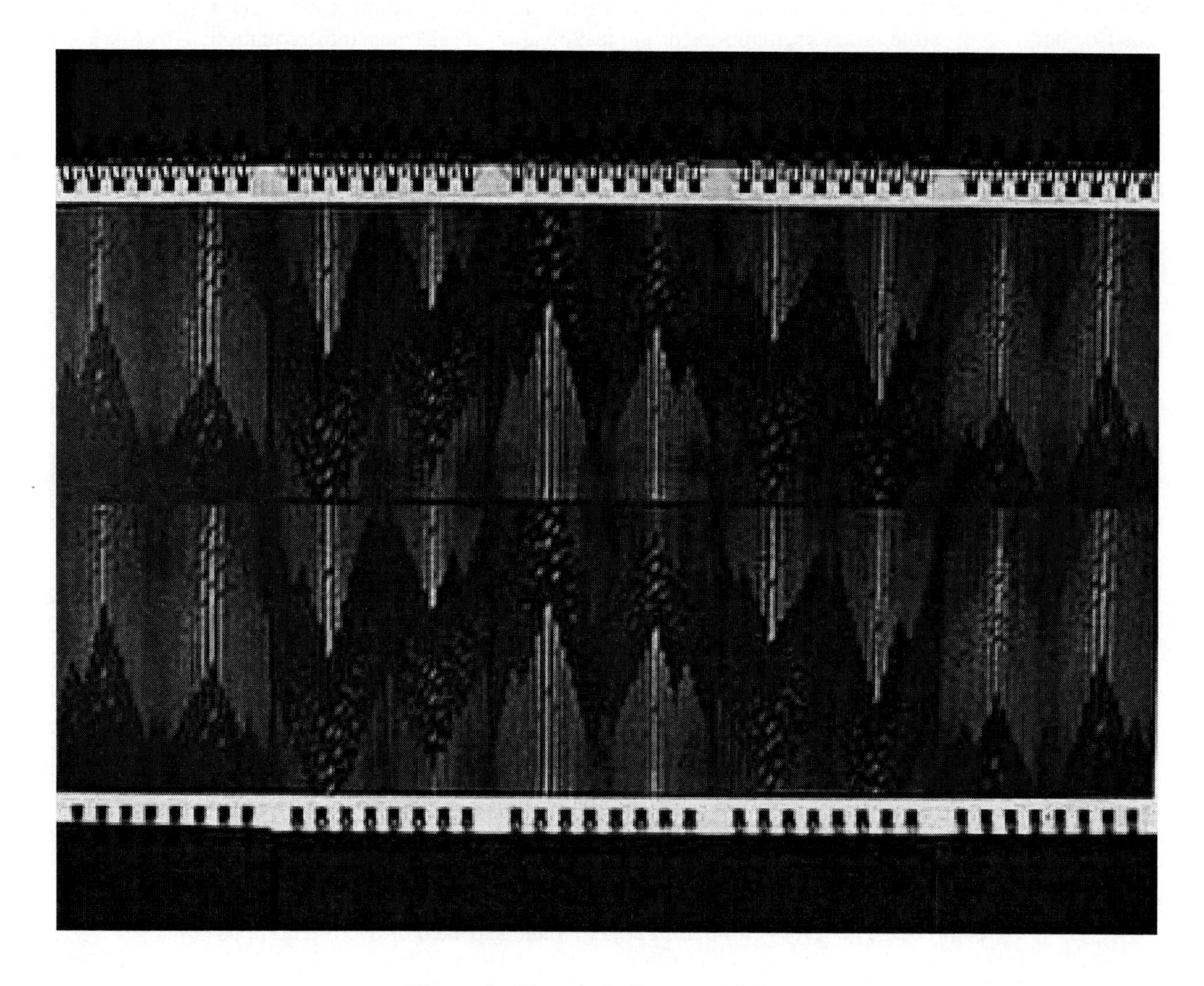

Figure 3: Chromatic Fantasy Flipbook

Chromatic Fantasy Video

After making the flipbook, which is a manual way of animating the images, the next step was to bring it into the world of technology by making a computer-animated video of the weavings dancing to Bach's 'Chromatic Fantasy'. I spent the last several months of my time in graduate school working in the computer graphics lab bringing this idea to life.

The first step was to record the music into the computer. By bringing the music into the program Macromind SoundEdit, I was able to see the music on the screen in wavelength form. I adjusted the size so that I could see two seconds of the music on the screen at a time, and then divided that up into fifteen partitions per second. I printed out the piece of music in this form, and then taped all of the pages together into one long wavelength pattern.

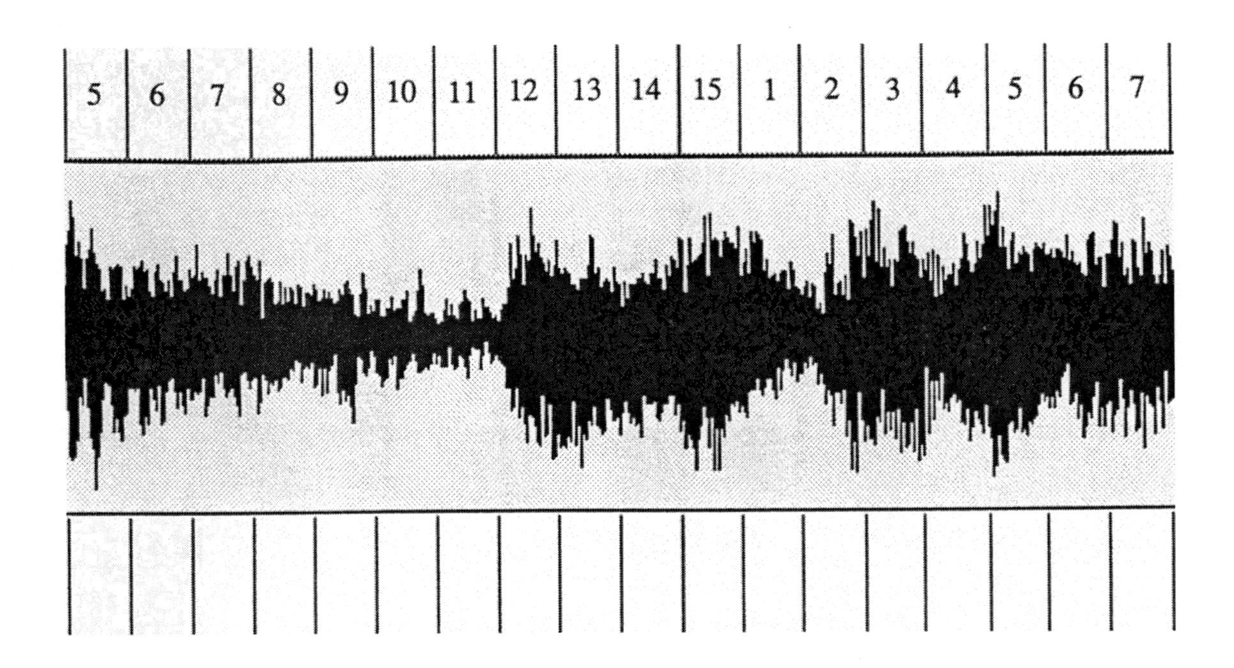

Figure 4: "Chromatic Fantasy" music shown in Macromind SoundEdit

Next I scanned the photograph of my woven panels into the computer and worked with the program Macromind Director to create the video. I entered each of the ten possible images of the weavings as individual cast members, along with a plain black background screen. The proportions of a computer screen allowed for three rows of five images in each row.

Using the pages I had printed out of the wavelength patterns of the music, I began to work out a screenplay of actions for the visual images. The fifteen partitions per second in the music corresponded to fifteen frame changes per second in the animation. By listening to the music while comparing the written score with the wavelength patterns, I could pinpoint where specific changes in the music were happening. While listening to the music I visualized the movements of the images and wrote down the sequences of actions I wanted them to take.

Returning to the computer, I placed the individual images on the screen in their chosen positions for each frame change. With the configuration on the computer screen of three rows of five images, I had one quadrillion possible combinations to choose from. In order to make the project manageable in the time frame I had available, I ended up working with the first two and a half minutes of the musical composition, which meant creating slightly over 2200 different frames. I used various special effects that were available in the animation program for transitions between different sections of the music. After all the frames were in place, quite a bit of fine-tuning was required to get the musical track and visual track running in synch with each other.

The end result was an animation of the images of the woven panels dancing around on the screen to the music of Bach which originally inspired their creation. The project came to a full circle with the completion of the video.

In the year 2000 the five woven panels were purchased along with a copy of the flipbook and the video by the Alberta Kimball Auditorium in Oshkosh, Wisconsin. People can view the woven panels and the flipbook in the lobby as they arrive to attend concerts.

References

David, Hans T. and Arthur Mendel, *The Bach Reader*, W.W. Norton & Co., 1945.

Garland, Trudi H. and Charity V. Kahn, *Math & Music: Harmonious Connections*, Dale Seymour Publications, 1995.

Godwin, Joscelyn, *Harmonies of Heaven and Earth*, Inner Traditions International, 1987.

Asymmetry vs. Symmetry in a New Class of Space-Filling Curves

Douglas M. McKenna • Mathemæsthetics, Inc.
PO Box 298 • Boulder • Colorado • 80306-0298
doug@mathemaesthetics.com

Abstract

A novel Peano curve construction technique shows how the self-referential interplay between symmetry and asymmetry based on the translation, rotation, scaling, and mirroring of a single angled line segment that traverses a square evinces rich visual beauty and optical intrigue.

Consider the patterns, or geometric motifs, labeled **C, D,** and **E** in Figure 1. Each is a sequence of connected line segments that differs slightly from its neighbor:

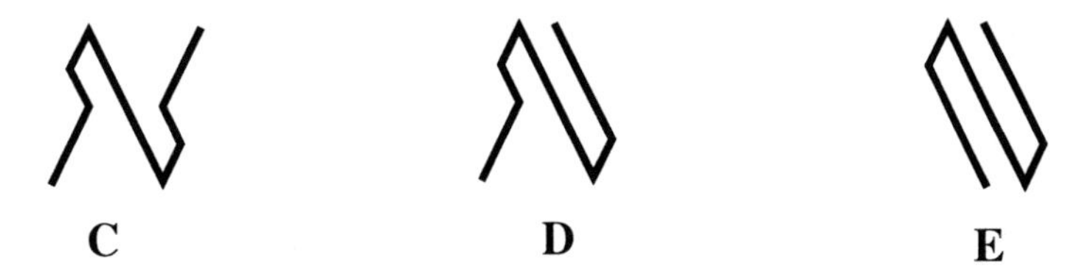

Figure 1: *Three patterns each differing from its neighbor by one line segment*

Before continuing, take a moment to think about the answers to the following aesthetic questions:

Which pattern, **C, D,** or **E**, is more *symmetric*?
And, more subjectively, which pattern is more *elegant, pleasing to the eye*, and/or *beautiful*?

Aesthetic questions and answers are of course subject to personal whim, but they still serve to illuminate the makings of interesting mathematical art. So we first analyze the geometry underlying patterns **C**, **D**, and **E** to show that the foregoing questions have an unexpected answer.

As Figure 2 shows, each of these three motifs is embedded within a square that is divided into a simple, rotationally symmetric arrangement of ten smaller squares, in two sizes, that tile (cover without gap or overlap) or dissect the original. All three patterns are the same length, and each traverses through the same set of smaller square tiles in the same order. Each contains line segments that are translationally parallel to other segments, which is a visually pleasing form of internal reference. Patterns **C** and **E** are rotationally symmetric 180° about the center, whereas **D** is not. None exhibits either vertical or horizontal bilateral symmetry. All three patterns exhibit an identical internal duplication: the upper-left quadrant (containing four smaller square tiles) of each is congruent with the lower right quadrant, rotated 180° about the center. Again, that internal central point reflection makes all three patterns visually interesting by virtue of the internal correlation and attendant partial rotational symmetry. Yet pattern **D** seems symmetrically awry, as if it were a poor compromise between **C** and **E**, both of which could be used in a simple, translationally repetitive frieze.

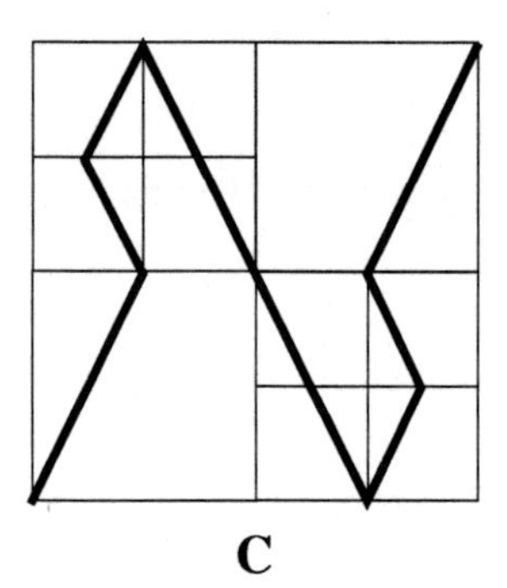

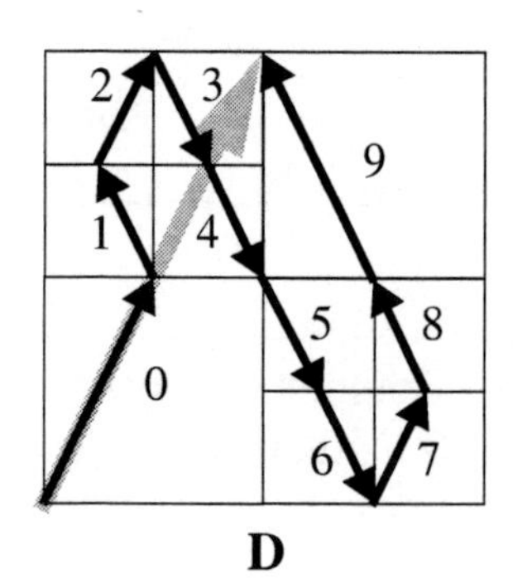

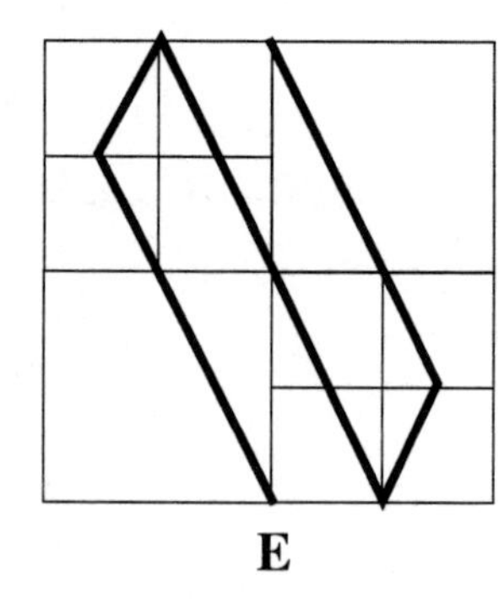

Figure 2: *Each pattern traverses the same 10 smaller squares that cover the original*

While patterns **C** and **E** exhibit more symmetry, thereby perhaps winning the aesthetic beauty contest, a closer look at **D** shows that it has a deeply elegant, geometric property that its neighbors are missing.

Pattern **D** is a path that traverses its enclosing square by travelling from a corner to the center of an opposite side (gray arrow, above). But the internal smaller square edges divide the pattern into segments, several of which are co-linear. Thus, if one takes **D** as composed not of six but of ten connected, directed line segments, each of these ten segments traverses—while maintaining piecewise continuity—exactly one of the ten sub-squares *in essentially the same geometric manner as **D** traverses its own square*. Each segment connects its smaller square's corner to the center of an opposite side, in either a right- or left-handed manner, and in either a forward or backward direction, and in various of four possible rotated orientations.

Thus, while all three patterns are essentially tile designs, pattern **D** is a spatially recursive "detour" that is composed of parts that each accomplish the same goal as the whole does, but at two smaller scales. Because each smaller square can itself be similarly subdivided, each leg of the **D**-tour can be replaced—after a simple linear transformation—by a smaller half- or quarter-size **D**-tour, yielding a **D**-**D**-tour (denoted $\mathbf{D}^2$), all without affecting the continuity of the overall tour The limit of the convergent sequence of $\mathbf{D}^n$-tours will be a space-filling Peano curve [1][2] that passes every point in the original bounding square region. The classic Hilbert and Peano Curves can be analysed using base 4 or 9 number systems [1], but in this new construction, simply label the appropriately ordered ten subsquares with the digits 0...9 (Figure 2, center). Any infinite decimal expansion of a point in the interval [0,1) maps—under the appropriate geometric ordering of subsquares and iterated linear transformations, one per digit, each scaling by at most 1/2—onto a point in the square.

Like Peano's original 3x3 (analytic) construction, each finite stage of the composition increases in length by a factor of 3. Unlike the original Peano Curve, here finite stages contain copies of previous, less-detailed stages at different sizes, more akin to Mandelbrot's "Snowflake Sweep" construction [3]. The multitude of different scales visually draws out the recursive quartering of the square and all its subsquares. Hence, I enjoy calling this space-filling curve construction the "Peano Quartet" (Figures 3 and 4):

Figure 3: *Stages 0 through 5 of Peano Quartet space-filling curve construction*

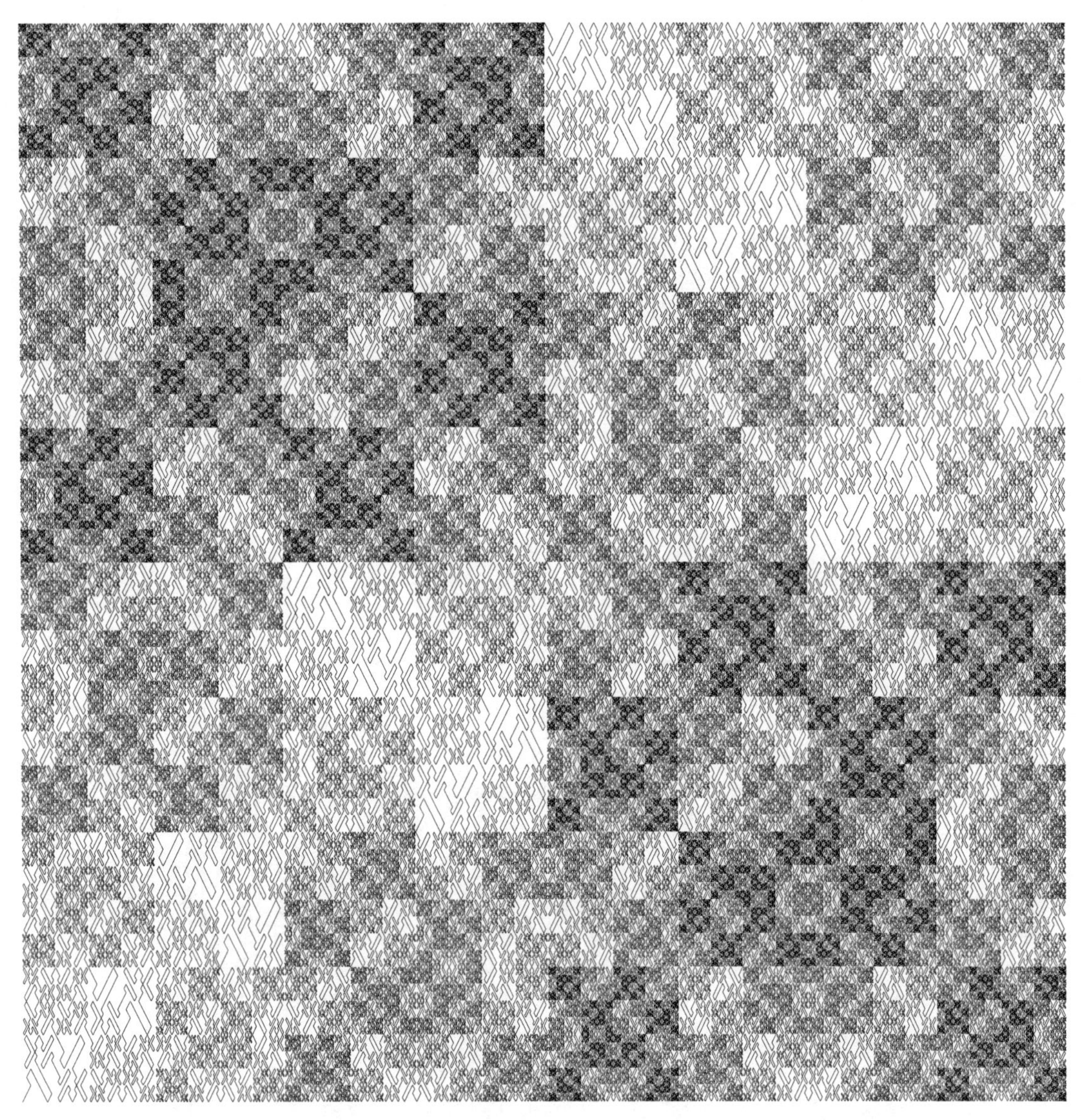

Figure 4: *Stage 6 of the Peano Quartet contains 10^6 line segments*

The original Peano Curve's highly regular approximation paths are self-contacting at every other lattice point [3]. But in the Quartet construction, finite approximations are nearly everywhere self-avoiding ($\mathbf{D}^2$ with 100 segments has only 2 points of self-contact, and $\mathbf{D}^3$ with 1000 segments has only 29). This is due primarily to the fact that, other than the lower left corner once, no finite $\mathbf{D}^n$-tour can ever contact the logical right or left side of its embedding square. Uniform, square-filling, edge-replacement Peano curves, such as McKenna's E-Curve and its relatives [2][4], cannot be completely self-avoiding (at finite stages) below order 5. In the limit, though, all space-filling curves are self-contacting, surjective mappings [1].

Because all finite $\mathbf{D}^n$-tours inherit $\mathbf{D}$'s essential asymmetry, but still have internal parts that are congruent, each finite stage displays a wonderful visual tension between high-level symmetry and low-level asymmetry. When thin lines with finite thickness illustrate the path, the parts drawn at the smallest scales will be darker. This results in a rich binary quilt (Figure 4) composed of repetitions of the initial diagonal line. The upper-left quadrant is congruent under 180° rotation about the center with the lower-right quadrant. The remaining two quadrants, however, are not congruent.

Self-avoidance of an approximation path is both visually important and combinatorially constraining. Each finite Peano Quartet approximation ($\mathbf{D}^n$) is a piecewise linear, one-dimensional curve that—regardless of self-contact at some subsquare corner—divides its world into two parts. When one colors one side of the path black and the other side white, in addition to handedness one can also have visually positive or negative subsquares that correspond to the direction the curve is travelling across that square. The nearly complete (finite) self-avoidance means that large white and black regions of the traversed square are topologically connected at more than just single points of self-contact. This hides the underlying geometry and leads the eye/brain to integrate larger scale structures and imbue connected areas with visual form and meaning. This is in striking contrast to the classic Peano Curve, whose approximations, when similarly colored, create nothing more than a uniform, diagonal checkerboard pattern [3]. But here, different sizes fool the eye into seeing distances and depth, foreground and background take on the same shapes, and an unexpected and intriguing visual property—an optical illusion—appears: the horizontal binary division lines are parallel, but rather unsettlingly do not look so (Figure 5):

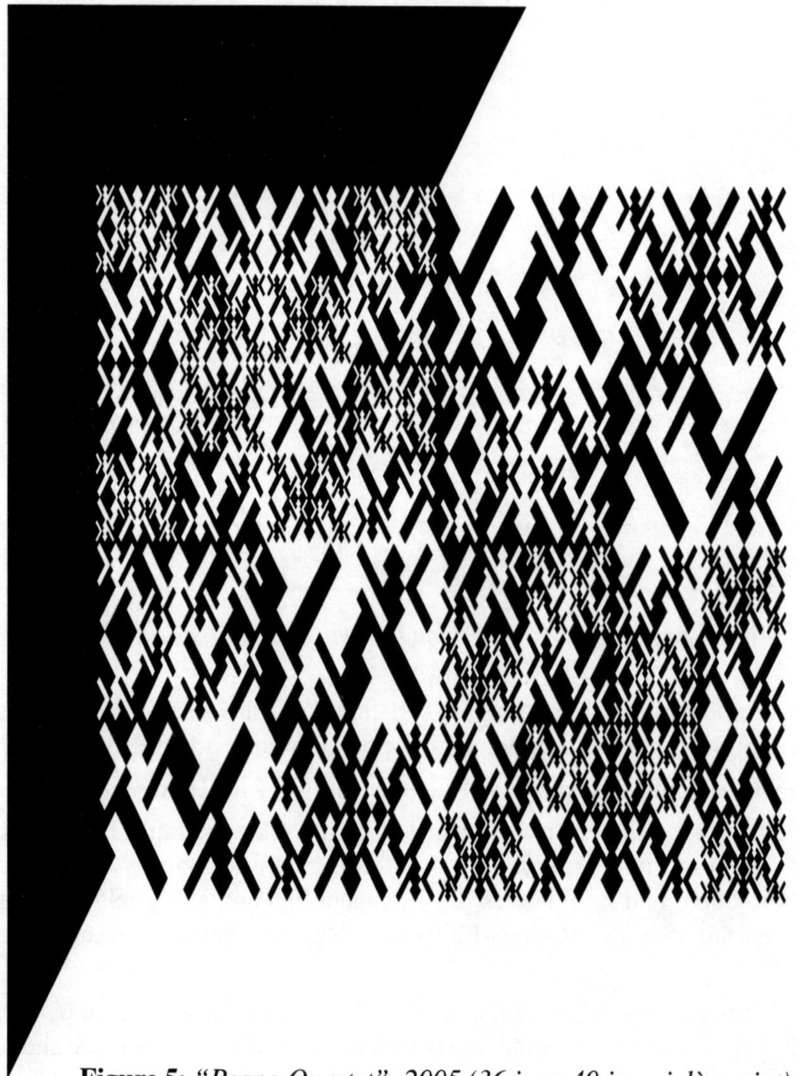

Figure 5: *"Peano Quartet", 2005 (36 in. x 40 in., giclèe print)*

The question arises as to the uniqueness of the construction. The combinatoric space is highly constrained, especially at the four corner squares where, regardless of their sizes, path connectivity from corner to top center requires that all four corner subsquares each have one of only two possible directed line segments traversing them (Figure 6). Pattern **D** is the only possible ten-segment, self-avoiding traversal from lower-

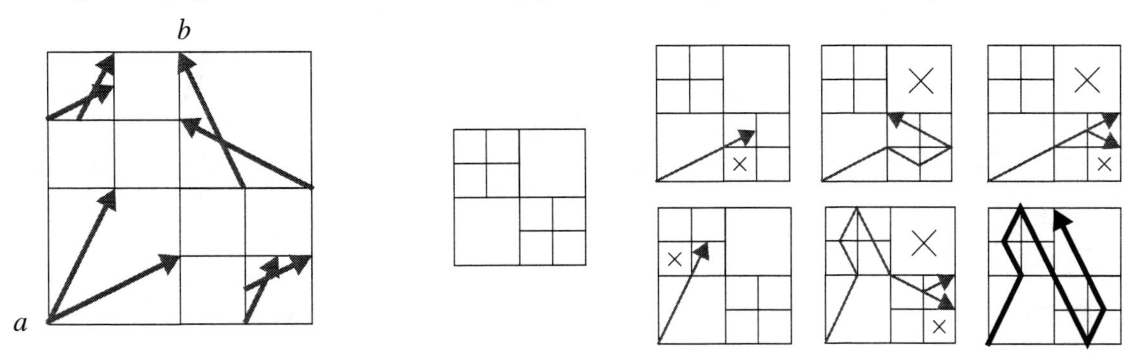

Figure 6:
Corner squares admit only two possible segment choices as part of any self-avoiding directed traversal from a to b

Figure 7:
*Pattern **D** is unique for given dissection into 10 subsquares. Squares for which it is impossible to continue building a solution are marked with an* X.

left corner to the center of the top side of the Peano Quartet's underlying dissection (Figure 7). Any other choices of path extension lead to contradictory situations. Because of the asymmetry of the corner-to-center traversal goal, even a simple rotation or reflection of the square's dissection arrangement can change the solvability of the problem. For instance, when the Quartet's underlying dissection is rotated 90°, there are no solutions (Figure 8).

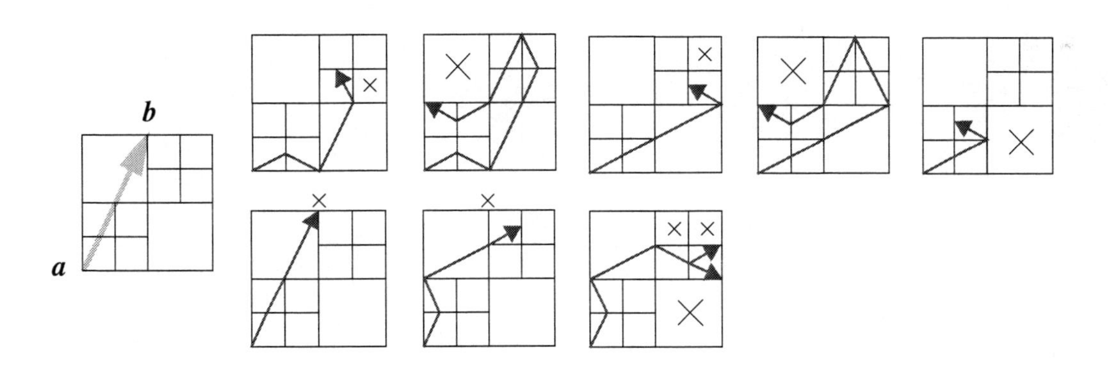

Figure 8: *No solutions from a to b when Quartet's dissection is rotated 90°. An* X *at point b means the path has prematurely arrived, or can never arrive given the current partial path.*

Among the 16 possible dissections (counting rotations and reflections) of a square into ten subsquares, two a quarter size and eight a sixteenth size (Figure 9), hand enumeration shows that there exist 10 space-filling curve generators that are locally self-avoiding (Figure 10). None of these generators is symmetric, and local self-avoidance is necessary, but not sufficient, for global self-avoidance of higher level stages. Nonetheless, many of them generate curve approximations that are largely without self-contact, leading to larger visual features with which the eye can play. Interestingly, one solution is the same shape as pattern **D**.

The recursive tiling patterns that these asymmetric generators self-referentially specify are more successful aesthetically when the underlying dissection has its own symmetries that more detailed stages inherit.

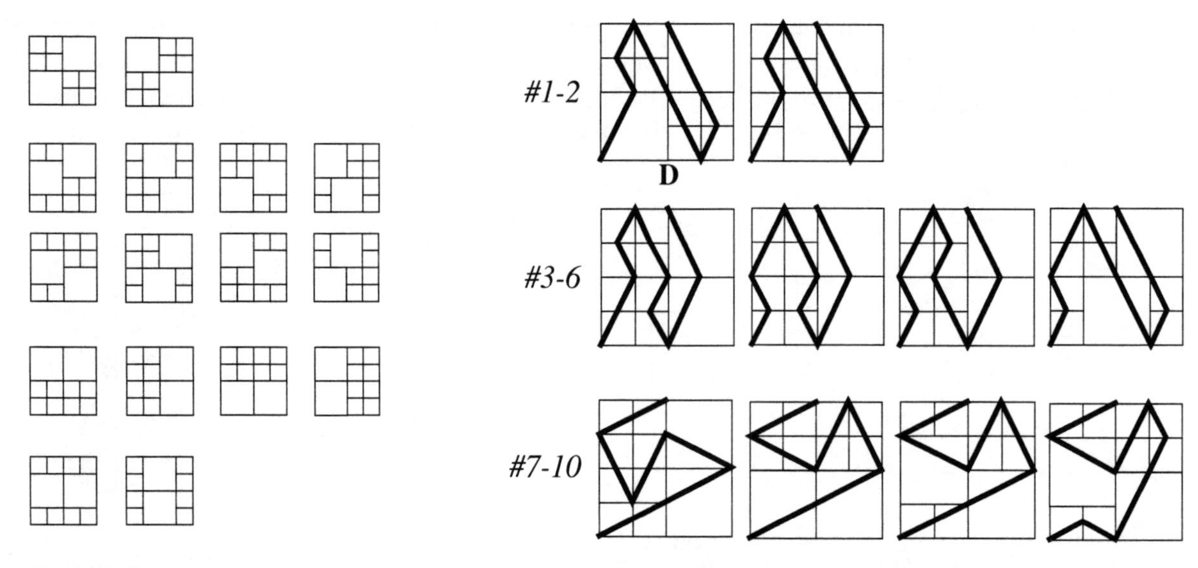

Figure 9: *16 dissections, counting rotations and reflections*

Figure 10: *10 space-filling curve generators. #7 is the basis for the print below*

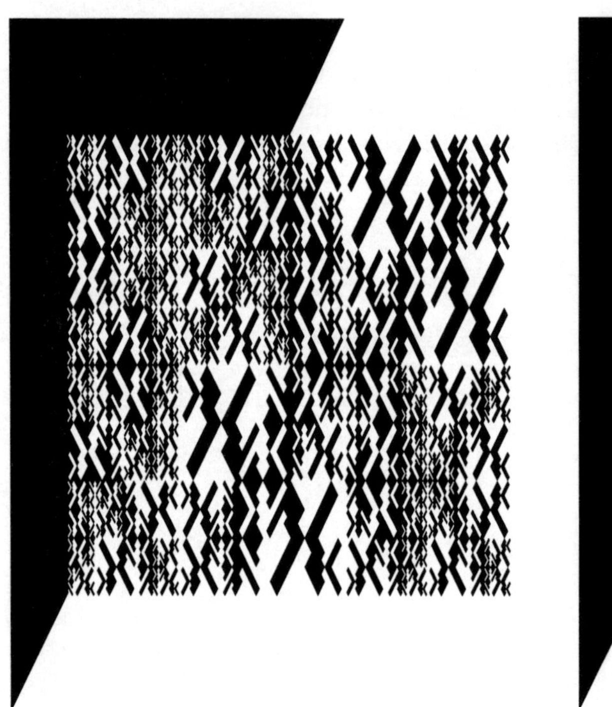

Figure 11: *The Peano Quartet's fraternal twin has fewer symmetries*

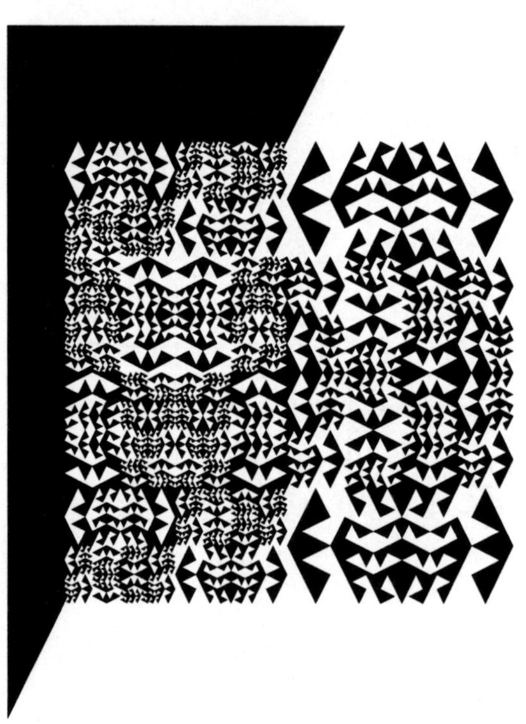

Figure 12: *"Platonic Dance" (giclée, 2006)*

Compare the more chaotic feeling of Figure 11 to Figure 12 ("Platonic Dance"), where the latter exhibits strong bilateral symmetries in many different scales and orientations, because the originating dissection does. Figure 13 ("Blade Rainer") is based on stage 4 of generator #4 from the above list, altered at the final drawing stage to be completely self-avoiding. Figure 14 is a self-avoiding variation based on generator #8. Both are wildly different in visual and even emotional impact and feel from each other (and from either of the above two), by virtue of how our mind and eye integrate and recognize connected forms.

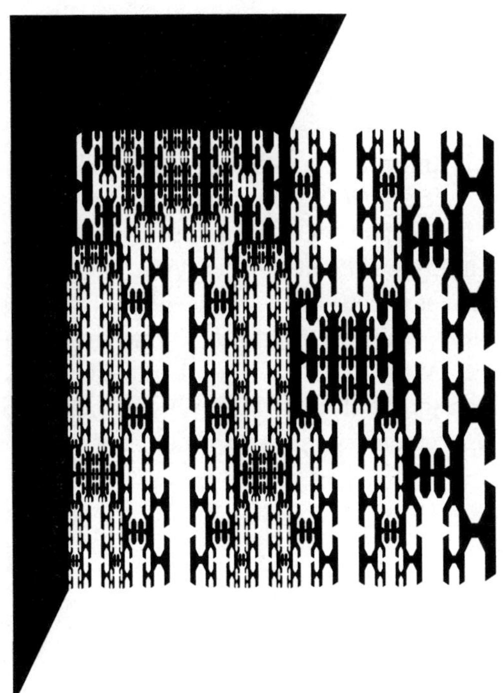

Figure 13: *"Blade Rainer" (2006)*

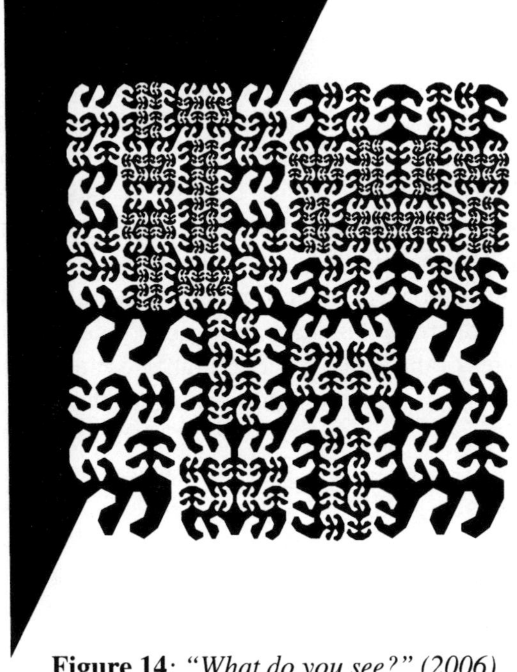

Figure 14: *"What do you see?" (2006)*

If one restricts generator patterns to be self-avoiding and only traverse equal-sized subsquares, there are several locally, but not globally, self-avoiding generators on the order 4 (4 x 4) subdivision). For example, patterns **A** and **B** each generate a more uniformly self-contacting, space-filling curve that, when colored as a tiling at, e.g., stage 3, is visually reminiscent of the textiles of some South American cultures (Figure 15).

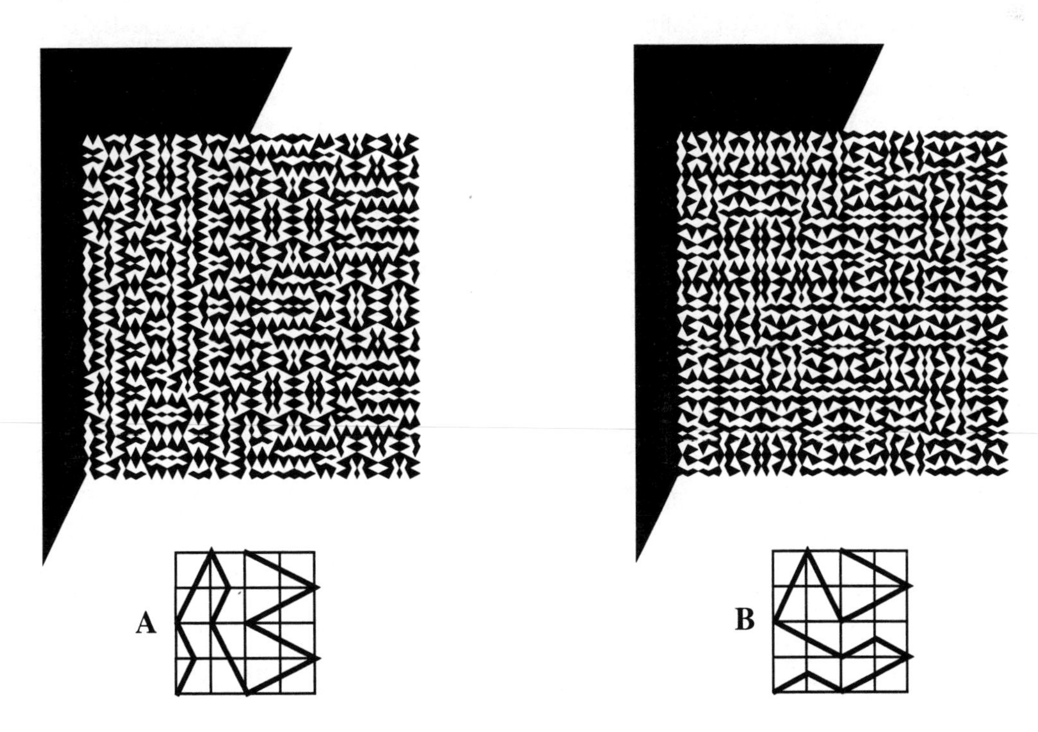

Figure 15: *Stage 3 of two space-filling generators **A** and **B**, within 4 x 4 tile designs*

Neither of these is as visually rich as the more exuberant Peano Quartet and its brethren, because the hierarchical nature of the $\mathbf{A}^n$ and $\mathbf{B}^n$ stages is hidden by the homogeneity of the subsquare scales at any given construction stage.

Whenever non-mathematical visitors view the large framed copy of the Peano Quartet (Figure 5) that I have on my wall, they are visually fascinated, and invariably think I'm making choices, following my muse. I tell them that the Peano Quartet is a composition, but that it is a mathematical one (all puns intended). Yet the crystalline, recursively raucous, connected collection of symmetries and asymmetries seems deeply important to treating these compositions as purely aesthetic objects with inherent visual intrigue and meaning to the viewer. Whether they are more platonically than artistically beautiful is an interesting and ancient question, because art, like math, is created/discovered through a person's choices and exploration within some constrained, combinatoric space. Which is why, when I printed one of my first copies of the Peano Quartet, I labeled it as both "A/P" (Artist's Proof) and "D/P" (Discoverer's Proof). The viewer can then interpret it mathemæsthetically according to their own philosophical slant.

Conclusion

The Peano Quartet's generator pattern **D** is one of 10 possible space-filling curve generator patterns that traverse a dissection of a 4 x 4 square into any of 16 spatial arrangements of two 2 x 2 subsquares and eight 1 x 1 subsquares. Each generator constitutes the asymmetric, self-referential, geometric code that builds a rich tapestry of threaded subsquares that cover an original square, in the limit as a space-filling curve. Finite approximations of these recursive constructions create combined curve and tiling patterns that are much more intriguing and dynamic to the eye than the more commonly known space-filling constructions based on line segments all of equal length.

Beauty in both math and art evinces itself more deeply and interestingly when there's a tension between symmetry and asymmetry. Pattern **D** and its fellow generators are just as symmetrically elegant in their own hierarchical scaling way as patterns **C** and **E** are rotationally. But recursive structure combined with heterogeneous sizes makes **D** and its fellow detours especially intriguing, both visually and combinatorially, more so than the merely pleasant patterns **C** and **E**, whose symmetry is self-evident but only skin-deep. The aesthetic value of these square-traversal patterns also trumps those created by **A** and **B**, where underlying self-referential, hierarchical structures are washed out by the drone of their "one-note" scales.

The author wishes to thank Oren Patashnik and the referees for their helpful suggestions and comments.

References

[1] H. Sagan, *Space-Filling Curves*, Springer-Verlag, ch. 1-3, 1994.

[2] D. M. McKenna, "SquaRecurves, E-Tours, Eddies, and Frenzies: Basic Families of Peano Curves on the Square Grid", *The Lighter Side of Mathematics: Proceedings of the Eugène Strens Memorial Conference on Recreational Mathematics and its History*, MAA, pp. 49-73, 1994.

[3] B. B. Mandelbrot, "Harnessing the Peano Monster Curves", *The Fractal Geometry of Nature*, Freeman, pp. 68-69, 1982 (illustrated in part by the present author, see p. 444).

[4] P. Prusinkiewicz and A. Lindenmayer, *The Algorithmic Beauty of Plants*, Springer-Verlag, pp. 12-14, 1990.

Modular Perspective and Vermeer's Room

Tomás García-Salgado
Faculty of Architecture, Autonomous National University of México
tgsalgado@perspectivegeometry.com

Abstract

The room's dimensions of the Music Lesson (ML), as deduced in my first perspective analysis [1], corroborate that the projected image on its back wall approximates the real size of the painting, as Steadman first pointed out. It seems unlikely that the tiled floors in Vermeer's paintings were done at random. Instead, some of them seem to have a consistent image formation of about 90° of aperture of visual field, which speaks on behalf of the use of the camera obscura. Steadman based his consistency analysis of the underlying tiled floor grids of Vermeer's paintings in the inverse perspective method [2], finding that about six of them seem to depict the very same room. Following this idea, but instead of deducing the room's plan and elevation as he did, I will proceed directly in perspective with the aid of my *Modular Perspective* method. Thus overlaying the floor grid of the ML to another painting's floor grid, I will prove if they are consistent or not. In addition, if they are so, the real size of the second floor grid will be deduced. As far as I know, such a perspective proof has never been attempted before.

1. The Riddle of the Back Wall as Perspective Plane

Steadman sustains the hypothesis that Vermeer painted six of his works in the same studio [3]. His accurate architectural reconstructions show simultaneously the vantage points location of each one of them and how their projected images lie on the back wall. I, instead, displaced the perspective plane (PP) onto the back wall position to catch the painting's projected image on it. For this reason the term *perspective plane of the back wall* (PPbw) is introduced. The term *observer point* will also be used instead of the *camera vantage point* to avoid confusions onto which plane —the PP or the PPbw— the image is projected. The interchange of these planes aroused an unexpected riddle, which can hardly be visualized without the help of Figures *1a* and *1b*.

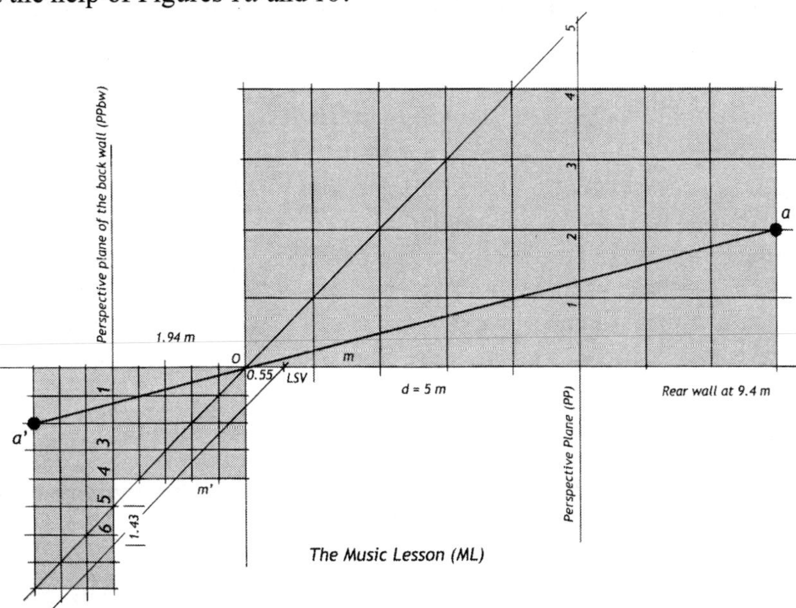

Figure 1a: *Relative positions, in the ML, of the PP and the PPbw from the observer.*

As it is evident in Fig. *1a*, the distance between the *observer point* and the PP is 5 *m* (modules), while being of about ≈ 2 *m* from the *observer point* to the PPbw. In other words, although both perspective planes have different modular distances, they remain proportional to one another. As we know, any given point within the observer's visual field relates to its identical within the camera's visual field. Notice in Fig. *1a*, how points *a* and *a'* have the same modulation in the PP and the PPbw respectively, while line *a-a'* encounters the visual of symmetry (*vs*) at the *observer point*.

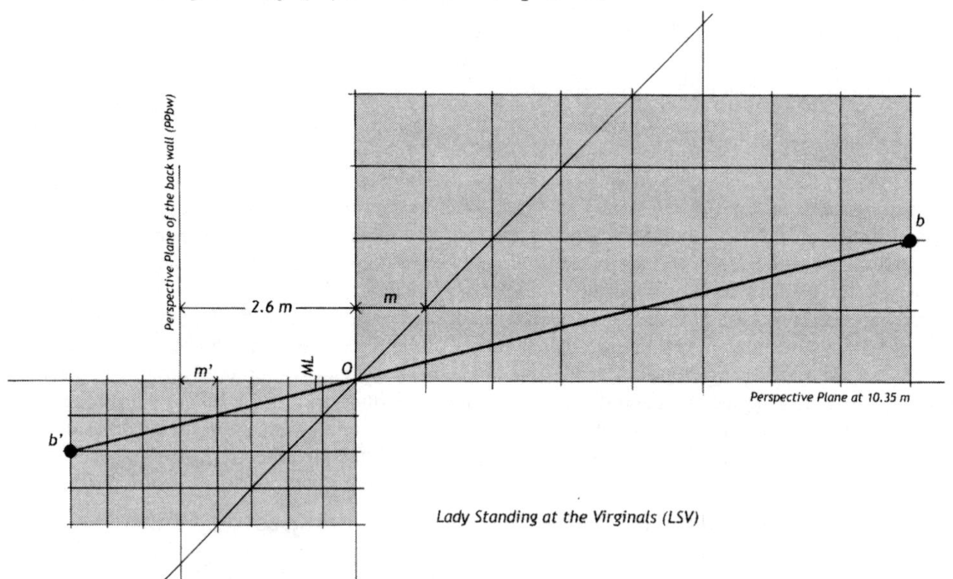

Figure 1b: *Relative positions, in LSV, of the PP and the PPbw from the observer.*

You may notice in Fig. *1b*, how the observer's distance increases in comparison to that of Fig. *1a*. Here, the distance between the PPbw and the *observer point* has changed to ≈ 2.5 *m*. When both figure *1a* and *1b* overlap at their correspondent PPbw, lines *a-a'* and *b-b'* run parallel one to another. On the contrary, when these figures overlap at their correspondent PP, then line *a-a'* and *b-b'* become as one. What these figures tell us is that the true size of the paintings corresponds to their projected image on the PPbw, whereas our perspective analysis takes place directly on the PP. Therefore, the riddle's mean is how the PPbw can be measured onto the PP.

2. Solving the Riddle

Based on the ML's real dimensions (≈ 64 x 73 cm), the size for other pictures will be deduced. A testing problem since the *observer point* is not common to all paintings. First, we need to prove that trimming a painting does not affect the absolute distance between *the vanishing point* (*vp*) and *the vanishing distance point* (*vdp*). Why should we do this? Because most of Vermeer's paintings show an asymmetric position of the *vp*, suggesting they were trimmed. For instance, Lady Standing at the Virginals (LSV) looks like a close up scene, emphasizing the female character without including completely the scene.

As it is evident through Figs. *2a-2e*, the *vdp* increases whereas the visible floor tiles decrease in depth, which means that the absolute value between the *vd* and the *vdp* remains constant in all figures. Therefore, adding the distance *vd-vdp* and the visible floor tiles in each figure, we have: Fig. *2a* (5.0 *m* + 9.4 *m* = 14.4 *m*), Fig. *2b* (5.4 *m* + 9.0 *m* = 14.4 *m*), Fig. *2c* (5.9 *m* + 8.5 *m* = 14.4 *m*), Fig. *2d* (6.4 *m* + 8.0 *m* = 14.4 *m*), and Fig. *2e* (9.9 *m* + 4.5 *m* = 14.4 *m*). Which proves that the image formation of a painting does not change when it is trimmed; it remains constant in any case. The ML and LSV have different observer's positions. The former is at 14.4 *m*, as we already know, and the second is at 13.85 *m*. Hence, the difference between both positions is: 14.4 *m* – 13.85 *m* = 0.55 *m* (see Fig. *3b*).

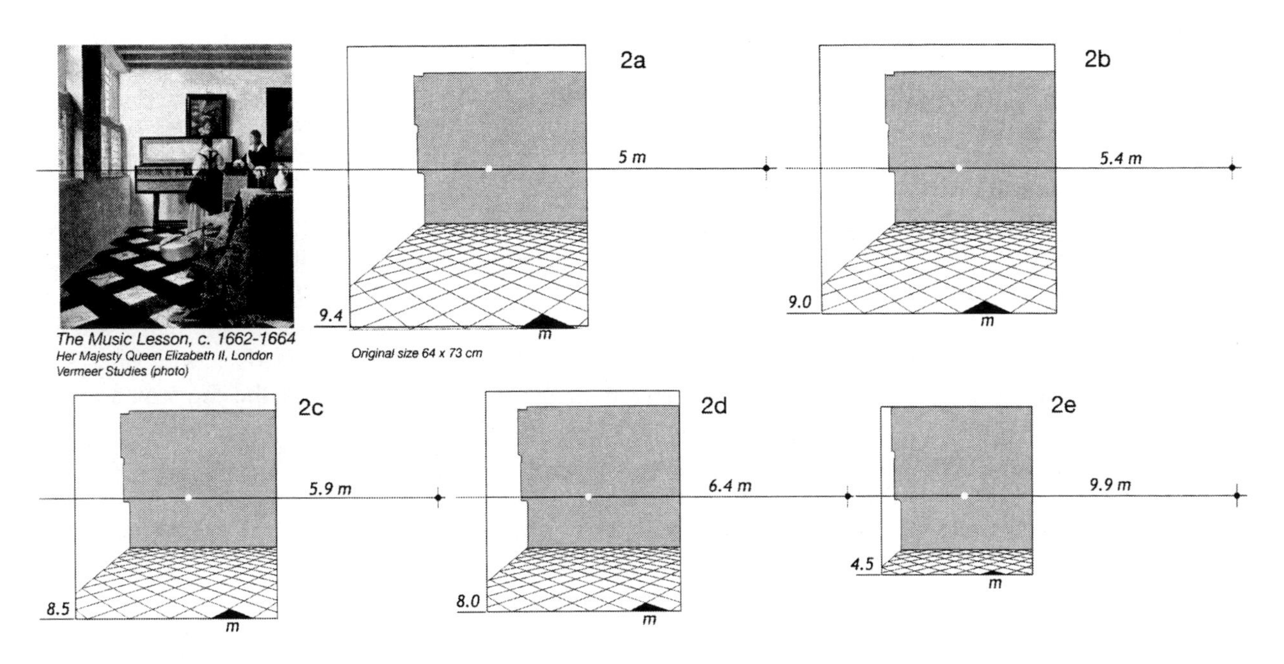

Figures 2a-2e: *Four trimming stages of the Music Lesson.*

Let's return to Fig. *1a* to interpret it's meaning on the PPbw. Comparing the modular measures between the observer's visual field and that of the PPbw, it is found that 1.93 *m* of the ML fits with 5 *m* of LSV. That is, any given measure within the observer visual field has to be multiplied by the factor $f = 2.6$, for it to lie onto the PPbw. So we have: 0.55 *m* x 2.6 = 1.43 *m*. This value is useful in finding out the *vdp* of LSV on the PP, as follow: 1.43 *m* + ML*vdp* = LSV *vdp* (on the PP) = 1.43 *m* + 5 *m* = 6.43 *m*.

3. The Music Lesson and Lady Standing at the Virginals

Our geometrical analysis begins by outlining both the ML and LSV in perspective [4]. Then after, both drawings are superimposed at their *vp*, as it is shown in Figure *3a*. In order to deduce the *dvp*'s position in each drawing, the diagonal-tiles of the floor are counted along the bottom row as modules. Thus, by carrying them over the *vh* we have: *dvp* = 5 *m*, in the ML, and *dvp* = 10.35 *m*, in LSV.

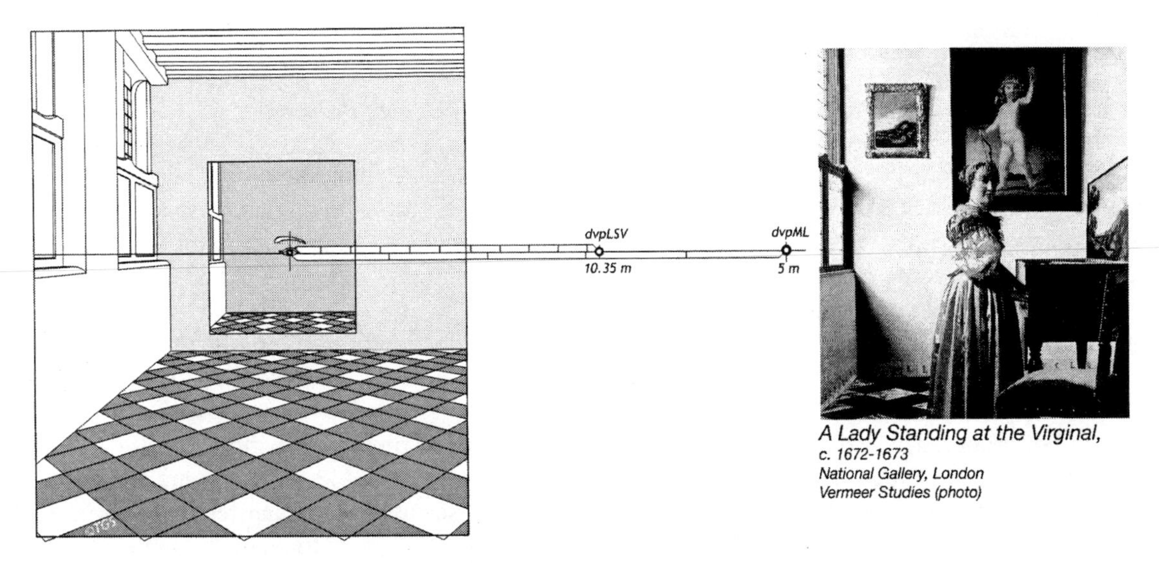

Figure 3a. *The ML and LSV overlapped at their vanishing points (vp).*

• Now, the *observer point* position for each painting is deduced as follows:
ML's *observer point*: 5 *m* (*dvp*) + 9.4 *m* (visible modules of the floor) =14.4*m*
LSV's *observer point*: 10.35 *m* (*dvp*) + 3.5 *m* (visible modules of the floor) = 13.85 *m*
• To find out the ML's true size, the distance from the *observer point* to the back wall should approximate 1.94 *m* (76 cm), which gives a diagonal-tile of 39.1 cm.
• The true size of LSV is deduced on the PP as follows: Locate point *c* by its modular perspective coordinates (X = 0, Y = -2.60 *m*, P = 0). This point lies exactly at the bottom of the visual field. Then drop the *c-vp* vertical visual ray by joining *c* with the *vp*. Next, mark on the *vh*: LSV2 *vdp* = 6.43 *m*, as deduced in section 2. This point lays on the PPbw, where the real image is projected. Therefore, sliding line *c*—LSV*dvp* until it meets point LSV2*vdp*, a new point *c'* is found where it crosses with line *c-vp*. This new point delimits the true size of LSV projected on the PPbw. Finally, after the drawing's visual field is completed, it can be attested that LSV measures ≈ 45 x 51 cm, as it is shown in Figure *3b*.
• Two points are required for the consistency proof: point *a* selected at random in LSV, and point *a'* deduced in the ML. The height (*vh*-floor) of point *a* needs to be adjusted from -2.60 *m* to -2.75 *m* to reach the ML's floor level, see Figure *3c*. This explains why point *a* appears lower than its actual position.

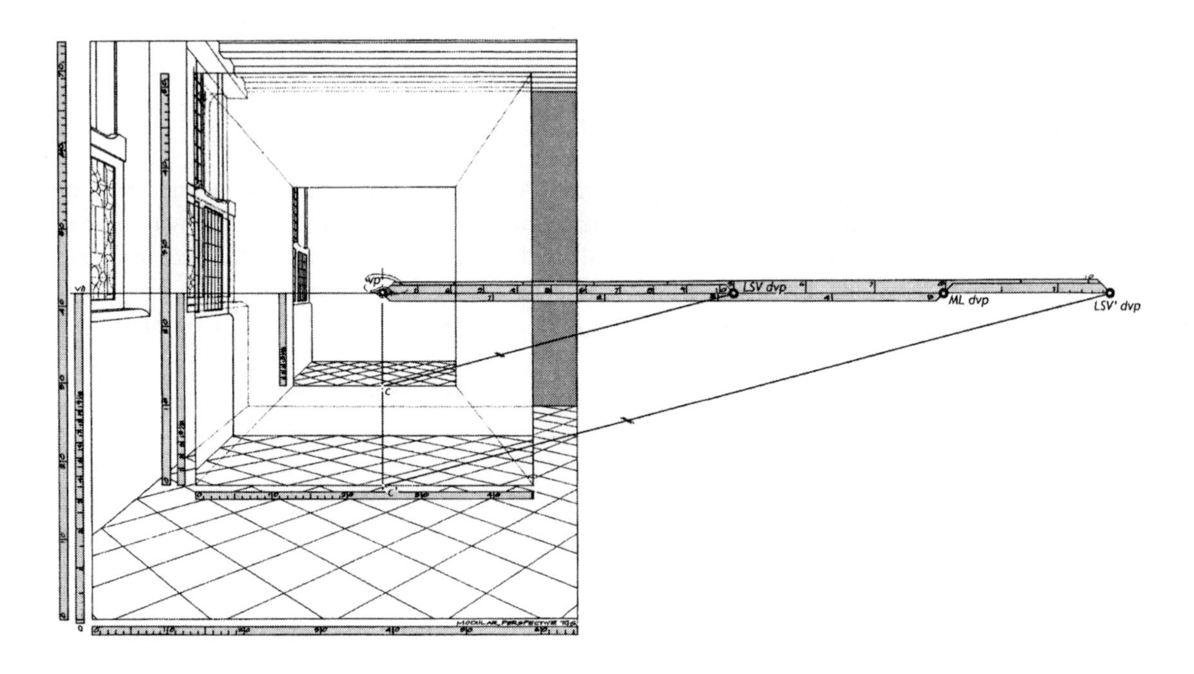

Figure 3b. *LSV measures, of≈ 45 x 51 cm, were deduced.*

• By diminishing from the rear wall 0.55 *m*, in the ML, point *a'* is found along the visual ray *a-vp* [5]. Thus, we can assure that both points *a* and *a'* occupy identical positions in the visual space. Finally, when line *a-dvp*LSV2 slides down until it meets line *a'-dvp*ML, it should result in them running parallel to each other, proving that points *a* and *a'* perfectly match point-to-point. By repeating the test for other pairs of points, the same result was obtained.
• The windows of LSV and the ML are of different types and the height of their sills also differ. This can be proven by carrying out a 45° diagonal line from each windowsill to the floor, having as a result 2.5 *m* in the ML, and 2.3 *m* in LSV.
• Despite both rooms seem to have the same dimensions and floor tiles, their windows do not match. It is unlikely that Vermeer intentionally decided to change both the windows' lead pattern and the sill's height, just for aesthetical reasons. His mindfully executed still life scenes rather manifest a pictorial realism. Whether the room depicted in LSV is that of the ML or not, they still seem to be of the same dimensions.

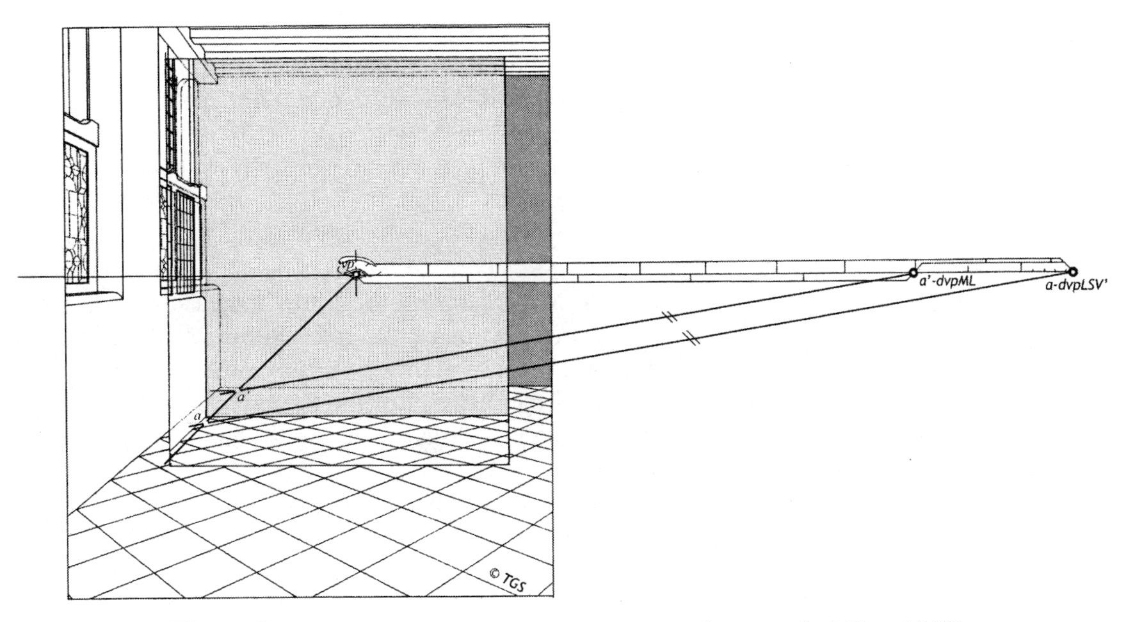

Figure 3c. *The consistency prove, point-to-point, between the ML and LSV.*

By 1672, when Vermeer painted LSV —after ten years the ML was painted—, it probably was not in the same room that the ML was, but most likely another room whose back wall approximates to it.

4. The Music Lesson and The Concert

It is impossible to make any assumption about the Concert's room identity since it has no windows. Once again, the tiled floor becomes the only reliable geometrical element available for the consistency proof.

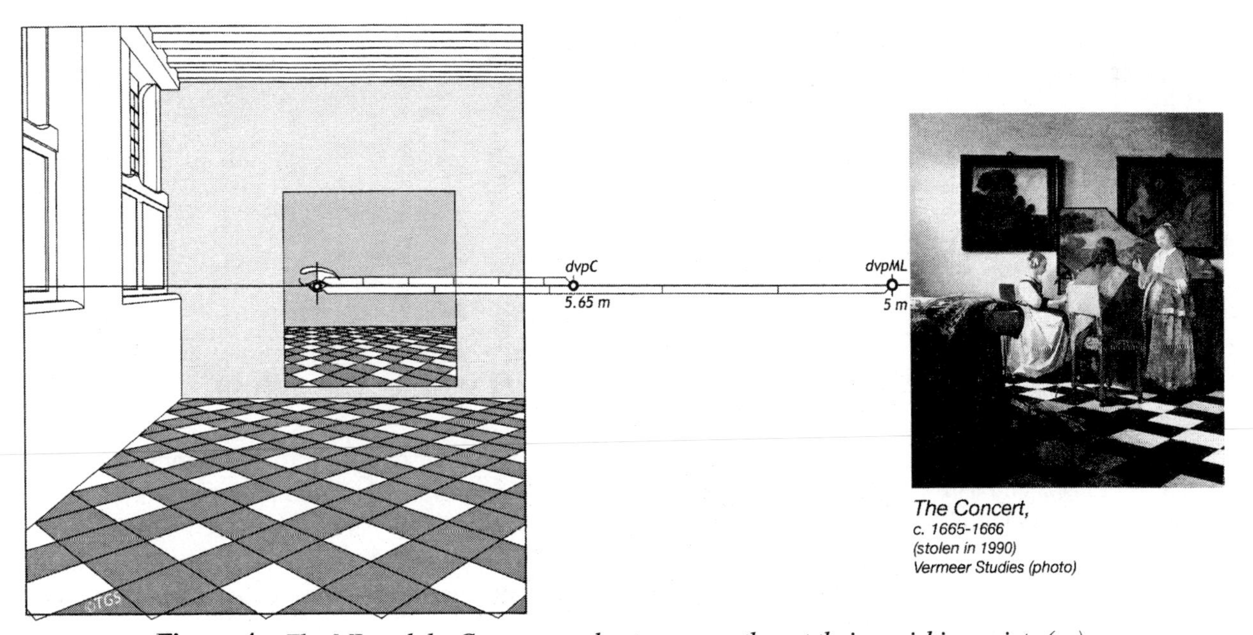

The Concert,
c. 1665-1666
(stolen in 1990)
Vermeer Studies (photo)

Figure 4a. *The ML and the Concert overlap to one another at their vanishing points (vp).*

• In Figure *4a*, both perspectives of the ML and the Concert overlay to one another at their *vp*. Each *dvp* was measured in modules at its correspondent limit of visual field, and carried over to the *vh*, as follows: *dvp* = 5 *m*, in the ML, and *dvp* = 5.65 *m*, in the Concert.

• Thus the *observer point* position for each painting is deduced as follows:
ML's *observer point*: 5 *m* (*dvp*) + 9.4 *m* (visible modules of the floor) =14.4 *m*.
The Concert's *observer point*: 5.65 *m* (*dvp*) + 8.20 *m* (visible modules of the floor) = 13.85 *m*.
• The real size of the Concert on the PP is deduced as follows: Locate point *c* by its modular perspective coordinates (X = 0, Y = -2.17, P = 0). This point lays exactly at the inferior limit of the visual field. Then drop a vertical visual ray *c-vp* by joining *c* with the *vp*.

Deducing the difference between both scenes' *observer point* we have: 14.4 – 13.85 = 0.55 *m*, thus the factor we are looking for is: 0.55 x 2.6 = 1.43 *m*. Now, carrying the PPbw's measures into the PP we have: 1.43 *m* + ML*vdp* = 1.43 + 5 = 6.43 *m* = C *vdp*. This point lays on the PPbw where its real image is projected. Sliding a parallel line from *c* – C *dvp* until it meets point C2 *vdp*, a new point *c'* is found where it crosses the vertical *c* – *vp*. This new point delimits the real size of the Concert projected on the PPbw. Finally, the whole visual field is completed by carrying visual rays from the *vp*, having as a result that the Concert measures ≈ 64.5 x 72.5 cm, as it is attested in Figure *4b*.

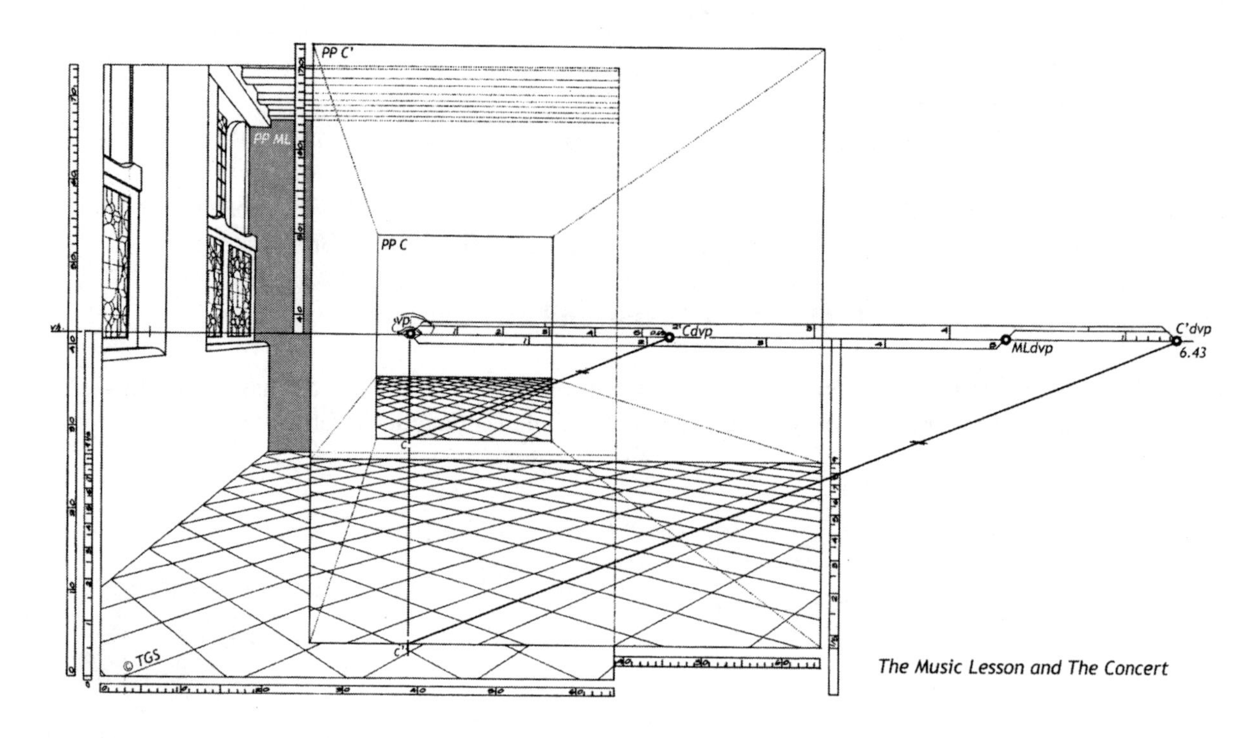

Figure 4b. *The Concert's measures, of ≈ 64.5 x 72.5 cm, were deduced.*

5. The Music Lesson and The Girl with a Wineglass

Based on Steadman's theory, of a ceramic tile been one-half of the marble's [6], the underlying floor grids of the ML and The Girl with a Wineglass (TGW) failed to match, since the aforementioned margin of error goes beyond of an acceptable tolerance. The Glass of Wine (GW) is also included here because it seems to depict the very same room of TGW, given that both appear to have the same window's leaden glass-pattern and floor tiles. To prove this, the missing part of the window was reconstructed in TGW, and then verifying its position with that of the GW.

As it can be seen in Figures *5a* and *5b*, the interval of the central window was found between the ≈ 10.4 *m* – 18 *m* rows for both grids [7], proving that they have the same embrasure span, while in Steadman's axonometric views this embrasure span commences at 9.5 *m* in GW and at 11.5 *m* in TGW.

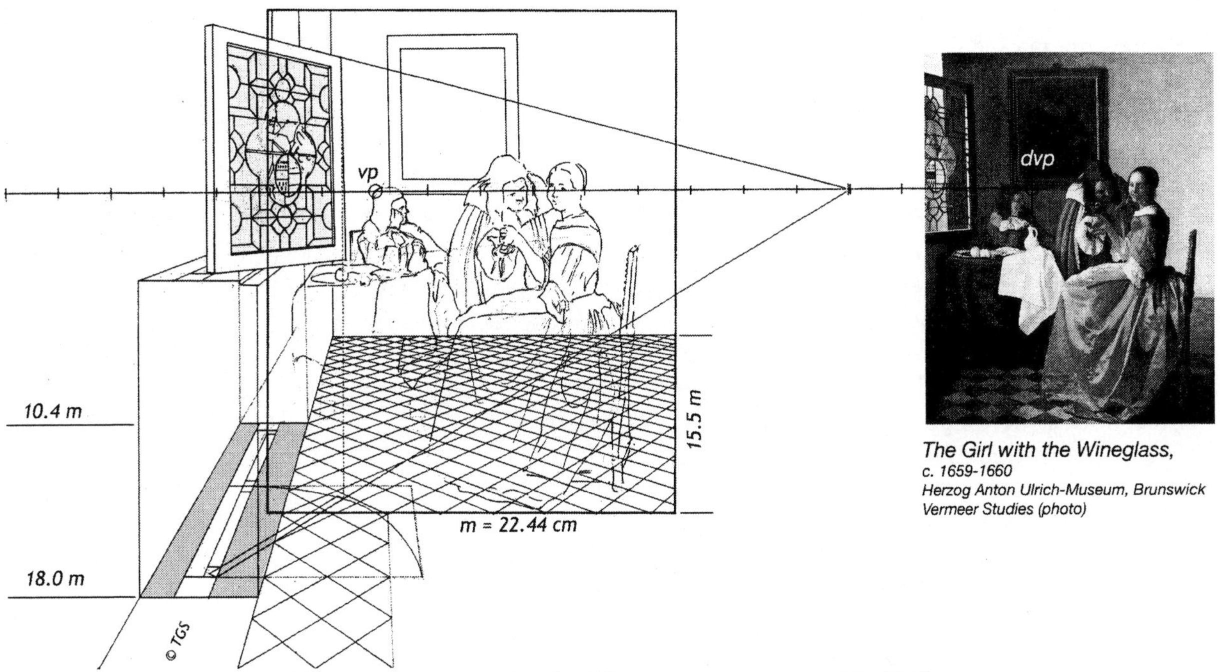

Figure 5a. *The Girl with a Wineglass reconstructed by TGS.*

If Steadman's assumptions were true, then the expected modular position for both windows must lay between the 12 *m* – 20.8 *m*, in accordance with the ML, which lies between the ≈ 6 *m* – 10.4 *m*. In addition, the window's height does not match: 2.60 *m* ≈ 2.25 *m* (the ML Vs. both TGW and WG respectively). Not only Steadman's measures are incompatible with the ML but also mine's. This discrepancy suggests that both floor tilling were not in 0.50:1 proportion. They instead seems to be in ≈ 0.575:1 proportion, as deduced through Figs. *5a* and *5b*, as follows:

TGW & GW window's embrasure: 10.4 *m* – 18 *m* = (6 x 1.74 – 10.4 x 1.74) ≈ 10.4 *m* – 18.1 *m*.
TGW & GW window's height: 4.5 *m* ≈ (4.5/1.74) ≈ 2.59 *m* ≈ ML window's height ≈ 2.60 *m*.

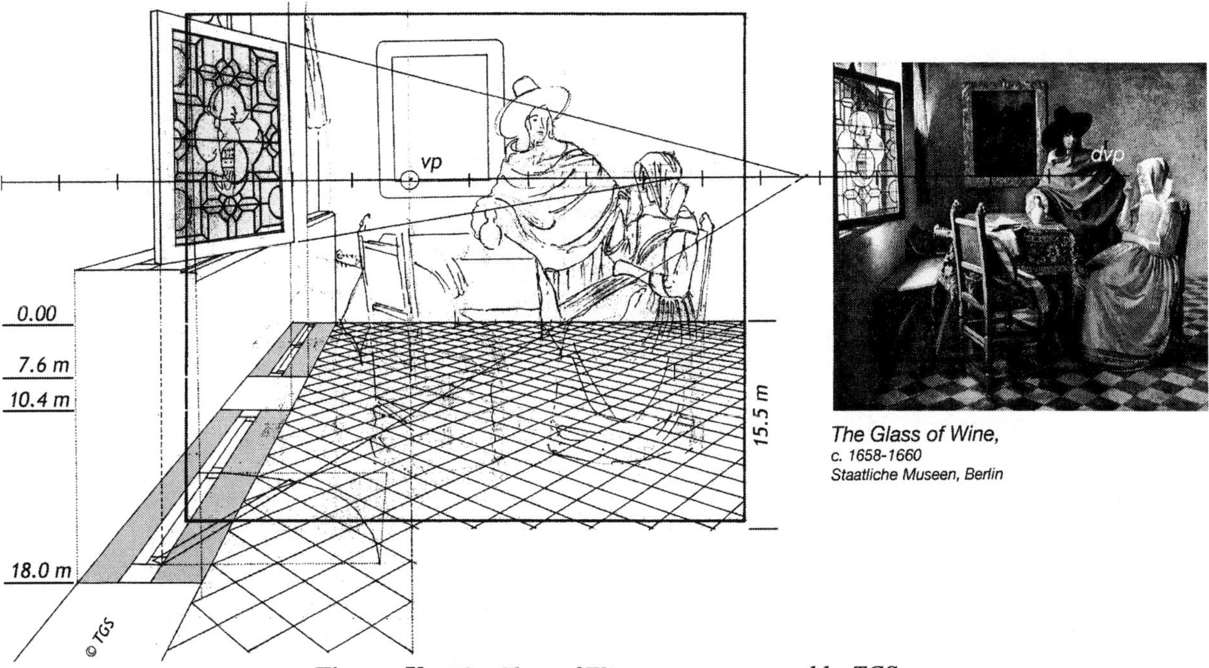

Figure 5b. *The Glass of Wine as reconstructed by TGS.*

So the awkward questions should be: how the ceramic floor grid could possible has served as a reference to deduce the marble floor under unmatched proportions? What if the marble flooring was in place at the time TGW was painted? At this point, we might conclude that the ceramic floor was captured at its current size by the camera, and then when it was renovated to the marble it was captured in the same way.

In both drawings, TGW and GW, the windows hold the clue to deduce where the *observer point* stands. In the TGW, the *observer point* is at: 12.2 *m* (*dvp*) + 15.5 *m* (visible modules of the floor) = 27.7 *m.* See Figs. *5a* and *5c*. Its position lays behind of the ML's but not too much to throw away the idea of it being the same room, since its real dimensions are not yet conclusive. If visual ray (*vr1*) were carrying out from the *observer point*, passing by tangentially to the left column, it would hit on the second column's lateral-face, as in the painting itself. Repeating the same proof for the visual ray (*vr2*), from Steadman's *observer point* estimation, it would fall outside of the window. This is the bottom question about the ceramic and marble floor's tile proportion. The same proof, as described for TGW, throw identical results for GW. Thus, my estimation of the room's width for both TGW and GW is about 7.63 m, instead of 6.46 m. This is why the consistency proofs between TGW and the ML's underlying grids is not practicable, since they seem to have different back wall positions, or even relate to different rooms.

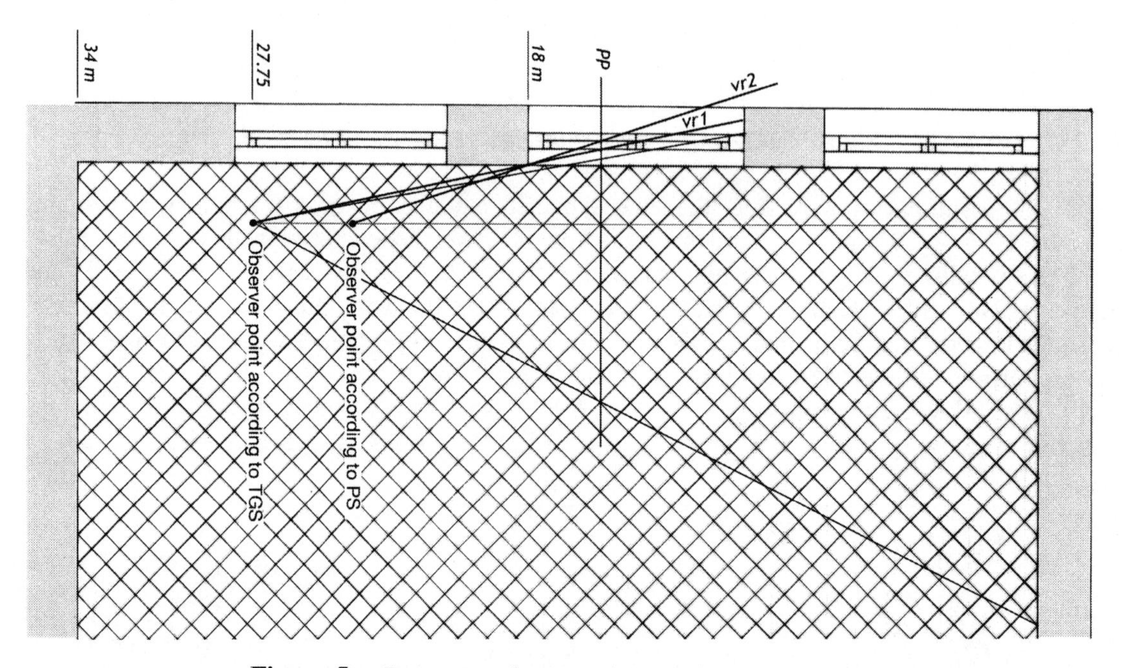

Figure 5c. *Comparison between the visual rays, vr1 and vr2.*

Notes and References

[1] Tomás García-Salgado, *"Some Perspective Considerations on Vermeer's The Music Lesson"*, *perspectivegeometry.com* e-journal, Number 1 (2003).
[2] Philip Steadman, *Vermeer's Camera, Uncovering the Truth Behind the Masterpieces* (UK: Oxford University Press, 2002), p. 73.
[3] Steadman [2], pp. 102-103.
[4] A margin of error of about ± 2 percent must be estimated since the photographs were not in true size.
[5] This value represents the difference between both observers in the room.
[6] Steadman [2], pp. 101, 114-115.
[7] Considering the diagonal of a ceramic tile as one *module* ($m \approx 31.7$ cm); counting from back to front.
Photo credits, as indicated in foot captions, are from: *Vermeer Studies*, Ivan Gaskell and Michiel Jonker, Eds. National Gallery of Art, Washington, 1998.

On the Bridging Powers of Geometry In the Study of Ancient Theatre Architecture

Zeynep Aktüre
Department of Architecture
Izmir Institute of Technology
Gülbahçe Campus, Urla
35430 Izmir, Turkey
E-mail: zeynepakture@iyte.edu.tr

Abstract

The on-going popularity of the Vitruvian layout for the Latin theatre is largely due to its capacity to bridge across several disciplines, which seems to appeal to a certain conception of material culture that assumes the existence of a plurality of formally similar structures of culture beyond surface phenomena. These tend to be not merely potent in their explanatory force but also gratifying aesthetically and ethically. Modern scholarship has forcefully promoted such a conjunction of truth, beauty, and goodness in the link between the Theatre in the Asklepieion at *Epidauros* and Pythagorean speculation. However, similar cognitively-significant structural or formal bridges would seem difficult to establish in all examples. In their absence, the search for a perfect geometry of perfect shapes beyond the extant remains may turn into a purely formalist exercise made possible by the capability of geometry to serve as an analytical tool through a reduction of the architectural code to a geometric code. This is a dilemma intrinsic in the difficult relation between architecture and geometry. In fact, Vitruvius seems to have noticed the problem long ago and tried to build a material bridge between his geometric assembly and the architectural project by recognizing the necessity to give up symmetry in the latter, wherever required by the nature of the site or the size of the project.

1. A Structural Bridge: The Vitruvian Layout for the Latin Theatre

David B. Small is one of the scholars who examined archaeological evidence for Roman period theatres to see that the majority were not constructed following the method described by the first century BC Roman architect Marcus Vitruvius Polio in the fifth book of his famous *De architectura*, which was dedicated to the Emperor Augustus [1]. This conclusion would appear to be in line with the critical opinion that Vitruvius' theory of architecture bears almost no relation to the realities of contemporary practice and had little influence on the near future [2]. His investigations led Small to devising, on the basis of the Vitruvian method, an alternative with greater explanatory power. His was one in a series of attempts including those of P.C. Hammond, F.B. Sear, Salvador Lara Ortega, and Luis Moranta Jaume [3, 4, 5, 6]. When taken together, these efforts attest the on-going popularity of the Vitruvian layout for the Latin theatre by taking it as a starting point, as do the eagerness of some archaeologists to adjust their evidence following Vitruvius' step-by-step method of design [1]. Examples for the latter would come from a wide geography, expanding from *Bilbilis Augusta* and *Segóbriga* in the Iberian to *Ancyra* in the Anatolian peninsulas and beyond [5, 6, 7, 8].

This paper will attempt to explain this on-going popularity of the Vitruvian layout by its capacity to bridge across several disciplines. To start with geometry, Vitruvius identifies the Latin theatre in the mirror of the Greek one, by the "*difference, that theatres designed from squares are meant to be used by Greeks, while Roman theatres are designed from equilateral triangles.*" [9] However, both design processes start with a circle inside which are inscribed, at equal distances apart and touching the boundary line of the circle, three squares in the Greek and four equilateral triangles in the Latin practice to produce

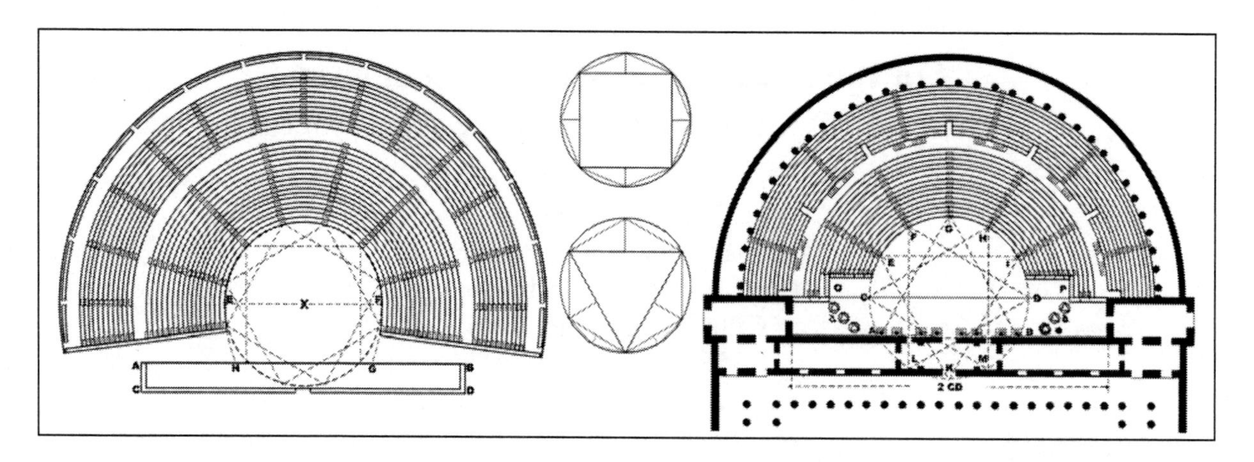

Figure 1: *The Vitruvian layouts for Greek (left) and the Latin theatres (right) and the quadrature of the circle according to Simplicius (centre top) and Themistius (centre bottom)* [after 9 and 10].

a circle of twelve points in both cases. The position of all the architectural components of both types is defined in the plan layout with respect to these initially-inscribed squares or triangles, in structural relation with each other. As to the design of the ascending rows of seats in the third dimension, Vitruvius argues for the convenience of the same dodecagonal layout for an arrangement with bronze sounding vessels that would satisfy perfectly the necessities of a physical phenomenon such as sound, on the basis of "*the canonical theory of the mathematicians and that of the musicians*".

Modern scholarship has demonstrated that the equivalence of the two resulting schemes in terms of mathematical logic would suggest an origin in the two apparently different but essentially analogous methods used by the Sophists to resolve the quadrature of the circle [10]. Their overlapping would provide an embryonic image of the *signifer circulus* formula for celestial harmony, which was developed on the basis of the zodiac circle. As explained by the Geminos of Rhodes (110-40 BC), it was the position of the signs in trigons on the sphere of the fixed and of those in quadrature that produced an accord with their opposition. The circle or the sphere were, by definition, generators of perfection and the inscription of a figure inside a circle or a solid inside a sphere was the proof, both for the ancient geometrician and for the ancient philosopher, of the accomplished character of that figure or object. The impossibility of inscribing any other perfect shape into a circle of twelve zodiacal points would suggest the idea of an innate binarism for the two Vitruvian schemes. This idea of binarism, appearing within the context of a pair of geometries that apparently bridges, as such, across the disciplines of geometry, architecture, physics, mathematics, music, logic, astronomy, philosophy, and taxonomy, would lead to a certain conception of material culture whose critical evaluation may help in a better understanding as to why such congruence would appear to provide a more satisfactory explanation of architectural phenomena than any other that would evaluate them as objects in and of themselves.

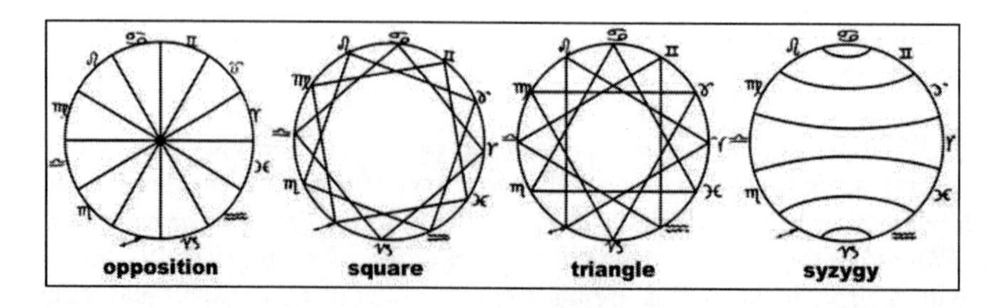

Figure 2: *The* signifer circulus *formula as explained by the Geminos of Rhodes* [after 10].

The reference here is to a model of *structuralisme* constructed by Ernest Gellner as a generative method consisting of a set of ideas or themes from which arguably follow the observed activities and positions of people normally described as structuralists [11]. These include a belief in the existence of a plurality of structures that "*formally resemble each other and are rooted in some generically shared structure of the human mind.*" In the analyses of material culture, the idea of a core set of elements (or structure), that are generally assumed to occur in a pair of polar opposites and constitute the limits of the world in question, comes to the fore as generating the system of signs. This would accord with Gellner's classification of structuralism as a form of emanation, which is one of the two main conceptions of causation, with the other being covering law. The two are parallel in their mutual presumption of a permanent core or another reality beyond surface phenomena. Emanationist explanations assume that the regularities discerned in the phenomena open to view emanate or flow from that permanent core that is normally (or permanently) hidden from view. Any attempt to bind phenomenal regularities under generalizations, including surface classification and prediction, is bound to remain superficial without an understanding of those inner forms that are, on the other hand, accessible only through their alleged manifestations in the phenomena. "*Those inner forms tend to be not merely potent in their explanatory force but also gratifying aesthetically and ethically. They reveal a moral as well as an ontological order; in fact these various orders converge. Truth, beauty, and goodness are one.*" [11] In this way, emanationist explanations provide, not only shorthand summaries of surface patterns, but also "*a deep, permanent, morally saturated, and satisfying reality, qualitatively different from and superior to the ephemeral and amoral connections observed on the surface of things.*" [11] The on-going appeal of the Vitruvian layout for the Latin theatre seems to be rooted in its capacity to reveal such a reality, in the face of (or perhaps due to) the fact that the corpus of ancient theatre remains is too amorphous to be equally gratifying in itself.

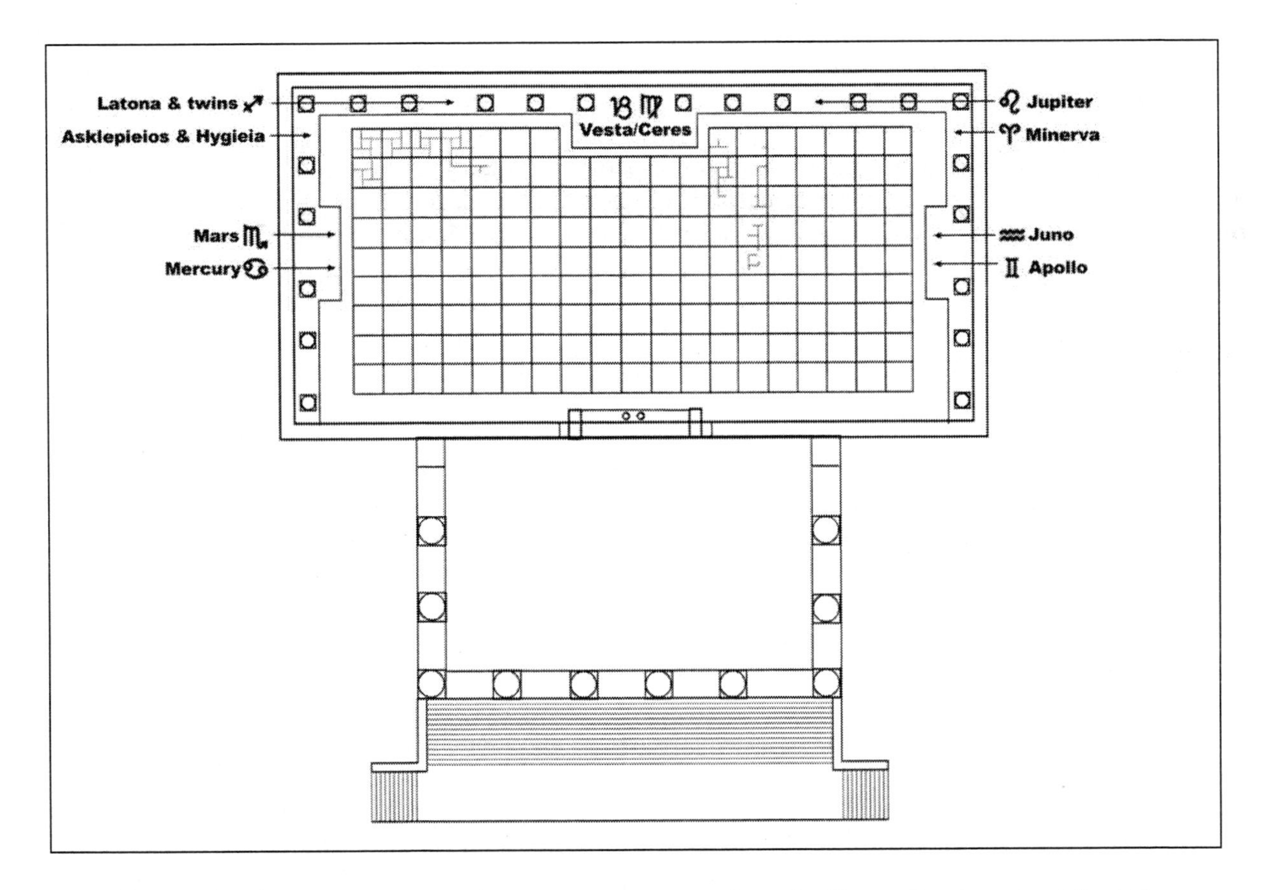

Figure 3: *The reconstruction of the statuary program at the aedes Concordiae Augustae* [after 13].

Architecture is a phenomenon beyond which the existence of another reality is, not presumed, but known to exist: the conscious will and performance of man. Beyond the Vitruvian layout for the Latin theatre, with all its cross-disciplinary associations, there would seem to exist the conscious will and performance of one man: the Emperor Augustus, who is portrayed by Paul Zanker as having structured an Imperial Roman culture using the power of images [12]. So, Pierre Gros would seem to have a strong point in suggesting that the *signifer circulus* formula for celestial harmony may have had a more direct application on the Vitruvian scheme for the Latin theatre than generally admitted [10]. A most direct reference is to be found in the sculptural program of the Temple of Concordia Augusta in Rome wherein the pairings of Juno and Apollo, Mars and Mercury, and Vesta and Ceres would have served to represent the harmonious world order that is being established by Augustus through the very principle of opposition in trigons and quadrature on the sphere of the fixed, as mentioned by the Geminos of Rhodes: "*Vesta rules Capricorn, Ceres, and Virgo, and Capricorn and Virgo are in the same trigon; Mars rules Scorpio, Mercury, and Cancer, and Scorpio and Cancer are members of another trigon; Juno rules Aquarius, Apollo, and Gemini, and Aquarius and Gemini are members of the third trigon. Significantly, Manilius points out that relations within a trigon were not always untroubled, but ultimately harmony prevailed and balance was maintained.*" [13] Vitruvius' dedication of a whole book to the Latin theatre would have indicated his awareness of the key role to be played by the theatrical edifice in the establishment of this difficult balance within the new world emerging under Augustus.

2. A Formal Bridge: The Theatre of the Asklepieion at *Epidauros*

As such, the Vitruvian layout for the Latin theatre would seem to owe its bridging powers to a careful structuring by the Roman architect of the current ideas of his time to serve the political agenda of the first Roman emperor to whom his treatise on architecture was dedicated. This oldest surviving treatise on architecture was rediscovered and printed for the first time in 1484 [14]. Publications documenting attempts at a restitution of ancient theatre remains in the Iberian Peninsula following Vitruvius' directions date back to as early as the sixteenth century, possibly due to the influence of the *Medidas del Romano* written by Diego de Sagredo, which appeared in Spanish in 1526 [15]. From that time on, the Vitruvian model has been so influential on our modern views about the Greek and Roman theatres that it served as a measure for architectural achievement in ancient theatres. A notable example in this respect is Gottfried Gruben's description of the renowned Hellenistic Theatre of the Asklepieion at *Epidauros*, on the basis of Wilhelm Dörpfeld's early plan of 1896, as the best among all surviving theatres in displaying the geometrically determined principles of design referred to by Vitruvius that manifested, according to the author, in the division of its *cavea* into twelve by taking as reference the corners of a dodecagon inscribed in the basic circle of the *orchestra* [16].

However, very detailed and precise measurements taken by Armin von Gerkan and Wolfgang Müller-Wiener in the period 1953-56 were soon to establish the basic organizing scheme possibly of the whole edifice to be the inner circle of its *orchestra* inscribed by a perfect pentagon [17]. Associated traditionally with Venus, the pentagon was apparently more potent in its explanatory force than the square or the equilateral triangle for the case of *Epidauros*, as it described a five-pointed star that symbolized *Hygieia*, daughter of Asklepieios and personification of health, for the Pythagoreans [18]. Around 450 BC, Greek thinkers had discovered the Golden Ratio from Pythagorean speculation and from the inscription of this very form inside a circle. There are echoes of this theorem in Plato's *Timaeus* (31b-33c), in the Treasury of *Cyrene* at *Delphi* designed as a demonstration and exemplification of mathematical wisdom, and in literary masterpieces including works of Vergil, Lucretius, Catullus, and Lucan, wherein passages are divided into line-totals that are in ratio to each other and to the whole passage or poem in Golden Section relations [19]. Jean Bousquet has illustrated the use of the Golden Section, and the Fibonacci Series intimately related to it, in the arrangement of the tiers in the *cavea* of the Theatre of *Epidauros* [20]. There are 21 rows in the upper and 34 in the lower *cavea*, which sum up to 55. All three are numbers in the Fibonacci Series and 21:34 is in ratio to 34:55. Additionally, these numbers constitute a Grand Tetractys. R.V. Schoder believes that, the architect's choice for precisely these numbers would have resulted from "*a deliberate desire to work into his design the perfections and mysterious interrelations of these numbers which contemporary Pythagorean speculation extolled.*" [19] So, here again, we seem to be faced with a layout that owes its bridging powers to a careful structuring of the current ideas of a certain period but this time the product is a work of architecture, instead of a verbal description of a design method in a treatise on architecture. This would seem to make a great difference with Schoder's evaluation of it as "*the most aesthetically pleasing arrangement in the history of theatre design*" [21]:

> True, an observer is not quite aware of all these mathematical correlations in precise equivalents and factors and multiples, but he is aware of the over-all harmonious symmetry, and pleased by it. [...] As in understanding a painting, so here one discovers the details and complex associations of individual elements only by close-up study and analysis, which seems mechanical and dehumanizing; but this necessary process of factual dissection and part-by-part examination leads to a better aesthetic appreciation also and is the sole means of discovering and adverting to—and thereby enjoying—some facets of the artistic accomplishment. [16]

With this statement, we seem to move one step closer to the structuralist idea of "*some generically shared structure of the human mind*" that is capable of appreciating the beauty of a work of art or architecture through an understanding of the underlying geometric forms and mathematical correlations. In this way, we also come closer to what corresponds to aesthetic formalism in the philosophy art. Aesthetic formalism may be described as a body of ideas according to which the aesthetic value of natural or manmade objects is determined by their form understood as opposed to matter, as opposed to content, as opposed to formlessness or, notably, as structure where the structure is constituted by a system of relations [22]. Theoreticians considered to be formalists have used the notion of form primarily to denote "*a certain* arrangement *of parts, a structure of elements, or a global* composition *of elements of a work or some other object*" and, therefore, in opposition to matter or substance [23]. At the origin of artistic formalism was a protest against the treatment of works of art as substitutes for politics, morality, or religion, as in Victorian expectations for an enhancement in the morality of the society through literature. This defense of art's autonomy by renowned theoreticians and critics working in diverse fields has had its share in the notion of the uniqueness of individual art forms and values exclusive to each one of them. However, this notion has had the risk of leading to an experience and appreciation works of art exclusively as objects in and of themselves rather than within a broader artistic and cultural context, without an acknowledgement of the cultural nature of art or the cognitive significance of individual works beyond their purely aesthetical properties. This has been marked as a major weakness of formalist approaches, together with their strong normativistic tendencies that reveal in an imposition of aesthetic experiences based on formal aspects as the only correct model of art reception and art evaluation. [23]

Reviewing the extant literature on the Theatre of *Epidauros* under the light of these criticisms would reveal the determining influence of the artistic and cultural context of the monument, as well as the cognitive significance of the design tools used in the shaping of its individual components over the resulting aesthetical experience. The building is thought to be part of an ambitious building program implemented during the fourth and third centuries BC in the Sanctuary of Asklepios that had become, by that time, the most celebrated healing centre of the ancient world [24]. A Polykleitos the Younger was employed for the design of the monument, who must have been an Argolid architect rather than the renowned fifth century sculptor [19]. Nevertheless, he is one of the very few ancient architects whose name has survived up to our times in connection with a theatre design, perhaps due to the significance of that very design itself. In the light of the previously reported analyses, Polykleitos would seem to have come up with an idea that is meaningful in (and probably only in) connection with the cult of Asklepios although the very same scheme would have been equally gratifying aesthetically in another context if it were to be judged "*only by a close-up study and analysis*", as suggested by Schoder, and not as part of a unique natural and cultural landscape. It apparently was the prosperity of the healing complex inserted in that landscape that made possible the application of that design idea in such precision that modern scholars could reveal the geometric and mathematical relations structuring it by working back from the extant remains and then discover the associations of those relations with Pythagorean speculation. Their analyses have provided a potent and gratifying explanation of the impressing architectural characteristics of the edifice.

3. As a Conclusion: A Material Bridge

The mechanism here seems to be the one described by Umberto Eco through a distinction between the denotations and connotations of architectural objects [25]. While the denoted meaning of architectural objects is the function they make possible, their formal characteristics may also "*refer to a certain conception of inhabitation and use; they may connote an overall ideology that has informed the architect's operation.*" In the semiotic mechanism described by Eco, connotations of architectural objects rest on the denotation of their primary function. In an attempt to move back from the complex and temporal connotations, it might be tempting "*to hypothesize for architecture something like the 'double articulation' found in verbal languages, and assume that the most basic level of articulation (that is, the units constituting the 'second' articulation) would be a matter of geometry.*" Accordingly, the Euclidean *stoicheia* (i.e. the elements of classical geometry such as the point, the angle, or various curves and the straight line) would combine into certain high-level spatial units called *choremes* (i.e. the square, the triangle, the parallelogram, the ellipse, and even rather complicated irregular figures that can be defined with geometric equations of some kinds) that belong to a "first articulation" and begin to be significant. The "second articulation" would be based on these units. A third level of articulation is suggested by solid geometry and other articulation possibilities may be assumed with non-Euclidean geometries.

The same code may be argued to lay behind some artistic phenomena, including abstract and representational art, on the premise of the long-held conviction that configurations in the latter can be reduced to a rather complex articulation of primordial geometric elements. In this way, the analytic possibilities offered by geometry enable a comparison of architectural phenomena with other type of phenomena by describing them in the same terms, as in the case of the Vitruvian schemes or the analyses of the Theatre of *Epidauros*. For Eco, this reveals the capability of geometry to serve as a *metalanguage* that might even be identified with a *gestaltic* code presiding over our perception of all such forms, as assumed by *structuralistes*. The problem is that, "*the fact that architecture can be described in terms of geometry does not indicate that architecture as such is founded on a geometric code.*" [25] An analogous example comes from language studies wherein the fact that both Chinese and Italian words are articulated in phonemes and, therefore, can be seen as a matter of amplitudes, frequencies, or wave forms does not indicate that "*Chinese and Italian rest on one and the same code; it simply shows that the languages admit of that type of analysis, that for certain purposes they can be reduced to a common system of transcription.*" [25] This would reveal a dilemma intrinsic in the relation between architecture and geometry in analytical studies on ancient theatre architecture, which is rooted in the fact that all ancient theatre buildings would have been laid out using a geometric assembly one some sort. Gros observes that Vitruvius had noticed the problem long ago and tried to establish a more direct relation between his geometric assembly and the architectural project with recommendations for giving up symmetry for the sake of utility in steps, curved cross-aisles, their parapets, the passages, stairways, stages, tribunals and the like, which need to have the same size both in a small and a large theatre [9, 10]. In the same paragraph, Vitruvius admits the impossibility of applying the proportional systems he proposes in all theatre constructions. "*Instead, it is up to the architect to note in which dimensions it will be necessary to pursue symmetry and in which to make adjustments according to the nature of the site or the size of the project.*" [9] And it is up to us, the modern investigators of ancient theatres, to decide in which theatres it would be meaningful to search for a perfect geometry of perfect shapes beyond the extant remains and in which to be contended with whatever remains we have in hand, without attempting to move back from them with the hope of finding a more satisfactory explanation in the connotations of such geometries.

It would seem that certain buildings such as the Theatre of *Epidauros*, with its highly symbolic religious connotations within the Sanctuary of Asklepios, would afford all the purposes required for such a reduction to a common system of transcription through a description in terms of geometry. Others that are located in major administrative centers of the Roman Empire such as *Ancyra*, the capital of the Roman province of *Galatia* in Anatolia, may motivate an analyses in reference to the Vitruvian layout for the Latin theatre, if by nothing else, by the direct link established with the Augustan project through the dominating presence of the nearby Temple of Augustus and Roma that features the monumental inscription panels of the complete text, in Latin and in Greek translation, of *res gestae divi Augusti*, the first and last report given by a Roman emperor, the Emperor Augustus, to his nation of his life and deeds. However, not all ancient theatres, Greek or Roman, may be expected to have equally strong connotations. Some, such as the Theatre of *Segóbriga*, were apparently constructed only to perform their denoted function and with a limited budget that could be allocated so late as the aftermath of Vespasian's grant of municipal status (*ius Latii*) to communities in the province of *Baetica* in Iberia. This should not be taken to mean the impossibility of their having been designed after an elaborate system of geometrical or mathematical relations beyond their material remains. Lara Ortega and Moranta Jaume have separately suggested two schemes for the monument featuring respectively four and three equilateral triangles inscribed in a circle, taking as a basis the Vitruvian layout for the Latin theatre [5, 6]. It should be taken, instead, as a warning against the irresistible charms of fabricating one such layout for the more deteriorated examples, such as the Theatre of *Bilbilis Augusta* [7].

References

[1] D.B. Small, 'Studies in Roman Theater Design', *American Journal of Archaeology* 87 (1983) pp.55-68.

[2] John Onions, *Bearers of Meaning - The Classical Orders in Antiquity, the Middle Ages, and the Renaissance* (Princeton, New Jersey: Princeton University Press, 1988) pp.40-41.

[3] P.C. Hammond, *The Excavation of the Main Theatre at Petra, 1961-1962 - Final Report* (London: Colt Archaeological Institute Publications, Bernard Quaritch Ltd., 1965).

[4] F.B. Sear, 'Vitruvius and Roman Theater Design', *American Journal of Archaeology* 94 (1990) pp.249-58.

[5] Salvador Lara Ortega, 'El trazado vitrubiano como mecanismo abierto de implantación y ampliación de los teatros romanos', *Archivo Español de Arqueología* 65 (1992) pp.151-79.

[6] Luis Moranta Jaume, *Hipóthesis de la Existencia de un Teatro Romano en Palma de Mallorca* (http://palma.infotelecom.es/~moranta, accessed on September 22, 2000).

[7] Martín Martín-Bueno and J. Núñez Marcén 'El teatro Municipium Augusta Bilbilis' in *Teatros Romanos de Hispania*, edited by S. Ramallo and F. Santiuste (Murcia: Universidad de Murcia y Colegio de Arquitectos de Murcia, 1993) pp. 119-32.

[8] Musa Kadıoğlu, 'Ankyra Tiyatrosu: Ön Rapor' in *I.-II. Ulusal Arkeolojik Araştırmalar Sempozyumu*, edited by Tunç Sipahi and Zeynep Çizmeli Öğün (Ankara: AÜDTCF Yayınları, 2004) pp.123-40.

[9] Vitruvius, *The Ten Books on Architecture*, translated by Morris Hicky Morgan (New York: Dover Publications, Inc., 1960; reprint of the 1914 Harvard University Press edition) p.153.

[10] Pierre Gros, 'Le schème Vitruvien du théâtre latin et sa signification dans le système normatif du *De Architectura*', *Revue Archéologique* (1/1994) pp.57-80.

[11] Ernest Gellner, 'What is structuralisme?', in *Theory and Explanation in Archaeology*, edited by Colin Renfrew, Michael Rowlands and B.A. Segraves. (New York: Academic Press, 1982), pp.97-124.

[12] Paul Zanker, *The Power of Images in the Age of Augustus*, translated into English by Alan Shapiro (Michigan, Ann Arbor: The University of Michigan Press, 1990).

[13] B.A. Kellum, 'The City Adorned: Programmatic Display at the *Aedes Concordiae Augustae*' in *Between Republic and Empire – Interpretations of Augustus and His Principate*, edited by Kurt A. Raaflaub and Mark Toher (Berkeley, Los Angeles & Oxford: University of California Press, 1990) p.295.

[14] Margarete Bieber *The History of the Greek and Roman Theater* (Princeton, New Jersey: Princeton University Press, 1961) p.255.

[15] Joseph Rykwert *The Dancing Column - On Order in Architecture* (Cambridge, Massachusetts; London, England: The MIT Press, 1996) p.401.

[16] Helmut Berve and Gottfried Gruben, *The Architecture of the Temples, Theatres and Shrines*, volume 2 (London: Thames and Hudson, 1963).
[17] Armin von Gerkan and Wolfgang Müller-Wiener, *Das Theater von Epidauros* (Stuttgart: W. Kohlhammer, 1961).

[18] Lutz Käppel, 'Das Theater von Epidauros. Die mathematische Grundidee des Gesamtwurfs und ihr möglicher Sinn', *Jahrbuch des Deutschen Archäologischen Instituts* 104 (1989) pp.83-106 cited in J.R. Green, 'Theatre Production: 1987-1995', *Lustrum* 37 (1995 [1998]) p.52.

[19] R.V. Schoder, 'II. The Theatre at Epidauros as a Work of Art', in *Greek Drama – A Collection of Festival Papers II*, edited by Grace Lucile Beede (Vermillion, South Dakota: The Dakota Press, the University of South Dakota, 1967) pp.13-39.

[20] Jean Bousquet, 'Harmonie au Théatre d'Epidaure', *Revue Archéologique* 41 (1953) pp.41-49.

[21] R.V. Schoder *Ancient Greece from the Air* (London: Thames and Hudson, 1974) p.66.

[22] David Pole, *Aesthetic, Form and Emotion* (London: Duckworth, 1983) cited in [23].

[23] Bohdan Dzemidok, 'Artistic formalism: its achievements and weaknesses', *The Journal of Aesthetics and Art Criticism* 51/2 (1993) pp.185-193.

[24] Juan Miguel Casillas and César Fornis 1995. 'Epidauro y el culto a Asclepio', *Revista de Arqueología* 173 (1995) pp.28-39.

[25] Umberto Eco, 'Function and Sign: The Semiotics of Architecture', in *Rethinking Architecture – A Reader in Cultural Theory*, edited by Neil Leach (London and New York: Routledge, 1997) pp.182-202.

The Gemini Family of Triangles

Alvin Swimmer
Barrett Honors College
Arizona State University
Tempe, AZ 85287-1612
aswimmer@asu.edu

Mary C. Williams
7736 E. Portland St
Scottsdale, AZ 85257
mary.c.williams@asu.edu

Abstract

There are a series of triangles in the pentagon/pentagram figure that can be used advantageously in quilting. We are going to investigate these triangles both mathematically and artistically.

Introduction

A regular pentagon P with its inscribed "star" (the pentagram constructed from the five diagonals of P) Figure 1 can be advantageously used to construct quilt designs that are aesthetically pleasing and mathematically fascinating. Unfortunately, the quilt program's [2] only five point star block was not properly aligned. In this paper we will discuss a correct mathematical construction of P and the related quilt designs.

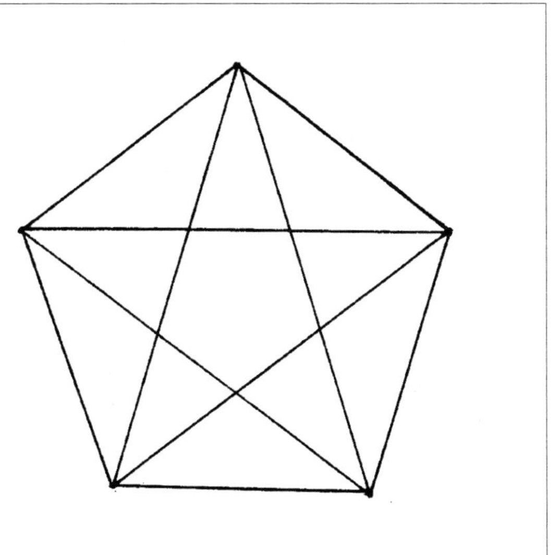

Figure 1: Regular Pentagon with Inscribed Pentagram

Mathematics

The first thing one should notice in figure 1 is that the star subdivides the area bounded by P into a small central regular pentagon and ten isosceles triangles circumscribing it. These ten triangles are of two distinct types and indeed all of the thirty-five triangles that can be formed from the sides and diagonals of P are of one or the other of these two types.

L. Gordon Plummer says “These triangles were held by the Greeks to be so important that they were named after the two stars in the sky associated with the constellation Gemini, the twins: Castor and Pollux.” [3]

A **Castor** Triangle is an isosceles triangle whose apex angle measures $\pi/5$ radians and consequently has base angles each measuring $2\pi/5$ radians.

A **Pollux** Triangle is an isosceles triangle whose apex angle measures $3\pi/5$ radians and consequently has base angles each measuring $\pi/5$ radians.

The construction of a Castor triangle is the key to inscribing a regular *decagon* (and hence also a regular *pentagon*) in a circle. For the central angle subtended by a side of a regular decagon, is $2\pi/10 = \pi/5$ and consequently the triangle formed by this side and the two radii terminating at the ends of this side is a Castor triangle. As a further consequence, the triangle formed by two diagonals of a regular *pentagon* emanating from the same vertex and the side of the regular *pentagon* that terminates these diagonals is also a Castor triangle [1].

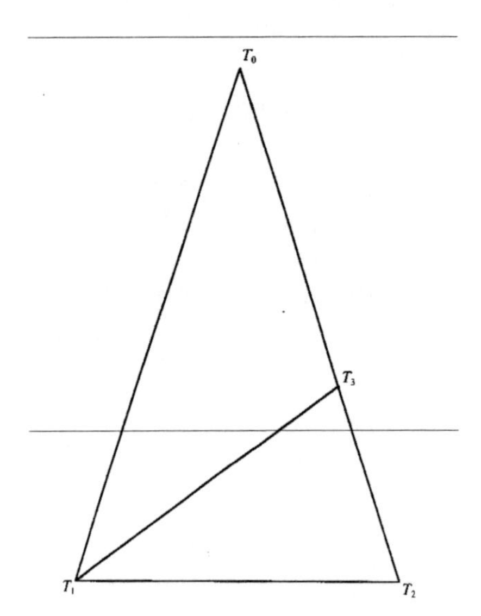

Figure 2: Mother Castor and Castor and Pollux twins

Let $[T_1,T_3]$ be the bisector of the base angle of $C_0 = [T0,T1,T2]$ with vertex T_1 and foot T_3 on $[T_2,T_0]$. This bisector subdivides C_0 into two isosceles triangles, $C_1 = [T_1, T_2, T_3]$ which is similar to C_0 and $P_1 = [T_3, T_0, T_1]$. The apex angle of P_1 measures $3\pi / 5$ radians and hence is a Pollux triangle. The common side $[T_1,T_3]$ of both triangles is equal to $[T_1,T_2]$ (which is the side of the decagon) and $[T_3,T_0]$. This configuration, consisting of two isosceles triangles, one a Castor and the other a Pollux, which share a common side with the apex angle of each contiguous with a base angle of the other, we call a pair of **Gemini Twins** whose **mother** is the Castor triangle from which they were derived and to which they return when their common side (the bisector) is erased.

Since C_0 and C_1 are similar and isosceles and P_1 is isosceles we can write
$\varphi = [T_0T_1]/[T_1T_2] = [T_1T_2]/[T_2T_3] = [T_2T_0]/[T_1T_2] = ([T_2T_3]+[T_3T_0])/[T_3T_0] = [T_2T_3]/[T_3T_0] + 1.$

In terms of the ratio **φ**, of the long side to the short side of the Castor triangles, this equation becomes
(*) $\varphi = 1 / \varphi + 1.$

Thus to construct a decagon and pentagon we must find the point T_3 on the radius $[T_0T_2]$ that divides it in the ratio φ. The length $[T_0T_3] = [T_3T_1] = [T_1T_2]$ is the length of each side of the decagon. [1] T_3 is not hard to construct because (*) the defining algebraic relation for φ yields numerically $\varphi = (\sqrt{5} + 1)/2$ and $1/\varphi = (\sqrt{5} -1)/2$ and also dividing (*) by φ yields $1 = 1/\varphi^2 + 1/\varphi$.

To see how Figure 2 can be constructed, given a circle C with center T_0 and radius $[T_0,T_2]$ as in Figure 3. We take the radius as our unit of length; $[T_0T_2] = 1$. Bisect $[T_0T_2]$ and label the midpoint M. Construct A_1 satisfying $[T_2A_1]$ perpendicular to $[T_0,T_2]$ and length $[T_2A_1] = [T_2M] = ½$. Then $[T_0,T_2,A_1]$ is a right triangle with hypotenuse length $[A_1T_0] = \sqrt{5}/2$. Let A_2 be on the hypotenuse satisfying $[A_1A_2] = [A_1 T_2] = ½$. Then $[A_2T_0] = (\sqrt{5}/2 - 1/2) = 1/\varphi$.

Next find T_3 on $[T_0,T_2]$ satisfying $[T_0T_3] = [A_2T_0] = 1/\varphi$. Then $[T_3T_2] = 1 - 1/\varphi = 1/\varphi^2$. Hence $[T_0T_3]/[T_3T_2] = (1/\varphi)/(1/\varphi^2) = \varphi$. Finally, find T_1 on C satisfying $[T_0T_3] = [T_3T_1] = [T_1T_2]$ which is the side of the decagon. From which it and the pentagon can be constructed.

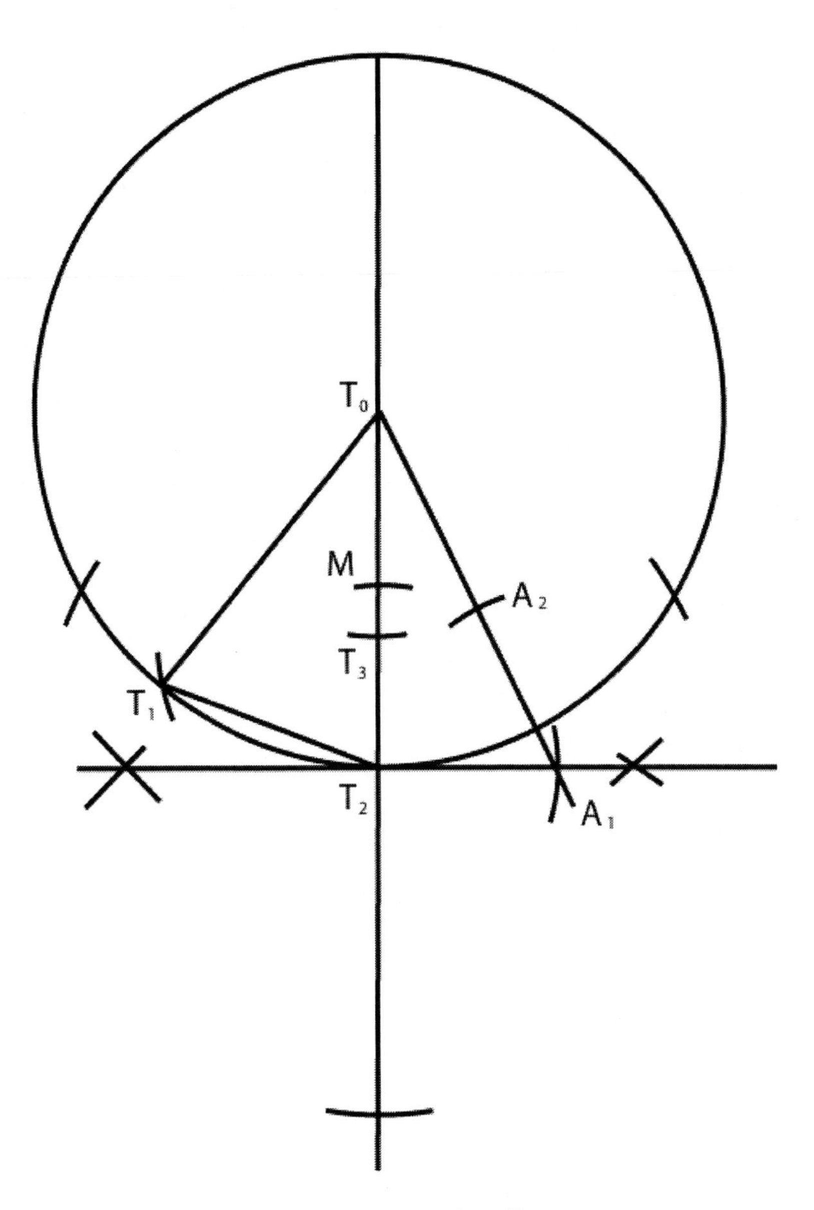

Figure 3: Construction diagram

Constructing the Quilt

Using this correct construction, we can now proceed with quilt design. We collaborated on several designs. They came up sequentially while trying to fix problems in each design. The first design was to illustrate all of the Gemini twins and mothers. It was not aesthetically pleasing, even though it did show all of the twin pairs. See Figure 4.

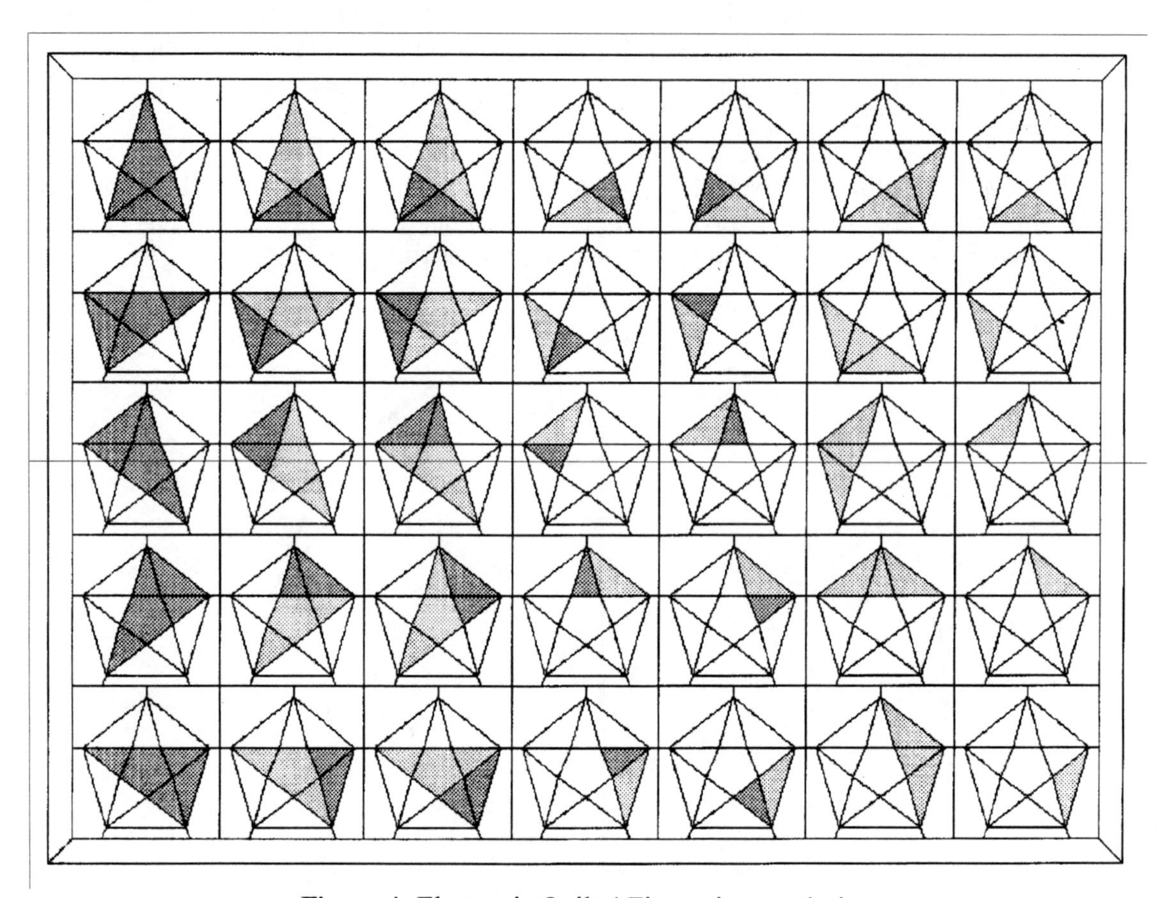

Figure 4: Electronic Quilt 4 Five point star design

The darker grey triangles are the Castors while the lighter grey triangles are the Pollux. This uses the misaligned block from EQ4 [2]. If you carefully lay a ruler along one side of the interior pentagon, you can see that it does not line up with the two Castor triangles that are the points of the star. Since the quilt program is supposed to print out the pattern pieces, I (Mary) would not use it for a real quilt. However, it is included because it shows all the triangles and orientations quite nicely.

It also demonstrates how to piece the block, since some of the interior lines extend to the side of the block, these are seam lines. There is more than one way to piece the block, but since it was unused, it is not necessary to say how.

I decided to inlay the four paired twins (columns 2 to 5 above) and the two solo Pollux (column 6 & 7 above) onto a larger Mother imbedded in a pentagon. Then by varying the orientation by rotation of the entire design one could see all the possible triangles. Figure 5.

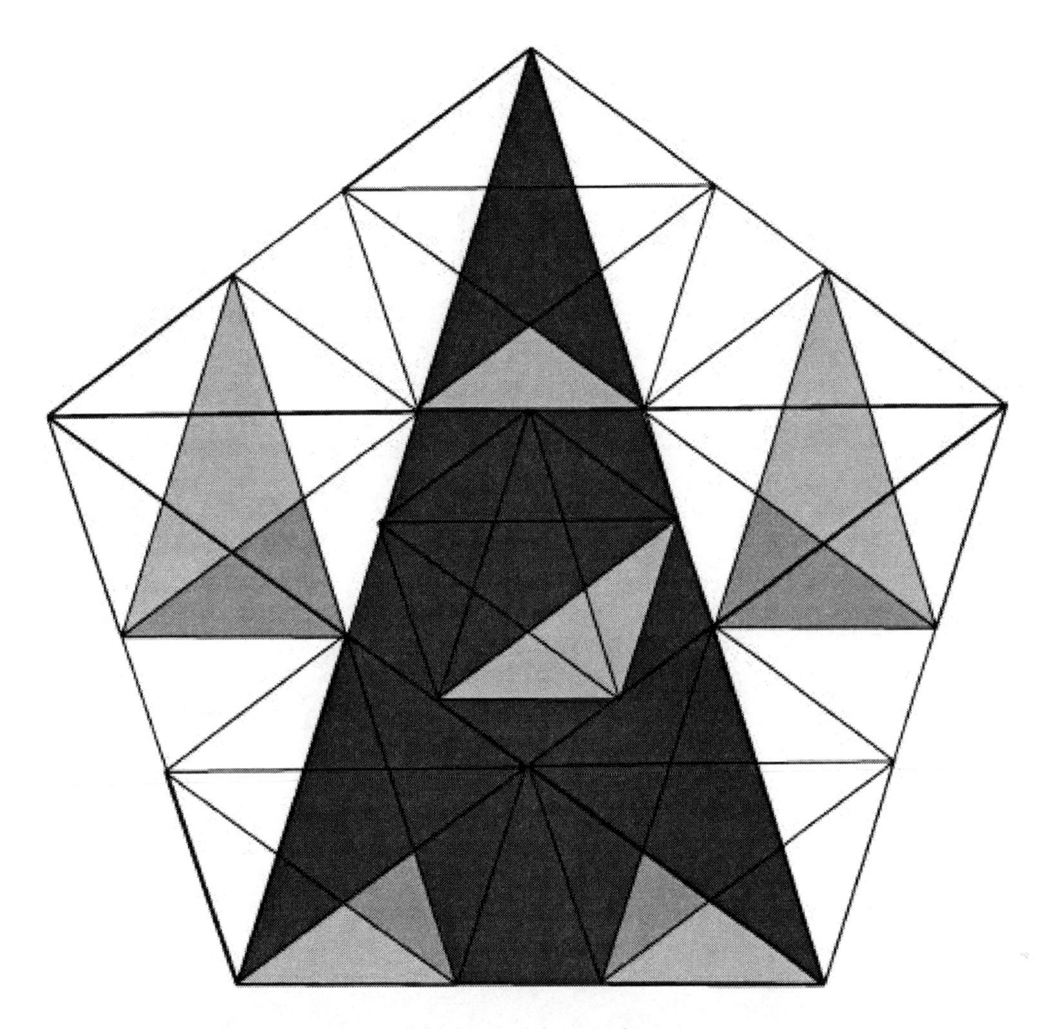

Figure 5: First design

When coloring the above design, I constantly lost the Mother Castor Triangle amid the colors when the largest Pollux and Castor pairs were in the lowest positions at the base of the Mother. I found that I needed two colors for the Castor triangles because if I had only one, how could it be seen against the larger Mother Castor? While it is not possible to have total symmetry in this design, a pleasant balance is desired, and we hope, achieved.

Seaming this design looked to be possible from the inside out, with modifications made to insure straight seams (highly desired by quilters). It is also possible to develop the design through coloring to emphasize the pentagrams and omit the twin aspects as given here. Or the pentagrams can be quilted and thus retained even when not colored.

So I further modified the design so the twin pairs and solo triangles are much smaller and finally all the triangles show up. However, this design would be very tedious to sew. Figure 6.

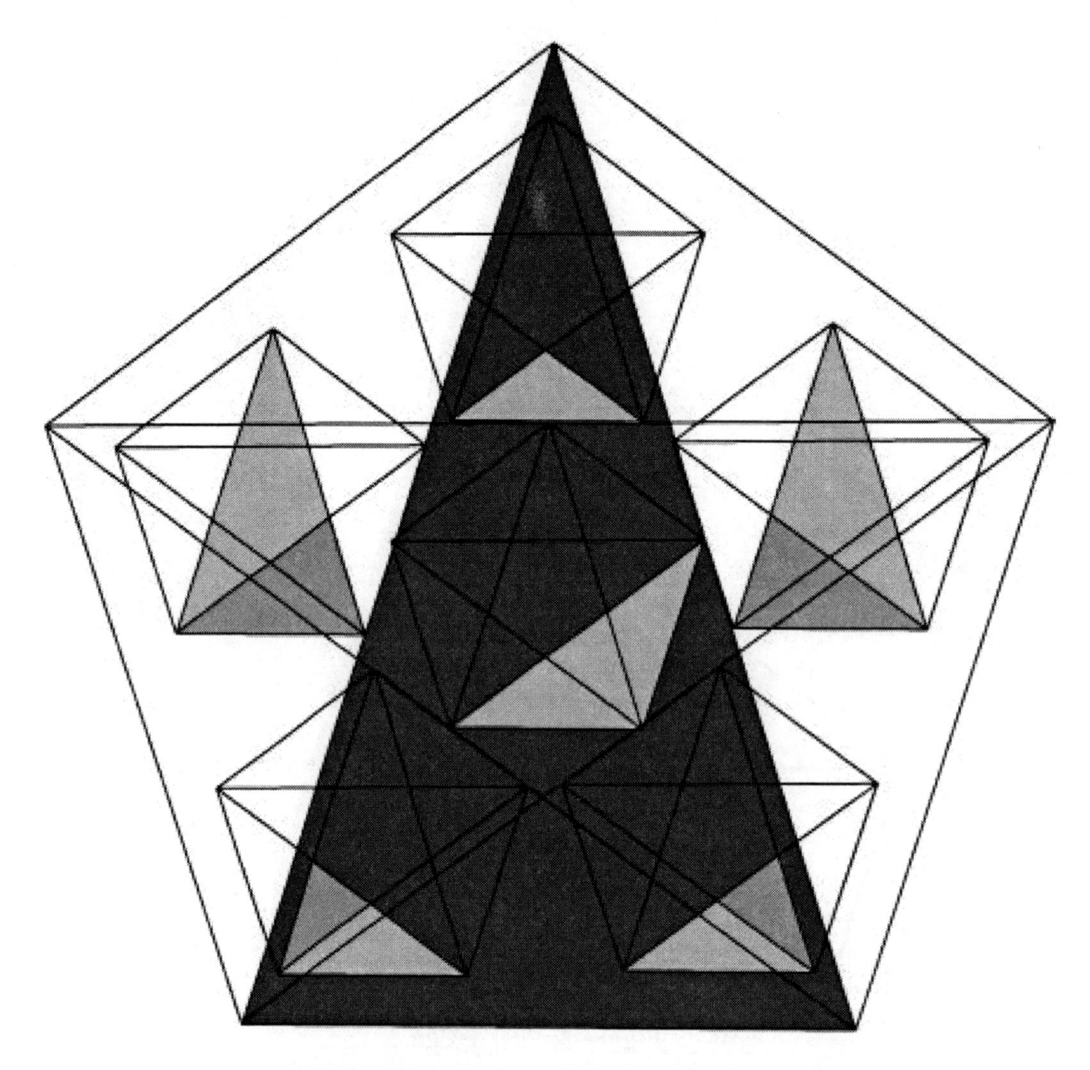

Figure 6: Adjustment of interior pentagons to be all one size.

While the main seam lines are retained from Figure 5, there would be more quilt lines to quilt and some confusion would result as questions could arise concerning design lines versus the 'is that a seam line or a quilt line or did you goof'. Then there are some nasty places at the corner of each one of the outer ring of pentagons where it approaches the corner of the outer pentagon. There is this $2\pi/5$ angle with the $8\pi/5$ angle that quilters just don't want to sew. It is possible to add a seam line to change the angles to a straight seam, but that could detract form the overall design. The problem of design line that cannot be incorporated as a seam line is an ongoing challenge for mathematical quilts.

But one last idea occurred to me of situating them as the seven sisters, of six hexagons around a central hexagon, but after consultation, we finally decided on the second design (Figure 5) with a twist. Figure 7.

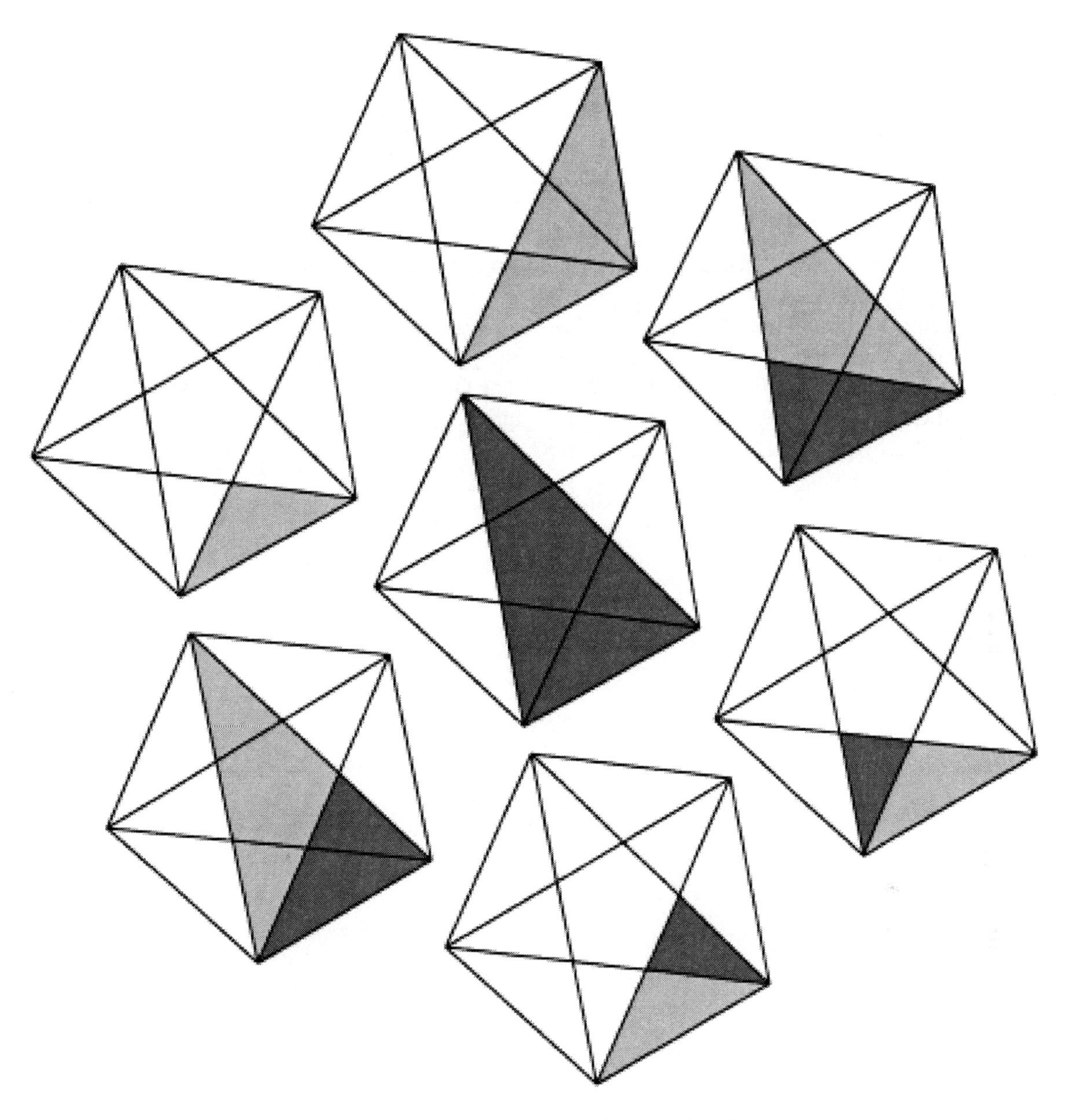

Figure 7: Seven Sisters variation of design

This clearly shows all of the twin pairs and solos, and can be rotated to show the proper orientation. The problems sewing here are the fact of sewing a pentagon within a hexagon. One could just sew fabric of the background color and then fit them with out a pattern for the background pieces. An alternative would be to construct the seven pentagons/pentagrams and suspend them with monofilament as a mobile. If this proves successful, we will show it off later.

References

[1] Euclid, Book IV Proposition 10 and 11 translations by Sir Thomas L. Heath
[2] Electronic Quilt 4, 1999
[3] L. Gordon Plummer, *The Mathematics of the Cosmic Mind* page 186, 1970

References not quoted but researched

H. S. M. Coxeter, Introduction to Geometry 2nd Edition pages 26 to 28 and 37
Jim Loy, The Regular Pentagon, www.jimloy.com/geometry/pentagon.html
Henry H. Ludlow, Geometric Construction of the Regular Decagon and Pentagon, 1904
Mathematical Countryside, www.literka.addr.com/mathcontry/pentagon.html
Jinney Beyer, Patchwork Puzzle Balls pages 76 to 83

Taitographs

Drawings made by machines

Jack Tait
Elm Cottage 4 Bronydd Villas
Clyro Herefordshire HR3 5RX
Email- jack.tait404@virgin.net
Website - taitographs.co.uk

Abstract

If a machine is instructed to make drawings and the results are viewed in the same way that a person's drawings are read, then speculation about the nature of creativity and art is not only possible but desirable. The decision making process becomes transparent because the maths, mechanics and after treatment are available for scrutiny, unlike the partially subconscious aspects of a person's drawing activity. It is proposed that the ideal way to meet the 'Bridges' aspirations is to follow Harold Cohen's exhortation that the most important task at the end of the 20th C (and beginning of the 21st) is to study how art works. My machines are electro-mechanical devices; from simple instructions they produce rich and complex images. Questions raised by machine drawings will be examined below.

Apposite limericks?

A scientist with maths in her heart
Set a sequence of numbers to start
Imagine her cries
Of joy and surprise
When her work was counted as ART

An artist who thought he could add
Made constructivist work, and was glad
But he sang a sad song
When his maths was proved wrong
Did this mean that his art was too bad?

1. Introduction

1.1. Questions. At a Bridges conference we have a gathering of able minds from science and art. In the past, undue prominence has perhaps been given to the 'nuts and bolts' of the activities and maybe insufficient attention paid to how the results, displayed as evidence of bridge activity, are assessed for provenance and meaning in terms of art objects. Whilst I am aware that "It's art because I say it is" has some currency, this will not do when scientists and artists meet to explore common ground. Given my art education background I feel able to pose what I hope are pertinent questions. If they have some resonance, then surely there is a chance that we might make some intelligent speculations about how art works. When sets of numbers are manipulated, the maths might be elegant algorithms, but how are we to judge whether the results 'add up' in terms of art? The converse is also true when artists flirt with numbers in the hope that the underpinning maths will give their work extra gravitas. We can all be vulnerable in this respect if we wish to attain any standing alongside more conventional areas of science and art. Evidence that we have examined the whole set of issues can only promote our cause and help us to be taken seriously. I am very aware that this point of view might be controversial but perhaps friendly argument and debate are the lifeblood of a good and rewarding conference.

2. Precedents

2.1. Art history. Precedents exist where the choice has been made to veer away from classical drawing and painting, so there is no need to make a special case when using machines to produce art. They are: -
(2.1a) Abstract drawings now have same validity and meaning as figurative work [1]
(2.1b) Mechanical aids, machines, cameras and computers are acceptable means of production. [2+3]

2.2. Variables. There remain only two points to address. As in photography [4] the main variables are WHERE you stand and WHEN you press the shutter; in using drawing machines they are WHAT they do and WHY the results might be regarded as art and their meaning evaluated. I will first deal with the 'WHAT' putting more emphasis on the questions raised by the process. Fine detail about my machines and their workings are given on my website. In examining the 'WHY' I want to use machine drawings to raise questions about our conceptions/misconceptions about art.

3. The WHAT.

3.1. Method Before describing my machines, I will deal with motivation and process. I prefer to call my activity 'Printmaking' in the broadest sense based on my experience in 'classical' litho, etching and silkscreen work[12]. I am not a mathematician and struggle to deal with some of the maths underpinning my machines. As well as making prints, I am also a photographer and engineer, skills which came together when I used to build cameras. I design, build, and work with drawing machines, who's output I scan into the computer, process in Photoshop and make final prints on an inkjet printer. Photoshop and the inkjet printer have simply taken the place of the litho press; the thinking is the same. The results are 'Taitographs'- hybrids; part drawing, photograph, print and digital image.

With machine-generated work, the steps in production provide concrete data and insights maybe derived into its workings. Curiosity is my starting point; design problem solving may stimulate intuitive insights. The ability to visualise the outcome of a set of actions is essential. How decision-making takes place, the relationship of calculation, intuition and random effects, the criteria for making choices and the choice of colour are all equally transparent steps. Finally, making wrong decisions and mistakes is an inevitable but ultimately valuable component of a design method. Whilst the above is helpful, what completes the array of data is the effect of the images on the viewer. (By viewer, I mean those with an art background, preferably other practitioners) Having the viewer's reaction [10] completes a feedback loop. At this point all the information is available with which to speculate. This theme is taken further in the 'WHY' section.

3.2. Questions arising. It is part of my strategy to test viewer's reactions frequently and if appropriate to act on them. This has proved very beneficial, but is perhaps more applicable when working with machines than in the case of hand/eye work. Two questions in particular have been raised and are outlined below. with a bearing on both the WHAT and the WHY.

3.3. To Colour or not? It has been suggested that adding colour to the images detracts from the power and impact of the drawings in their 'raw' state. I have some sympathy with and understanding of this view, but my primary concern is that the viewer is first drawn to look at the image. Colour is a powerful expressive tool when used subjectively and a supportive component of the drawing particularly when it is rule based as many constructivist strategies are. **Fig 2.** Colour attracts a wider variety of viewers than would a monochromatic image and once the viewer is captivated, then I hold that the essential qualities of the drawing will insinuate themselves into the viewer's mind, perhaps without them realising it. There is no doubt that the drawing qualities are paramount, but not all viewers bring the same experience to looking at images, **4.4** below. This is also part of communication theory **4.2** below, in that the message of the drawing is received with as little 'noise' as possible and just as important, is more likely to be retained. I am supported in my view by Harold Cohen[5] who also chose to add colour to his computer-generated drawings. See **Figs 1+2**, identical drawings with and without colour.

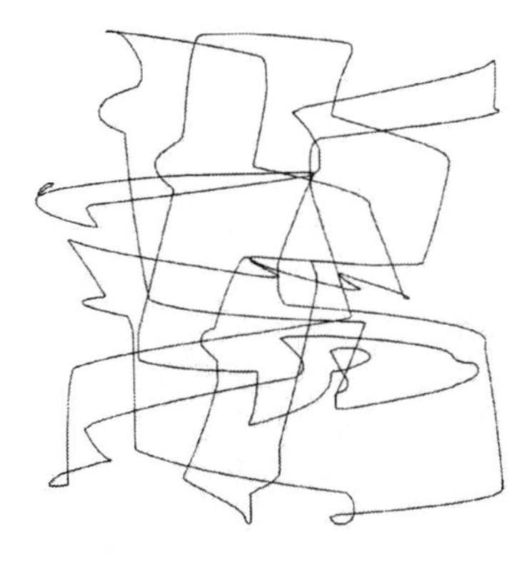

Figures 1+2 *Green shoots in the desert*

3.4. Digital versus Analogue If the question were that simple then computer programming would be the first choice, but the two are not comparable. Building analogue machines is significantly different from writing a computer programme. With analogue machines, I do not know the exact outcome and I am often surprised by the results, although I treat my work as a design process. This is associated with the expression "Look what **It** has just done!" There is only a vague feeling that a particular avenue will be fruitful. When I write a computer programme I have a more concrete notion of where the process is going. For instance, if I explore sine waves, I need formulas in place, have a clear idea of the end result and limit the range of decision making. Programme writing is exact and unforgiving; it runs or it does not. Admittedly, occasional mistakes may be intriguing, but this seems to be rare. Each method is good and there is a case for using both. **Figure 3** is a computer-generated image which would have been difficult to do by an analogue machine, whilst **Figure 4** is unlikely to have been programmed. Until it was drawn, I did not know the outcome, but a product of intelligent exploration and programming not in any way a 'happy accident'.

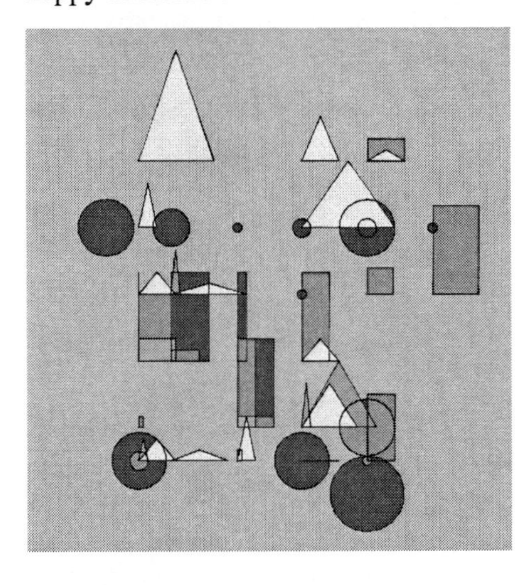

Figure 3 - *Homage to Kandinsky*

Figure 4 - *Rotor 6*

3.5. Electro-mechanical devices Most of my machine images are of this type. There are two. The first set has the 'programming' integral (by this I mean adjustment of the settings - see below) and the images are changed by means of the relationship of linkages, drive wheels and ratios. Both are wholly deterministic and have no random element in their sequencing. Figure 5 shows the Linkogram and figure 6 the Sinewave machine.

Figure 5 - *Linkogram machine*

Figure 6 - *Sinewave machine*

3.6. Set 1 - Deterministic machines. The Linkogram is 'programmed' by the setting position of the clock face wheels together with the ratios of the X wheel set (top left and bottom right) to the Y set (top right and bottom left) and to the turntable speed. A typical setting might be X=2, Y=3, Turntable =2. The start position of the pen also alters the character of the drawing. Lifting and lowering the pen in synchronisation with the Y set of wheels afford further variety. **See figures 8 and 9.** The organic growth of the lines is controlled by the very slight variation of the wheels' speeds i.e. the X and Y sets differ by approximately 1:1.05 whilst the X:Y speed difference is 1:1.03. This means that the positions of the clock settings varies as the machine draws and causes the line growth to continue at around x 1 to x 1.5 line widths on each revolution. The settings are sufficiently accurate to allow almost identical repeats. A light pen is available which enables the drawing to be made on photographic paper (see I below right).

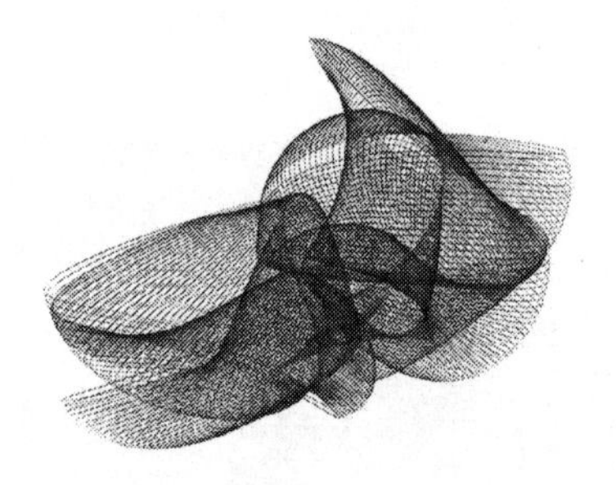

Figure 7 – *Linkogram*

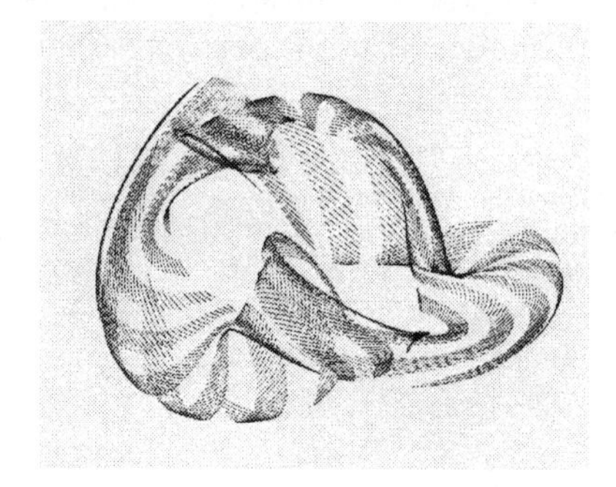

Figure 8 - *Blue line wih pen lift*

Figure 9 - *'P'- light pen*

3.7. Linkogram. Figure 7 is a simple linkogram image allowed to run for an extended time, **Figure 8** had the pen lifted and lowered and **Figure 9** had the light pen lifted in and out of focus, where a light pen 'wrote' onto photographic paper.

3.8. The Sinewave machine. Figure 6 is 'programmed' via switches to control the line spacing and the wave growth speed. The number of waves is governed by the gearing to the main sinewave gearbox shown on the bottom of the picture. Altering these gears can change the number of waves. **See Figure 10.**

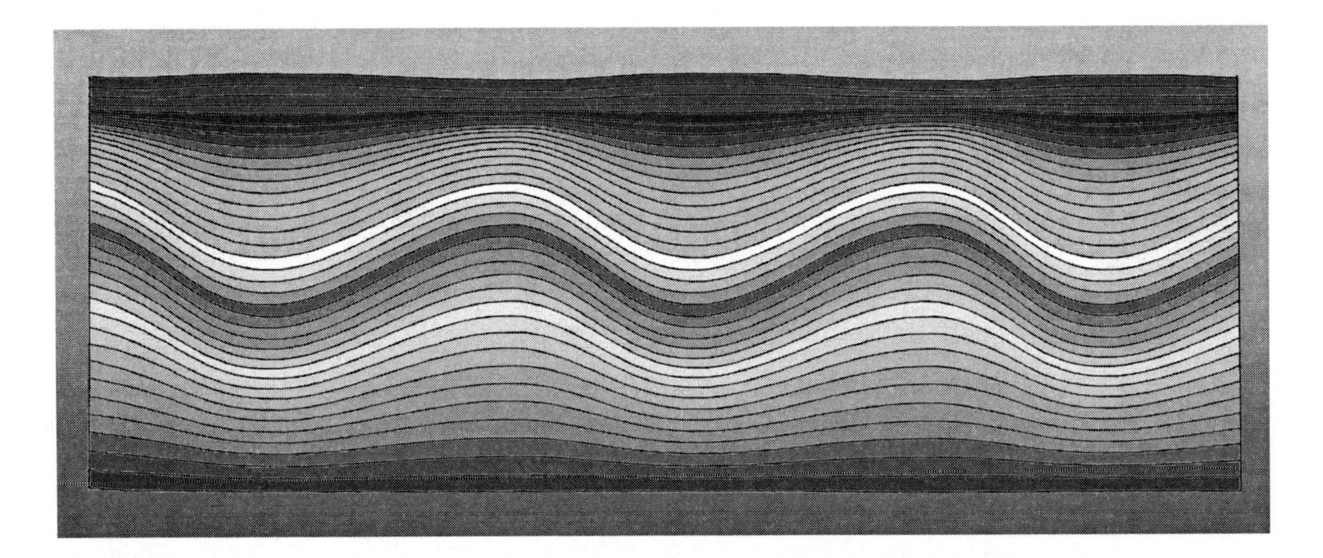

Figure 10 - *Seascape*

3.9. Set two – Programmed machines The second set of machines relies on the input from sequential time programmers to execute the drawing. There are two such sequential programmers at present; each has a quasi-random element built in and can be used to control a number of machines. By quasi-random I mean that sets of switches are driven by gears whose numbers of teeth are slightly different and take time before they repeat. Added to this the sequential timers and reversing switches may be applied to each of four different motors; the practical outcome is a large number of permutations.

Figure 11 - *NSEW*

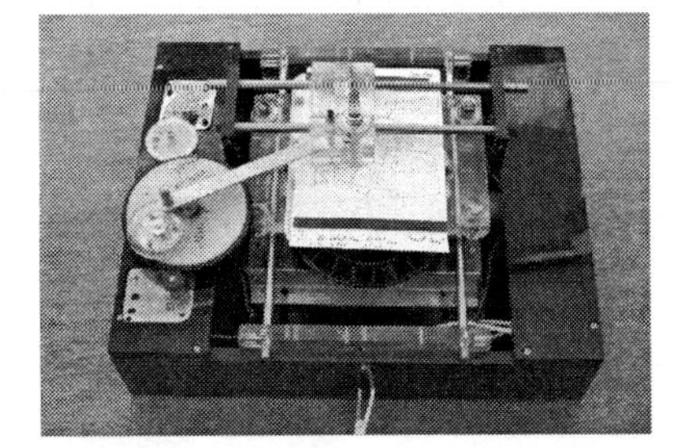

Figure 12 - *NSEWsp*

Figure 13 - *Turntable*

3.10. NSEW - In this machine **Figure 11** , simple X and Y axis movements are driven by d.c. motors; voltage variation gives different speeds. Underneath there is a reversing relay box to automatically reverse the direction at each end of the travel. This is in addition to any reversing done by the programmer and is solely to prevent overrun. This simple range of right angle movements in North, South, East, West directions (hence the NSEW name) is able to offer a wider range of drawings than might be expected.

3.11. NSEWsp - **Figure 12** shows a more complex version of the above, having a sun and planet gear system on each axis, with two motors to each, and offers greater richness and variety. This set up can produce lines which are not confined to right angles as in the first NSEW machine. With my elementary maths, I calculate that the theoretical extent of possible images from the NSEWsp model is in the order of 37 million.

3.12. Turntable - The third machine in **Figure 13** is a turntable unit coupled with a rotating pen and a simple harmonic linkage. It has three separate motors and the combinations available to either programmer are large. The sequential programmer can just be seen at the top right of the picture.

4. The WHY

4.1. The Influences. When dealing with the WHY, I want to start by recognising the influences which have a bearing on my work. At art college and afterwards, I immersed myself in the range of work listed below. In cases, such as Cezanne, this had a direct bearing on my still life photography. Others were indirect and include Maholy Nagy, Man Ray, Paul Klee, Kandinsky, Mondrian, Cezanne, Matisse, Kenneth Martin, Brigit Riley, Vassarely, Tanguely, Harold Cohen, Constructivism, Kinetic art, Cybernetics, and Artificial Intelligence. The influences did not lead to machine building but developed alongside it. It was never my intention to 'mimic' the constructivist pictures and if there are obvious parallels then these are done as 'homage' to the painters and are an attempt to understand their preoccupations more fully [6,7+8] If similarities do exist then this has no relevance to the standing of my drawings; they derive nothing from any likeness. I am only interested in the significance of what machines can do. If one movement in art were to be selected as pivotal, it would be the Bauhaus and in particular the philosophy of Moholy Nagy, Man Ray, Klee and Kandinsky. If I have to classify my work, then it paddles in a backwater of constructivism with a bit of cybernetics thrown in for good measure!

4.2. Design process [9] **and Communication theory**

In trying to learn more, we need a framework in which to place data referred to above in motivation and method. The Bauhaus teaching set the path, in particular Moholy Nagy's telephone pictures. This was further encouraged by the Harold Cohen's statement quoted in the Abstract. Design process may help and two questions may be borrowed from it: By what criteria are alternative solutions rank ordered? How do we evaluate the results?

4.3. The Philosophy

In conventional art practice, the underpinning philosophy is very complex and outside the remit of this paper. Our prime concern is how is meaning derived from marks on paper. Using machines to make drawings simplifies things. Clear instructions must be given and 'allowable decisions' chosen. The process then becomes 'outside our minds' and evaluation can happen 'as if the images were not ours.' The response "I would not have thought of doing that" *(had I been drawing in the conventional way)* becomes significant. The nuts and bolts are accessible, the viewers' reaction generates feedback and the process is complete.

4.4. The Beholders Share [10]

Gombrich stresses the vital role played by the viewer's experiences in governing their evaluation. From this I developed, in my design teaching, a way of differentiating between a **design object** and an **art object**. If a design object is evaluated by a number of viewers, they all respond to it in a similar way, i.e. a chair will not be mistaken for a table. Their previous experiences are unlikely to condition their responses to any great extent; they all get the same message. When an art object is evaluated, their response to it is likely to be different in each person because they have brought to it a wholly different background experiences. The message received is almost certain to be individual. I propose that this is a workable benchmark for identifying whether or not an artwork 'adds up'.

4.5. The Turing test [11]

Recentl machines have complicated linkages and programmers containing quasi-random elements. **See figures 12+13.** Earlier images had deterministic characteristics and might not have 'added up to art' ref. **4.4** above. Now, the complexity and richness of images shown in **Figures 4, 14 and 15** take on ambiguous qualities; they ceased to shout "machine image". Many viewers have felt that they exhibit characteristics more akin to hand-drawn abstract paintings. Given that a concensus exists, then a Turing type test can now be proposed. Can the viewer tell the difference between an image made by hand and one done by machine? If they create the same viewer response then perhaps no difference exists! From this, can we begin to speculate how art works? Recent work, using light pens fed into a digital camera, suggests more potential as the images look wholly photographic.

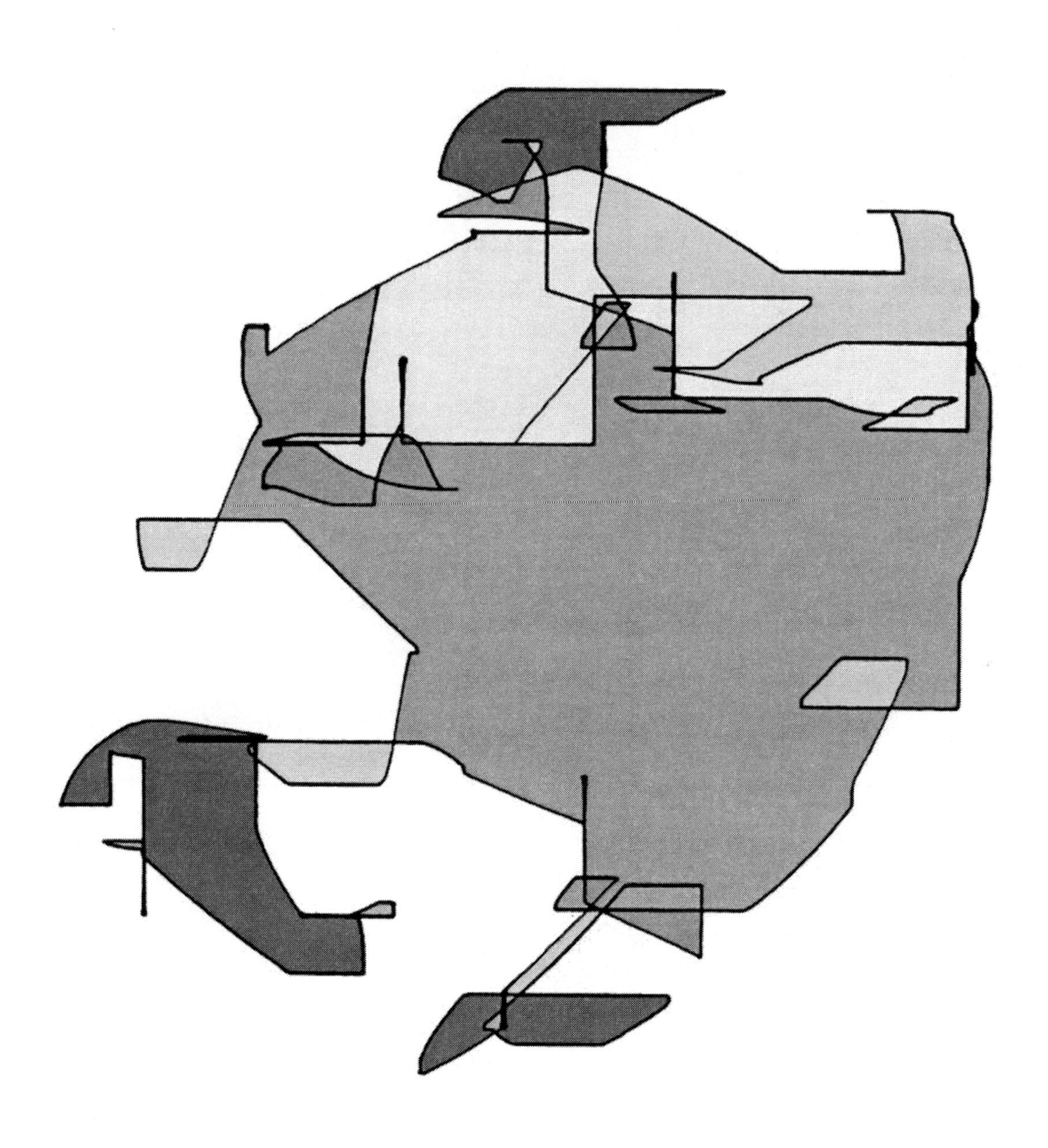

Figure 14 - *Dancer*

5. Conclusion

5.1. Sharing questions. In showing machine drawings I have put intriguing questions. I accept that it is impossible to rule out subjectivity in picture-making activity; our subconscious minds always insinuate themselves into every corner. However, speculation that the machine drawing activity might unravel some threads of the creative act, has motivated and enriched my printmaking. With luck, something useful might have been transmitted and even received. In the final instance, I hope that my images will be more eloquent than I have been.

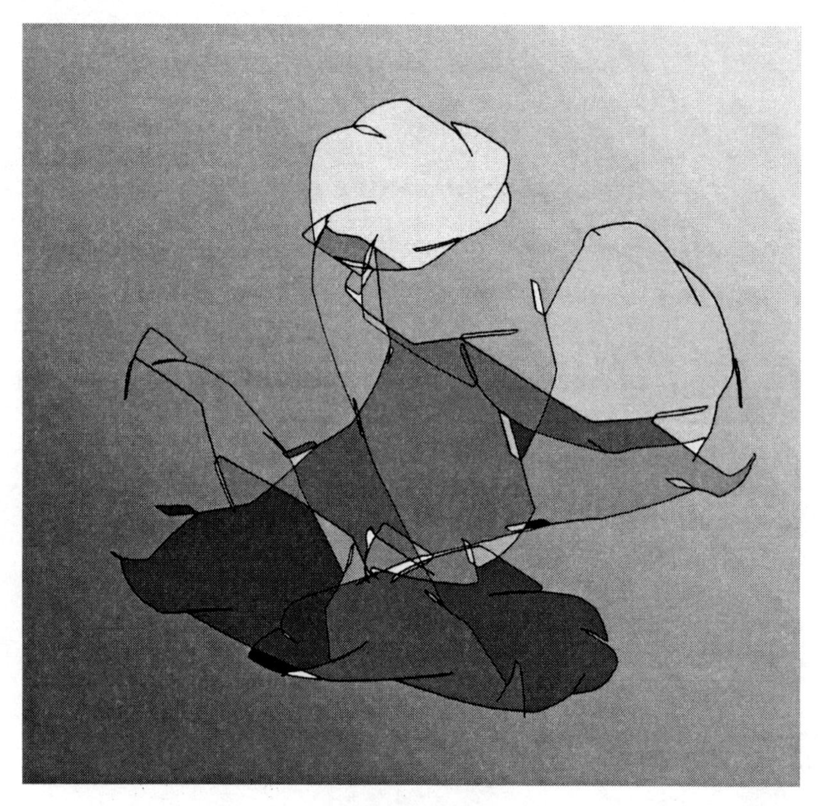

Figure 15 – *Figures*

6. References

[1] Royal Academy of Arts London Bauhaus - German Exhibition - Catalogue 1968
[2] Tate Gallery - Tinguley Exhibition catalogue - 1982
[3] Malina F - Kinetic Art Theory and Practice - Dover Publications New York 1973
[4] David Hurn/Bill Jay 'On being a photographer'- 3rd ed.- Lenswork USA 2001
[5] Tate Gallery - Harold Cohen - 1983
[6] Stoichita V I - Mondrian - Meridiane Publishing House Bucharest 1979
[7] Tate Gallery - The Moderns and their World - Phoenix House London 1957
[8] Tate Gallery - Kenneth Martin -1975
[9] Sebastian Lera - Synopses of some recent studies of design process and behaviour- RCA 1983.
[10] Gombrich E H - Art and Illusion - Phaidon Press - London 1977
[11] Alan Turing - Computing machinery and intelligence - Mind 1950
[12] Jack Tait - Beyond Photography - Focal Press - 1977
Other publications which have had a general bearing on the above are: - Koestler A -The Act of Creation - . Hutchinson+Co 1964, and Collingwood R G - The Principles of Art - OUP 1979.

Photography and the Understanding of Mathematics

Richard Phillips
Self-employed writer & photographer
4 High Street, Badsey
Evesham WR11 7EW, UK
E-mail: Phillips@badsey.net

Abstract

This paper considers ways in which photographs help our understanding and teaching of mathematics. Some historical landmarks are considered from Muybridge's galloping horses to mathematics trails snapped with mobile phones. The possibilities have always been limited by the available technology and have been shaped by changing attitudes to mathematics teaching. It is argued that in mathematics teaching, photographs are not just for illustration. They provoke discussion, pose problems and provide data. We can measure them and model them with graphs. The approach adopted for developing the *Problem Pictures* calendars and CD-ROMs is described together with some of the ways these resources are used.

1. A Little History

From the earliest days of photography, the medium has had mathematical possibilities. A simple frontal photograph of a building could be measured to extract angles and ratios with an accuracy that had not been possible when working from a drawing or painting. The first photographers may not have recognized the potential but it has always been there. [1]

In 1878, Eadweard Muybridge took a famous series of photographs of a trotting horse. Its owner, Leland Stanford, had proposed, that during a horse's running stride, there is a moment of suspension where no hooves are touching the ground, and Muybridge's pictures proved that he was right. Today Muybridge's work is seen as a landmark in motion photography, and by biologists, as pioneering work in understanding animal movement. To ask 'In what order does an animal place its feet on the ground?' is also a mathematical question, demanding analysis and modeling of a complex situation.

In the 1930s, Harold Edgerton's experiments with multiple-flash photography created stunning images that seem to pose their own mathematical questions. Edgerton's picture of the golfer Dennis Shute was taken at 100 flashes per second. How fast was the club moving when it struck the ball? How fast does the ball move? How long does the golf shot last? How close is the motion to a circle?

Figure 1: *Harold Edgerton's 'Multiple-flash photographs of the golfer Dennis Shute', about 1935.*

It has taken a long time for photographs to begin appearing regularly in mathematics textbooks and journals. One imagines that it was easy for an author to convince a publisher of the value of photographs in geography textbooks but the case for mathematics was harder to make. A small number of photographs were used to good effect in *Mathematics: A Human Endeavor* published in 1970 [2]. In 1980, the classic Dutch publication *Schaduw en Diepte (Shadow and Depth)* [3] is an early example of teaching material using photographs to pose mathematical questions. The book explores ideas about light, space and perspective.

In the UK an important landmark year was 1983 when Ray Hemmings and Dick Tahta took over the editorship of the journal Mathematics Teaching [4]. Their impact has been considerable and introduced many changes that have lasted to this day. They chose to fill their journal with photographs. Interspersed among the articles are images of coal hole covers, scaffolding, balustrades and more, mostly in grainy black and white. There is a double page spread of car wheels contributed by Marion Walter. Very few of the images had captions, and those that did are very minimal. The photographs were not accompanied by explanations, questions or activities. These were left for the reader to invent. Nothing was written to justify the pictures as being in any way mathematical but there was no doubt that they were.

Figure 2: *Use of photographs in the UK journal Mathematics Teaching between 1983 and 1985.*

This generous use of photographs became possible because of changes in printing costs. In the 1970s a separate printing plate had to be made for every photograph at considerable expense. By the 1990s, with the arrival of computer filmsetting, it made almost no difference whether a page had one photograph or a hundred photographs. When Laurinda and Tony Brown took over the editorship of Mathematics Teaching in 1987 they enthusiastically continued the imaginative use of photographs including some playful juxtaposition of text and image. Photographs started to appear more often in other mathematical journals and books. Also in 1987 the Mathematical Intelligencer's Mathematical Tourist column began publishing photographs of mathematical significance from around the world. In the same year the NCTM published 'Geometry in Our World,' a collection of colour slides as a classroom resource. In 1988, *Geometric Patterns from Roman Mosaics* was the first of a series of photo books by Robert Field from Tarquin Publications.

In more recent years, a different type of photograph has become more common in several of the mathematics teaching journals. These could be described as 'role model' photographs. They show teachers and students working at mathematical activities in a wide variety of ways. Often the pictures show activities away from the classroom or using practical apparatus. Editors seem to use these photographs to convey several types of message, such as:

- There are many ways of learning mathematics.
- Mathematics can be fun.

- Mathematics can be engrossing.
- You can do your own mathematical investigations.

Some editors in the late 1980s were cautious about using pictures of this kind because they had received frequent criticisms of the kind 'Why are there more boys than girls in this picture?' or 'Why are there more white faces than black faces in this picture?' Fortunately this has not seriously inhibited the use of role model photographs, which are now an established way of communicating about ways of learning mathematics.

Figure 3: *A 'role model' photograph from cover of Mathematics Teaching, 2004.*

A recent development in print publications is the Mathematical Lens column in NCTM's Mathematics Teacher. This is edited by Ron Lancaster from Ontario, an enthusiastic user of photographs for mathematics teaching. Each article typically consists of a single photograph which is used to pose numerous questions. Some questions are quite simple and others require considerable work, sometimes using resources like interactive geometry software. The questions are followed by answers and a commentary.

About 1990 the 'digital multimedia' revolution made it much easier to work with photographs, sound, and video on a computer. A range of technological advances opened up new possibilities including the first affordable digital cameras and the arrival of CD-ROM and other low-cost, high capacity storage media. The world wide web followed soon after. The revolution continues to this day embracing new technologies along the way, such as mobile phones that take photographs.

"World of Number" was an early UK multimedia project using interactive video discs under computer control to deliver a variety of classroom resources. In theory it was possible to put 45,000 photographs on one side of a laser vision disc. The project's Picture Gallery offered a large collection of photographs, mostly taken frontal parallel to make it easy to measure them with the toolbox software provided. [5]

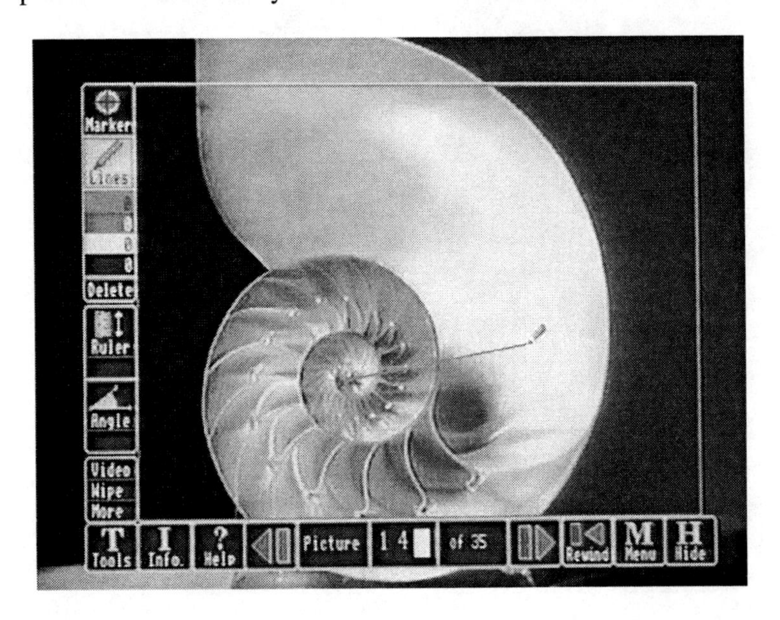

Figure 4: *Angle measurements being made on a photograph using the 'World of Number' software toolbox.*

As well as multimedia publications, the revolution opened the way for students and teachers to take their own photographs. Primary age students at Roch School in Wales have developed a photographic mathematics trail around their village and published it on the internet [6]. A teacher at another primary school arranged her class into a 'living bar chart'. Only by taking a photograph could she show the class the pattern that they had made.

Figure 5: *Students make a human barchart to show their favourite fruit during a Maths Day at Mosley Primary School, Staffordshire.*

2. Problem Pictures

Over the last 25 years photography has grown significantly more important in understanding mathematics. There have been many different approaches to using the medium and its impact has touched teachers and students at every level. Building on this tradition I have developed an approach to publishing photographic resources for mathematics which I call *Problem Pictures.*

The photograph of the patchwork quilt and the accompanying text is an example.

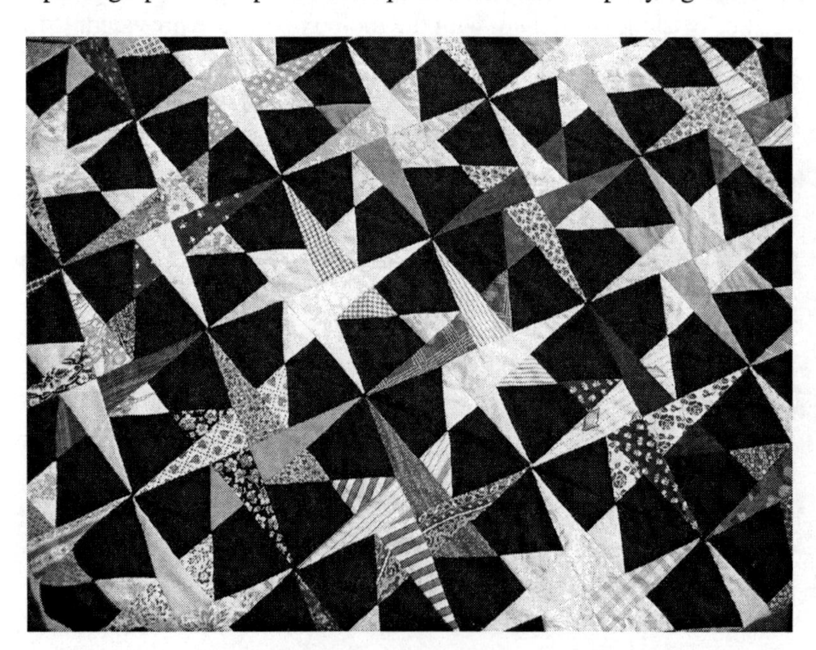

This beautiful patchwork quilt has a complex design.
Try to see the basic units from which it is constructed.
Can you see any right angles here?
If the whole quilt has 100 black quadrilaterals, how many triangles does it have?

Figure 6: *Example of Problem Pictures photograph and text.*

The choice of the text that accompanies the photograph is of some importance. Because it is quite short (just four sentences) it is likely to be read. The language is chosen to be readable and informal. It suggests questions and activities, but the reader is quite at liberty to do something different with the photograph.

The instruction 'Try to see the basic units' is ambiguous. The basic units could be the pieces of cloth from which the quilt is constructed, or the names for their geometric shapes. It could also mean the 'repeat' unit within the pattern. Such ambiguity would be unacceptable in a task that was designed for assessing students. But the game is not to measure students but to engage their interest. By being ambiguous the phrase works for a very wide range of ages and abilities. 'Try to see the basic units' could be tackled by a seven year old, but could also be quite challenging to readers of this paper.

'Can you see any right angles here?' is a more closed question. It is also a kind of hint, helping the observer to structure their viewing of the pattern.

The final question about the number of triangles is more difficult. It is reasonable that the questions asked should get progressively more difficult. But this question can also be approached in different ways and is therefore open to students with different degrees of knowledge. Primary age students should be able to estimate an approximate answer.

This pyramid is built entirely from pieces of Turkish delight. The base of the pyramid is made from a square 16 by 16, using 256 pieces of Turkish delight. On top of this is a second square 15 by 15, a third square 14 by 14, and so on, up to the single piece of Turkish delight at the top.
How many pieces are there in the top three layers?
How many pieces were used altogether?

Figure 7: *A second Problem Pictures photograph.*

A second example shows a pyramid built from pieces of Turkish delight. The questions are about number patterns. Although no algebra is needed, it could be used. Students who know the formula for summing squares or the formula for the volume of a pyramid might find these useful. It is the kind of problem where you bring to it whatever knowledge and skills you have.

Most Problem Pictures photographs come with about four sentences of text. These offer questions and activities that work at a wide range of ability levels. The materials are designed to be as open and inclusive as possible.

It is useful to contrast this with the approach taken by Ray Hemmings and Dick Tahta in the 1980s with Mathematics Teaching. Their photographs generally came without captions or questions. But their intention was much more than to decorate the empty spaces in their journal. We know from other aspects of their work that they were committed to a very open approach to education where teachers ask very few questions and, as much as possible, the responsibility for learning is taken by the students. A photograph without a caption gives the learner the widest scope to pose their own questions. But it also greatly increases the chances that the photograph will simply be ignored.

Different again, Ron Lancaster in his Mathematical Lens column, accompanies one photograph with a whole article. With so many words, it is possible to explore in much more depth. There is considerable ingenuity in devising a wide range of interesting questions that emerge from one picture. The drawback here is that the reader is less likely to pose their own question. With so many words and so many possibilities, the ownership of the problems belongs more to the author than the reader.

Problem Pictures limits the text to about four sentences. This formula also suits the media in which the photographs are published – as CD-ROMs [7] and as calendars and posters from AAMT [8]. The open character of the material means that the photographs have been used by teachers in several different ways. One possibility is to lead a whole class discussion with photographs displayed on an electronic whiteboard, usually to begin a lesson. Some teachers use the photographs to prepare their own worksheets. Most work is with students aged between 9 and 16 years, but there are examples of use with older and younger students too.

Peter Finch is a teacher from South Australia. He assigned his year 8 students "research tasks" as project work. They had to choose one of seven photographs on Problem Pictures CD-ROM and answer the questions in the form of an A4 poster that would be understandable by someone who knows nothing about the task (see Figure 8).

Figure 8: *One of Peter Finch's students produced this poster about triangular numbers.*

Adrian Oldknow [9] has developed some activities that involve modelling photographs with graphs. A photograph can be copied from a webpage and pasted into software that allows a graph to be superimposed. Recent versions of The Geometer's Sketchpad and Cabri Geometry are suitable for this. Three sections of the Problem Pictures Themes CD-ROM have photographs specially taken to be used in this way. This modelling activity offers a learning environment that encourages experimentation.

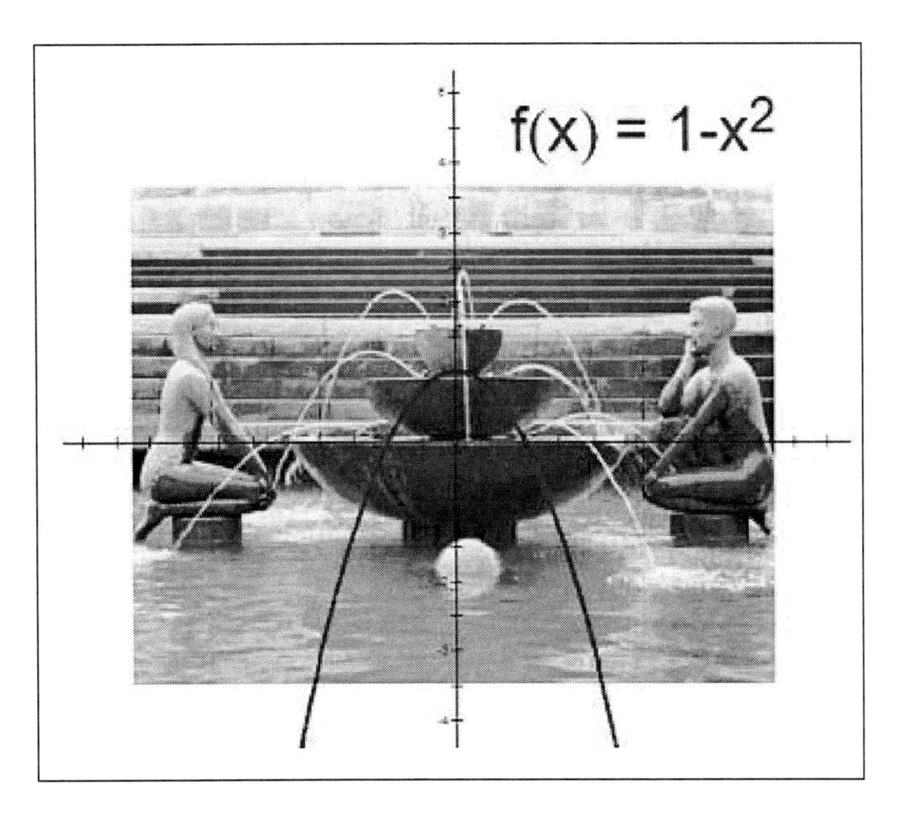

Figure 9: *Modelling the trajectory of water from a fountain.*

A workman has just finished painting the word STOP on the road. Four stencils were used to paint the letters.

Imagine the stencils are turned over. The letters T and O would still look like a T and an O. But what would happen to the other letters?

Suppose you had stencils for all 26 capital letters. How many of these could be turned over without changing the shape of the letter?

Figure 10: *A visualization task.*

3. Conclusions

Photographs are an excellent way of building bridges between mathematics and the real world. They bring relevance and interest to a subject that is often see as abstract and remote. There are many ways of using photographs in teaching mathematics, with exciting possibilities for students of all ages.

In the past, the cost of the technology has often limited the educational possibilities of photography. Today, this obstacle has been removed. There has never been a better time to use photography to improve the understanding of mathematics than now.

Acknowledgements

I am grateful to Will Moroney and the AAMT team for their support and encouragement of these activities. Thanks also to Laurinda Brown for discussing the use of photographs in the journal *Mathematics Teaching.*

References

[1] H. & A. Gernsheim, *A Concise History of Photography*, Thames and Hudson, London, 1965.

[2] H. A. Jacobs, *Mathematics: A Human Endeavor*, Freeman, San Francisco, 1970.

[3] A. Goddijn, *Schaduw en Diepte (Shadow and Depth)* Rijksuniversiteit Utrecht, the Netherlands, 1980.

[4] The journal *Mathematics Teaching* is published by the Association of Teaching of Mathematics, Derby, UK. See also: www.atm.org.uk

[5] *World of Number*, interactive video discs, Shell Centre & New Media, UK, 1992.

[6] See www.ngfl-cymru.org.uk/vtc/ngfl/eng/ks2/mathematics/number/maths_trail_roch

[7] *Problem Pictures* and *Problem Pictures Themes*, collections of colour photographs with activities for mathematics teaching on CD-ROM, Badsey Publication, Badsey, UK, 2001, 2005. An American edition titled *Mathematical World* is published by Key Curriumum Press. See www.problempictures.co.uk

[8] *Problem Pictures* calendars have been published annually since 2003. There is also a set of four posters. Australian Association of Mathematics Teachers, Adelaide. See also www.problempictures.co.uk

[9] A. Oldknow, ICT - bringing maths to life and vice versa, *Micromath* 21(2), 2005. See also www.adrianoldknow.org.uk

Inference and Design in Kuba and Zillij Art with Shape Grammars

Ramgopal Rajagopalan[1], Eric Hortop[3], Dania El-Khechen[1], Cheryl Kolak Dudek[2], Lydia Sharman[2], Fred Szabo[3], Thomas Fevens[1], Sudhir Mudur[1]
{Computer Science and Software Engineering}[1], {Fine Arts}[2], { Mathematics Department }[3],
Concordia University,
1455 de Maisonneuve Blvd. W., Montreal, Quebec H3G 1M8, Canada
{r_rajago, d_elkhec, fevens, mudur}@cs.concordia.ca, cdudek@sympatico.ca,
lydia@alcor.concordia.ca, fredszabo@gmail.com, ehortop@metawidget.net

Abstract

We present a simple method for structural inference in African Kuba cloth, and Moorish zillij mosaics. Our work is based on Stiny and Gips' formulation of "*Shape Grammars*". It provides art scholars and geometers with a simple yet powerful medium to perform analysis of existing art and draw inspiration for new designs. The analysis involves studying an artwork and capturing its structure as simple shape grammar rules. We then show how interesting families of artworks could be generated using simple variations in their corresponding grammars.

1. Introduction

Generative specification [1] of paintings has long been a topic of interest. Stiny and Gips' seminal works [2, 3] well demonstrates the potential and usefulness of this model of generating art forms. In this paper we illustrate the application of a generative model to enable us interpret the geometric structure inherent in Kuba[1] textile and Moroccan zillij type of art forms (see Figure 1).

1.1. Kuba and zillij Art. Kuba textile designs can best be described as compositions of "simultaneous diversity."[4]; women artisans incorporate spontaneity and improvisation in their designs to achieve uniqueness and individuality, part of their African aesthetics [5]. They are most often characterized by semi-symmetry. The semi-symmetry in Kuba weaving is achieved by the juxtaposition of distinct geometric motifs and by controlled variations in texture, scale, shape, orientation, and/or colour. (see Figure 1) [5]. On the other hand traditional zillij mosaics depict rigorous symmetry. They are characterized by their intricate interlace pattern and periodic tiling [6]. While their existence has been traced back to the twelfth century, they reached a zenith in Fez in the seventeenth century [7]. Some of the best examples can be found in the *madrasahs* (universities) in the old Medina.

These artworks are an aesthetic expression of culture, social context, and history of a region. Unfortunately not enough knowledge about this expression survives today. As noted by Kaplan et. al. [8], with their original design techniques for such art lost in the sands of time, analysis of these art forms generate considerable interests in the contemporary design and arts community.

1.2. Structural Inference. While Moroccan zillij art exhibits rigorous symmetry, Congolese Kuba cloth's complexity derives from semi-symmetric compositions of elementary geometric motifs. Although the richness of these two art forms is distinctly different, the challenge of inferring the structure of geometric patterns is present in both. In the context of this paper, by *structural inference* we mean classifying the structural composition of these art forms using computational tools. The desired classification should help

[1] African Kuba textiles are also referred to as BaKuba.

an artist, designer or historian understand the underlying generative process involved in creating the design. The emphasis is on finding simple geometric classification techniques so that the knowledge could be easily codified and conveyed to others. From our perspective, this objective presents many analytical challenges which are characteristic of classification and generation problems in pattern design. While continuing to experiment with various techniques [14, 15] like group theory, neural nets and pattern recognition, we have found that shape grammars provide a powerful paradigm both for representing the structural complexity in existing designs and for creating newer ones.

Figure 1: *Top row illustrates* ***inference (analysis)****; the image on the left shows an example of an African Kuba cloth design, the middle image has identical geometric motifs overlaid on the design and the right image demonstrates all the five different types of motifs inferred from the design overlaid on top. The overlaid geometric pattern of the motifs on the right image is generated using a simple shape grammar as discussed in section 3. Bottom Row illustrates* ***new designs (generation)****; the image on the left shows an example of a Moroccan zillij mosaic with an overlaid motif imitating the underlying strand shape. The middle image retains the underlying structure and markers while modifying terminal symbols. The rightmost design simplifies the rule application sequence while changing a terminal and adding scaling to create a radically different design. These images demonstrate new family of art inspired by simple variations in the grammar of the zillij mosaic (see section 5).*

1.3. Previous Work. Earlier research attempts addressing the structural inference problem in Kuba and zillij art are inspiring. Moser and Coxter [17] give a comprehensive treatment to the 17 wallpaper groups employing group theory. Grunbaum and Shephard [6] classified periodic Islamic patterns like zillij using tiling theory. They obtain a fundamental region in a given pattern and infer its symmetry group by observing the strand's movements in it. Abbas and Salman [9] confirmed the theoretic analysis on a large collection of historical designs. For Kuba textiles, Washburn and Crowe [10] confirmed that while 12 of the 17 possible ways a design can be symmetrically varied on a surface are found in these textiles, they are most often characterized by semi-symmetry. Most recently, Grünbaum [18] has developed symmetry schemes for pattern variations on a two dimensional plane.

In design creation, Ostromoukhov [11] proposed steps for Islamic art generation by extending Grünbaum's group theory analysis [6], Dewdney [12] outlined a method for constructing designs based on reflecting lines through a regular arrangement of circles. Kaplan and Salesin [8, 13] presented a software tool, Taprats, which carries out a tile-based construction. It generates the pattern over a fundamental region of the tiling and then subsequently replicates it over the entire plane using the symmetry of the tiling. Taprats allows users to specify the details of the different regular polygons of the tiling using stars and rosettes. For irregular polygons the software infers the pattern based on the existing specification of the regular ones. For Kuba art, Washburn and Crowe [10] present interesting analytical results, but there is little suggestion of how new Kuba like designs might be constructed.

1.4. Our Contribution. The focus of our research is to develop a single, powerful computational system for analyzing the symmetric zillij mosaics, as well as the semi-symmetric African Kuba cloths. This approach has the advantage over traditional methods of providing a uniform analytical approach to pattern analysis, independent of the presence of symmetry. In addition, the generative design mechanism helps in capturing the designs uniquely.

The rest of the paper is organized as follows. Section 2 presents the needed background on shape grammars for specifying the structure. Section 3 shows how grammar rules help in structural inference and subsequently generating an entire family of new designs by simple variations in the parent design's grammar rules. Section 4 outlines the essential components of our Generative Design System. Techniques for creating new designs inheriting a parent's structural properties are given in Section 5. The paper concludes in Section 6 by exploring some opportunities for future work.

2. Shape Grammars

2.1. Pictorial Specification. If we consider the proposition that an artist executes a finite set of shapes and number of transformations in creating an artwork then the resulting painting, sketch or engineering plan could be defined using a *pictorial specification.* Stiny defines a pictorial specification [2] as a set of drawing instructions which can be implemented on a finite plane, in finite time. At the heart of a pictorial specification is an *arrangement of finite lines* (both curved and straight). Any *shape* is characterized by using a particular arrangement. Arrangements are formally captured as a finite sequence of *Euclidean transformations* (which are translation, rotation, scale and reflection). Next in turn we could use shapes as primitives and apply transformations to them to develop patterns based on geometry. This process could be carried out recursively till the desired picture is achieved. Colouring techniques in the artwork could associate as properties of basic shapes (closed ones). Texturing, blending, and fading are the primary coloring techniques employed. But these are not considered as part of a pictorial specification. Hence a pictorial specification should capture the employed shapes and the sequence of transformations on them.

2.2. Formal Grammar and Shapes. The concept of using a grammar to describe a family of artworks has its origins in linguistics. Noam Chomsky [19] defines a formal grammar to be an abstract structure that describes a formal language precisely: i.e., a set of rules that mathematically delineates a (usually infinite) set of finite-length strings over a (usually finite) alphabet. Formal grammars are so named by analogy to grammar in natural language. Formal grammars fall into two main categories: *generative* and *analytic.* A *generative grammar*, the most well-known kind, is a set of rules by which all possible string(s) in the language to be described can be generated by successively rewriting strings starting from a designated start symbol. A generative grammar in effect formalizes an algorithm that generates strings in the language. An *analytic grammar*, in contrast, is a set of rules that assume an arbitrary string to be given as input, and which successively reduce or analyze that input string and yields a final Boolean result indicating whether or not the input string is a member of the language described by the grammar. An analytic grammar in effect formally describes a parser for a language.

In short, an analytic grammar describes how to read a language, whereas a generative grammar describes how to write it. A *shape grammar* by definition is a generative grammar used to capture a family of pictorial specifications. It provides a structured mechanism to capture the recursion inherent in the artwork.

2.3. Definition of a Shape Grammar. Gips [2] and Stiny [3] define a shape grammar as a quadruple: $SG = <V_T, V_M, R, I>$ where –

- V_T is a finite set of shapes. A finite arrangement of an element or elements of V_T in which any element of V_T can be used multiple times with the desired Euclidean transformations (translation, rotations, scale and reflection), gives the set V^*_T. Also we define $V^+_T = V^*_T - \phi$.
- V_M is a finite set of shapes such that shapes in the nonempty sets V^+_T and V^+_M are distinguishable.
- R is a finite set of shape rules of the form $u \rightarrow v$, where u and v are shapes formed by the shape union of shapes in V^*_T and V^*_M. The shape u must have at least one sub-shape that is a shape in V^+_M. The shape v may be the empty shape.
- I is the shape formed by the shape union of shapes in the power sets V^*_T and V^*_M. I must have at least one sub-shape that is a shape in V^+_M.

Shapes in the sets V^*_T or V^+_T are called terminal shapes (or *terminals*). Shapes in the sets V^+_M or V^*_M are called non-terminals (or *markers*). Terminals and markers are distinguishable, i.e., given a shape formed by the shape union of terminals and markers, the terminals occurring in the shape can be uniquely separated from the markers occurring in the shape. No shape in V^+_T is a sub-shape of any shape in V^+_M and vice versa.

For the *shape rule* $u \rightarrow v$, u is called the *left side* and v the *right side* of the shape rule. The shape consisting of all the terminals in the left side of the shape rule is called the *left terminal*. The shape consisting of all the markers in the left side of a shape rule is called the *left marker*. (The left marker has at least one sub-shape that is a marker.) *Right terminal* and *right marker* are defined similarly. The shapes u and v are represented in identical sized canvas to show the correspondence between them (in terms of their Euclidian transformations). I is called the initial shape. The shape consisting of all the markers in I has at least one sub-shape that is a marker. Euclidean transformations could be applied to any of the above shapes to facilitate creation and orientation. But only those that are applied to markers in a rule are applied to the subsequent generations employing that marker.

A shape is generated from a shape grammar by beginning with the initial shape and recursively applying the shape rules. The shape generation process is terminated when no shape rule in the shape grammar can be applied. The *sentential set* of a shape grammar, *SS(SG)*, is the set of shapes (sentential shapes) which contains the initial shape and all shapes which can be generated from the initial shape using the shape rules. The *language* of a shape grammar *L(SG)*, is the set of sentential shapes that contains only terminals. In other words, the language defined by a shape grammar L(SG) is the set of shapes generated by the grammar that do not have any sub-shapes which are markers. The language of a shape grammar may be finite or infinite set of shapes. The next section demonstrates examples.

3. Inference and new design

3.1. Developing Shape Grammars. We shall illustrate the application of shape grammars using the Kuba design from Figure 1. Kuba designs present a challenge to shape grammar based interpretation because the set of rules need specificity and detail to describe their complex compositions. Our aim is to enable development of simple shape grammars that do not have an overwhelming set of rules and yet have the full power to capture the structure of the design and its creational process. Figure 2 shows a simplified subset of V_T, V_M, and I of a shape grammar SG1 for the Kuba design in Figure 1. V_T contains a single element, a geometric motif traced out from the Kuba design presented in Figure 1. V_M contains an arrow shape as its only element.

The pattern generated using the four grammar rules of SG1 is shown in Figure 2. Application of rule 1 to initial shape results in the creation of a geometric motif (terminal) and an arrow (marker). Application of rule 2 and 3 causes a translation by x and $-x$ units respectively. Rule 3 causes a horizontal flip. The availability of a marker on the right side forces the continuation of the generation process till we apply a rule in which the marker is dropped from the right. With no marker in the canvas the generation process halts. The desired placement of our terminal symbols (as shown in Figure 1, middle) is achieved by applying rule 2 three times, followed by rule 3 twice, and finally rule 4 twice. Please note that the above generation is parallel in nature, i.e., whenever a shape rule is used it is applied simultaneously to every part of the shape to which it is applicable. Examples of new designs by modifying the parent's grammar are discussed in section 5.

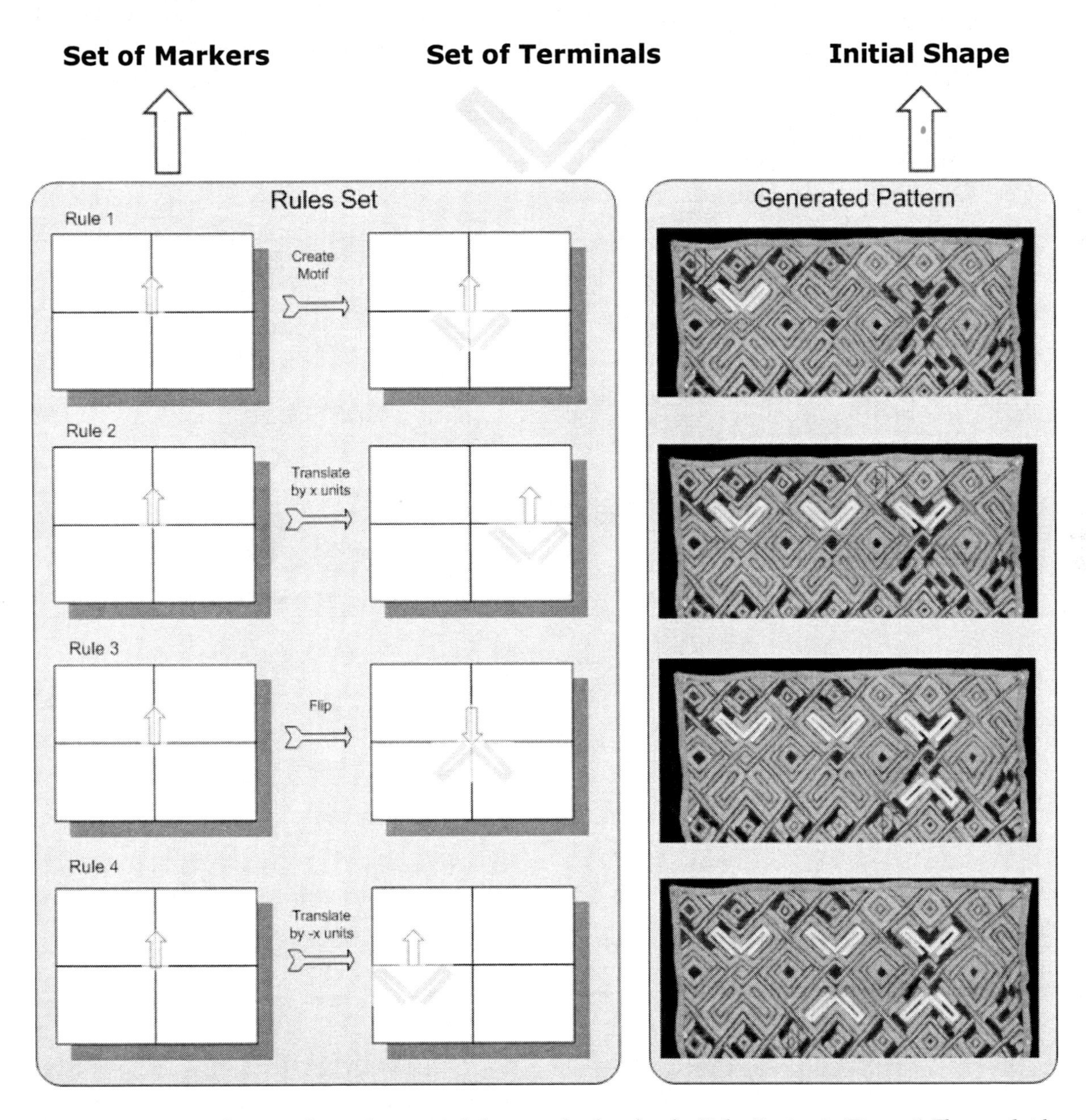

Figure 2: *SG1; Set of terminals, markers initial shape and rules for the Kuba Design in Figure 1.The overlaid pattern on the right is generated by the corresponding left hand side rule application on the design.*

4. Components of the Generative Design System

4.1. XML-based Specification. Computer implementation-wise, we capture the shape grammar for any design in XML with the future intent of being able to create further tools for structure analysis. Our XML schema for shape grammar specification is simple. It consists of three parts: *ShapeList*, *RuleList* and *SequenceList*. ShapeList captures the terminal and marker shapes in the design i.e. the sets V^*_T and V^*_M. RuleList captures the rule set *R*, of the grammar. For simplicity we use a unique marker as the start symbol. This facilitates modeling the set of initial shapes *I*, as additional rules. The SequenceList captures the sequence of rules applied to generate the design. Figure 3 lists the shape grammar SG1 (from Figure 2) in our XML. Please note that Euclidean transformations (translation, rotation, scale and reflection) are specified as an inherent property of the rule. This is in complete compliance to the shape grammar specification outlined in section 2. Colors could be associated with each shape. Transparency of the shapes can also be set. Transparency is a useful characteristic when it comes to blending shapes and creating new designs (refer to Figures 4 and 5).

4.2. Grammar Interpreter. The grammar interpreter takes a shape grammar in XML as input and creates a tree based representation of the pictorial specification. Depending upon the sequence of shape rules provided it accordingly creates the design on the canvas.

```
<?xml version="1.0" encoding="UTF-8"?>
<!DOCTYPE ShapeGrammar SYSTEM "sg_ver_1.01.dtd">
<ShapeGrammar>
        <ShapeList>
                <Shape type="marker"  shapeID="ARROW">
                        <Polygon color="150, 0, 0" opacity="0.0" stroke="2.0">
                                <Point x="320.0" y = "300.0" />
                                <Point x="300.0" y = "320.0" />
                                <Point x="280.0" y = "300.0" />
                                <Point x="290.0" y = "300.0" />
                                <Point x="290.0" y = "250.0" />
                                <Point x="310.0" y = "250.0" />
                                <Point x="310.0" y = "300.0" />
                        </Polygon>
                </Shape>
                <Shape type="terminal"  shapeID="MOTIF">
                        <Polyline color="251,228,140" opacity="0.0" stroke="8.0">
                                <Point x="364.0" y = "361.0" />
                                <Point x="327.0" y = "324.0" />
                                <Point x="335.0" y = "317.0" />
                                <Point x="380.0" y = "357.0" />
                                <Point x="422.0" y = "316.0" />
                                <Point x="431.0" y = "325.0" />
                                <Point x="396.0" y = "362.0" />
                        </Polyline>
                </Shape>
        </ShapeList>
        <RuleList>
                <Rule ruleID="INITIAL">
                        <Left>
                                <CompositeShape><RuleShape shapeRef= "ARROW" /> </CompositeShape>
                        </Left>
                        <Right>
                                <CompositeShape>
                                        <RuleShape shapeRef= "ARROW" />
                                </CompositeShape>
                        </Right>
                </Rule>
                <Rule ruleID="TRANSLATE_ON_X">
                        <Left>
                                <CompositeShape><RuleShape shapeRef= "ARROW" /> </CompositeShape>
                        </Left>
                        <Right>
```

```
                                        <CompositeShape transformation="translate(180, 0)">
                                                <RuleShape shapeRef= "ARROW" />
                                                <RuleShape shapeRef= "MOTIF" />
                                        </CompositeShape>
                                </Right>
                        </Rule>
                        <Rule ruleID="TRANSLATE_ON_-X">
                                <Left>
                                        <CompositeShape><RuleShape shapeRef= "ARROW" /> </CompositeShape>
                                </Left>
                                <Right>
                                        <CompositeShape transformation="translate(-180, 0)">
                                                <RuleShape shapeRef= "ARROW" />
                                                <RuleShape shapeRef= "MOTIF" />
                                        </CompositeShape>
                                </Right>
                        </Rule>
                        <Rule ruleID=" FLIP_ABOUT_X">
                                <Left>
                                        <CompositeShape><RuleShape shapeRef= "ARROW" /> </CompositeShape>
                                </Left>
                                <Right>
                                        <CompositeShape transformation="translate(0, 800); scale (1,-1)">
                                                <RuleShape shapeRef= "ARROW" />
                                                <RuleShape shapeRef= "MOTIF" />
                                        </CompositeShape>
                                </Right>
                        </Rule>
                </RuleList>
                <SequenceList>
                        <Sequence ruleRef="INITIAL"/>
                        <Sequence ruleRef="TRANSLATE_ON_X" count="3"/>
                        <Sequence ruleRef="FLIP_ABOUT_X"/>
                        <Sequence ruleRef="TRANSLATE_ON_-X" count="2"/>
                </SequenceList>
</ShapeGrammar>
```

Figure 3: *XML based specification of shape grammars. Please note the correspondence of the marker and terminal shape listed in the <Shape> tags with the sets V^*_T and V^*_M of SG1. Also note the four rules listed in the <Rule> tags and the sequences under <Sequence> tags correspond to the rule set and sequence of application respectively, in Figure 2.*

5. Inheritance and New Designs

5.1. Creating a Family of New Designs. As mentioned in section 1.4, once we capture the structure of a base design using shape grammars we can generate interesting families of artworks by simple variations in its grammar rules.

Figure 4 (first row) demonstrates the grammar rules for the zillij mosaic from Figure 1 (Being evident in the rules, the marker and terminal shapes are not shown separately). The rest of the designs in Figures 4 portray the results of simple variations in the shapes, rules and rule sequences of the grammar.

A grammar can generate related designs within its language, or the grammar can be modified. A modified grammar can inherit terminal shapes from a parent source and rearrange them (Figure 4, row 2; column 2), perhaps changing its symmetry group (Figure 4, row 4; columns 2 & 3), adding details, transforming terminals or modifying their arrangement (Figure 4, row 4; column 1), or retain some or all of its arrangement of markers but change terminals to build a new design with the same fundamental structure (Figure 4, row 3). A child grammar may even retain the exact same pictorial specification but modify colouring rules, yielding a new design through different blending, line traits or colour assignments (Figure 4, row 2; columns 2 & 3).

By traversing grammars' languages or reworking grammars, artists, designers and historians can wander freely in the "neighbourhood" of an existing design, exploring variations and discovering what elements, for them, define the character of an existing pattern. By modifying a pattern to replicate another existing design, the artist or researcher can get a hands-on feel for the similarities and differences between them. The end product of such work may be new patterns, or simply a better understanding of the ones under examination.

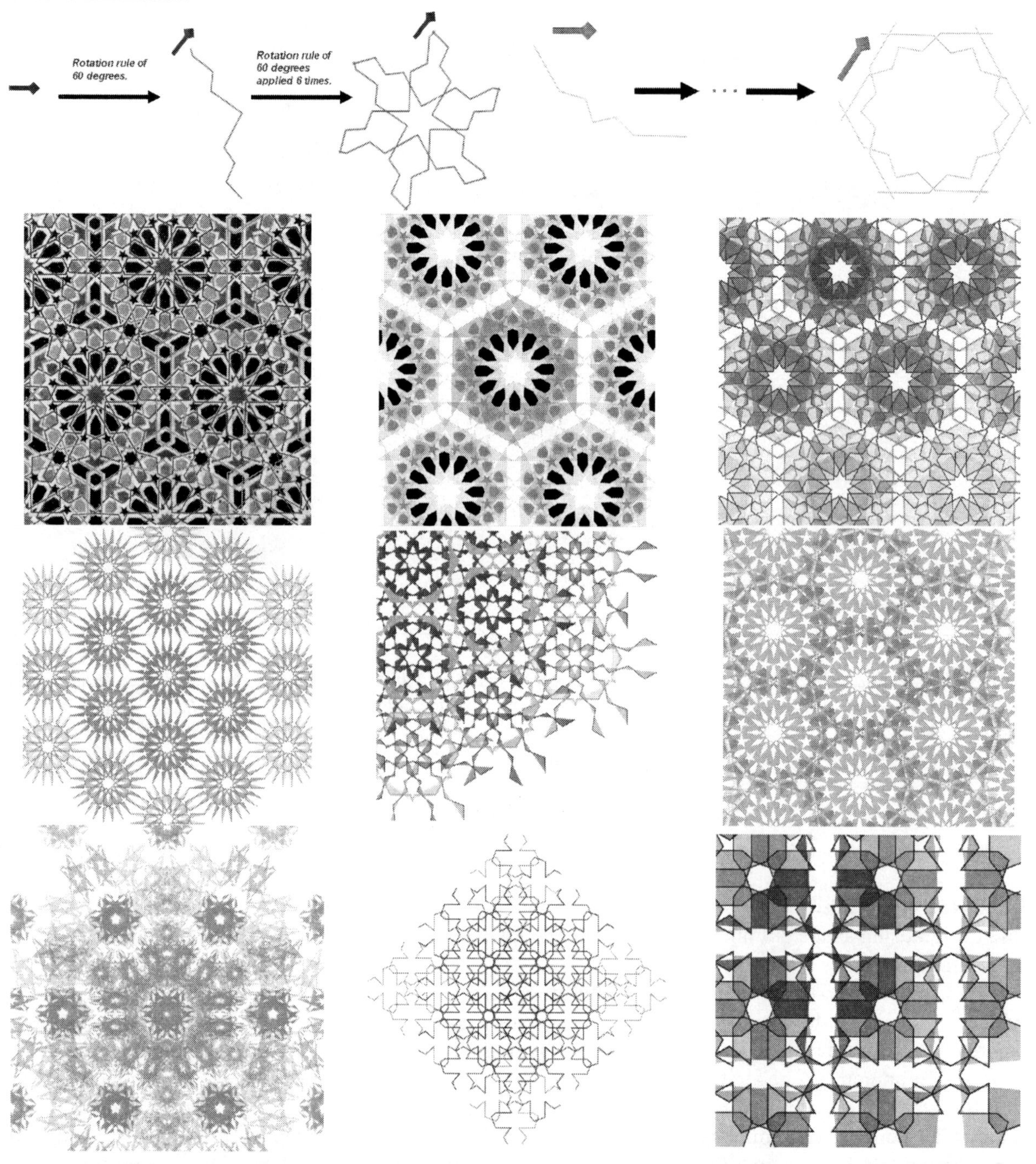

Figure 4: *Row 1: It illustrates the two different strands (terminal shapes) of the zillij mosaic from Figure 1 and their respective shape grammar rules. Row 2: the first image illustrates the recursive application of the rules to achieve the structural inference; the next two images show simple changes in colouring schemes and use of additional motifs adhering to the same grammar rules. Row 3: these designs retain the rules and sequences described in the row 1 but use different terminal shapes (strands). Row 4: the first image has one of the strands of the original mosaic with a non-uniform scale in the rule, generating a non-periodic design; the next two designs are produced by replacing the 6-fold dihedral symmetry in one rule with a 4-fold one, and modifying a terminal symbol.*

6. Conclusion and Future work

Kuba cloth and zillij mosaics seem to be on opposite ends of a design spectrum from the apparently arbitrary deviations within Kuba motifs to the characteristic repetitions of a zillij pattern. These designs present a challenge to shape grammar based interpretation because the set of rules need specificity and detail to describe their complex compositions. Our aim is to enable development of simple shape grammars that do not have an overwhelming set of rules and yet have the full power to capture the structure of the design and its creational process.

The comparison between the periodic, elegant designs of the zillij mosaics and the seemingly arbitrary, improvisational Kuba cloth suggests opportunities for generating an entire class of artworks by simple variations in the grammar rules. As illustrated in Figures 4 and 5 the concept of inheritance from parent design is an interesting contribution of our work. It gives rise to many questions yet to be answered such as what extent of change to the grammar maintains a clear inheritance relationship to the parent design? Radical changes in symmetry group and design elements are supported in our system: how should those changes be considered to support contemporary design practice? How close is the mapping between our grammar process and the development of designs by traditional artisans?

Currently we are exploring the following areas:

- Explorations are being undertaken to create variations, possibly by using heuristic based approaches in the zillij rule set and create an analogue for understanding the geometry of the Kuba cloth patterns.
- Our current efforts are also geared towards building an "artist-friendly" system [16] that facilitates the structural analysis of these artworks through the interactive processes of creating, modifying, comparing and searching for descriptive shape grammars. While the back end of such a system is a databank of artworks in different states of analysis, the front end is a simple to use graphical user interface that provides intuitive interaction techniques for typical shape grammar tasks such as grammar based retrieval, production rule specification or modification, visualizing the evolution of the design as the production rules are applied or modified and structural similarity checking through shape grammar comparisons.
- As our generative system does not assume a regular underlying tiling, interlacing of strands cannot be performed on fundamental regions as proposed in [6]. We are still trying to arrive at intelligent ways to perform interlacing based on the information provided in the grammars.
- We are also looking at the possibility of deriving the tiling of a periodic generative specification. We believe that the shape grammar, if well-defined, has the flexibility of predicting the symmetry group of the generated pattern and hence it inherently contains the classification of the tiling in one of the 17 wallpaper patterns [17]. This classification is useful since it allows one kind of comparison between existing and newly generated patterns. This is currently one of our main research foci.
- We are exploring the range of variation in grammar rules and its effect on the inheritance relationship with the parent design. It is not certain where culturally or structurally clear inheritance relationships end and novel designs borrowing from a parent artefact begin. Nevertheless, the inheritance of shape grammar elements can be applied in elegant and novel ways to build designs which draw on an existing, rich heritage.

7. Acknowledgements

This work is supported by grants from NSERC (Science and Engineering Research Council, Canada) Discovery grant programme, FQRSC (Fonds Quebecois de la recherché sur la société et la culture) Appui à la recherche-création programme, SSHRC (Social Sciences and Health Research Council, Canada) Research/Creation in Fine Arts programme, Hexagram and Concordia University Faculty Research grants.

References

[1] Stiny G, Gips J, 1972, "*Shape Grammars and the Generative Specification of Painting and Sculpture*" in C V Freiman (ed) Information Processing 71 (Amsterdam: North-Holland) 1460-1465. Republished in Petrocelli O R (ed) 1972 The Best Computer Papers of 1971: Auerbach, Philadelphia 125-135.
[2] Stiny, George. "*Pictoral and Formal Aspects of Shape and Shape Grammars.*" Basel, Switzerland: Birkhäuser Verlag, 1975, pp 28–39; 122–126.
[3] Gips J, 1975 "*Shape Grammars and Their Uses: Artificial Perception, Shape Generation and Computer Aesthetics*" (Birkhaüser, Basel:, Switzerland)
[4] Adams, Monni, "*Beyond Symmetry In Middle African Design.*" In: African Arts 23, 1: 35-43, 102-3, 1989. pg.35.
[5] Ibid, pg 37, 41.
[6] Branko Grünbaum and G.C. Shephard. *Interlace patterns in Islamic and Moorish art.* Leonardo, 25:331-339, 1992.
[7] El-Said, Issam and Ayse Parman. *Geometric Concepts in Islamic Art.* London: Scorpion Publishing, 1976.
[8] Craig S. Kaplan and David H. Salesin. *Islamic star patterns in absolute geometry.* ACM Trans. Graph., 23(2): 97–119, 2004.
[9] Abas, S., Salman, A. *Geometric and group-theoretic methods for computer graphics studies of Islamic symmetric patterns*. Comput. Graph. For. 11, 1, 43–53, 1992
[10] D. W. Crowe and D. K. Washburn, *Symmetries of Culture: Theory and Practice of Plane Pattern Analysis*, Seattle and London: University of Washington Press, 1988.
[11] Ostromoukhov, V.. *Mathematical tools for computer-generated ornamental patterns.* In Electronic Publishing, Artistic Imaging and Digital Typography. In Lecture Notes in Computer Science, vol. 1375. Springer-Verlag, New York, 193–223, 1998
[12] Dewdney, A.. *The Tinkertoy Computer and Other Machinations*. W. H. Freeman, 222–230, 1993
[13] Kaplan, C. S.. *Taprats*. http://www.cgl.uwaterloo.ca/~csk/washington/taprats/. , 2000
[14] Ramgopal R and Miguel A, "*Generative Design System*", Technical Report, 25th Sept 2004, URL (http://www.cs.concordia.ca/~r_rajago/GDS/researchReport-0.5.pdf)
[15] Cheryl Kolak Dudek, Lydia Sharman, Fred E. Szabo, Sushil Bhakar, Eric Hortop, Yun Li, Wumo Pan, *From Ethno-mathematics to Generative Design: Metapatterns and Interactive Methods for the Creation of Decorative Art.*, 8th International Conference on Information Visualisation (IV 04), London, 2004
[16] Yojana Joshi, Ramgopal Rajagopalan, Cheryl Kolak Dudek, Lydia Sharman, Sudhir P Mudur, "*Designing a user interface for Generative Analysis of African Kuba Textiles*", Third Iteration Conference on Generative Systems in the Electronic Arts, Nov 30th – Dec 2nd, 2005, Melbourne, Australia.
[17] H. S. M. Coxeter, W. O. J. Moser, "*Generators and relations for discrete groups*", 3rd Ed, Berlin; New York : Springer-Verlag, 1972.
[18] Grunbaum, Branko. "Periodic Ornamentation of the Fabric Plane: Lessons from Peruvian Fabrics," Chap. 2 in Symmetry Comes of Age: The Role of Pattern In Culture. Seattle and London: University of Washington Press, 2004.
[19] Noam Chomsky: *Three models for the description of language*, IRE Transactions on Information Theory, 2 (1956), pages 113-124

Green Quaternions, Tenacious Symmetry, and Octahedral Zome

David A. Richter
Department of Mathematics, MS 5248
Western Michigan University
Kalamazoo, MI 49008, USA
E-mail: david.richter@wmich.edu
Website: http://homepages.wmich.edu/~drichter/

Scott Vorthmann
1317 Lacresta Dr
Freeport, IL 61032, USA
E-mail: scott@vorthmann.org
Website: http://www.vorthmann.org/zome/

Abstract

We describe a new Zome-like system that exhibits octahedral rather than icosahedral symmetry, and illustrate its application to 3-dimensional projections of 4-dimensional regular polychora. Furthermore, we explain the existence of that system, as well as an infinite family of related systems, in terms of Hamilton's quaternions and the binary icosahedral group. Finally, we describe a remarkably tenacious aspect of H_4 symmetry that "survives" projection down to three dimensions, reappearing only in 2-dimensional projections.

1. Introduction

This is a report on a journey of discovery we have shared over the past year, as we endeavored to gain deeper insights into the mathematics surrounding the Zome System of Zometool, Inc. This collaboration has borne some intriguing fruit, specifically the results we describe in this paper concerning generalizations of Zome, and a surprising aspect of symmetry and projection.

Our interest in Zome centers on the way that it hints at deep underlying mathematics: serendipity becomes commonplace, and startling coincindences come to be expected. The "octahedral Zome" and "green quaternions" of our title were quite surprising initially, but turn out to have a very straightforward explanation, detailed below. Nonetheless, the explanation has opened up new vistas, by giving us the ability to characterize some generalizations of Zome. The mechanism of "tenacious symmetry" is not so easy to explain, so we must content ourselves with merely describing its characteristics and some examples.

Our collaboration was also fruitful as a synergy between pure mathematics and the applied science of visualization and modelling software. Scott's Zome modelling software, "vZome", gained a number of rich capabilities as he learned more mathematics from David, and David gained a deeper insight, and even learned some geometry, by working with and talking about vZome. A highly productive feedback loop thus developed. Since this is as much a story of a collaboration as it is about beautiful mathematics, we have decided to tell it in roughly chronological order.

Along the way, we have seen many novel views of some classical polytopes, and here we present a few of these as well. It is important to stress that, with all of our static art being 3-dimensional, it is notoriously difficult to visualize the complicated and beautiful objects which exist in higher dimensions. Zome and vZome have allowed us to see some fantastic properties of these objects.

2. Octahedral Serendipity

The journey began with David's mention of the fact that one may inscribe five copies of the 600-cell in the vertices of the 120-cell, [1]. One may see this in the usual three-dimensional Zome projection of the 120-cell: Starting with any one of the projected dodecahedra, one selects any one of the ten regular tetrahedra inscribed in its vertices. The faces of this tetrahedron are shared by four more tetrahedra, and one can proceed in this way to construct a projection of all the tetrahedra in the 600-cell. Since the original 120-cell has a dodecahedral cell at the center, rather than a face, edge, or vertex, it is a "cell-first" projection, and hence the new 600-cell projection is necessarily also a cell-first projection. This projection of the 600-cell had not been seen by either of us, but is certainly not unknown, [1, 7]. The figure below shows a "4D-cutaway" view; it is halved in four dimensions to omit the central involution, then almost-halved again in three dimensions. The left-hand pair is a parallel binocular stereo view, and the right-hand pair is a cross-eyed binocular stereo view.

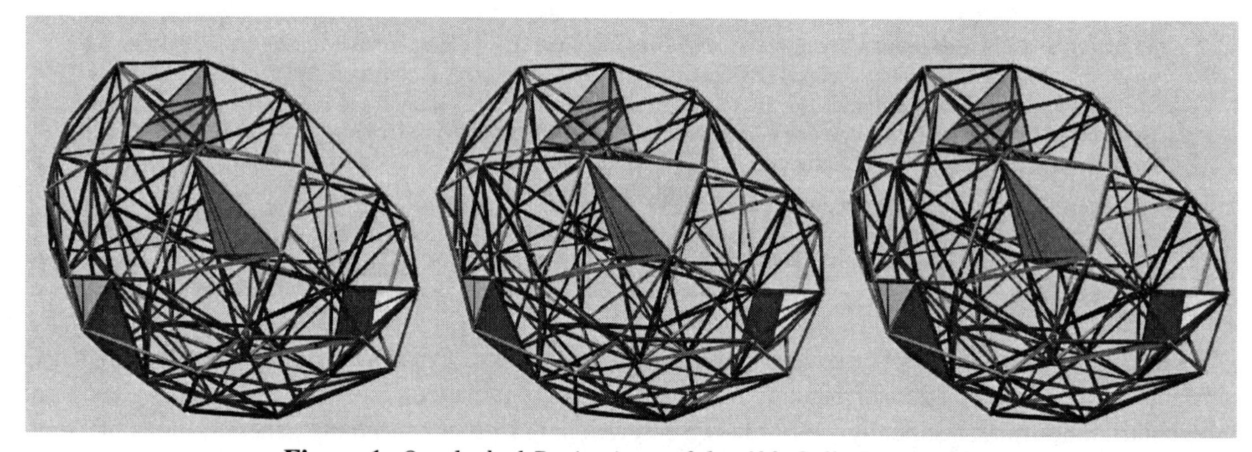

Figure 1. *Octahedral Projections of the 600-Cell, Quartered.*

It turns out that most of the edges of these tetrahedra do not correspond to any existing Zome struts. The vZome program, however, can construct a virtual strut between any two connector balls. When constructing such an "unknown" strut, vZome assigns a color in the same pattern as real Zome: All directions ("zones") that are equivalent under icosahedral symmetry are assigned the same color. This has the effect of highlighting the symmetry (or asymmetry) of models. After performing the construction in vZome, when we finally cleared away all the 120-cell "scaffolding" to see the 600-cell, we realized that the model used some of the blue, yellow, and green zones from the original Zome System, and three new zones. We have since christened these new zones with the colors maroon, olive, and lavender.

Although our original construction appeared to have tetrahedral symmetry overall, the full projection actually has "pyritohedral" symmetry, with two tetrahedra combined as in Kepler's stella octangula at the center. The pyritohedral symmetry group is the symmetry group of an idealized pyrite crystal. Geometrically, this is the direct product of the group of 12 orientation-preserving symmetries of the tetrahedron with the 2-element group generated by the "central involution" or "antipode" map

$$(x, y, z) \mapsto (-x, -y, -z).$$

This group arises in the context of Zome when one attempts to build a model with octahedral symmetry; it is actually impossible to build a Zome model with ideal octahedral symmetry (when one includes the symmetry of the Zome parts themselves rather than treating them as idealized points and line segments). The symmetry group of every Zome model is necessarily a subgroup of the full group of symmetries of the icosahedron, but no subgroup of the icosahedral group is isomorphic to the octahedral group. The largest subgroup of the icosahedral group which is also a subgroup of the octahedral group is the pyritohedral group.

Examination of the cell-first 600-cell model reveals that it actually has full octahedral symmetry in every sense except in the precise geometry of the real Zome balls and struts, for the reason explained above. Observing this and the fact that only three new directions were necessary for the whole model, we realized that we had stumbled on a novel Zome-like system with octahedral symmetry rather than icosahedral symmetry. This system shares all the other characteristics with the original Zome system: the ability to scale by powers of the golden ratio τ, strut coloring by symmetry group equivalence, and a surprising variety of constructible triangles and tetrahedra from a small set of zones. Although we have not yet proposed a formal mathematical definition, we consider this new Zome-like system as an example of a "zoning system". Such "zoning systems", generally, should possess properties like those outlined above, and indeed we will show that there are many systems that fit within such a framework.

With our discovery of the octahedral system, Scott immediately set to work to make vZome support it, both in the rendering of the virtual parts, and in the available symmetry operations. Scott soon realized that he could construct a 120-cell projection with the same symmetry by constructing a vertex at the center of each tetrahedron in the 600-cell projection, then connecting each vertex to its four nearest neighbors. After adding a centroid command to vZome, we quickly produced a vertex-first, octahedrally symmetric projection of the 120-cell. Although a black-and-white figure cannot readily convey it, this projection is entirely constructible within our new octahedral Zome-like system. As above, this figure shows a 4D-cutaway view in two stereo pairs.

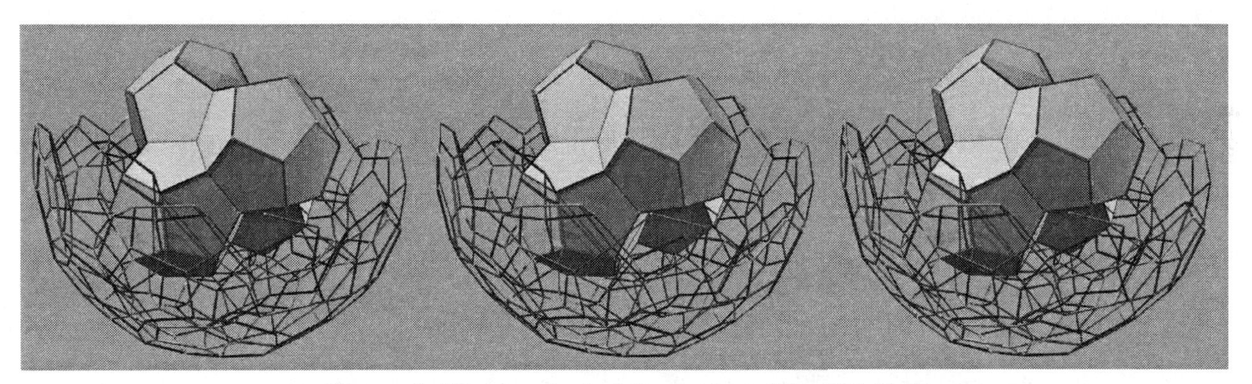

Figure 2. *The Octahedral Projection of the 120-Cell.*

3. What Color Is Your Quaternion?

Having discovered one new zoning system and being interested in finding more, we set out to explore how and why the new system worked. Although we constructed the octahedral 600-cell projection in three dimensions, David pointed out that as a faithful projection of the 4-dimensional object, it could equally well be produced by rotating the 600-cell in 4-dimensional space before projecting to three dimensions, and that such rotation could be accomplished by quaternion multiplication. Hamilton's quaternions $\mathbb{H}$ are intimately related to 4-dimensional geometry [3]. Furthermore, the original blue-yellow-red Zome System of Zometool is intimately related to the binary icosahedral group I and its embedding in the group S^3 of unit quaternions.

There are infinitely many different ways to embed the group $I \hookrightarrow S^3$, but one in particular leads to a set of easily expressed vectors that correspond to the Zome System. First, denote $\tau = \frac{1}{2}(1+\sqrt{5})$ and $\sigma = \frac{1}{2}(1-\sqrt{5})$, so that τ is the golden ratio and σ is its conjugate. One can list all 120 vectors quickly after noticing that they only take on three different "shapes". These shapes are representatives of orbits in the 192-element group which contains arbitary sign changes and all even permutations on the four coordinates. The shapes of vectors in this embedding of I are

$$(1,0,0,0),\ \frac{1}{2}(1,1,1,1),\ \text{and}\ \frac{1}{2}(0,\sigma,1,\tau),$$

of which there are respectively 8, 16, and 96 vectors. (We often use the two notations $w+ix+jy+kz$ and (w,x,y,z) for quaternions interchangebly. As a rule, we use (w,x,y,z) when we want to denote points or vectors and $w+ix+jy+kz$ when we want to denote a multiplication operator.) Notice that, although here we have presented three different shapes of vectors, this set of 120 vectors constitutes a single orbit under the H_4 symmetry group, of which this 192-element group is a subgroup.

One obtains the Zome System by considering $\pi(2I)$, where π is the projection from 4-dimensional to 3-dimensional space that maps according according to the formula

$$\pi : (w,x,y,z) \mapsto (x,y,z).$$

(We use $2I$ to indicate that we pre-multiply each element of I to avoid dealing with $\frac{1}{2}$ factors.) One assigns a color to each vector in $\pi(2I)$ according to its shape. Blue vectors have the shape $(2,0,0)$ or $(\sigma,1,\tau)$, yellow vectors have the shape $(1,1,1)$ or $(\tau,0,\sigma)$, and red vectors have the shape $(1,\sigma,0)$ or $(0,\tau,1)$. Here "shape" refers to orbits on 3-space under the pyritohedral group, as described earlier. Naturally, Zome also allows that every multiple of these vectors by a power of the golden ratio τ be a standard strut. Indeed, one should notice, for example, that $\tau(1,\sigma,0) = (\tau,-1,0)$, and this is the longer of the two types of red struts described above. Every blue strut has length $\tau^n \cdot 2$, every yellow strut has length $\tau^n \cdot \sqrt{3}$, and every red strut has length $\tau^n \cdot \sqrt{1+\sigma^2}$.

Recall that these are all projections of a single orbit of vectors equivalent under H_4 symmetry. If one imagines a four-dimensional analogue of Zome, the actual 600-cell and 120-cell could be entirely constructed using "blue hyper-struts" based on the vectors in I. In fact, all 15 convex uniform polychora with H_4 symmetry could be constructed with them, and therefore 3-dimensional π-projections of them can be constructed with the Zome system. The current existence of the Zome system itself is directly related to Marc Pelletier's recognition of this fact (when he was just 17).

If the set I can be considered the "blue hyper-struts" in four dimensions, the set $(1+i)I$ can be considered the "green hyper-struts". Note that the two sets are related to each other by a quaternion multiplication that composes a dilation by $\sqrt{2}$ with a 4-dimensional "rotation", and an object with H_4 symmetry can be constructed with either set alone. However, if the two constructions are both projected to a hyperplane perpendicular to a "blue" vector, the resulting three-dimensional projections are very different, as one might expect. Remarkably though, that difference is manifested in the fact that one projection has icosahedral symmetry and the other has octahedral symmetry.

We obtain our 6-color octahedral zoning system by considering set $\pi(2(1+i)I)$. As is the case with the elements of I, one notices that the vectors in the set $2(1+i)I$ also take on a small number of shapes. However, for reasons which one will find in [2], the group that we need is slightly restricted. Instead of allowing arbitrary sign changes, we only allow an even number of sign changes of the 4 coordinates. Since we still allow even permutations of the coordinates, we have a group of order 96 that preserves the shape of the vectors. Explicitly, there are four shapes

$$(-\tau^2, \sigma, \sigma, \sigma),\ (\sigma - \tau, 1, 1, 1),\ (\sigma^2, \tau, \tau, \tau),\ \text{and}\ (2, 2, 0, 0),$$

with corresponding orbit sizes 32, 32, 32, and 24. Projecting down to 3 dimensions, we notice that the number of signs is no longer relevant. For example, notice that (w, x, y, z) and $(-w, -x, y, z)$ agree except for an even number of sign changes, whereas their images $\pi(w, x, y, z) = (x, y, z)$ and $\pi(-w, -x, y, z) = (-x, y, z)$ agree except for an odd number of sign changes. Indeed, again we use the pyritohedral group to obtain the entire orbit of directions, and one may quickly tabulate the 6 shapes of vectors.

Shape	Number of Zones	Color	Shape	Number of Zones	Color
$(0,0,2)$	3	blue	$(1,1,\tau-\sigma)$	12	maroon
$(2,2,0)$	6	green	(σ,σ,τ^2)	12	lavender
$(1,1,1)$	4	yellow	(τ,τ,σ^2)	12	olive

Table 1. *The Octahedral Zoning System.*

Note that for any vector in this zoning system, at least two of the coordinates are equal in magnitude. This is equivalent to saying that all these struts lie in a plane orthogonal to some green strut. Also notice that a subset of the original blue, yellow, and green zones appear naturally in this system. However, if we want this new zoning system to accommodate the 600-cell, we must include the new maroon, lavender, and olive zones as well.

Armed with some of this knowledge, Scott was able to quickly implement quaternion multiplication in vZome when importing a four-dimensional data set or when generating one of the H_4 polychora. In vZome, one specifies a quaternion by selecting an existing strut. Although the strut is specified by a vector (x, y, z) with only 3 coordinates, the above analysis shows that the fourth coordinate w is determined up to a sign by the zone in which it lies. Selecting any green strut results in a quaternion having the same shape as $1 + i = (1, 1, 0, 0)$, and all such quaternions have effect equivalent to mapping $2I \to 2(1 + i)I$. In short, achieving the octahedral projections of the 120-cell or 600-cell is now as easy as three clicks, a vast improvement over our original manual derivation.

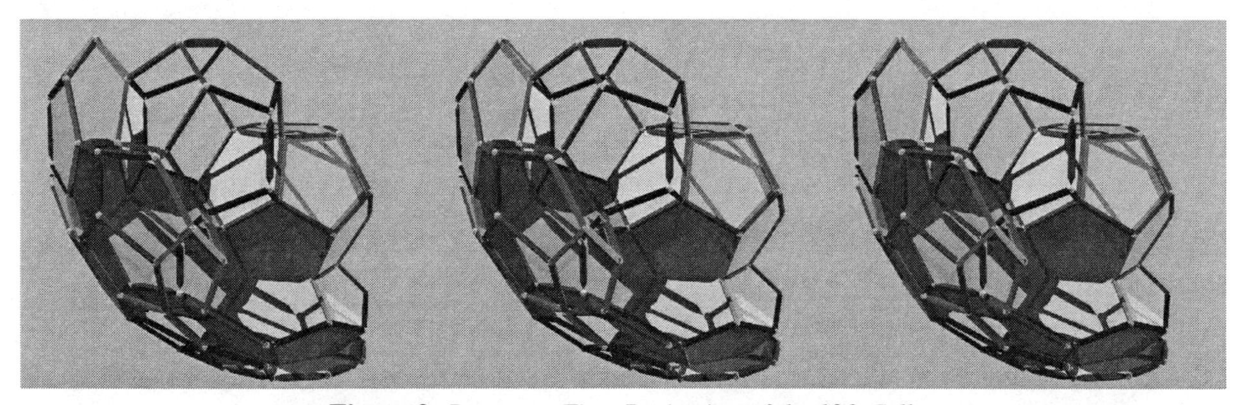

Figure 3. *Pentagon-First Projection of the 120-Cell.*

Availability of quaternion multiplication in vZome immediately begged the question: What is the effect of using the various "colors" of quaternions? First, since blue struts are all in the orbit of the quaternion $(2,0,0,0)$, which is an axis of symmetry in H_4, using a blue quaternion has no effect modulo a dilation. For other colors, the answer proved serendipitous in the way we have come to expect of Zome. Applying a "red" quaternion to an object with H_4 symmetry yields a 3-dimensional object with the symmetry of a pentagonal antiprism; the projection is symmetric around the red strut used as the quaternion. This results in a face-first projection of the 120-cell with two overlapping pentagons in the center, and an edge-first projection of the 600-cell. The figure above is a cutaway that shows one example of each of the 9 different dodecahedral cell shapes in the former model, with some faces present for clarity.

Naturally, applying a "yellow" quaternion produces projections symmetric around that yellow strut. This yields a face-first 600-cell, centered on overlapping triangles, and an edge-first 120-cell. Both red and yellow quaternion multiplications yield new zoning systems analogous to the octahedral system we have described, but with different symmetries. In both cases, the vectors of I are mapped to a small number of 3-dimensional shapes, although more shapes (thus colors) are required than for the octahedral system. Indeed, any quaternion multiplication of I yields a zoning system capable of rendering orthogonal projections of H_4 polychora.

4. Tenacious Symmetry

The group H_4, with 14,400 elements, is moderately large and exotic, compared to the sizes and variety of the finite groups which act in 4-dimensional space. Having so much symmetry, some unusual traces of this symmetry remain when these objects are projected down to three and two dimensions. We refer to this as "tenacious symmetry": Much symmetry is necessarily lost in a projection to lower dimension, but it somehow refuses to be totally eradicated. Moreover, due to the observation in [8], we see that the original Zome System is directly related to the famous E_8 "Gosset" lattice, whose point symmetries comprise a group with nearly 700 million elements. Certainly we will see a wide variety of highly-symmetrical objects by considering different views of this amazing object.

Figure 4. *Two Views of the Compound of Fifteen 16-Cells.*

An easily-accessed example of this phenomenon is provided by the Van Oss projection of the 600-cell, which appears as the frontispiece in [1]. The symmetry of this figure is the dihedral group with 60-elements, i.e., the symmetry group of a regular 30-sided polygon. This projection reveals the fact that this dihedral group, while it may not be isomorphic to any subgroup of H_4, is still closely related; indeed, H_4 contains elements of order 30.

Anyone who has worked on a Zome model of an H_4 polychoron is familiar with the obvious symmetries visible when looking through such a model. Few are surprised, moreover, that an object with icosahedral symmetry can exhibit 10-fold rotational symmetry in a parallel 2-dimensional projection. When David undertook to build a Zome model of the regular compound of fifteen 16-cells, using vZome as an aid for visualization, he began to see more surprising symmetries. Scott and other Zome enthusiasts experienced this firsthand when David led construction of the latter model, pictured above, at a meeting in Chicago in September 2005. This model has pyritohedral symmetry, and, in particular, its symmetry group has no elements of order 5. Nevertheless, as the photo on the right shows, the projection of the model onto a plane perpendicular to a red zone has pentagonal symmetry. This is a remnant of the symmetry group of the corresponding 4-dimensional polytope, which does have elements of order 5.

The vZome program is ideally suited to exploring this phenomenon. Consider the red-quaternion projection of the 120-cell; that projection has the same symmetry as a pentagonal antiprism. Although that symmetry group has no elements of order three, there are ten distinct orthographic projections of this object to two dimensions that exhibit six-fold rotational symmetry. Five of those projections produce a figure as shown below in the first two images, first showing the strut colors and second in a simple wireframe. The other five produce a figure as shown in the third image.

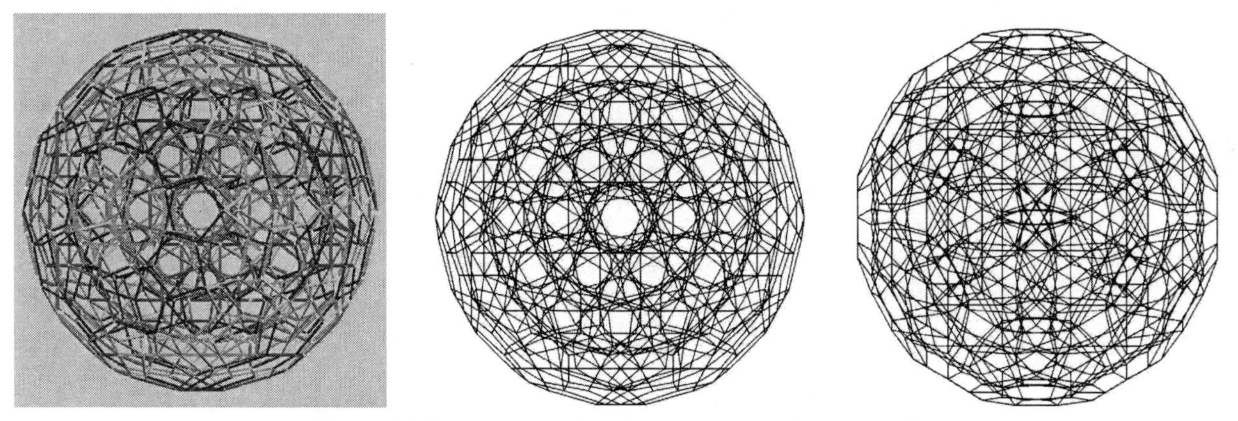

Figure 5. *Red-Quaternion Projections of the 120-Cell.*

Tenacious symmetry is in fact displayed in all Zome-axis projections of any H_4 polychoron, and similarly if a yellow quaternion is applied rather than a red one. In other words, although most of the rich symmetry of H_4 is inevitably lost in these 3-dimensional projections, that symmetry is too "tenacious" to be completely eradicated – traces of it remain when one projects again down to two dimensions along particular axes.

5. Conclusion

To reiterate, there are in fact an infinity of Zome-like systems based on the binary icosahedral group, one for each unit quaternion. However, they get progressively less interesting as the elements of I are mapped to more generic elements with respect to the group H_4, and less and less symmetry is preserved in the three-dimensional projection. With icosahedral symmetry, the original blue-yellow-red Zome System is clearly the most symmetric of this infinite set. As in the case of the octahedral system based on green-quaternion multiplication, the red and yellow quaternions similarly generate Zome-like systems with a small set of vector shapes. All of these systems share the useful property of arbitrary scalability by powers of τ, without requiring additional, longer strut lengths. These scalability and "small inventory" properties make all such systems potentially interesting to artists and engineers alike.

We are proposing to generalize and study a wide class of such "zoning systems", such as those that have been discussed here. Here is a review some of the critical properties shared by all of the zoning systems we have seen. First, all of these systems are based on connector balls and struts. Second, there is a group G which acts on the ambient space and for which the symmetry group of every model is necessarily a subgroup of G; the symmetry of the connector ball is G. Third, the original 4-dimensional orbit of 120 elements in I maps to a small set of orbits under G, with the size of that set dependent on the quaternion q used to map I to $\pi((2q)I)$.

As has been observed in [8], another characteristic of the Zome System is that it is closely related to a genuine lattice in 8 dimensions. One may describe this lattice quickly as a subset of $\mathbb{R}^8$, but is is also embedded naturally in $\mathbb{R}^4$ in such a way that the lattice points may be positioned arbitrarily close to each other. More precisely, the lattice points comprise a dense subset of $\mathbb{R}^4$ under the usual norm. This property leads to a fourth characteristic of all these zoning systems: The idealized locations of the connector balls comprise a dense subset of the ambient space. From the perspective of one who wishes to create an interesting model, we regard this property as critical; in a theoretical sense, it provides the user with the liberty to place objects in virtually any location s/he desires.

References

[1] H. S. M. Coxeter. *Regular Polytopes.* 3rd ed. Dover Publications Inc., New York, 1973.

[2] John H. Conway and Derek A. Smith. *On Quaternions and Octonions: Their Geometry, Arithmetic, and Symmetry.* A K Peters Ltd, Wellesly, Massachusetts, 2003

[3] P. Du Val. *Homographies, Quaternions and Rotations.* Oxford University Press, 1964.

[4] Veit Elser and Neil J. A. Sloane. A highly symmetric four dimensional quasicrystal. *J. Phys. A: Math. Gen.* **20** (1987), 6161-6167.

[5] Thorold Gosset. On the regular and semi-regular figures in space of n dimensions. *Messenger of Math.* **29** (1900), 43-48.

[6] George W. Hart and Henri Picciotto. *Zome Geometry: Hands on Learning With Zome Models.* Key Curriculum Press, Emeryville, California, 2001.

[7] Koji Miyazaki. *Higher-Dimensional Geometry.* Kyoto University Academic Publishing Association, 2005

[8] David A. Richter. Two results concerning the Zome model of the 600-cell. Renaissance Banff/Bridges Conference Proceedings, 2005.

Mathematics and Music – Models and Morals

Meurig Beynon
Computer Science Department
The University of Warwick
Coventry CV4 7AL, UK
wmb@dcs.warwick.ac.uk

Abstract

The intimate association between mathematics and music can be traced to the Greek culture. It is well-represented in the prevailing Western musical culture of the 18th and 19th centuries, where the traditional cycle of fifths provides a mathematical model for classical harmony that originated with the well-tempered, and later the equal-tempered, keyboard. Equal-temperament gives equivalent status to all twelve tonal centres in the chromatic scale, leading to a high degree of symmetry and an underlying group structure. This connection seems to endorse the Pythagorean concept of music as exemplifying an ideal mathematical harmony. This paper examines the relationship between abstract mathematics and music more critically, challenging the idealized view of music as rooted in pure mathematical relations and instead highlighting the significance of music as an association between form and meaning that is negotiated and pragmatic in nature. In passing, it illustrates how the complex and subtle relationship between mathematics and music can be investigated effectively using principles and techniques for interactive computer-based modelling [17] that in themselves may be seen as relating mathematics to the *art* of computing – a theme that is developed in a companion paper [2].

1. Mathematics and music in harmony

1.1. The music of the spheres. The notion of an intimate relationship between mathematics and music has a long history. The correlation between vibrating strings of different lengths and musical pitch established a strong connection between numerical and aural relationships, and supplied the basic intervals upon which the Western musical tradition is based. The vibration of a string could be identified as made up of many concurrent basic vibrations, comprising the fundamental note and many harmonics. The qualities of notes as played by different instruments could be analysed with respect to the *harmonics* they generated. The harmonic constituents of sounds acquired such fundamental importance in thinking about music that they later became the point of departure for many music theorists. Schenker (1868-1935) [11], for instance, set out to demonstrate that the tonal system had its roots in nature, and – building on this basis – even went so far as to attribute a primary, distinguished status to musical traditions based on tonality, and to their underlying aesthetics. In this view, the correspondence between the findings of natural science, mathematical models and human psychology gave a special significance to the classical music culture of the West.

1.2. The cycle of fifths. The advent and development of keyboard instruments led to a rationalization that had enormous ramifications for music. Once the range of a keyboard was sufficient to allow a single interval to be spanned both by a sequence of octaves and "perfect fifths", the need for some compromise in tuning each octave span of the instrument became apparent. This led to a mode of tuning known as "well-temperament", first widely used in the Baroque period, that had the liberating side-effect of allowing any note on the keyboard to be the tonal centre of a musical composition – a development celebrated to great effect by J S Bach in his 48 Preludes and Fugues for the Well-Tempered Clavier. The influence of composers such as Bach on musical composition was profound. It opened the way to a full exploration of tonal possibilities associated with all twelve notes of the chromatic scale. Where a mathematical perspective on music is concerned, there is a subtle but significant distinction between *well* temperament and *equal* temperament, as practiced from the middle of the nineteenth century. To quote Loy's account of tempering [8]: with well-temperament: "None of the scales or chords sounded bad. In fact every major and minor key sounded different. C sounded placid and fairly uninteresting. The more distant keys sounded more interesting. You might call some keys harsh, or agitated, or tense. And so, music could be written to suit the mood (or color) of each key". With equal temperament, in contrast: "No key sounds bad. No key sounds pure. All keys are somewhat interesting. But, they are all the same".

In effect, the emancipation of all twelve major and minor keys made it possible to conceive key relationships in an abstract mathematical fashion. The most widely cited form of this structure is that based on the cycle of fifths, as depicted in the model shown at the left in Figure 1. [The figures in this paper have been extracted from a poster for which a full colour image can be accessed and downloaded from http://empublic.dcs.warwick.ac.uk/projects/kaleidoscopeBeynon2005/posters/erlkonigPoster.pdf] In the cycle of fifths, adjacency of keys is associated with similarity of key signature. This connects keys whose tonic notes differ by a perfect fifth (cf. the red edges making up the two circuits in Figure 1) and connects a major key with its "relative minor" key with which it shares the same key signature (cf. the bi-directed green edges that define the spokes). As in Figure 1, capitalized and lower case letters will be used to refer to major and minor keys respectively.

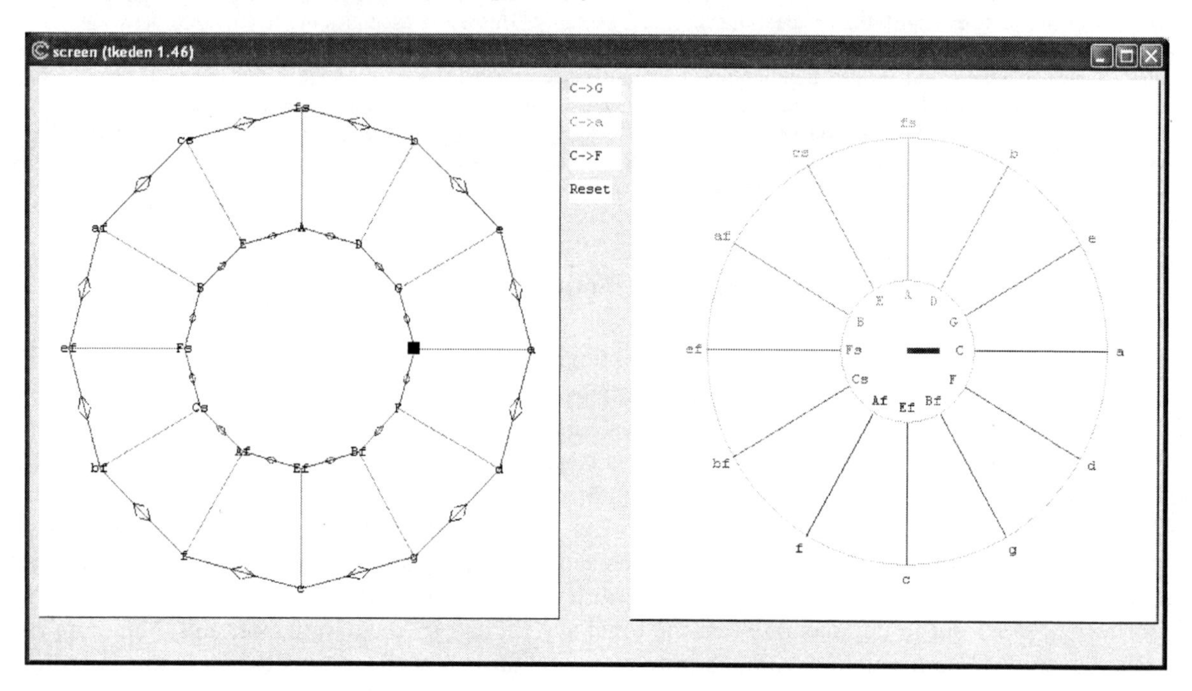

Figure 1: *The cycle of keys and an associated colour wheel*

1.3. Modelling the cycle of keys. Figure 1 depicts an interactive computer model of the cycle of keys that has been constructed by the author using modelling principles and tools that have been described in detail elsewhere [2,17]. The model comprises two connected visual components to represent two different ways in which the notion of current key is perceived in the mainstream tradition of tonal music. The diagram on the left (a "group graph") is the kind of model that underlies the basic music theory that is required to learn skills such as playing scales on an instrument. It is also used for harmonic analysis in music, as it might be applied to the works (e.g.) of Bach, Handel, Haydn, Mozart, Beethoven, Schubert, to traditional hymn tunes and to certain idioms of popular music. It is a framework that is harder to apply directly to music from the so-called 'romantic' tradition of the nineteenth century, as represented in the music of (e.g.) Chopin, Wagner, Liszt and – to a lesser extent – Schumann or Brahms, because of the greater complexity of their harmonies, and is a framework that does not apply to music of earlier traditions (such as that of Monteverdi or Palestrina) or to 20th century music that is atonal or polytonal.

At the heart of the harmonic system is the cycle of keys depicted in the group graph on the LHS of Figure 1. There are twelve such keys, corresponding to the 12 essentially distinct notes C, $C^{\#}$, D, $D^{\#}$, E, F, $F^{\#}$, G, $G^{\#}$, A, $A^{\#}$, B on a piano keyboard. In the group graph, each node on the inner cycle represents the tonic note in a specific major key, and this is connected by a green bi-directed edge to the submediant – the tonic note of its relative minor, by a red edge to the dominant, and by a blue edge to the subdominant. In this context, the submediant, dominant and subdominant respectively refer to the notes that lie at intervals of a minor third below, a fifth above and a fourth above the tonic note respectively. In Figure 1, the current tonic note is indicated by a small black square, and corresponds to the key of C, with relative minor a, dominant G and subdominant F. The current key is also indicated in the display on the RHS of Figure 1, in which the keys are disposed on the spokes of a wheel and coloured so as to suggest the different ways in which a musician experiences keys. As a mathematical model, the group graph in Figure 1 is associated with a particular Abelian product of cyclic groups, namely $C_2 * C_{12}$. The symmetry of the group graph reflects the sense in which an experienced musician maintains a sense of the global tonal framework whatever the current key.

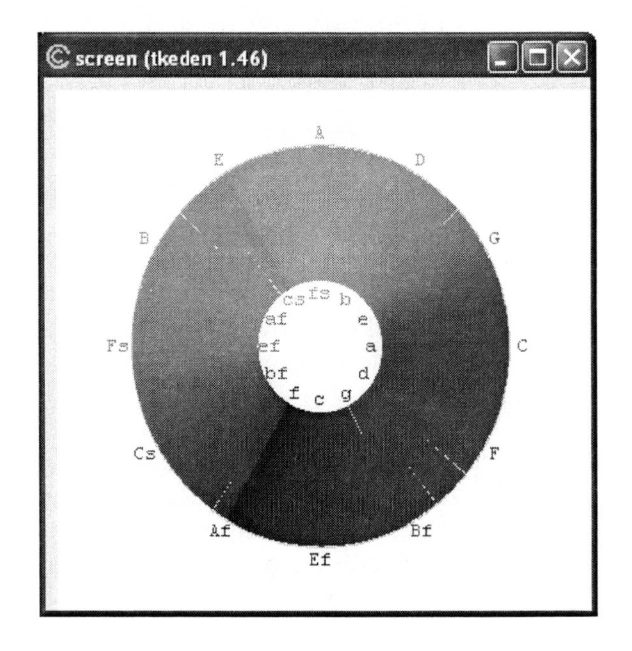


```
func colourwheel {
        para angle;
        auto r0, g0, b0;
        if ((4*PI/3<=angle)&&(angle<2*PI)) {
                angle = angle - 4*PI/3;
                r0 = int(255*sin(2.0*angle/3.0));
                g0 = 0;
                b0 = int(255*cos(2.0*angle/3.0));
        }
        else if ((2*PI/3<=angle) && (angle<4*PI/3)) {
                angle = angle - 2*PI/3;
                r0 = 0;
                g0 = int(255*cos(2.0*angle/3.0));
                b0 = int(255*sin(2.0*angle/3.0));
        }
        else {
                r0 = int(255*cos(2.0*angle/3.0));
                g0 = int(255*sin(2.0*angle/3.0));
                b0 = 0;
        }
        return rgb2colour(r0,g0,b0);
}
```

Figure 2. *The colourwheel from which key colours have been extracted, together with its functional definition*

1.4. Modelling keys by colours. The component on the right-hand side of the model (the colourwheel) reflects a complementary aspect of musical experience: the sense that each key has its own distinct character and 'colour'. Notwithstanding the symmetry of the cycle of keys, it would (for example) be regarded as a significant act in musical terms to rewrite Beethoven's Choral Symphony (written in D minor) in C minor. One aspect of this is that the choice of key influences the absolute pitch and impacts on the technical difficulty and even feasibility of the instrumental parts. Over and above this, the associations of different keys are deeply embedded in the classical musical tradition, and appear to exercise an important influence over the kind of music that a particular composer writes.

The model depicted in Figure 1 exploits definitions similar to those underlying the cells of a spreadsheet to maintain dependencies between the values of observables. As an example of such a dependency, the colours associated with keys on the RHS of Figure 1 are drawn from the colourwheel in Figure 2, where the association between the colour of a ray and its orientation is defined by the piecewise function – based on interpolating between the red, green and blue colour constituents – that is encoded on the right of Figure 2. Another dependency is used to link the abstract change of key in the group graph (as typically recorded cerebrally by the musician in the role of musical analyst) to the associated experiential change in tonal colour (as typically registered as a 'felt experience' by the performer). In this way, changes of key – and to some extent ambiguity about current key – may be represented by the motion of a 'colour wheel' in which each direction corresponds to a different colour.

2. Some hints of discord

2.1. The status of mathematical models of music. Mathematical models of musical phenomena may appear to endorse the absolute, transcendental nature of the relation between mathematics and music. It is perhaps fitting that some of best illustrations of modulation conforming closely to the cycle of keys are to be found in the hymnals of the 19th century. But, whilst the cycle of keys and the colourwheel can be interpreted as concretizations of precise abstract mathematical concept in the spirit of Turkle and Papert [15], the notion that they embody an absolute mathematical theory of music is suspect. From a detached musical standpoint, the colourwheel and group graph of Figure 1 may be seen as contrived mathematical abstractions within a concrete world of fuzzy experience.

This is most clearly evident in respect of the colourwheel. There is no accepted rationale to justify the choice of a particular key-colour association (whether or not a musician consciously associates keys with colours – as in instances of synaesthesia, to which composers such as Scriabin may have been subject [5]). In an extended discussion of this issue, Tovey [14:8,9] observes that "Beethoven ... when setting Scottish melodies, wrote to his English publisher, Thomson, that the key of A flat did not fit a certain tune that was sent him, inasmuch as that tune was marked *amoroso*, whereas the key of A flat should be called *barbaresco*", but goes on to deride the notion that "transposition would be equivalent to altering all the colours of a picture" as "a favourable example of the kind of fantasy which many learned musicians still fail to confine to its proper place among psychological obscurities".

Similar reservations apply to the group graph. In the first place, in aural terms, the cycle of keys is associated with the accidental approximation in pitch that underlies the notion of equal-temperament. Specifically, the 'cycle' is based on the false premise that 12 intervals of a perfect fifth (such as one might trace on a piano keyboard) generates exactly the same difference in pitch as a sequence of seven octaves. In reality, moving up a perfect fifth raises the frequency of a note by a factor of 1.5, whilst moving up an octave raises its frequency by a factor of 2, and $(1.5)^{12} = 129.746337890625$ merely approximates 2^7. On this basis, the formal model is an inexact abstraction from experience, in contrast to (say) the Platonic lines and points of Euclidean geometry. (Compare the rhythmic device used by Schumann in the second of the Drei Stücklein of his Bunte Blätter op. 99, where the 4th and 6th notes in a bar comprising 8 demisemiquavers carry a melody to be played in triplet time – a mathematical impossibility, but plausible in the specified *Sehr Rasch* tempo since 8/3 and 16/3 approximate 3 and 5 respectively.)

Secondly, the correspondence between the abstract harmonic framework depicted in Figure 1 and music within a particular tradition cannot be formalized. The closest correspondence is typically to be found in the least sophisticated music of Haydn or Mozart, where the harmonic structure is traced beat-by-beat through the chords of an accompaniment, and the changes of key such as are represented in the group graph predominate. In contrast, though Bach's music (and the 48 Preludes and Fugues for the Well-Tempered Clavier in particular) is conceived within the framework, the characteristic texture of his music does not for the most part lend itself to a simple association between simultaneously sounding notes and tonal harmonies. And whilst most of the music of Beethoven and Schubert relies upon clear harmonic textures, it exploits tonal juxtapositions quite unlike those represented by the edges of the group graph in Figure 1. The problematic aspects of the model of key relationships in Figure 1 are exemplified in the distant relationship it establishes between tonic major and minor, of which Tovey [14:11] writes: "First, let us be quite clear about the contrast between tonic minor and tonic major. Remember that the contrast is not a 'modulation' or change of key at all: it is a change of outlook while we stay at home."

2.2. Alternative models of tonality. In his analysis of tonality in Beethoven [14], Tovey advocates an alternative model of tonality. The principle behind his analysis is that the keys that are most closely related harmonically to a major key are those whose triads can be found in its major scale. For example, applying this criterion, the key of C is directly related to d, e, F, G and a. A similar criterion can be applied to minor keys: for instance, the key of c is directly related to Eb, f, g, Ab and Bb. This supplies an alternative to the algebraic model of tonal relationships in the cycle of keys – a looser concept of neighbourliness of keys that can be expressed geometrically by laying out all 24 major and minor keys on the surface of a torus. Such a geometric framework was devised by Schoenberg [12, 10], and set out as a chart of harmonic regions as depicted in Figure 3.

			d#	F#	f#	A	a	C
			g#	B	b	D	d	F
			c#	E	e	G	g	Bb
	d#	F#	f#	A	a	C	c	Eb
	g#	B	b	D	d	F	f	Ab
	c#	E	e	G	g	Bb	bb	Db
F#	f#	A	a	C	c	Eb	eb	Gb
B	b	D	d	F	f	Ab		
E	e	G	g	Bb	bb	Db		
A	a	C	c	Eb	eb	Gb		
D	d	F	f	Ab				
G	g	Bb	bb	Db				
C	c	Eb	eb	Gb				

Figure 3. *The chart of the regions, from Arnold Schoenberg: Structural Functions of Harmony*

As explained in detail in [14], Tovey's treatment of tonality has a quite different focus from the cycle of keys model in Figure 1. Where the edges of the group graph purport to specify the particular modulations that normally arise in classical harmonic sequences, the emphasis in Tovey's analysis is rather on a more topological notion of neighbourliness of keys. For instance, whereas the group graph distinguishes modulations to the local relative major/minor, dominant and sub-dominant, Tovey maintains that these keys are no more closely related to the tonic than the rest of the five related keys. This accords with Tovey's central thesis about Beethoven's music – that it is not so much distinguished by novel local harmonic progressions, but by its contribution to "the long-range power of handling tonality [that] is in its earlier stages downright incompatible with concentration upon new chords and new progressions".

Further analysis of the music of the later classical period suggests other ways of elaborating on standard models of tonality. In addition to modulation to neighbouring keys in Figure 3, Tovey [14] also identifies Neapolitan transitions as characteristic of Beethoven's harmonic idiom – as exemplified in the initial bars of the Appassionata Piano Sonata op. 57, where the opening phrase is immediately repeated a semitone higher. The music of Schubert is also recognized as exhibiting novel harmonic effects, characterized, especially in the context of his finest songs, by unexpected dramatic progressions. The brief discussion that follows is elaborated in [2].

At one level, Schubert's music is characterized by very strongly defined local harmonic progressions, more in keeping with an algebraic than a topological model of structure. The simple ballad *Der König in Thule* is an excellent example of a musical composition that exemplifies structure within the traditional cycle of fifths framework. It not only exhibits direct modulations between closely related keys, but has the remarkable feature of being based solely on chords that are in first position – that is, that have their tonic note in the bass. Of Schubert, Tovey writes [14:30]: "He came only gradually under Beethoven's spell after his sense of tonality had developed quite independently; but his tonality is exactly Beethoven's in its fullest range, and is intensified by concentration in lyric forms to which Beethoven contributed little". Schubert's "independent sense of tonality" arguably put more primary emphasis on the traditionally most commonplace modulations (to the dominant, subdominant and relative minor), and in lyric forms focused on harmonic tension and argument rather than on loose and exploratory juxtaposition of keys. Through reinforcement of conventional expectation, such an idiom gives greater scope for immediate local harmonic surprise. To some listeners, it also conveys an unwelcome impression of naivety – witness George Bernard Shaw's scathing critique of the Great C Major Symphony: 'a more exasperatingly brainless composition was never set on paper' [13:99].

Because of the underlying orientation of Schubert's harmonic idiom, the restricted model of tonality afforded by the group graph in Figure 1 is a promising basis for the analysis of his music, but something further is required. Attempting to apply the simple techniques that apply to the *Der König in Thule* to more sophisticated compositions poses problems. One of Schubert's trademark harmonic innovations is found in his treatment of tonic major and minor. Much more is involved here than Tovey's "change of outlook whilst we stay at home". In the extended ballad *Erlkönig*, one of Schubert's most ambitious songs, it is not simply that the music visits tonic major and minor keys in quick succession, but that tonic major and minor are conflated in passages in such a way that the whole tonality of the music is called into question. Figure 4 depicts two experimental models, each constructed by modifying the group graph in Figure 1, that offer alternative ways of expressing Schubert's use of tonic major-minor.

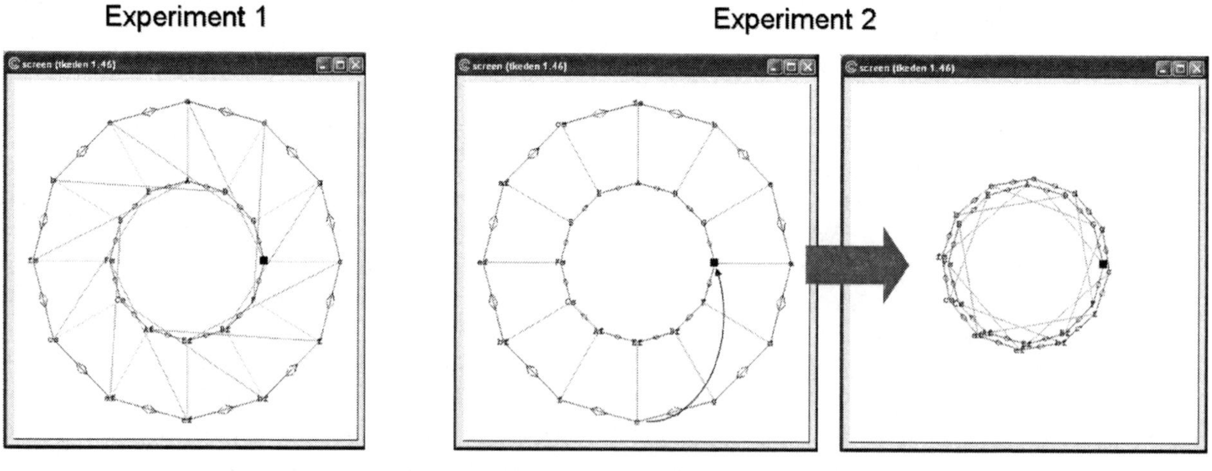

Experiment 1

Experiment 2

Figure 4. *Two experiments in modelling Schubert's conflation of tonic major and minor*

In Experiment 1, a new generator is introduced to represent tonic major-minor modulation. For this generator to be its own inverse, it is necessary to reappraise the structure of the group. This entails reversing the directed edges on the outer circuit. As is explored by Waller [16], this identifies the resulting group graph with the automorphism group of the toroidal space of keys depicted in Figure 3. In Experiment 2, the group graph in Figure 1 undergoes a different transformation, corresponding to being mapped on a homomorphic image in which tonic major and minor tonalities become conflated. This better reflects the character of Schubert's harmonic innovation.

3. Resolution through model making

3.1. Making mathematical models of music. The above discussion illustrates something of the range and diversity of issues that are involved in making models to assist musical analysis. It highlights the contrast between alternative ways of interpreting group graphs – as encoding distinguished transitions (as in the classical cycle of keys), or as defining a set of transformations in a space of tonality. These alternative views relate to algebraic and topological perspectives on tonal relationships. Waller's model [16], a variant of the group graph on the left of Figure 4 in which the group generators are the three involutions exemplified by C↔a, C↔c, C↔e, illustrates a further nuance: these three generators are implausible as distinguished modulations in a standard musical idiom, and have no special significance as generators of the automorphism group of the space of keys, but are informative in relation to symmetries and musical chords such as augmented triads and diminished sevenths. Equally important is the nature of the interpretation of mathematical diagrams involved. The virtue of Experiment 2 as a representation is that it accommodates the harmonic excursions of *Erlkönig* without dispensing with the cycle of fifths model of tonal relationships. In using this model in context (as when tracking the harmonic progressions as the song is played [2]), it is not appropriate to make sudden transitions between the full group graph and its homomorphic image. The gradual transformation of one group graph into the other suggested by the black arrow in Figure 4 is more convincing, and could be appropriately synchronised with the music so as to suggest the presentiment of encroaching harmonic disintegration that an appreciative and familiar listener experiences. Significantly, such an interpretation exploits a serendipitous informal extension of the semantics of the diagram in Figure 4 whereby the locality of the nodes of the abstract graph is taken into account. Taken together, such model-making exercises suggest quite different priorities from those normally associated with mathematical modelling. A strict and formal mode of interpretation may suit a Pythagorean or Schenkerian view of music, but does not give the scope for ambiguity that is required in creative musical analysis. It is not only helpful to be able to make mathematical models of many different varieties, but also to explore the semantic relationships between them, and to study them in relation to the discourse surrounding them and the processes that mediate their evolution.

3.2. Model-making from a musical perspective. Though the term "model-making" is not prominently used in musicology, it can be seen as an essential aspect of musical theory and analysis. The range of modelling activities involved is broad, and encompasses far less formal and mathematical approaches than are represented in this paper, but the semantic challenges are similar. The trend in musicological practice has been from an idealized to a pragmatic interpretation of music theory: witness how Caplin [3] restricts his theory of formal functions to *instrumental music of specific composers between 1780 and 1810* [6], and the systematic but *informal* nature of Cooke's attempt to describe the semantics of the language of music [4]. The focus has also shifted to exploratory and culturally mediated activities shaping musical meaning. As Kofi Agawu [1] observes: "... when foundations [of Music] are enshrined in Theory and deployed in something called Theory-based analysis, they do nothing but reproduce themselves. It would be wise, then, in seeking to escape the circularity of theory-based analysis, to return to an issue that exercised several minds as far back as the 1960s, and to consider detaching theory from analysis. Theory is closed, analysis open. A theory-based analysis does not push at frontiers in the way that a non-theory-based one does. Theory is foundational, analysis non-foundational. But analysis is also performance, and its claims are to a different order of knowledge than, say, historical or archival knowledge."

Agawu's agenda calls for model-making that is eclectic, open-ended and embraces informal and experientially mediated semantics. The flexibility of the observational and computational framework required for this purpose is illustrated by Huron's Humdrum Kit [7]. As explained in [2], the models devised for this paper take the form of linked *construals* [17], rather than discrete programs, and give support to match McCarty's dictum [9:27]: "computational models, however finely perfected, are better understood as *temporary states in a process of coming to know* rather than fixed structures of knowledge".

3.3. Morals for the musical analyst and model-builder. The idealized concept of music, inherited from the ancient Greeks, suggests a relationship between mathematics and music that has a transcendent quality. Modern computing technology may, through its apparent emphasis on formal musical syntax and the digitization of sound, promote an abstract view of music as comprehensively captured by a mathematical specification. In reality, mathematical representations of music illustrate a relationship between form and content that has great complexity and subtlety. Computing technology is most significant because it enables us to model this rich and ever evolving relationship more vividly and dynamically than was possible in the past. In our enthusiasm for interpreting music in mathematical terms, we should not lose sight of the pragmatic considerations that enable us to match musical compositions to abstract structures (such as the coincident approximate equality of 3/2 to the power 12 and 2 to the power 7, and the need to discount the creative impulses that inspire composers to deviate from conventional relationships). Nor should we be misled into supposing that model-building with computers commits us to imposing formal mathematical models on music. Mathematics, music and modelling are all activities that derive their creative essence from relationships between formal patterns and experiences that are always fluid and open to negotiation. This flexibility and openness should be reflected in the educational and technological support that is provided for them.

References

[1] Kofi Agawu, Analyzing music under the new musicological regime, Music Theory Online 2(4), 1996

[2] Meurig Beynon, Steve Russ, Willard McCarty, Human Computing: Modelling with Meaning, The Journal of Literary and Linguistic Computing, 2006 (to appear)

[3] William E. Caplin, *Classical Form: A Theory of Formal Functions for the Instrumental Music of Haydn, Mozart, and Beethoven*, New York and Oxford: Oxford University Press, 1998

[4] Derycke Cooke, *The Language of Music*, Oxford: Oxford University Press, 1959

[5] Richard Cytowic, Synesthesia: Phenomenology & Neuropsychology, Psyche, 2(10), July 1995, `http://psyche.cs.monash.edu.au/v2/psyche-2-10-cytowic.html`

[6] Floyd Grave, Review of William E. Caplin, Classical Form, Music Theory Online 4(6), 1998

[7] David Huron, The New Empiricism: Systematic Musicology in a Postmodern Age, `http://dactyl.som.ohio-state.edu/Music220/Bloch.lectures/3.Methodology`

[8] Jim Loy, `http://www.jimloy.com/physics/scale.htm`

[9] Willard McCarty, *Humanities Computing*, Houndmills, Basingstoke: Palgrave Macmillan, 2005

[10] Andy Milne, The Tonal Centre, `http://www.andymilne.dial.pipex.com/index.shtml`

[11] Heinrich Schenker, *Free Composition: Vol. 3 of New Musical Theories and Fantasies*, Ernst Oster (Ed.) Pendragon Press, 2001

[12] Arnold Schoenberg, *Structural Functions of Harmony* (New York: Norton, 1954); rev. ed., ed. Leonard Stein (New York: Norton, 1969)

[13] George Bernard Shaw, *G.B.S. on Music*, Harmondsworth, Middlesex: Penguin Books Ltd., 1962

[14] Donald Tovey, *Beethoven*, Oxford University Press, London, 1944

[15] Sherry Turkle, Seymour Papert, Epistemological Pluralism: Styles and Voices within the Computer Culture. In Harel, I., Papert, S. (Eds.) *Constructionism*. Norwood, N.J. Ablex Pub. Corp, 161-191, 1991.

[16] Derek Waller, Some combinatorial aspects of musical chords. In *Mathematical Gazette* (March), The Mathematical Association: 12-15, 1978.

[17] The EM website at `http//www.dcs.warwick.ac.uk/modelling`

Teaching Design Science: An Exploration of Geometric Structures

Carl Fasano
Department of Foundation Studies
Rhode Island School of Design
2 College Street
Providence, RI 02903
e-mail: fasano@gmail.com

Abstract

The late Dr. Arthur Loeb, professor in the Department of Visual and Environmental Studies at Harvard University, developed and taught Design Science/Synergetics, an exploration of three-dimensional space, and Visual Mathematics, which explored the parameters of structure in two and three dimensions for more than two decades. The main foci of design science were geometry, mathematics, design and the beauty that resulted from this marriage. Dr. Loeb's widow, Charlotte Loeb, donated the Design Science Teaching Collection to the Edna Lawrence Nature Lab at Rhode Island School of Design in 2003. In its new environment, the Teaching Collection is inspiring both faculty and students. This paper includes examples of models made by RISD students in response to questions arising from the study of geometry and design science. I will also discuss how a heuristic approach to teaching design science meshes synergistically with an art and design curriculum.

1. Introduction

The cultivation of creativity requires fine balance among the teaching of technique, the stimulation of curiosity, guidance, group dynamics and respect for an individualistic sense of the creative act. In the education of the artist these countervailing forces serve a multifaceted role: a.) guiding young minds without squelching creativity and offering paradigms for further investigations, b.) encouraging a sense of participation within a creative community or tradition while respecting the privacy and introspection necessary for the creative act.

Learning geometry requires the acquaintance with and understanding of a necessary body of core concepts and mathematical techniques. How these concepts and techniques are arrived at is crucial to the ability of students to manipulate them creatively and to feel enabled to further explore.

In drawing or sculpting, for example, simply moving a pencil or molding a piece of clay is considered an appropriate and indeed fundamental component of the larger diet of artistic exploration. The recognition that activity on such a primary level is as central to maintaining and developing artistry as planning and executing a more technically ambitious project such as a large scale oil painting or casting a sculpture in bronze shows a deep self-awareness and understanding of the nature of creativity. With this understanding, an artist realizes that there is no set hierarchy of values attached to a given creative activity - only as I mentioned above, "a diet" based on openness to possibilities that arise via various processes.

In mathematics, in this case geometry, a student's notion of creative manipulation often seems to be the privilege of a gifted few. There is a perception that technique and conceptual apprehension must precede any creativity; that the prodigy who solves a problem or ventures a novel theorem is born with an innate ability that is denied to the vast majority of people. This "perception" per se is open to many debates concerning the role of innate abilities. But what stands in undeniably daunting truth is that this perception is at the root of one of the greatest obstacles to the study and enjoyment of mathematics:

the fear of not knowing something specific, a correct answer for example, can feel as terrifying as trying to land a plane on an island at night in dense fog.

And yet we know that one can have levels of understanding of mathematical ideas. An initial glimpse into the spirit of duality or instantaneous rate of change or the possible implications of the Fibonacci series...and so forth is not simply an incorrect understanding because it seems incomplete but an exciting, stimulating revelation that energizes one to look deeper - like the quick gestural sketch that helps an artist to bring life to a larger, more worked piece.

2. The Heuristic Approach and Results

By taking a hands-on approach to studying geometry, the initial glimpse can be literally tangible such as a paper model of a dodecahedron or the construction with compass and straight edge of a pentagon. In this sort of educational environment, that which is "correct" is also fascinatingly beautiful.

Below are some related images of models made by RISD students:

Figure 1: *Icosahedron*

Figure2: *Model shown in figure 1, unfolded to show a stellated pentagonal dodecahedron embedded in it*

Figure 3: *Detail: interior surface of model Shown in figure 1.*

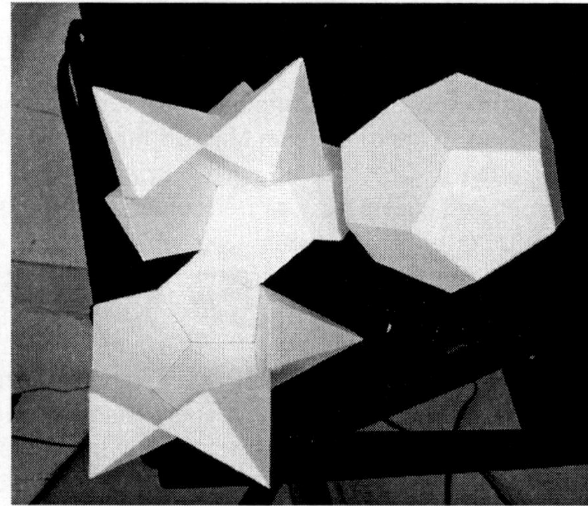

Figure 4: *pentagonal dodecahedron embedded in the stellated dodecahedron shown in figure 2.*

Figure 5: *Polyhedron obtained by turning the model shown in figure 1 inside out.*

Figure 6: *pentagonal dodecahedron with five wire tetrahedra embedded in it. Models shown above by Andrew Goett, freshman.*

Figure 7

Figure 8

Figures 7 and 8: *Compound of five tetrahedra with pattern that demonstrates a uniform organization of five and three-fold symmetries. Models by photography student Tom Prado, '07*

Some other course topics and activities are: Platonic solids, duality, Euler-Schlafli equation, stellation, truncation, Archimedean solids, rigorous perspective drafting, Schlegel diagrams, space filling: rhombic dodecahedron, octahedron/tetrahedron ratios, introduction to group theory space filling form exercise using either projection drawing or model-making, 5-fold forms: icosahedron, pentagonal dodecahedron, triacontahedron etc, degrees of freedom, stability, jitterbugs, basic tensegrity structures and dome systems, building tensegrity forms, preferably with an eye toward eventual applications to sculpture, furniture, architecture/shelter, etc.

3. The RISD Context: Open and Conducive to Sharing

At the end of the first term, Caryn Johnson (Dr. Loeb's Teaching assistant of ten years who was my tutor for design science) and I were discussing the course with a few students. One remarked, "it will be interesting to see how this course affects the (study of sic.) design at RISD". At first I thought that this was a lovely way to make a compliment. On further reflection, I realized that the student was addressing a unique characteristic of the RISD environment (in fact, art and design schools in general). Students make work in clear sight of classmates, and often leave their work in studios that are used by many students. There is an emphasis at RISD on sharing; it is understood from the beginning that ideas must not be considered as proprietary products. When Picasso said, "I do not borrow ideas from other artists, I steal them" (paraphrase), he was not expressing a sociopathic tendency, he was simply being honest. When an artist "steals" an idea, it is to develop it, to incorporate it into his/her own vision. And so the ideas, expressed in models and drawings as a result of the design science course have been "stolen", passed around, expanded upon, integrated into other design studies and influenced students' work in many areas of study.

The ability to visualize, draw, build, to think with one's hands is a huge benefit in terms of being prepared to study design science. During the foundation year (first year of study) and during the course of subsequent years spent in their respective majors, students are primed with strong intuitive and practical understandings of the laws of space and form. Most aspects of this knowledge are defined in a clear, sophisticated and structured manner.

On the other hand, there are large gaps left in design terminology from the standpoint of a lexicon that is common across disciplines. There is also a dearth of understanding about the nature of polyhedra and polygons - their extent, categorization and structural properties. From the point of view of studio practice, this is clearly not an impediment; the RISD student often goes on to create wonderful things and to be successful in his/her field. Students who have taken a term of design science have recognized that it has opened their eyes and minds to another world of possibilities.

4. The Power of Geometric Models for Artists

The discomfort with looking more deeply at art, via geometry, or design science in this case, may ultimately result in an unhealthy emphasis being placed on the product of creative activity. This discomfort may have a paralyzing effect on the ability to see new solutions. Often during the creative process things are discovered which are not sought. One is reminded of the Zen homily which cautions," If you only focus on the goal, you will not see the dangers and pleasures of the path, whereas if you attend to the path, you realize that the path is the goal and the goal the path. If we know what we are seeking, discovery is redundant. So emphasis is placed on enjoying exploration in this design science class.

Artists are inclined by training to be cognizant of the fact that there are ideas attendant to an artwork. These ideas, primarily of an historical/cultural context, become implicit components of artistic intention. By contrast, pieces are not usually undertaken that are meant to demonstrate and elucidate concepts while at the same time embodying those concepts. To put it another way, the geometric models are not denotive or connotative as an allegorical painting or a work that contains religious symbols might be. The models are the physical equivalent of a verbal or written description of the geometric ideas under scrutiny. At the same time, the variations made possible by introducing a physical, materially rich study that utilizes model-making often indicates the breadth of nuances potentially left to be discovered. This is all the more heartening when we reflect on the fact that these are nuances of mathematically precise ideas. So we learn to see that mathematical objects may be as multi-dimensional as a person that we know or a character in a play. We can replace the fearful monolith of "that which is correct and only this" with a sense of playful negotiation; an invitation to muse.

The arts give us joy to the extent that they reveal new ways of thinking and perceiving. This generates optimism that the world might have infinite facets -a most beautifully cut diamond. In that same sense, mathematics is like poetry. Or more precisely, mathematics and poetry are both forms of expression that seek to discover using the tools of form. In both, content is pretext for structural investigations. In each, the structural investigation reveals and abstracts universal principles purely for the sake of this activity. They are both processes of making metaphorical statements and therefore involve a reorganization of information. Each structure that is revealed suggests new paths specific to it and paths towards new structures.

By way of illustration, Sara Kudra, a sophomore in architectural studies writes: "Design Science has allowed me to understand shapes in a new context that I had never considered before. The class has opened my thoughts about simple things; the way solids are formed, their relationship to each other and the mathematics that fundamentally drives them all.

On the first day of class my preconceptions were contradicted, and I was astonished to be seeing new relationships in everyday things. We were asked to slice a cube along a plane in such a manner that we would get a regular hexagon. The solution, it turns out is a rather simple one that proves the old adage, 'think outside of the box'; but although the answer was not mind bending like some of the conversations about the fourth dimension have been, it still was as astonishing. The reason it such an amazing experience was because it had been in front of my eyes the whole time; I only needed something to open them to it.

Figure 9: Cube sliced perpendicular to its three-fold axis of symmetry, at the midpoint of that axis.

This awakening is a great benefit of Design Science. There are relationships between all geometry that we use, and to be able to enlist that knowledge to manipulate and inform your work is a wonderful understanding to have. Students do not just pass through this kind of class; although it never feels like work, there is a much greater depth of understanding when you use your hands to make these relationships for yourself.

The advantage of having this kind of class in a studio art and design environment is that the pupils that are engaged in the work are able to find things out on their own by building models and through observation. If it looks like there may be a relationship between forms, or it appears as though one shape is derived form another, then students are encouraged to find that out on their own by constructing the form. I can honestly say that there is no better way, as a student, to understand what I am working with than by being challenged to make it on my own, and draw my own conclusions without being handed the answers for the sake of information. When students are told a solution in order to memorize it and accept it then they do not actually understand it or know why it is so; but when a student is told to make it (specifically in an art setting) he/she comes away from it not only with a greater depth of understanding, but can also show you all sorts of relationships that were previously unseen and they will tell you all about them.

Having gone through this class has changed my view of simple things; I no longer can assume relationships in design without thinking how they have come to be. As I continue into my next semester of school I will take with me a greater appreciation about the world of shapes and forms, and as an architecture student I will use those thoughts to my advantage as I design in the future. The extent of design Science is not limited to polyhedra, but is something that reaches much further than just that. It is a way of understanding- relationships, orders, structures, and reasoning that can be applied to any work; and has an especially strong relation and contribution to the art world. For the select few students at RISD who are not timid about embracing their interest for geometry this class will change them for the better, and they will be able to encourage the next generation to engage and be educated about the ways of Design Science. " -Sara Kudra, architecture student '08.

An artistic masterpiece might safely be described as something composed of only necessary parts - and beyond that the composition transports us to its own world - a world of order. It might suggest disorder but the vehicle of the message must itself be orderly - it is from this that expression derives power. In speaking of mathematics Bertrand Russell might well be speaking of the fine arts, "Remote from human passions, remote even from the pitiful facts of nature, the generations have gradually created an ordered cosmos, where pure thought can dwell as in its natural home and where one, at least, of our nobler impulses can escape from the dreary exile of the actual world." [1]

5. Conclusion

We sense an ordered fabric to physical reality but we have only occasional, tantalizing glimpses of it. With the study of design science, students learn that the boundaries between internal and external realms of discovery are fluid. The difference between art and mathematics is that for art this beauty and clarity seems to gestate internally and brings about the birth of living expression. In mathematics this wonder appears all around us. We see patterns again and again in a surprising range of places, objectively distant from our selves.

The great resonant ideas have been voiced repeatedly, not simply for the sake of reiteration or bombastic pride, but in the desire to clarify or even purify them. In the visual arts this is fairly easy to accept; a portrait from ancient Rome and a portrait by Alberto Giacometti may seem equally necessary to the tradition of painting. But a model of a dodecahedron helps students to see that what they know about the power of material exploration to bring new ideas to life applies to geometry as well... and then, what else?

This course in geometry enables students to see that it is the form of these ideas that changes. This is far from meaning that there is a paucity of wisdom. Form is not necessarily the superficial aspect; in fact, the idea resides within the form much as the spirit resides within the body. Without a form to animate, an idea is without voice.

References

[1] **Bertrand Russell,** *from The Study of Mathematics: Philosophical Essays*

More "Circle Limit III" Patterns

Douglas Dunham
Department of Computer Science
University of Minnesota, Duluth
Duluth, MN 55812-2496, USA
E-mail: ddunham@d.umn.edu
Web Site: http://www.d.umn.edu/~ddunham/

Abstract

M.C. Escher used the Poincaré model of hyperbolic geometry when he created his four "Circle Limit" patterns. The third one of this series, *Circle Limit III*, is usually considered to be the most attractive of the four. In *Circle Limit III*, four fish meet at right fin tips, three fish meet at left fin tips, and three fish meet at their noses. In this paper, we show patterns with other numbers of fish that meet at those points, and describe some of the theory of such patterns.

1. Introduction

Figures 1 and 2 below show computer renditions of the Dutch artist M.C. Escher's hyperbolic patterns *Circle Limit I* and *Circle Limit III* respectively. Escher made criticisms of his first attempt at creating hyperbolic art,

Figure 1: A rendition of Escher's *Circle Limit I.* **Figure 2:** A rendition of Escher's *Circle Limit III.*

Circle Limit I. However, he later redressed those deficiencies in *Circle Limit III.* In a letter to the Canadian mathematician H.S.M. Coxeter, Escher wrote:

> *Circle Limit I*, being a first attempt, displays all sorts of shortcomings... There is no continuity, no "traffic flow," nor unity of colour in each row... In the coloured woodcut *Circle Limit III*, the shortcomings of *Circle Limit I* are largely eliminated. We now have none but "through traffic"

series, and all the fish belonging to one series have the same colour and swim after each other head to tail along a circular route from edge to edge... Four colours are needed so that each row can be in complete contrast to its surroundings. ([6], pp. 108-109, reprinted in [4])

Escher had been inspired to create his "Circle Limit" patterns by a figure in one of Coxeter's papers [2]. That figure "gave me quite a shock" according to Escher in a letter to Coxeter, since the figure showed Escher how to make "circle-limit" patterns. Some of this important Coxeter-Escher interaction is recounted in [3].

In turn, Coxeter, was later inspired to write two papers explaining the mathematics behind Escher's *Circle Limit III* [3, 4]. In the same issue of *The Mathematical Intelligencer* containing Coxeter's second paper, an anonymous editor wrote the following caption for the cover of that issue, which showed Escher's *Circle Limit III*:

> Coxeter's enthusiasm for the gift M.C. Escher gave him, a print of Circle Limit III, is understandable. So is his continuing curiosity. See the articles on pp. 35–46. He has not, however said of what general theory this pattern is a special case. Not as yet. [1]

It seems that Coxeter did not describe such a general theory, or at least did not publish it. The goals of this paper are to provide the beginnings of a general theory and to show some sample patterns. Before developing our theory, we start with a short review of some hyperbolic geometry. With this background, we can then present a general theory of "Circle Limit III" fish patterns. Next, we examine special cases that are more amenable to calculation, showing sample patterns along the way. Finally, we indicate directions of further research.

2. Hyperbolic Geometry

Escher used the *Poincaré disk model* of hyperbolic geometry for his "Circle Limit" patterns. In this model, Euclidean objects are used to represent objects in hyperbolic geometry. The points of hyperbolic geometry in this model are just the (Euclidean) points within a Euclidean bounding circle. The hyperbolic lines are represented by circular arcs orthogonal to the bounding circle (including diameters). For example, the backbone lines and other features of the fish lie along hyperbolic lines in Figure 1. The hyperbolic measure of an angle is the same as its Euclidean measure in the disk model — we say such a model is *conformal* — so all fish in a "Circle Limit III" pattern have roughly the same Euclidean shape. We note that equal hyperbolic distances correspond to ever smaller Euclidean distances toward the edge of the disk. For example, all the black fish in Figure 1 are hyperbolically the same size, as are the white fish; all the fish in *Circle Limit III* are the same (hyperbolic) size. The Poincaré disk model was appealing to Escher since an infinitely repeating pattern could be shown in a bounded area and shapes remained recognizable even for small copies of the motif, as a consequence of conformality. Escher was more interested in the Euclidean properties of the disk model than the fact that it could be interpreted as hyperbolic geometry.

One might guess that the backbone arcs of the fish in *Circle Limit III* are also hyperbolic lines, but this is not the case. Even Escher believed the backbone arcs were orthogonal to the bounding circle, but he accurately drew them as non-orthogonal circular arcs. They are *equidistant curves* in hyperbolic geometry: curves at a constant hyperbolic distance from the hyperbolic line with the same endpoints on the bounding circle. For each hyperbolic line and a given distance, there are two equidistant curves, called *branches*, at that distance from the line, one each side of the line. In the Poincaré disk model, those two branches are represented by circular arcs making the same (non-right) angle with the bounding circle and having the same endpoints as the corresponding hyperbolic line. Equidistant curves are the hyperbolic analog of small circles in spherical geometry: a small circle of latitude in the northern hemisphere is equidistant from the equator (a great circle or "line" in spherical geometry), and has a corresponding small circle of latitude in the southern hemisphere the same distance from the equator. We usually do not use the term "equidistant curve" in Euclidean geometry since parallel lines have that property (and are not curved).

There is a *regular tessellation,* $\{m, n\}$ of the hyperbolic plane by regular m-sided polygons meeting n at a vertex provided $(m-2)(n-2) > 4$. Escher used the regular tessellations $\{6, 4\}$ and $\{8, 3\}$ as the basis of his Circle Limit patterns ($\{6, 4\}$ for *Circle Limit I* and *IV*, and $\{8, 3\}$ for *Circle Limit II* and *III*). Figure 3 shows the tessellation $\{8, 3\}$ (heavy lines) superimposed on the *Circle Limit III* pattern. As one traverses edges of this tessellation, alternately going left, then right at each vertex, one obtains a zig-zagging path called a *Petrie polygon.* The midpoints of the edges of a Petrie polygon lie on a hyperbolic line by symmetry. The vertices of the Petrie polygon lie alternately on each of the two equidistant curve branches associated to that line — this is shown in Figure 4 with the Petrie polygon drawn with thick lines, the "midpoint" line and the equidistant curves drawn in a medium line, all superimposed on the $\{8, 3\}$ tessellation (lightest lines). Escher only used one branch for fish backbones from each pair of equidistant curves in *Circle Limit III.* If he had consistently used the other branch, the pattern would have been rotated about the center by 45 degrees.

Figure 3: The tessellation $\{8, 3\}$ underlying the *Circle Limit III* pattern.

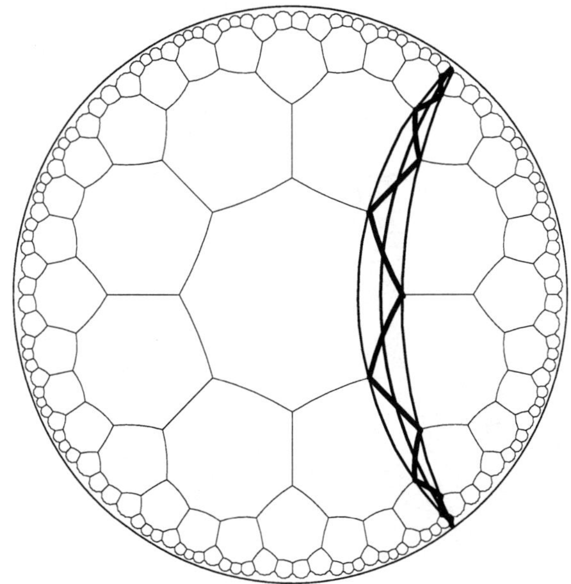

Figure 4: A Petrie polygon (heavy), a hyperbolic line and two equidistant curves (medium) associated to the $\{8, 3\}$ tessellation (lightest lines).

3. The General Theory of "Circle Limit III" Patterns

If we examine *Circle Limit III*, we see that four fish meet at right fin tips, three meet at left fin tips, and three meet at their noses (and tails). We generalize these numbers to patterns of fish with p fish meeting at right fins, q fish meeting at left fins, and r fish meeting at their noses. We will label such a pattern (p, q, r). So *Circle Limit III* would be called $(4, 3, 3)$ in this notation. One could conceptually "reflect" all the fish of a (p, q, r) pattern across their backbone lines to obtain a (q, p, r) pattern, but this is a true hyperbolic reflection only when $p = q$.

One restriction that we make is that r be odd so that the fish swim head-to-tail, in order to achieve "traffic flow." Also, by examining the left fins of *Circle Limit III*, we also require that p and q be at least 3, since two fins could not have tips that meet. And of course r must be at least 3 too (since r is odd and greater than 1). Consequently, the "smallest" example of such a pattern is $(3, 3, 3)$, realized by Escher in his *Notebook Drawing 123* [8, Page 216]. This $(3, 3, 3)$ pattern is based on the regular tessellation $\{3,6\}$ of the Euclidean plane by equilateral triangles, each triangle containing three half-fish. It is interesting that this simpler drawing is dated several years after the much more complicated *Circle Limit III*. We also note that we do not consider *Notebook Drawing 122* to be a valid "Circle Limit III" pattern, since, as in *Circle Limit I,* fish meet "head-on", not head-to-tail. This pattern is based on the Euclidean tessellation of squares, each square containing four half-fish, and would be denoted $(4, 4, 2)$ if we allowed r to be even.

There is another natural tessellation that we can associate with the *Circle Limit III* pattern, obtained by dividing the octagons in Figure 3 into four "kites" — convex quadrilaterals with two pairs of adjacent equal sides. Each kite can serve as a *fundamental region* for the pattern since it contains exactly the right fish pieces to assemble one complete fish. Figure 5 shows this kite tessellation superimposed on the *Circle Limit III* pattern. In general for a (p, q, r) pattern, one can use a kite-shaped fundamental region with vertex angles $\frac{2\pi}{p}, \frac{\pi}{r}, \frac{2\pi}{q}$, and $\frac{\pi}{r}$. We note that a quadrilateral is hyperbolic precisely when the sum of its interior angles is less than 2π, which translates to the following inequality for our kites: $\frac{2\pi}{p} + \frac{\pi}{r} + \frac{2\pi}{q} + \frac{\pi}{r} < 2\pi$ or in other words: $\frac{1}{p} + \frac{1}{q} + \frac{1}{r} < 1$. Figure 6 shows the kite tessellation corresponding to the (4,3,5) pattern and containing fish backbones along equidistant curves.

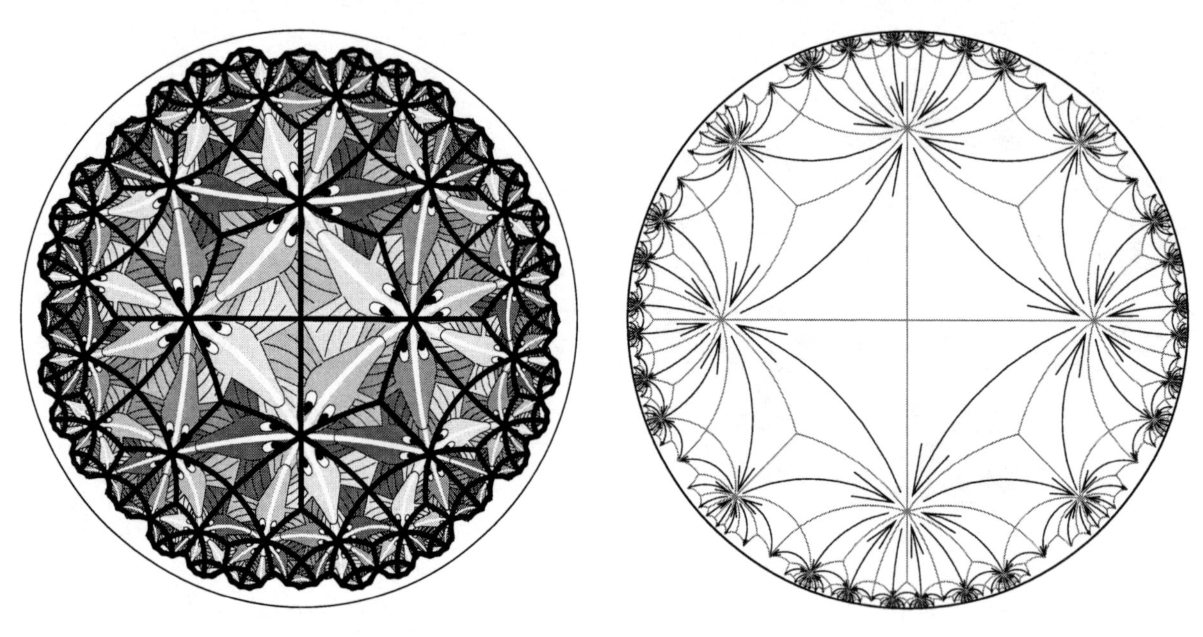

Figure 5: The kite tessellation superimposed on the *Circle Limit III* pattern.

Figure 6: The kite tessellation (lighter lines) of the (4,3,5) pattern, with fish backbones (darker arrows) along equidistant curves.

4. Special Cases

There are two special cases that can be analyzed in more detail. The first case we consider is when $p = q$, so that the fish are symmetric. In this case the backbone curves are (hyperbolic) lines. When $p \neq q$ and the fish are not symmetric, their backbone lines bend away from the side with the larger number of fish meeting at a fin tip. For the second special case, we assume that q and r are both 3.

In the first case, the fish are symmetric by reflection across the hyperbolic backbone lines. Thus we can use half a fish for the motif and an isosceles triangle that is half of a kite (with angles $\frac{2\pi}{p}, \frac{\pi}{2r}$, and $\frac{\pi}{2r}$) for the corresponding fundamental region. Figure 7 below shows a $(4, 4, 3)$ pattern, but with angular fish in the style of *Circle Limit I*. Escher's *Notebook Drawing 123*, mentioned above, is the "smallest" example of this special case, since it is a $(3, 3, 3)$ pattern.

Figure 8 is a fin-centered version of Figure 7 that answers most of Escher's criticisms of his *Circle Limit I* pattern. This pattern was obtained by the following sequence of steps. First, two of the noses of the white fish (and tails of the black fish) were made narrower so that three white fish could also meet nose-to-nose. This solved the "traffic flow" problem (and made the white fish congruent to the black fish). The fish would now all swim the same direction along a backbone line, but would be alternately colored black and white. To solve this "unity of colour" problem we use the minimum number of colors, three, to re-color the fish, yielding the pattern of Figure 7. Finally, Figure 8 is derived from Figure 7 by hyperbolically translating a 4-fold fin meeting point to the center of the bounding circle. Fortunately Escher did not follow this sequence, so that we have the beautiful *Circle Limit III* pattern instead.

Figure 7: A (4,4,3) pattern derived from the *Circle Limit I* pattern.

Figure 8: A (4,4,3) pattern in the style of *Circle Limit I* (derived from Figure 7).

In the second special case, when $q = r = 3$, we can calculate the angle, ω, that the "backbone" equidistant curves make with the bounding circle. Again, Escher's *Notebook Drawing 123*, a $(3, 3, 3)$ pattern, is the "smallest" example of this special case too. Coxeter computed ω for *Circle Limit III* ($p = 4$) in two ways: first by using hyperbolic trigonometry [3], and later by using Euclidean techniques [4]. We follow Coxeter's first method to calculate ω in terms of p. First, we note that the regular tessellation $\{2p, 3\}$ can be superimposed on a $(p, 3, 3)$ pattern just as in Figure 3 (with nose and left fin points at alternate vertices of the $2p$-gons). We next make additions to Figure 4 to obtain Figure 9: we add lines ON, OM, and NL, and label points L, M, N, and O. Angles $\angle OMN$ and $\angle MLN$ are right angles.

We wish to calculate the distance $\overline{NL}$ (an overline above a line segment denotes its hyperbolic length) from the left equidistant curve to hyperbolic line passing through L and M. This distance is related to *angle of parallelism*, the angle between NL and the hyperbolic line $N\infty$ (not shown — it is different than the equidistant curve going through N and ∞ which is shown and clearly makes a right angle with NL). If α denotes the angle of parallelism, this important relation in hyperbolic geometry is: $\cos\alpha = \tanh\overline{NL}$ [7, Page 402]. It turns out that angle of parallelism α is the same as the angle of intersection ω of the bounding circle with the equidistant curve at that distance from its hyperbolic line. Thus we have:

$$\cos\omega = \tanh \overline{NL}$$

We can calculate $\overline{NL}$ by solving the two right triangles $\triangle OMN$ and $\triangle MLN$ using standard formulas from hyperbolic trigonometry [7, Page 403, Theorem 10.3]. First, we use $\triangle OMN$ to compute $\overline{MN}$ by:

$$\cosh \overline{MN} = \cos(\frac{\pi}{2p})/\sin(\frac{\pi}{3}) = \frac{2}{\sqrt{3}}\cos(\frac{\pi}{2p})$$

Next, we use $\triangle MLN$ to compute $\overline{LN}$ from $\overline{MN}$ by:

$$\tanh \overline{LN} = \cos(\frac{\pi}{3})\tanh \overline{MN} = \frac{1}{2}\tanh \overline{MN} = \frac{1}{2}\sqrt{1 - 1/\cosh^2(\overline{MN})}$$

using the relation $\tanh^2 = 1 - 1/\cosh^2$. Finally, can combine these equations to obtain:

$$\cos\omega = \frac{1}{2}\sqrt{1 - 3/4\cos^2(\frac{\pi}{2p})}$$

When $p = 4$, $\cos\frac{\pi}{8} = \sqrt{\frac{2+\sqrt{2}}{4}}$, consequently $\cos\omega = \sqrt{\frac{3\sqrt{2}-4}{8}}$ and ω is approximately $79.97°$. Coxeter obtained a different, but equivalent expression for ω. Similarly, when $p = 5$, $\cos\frac{\pi}{10} = \sqrt{\frac{5+\sqrt{5}}{8}}$, $\cos\omega = \sqrt{\frac{3\sqrt{5}-5}{40}}$, and $\omega \approx 78.07°$. Figure 10 shows a $(5, 3, 3)$ pattern which was used as the basis for the 2003 Mathematics Awareness Month poster and whose background is described in [5].

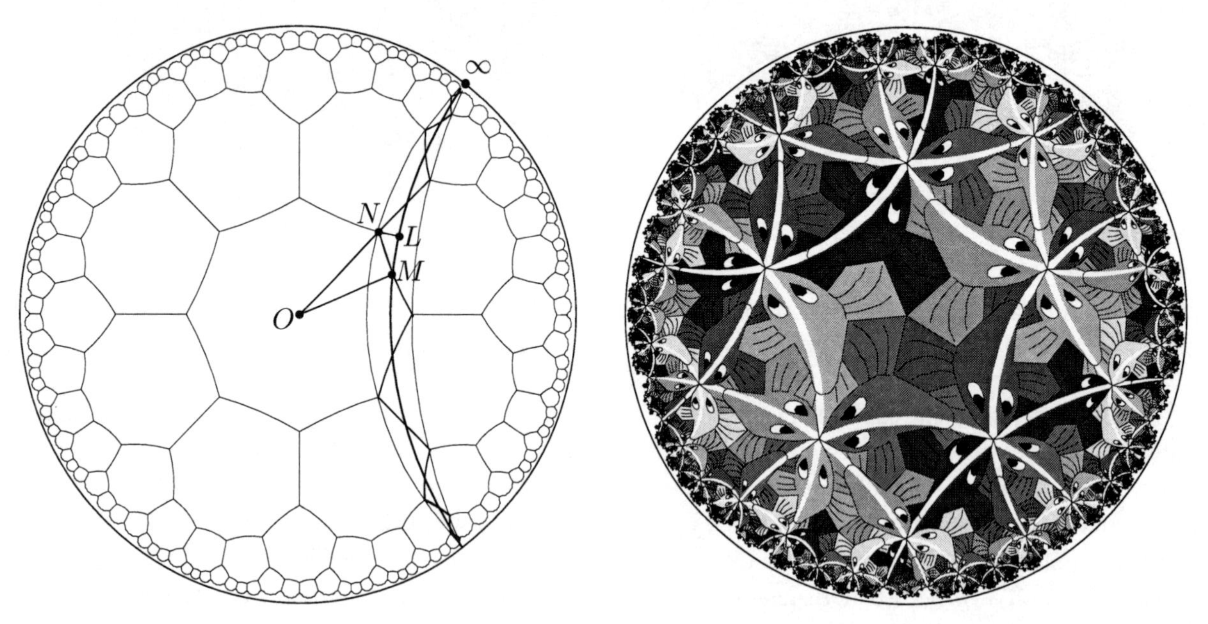

Figure 9: Two right triangles MON and MLN added to Figure 4 used to calculate ω.

Figure 10: A $(5, 3, 3)$ fish pattern.

We can also take the limit as p goes to infinity and obtain the limiting equation $\cos\omega = 1/4$ or $\omega \approx 75.52°$. Thus ω lies in the interval $(\cos^{-1}(\frac{1}{4}), \frac{\pi}{2}]$ for finite p, q, and r. Of course we cannot actually draw a pattern with an infinite number of fish meeting in the center of the disk (but an infinite number of fish fins *can* meet on the bounding circle, as indicated below).

The concept of (p, q, r) is symmetric in p and q, of course. Figures 11 and 12 below show examples in which $p = r = 3$. We have put the right fin tips at the center of the disk in agreement with Escher's *Circle Limit III*. Figure 11 shows a $(3, 4, 3)$ pattern related to *Circle Limit III* except that the numbers of fish meeting at left and right fin tips have been switched. Figure 12 shows a $(3, 5, 3)$ pattern that bears the same relationship to our Figure 10 above. Note that these figures are not just translations (or reflections) of *Circle Limit III* and Figure 10, since the number of fish meeting at right (and left) fins is different.

There are two interesting aspects of $(3, q, 3)$ patterns. First, the three backbone lines closest to the center have "straightened out" into chords of the bounding circle, resulting in a Euclidean equilateral triangle of backbones in the center. When I first created such a pattern (by translating to the origin a left fin meeting point of the *Circle Limit III* pattern) about 20 years ago, I told Coxeter that I was astonished by this phenomenon. He replied (words to the effect):

> Well, you shouldn't have been. Any triangle made up of three congruent circular arcs meeting at 60-degree angles must obviously be a Euclidean equilateral triangle.

It may be possible to exploit the simple geometry of these $(3, q, 3)$ patterns to compute ω more easily.

Figure 11: A (3,4,3) pattern related to Escher's *Circle Limit III.*

Figure 12: A (3,5,3) pattern related to our pattern of Figure 10.

A second aspect of $(3, q, 3)$ patterns is that we can conceive of taking the limit of them as q tends to infinity. As q goes to infinity, the left fin tips would get farther and farther from the center of the disk, until in the limit, they would be on the bounding circle. It would theoretically be possible to draw such a pattern, but our current software can only handle finite values of p, q and r.

5. Conclusions and Future Work

We have described a general theory of (p, q, r) "Circle Limit III" patterns and shown some examples. It would seem to be a worthy goal to find a general formula for ω in terms of $p, q,$ and r. With current software, (p, q, r) patterns can only be created one at a time. It would certainly be useful to have a program that could automatically create a new (p, q, r) pattern with different values of $p, q,$ and r from an existing pattern. Another interesting direction would be to investigate (and draw) (p, q, r) patterns with one of q, or r being infinity (patterns may also exist with both q and r being infinity). A seemingly difficult problem is to automate the process of determining a coloring for a (p, q, r) pattern that has the same color along any line of fish and adheres to the map-coloring principle that adjacent fish have different colors.

Acknowledgments

I would like to thank Lisa Fitzpatrick and the staff of the Visualization and Digital Imaging Lab (VDIL) at the University of Minnesota Duluth. This work was also supported by a Summer 2005 VDIL Research grant. I would also like to thank the reviewers for a number of helpful suggestions.

References

[1] Anonymous, *On the Cover,* Mathematical Intelligencer, **18**, No. 4 (1996), p. 1.

[2] H.S.M. Coxeter, *Crystal symmetry and its generalizations,* Royal Society of Canada, (3), 51 (1957), pp. 1–13.

[3] H.S.M. Coxeter, *The Non-Euclidean Symmetry of Escher's Picture 'Circle Limit III',* Leonardo, **12** (1979), pp. 19–25.

[4] H.S.M. Coxeter. *The trigonometry of Escher's woodcut "Circle Limit III",* Mathematical Intelligencer, **18**, No. 4 (1996), pp. 42–46. This his been reprinted by the American Mathematical Society at: `http://www.ams.org/featurecolumn/archive/circle_limit_iii.html` and also in *M.C. Escher's Legacy: A Centennial Celebration,* D. Schattschneider and M. Emmer editors, Springer Verlag, New York, 2003, pp. 297–304.

[5] D. Dunham, Hyperbolic Art and the Poster Pattern:
`http://www.mathaware.org/mam/03/essay1.html`, on the Mathematics Awareness Month 2003 web site: `http://www.mathaware.org/mam/03/`.

[6] B. Ernst, *The Magic Mirror of M.C. Escher,* Benedikt Taschen Verlag, Cologne, Germany, 1995. ISBN 1886155003

[7] M. Greenberg, *Euclidean & Non-Euclidean Geometry, Third Edition: Development and History,* 3nd Ed., W. H. Freeman, Inc., New York, 1993. ISBN 0716724464

[8] D. Schattschneider, *M.C. Escher: Visions of Symmetry,* 2nd Ed., Harry N. Abrams, Inc., New York, 2004. ISBN 0-8109-4308-5

J-F. Niceron's *La Perspective Curieuse* Revisited

J. L. Hunt
Department. of Physics
University of Guelph
Guelph, ON N1G2G8 Canada
e-mail: phyjlh@physics.uoguelph.ca

Abstract

J-F Niceron's well known work on the mathematics of anamorphism *La Perspective Curieuse* is a much quoted but perhaps less read classic. In particular the templates he provides for various transformations are commonly used as a starting point by those artists who occasionally practise the anamorphic art. Some of these templates are known to be approximations and some are exact. In the process of casting the mathematical descriptions of these templates into modern notation suitable for computation, a peculiar error has been found in Niceron's analysis of transformations onto the surface of a cone or pyramid. The correct relationships are presented and possible reasons for the error are discussed.

1. Introduction

Anamorphic Art still engenders a response from the general public on the rare occasion when it is placed on exhibit. A few artists still practise it and produce anamorphoses usually of the most familiar type viz. that resolved in a cylindrical mirror. To make the anamorphic original it is common to use the templates provided by J-F. Niceron in his book *La Perspective Curieuse* [1] first published in 1638. Every person interested in the field of anamorphism knows and consults this book [2], principally for the working templates. Practitioners probably do not read the text carefully and perhaps do not know where Niceron's prescriptions are exact and where approximate. This difficulty is compounded by the fact that the book has not been translated in its entirety into English (although there is an Italian translation of the second and third volumes [3]) and Niceron's 17th century prose style is not exactly easy.

Whether or not the templates are exact or approximations is probably moot to the practising artists who may use them only for a beginning and make subsequent alterations in the finished product to suit their taste and eye. However, in the computer era, when the manipulation of bitmaps is routine, knowing the correct analytical prescription for the various forms of anamorphic transformation is desirable [4]. If the exact analysis is intractable then the effect and magnitude of any approximations must be known.

For example, we know from his own words that Niceron's template for the cylindrical mirror anamorph is a somewhat crude approximation, whereas his geometrical analysis of the tilted plane viewed from a finite distance is exact [4]. This raises the question then as to which other analyses in Niceron's classic book are exact and which are approximations since he only rarely comments on this fact. In the course of producing an English translation of the work it has been discovered that most of the matter of the second volume is not only approximate but incorrect in a manner that is very peculiar considering that Niceron was a skilled mathematician and, moreover, under the tutelage of an expert like Mersenne.

2. Anamorphic Transformation on to the Surface of a Right Circular Cone.

Having produced his version of the standard theory of perspective in the first volume, in the second volume he gives an exact prescription for the plane anamorph viewed from a finite distance. He then begins a lengthy consideration of the perspectively-distorted image on the surface of a right circular cone and right pyramids of various order. Virtually without any preamble and no justification he presents a template and scheme to construct an anamorphic image on the exterior surface of a cone and shows that it is related to the table of tangents. The image is to be resolved by observation along the axis from some previously chosen point above the apex. As Niceron proves nothing in the book but presents everything as a series of assertions it perhaps can go unnoticed that the template is, in fact, incorrect. The proposed scheme is shown in Figure 1 adapted from Plate 14 of the work. The construction is effected in XL by dividing 45° of the left hand sector into N equal parts (here $N = 8$). The equiangular radii are extended onto a line AB which they cut in a manner that produces lengths proportional to the tangent of a series of angles; in this case $45/n$, where $N \geq n \geq 1$. The number N is the number of radial and azimuthal divisions into which the original image is divided in the cone's base (in this case therefore, 8 as shown in XLI). The right hand sector in XL is the actual 90° sector to which the anamorph will be transformed. It is divided into equiangular segments whose width increases radially to account for the perspective. The sector is then rolled into a cone (apex A and lines AC and AB congruent) and observed along the axis from a point which is as much elevated above the apex as the apex is above the base. Niceron does not aim for generality as the method only applies to a right sector which, when rolled into a cone, has a total apex angle of 28°.

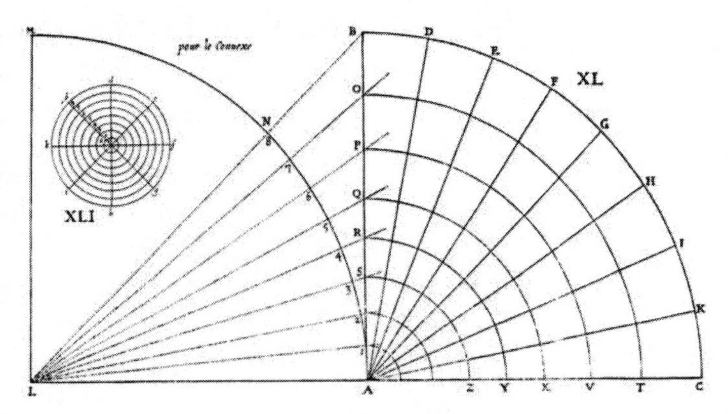

Figure 1: *Niceron's graphical prescription for constructing an anamorphic image on the exterior surface of a cone.*

The correct transformation is shown in Figure 2 where the eye at $2h$ observes a point on the conical surface at m' projected at ρ in the base of radius R. Similar triangles show that the relationship between m' and ρ is

$$m' = \frac{\rho}{2R-\rho}\sqrt{R^2+h^2} = \frac{\rho r}{2R-\rho} \quad (1)$$

or, if ρ is a fraction f of R then,

$$m' = \frac{f}{2-f}r \quad (2)$$

Niceron considers only the case of a 90° sector cone and observation at $2h$ but both of these can be easily included in a more generalized formulation (See Appendix). The kernel of this transformation $f/(2\text{-}f)$ is to be compared with Niceron's kernel: $\tan(f\times 45°)$. The anamorphic transformation of a simple square grid onto the surface of a cone obtained by applying Equations 1, 2 is shown in Figure 3.

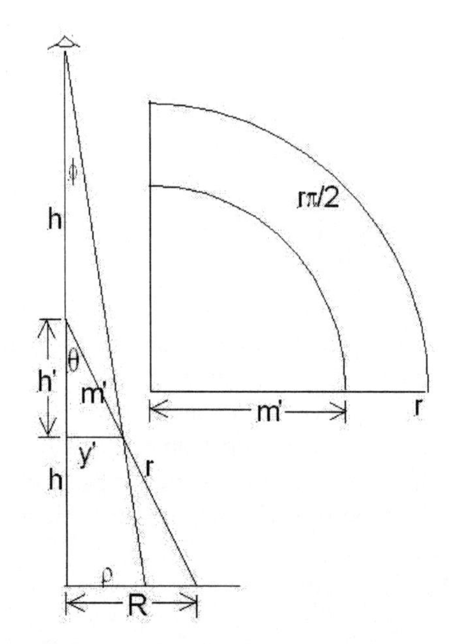

Figure 2: *The geometry of the anamorphic transformation to the exterior surface of a cone.*

The mystery is to understand how Niceron missed this simple transformation. It is all the more mysterious since the correct transformation was actually at hand in the immediately previous subject where he presented a well known and rigorous template for the transformation of a square tessellation as shown in Figure 4. He apparently failed to recognize that the transformation of the centre line of this figure is precisely the required transformation (Equation. (1) and (2)) if h is set equal to d, i.e. observation takes place at $2h$.

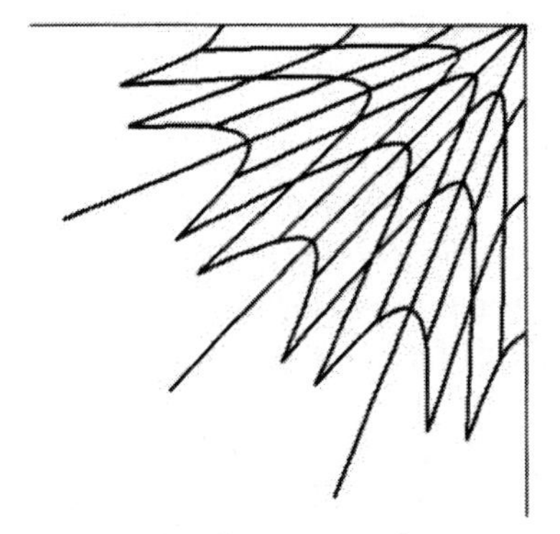

Figure 3: *The anamorphic transformation of an 8X8 square grid. The reconstruction is shown in Figure 5.*

The differences between the two formulations on the cone are subtle but observable, for a regular array for example, as shown in Figure. 5. The grid on the left has been calculated using Niceron's kernel and that on the right by Equations. (1), (2). A paper cone has been prepared and the reconstruction photo-graphed under identical conditions from height $2h$. The irregularities produced by the imperfect closing of the seam are evident, but the figure on the right has straighter lines and tiles which are more accurately square. Similar tests performed with concentric circles yield identical results.

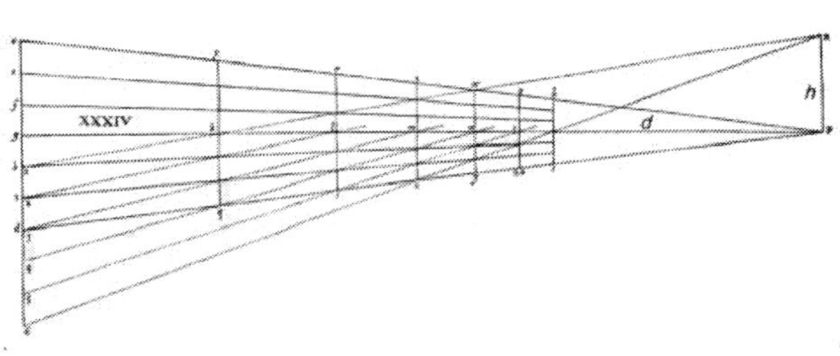

Figure 4: *Niceron's exact prescription for the perspective of a square tesselation (The letters d and h have been added.)*

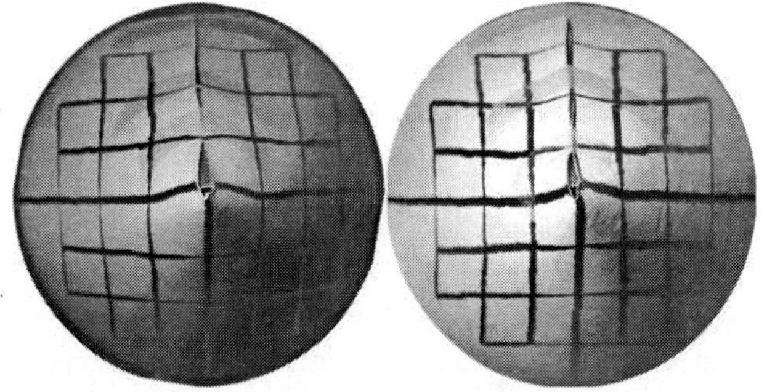

Figure 5: *Critical comparison of a square grid transformed according to the kernel of Niceron (left) and Equations 1 and 2 (right).*

The next topic in his book is the related one of the anamorphic image on the interior surface of the cone and, using a similar incorrect recipe, he comes to the correct conclusion that the same numerical results apply except in reverse order: widest segments nearest the apex and narrowest at the base.

3. Anamorphic Transformation on to the Surface of a Right Pyramid of any Order.

Niceron next turns his attention to the problem of a pyramid with a square base and also observed along the axis. Again he confines his prescriptions to those pyramids which unfold into a 90° sector and as before the more general equations are easy to derive; the recipe is given in Figure 6, (his XLIX and LI). The recipe is absolutely identical to the cone except for the segmentation of the pyramid base, where the resolved image appears, is even more complex and difficult to use in practise. For the pyramid surface, again it is divided into sectors of equal angle and divided radially using the tangent formulation given previously.

Now the question really arises as to why he did not see the error? Why did he not recognize that each face of this pyramid is exactly the case of the tessellation shown above in Figure 4? He seems to have been blind to everything except the apparent cleverness of his first guess using tangents, the error in which is unlikely to be revealed by transforming artistic images; it requires simple geometric shapes to do that.

For calculation purposes it is simplest to reduce the pyramid case to that of the cone. How this can be done is illustrated in Figure 7 for the case of a 4-sided pyramid. Again the reconstruction is to be viewed from $2h$ along the pyramidal axis and the reconstructed image is taken to be in the pyramid base. The pyramid is characterized by a base element b and an axial height h; only pyramids that unfold to a right-angle sector are considered. Imagine a vertical, axial slice through the pyramid creating the vertical plane ACE. This is the same as a section through a cone of base radius R and height h. A point P in the base has coordinates x, y or ρ, α; and as before $f = \rho / R$.. Therefore the point P can be transformed from x, y (ρ,α) to x', y' $(m'$, $\phi)$ using Equations (1), (2). As was done for the cone in Figure 5, the transforms are compared for a 4-sided pyramid in Figure 8. It is even more apparent that the tangent kernel is incorrect as it gives a pattern with converging lines and rectangular rather than square tiles.

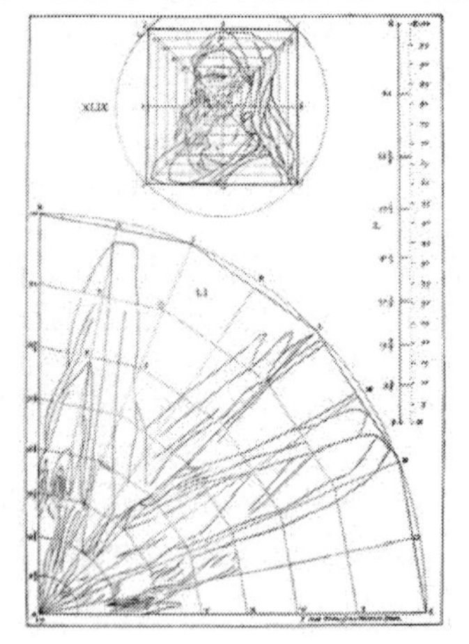

Figure 6: *Niceron's template for putting an anamorphic figure on the exterior surface of a 4-sided pyramid. XLIX-the original drawing, XLI-the transformed drawing, XL-the table* of tangents.

A topic which Niceron does not address is that of the continuity of lines across pyramid boundaries. In performing the anamorphic transform how does one insure that a line which crosses the boundary between two faces does not show a break or other discontinuity when observed? In the case of the 4-sided pyramid the condition is particularly trivial, which may be the reason he ignores it. From Figure 8 it is obvious that a vertical line in one face becomes a horizontal line when crossing to an adjacent face and vice versa. In Figure 9 is shown the reconstruction of a circle which intrudes on two faces adjacent to the central one.

Figure 8: *Critical comparison of a square grid transformed according to the kernel of Niceron (left) and Equations 1 and 2 (right)*

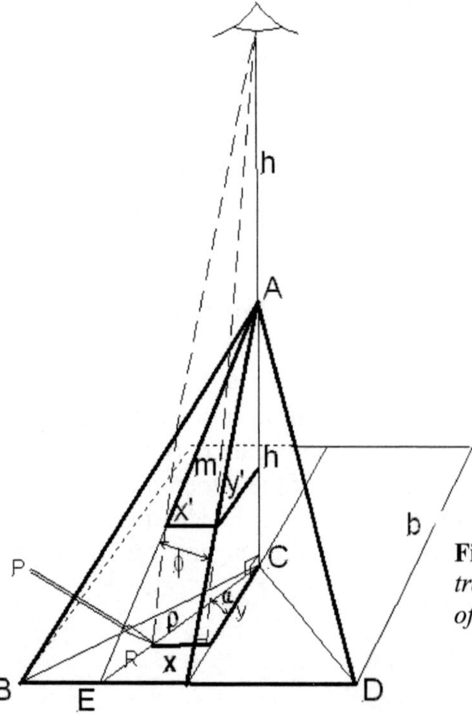

Figure 7: *Geometry of the transformation on the exterior surface of a 4-sided pyramid.*

Niceron mentions pyramids of other orders (3-sided, 5-sided etc.) but shows no examples. In these cases the continuity conditions are not so trivial. They can, however, be avoided (or rather, automatically applied) in every case by using the following recipe:

1. Divide the base into as many equal sectors as the base has sides.
2. Place the image to be transformed in the base and choose one of the sectors as the "calculation" sector.
3. When all the elements in the calculation sector have been transformed, rotate the entire image by one sector and transform the new elements as before.

This procedure guarantees the continuity of the image elements as the conditions are now automatically imposed by the proper rotation of the coordinates. In this way anamorphic pyramids of any order can be generated.

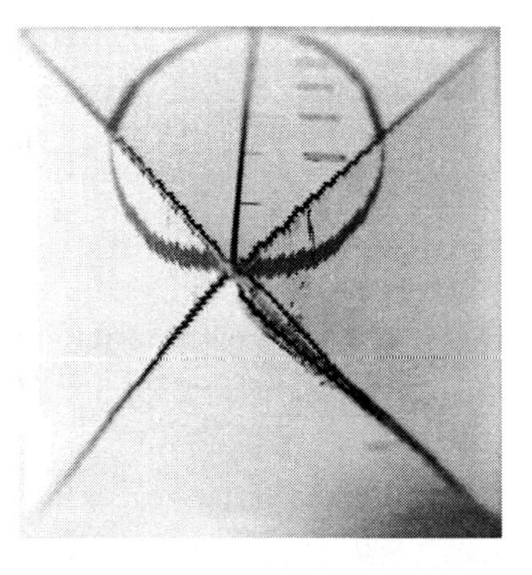

Figure 9: *A circle residing on three faces of a 4-sided pyramid with the proper continuity.*

4. Discussion of Niceron's Prescription

In trying to understand Niceron's construction of Figure 1 and why it is incorrect, it helps to see how it connects to the geometry of the cone or pyramid. This is attempted in Figure 10 where the capital letters indicate points common with Figure 1. The half of the real cone is shown in bold lines as AxC and the base radius R is one quarter the slant side r; this is necessary to produce a right-angle sector ABC which rolls up to form the cone (apex A, centre x).

The eye point y is placed so that yA = Ax (observation at $2h$). The construction of Figure 1 is obtained by drawing a triangle on AB as ABL with the angle at L and B equal to 45°. Looking at the sight-lines from y we have:

1. yA is directed to the apex ($\theta = 0$, $\tan 0 = 0$) and thus to the centre of the base as required.
2. If $\theta = 45°$ then this determines the point B which tracks by means of the arc BC to the periphery of the cone's base and thus determines the limiting sight-line yC.

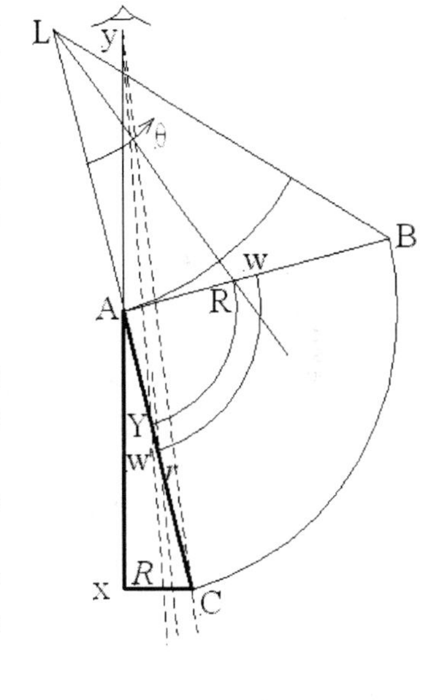

Figure 10: *Analysis of Niceron's tangent prescription for the cone. Dashed lines are sight-lines, capital letters correspond to Figure 1 and italics are dimensions.*

An example of Niceron's prescription is the following: Find where the sight-line to the point half-way between the centre and the periphery of the base, cuts the surface of the cone. Bisect the angle BLA and draw an arc from where the bisector cuts the line AB at R and track that to the cone's surface at Y. Since $\tan 22.5° = 0.414$, then the point Y is 41.4% distant from the apex along AC. Niceron infers then that the sight-line from y through Y will fall half way along the radius R, which is manifestly untrue.

He would have been aware of the fact that bisecting the line AB to produce the point w, which tracks to w′, produces a sight-line too near the periphery and that a mechanism had to be found to move that point back toward the centre, which the tangent calculation accomplishes but not sufficiently.

Whatever mechanism is used it must give a value of 1 at 45° and 0 at 0° which the tangent does (as does $f/(2-f)$ of Equation. 2 for $f = 0$ and 1). It is difficult to think that Niceron could not do the simple calculation of Equation 1 but perhaps he felt constrained to find the best approximation that could be carried out with ruler, compass and pencil alone. Indeed throughout the book he reminds us that he is speaking to the practitioners who, presumably, he did not expect to do calculations and make detailed measurements. However, nowhere in Volume 2 does he indicate that this is an approximation.

Appendix: The Generalized Equations for the Conical Transformation.

Niceron deals only with the special case where observation is from a point at h above the apex but it can be generalized. If the observation point is at $h + qh$, where q is a dimensionless number then the transform equation (analogous to Equation 2) is,

$$m' = \frac{qf}{1-f+q} r . \qquad (3)$$

Niceron also considers only a cone rolled from a right sector, i.e. in Figure 10 angle BAC ($= \theta$) = $\pi/2$ radians. To further generalize, the cone can be rolled from a sector of any angle, although, clearly very flat anamorphs produce unspectacular results. If a fraction ε of the circumference $2\pi r$ is used then,

$$\theta = \sin^{-1}\varepsilon, \qquad (4)$$

and

$$h = \left(R/\varepsilon\right)\sqrt{1-\varepsilon^2} . \qquad (5)$$

Using Equations (3) to (5) the anamorph of any image can be constructed.

References

1. Jean-Francois Niceron, *La Perspective Curieuse,* Chez Pierre Billaine, Paris, (1638)
2. Jurgis Baltrusaitis, (Translated by W.J. Strachan), *Anamorphic Art,* Abrams, New York (1977)
3. www.artetoma.it/anamorfosi/niceron.PDF
4. Hunt, J.L., Nickel, B.G., Gigault, Christian. *Anamorphic Images,* Am. J. Phys. **68**, pp 232- 237 (2000).

A meditation on Kepler's Aa

Craig S. Kaplan
David R. Cheriton School of Computer Science
University of Waterloo
csk@cgl.uwaterloo.ca

Abstract

Kepler's *Harmonice Mundi* includes a mysterious arrangement of polygons labeled Aa, in which many of the polygons have fivefold symmetry. In the twentieth century, solutions were proposed for how Aa might be continued in a natural way to tile the whole plane. I present a collection of variations on Aa, and show how it forms one step in a sequence of derivations starting from a simpler tiling. I present alternate arrangements of the tilings based on spirals and substitution systems. Finally, I show some Islamic star patterns that can be derived from Kepler-like tilings.

1. Introduction

Can a tiling of the plane be produced in which every tile has fivefold rotational symmetry? The question seems first to have been explored by Johannes Kepler in *Harmonice Mundi*. Although he did not resolve the question one way or the other, he did produce some remarkable tilings in the process of seeking an answer. They are reproduced in the frontispiece of Grünbaum and Shephard's *Tilings and Patterns* [3]. The general problem remains unsolved to this day [1, 2, 7], although solutions exist in which no bound is placed on the sizes of tiles.

Every finite drawing of a tiling requires the viewer to make a tacit assumption that the drawing could in theory be continued in an obvious way to cover the whole plane. In each case, that assumption may conceal a mathematical problem of lesser or greater difficulty. In many of Kepler's simpler drawings, the manner of continuation may indeed be considered obvious. But the same is not true for his arrangements that include shapes with fivefold symmetry.

Of these arrangements, the drawing labeled Aa is perhaps the most intriguing. It contains more tiles than its brothers Z and Bb; it seems as if Kepler was more confident of the continuation of this tiling. Indeed, we now know that Aa can be extended to a tiling of the plane. It is far less obvious what Kepler had in mind for Z and Bb (though Grünbaum and Shephard mention a possible solution for Bb).

One possible method of tiling the plane as an extension of Kepler's Aa is attributed to Dessecker and described by Grünbaum and Shephard [3, Section 2.5]. Consider the $36°$ rhomb in Figure 1(a). The interior of the rhomb can be marked with whole and partial polygons. It turns out that when laid out edge-to-edge in a periodic pattern, copies of this rhomb generate a periodic tiling by pentagons, pentacles, decagons, and "fused decagon pairs" that Kepler called "monsters". Remarkably, this rhomb can also fit snugly with a rotated copy of itself staggered by one half of a rhomb edge (and not by the golden ratio, as Grünbaum and Shephard claim). Moreover, the rhombs can then be put into an arrangement with finite symmetry group d_5 in such a way that the markings on them form a non-periodic tiling of the plane. This new non-periodic tiling contains Kepler's Aa patch as a subset. The rhomb tiling and Kepler tiling are given in Figure 1(d) and (e).

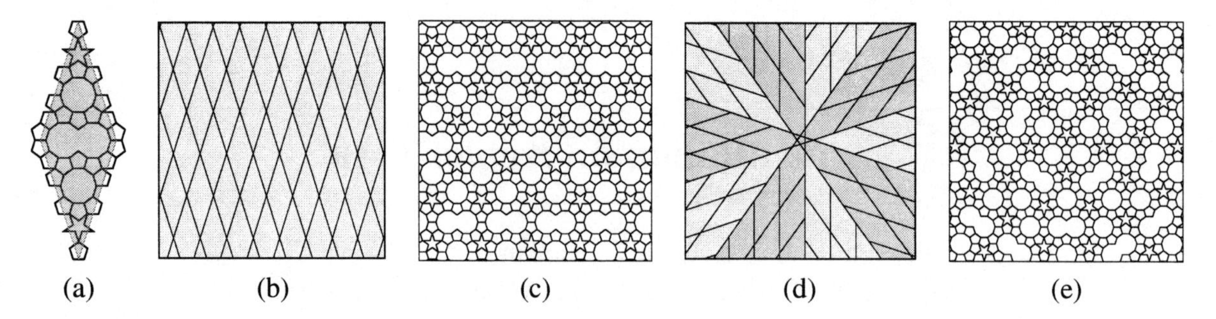

Figure 1 A demonstration of the construction of Kepler's Aa tiling. The rhomb in (a) forms a periodic tiling in (b), and its markings form a corresponding periodic tiling in (c). Because this rhomb can also meet a copy of itself staggered by half an edge length, the arragement of rhombs in (d) also produces a valid tiling, as shown in (e).

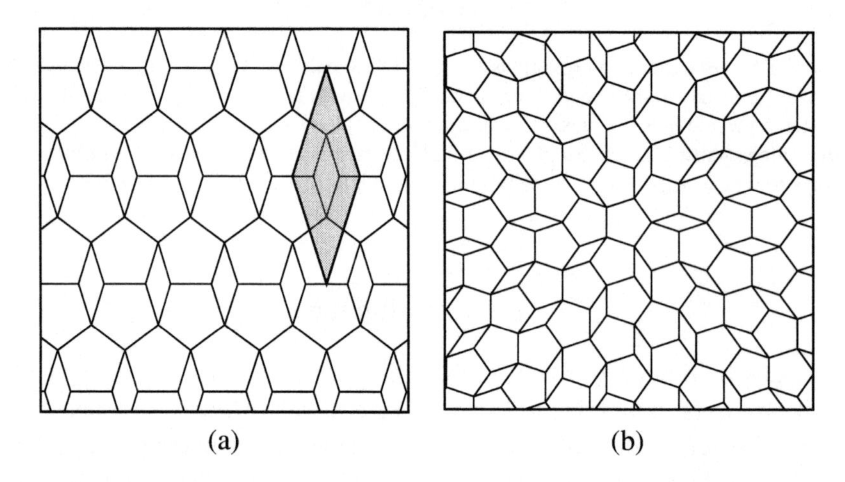

Figure 2 A simple tiling by pentagons and rhombs, with one possible translational unit shaded. In (b), the translational units are rearranged into a d_5-symmetric tiling in the spirit of Kepler's Aa.

In this paper, I use Kepler's Aa as a starting point for a process of exploration. First, I show how the Aa tiling is but one step in a larger progression of related tilings based on $36°$ rhombs. I then show how to create spiral arrangements of rhombs, giving attractive new tilings. I show how these rhombs can be used as a basis for developing substitution systems, producing additional related tilings without being tied to the underlying rhombs. Finally, I explore some of the decorative possibilities that arise when these tilings are used as templates for the construction of Islamic star patterns.

2. Other Kepler rhombs

Kepler's rhomb in Figure 1(a) can be derived in a fairly natural way from a simpler tiling. In fact, this derivation produces an infinite sequence of marked rhombs, each of which can then generate tilings of the plane in the same ways as Kepler's. In this section, I explain the derivation and demonstrate the first few members of the sequence.

We begin with a well-known periodic tiling by regular pentagons and $36°$ rhombs, as shown in Figure 2(a). One natural translational unit for this tiling is a larger rhomb centred on a rhombic tile. This translational unit also contains fragments of pentagonal tiles adding up to two whole pentagons. We refer to this marked rhomb as $k1$. Observe that copies of $k1$ can meet in the same ways as the rhomb used in Kepler's Aa, and indeed can produce a similar d_5-symmetric tiling, as shown in Figure 2(b).

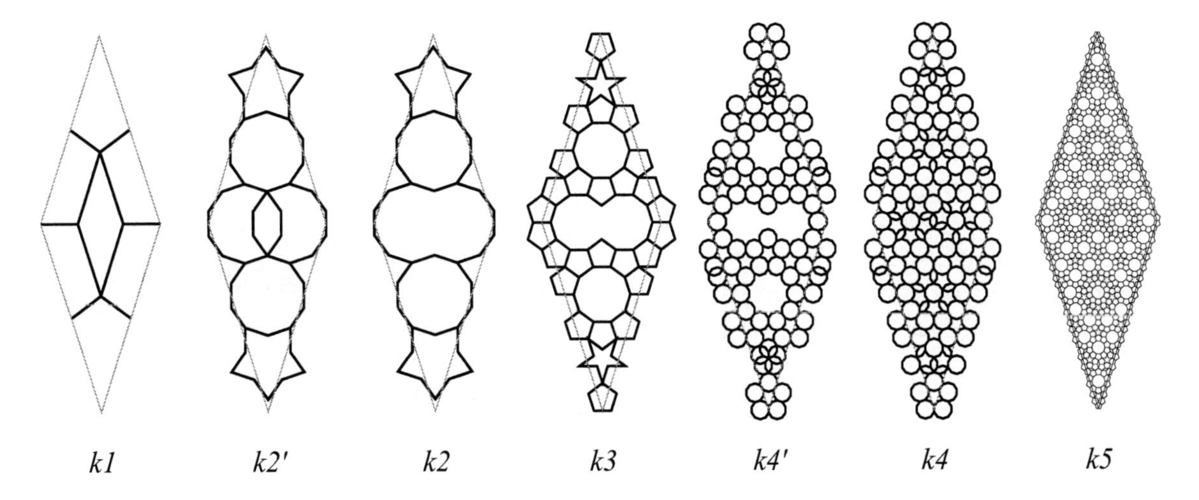

Figure 3 A sequence of marked rhombs, all of which lead to Kepler-like periodic and non-periodic tilings.

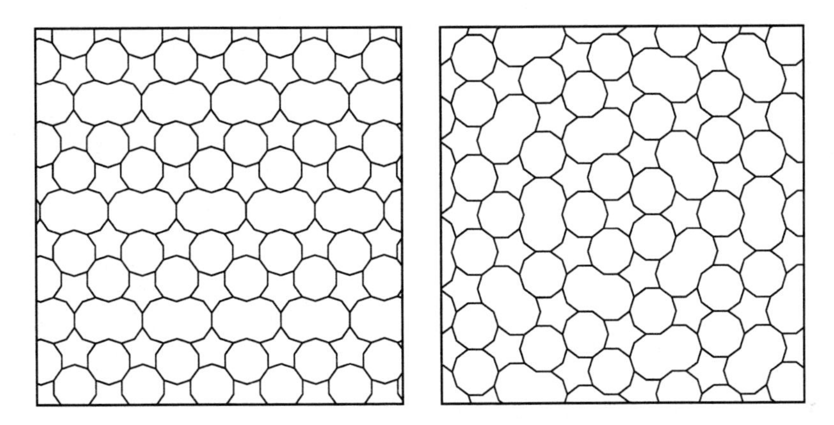

Figure 4 Periodic and non-periodic tilings by marked rhomb $k2$.

In every tiling generated by the markings on $k1$, the angles between adjacent edges at a vertex are multiples of $36°$. It is therefore possible to surround every vertex by a regular decagon in such a way that the vertex's edges are perpendicular bisectors of the decagon edges. Thus we place a decagon at every vertex, scaled so that they meet edge-to-edge when the underlying vertices are adjacent. This substitution yields the marked rhomb $k2'$ in Figure 3. This rhomb produces a tiling with decagons (some of which overlap) and five-pointed stars. We can produce a simpler design, $k2$, by deleting the regions where decagons overlap, resulting in a single large shape that is the union of two decagons. Note that the decagons overlap differently than in the arrangement of Figure 1(a). Inspired by Kepler's use of the name "monster", in previous work I classified Kepler's shape as a $(10, 2)$-monster and the new shape in $k2$ as a $(10, 3)$-monster [4, Section 3.10]. Periodic and d_5-symmetric tilings by $k2$ are shown in Figure 4.

In $k2$ and $k2'$, the angles at every vertex are multiples of 72 degrees. We can therefore perform a similar process as above, this time placing scaled regular pentagons around every vertex. Performing this substitution on $k2$ yields the rhomb $k3$, which is none other than the generator of Kepler's Aa.

We can continue this process indefinitely, alternately placing decagons or pentagons at vertices to produce finer subdivisions of a $36°$ rhomb. At every step, we get a new source of Kepler-like tilings. Figure 3 shows the first few steps of this substitution sequence, ending in $k5$.

Occasionally we will want to adjust the outcome of the substitution. In some cases I replace overlap-

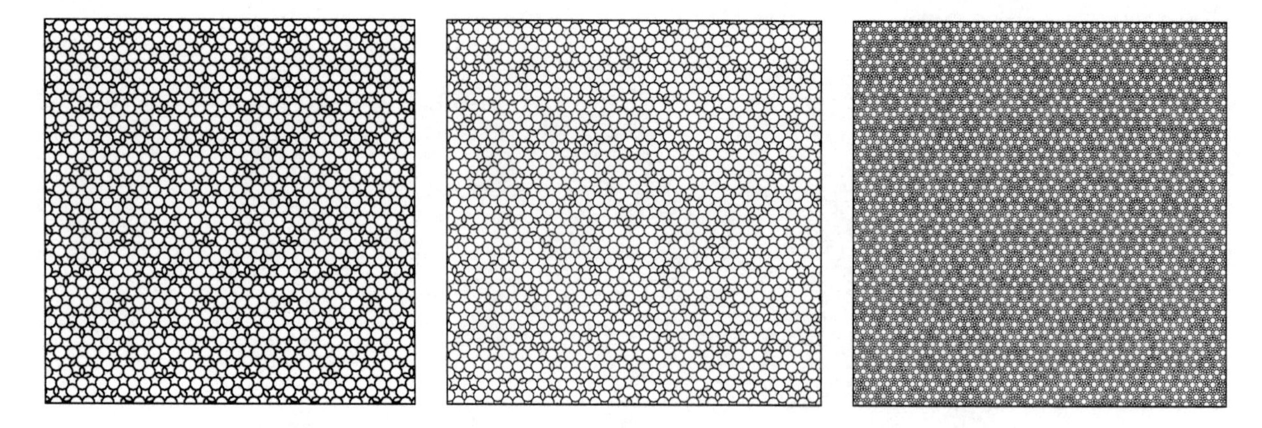

Figure 5 Periodic and non-periodic tilings by marked rhombs $k4$ and $k5$.

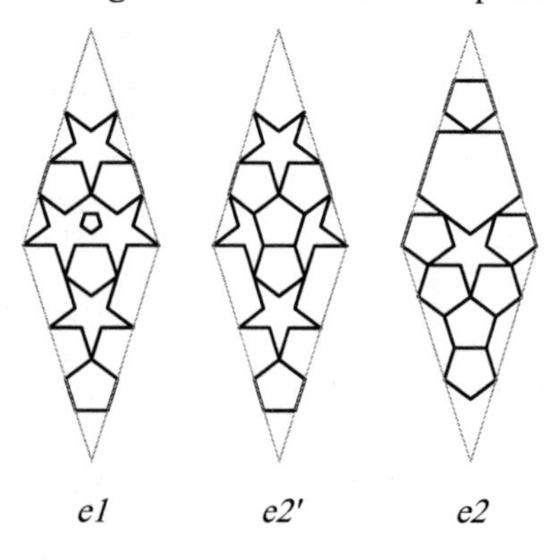

Figure 6 Other marked rhombs that yield interesting tilings. The rhomb labeled $e1$ is due to Eberhart. Rhomb $e2'$ is a simple modification suggested by Grünbaum and Shephard. On the right, $e2$ is the same as $e2'$ but with the enclosing rhomb shifted slightly relative to the underlying periodic tiling.

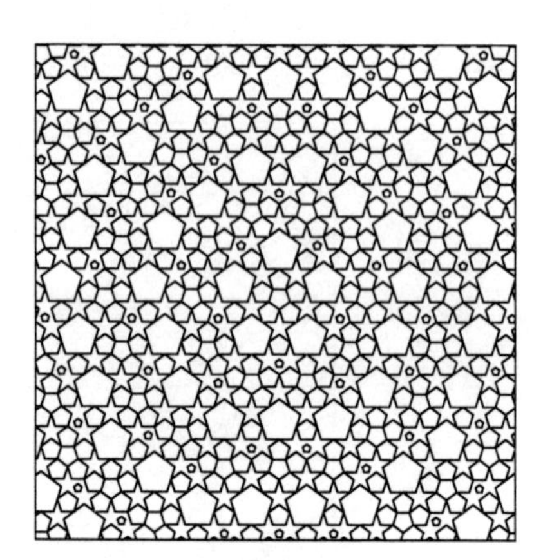

Figure 7 A d_5-symmetric tiling constructed using copies of $e1$.

ping decagons with an alternate arrangement of smaller tiles; in others I fill large empty regions with tiles. Both adjustments can be seen in the transition from $k4'$ to $k4$ in Figure 3. Tilings by $k4$ and $k5$ appear in Figure 5.

Grünbaum and Shephard also present an alternate rhomb, due to Stephen Eberhart. It is labeled $e1$ in Figure 6. This rhomb generates two sizes of regular pentagon, and uses a fused pair of pentacles. Because $e1$ is not centrally symmetric, it cannot be derived in an obvious way from $k1$. Unlike $k3$, Eberhart's rhomb really does fit with copies of itself staggered by the golden ratio. Note that a slight variation of $e1$ is also possible in which the fused pentacles are replaced by two half-pentacles and a pentagon. This attractive alternative is labeled $e2'$ in Figure 6. Figure 7 shows a d_5-symmetric tiling generated by copies of $e1$.

3. Spiral arrangements

In all of the cases above, rhombs can meet their neighbours either edge-to-edge or staggered by some amount. These possibilities lead naturally to periodic tilings or tilings with finite symmetry d_5. The staggered layout suggests that a third arragement of rhombs should also be possible, one where rhombs spiral outward from a centre. Figure 8(a) shows a spiral arrangement of $36°$ rhombs that can meet their neighbours staggered by half an edge length. The only additional question is how to fill the leftover space in the middle

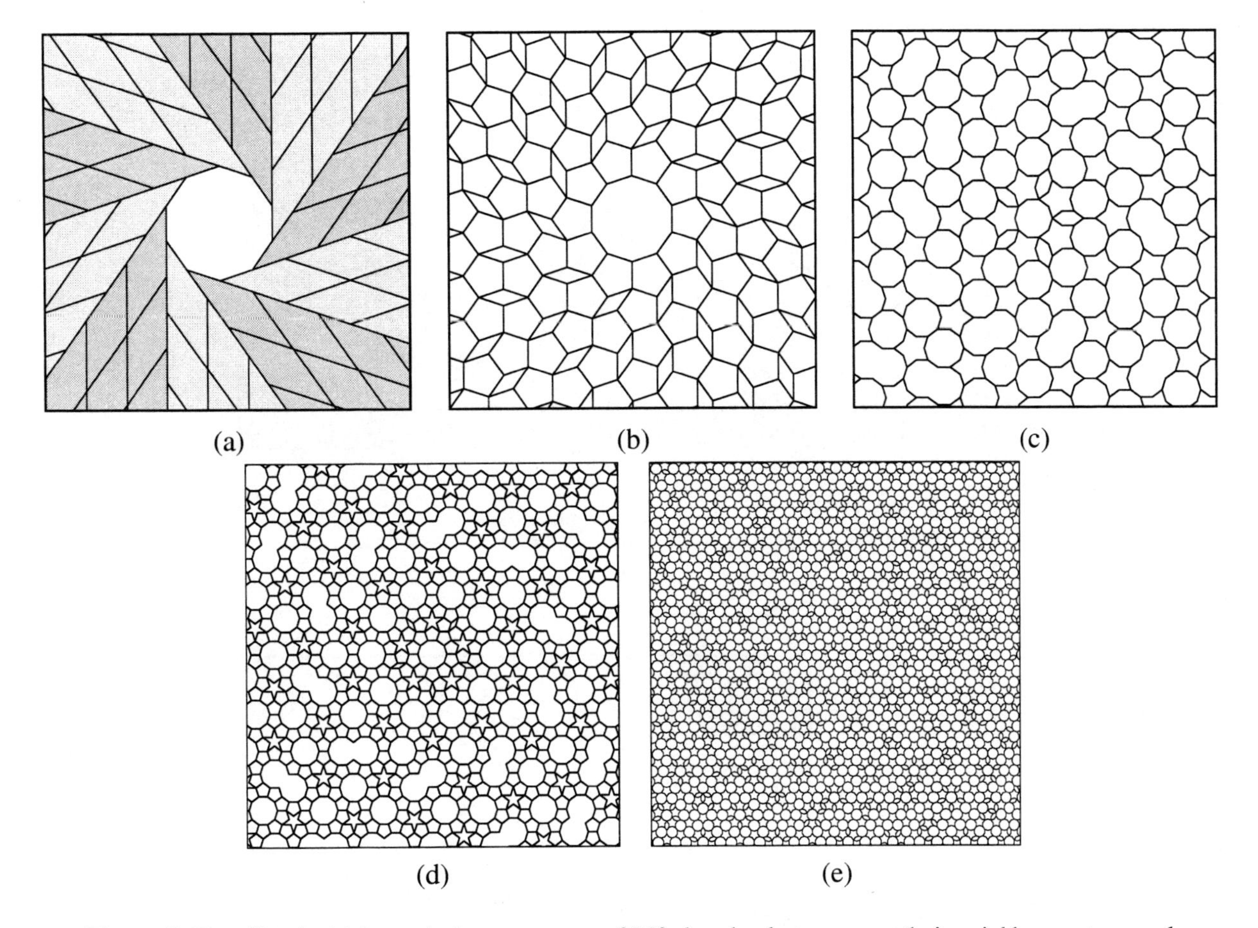

Figure 8 The tiling in (a) is a spiral arrangement of $36°$ rhombs that can meet their neighbours staggered by half an edge length. This arrangement leaves an unfilled regular decagon in the middle of the tiling. Tilings (b), (c), (d), (e) are examples of spiral arrangements using rhombs $k1$, $k2$, $k3$, and $k4$ respectively.

of the tiling. I have developed plausible arrangements of tiles through experimentation. In every case, I have tried to avoid introducing new tile shapes. Examples of spiral tilings based on $k1$, $k2$, $k3$, and $k4$ are given in Figure 8.

Figure 9(a) shows a spiral arrangement based on $e1$. Because $e1$ meets itself staggered by the golden ratio, the central decagon is not regular, but the region can still be filled by introducing half pentacles. The spiral tiling by $e2'$ is evidently similar in structure. Interestingly, the arrangement of tiles in $e2'$ can be translated relative to the surrounding rhomb to produce a new rhomb $e2$. This new rhomb generates the same periodic and d_5 tilings of the plane as $e2'$. But the resulting spiral tiling is different, as shown in Figure 9(b).

4. Related substitution tilings

If we superimpose any of the rhombs of Figure 3 on top of any other, a recursive structure suggests itself. For example, every tile from $k3$ can be seen as filled by smaller tiles from $k5$. Any of these superpositions leads to a substitution system that produces new tilings from old ones. These systems are reminiscent of Penrose's early experiments that ultimately led to his discovery of the aperiodic tile set that Grünbaum and Shephard refer to as $P1$ [6]. Two examples of substitution systems inspired by the Kepler rhombs are given in Figure 10. In Figure 11, a related tiling is coloured in.

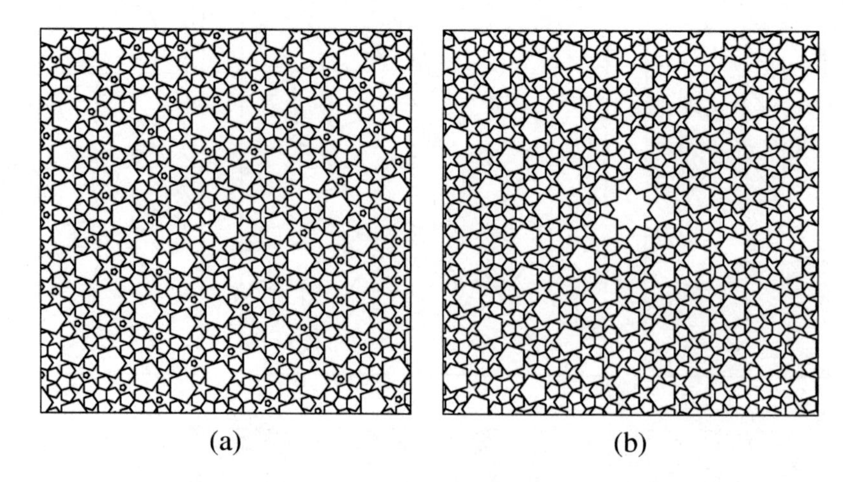

Figure 9 Examples of spiral tilings based on rhombs $e1$ and $e2$.

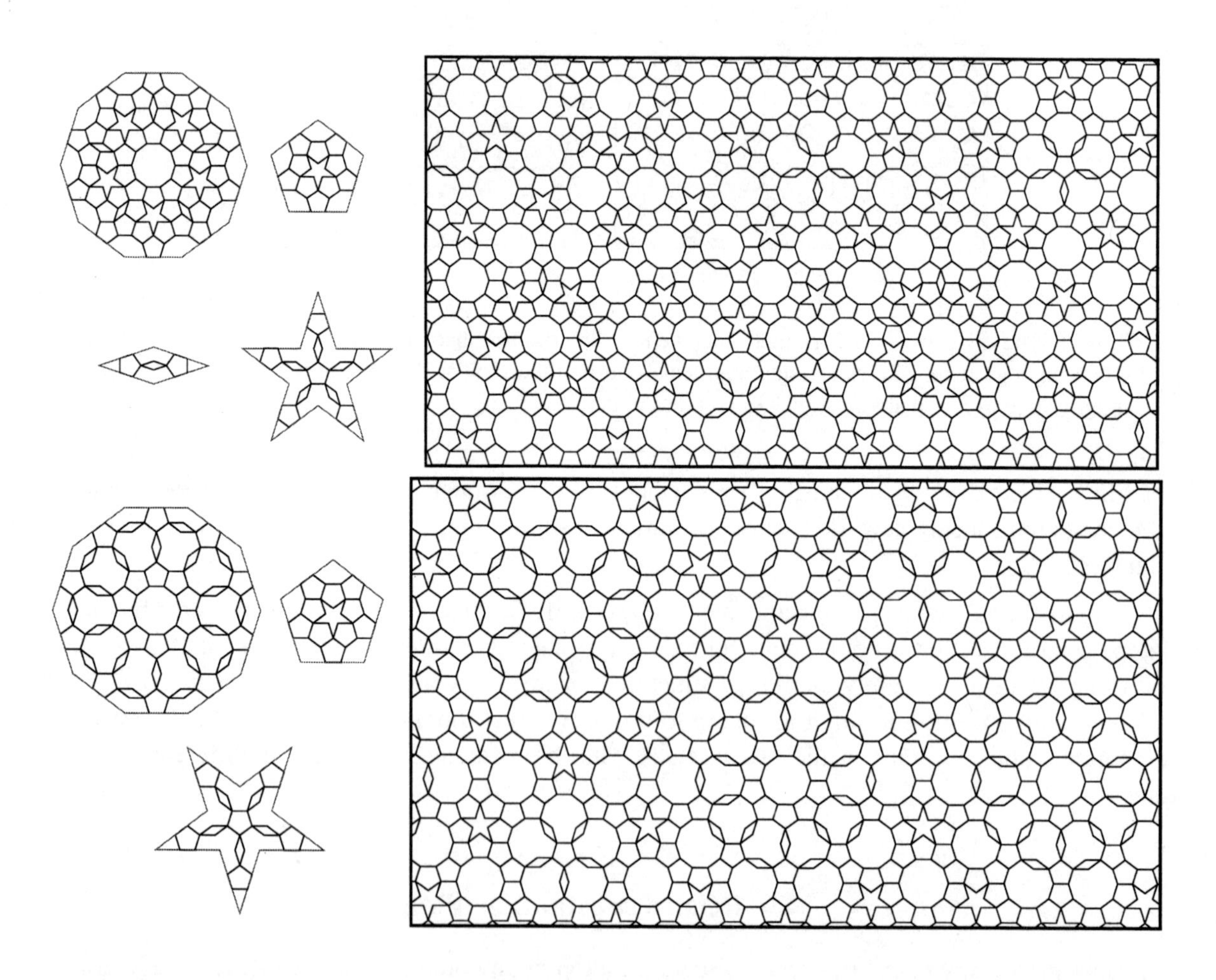

Figure 10 Examples of substitution tilings inspired by examination of the Kepler rhombs $k1$, $k3$, and $k5$. The top diagram shows a simple substitution system that permits decagons to overlap in $36°$ rhombs. In the bottom diagram, this overlapping is extended to the point where explicit rhombs are unnecessary, and the entire tiling can be regarded as composed of pentagons, pentacles, and overlapping decagons.

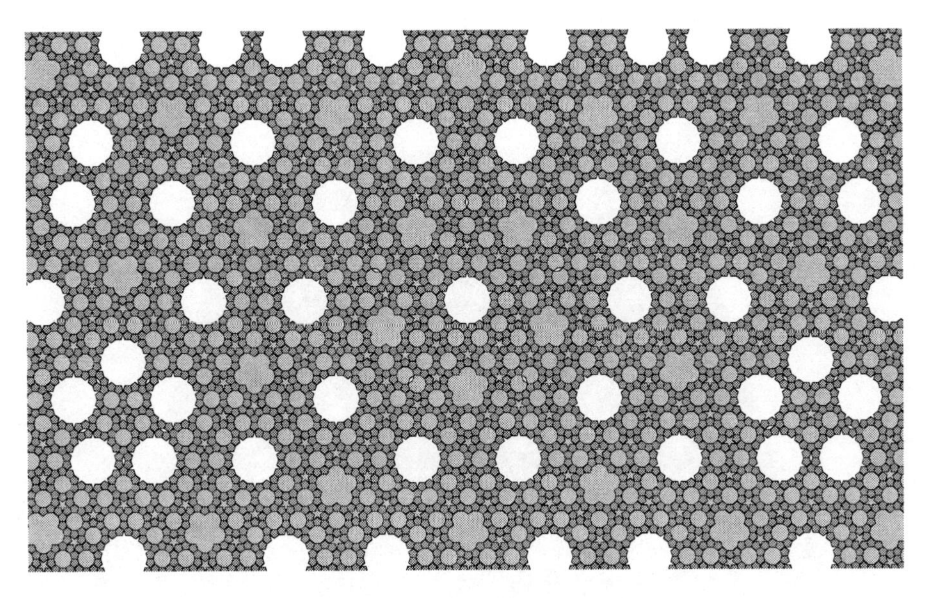

Figure 11 A colourful example of a substitution tiling. This example uses the substitutions defined in the top diagram of Figure 10, but includes a final step that replaces some repeated arrangements of smaller tiles with large symmetric ones.

5. Kepler star patterns

Because these tilings contain many regular polygons, it is natural to consider them as templates for producing Islamic star patterns via the polygons-in-contact method [5]. Figure 12 gives two examples of Kepler-based star patterns. The first is derived from the original Kepler Aa tiling. The second is based on the substitution system at the bottom of Figure 10. In both cases, small adjustments must be made so that attractive motifs can be found for pentacles. Novel motifs must also be supplied for overlapping decagons (or, equivalently, monsters). Details on both adjustments are given in previous work [4, Section 3.10].

The use of Kepler's tilings as a base for constructing Islamic star patterns is a satisfying mix of art and mathematics from across different centuries and cultures.

References

[1] Hallard T. Croft, Kenneth J. Falconer, and Richard K. Guy. *Unsolved Problems in Geometry*. Springer-Verlag, 1991.

[2] L. Danzer, B. Grünbaum, and G. C. Shephard. Can all tiles of a tiling have five-fold symmetry? *American Mathematical Monthly*, 89:568–585, 1982.

[3] Branko Grünbaum and G. C. Shephard. *Tilings and Patterns*. W. H. Freeman, 1987.

[4] Craig S. Kaplan. *Computer Graphics and Geometric Ornamental Design*. PhD thesis, Department of Computer Science & Engineering, University of Washington, 2002.

[5] Craig S. Kaplan. Islamic star patterns from polygons in contact. In *Proceedings of the 2005 conference on graphics interface*. Canadian computer human communication society, 2005.

[6] Roger Penrose. Pentaplexity. *Mathematical Intelligencer*, 2:32–37, 1979/80.

[7] John Savard. Pentagonal tilings. `http://www.quadibloc.com/math/penint.htm`.

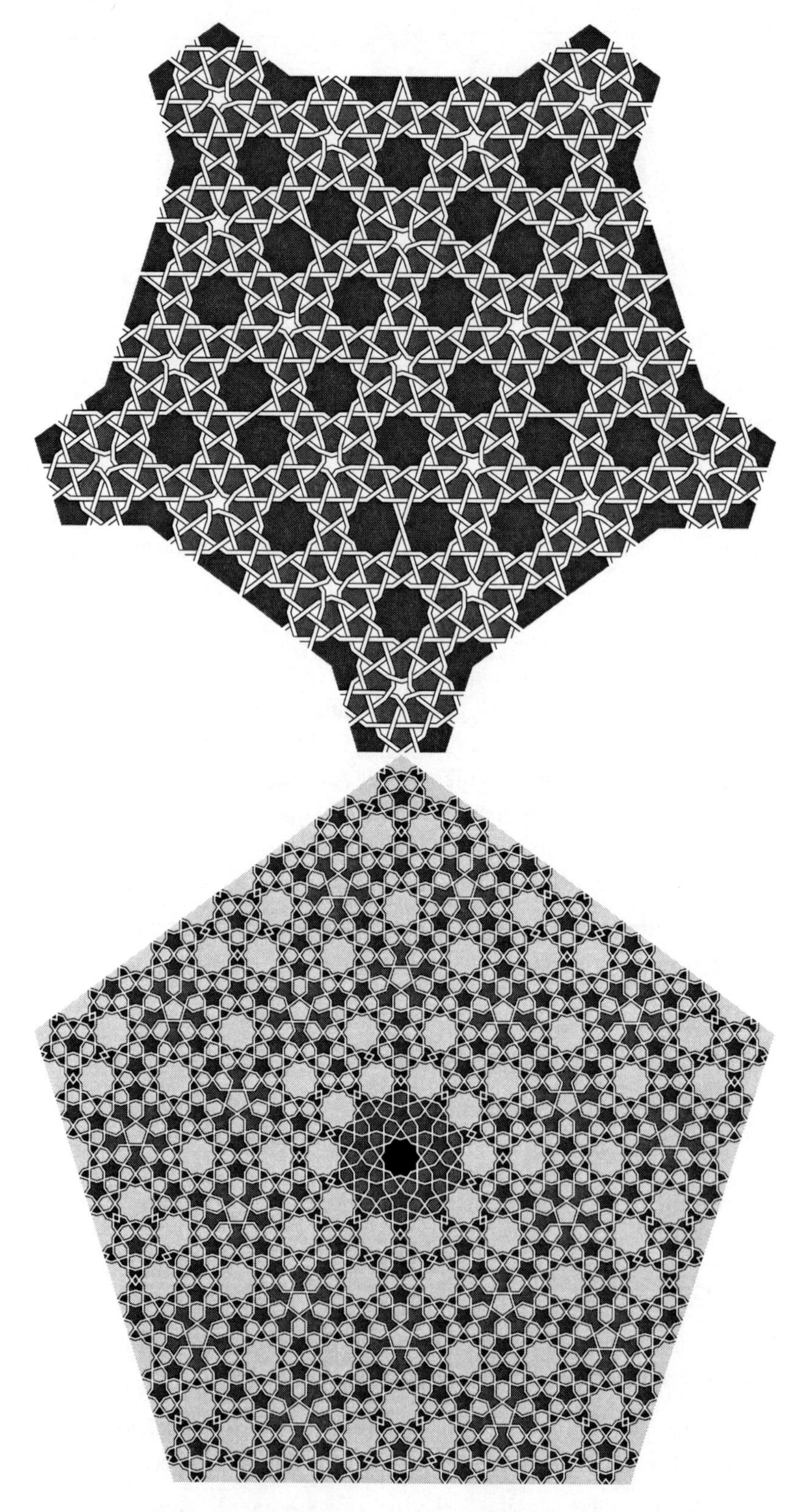

Figure 12 Rendered star patterns that use Kepler tilings as templates.

Approximating Mathematical Surfaces with Spline Modelers

Stephen Luecking
School of Computer Science, Telecommunications and Information Systems
DePaul University
243 South Wabash
Chicago, IL 60604
sluecking@cs.depaul.edu

Abstract

Computer modeling permits the creation and editing of mathematical surfaces with only an intuitive understanding of such forms. B-splines used in most commercial modeling packages permit the approximation of a wide variety of mathematical surfaces. Such programs may contain tools for aiding in the production of these surfaces as physical sculptures. We outline some techniques for non-mathematical designers and sculptors to produce these objects with conventional modeling.

Mathematical sculptures are premised on the inherent beauty discovered again and again in mathematical objects. Among the most elegant and bewitching of these objects are the multitudes of surfaces that twine through 3D and 4D spaces. In industry-standard programs, intended for the designer and not the mathematician, no equations or formulations exist to generate such forms; rather they employ a general, all-purpose geometry for fashioning 3D objects from the designer's vision.

This geometry is the "material" of 3D graphics, specifically geometry that is computable. Surfaces are commonly computed by defining the surface with points in a coordinate space. The surface results from one of two methods of interpolation among these points: one method treats the points as vertices of polygons; the other uses a fabric of spline curves, known as B-splines, that pass through these points. The former is familiar to users of the Mathematica software and the latter to computer artists and designers seeking more natural representations of design products, characters, and scenes. Since both cases require interpolation, the results are approximate, although close enough in practice for almost all applications. What the computer produces, then, are representations or models of the surface.

In the hands of most artists, computer modeling is not at all mathematical, even though the artists are manipulating geometry at all stages. The user interfaces of all popular modeling programs work at making this manipulation intuitive and as close to drawing and design traditions as possible. Nevertheless, the best of these programs also boast ample tools for the mathematically inclined artist to build surfaces expressing mathematical elegance. Because such models are always approximations used to represent a surface they are closer to well-crafted plaster carvings, formed metal, or string models of surfaces than they are to the actual mathematical surface.

Developable Surfaces

Most modeler's represent ruled surfaces by sweeping straight lines through space. Many ruled surfaces are developable and can be crafted from sheet materials, as is the case of Naum Gabo's sculpture, "Spheric Theme" (Figure 2). Gabo disavowed the mathematical nature of his work and avowed, instead, that his surfaces were the natural outcome of experimentation with sheet metal's ease in fashioning developable surfaces. "Spheric Theme" originated, for example, from the joining of two split annuli cut from paper and joined at their split edges into one continuous surface. There is a mathematical spirit to such experiments, even if mathematics was never consciously a part of the process. The fact that the sweeping arcs of Gabo's sculpture began as flat sheets of bronze qualifies it as a developable surface. By definition a developable surface can be mapped – or unrolled – to a flat surface with no discernible distortions or projections.

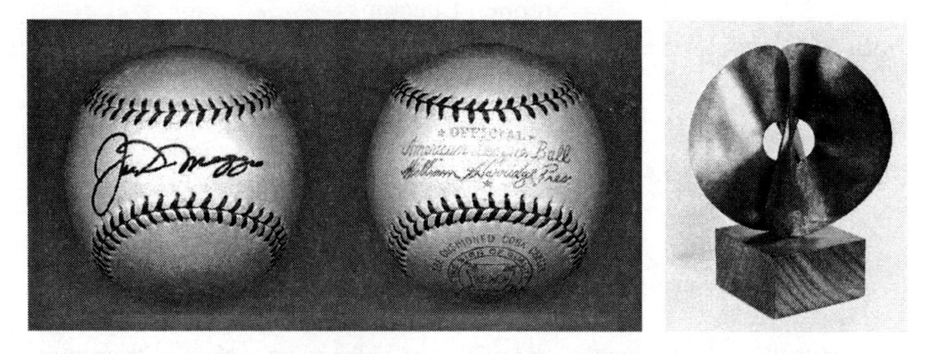

Figure 1 *Joe DiMaggio autographed baseball.* **Figure 2** *Naum Gabo, "Spheric Theme"*

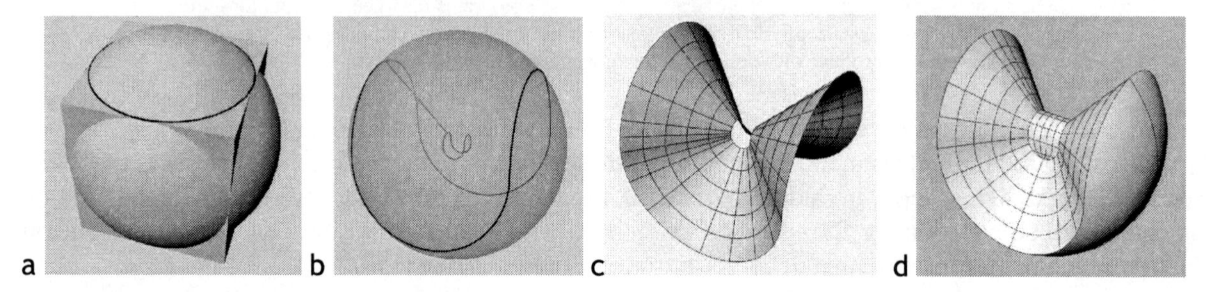

Figure 3a-c *Developing surface in a sphere* **3d** *Stitched polysurface*

Like the symmetry of a baseball's seam, Gabo's surface was the result of minimally crafting a sphere in space. Procuring form in a modeling program often follows such mathematically-spirited crafting rather than calculations and formulas. The bottom sequence in Figure 3 demonstrates such an approach. It begins with the intersection of a sphere and a cube whose edges are tangent to the sphere (Figure 3a). From the six tangent circles that ensue, four semi-circles join into a single curve dividing the surface of the sphere (Figure 3b). A smaller version of this curve is copied to the center to act as a second rail along which to sweep a surface (Figure 3c). The modeler affixes this surface to one half of the divided sphere and another surface segment at the center (Figure 3d).

Booleans and Surfaces

Once all of the surfaces have been stitched into a *closed polysurface*, the program treats it as a solid and it then becomes subject to Boolean edits. The Boolean operations – *union*, *difference* and *intersection* – allow surface models to imitate the edits of constructive solid geometry (*CSG*).

When two closed polysurfaces interpenetrate they can affect one another's shape by applying the Boolean operations. Union (Figure 4a) trims those surface portions enclosed within the penetration. The exterior surfaces remain to represent a single solid. Difference (Figure 4b) rejects the exterior surface segment of one shape and the enclosed surface segment of the other. The remainder appears as a solid with an extraction shaped like the subtracted solid. Intersection (Figure 4c) retains the interior surface segments while discarding the exterior. The new solid includes only that region of space common to both solids. The polysurfaces created by a Boolean operation always comprise portions of at least two surfaces.

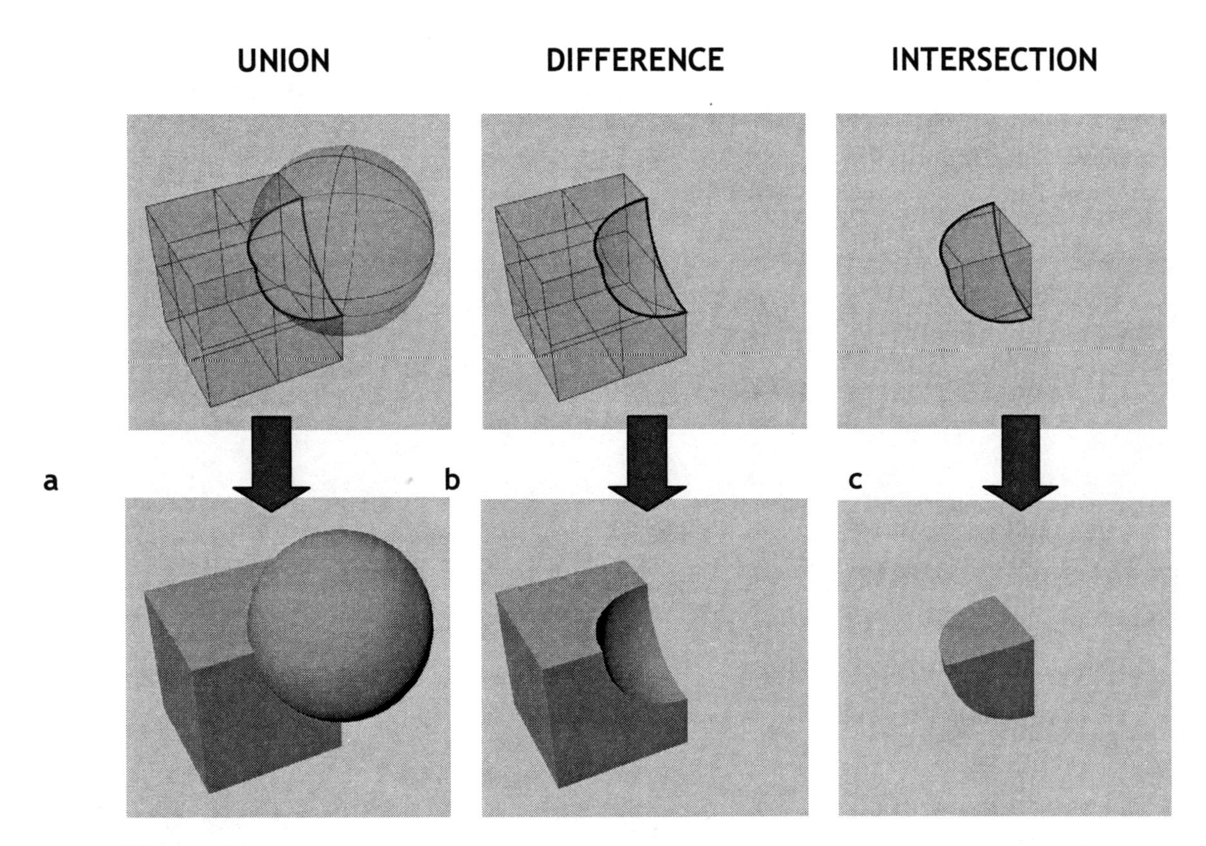

Figure 4 *Boolean operations.*

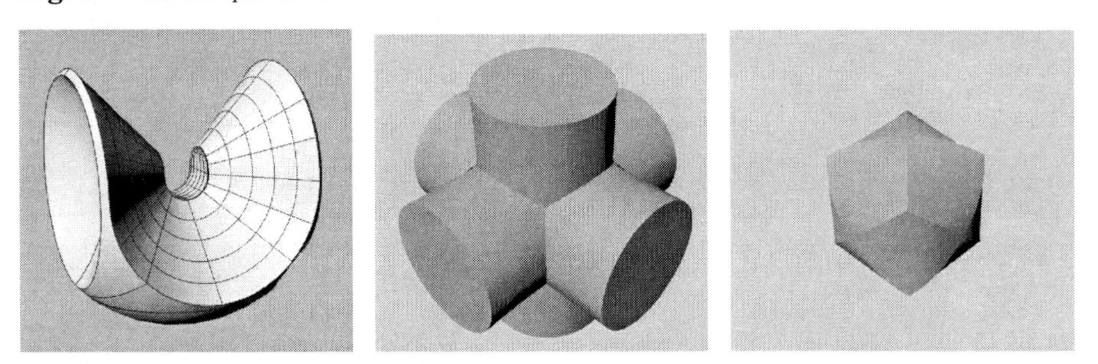

Figure 5 *Previous polysurface with spherical concavities.* **Figure 6a** *Orthogonally intersecting cylinders.* **Figure 6b** *Boolean intersection of three cylinders.*

In most design modelers, CSG modeling relies on sets of geometric *primitives*. The primitives are a set of commonly used geometric solids that the designer can quickly and easily insert into the modeling space. These may include boxes, cones, cylinders, spheres, toruses, ellipsoids and paraboloids. The last three provide a stock of ready-made quadratic surfaces, the spheroids. Other related, but less-used surfaces like the hyperboloid below can generate easily in a spline environment.

Geometric "Grain"

Gaspard Monge invented string models to instruct his students on the nature of ruled surfaces. One of the best known is the hyperboloid, depicted in Figure 7a as model by Theodore Olivier, one of Monge's most famous students. This model begins as a string model of a cylinder. The top circle then rotates to skew the string lines to form the hyperboloid. This can be generated by sweeping a straight line, thereby creating the effect of a series of canted strings connecting two end circles (Figure 7b,c). The same form may also be treated as a *surface of revolution*, created by revolving a hyperbola around an axis. Each method of construction will yield the same shape, but with differing internal geometry. Different geometries will have a significant effect on how the surface may be edited or manufactured.

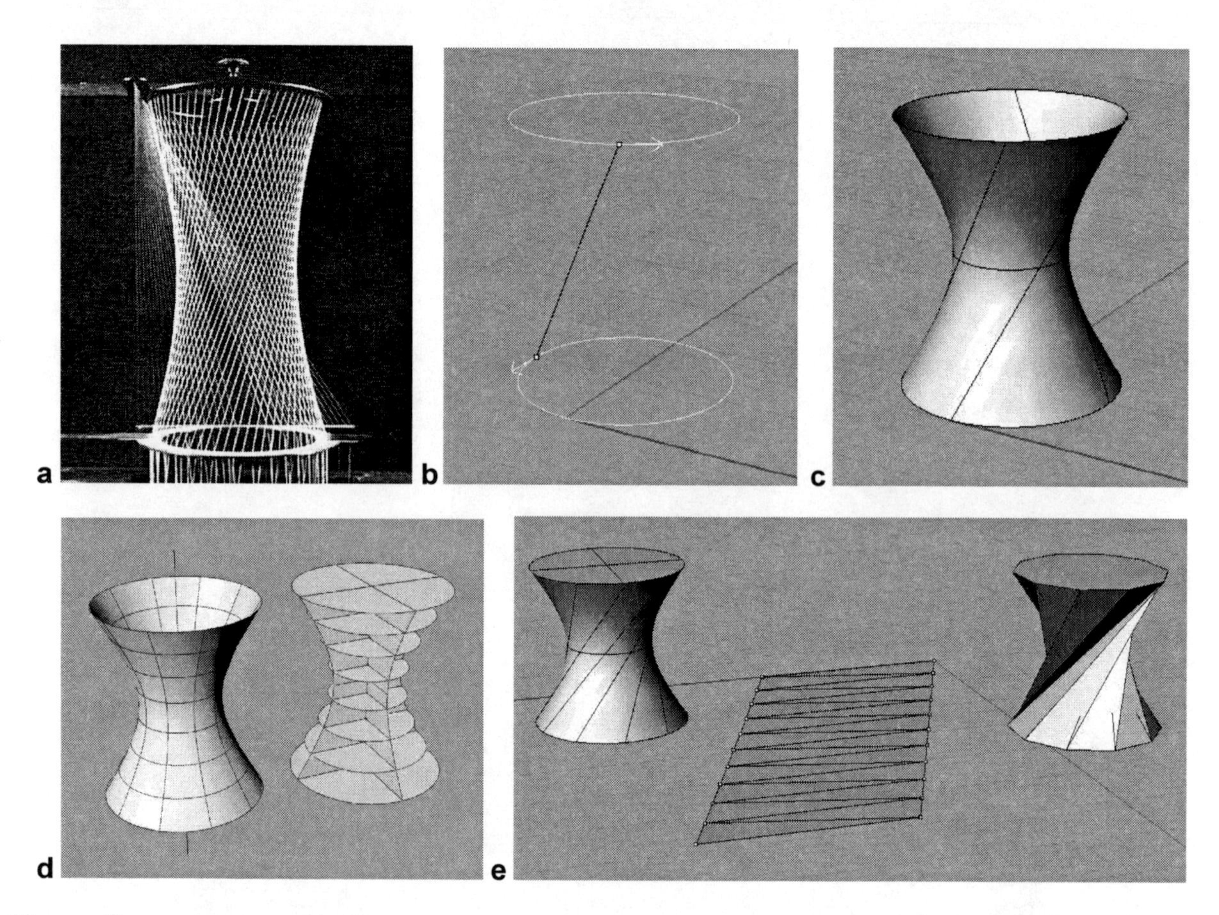

Figure 7a *String hyperboloid.* **b,c** *Ruled sweeping of the hyperboloid.* **d** *Sliced hyperboloid.* **e** *Folded hyperboloid.*

Much as the grain of a chosen material affects the crafting of objects from that material, so the fabric of the geometry used to craft a model determines much of the disposition of the model. According to the program's build geometry, the ruled version of the hyperboloid can unroll into a pattern, while the revolved version cannot unroll. This can mislead, since a prompt from the program also cautions that the unrolled surface pictured is 25% larger in area than the original surface. In truth the hyperboloid will not unroll, however, the program treats any ruled surface as developable.

The program recognizes that the revolved surface possesses double curvature: the hyperbola begins as a curve and revolves in a circle, yielding the second curve. This can be replicated through a structure of cross-sections (Figure 7d). The cross-sectional structure has the added benefit of providing an extremely strong armature for a sculpture. Crafted in thin plywood and filled with conventional auto body repair putty, a sculpture of great durability and permanence results.

The folded version (Figure 7e) is in fact a rotationally skewed, concave anti-prism. Its pattern was crafted by dividing the "unrolled" surface into right triangles. The 25% overage of the unrolling process disappears into the concave folds and the convex edges follow the actual surface of the hyperboloid.

Minimal Surfaces

Splines were thin strips of cedar or metal that early draftsmen/shipwrights could flex into a variety of non-circular and streamlined curves for tracing onto their design drawings. The spline would curve according to the natural effect of stress on the material. When mathematicians developed spline geometry they assumed that the spline behaved like an extremely thin beam and adapted Euler's formula for a bending beam to describe spline curves. The spline surface is described by a net of such curves that behave remarkably like sheets of highly flexible material under stress.

Spline surfaces can often take on characteristics of minimal surfaces, responding somewhat like soap films or shrink wrap when applied to an armature of curves. The top sequence of screenshots in Figure 8 illustrates the editing of a circle into an armature for the edges of a monkey saddle surface. Instructing the program to build a patch inside the curve will cause it to seek the most efficient surface configuration within the program's parameters. This will sometimes yield a strong approximation of a minimum surface, as is the case with the monkey surface, or it may yield the simplest executable construction.

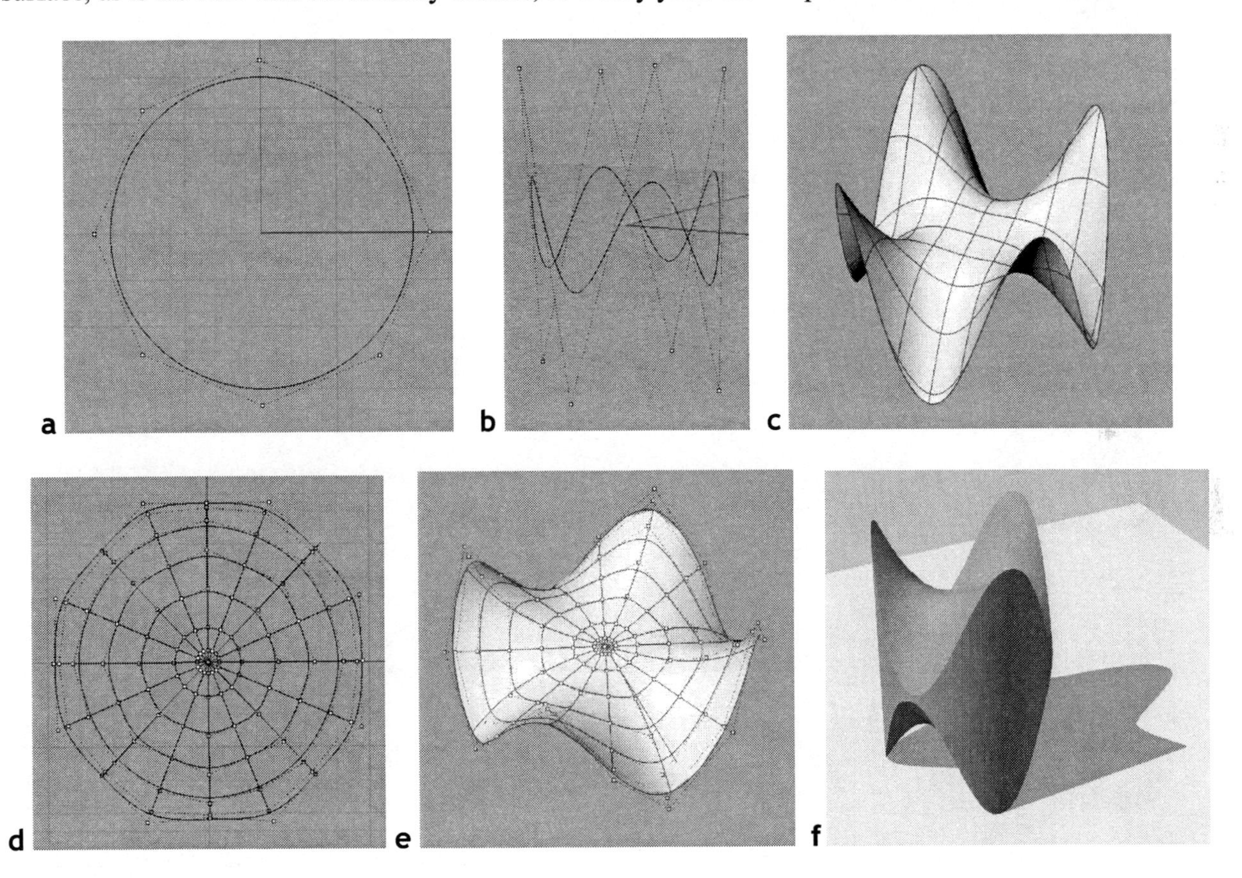

Figure 8a,b *editing a curve armature for the monkey saddle.* **c** *patch surface for monkey saddle.*
d.e *monkey saddle edited from a revolved disk* **e** *composite of cylinder and saddles*

The alternative construction from a revolved disk has its net aligned in a radial pattern from a center singularity. This version of the surface is stronger as a visual build than as a mathematical approximation. It results from a controlled hand edit of surface points rather than a solely computational execution. Its radial grain, however, is more in keeping with the symmetry of the surface and offers the opportunity to edit along the lines of that symmetry. The surface below is an edit of this surface with the center drawn out and radial veining added.

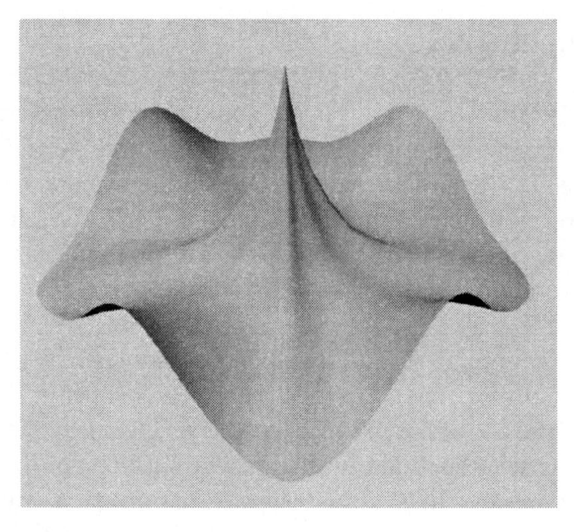
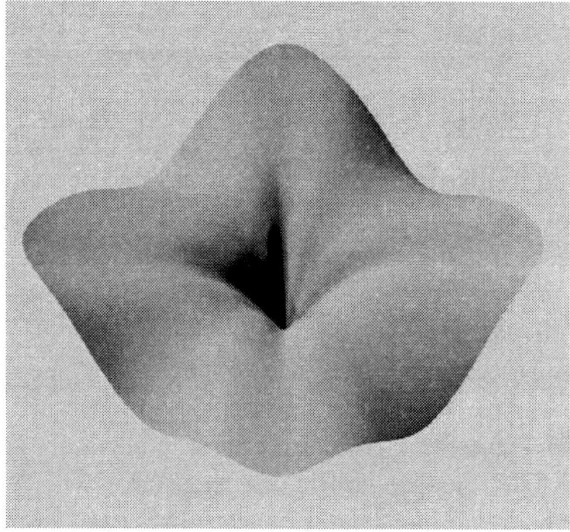

Figure 9 *floral surface from an edit of the radial monkey saddle.*

The hyperbolic paraboloid is a minimal surface that, like the hyperboloid, a modeling program can generate by sweeping a straight line. Such ruled surfaces, which can be modeled with taut string, can also be built at architectural scales with stressed steel cable. During the 1950's architectural engineers began applying this method to form roof shells of pre-stressed reinforced concrete – roofs whose double curvature provided strengths similar to domed roofs.

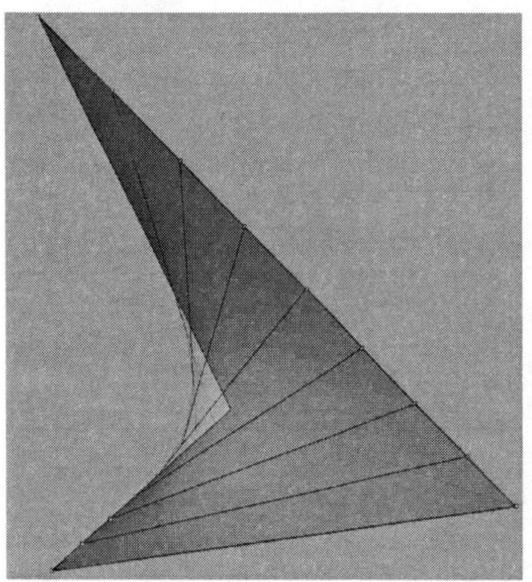

Figure 10 *hyperbolic paraboloid from swept line*

Figure 11 *Félix Candela's Iglesia de San José Obrero, Mexico.*

Shell Surfaces

As judged from their natural counterparts, shell surfaces are an especially beautiful category of form. The surface is essentially a circle swept along a spiral, flat or helical. In modeling programs, the spiral options are typically limited to Archimedean spirals, whereas natural shells configure more closely to equiangular, logarithmic, spirals. These spirals are also termed growth spirals; they possess scalar symmetry, in which equal radial divisions of the spiral increase by a fixed ratio.

Fortunately, growth spirals are easily constructed from a set of regularly-spaced radials and a connected series of perpendiculars drawn on those lines as in Figure 12 below. By increasing or decreasing the number of radials, the spiral will follow a more open or more closed path. The series of perpendiculars will interpolate into the spiral curve that closely approximates the growth pattern of natural shells. Constructing the actual shell surface is more difficult in practice than in theory. In almost all standard programs, attempts to sweep a circle along the spiral are frequently thwarted, yielding unpredictable effects.

A good approach is to "flow" a conical surface along the spiral (Figure 14). This cone should have an altitude equal to the length of the curve and should be rebuilt into a net comprising a larger array of splines to permit greater flexibility. Figure 13 illustrates rebuilding the cone's surface geometry: cone **A** is a cone as created by the modeling program and cone **B** is a rebuilding of that cone. The first cone is a true cone as reproduced by spline geometry; however, it has no control points to permit it to flex. A control point carries a mathematical weight, termed its *rho* value, that pulls the surface splines to curve toward that weight. Adding more control points permits finer curvature of the surface. Rebuilding adds more surface splines and therefore more control points. The rebuilt cone is no longer a true cone, but it is an admirably sufficient approximation.

A second approach is a little more complex, but offers more control. Create a triangular section of the cone. This shape will have only three edit points, one at each vertex and so it, too, requires rebuilding in order to smoothly flow along the spiral. The base edge of this triangular surface denotes the diameter and the side edges serve as two rails along which to sweep the circle. A two-railed sweep (Figure 15) is usually more stable and predictable than a sweep on a single rail. Figure 16 pairs shells to mimic the snail minimal surface. This surface is then capped by two hemispheres to create the sculpture in Figure 17.

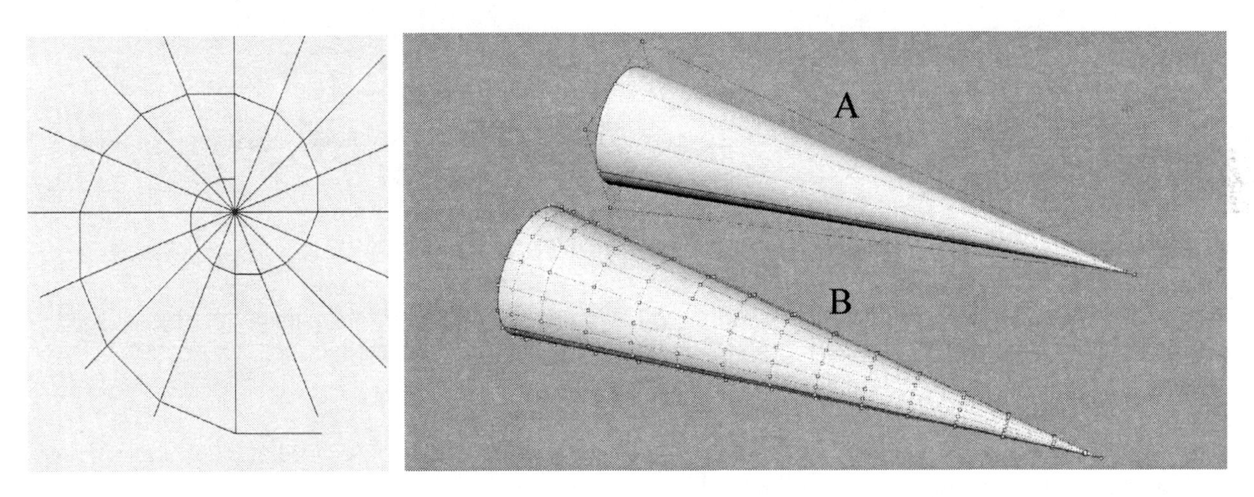

Figure 12 *Equiangular spiral.* **Figure 13a** *Original conical surface.* **b** *Cone rebuilt into editable surface.*

Virtual Sculpting

Most sculptors have an intuition for form, which in another life may have steered them into mathematics. Still, in this life, they have an appreciation for the elegance of mathematical surfaces that parallels that of the mathematician. As more and more sculptors integrate computer modeling into their tool set, the potential for them to create satisfactory replications of these surfaces becomes more and more accessible. By sculpting in the virtual materials of geometry, the intuition and craft of the sculptor can open up for them a new vocabulary of visual form.

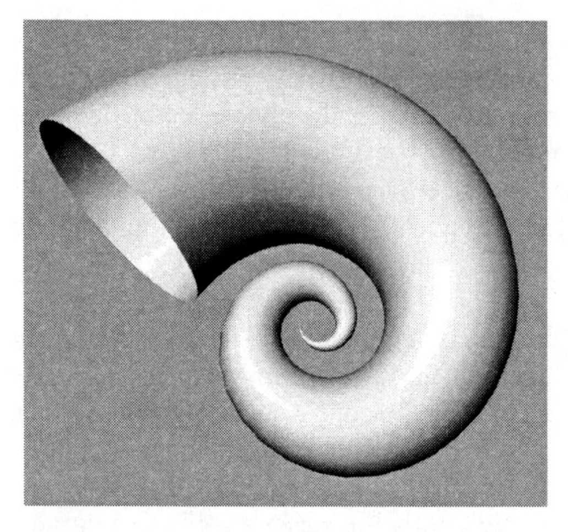

Figure 14 *Shell from flowed cone.*

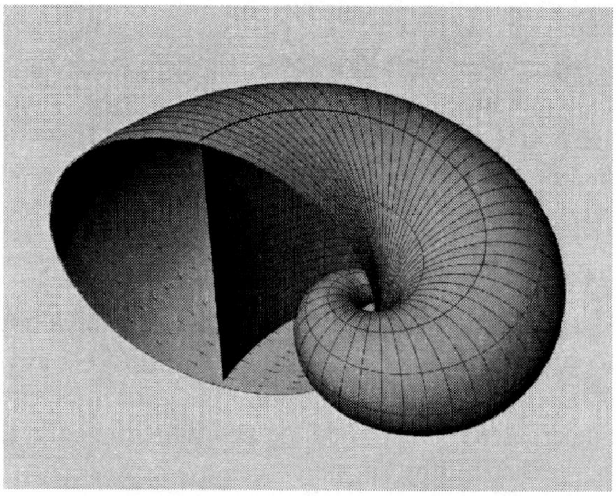

Figure 15 *Shell swept on two rails.*

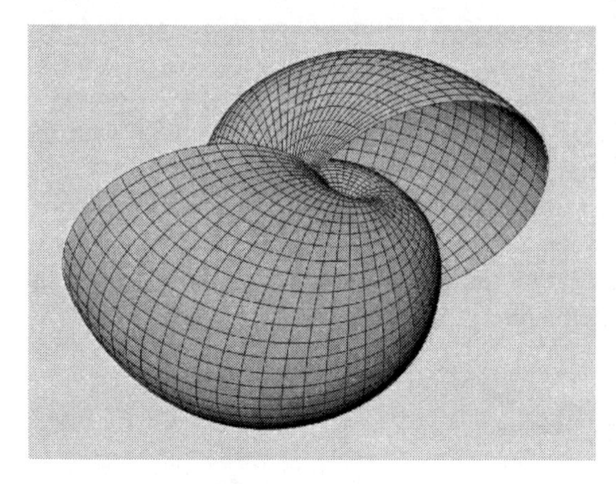

Figure 16 *Approximation of the snail surface.*

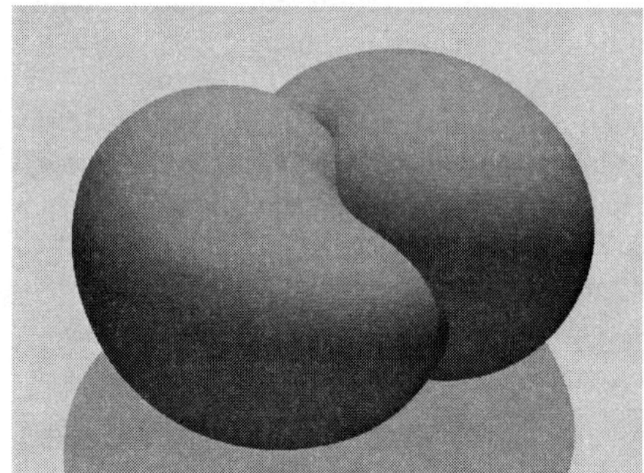

Figure 17 *Model for "Wrap", 2006.*

Bibliographic Notes

The software used to create the models is Rhino 3D, a spline modeling program developed by McNeel Associates founded by Robert McNeel. Mathematica, mentioned in the introduction, is a graphic software product of Wolfram Reasearch, Inc. founded by Stephen Wolfram.

The baseball in Figure 1, which bears Joe DiMaggio's autograph, can be purchased at www.sportsartifacts.com/ autographs.html.

The copyright on the image Naum Gabo's "Spheric Theme" (Figure 2) from an exhibition at the Tate Gallery in London is held by the © Courtauld Institute of Art.

The images of the string model of the hyperboloid (Figure 7a) and that of the hyperbolic paraboloid (Figure 11) are reproduced from a small, but fine, article by William C. Stone on Theodore Oliver's models held in the collection of Union College in Schnectady, NY and reproduced with their permission. The image of Félix Candela's Iglesia de San José Obrero is also reproduced by permission of Union College. http://www.union.edu/Academics/Special/Olivier/stone.php

The Lost Harmonic Law of the Bible

Jay Kappraff
New Jersey Institute of Technology
Newark, NJ 07102
Email: kappraff@verizon.net

Abstract

The ethnomusicologist Ernest McClain has shown that metaphors based on the musical scale appear throughout the great sacred and philosophical works of the ancient world. This paper will present an introduction to McClain's harmonic system and how it sheds light on the Old Testament.

1. Introduction

Forty years ago the ethnomusicologist Ernest McClain began to study musical metaphors that appeared in the great sacred and philosophical works of the ancient world. These included the Rg Veda, the dialogues of Plato, and most recently, the Old and New Testaments. I have described his harmonic system and referred to many of his papers and books in my book, *Beyond Measure* (World Scientific; 2001). Apart from its value in providing new meaning to ancient texts, McClain's harmonic analysis provides valuable insight into musical theory and mathematics both ancient and modern.

2. Musical Fundamentals

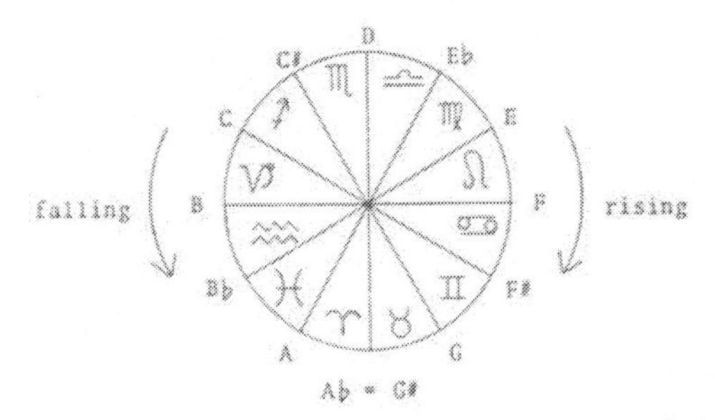

Figure 1. *Tone circle as a Single-wheeled Chariot of the Sun (Rg Veda)*

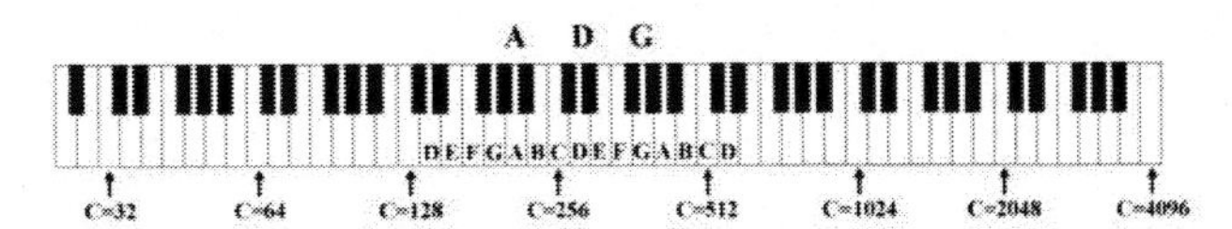

Figure 2. *The piano has 88 keys spanning seven octaves and twelve musical fifths.*

The chromatic musical scale has twelve tones, or semitone intervals, which may be pictured on the face of a clock or along the zodiac referred to in the Rg Veda as the "Single-wheeled Chariot of the Sun." shown in Fig. 1, with the fundamental tone placed atop the tone circle and associated in ancient sacred texts with "Deity." The tones are denoted by the first seven letters of the alphabet augmented and diminished by and sharps (♯) and flats (♭). For reasons that will become evident, we will choose D as the fundamental although the musical scale is indifferent to which tone is taken. If the fundamental is assigned a relative frequency of 1 unit then proceeding clockwise around the tone circle one arrives at the identical tone one octave higher with a relative frequency of 2. A revolution in the counterclockwise direction results in a tone one octave lower at a relative frequency of ½. It is the miracle of music that if the ratio of frequencies of two tones is a multiple of 2, the tones are perceived by the ear to be identical. As a result each tone will be considered to be a member of a pitch class of tones differing in frequency by a multiple of 2. The norm in the Old Testament is to use a double octave; i.e., 1:2::2:4, which also corresponds to Hebrews gematria for "Eden," 124, where the Hebrew letters were given numerical values. The chromatic scale in Fig. 1 equally divides the tone circle and as a result is referred to as an equal-tempered scale. Each tone is given a value of 100 cents on a logarithmic scale with 1200 cents to the octave. At the bottom of the tone circle, at a relative frequency of $\sqrt{2}$ is the tritone, the most dissonant interval of the chromatic scale.

Relative to the fundamental, the interval of the musical fifth (DEFGA) is found at 7 o'clock spanning 7 semitones. Its complement in the octave, 5 semitones, is called the musical fourth (DEFG) which arrives at the same pitch class as the falling musical fifth. The interval of the major third has a relative frequency of $\sqrt[3]{2}$ and is found at 4 o'clock. Its complement is the minor 6th at 8 o'clock, the position of the falling major 3rd. The remaining interval of importance to Western music is the minor 3rd found at 3 o'clock and its complement at 9 o'clock, the major 6th. The structure of music is built around chords consisting of the fundamental, major 3rd, and 5th known as the major triad and the fundamental, minor 3rd and 5th , the minor triad. The unison (1:1 ratio), octave, 4th, 5th, major and minor 3rds, and major and minor 6ths are the only consonant intervals of the chromatic scale.

The piano has 88 keys spanning seven octaves and 12 fifths as shown in Fig. 2 where, the piano, beginning with any tone, after 12 musical fifths one ends on the same tone seven octaves higher as shown in Fig. 3. Seven consecutive tones on the white keys gives rise to the heptatonic scale. A double octave is shown in Fig. 2 beginning on D; the double octave is the basic unit of the Bible. McClain refers to this mode as being the "menorah model" since it suggests the seven-branched candlestick referred to in Exodus 25:31-40 and found today in all synagogues. However, there are seven different modes of this scale depending on the choice of fundamental. The famous do re mi ... scale begins on C. This leaves the blacks positioned to sound the pentatonic patterns by themselves.

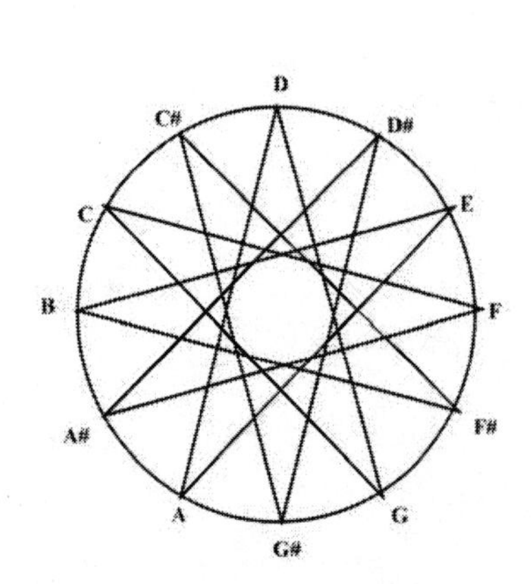

Figure 3. *Circle of fifths*

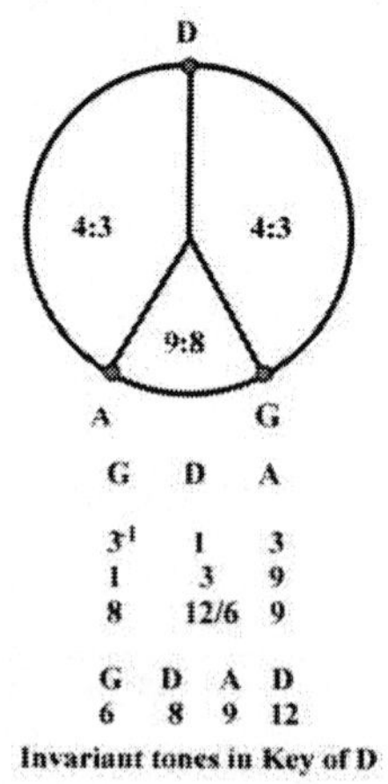

Figure 4. *Rising and falling fifths as twin tones.*

In the ancient world, the musical scale was defined by ratio of string lengths. Beginning with a length of string, placing the bridge at the midpoint of the string and plucking the remaining ½, one gets a tone one octave higher. Moving the bridge to a position where 2/3 of the string is plucked results in to a fifth, ¾ is a fourth. These were the principle intervals of Greek harmonic law making up the tetrachord of Pythagoras immortalized by Raphael in his painting Birth of Athens. The Just scale also uses 4/5 a major 3rd, and 5/6 a minor 3rd. Increasing the length of the strings by the inverse of these ratios give rise to the same intervals but in a falling direction. Therefore, 5/4 is a falling major 3rd and raising it an octave gives rise to 5/8, its complement the minor 6th. Likewise 6/5 is a falling minor 3rd while its complement is the major 6th a ratio of 3/5. The inverse of the relative string length is the relative frequency. A summary of the consonant intervals of Western music is given as:

(interval, No. of semitones (s), ratio): (unison, 0s, 1/1), (minor 3rd, 3s, 6/5), (major 3rd, 4s, 5/4), (fourth, 5s, 4/3), (fifth, 7s, 3/2), (minor 6th, 8s, 8/5), (major 6th, 9s, 5/3), (octave, 12s, 2/1)

When the tones are defined by rational numbers the tone circle is no longer equally divided. . On a monochord, 12 successive fifths no longer results in a tone in the same pitch class but to a slightly altered tone differing by approximately one-quarter of a semitone (24 cents), called the Pythagorean comma. The Pythagorean comma is a rich source of metaphor in ancient texts.

3. Twin tones

Consider Deity at D and the twin tones A and G shown in Fig. 4 serving Plato as harmonic and arithmetic means in the octave and achieved by rising and falling fifths. In Bible mythology, these symmetrical tones represent either twins or archangels found in the Bible. We will associate it in Sec.8 with Isaac's twin sons, Jacob and Esau. If D is given the value of 1 then A and G, being rising and falling musical fifths are 3/2 and 2/3 respectively, where we are taking the ratios to mean relative frequency (the inverse of string length), and D is the geometric mean of A and G. Notice that the octave interval subdivides into two musical fourths (5 semitones each) and a wholetone (two semitones). This also demonstrates the most fundamental principal of harmonic law, that intervals add while frequency ratios multiply, i.e., $5+2+5= 12$ while $4/3 \times 9/8 \times 4/3 = 2$ also $4/3 \times 9/8 = 3/2$ or $5 + 2 = 7$.

The integer 2 was considered to be the female number which can give birth to no new tones without the participation of the male number 3. Since tones differing by a multiple of 2 are in the same pitch class and are considered to be identical, multiples of 2 are suppressed and this triple of tones is represented by, 3^{-1}, 1, 3 as shown in Fig. 4. The second row of numbers in Fig. 4 is derived by multiplying the first row by the common denominator 3. Since A is the largest integer at 9, the fundamental D is multiplied by the smallest power of 2 to enclose 9. In other words an octave 12/6 is created between D and D' an octave higher that encloses 9. The tones G and D must be multiplied by powers of 2 to "seal" them in the 12/6 octave. The integers 8 and 9 are the harmonic and arithmetic means, respectively, of 6 and 12. In Epinomis (991) Plato states that "in the potency of the mean between these terms (6,12) with its double sense, we have a gift from the blessed choir of Muses to which mankind owes the boon of the play of consonance and measure, with all they contribute to rhythm and melody."

Opposite Deity is the dreaded tritone, "diabolus in musica" and represents the worst possible dissonance, but more important, it betrays a fundamental asymmetry in the middle of the tone circle as we shall see in the Sec.5. Its relative frequency is represented by $\sqrt{2}$ on the equal-tempered scale. Without the concept of irrational number, ancient cultures found ways to accurately approximate $\sqrt{2}$ by rational numbers. It is from the portion of the tone circle surrounding the tritone that, in metaphoric terms, the savior will be born.

Fig. 4 has been hypothesized by McClain to represent the heroic Jewish figure, David, since David's name in Hebrew gematria is 464 where 4:6 and 6:4 are increasing and decreasing fifths. McClain suggests that David's sling which he so deftly used to kill Goliath is inverted in Fig. 4.

4. Pentatonic scale

The pentatonic scale shown in Fig 5 is generated from its center at D by two successive rising and falling musical fifths as follows:

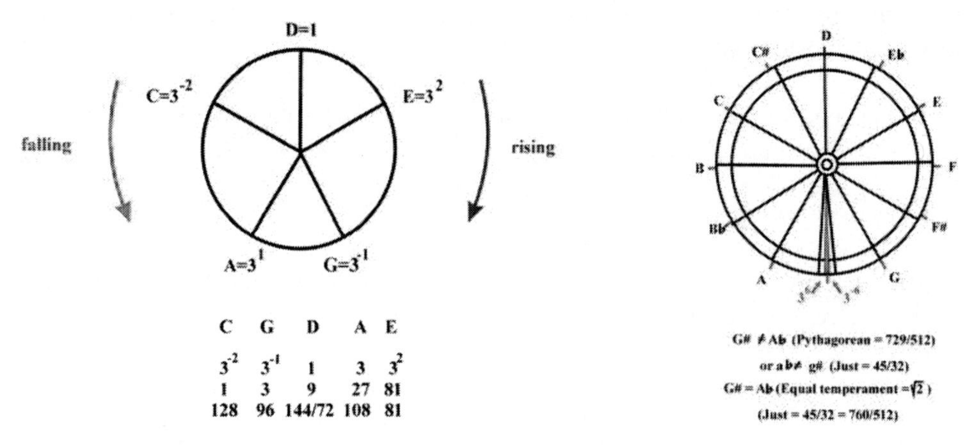

Figure 5. *Pentatonic scale*

Figure 6. *The Pythagorean comma*

Again, the first row of numbers is multiplied by the common denominator 9 to get the second row after which the terms are multiplied by powers of 2 to place them in the 144/72 octave which seals the set symmetrically with DEGACD as the scale order. Notice that the tones of the pentatonic scale occupy symmetric positions around the tone circle. Symmetric scales result from the choice of D as fundamental. Semitone intervals are dissonant or harsh to the ear. Since there are no semitone intervals in the pentatonic scale, any combination of tones has a pleasing sound. The pentatonic double octave is: 72:144::144:288 where 288 represents David's " 288 trained singers, sons of Asaph, Jeduthun, and Heman" (1 Chron. 25:1-8).

A similar analysis can be made for the heptatonic scale with three rising and falling fifths.

5. Pythagorean Comma

Six rising and falling fifths result in two representations of the tritone leaving a small gap between them, the Pythagorean comma, enclosing the equal-tempered value at $\sqrt{2}$ as shown in Fig. 6. With D as the fundamental, these rising and falling fifths are symmetrically placed around the tone circle and make up 11 of the 12 tones of the "Pythagorean scale." The last pair produces an asymmetric 13th tone in the middle of the tone circle which is the source of many metaphors in the Bible and other ancient texts. If D is the fundamental, then the pair of tritone approximations are G ♯ and A_b, two tones that are identical on the piano. These three tones are represented by integers as:

Figure 6a	Gsharp	D	Aflat
	3^{-6}	1	3^{6}
	1	729	$531441 = 3^{12}$

If "1" in the second row is multiplied by $2^{19} = 524288$ with a tail of David's 288 "trained singers", the ratio 531441: 524288 is the Pythagorean comma and represents the amount by which twelve musical fifths deviate from seven octaves ($2^7 \approx (\frac{3}{2})^{12}$ or $2^{19} \approx 3^{12}$). Clearly, no power of 2 can equal a power of 3. The savior will be born in this small gap between G ♯ and A_b. This gives rise to a wealth of sexual imagery in the Rg Veda in which this region is pictured as a "vagina" while the wedge to the center of the circle is a sepah or "penis." The whole construction arises from an interplay of "male" and "female" in which the commas that arise are a kind of "genital friction."

The savior is born in this womb. In fact, notice that G sharp at "1" can be made to represent Deity by multiplying it by 512 (nine powers of 2) and then switching places with 729 at D so that 512 is the fundamental and 729 now represents the tritone as shown in Fig. 6a. But $512 = 8^3$ while $729 = 3^6$ where

729/512 = 1.4238, slightly greater than $\sqrt{2}$. Is it too much to suggest, as McClain does, that $8^3 \equiv 888$, the number assigned the savior, in Biblical times Isaac and in later times Jesus Christ, while $3^6 \equiv 666$ is the "devil's number?" Pure doublings point to a misunderstood metaphor for "Israel" as YHWH's wife. Since the male number 3 is responsible for the small overlap between the first and thirteenth tones, McClain hypothesizes that this is the origin of the law requiring circumcision of males.

The twelfth tone of the chromatic scale is represented by the six digit number 531441. Clearly the ear can only perceive the first two digits. McClain feels that the "1" in the tail of the number represents the "forgotten cornerstone" while 1440 is the gematria for Adam. The tone circle from D to D' is made up of approximately 53 Pythagorean commas.

6. The Hebrew octave

Old Testament arithmetic is based on the numbers 1,2,3,4,5,6,7. In Genesis the first six days are used for the work of constructing the world while the seventh day is reserved for God. To construct tunings using musical fifths, only the numbers 1,2,3,4 and 6 were used, but we saw in the last section that the chromatic scale requires six digit numbers. There are references in the Bible to the "Anakim" or "giants" from whom the Chosen win the "Holy Land"(Deut. 1:28) with another tuning represented by smaller numbers. These Anakim correspond to the huge numbers and slightly excessive spiral fifths. In Greek mythology, 5 is the number assigned to humans; humans are "fivers." This number will reduce the numerosity of the chromatic scale from six digits to a mere two or three digits.

In the Bible, God's number seven is introduced since the Jews are God's people. The Bible octave for the "gestatation of the Savior" (Rev. 11-12) is:

$$35:49::50:70$$

where $49/35 = 70/50 = 7/5 \approx \sqrt{2}$, the simplest approximation of the square root of 2 using small integers. Therefore 49 and 50 enclose the tritone and all approximations to $\sqrt{2}$ occur within this tolerance on either side of the tritone. For example, the Pythagorean comma is included within this interval. But "seven Sabbaths of years" comprising 49 years constitutes the Jewish calendar cycle while the 50th year is the "Jubilee year" in which the land lies fallow (Lev. 25). The distance between 49 and 50 on the tone circle (34 cents) is 1% on either side of the tritone prompting McClain to associate this with New Testament references to a "tolerant God" in Matthew 18:12-14 and repeated in Luke 15:3-6. "Suppose a man has 100 sheep. If one of them strays, does he not leave the other 99 on the hillside and go in search of the one strayed." All approximations to the tritone occur within the 49-50 tolerance. This small errancy permits alternative metaphoric interpretations. It also points to God as a "craftsman god" as opposed to a god "scientific precision."

The number 70 is one of the most frequent numbers in the bible and generally reserved for God. At three places in the Bible 70 men are slain. For example, by the order of God, Ahab's 70 sons were killed, and their heads placed in baskets (Kings 2 10:1-7. In fact the Bible is the story of how the many gods of ancient civilizations were replaced by the "One God" and how a people, the Jews, were created to worship that God.

7. The Matrix

McClain's story of musical metaphor continues with the introduction of a matrix of integers based on multiples of the numbers 3 and 5. The bottom row of the matrix are powers of 3 while the rising edge are powers of 5. Elements within the matrix are multiples of both 3 and 5 as shown in Fig.7 beginning with 1 in the lower left-hand corner of the matrix and referred to by McClain as the "cornerstone" which will become the savior of the system.

The first twelve numbers of this matrix are: 1,3,5,9,15,25,27,45,75,81,125,135 which McClain assigns to The twelve sons of Jacob. Notice that the first four: 1,3,5,9, being single digit
are the sons of the "weak-eyed Leah," Jacob's first wife, while 125 is

Joseph's number and 135 belongs to Benjamin, the youngest son. The numbers 225 and 625 are assigned to Manasseh and Ephraim, the sons of Joseph. There are no more than twelve sons because Jacob dies at age 147.

125
25 75 225 625
5 15 45 135
1 3 9 27 81

Figure 7. *The matrix representing Jacob's family*

The 3,4,5- or 4,5,6-nature of the musical scale is revealed by the first triple of numbers 1,3,5 which is illustrated by

5   5
4 3   4 6

where the tone representing 1 has been inflated to 4 in the first triangle and the 3 is inflated to 6 in the second. Plato states in the Republic that "4,3 mated with 5, thrice increased, provides two harmonies (Republic 546a-d)" All the tones of the scale can be derived by a system of vectors from this 3,4,5-relation as shown in [4]. Consider the 4,5,6-triangle. It defines three intervals: 3:2, 5:4, and 6:5, a rising fifth, a rising major third, and rising minor third major third, the intervals that make up the major and minor triads of Western music.

Notice in Fig. 7 that each integer (frequency) in the matrix is the geometric mean of any pair of symmetrically placed integers, e.g., 15 is the geometric mean of 9 and 25, 3 and 75, and 1 and 225. As a result, the pairs may be thought of as inverses with respect to the central tone, e.g, 25/15 = 5/3 while 9/15 = 3/5. In this sense the matrix is able to define inverses in terms of integers.

8. The First Eight Sons of Jacob

Consider the first eight family members: 1,3,5,9,15,25,27,45. The next number in this sequence is 75, the age of Abram when he left Haran for the Holy Land (Gen. 12:5). Since every tribe presents to the tabernacle with "one silver plate whose worth was a hundred and thirty shekels" (Numb. 7), McClain suggests that Abram's tent was also worth 130 shekels, the sum of the first eight family members, each < < 60 in value.

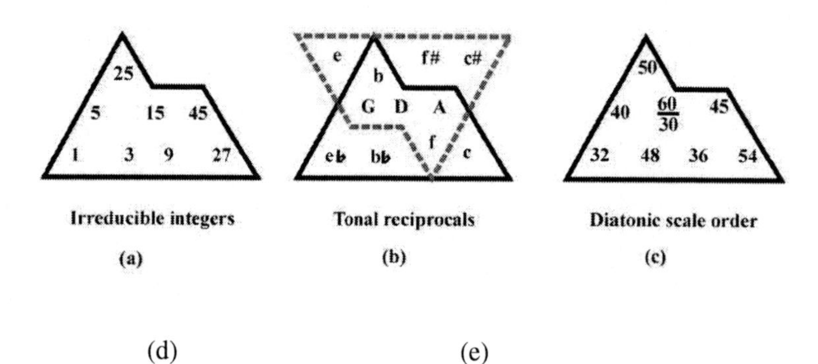

Figure 8. *The first eight members of Jacob's family create the "Davidic Tuning." The matrix in (a) is multiplied by powers of 2 into the 60/30 octave in (c) resulting in the tones in (b). The tones of (b) divide into a pair of plinths in (d) representing upward and down ward heptatonic scales. The tones fit within the Star of David.*

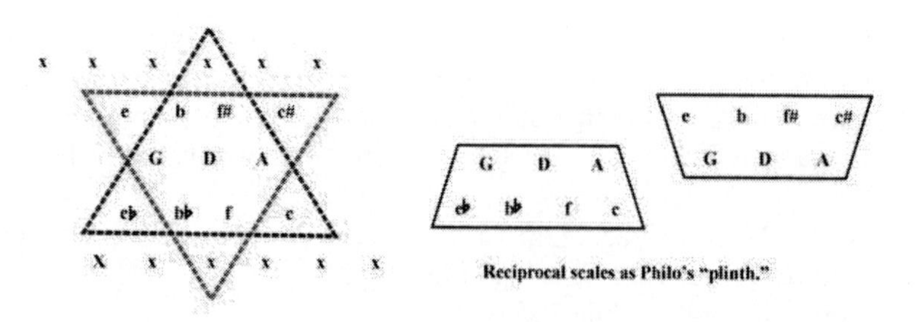

These eight matrix members are shown in Fig. 8a in a tent-like configuration suggestive to McClain of Abram's tent. In Fig.8c the numbers are multiplied by powers of 2 to place them in the

60/30 octave sealing 45, a computation shared in ancient Sumeria by base 60 sexagesimal arithmetic. Deity at D in the midst of the twin tones A and G are found along the central axis. In this matrix we see that, with respect to Deity, the same relationships hold as for the 4,5,6-triangle above. For example, 45:30 = 3:2 while 40:60 = 2:3. Likewise 50:30 = 5:3 while 36:60 = 3:5. Since (36, 50) and (40,45) are the only two symmetric pairs with respect to Deity in this matrix they have special importance. Also note that 50 is the arithmetic mean of 40 and 60 that brace it from below forming the major triad G, b, D, while 48 is the harmonic mean of 40 and 60 that brace it from above, forming the minor triad G, b flat, D. Similar relationships abound throughout the matrix.

The Hebrew double octave is: 30:60::60:120 where "the sons of Asaph, Jeduthun, and Heman—with cymbals, harps and lyres, stood east of the altar with 120 priests who were trumpeters. (2Chron. 5:12)" The age of 120 is the natural lifespan of humans set out in the Bible (Gen. 6:3) and was the lifetime of Moses. In Sumerian mythology, 60 is the god number of Anu, sky god and father of the Pantheon, 40 is Enki, the god of sweet water, and 50 is Enlil, the mountain god and active ruler. It is well known that the Sumerian Gilgamesh legend has many similarities to the story of Genesis transformed by the authors of the Old Testament to accommodate Jewish meanings.

In Fig. 8c, the relative frequencies of Fig. 8b are related to tones from what McClain refers to as the Davidic tuning for reasons that will soon become clear. Upper case letters are reserved for the tones defined using only the integers 2 and 3 as before. Lower case letters are the tones created by the integer 5. Tones from the circle of fifths lie on the central axis of the matrix and are often referred to as the "Pythagorean scale." The tones from the central axis and the two adjacent rows make up the "Just scale." Major triads of Western music appear in every upright triangle of nearest neighbors (above or below), and vice versa (e.g., GbD). Triangles of tones from the central axis and the row below results in minor triads (e.g., Gb_bD). There is a reference in the Gilgamesh legend to the central axis with its two neighboring rows as a "three-ply rope that is hard to break."

You will notice that McClain has introduced a pair of matrices that he refers to by the Hindu expression, "yantra." The upward yantra is the one in Figs. 8a and 8c while the downward yantra in Fig. 8b is turned at 180 deg. and its tones are "inverse" of the frequencies in Fig. 8c. All tones in the intersection of upward and downward yantras have inverses that can be expressed in terms of integers, i.e., each integer has a symmetric twin within the intersection. Although McClain has never found direct evidence for his yantras, the Hebrews' journey from Sinai to Edom with YHWH in the midst of "a pillar of cloud by day and in a pillar of fire at night (Numb. 14)" Could this refer to the downward yantra and upward yantras?

Fig. 8b is divided into a pair of upward and downward pointed plinths in Fig. 8d. These plinths comprise the rising and falling ancient Dorian scale with D as the fundamental having all the intervals of the do re mi...scale but in reverse order. The rising scales fit within the 60/30 octave and is represented as Davidic tunings since the tones fit easily into the star David as shown in Fig. 8e. As double octaves these tunings extend to 120 and 288 (the "120 trumpeters and 288 trained singers").

The mode of the Pentateuch and its symmetric opposite.

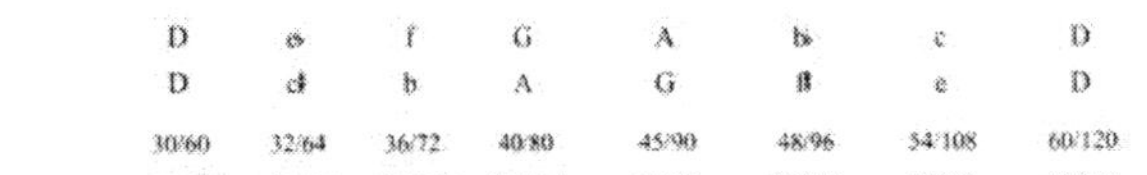

D	e♭	f	G	A	b♭	c	D
D	c♯	b	A	G	f♯	e	D
30/60	32/64	36/72	40/80	45/90	48/96	54/108	60/120
144/288	135/270	120/240	108/216	96/192	90/180	80/160	72/144

16/15 9/8 10/9 9/8 16/15 9/8 10/9

Figure 9. *The upward and downward heptatonic Dorian modes lie in the 60/30 and 144/72 octaves respectively. The double octaves have limits of 120 and 288, "David's singers" and the "priests who played trumpets."*

Along with the self-symmetric Phrygian mode, the Dorian scale comprises the two revered modes of ancient music.

McClain's matrices are very versatile and are used and reused to tell many stories. For example, consider 40 and 45 straddling Deity at 60/30. As mentioned above, 40 and 45 are the harmonic and arithmetic means of 30 and 60, respectively. They are also the twin tones mentioned in Sec. 3. Rebecca gave birth to the twins, Jacob and Esau when Isaac was 60 years old. Since Jacob was the younger, he gets the larger number 45 while Esau is 40. But according to the famous Bible story, Jacob stole Esau's birthright. This identity theft was necessary for Jacob, renamed Israel, and become the father of the Hebrews. We shall see in the next section that the number 45 is needed for this purpose. The number 15 from the matrix inflates to the Sumerian god number 60, symbolic of the "old" system. Abram at age 75 emerges from this yantra whose numbers sum to 130 as if the "triangle" and its shoulder (see Fig. 8b) were the family tent (McClain's metaphor). The number 15 will be replaced by the next number of the matrix 45 symbolic of the "new" system and representative of the patriarchs Isaac and Jacob. Abram will also make another appearance in this new yantra.

9. The Patriarch's Matrix

Consider Jacob's family matrix shown in Fig. 10. Using the same symmetry argument, there are 5 pairs of integers in symmetry around Deity at 45 (Jacob's number): (15,135), (27,75), (9,225), (3,625), and (25,81), the five pairs in inverse symmetry with respect to 45 in the family matrix of Fig. 6.

Figure 10. *The calendar matrix or "matrix of the Patriarchs" seals 11 of the 12 tones of the chromatic scale in the 720/360 octave with inverse symmetry. A hexagon of tones fit into "star of David."*

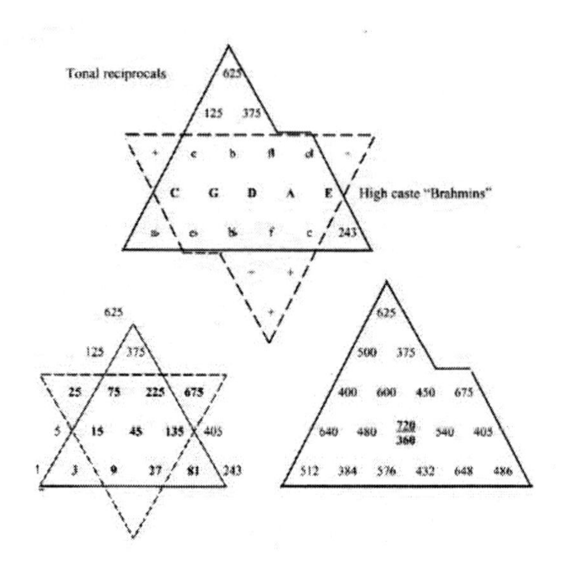

Jacob's number 45 is transformed as:
45 → 90 → 180 → 360 → 720 into the 720/360 octave which is the least common multiple of 72 and 60, the octave limits of the Davidic tunings of Fig. 9. Sarah was 90 when Isaac was born while Isaac died at 180. All twelve tones of Jacob's family matrix are sealed by the 180/90 octave. Within the 720/360 octave, eleven of the twelve tones are sealed with inverse symmetry as shown in Fig. 10b in the intersection of the upward and downward yantras and referred to by Plato in "Laws" as Poseiden's five twin sons. They are the eleven tones of the Just scale that can be expressed as the ratio of small integers, with the remaining tritone expressible as an awkward approximation to $\sqrt{2}$ under a "tolerant" Deity accustomed to such approximations. The base 60 system has now been transformed to a base 10 system inspiring the Hebrew importance of ten-ness. Notice that the pentatonic scale: CGDAE appears on the central axis defined again from Deity at the center. Tone pairs, such as c,C and e,E differ from each other by the ratio 81:80 (approximately 22 cents), known as the syntonic comma.

The hexagon in the midst of Fig 10b illustrates the rotational symmetry of the tone circle. The nor 3rd at f rotates to its complement, the major 6th at b; the major 3rd at f sharp rotates to its complement, the minor 6th at b flat; the fifth at A rotates to it complement, the 4th at G. Also the major triad (D, f ♯, A) of the rising scale maps to the major triad of the falling scale (D, b_b, G) while the minor triad of the rising scale (D, f, A) maps to the minor triad of the falling scale (D, b, G). Notice that hexagon as shown also in Fig. 8d and 10c with its Magen David triangles eliminates all elements lacking symmetric opposites. We see that the "glory of the Shekinah" in Fig 10c excludes Plato's syntonic commas at 400:405 and 640:648 to reveal the supreme elegance of YHWH's reductionism.

This matrix is considered to be the matrix of the "patriarchs" or the "calendar" matrix. In ancient cultures, 360 was considered to be the "canonical year" even though these societies were quite aware that this was not the length of either the solar or lunar years. In fact the ratio of 360 days to the lunar year of 354 days and the solar year of 365.25 days is remarkably close to the Pythagorean comma. The thirteen tones with inverse symmetry within the Patriarch's matrix and its inverse occupy symmetric positions around the tone circle shown in Fig. 11. Notice that the cornerstone at 512 makes a good candidate for the twelfth tone, G sharp, approximating the tritone at $\sqrt{2} \approx 720/512 = 45/32 = 1.406\ldots$ In a sense it "saves" the octave. In Sec. 5 we witnessed its birth in the region around the tritone .

10. The Marduk Matrix

Note in Sec. 8 and 9 that the six figure numerosity of the chromatic scale (see Sec. 5) has been reduced to a limit of 60/30 or 720/360. In Fig. 12 McClain introduces a matrix, the Marduk/Elohim matrix, with the ferocious numerosity, 8,640,000,000/4,320,000,000 which he attributes to Marduk (c.2000 BC) the head of the Babylonian Pantheon, whose only invariances under rotation are the "seven-headed dragon" as menorah scale, and it contains the Patriarch's yantra surrounding Deity. It is 15 rows high in perfect inverse symmetry on its

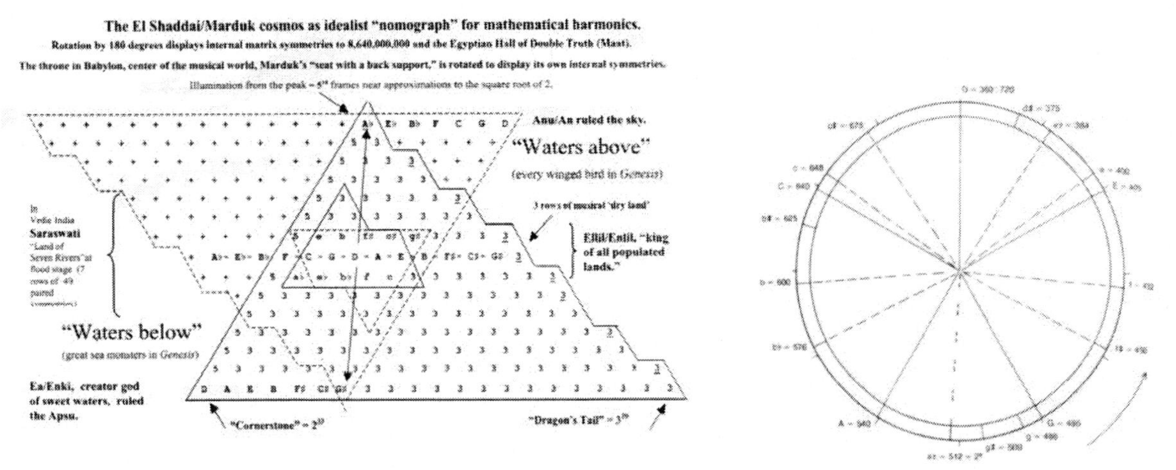

Figure 11. The calendar matrix defines six pairs of tones with inverse symmetry with respect to the fundamental D.
Figure 12. The Marduk/Elohim matrix.

central axis. The ratio of the highest and lowest tones, A flat and G sharp, to the fundamental give three decimal point approximations to $\sqrt{2}$ with the gap between A flat and G sharp now a mere 4 cents. McClain associates this matrix with the Biblical "flood" which rose 15 cubits in depth (Gen. 7:21). The Patriarch's matrix has been lifted seven rows in the air. Seven rows of this matrix have exactly seven counters, sharing reciprocity, reflecting the "seven sabbaths of years, that is seven times seven years" described in Sec 6. Notice the rows of 21 tones along the bottom of the matrix and 21 tones facing them along the top row of the inverted yantra. In Babylonian mythology, the dragon of the deep, Tiamat, has 20 dragon offspring. But the matrix is also important in Hindu mythology where one thousand Maha Yugas consitute a Kalpa of 4,320,000,000 "years;"4,320,000,000 and its double is Brahma - the Immense being - a cycle of 8,640,000,000 years. A Hebrew myth describes the digging of the Temple foundation to a depth of fifteen "cubits" and fifteen psalms are labeled "a song of ascents." In Egyptian mythology, fourteen steps lead upwards to the throne of Osiris—while forty-two judges look on from either side in the Hall of Double Truth: occupying 21 seats along opposite rows as in our matrix.

11. The YHWH Matrix

The enormous numerosity of the Marduk matrix can be reduced to exactly 9% of its value while at the same time accommodating more tones with inverse symmetry including the eleven tones of the octave in perfect inverse symmetry along the central axis shown in Fig. 13.

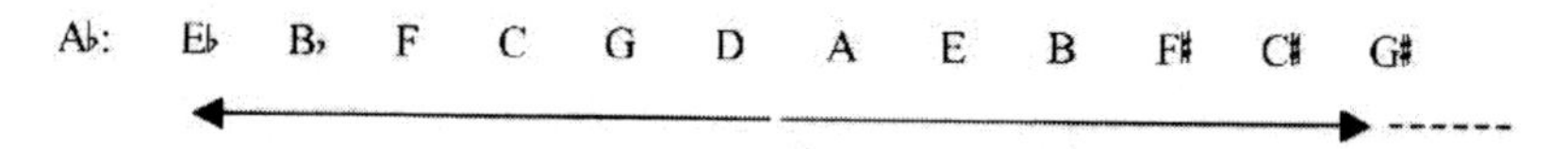

Figure 13. *The central axis of the YHWH matrix.*
Five tones are in inverse symmetry about the fundamental D (between the arrows). Twelve tones are sealed by 777,600,000 (outer brackets). The 1st and 13th tones, A flat and G sharp differ by the Pythagorean comma.

Consider the gematria value of YHWH which is: 10.5.6.5. McClain interprets this as $10^5 \times 6^5 =$ 777,600,000 and assigned to D (Deity); God's number 7 is repeated thrice while 600,000 are the number of bricklayers who fled Egypt with Moses. Fig. 13 shows God surrounded by his minion of ten men along the central axis in perfect inverse symmetry (inner brackets) with 12 tones sealed in the octave (outer brackets) and all thirteen tones along the axis. The three ply rope now extends to the entire circle of fifths and now includes major and minor triads in each of the twelve keys. When placed within a single octave the twelve tones form a "serpent tuning" in which the 1st and 13th tones differ only by the Pythagorean comma suggestive of the "ouribus" figure in which a snake eats its own tail. According to Isaiah 27: "On that day the Lord with his cruel and great and strong sword will punish Leviathan the fleeing serpent, Leviathan the twisting serpent and he will kill the dragon that is in the sea." McClain feels that all serpent myths depict this "serpent tuning" shown in Fig. 14, and that the 21 elements in the figure are an extension of the foundation in Fig. 4 ad account for the 21 elements in the base of the Marduk matrix in Fig. 11 and inspire the 42 gods of the Egyptians.

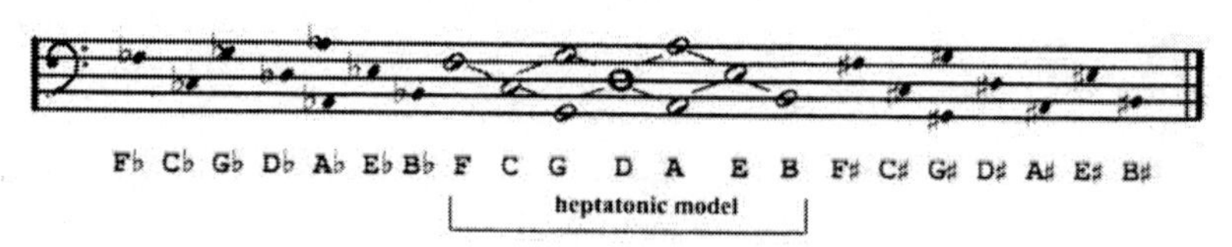

Figure 14. *The serpent tuning. All tones of the chromatic scale are places in scale order. A flat and G sharp differ by the Pythagorean comma giving rise to the metaphor of the "snake eating its own tail."*

Finally, the number 243, the last number in the bottom row of Fig. 10a, is the gematria sum of the letters of Abram, the father of the Jewish people before his name is changed to Abraham. But 243 x 32= 7776, the lead digits of YHWH. In Numbers 31, 32,000 virgins, with its lead digits 32, are among the spoils in the genocidal victory of the Israelites over the Midianite Kingdom. We see here Abraham lifted to the throne of God in order to identify God with his "people." A bloody and senseless episode of the Bible is shown in its true colors to be a brilliant piece of story-telling in the service of creating God's people.

12. Conclusion

This paper is meant as an introduction to Ernest McClain's Bible studies. McClain has produced impressive evidence to show that a lost knowledge of harmonics was used to produce numerous metaphors in the Bible with each metaphor having multiple meanings. An understanding of these hidden connections gives greater meaning to this great book while placing it squarely in the context of other ancient civilizations that also drew upon these ideas.

New ways in symmetry

María Francisca Blanco Martín[1] y Elena Elvira Nieto[2]
[1] Universidad de Valladolid, ESPAÑA
[2] Universidad del Nordeste, ARGENTINA
Colaborators: Raúl Capretini, Pablo Estévez
E-mail: fblanco@maf.uva.es, elenanieto14@yahoo.com

Abstract

This proposal presents the continuation of the task assumed some years ago by this interdisciplinary research team about the relations between Mathematics and Design.

The basic objectives in this proposal are:

1. To research about the syntactic, generative and methodological possibilities of mathematical models and fundamentally, geometric structures, as a base for the morphologycal definition of the objects, in their widest significance.
2. To study the transference of these knowledges to the educational level, through the implementation of learning situations that imply not only to offer the model, but also the ways of manipulation, extracting from it all its compositive possibilities. The idea is to establish a work methodology that can be applied to different situations, moving the students to be involved in each possible stage of the search.
3. To develop a systemic approach that allows the use of different informatical programs to promote creative development of students in the teaching- learning tasks.

Introduction

In this paper we present a synthesis of the work carried out on the possibilities of basic figures partitions arisen from their own geometric structure, using as a "design tool" the symmetry in connection to other such as, *tesselations*, *crystallographic groups*, and *regular divisions of the plane*, in order to develop different periodic mosaics.

About the teaching, we adopt a constructivist methodology: to present a design problem (containing something of game, something that implies creativity) that the student should solve making use of mathematical models in a quasi intuitive way, to explain rigorously the scientific justification, in each case.

As a conclusion, we believe it is possible to demonstrate the validity of this educational work methodology, in which the student is the main actor and can be motivated to inquire into the ways that geometry and symmetry offer, penetrating in the dialectical game of its rules, discovering relationships and generating forms in order to incentivate and potenciate the creative act.

1. Development

This work was previously started in the universities involved by putting into practice some guidelines related to the implementation of a body of learning situations which allow the student to develop from a mathematic model, different morphologic interpretations tending towards a deeper conceptualization of the design problems, both for the Architecture and Graphic Design courses of studies.

Our purpose is to complete this proposal in order to enable its realization, always under assessment, in the different courses involved. This implies the systematization of the integrated theoretical contents for the courses of Mathematics and Morphology of the Design which are ordered according to the increasing complexity of the theoretical and practical developments.

Since the current computer programmes allow different degrees of formalization in the graphic systems, it is our purpose to make the best of the possibilities offered, including the animation as the incorporation of movement, used from the perspective of the didactic strategies as well as the expression of the generation of the bi and tri dimensional form.

The first part of this work consisted in relating the morphologic operations of subtraction and transposition in the generation of the emerging forms to the geometric transformations corresponding to a given figure, mathematically formalizing such operations.

Analysing the possibilities of the square, as one of the regular polygons that tiling the plane, we chose it as the initial figure and, from its axis of symmetry, we performed the various partitions generated by the two axis of reflection. (**[3]**)

2. Partitions of the square

The set of movements of the plane that leave the square fixed, its group of symmetry, is a dihedral group of order 4, D_4 which is a Leonardo's group or group of rosettes:

$$D_4=\{ s_1, s_2, s_3, s_2, g_{O,\pi/2}, g_{O,\pi}, g_{O,3\pi/2}, g_{O,2\pi} \}$$

In which s_1, s_2, s_3, s_4, are the reflections about the lines M_1, M_2, M_3, y M_4, y $g_{O,\square/2}$, $g_{O,\pi}$, $g_{O,3\pi/2}$, the rotations of centre in the centre of the square and angles $\pi/2$, π, $3\pi/2$ y 2π.. A set of generators of this group is formed by $\{s_3, s_2\}$.

From the axis of symmetry of the square M_1, M_2, M_3 and M_4 we proved that it has only four different partitions, those generated by M_1 and M_2 (I); M_1y M_3 (II); M_4 y M_2 (III) y M_1 y M_4 (IV). (Fig. 1)

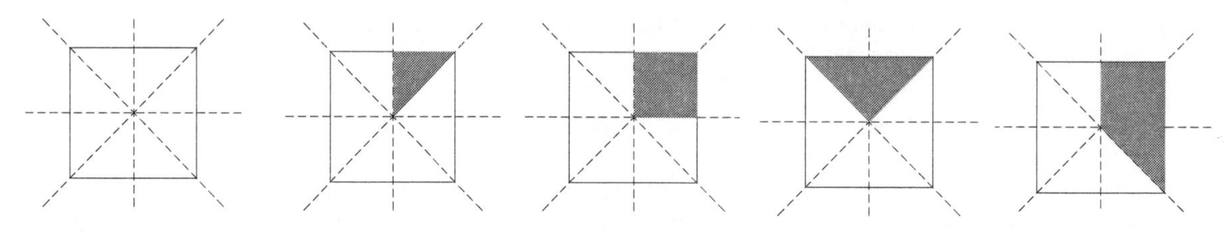

Figure 1

The ordered application of the morphological operations of subtraction and transposition to each of this partitions results in emerging figures that have the same area as the square but a different perimeter.

These morphological operations, geometrically correspond to the application of a movement to the parts taken from the square in such a way that some vertex of the partition coincide with some vertex of the square, having a side in common.

3. Emerging Figures

Each of the partitions generates the following non- isomorphic figures:

a) The partition I generates 14 figures: In (3) we studied this partition generating 14 emerging figures that tile the plane (except for I.3) and classifying the corresponding crystallographic group.
b) The partition II generates only 3 non- isomorphic figures (Fig. 2)
c) The partition III generates only 2 new different figures (Fig.3)
d) The partition IV generates 24 new different figures, which study will be the object of a forthcoming paper.

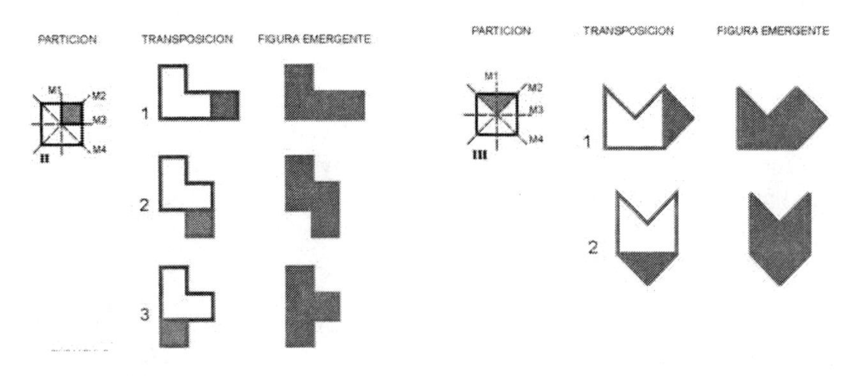

Figure 2 **Figure 3**

4. Geometric Generation of the Emerging Figures

We consider $\{O; \mathbf{u}, \mathbf{v}\}$, $\mathbf{u} = (1,0) = OQ$ y $\mathbf{v} = (0,1) = ON$, as a system of reference, bound to the square.

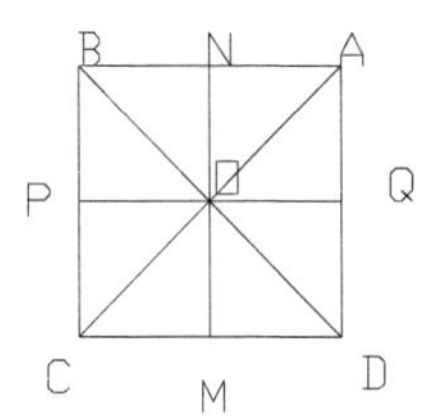

The emerging figures corresponding to the partition II (Fig.2) are obtained from the following movements:

1: Rotation of centre Q and angle π
2: Translation of vector $-2\mathbf{v}$
3: Rotation of centre P and angle $3\pi/2$

The emerging figures corresponding to the partition III (Fig.3) are obtained from the following transformation:

1: Rotation of centre A and angle $-\pi/2$
2: Translation of vector $-2\mathbf{v}$

5. Regular division of the plane

5.1.–The following step of this work consisted in taking each of the figures that are called emerging figures and verifying what type of movements of the plane (rotation, translation, reflection or reflection with displacement) is required by each of them to cover the plane by repetition and which ones allow that possibility.

When studying the partition II, we found that the three generated figures have the capacity of tiling the plane by using operations of symmetry

II.1 The tiling of II.1 is a crystallographic group type p1, the tile has no symmetry. The group is generated by the translations of the vectors $\mathbf{a} = \mathbf{u} + \mathbf{v}$ (diagonal of the square) and $\mathbf{b} = 4\mathbf{u}$. (Fig. 6).

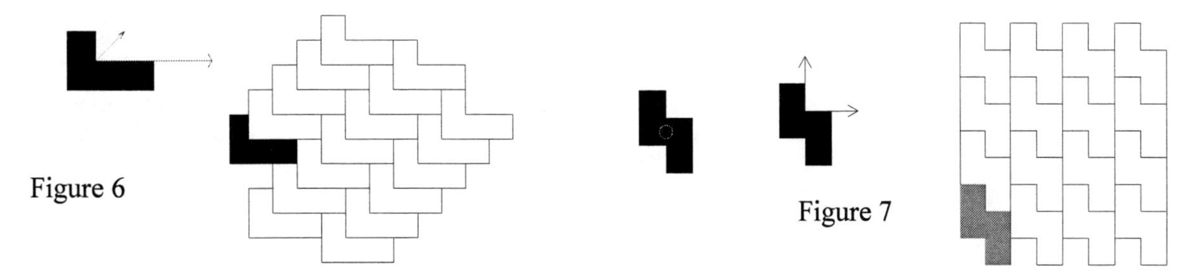

Figure 6

Figure 7

II.2. The tile corresponding to this partition is invariable by a rotation of angle 180° and centre of the midpoint of OM.
This partition gives rise to different tilings:

The tilings of figure 7 is a crystallographic group type p2, generated by the rotation previously mentioned and the translations of the vectors: **a** = 2**u** y **b** = 2**v.**

The tiling of the figure 8 is a group of type p2: generated by the rotation of angle 180° and centre of the midpoint OQ and the translations of the vectors: **a** = 2**u** y **b** = 2**v.**

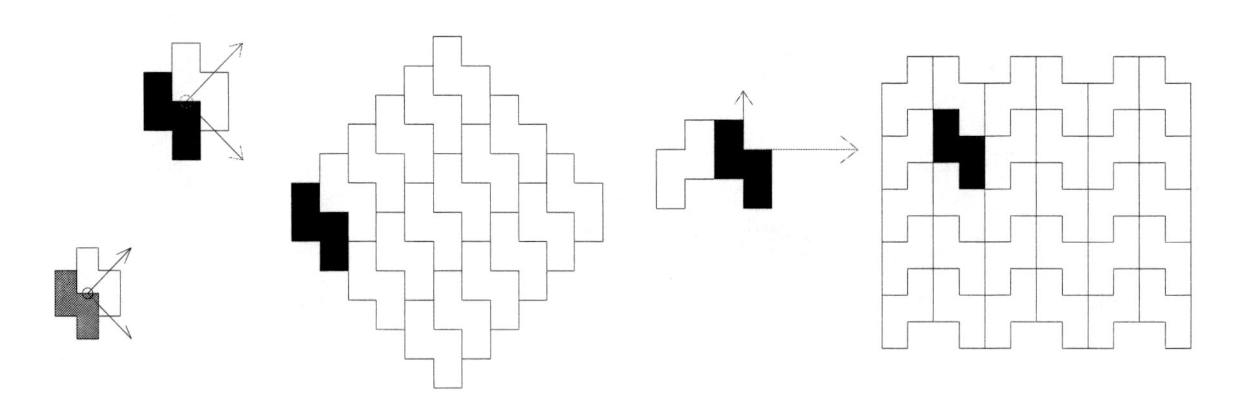

Figure 8 **Figure 9**

In the tiling of the figure 9, there is a symmetry with respect to the axis BC, (which) it is a group of type pmg: generated by the rotation of 180° and centre of the midpoint of OQ, the symmetry of the axis BC and the translations of vectors : **a** = 2**v** and **b** = 4**u.**

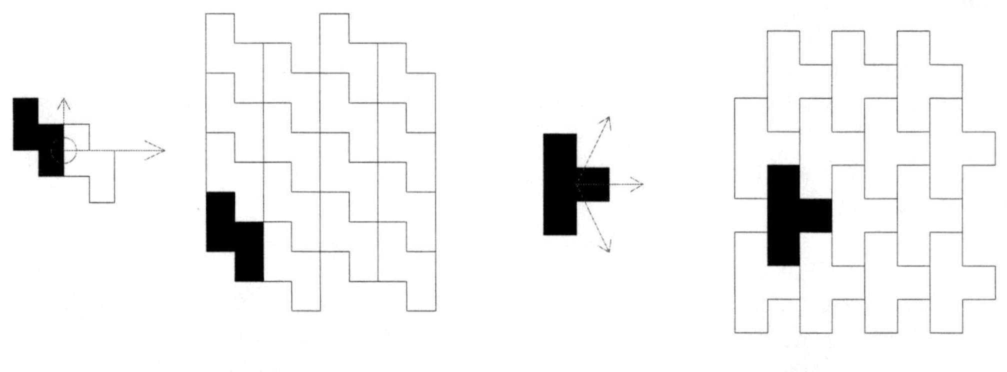

Figure 10 **Figure 11**

The tiling of the figure 10 is a group p2, generated by the rotation of centre D and angle π and the translations of vectors: **a** = 2**v** y **b** = 4**u.**

II.3 The tile corresponding to this partition is symmetric respecting to the straight line *r* that joins the midpoints of CP y DQ. Two different tiling with this tile are shown as follows.

The tiling of figure 11 is a crystallographic group of type cm, generated by the symmetry previously indicated and the translations of vectors: **a** = 2**u** y **b** = **u** + 2**v.**

The tiling of figure 12 is a group of type p4: generated by the rotation of the centre Q and order 4 and the translations of the vectors: **a** = 4**u** and **b** = 4**v.**

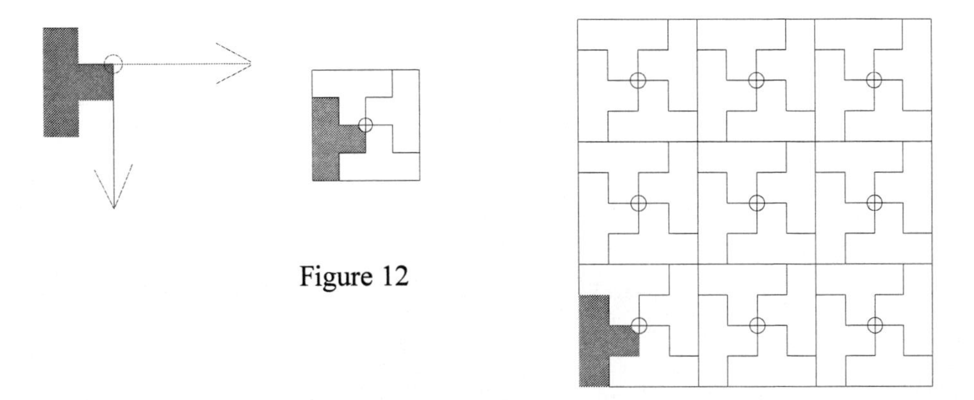

Figure 12

Partition III

The partition III gives rise to two emerging figures, the second one with an axis of symmetry.
III.1 The partition III.1 tiles the plane, figure 13, forming a crystallographic group of type p4, generated by a rotation of order 4 and centre shown in the figure and translations of vectors: **a** = 4**u** and **b** = 4**v.**

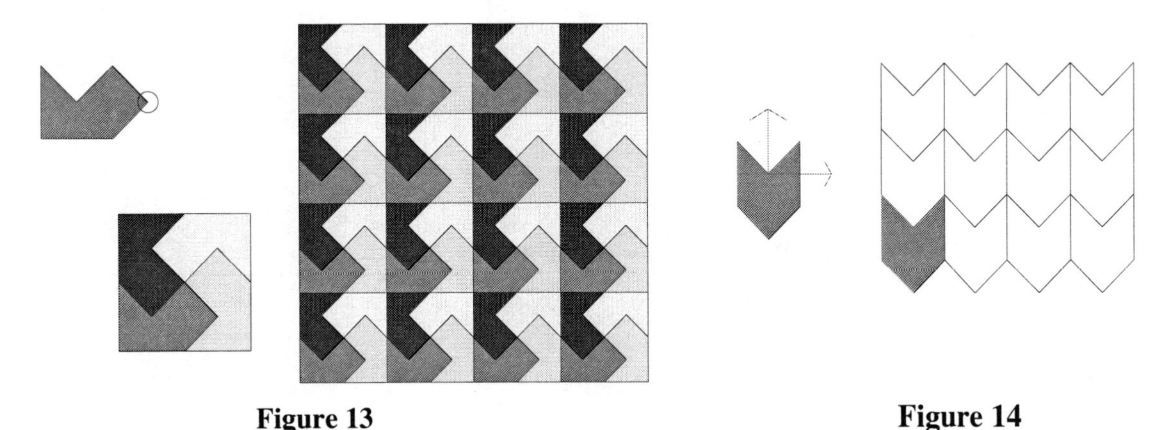

Figure 13 **Figure 14**

III.2 The emerging figure of this partition has a reflection of axis, the straight line MN.
The tiling of the plane is a crystallographic group of type pm, generated by the symmetry with respect to the axis MN and the translations of the vectors: **a** = 2**u** and **b** = 2**v.** (Fig. 14).

6 Conclusion

As it was already mentioned, we are presenting another stage of an ongoing research work. Considering a probable extension of this work, we submit it for its validity as a didactic tool before finishing each stage. At this respect and taking into account the various formal possibilities that these geometrical structures generate, as well as the inclusion of the colour as a problem to be worked out, we are presenting only two of the results obtained by the students of Graphic Design, in the Universidad Nacional del Nordeste, Argentina.(Figure 15) In this case, the students worked with different regular figures, as the hexagon, tiling the plane and using different schemes of colour in the grid generated using the same methodology as for the square. As we already said, this is a proposal to invite you to follow this new way of the symmetry to enjoy the design.

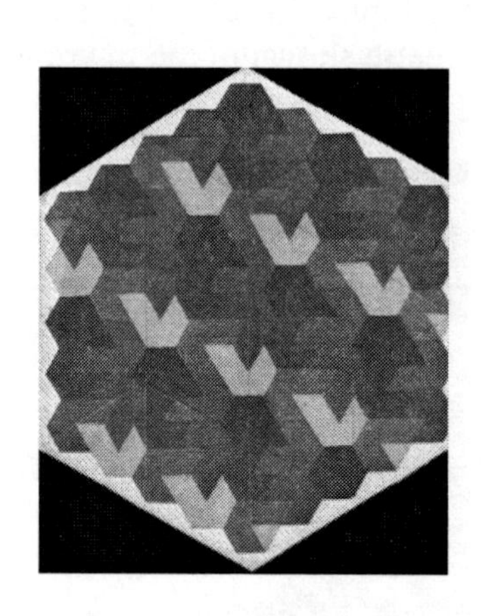

Figure 15

References

[1] BLANCO, MA. FRANCISCA: *Movimientos y Simetrías*. Ed. Publicaciones de la Universidad de Valladolid, Valladolid, 1994. ISBN 84-7762-413-5

[2] BLANCO, Mª. F. NIETO, E.: *Symmetry: A tool of Design. Mathematics & Design 2001*, Deakin University. Geelong, Australia, ISBN 0 7300 2526 8, pp 39-45.pp: 39-45.

[3] BLANCO, Mª. F. NIETO, E.: *The way of the symmetry. Mathematics & Design 2004*. ISBN 950-29-0823-6 Journal of Mathematics & Design. Volume 4, Nº 1, ISSN 1515-7881

[4] COSTA GONZÁLEZ, ANTONIO: *Arabescos y geometría.* (vídeo) CEMAV. Vicerrectorado de Metodología, Medios y Tecnología. UNED, 1995.

[5] COXETER, H. S. M.: *Fundamentos de geometría*. Ed. Limusa, Méjico, 1984.

[6] GHYKA, MATILA: *The Geometry of Art and Life*. Ed. Dover Publications, N. York, 1977.

[7] HAMBRIDGE, JAY: *The Elements of Dynamic Symmetry*. Dover Publications, Inc.- New York, 1967.

[8] KAPPRAFF, JAY: *Connections. The geometric bridge between art and science*. Mc. Graw –Hill, Inc. 1991.

[9] PEDOE, D.: *La Geometría en el Arte*. Ed. G. Gili, 1979

Linkages to Op-Art

John Sharp
20 The Glebe
Watford, Herts England, WD25 0LR
E-mail: sliceforms@compuserve.com

Abstract

Many artists using mathematical curves to generate lines in their work use Lissajous figures or cycloids. There are many other curves which can be drawn "mechanically" and linkages do not appear to have been used in an obvious way. In my op-art period many years ago, I used a simple linkage and I have resurrected this to create some new ideas following a particular interest in the lemniscate.

Introduction

The mechanical generation of curves has been a feature of many forms of art. Harmonographs, based on swinging pendulums were invented in the nineteenth century and computer versions have been used by Bob Brill in his work on Lissajous figures [1]. There were many lathe or gearing mechanisms which also developed in the nineteenth century giving rise to Rose Turning machines which are essentially cycloidal mechanisms. More complex versions of these were used to engrave plates for banknotes, although this is now done using computer programs. My work, described here, is based on a simple three bar linkage. The mechanics of such systems were developed to control machinery in the Industrial Revolution. A. B. Kempe's book of 1877 [2] on how to draw a straight line is the classic book, and mathematicians like Sylvester and Cayley were also bitten by the linkage bug. Sylvester's notable contribution was the pantograph for enlarging drawings. Kempe is also famous for his "proof" of the four colour theorem. The straight line of the title is important since it enabled linear motion to be turned into rotary motion and vice versa. James Watt's linkage of 1784 was fundamental to obtaining motion from beam engines.

The lemniscate from a linkage

The simplest linkage might be said to be a bar compass. One point of the bar rotates about a pivot and a pen is placed at another position on the bar. Kempe [2 p 4-5] describes how to build up complexity:

> Turning to that apparatus, we notice that all that is requisite to draw with accuracy a circle of any given radius is to have the distance between the pivot and the tracer properly determined, and if I pivot a second "piece" to the fixed surface at a second point having a tracer as the first piece has, by properly determining the distance between the second tracer and pivot, I can describe a second circle whose radius bears any proportion I please to that of the first circle. Now, removing the tracers, let me pivot a third piece to these two radial pieces, as I may call them, at the points where the tracers were, and let me fix a tracer at any point on this third or *traversing* piece. You will see that if the radial pieces were big enough the tracer would describe circles or portions of circles on though they are in motion, with the same ease and accuracy as in the case of the simple circle-drawing apparatus; the tracer will not however describe a circle on the fixed surface, but a complicated curve.

His figure is shown in figure 1. However, it is simpler to show linkages diagrammatically in as figure 2. The two end pivots or pins are fixed and the others can rotate. The freedom of the motion is limited and each point on the bars or links trace a curve. The linkage in figure 2 is sometimes called a three-bar and sometimes a four-bar linkage. The fourth bar is the fixed bar between the outer pins shown here. The central point of the middle bar traces a figure of eight curve known as a lemniscate when the distance between the end pins and the lengths are prescribed. If the short bars are of length √2, the long bar is 2, as is the distance between the foci, then the linkage generates a curve known as Bernouilli's lemniscate [3].

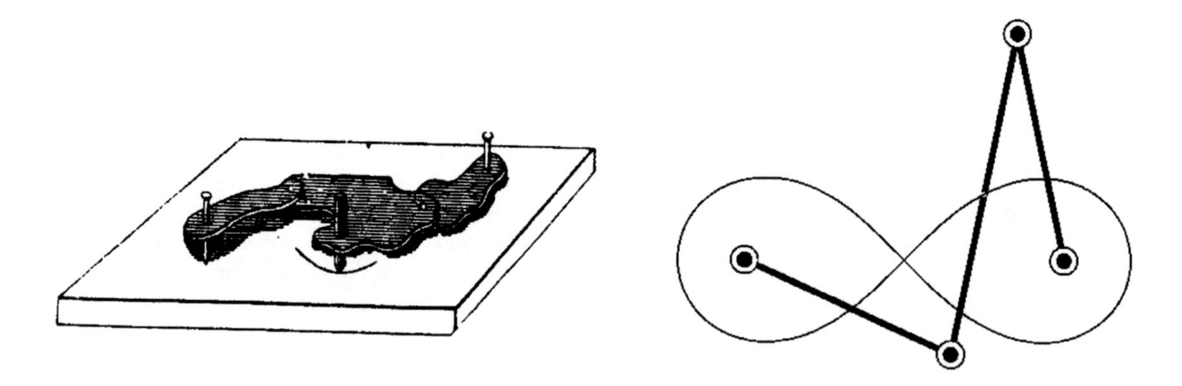

Figure 1, Kempe's figure

Figure 2, an "instrument" for drawing a lemniscate

The linkage is one of the simplest to make and a cardboard model is shown in figure 3. The bars of the linkage are three pieces of cardboard two of the same length and a longer central one with a hole in its centre for a pencil point. The pins or pivots are made using paper fasteners. The fixed, end pivots are drawing pins. Note how the lemniscate is incomplete. This is because the pencil hits the bars.

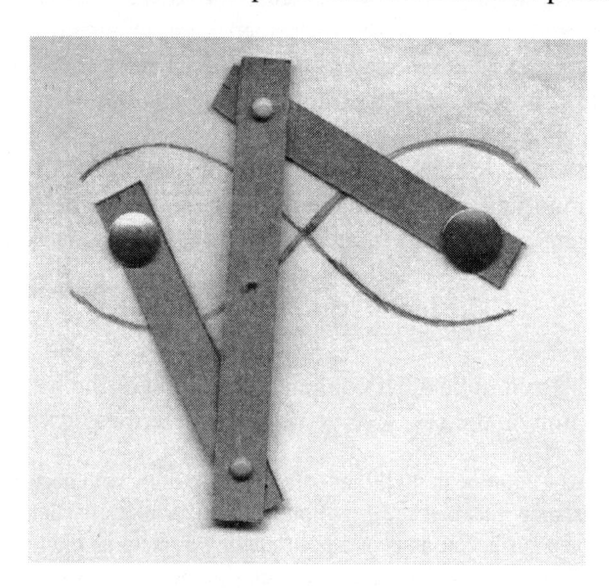

Figure 3, model of the lemniscate linkage

You can have a lot of fun by seeing what happens when you vary the distance between the positions of the fixed points. Wider distances than those for the standard lemniscate allow you to complete the curve. Figure 4 shows some examples of what happens. You could also change the relative lengths of the bars and obtain other curves or you could explore the family of curves that results from moving the pencil point along the central bar.

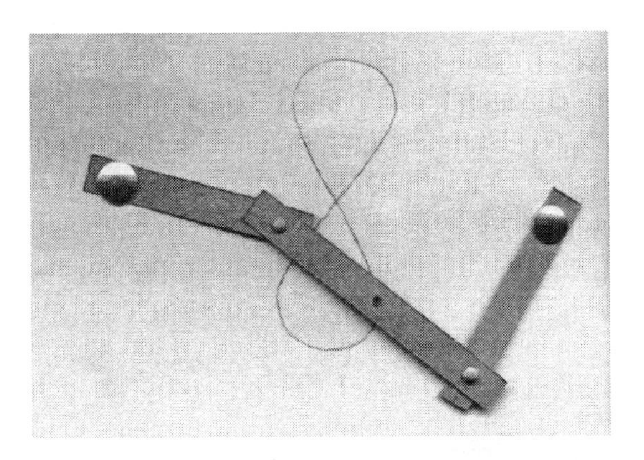

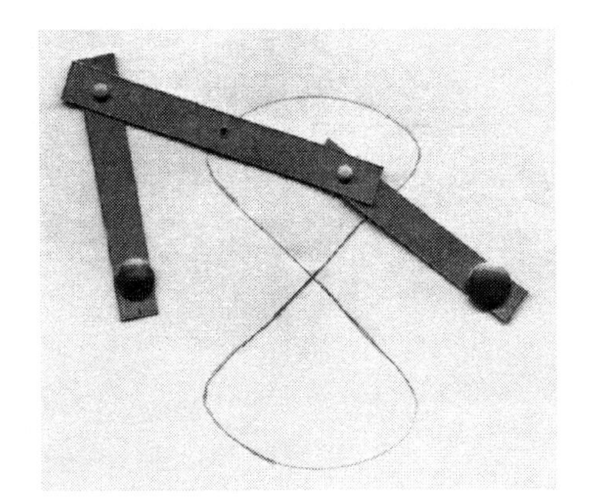

Figure 4, results of varying the fixed points

Some 25 years ago this is how I drew the curves for op-art pieces. It is now easier to work on the computer, plotting the equation for the lemniscate which in polar terms is:

$$r^2 = a^2 - [c\sin\theta \pm\sqrt{b^2 - c^2\cos^2(\theta)}]^2$$

where the origin is in the centre, the length of the two short bars is a, the central bar $2b$ and the distance between the outer pins is $2c$. Figure 7 shows the curves for various values of a, b and c turned into op-art.

Dynamic Geometry output

With the appearance of dynamic geometry packages like Geometer's Sketchpad and Cinderella [4], it is possible to create the curves as loci by constructing the linkage in the computer. The latter is very good for producing output that can be imported into drawing packages. Using these packages has the added benefit that you can vary the positions of various parameters and construct the locus in real time. This gives much more of an artist's control compared with manipulating equations.

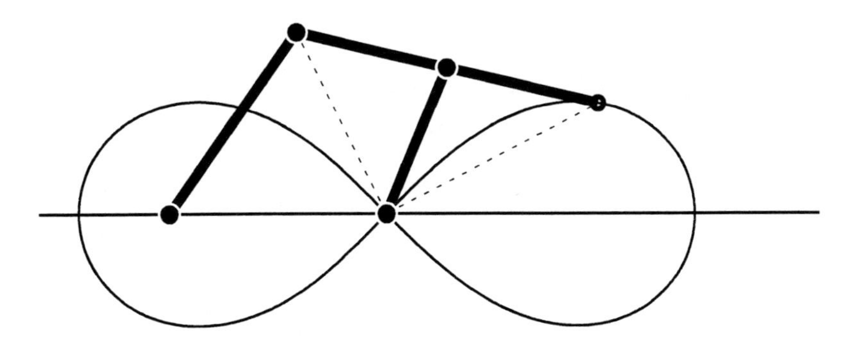

Figure 5, another lemniscate linkage

An alternative linkage for the lemniscate is shown in figure 5. A set of curves can be created by taking different positions on the top bar so that the lemniscate is transformed (figure 6). A set of further variations can be achieved by erecting a triangle on the top bar (figure 7). The results were used to create figure 9.

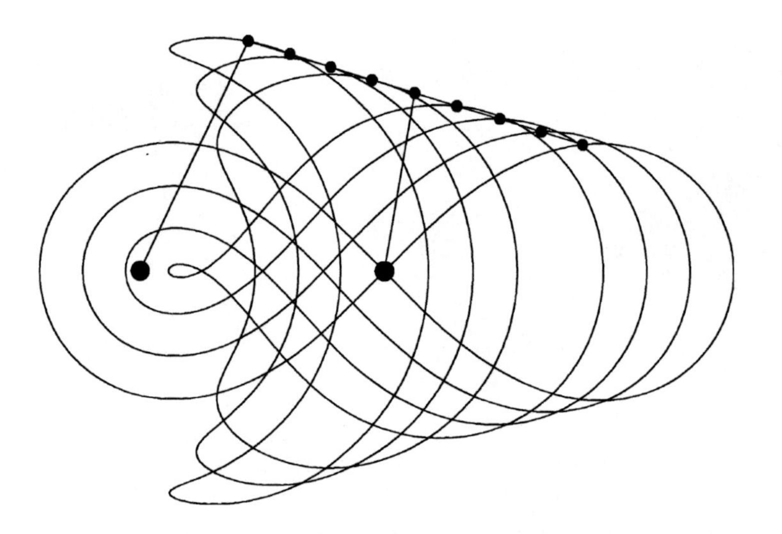

Figure 6, loci of different points on the top bar

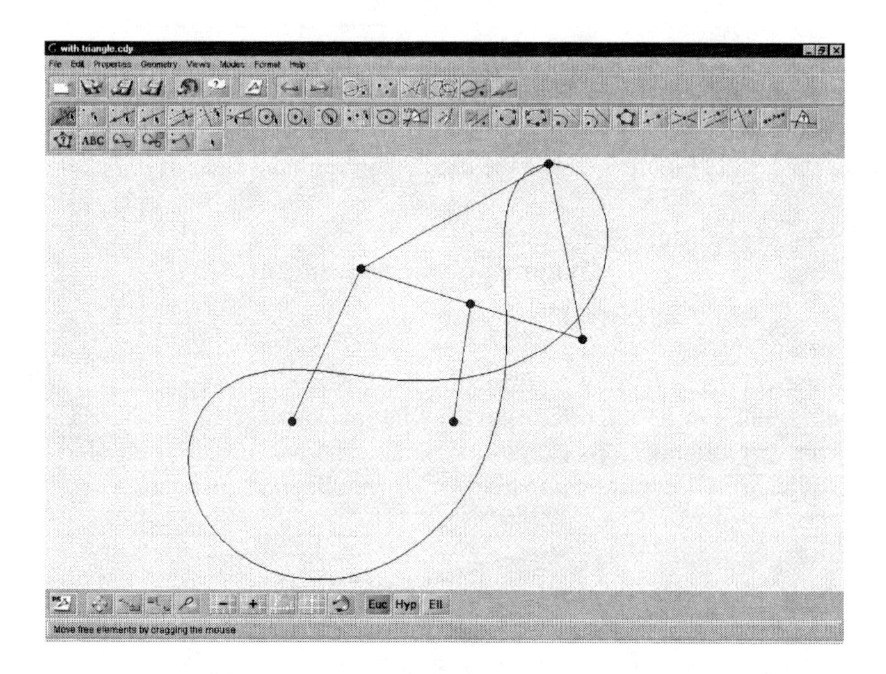

Figure 7, changing the locus (example showing Cinderella screen)

Checker-boarding

Combining and overlaying a set of curves then forms an ideal starting point for checker-boarding in black and white to form op-art designs. This is much easier to do on a computer than with pen and ink or painting. Figure 8 shows the result of using the results from equation 1. Figures 9 and 10 show the result of using output from Cinderella. The curves are exported as a PostScript file which is then imported into a graphics program and shaded by filling. This is far easier than using paint or pen and ink and is a boon to the artist as painting fine detail becomes harder with age as one gets more long sighted.

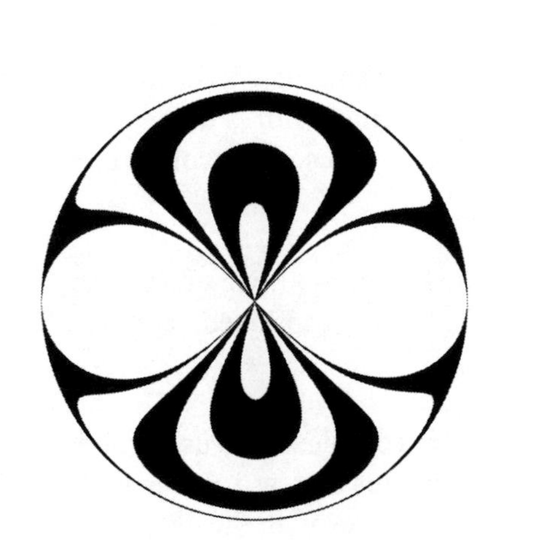

Figure 8, "Pinch" and "Hourglass"

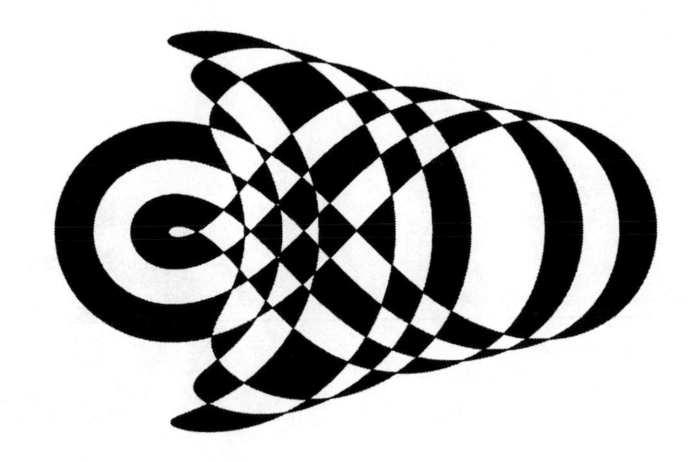

Figure 9, "Shockwave"

Figure 10, "Dancer" and "Falling Water"

Further possibilities

Computer output has the further advantage that it can be manipulated, for example with various filters. Figure 10 of using a bas relief filter and an impressionist filter with “Falling Water”. More could be done with colour and varying the line thickness of the output. Because I am colour blind, I tend to be happier with black and white.

Another option I mean to explore is that each of these curves is a slice of a surface. With many more linkages than I have described here and the variations I have mentioned, there is a lot of potential for many forms of art.

The linkages I have described here are very basic. More complex ones use sliding bars and many more linkages and pins. With increasing complexity, the variables increase, so the scope is enormous. Many books on linkages and mechanisms do not concern themselves with the curves. References 4-6, however are particularly good for the mathematics.

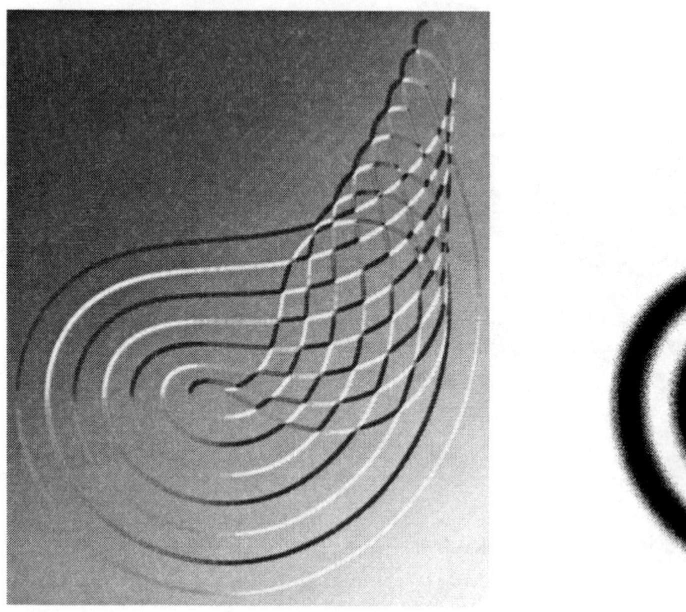

Figure 11, “Falling Water” with bas-relief and impressionist manipulation

References

[1] Bob Brill, “The Endless Wave”, Bridges Proceedings 2002, p 56.
[2] A B Kempe, “How to draw a straight line: A lecture on Linkages”, London 1877
[3] John Sharp, “The infinite Lemniscate”, *Infinity* 2006/4.
[4] Details of the Cinderella package can be found at `www.cinderella.de`
[5] R C Yates, “A Handbook of curves and their properties”;
[6] R C Yates, “Tools, A Mathematical Sketch and Model Book”, LSU Press 1941
[7] Eugene V Shikin, “Handbook and atlas of Curves”; CRC Press 1995

D-Forms: 3D forms from two 2D sheets

Tony Wills
Wills Watson + Associates
10 Greenham Road
London N10 1LP UK
W: www.wills-watson.co.uk
E: tony@wills-watson.co.uk

Abstract

Is there a significant branch of geometry that has been overlooked? Unlikely as it may seem, D-Form geometry provides designers, architects, sculptors and artists with a vast, new vocabulary of three-dimensional forms that are easy to play with and make. Easy as they are to fabricate, D-Forms are proving equally hard to predict with computing. This geometry exploits some interesting properties of developable surfaces that, among other things, will enable you to 'square the circle'.

Introduction

D-Forms are similar to forms used by sculptors such as Constantin Brancusi, Barbara Hepworth, Alexander Calder and Isamu Noguchi. At the same time, D-Forms are created with the seemingly simple mathematics of developable surfaces.

Peculiar as it may seem, I had the initial idea for D-Form geometry in a dream. Fortunately I had a pen and paper by my bedside and jotted down the details. Given that the structure of the benzene ring appeared to Friedrich August Von Kekule in a dream, then this might be good company to be in and not so strange after all.

We all have a vocabulary of forms in our heads and, as a product designer, I constantly update and refer to mine in the course of my work. So, the idea that there exists a large (possibly infinitely large) family of simple forms out there that I had never before encountered was something of a surprise to me – as I hope it will be to you. I have noticed that some of the people with whom I have shared this concept have developed a kind of D-Form obsession so maybe this paper should come with a mental health warning.

What is a D-Form?

Firstly, a D-Form is a kind of three dimensional equation: joining the edges of flat surface S1 plus flat surface S2 (which must have identical perimeter lengths) at initial points of contact A and B1 forms (equals) a specific 3D shape. For example, referring to figure 1a, by taking a pair of ellipses and joining point A on one to B1, B2, B3, B4 and B5 you will produce a series of 3D shapes that progressively change their form from one to the other. Depending on the ratio of the minor and major axes together and where the A and B points are placed, a whole range of D-Forms will appear. Five variations are shown in figure 1b.

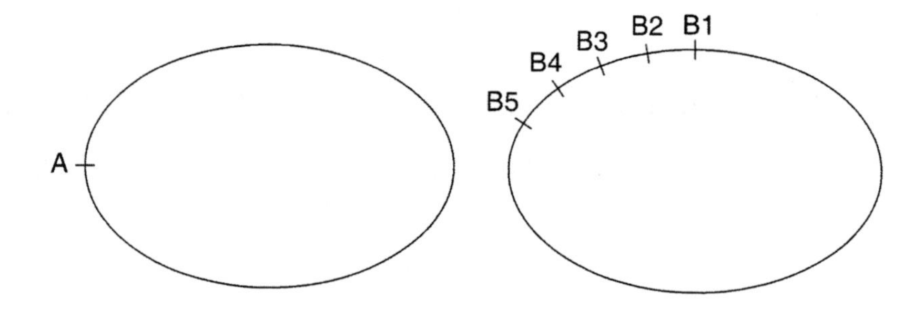

Figure 1a, two identical ellipses. Start joining A to B5, or B4, or B3 or B2 or B1 and different D-Form 'solids' will result as in Figure 1b

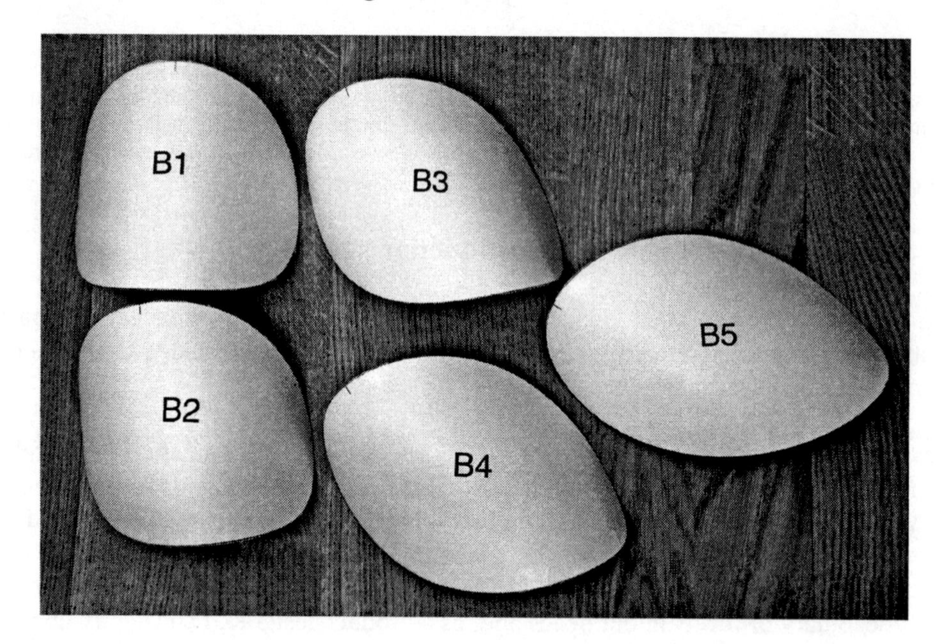

Figure 1b

Making a basic D-Forms

The first D-Form I stumbled across was simply this combination of two identical ellipses. When the initial joining points are the ends of the axes (major on one ellipse joined to the minor on the other) the result has a D-shaped cross section. To make a D-Form like this see figure 1a and join points A and B1. Draw the ellipses on a computer or construct them in the normal way. You can even draw a freehand shape like an ellipse. Cut them out from stiff paper or thin card. **Tip:** when identical shapes are being used, draw one and cut out two sheets at once thus ensuring identical shapes and therefore perimeter lengths. Choose a point A on one of the cut outs and a point B on the other but make sure A and B are in different places otherwise the form will lack and three dimension shape. Start taping the two perimeters of the cut outs together using short lengths of tape. **Tip:** I use a heavy tape dispenser and 'invisible' tape making it easy to tear off short lengths with one hand whilst the other manipulates the emerging D-Form. After a few minutes of taping you will see that each cut out is 'telling' the other what to do. It is as if this D-Form is a kind of three-dimensional equation where the sum of the two cut outs being joined, starting from the selected A and B points, equals a specific D-Form.

Materials

For practical purposes, sheets of relatively inelastic material like card, thin plywood, steel or aluminium are most likely to achieve a satisfactory result. Material stiffness and its relative thickness have a direct result on the scale of a D-Form that can be made and the scale is also affected by the tightness of curves required in the final form. This is a matter of experience and I have built D-Forms from a few centimetres across to over 4 metres high in many different materials. Mistakes I made include laminate that was too floppy and collapsed under its own weight to 8mm plywood that took ten men to force into shape. Steel sheet can be used and I have had quite large sheets cut with a laser so that tabs are added to the edges enabling them to be folded over during construction much like an old fashion tin toy. I have tried photo-etching thin sheets of brass and this works very well. A silversmith has made silver soldered D-Forms in bronze sheet for me but this can be expensive. Mesh or woven surfaces that can distort from rectangles to parallelograms are not 'D-Formable' although some interesting forms might still result as some fabric designers have found. Transparent drafting tape in a dispenser is best for joining card and I have successfully used a glue gun to internally join aero-ply D-Forms. If polypropylene is used as a skin on truncated D-Forms, filled resin can then be poured into the cavity and a cast made. This last method is particularly effective for maquettes of larger sculptures.

Variations

Some D-Forms where one or both of the sheets have vertices, will require creasing so they appear to have more than one developable surface. The 'Squaricle' (figure 4) is an example of this. The templates for making it are shown in figures 3.

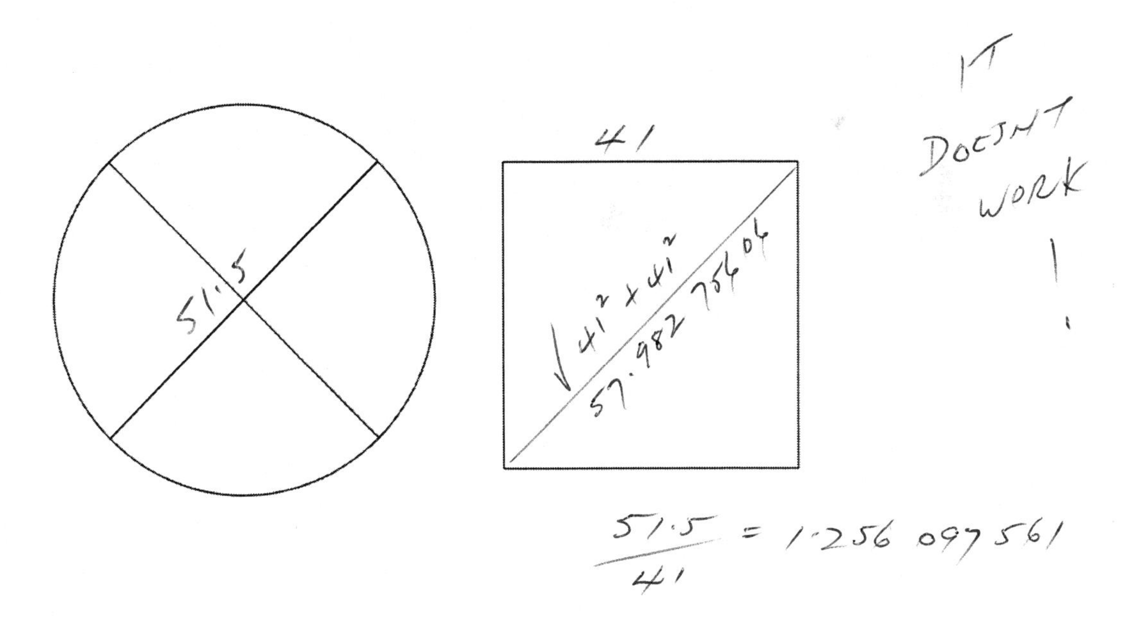

Figure 3, Squaricle templates Both lines on the circle are creased as 'crests' and the ends of the creases are joined to the corners of the square.

An alternative solution discovered by Prof. Tom Banchoff Brown University requires that the creases are made on a square drawn within the circle touching its perimeter. This gives a much flatter form. Other solutions are no doubt possible. Ellipses and rectangles are obvious contenders for variations on this theme.

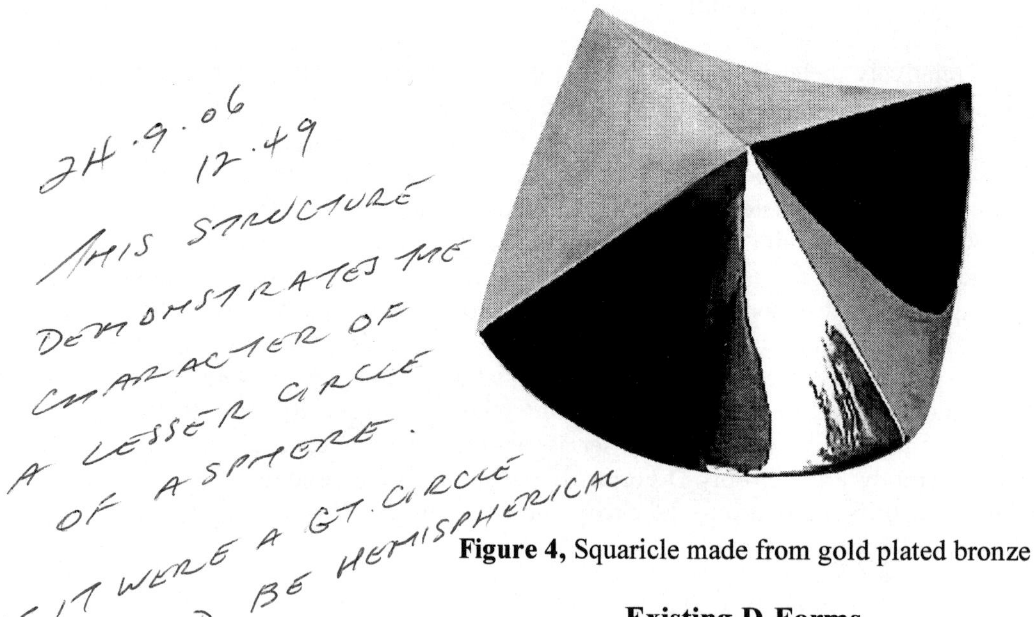

24.9.06
12.49
THIS STRUCTURE
DEMONSTRATES THE
CHARACTER OF
A LESSER CIRCLE
OF A SPHERE.
IF IT WERE A GT. CIRCLE
IT WOULD BE HEMISPHERICAL

Figure 4, Squaricle made from gold plated bronze

Existing D-Forms

There are some existing D-Forms in the world. For example, I recently saw a French street light with a D-Form luminaire made from a circle and an ellipse. Leather workers sometimes use basic D-Form geometry to make bags and upholsters make forms from just two surfaces of fabric but these tend to stretch depending on the warp and weft. In general, I have found that these existing shapes tend to be symmetrical about at least one axis. Taken to its extreme, D-Forming could be said to include a cube since by joining two rectangles (ratio 1:3 with creases a third the way along) then the cube can be made with one joined edge. However, it is the uncharted territory, particularly of twisted D-Forms, that interests me. The D-Form family is clearly very large and is a little like discovering a new continent, some coastal parts are already inhabited but the hinterland is virgin territory.

One might think that nature has used D-Forms, for example as seed-pods, but this does not appear to be the case. The closest natural phenomenon might be the structures made by weaver ants or tailor-birds where two leaves are sewn together to form nests.

Extending the concept

As a rule, the more simple the concept the more potential it contains. D-Forming is a very simple idea and, with the input of my colleagues, is living up to this observation. In my experience, ideas arise and are subsequently developed out of a context that usually includes respected collaborators who ask well-defined questions. The D-Form concept has been developed in this way. I took the first D-Form to one of my engineering colleagues Dr Philip Davies, who challenged me to try combining not two identical shapes but two entirely different ones. To our surprise, although some were more difficult to form than others, all the shapes and combinations we tried produced elegant D-Forms. One of these early tests was the Squaricle, created from the combination of a square and a circle, see figures 3 and 4. Whilst I don't claim to have 'squared the circle', it is an interesting form with several possible outcomes depending on how the circle is creased. Many other D-Forms followed from this simple experiment.

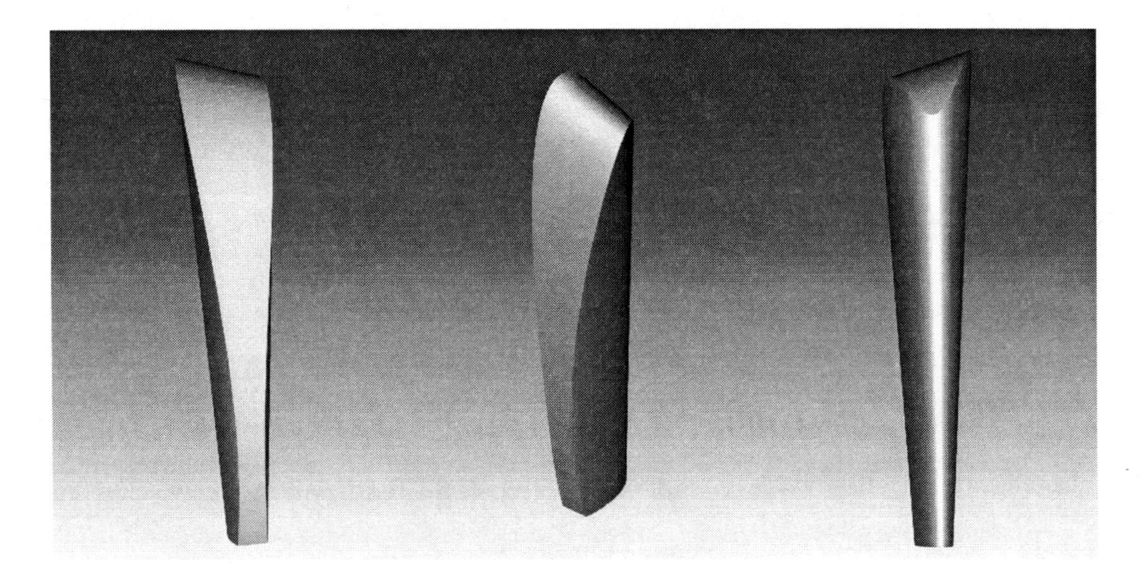

Figure 5, example of a truncated D-Form seen from various angles. This is a sketch for a stainless steel sculpture.

Anti D-Forms

Anti D-Forms came out of my collaboration with John Sharp. In one of our working sessions, rather than work with pairs of surfaces, we decided to try to join two holes with equal perimeters. Not only did this work but we found we could take the surfaces that we had removed to make the holes, construct the 'positive' D-Form from them and insert it precisely within the anti D-Form. After this, we tried half positive and half negative D-Forms. You can try this by cutting an ellipse out of a sheet and reintroducing it at an angle. We have recently asked what would happen if we didn't cut out the shapes at all, just joined two lines of equal length that are drawn on two surfaces. At the time of writing results are inconclusive but looks as if some D-Forms are possible using this method. Figure 6 shows an anti-D-Form.

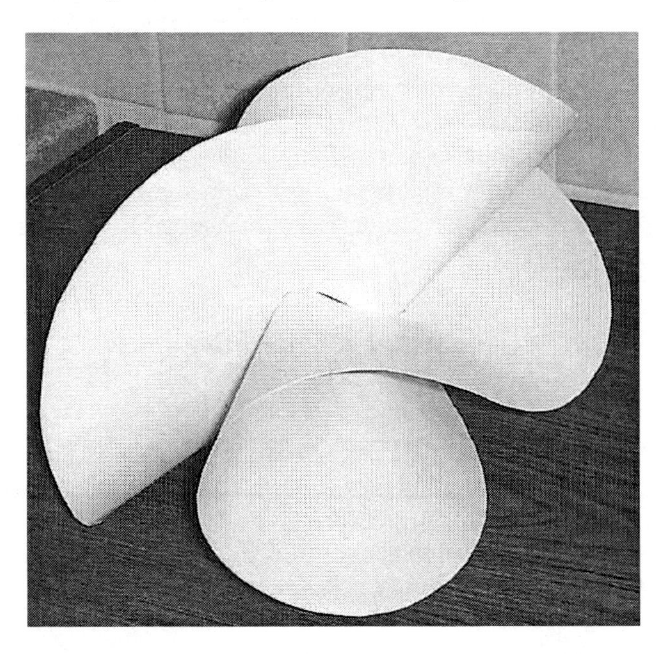

Figure 6, a pair of ellipses with elliptical holes that are joined together making an anti-D-Form

A general rule?

If D-Forms only worked with ellipses then a 'special rule' would be at work here. I am not a mathematician so I don't know how to about proving that a 'general rule' applies. However, it appears that there is at least a very large number of pairs of surfaces (with the same perimeter length) that, in combination and with different starting points, can create an even larger number of unique D-Forms. I have shown the D-Forms to amongst others, Prof Helmut Pottmann at the Institut Fuer Geometrie, Vienna. He describes D-Forms from a geometer's view point as an 'open problem' and describes them like this:

Let c1 and c2 be two smooth, closed, convex, planar curves of the same length, each bounding a flat piece of paper. Choose a point p1on c1 and a point p2 on c2, and glue the two curves to each other (according to arc length) starting with p1 glued to p2. The resulting single piece of paper forms a shape in space called a D-Form by Helmut Pottmann, Johannes Wallner, and Tony Wills. The curves c1 and c2join to form a closed space curve c bounding two developable surfaces S1and S2 These authors ask two questions:

1. It is not clear under what conditions a D-Form is the convex hull of a space curve.
2. After some experiments we found that, surprisingly, both S1and S2 were free of creases, but we do not know whether this will be so in all cases.

Advanced D-Forming

To construct more complex D-Forms it is important to have access to a computer program that allows the user to draw the developed surfaces and measure the perimeters, resizing as required. For D-Forms and my product design work I use Concepts Unlimited, a high end 3D modelling programme from Cadsoft Solutions, but this is somewhat over specified for the two dimensional aspect of drawing and resizing the developed surfaces. In common with even the most sophisticated design programmes, CU can only draw the most basic three dimensional D-Forms consisting of two intersecting cylinders. There seems to be no software program that could process two equal perimeter surfaces and emulate the building of a D-Form from them.

Our experiments show that, provided the perimeters of the two surfaces are continually convex, using any pair of ***different*** start points will make a unique D-Form. An obvious exception is that that joining two circles only produces a flat disk. An interesting experiment is to divide one of the perimeters up into say ten equal lengths and use these divisions as starting points so that nine D-Forms in different stages of rotation are created. It gets more interesting and difficult if concave curves are included in one or both of the perimeters. Many D-Forms are still possible with concave edges but there seems to be some kind of poorly defined boundary beyond which distortion starts to happen as if the surfaces are tying to enter 'negative space'.

Personal and collaborative work

As a product designer interested in geometry, I have occasionally used D-Forming in my work. It is very satisfying to create a D-Form as a mould and cast a perfectly visually balanced object from it. It is the 'rightness' that D-Forms have that particularly interests me. Since I originally conceived the idea some years ago I have developed a reasonably good eye for how to predict a D-Form from its developed surface but they still have the power to surprise me. Figure 7 shows me holding two steel D-Forms.

Figure 7, D-Forms made from sheet steel

Figure 8 shows how D-Forms have been used as part of a bench. The ends of the bench are cast in concrete using a D-Form.

Figure 8, left: street furniture made using D-Forms moulds and right: concrete castings

Intellectual Property

Rather than protect the Intellectual Property inherent in D-Forms, I want the concept to be 'open source'. Please feel free develop the concept independently but all I ask is that you acknowledge me as your source and inform me of your progress so that ideas can be shared.

Summary

Although I discovered D-Forms, the mathematics of them remain much of a mystery to me - if you have some insights do please contact me. In particular, I would very much like to be able to draw the developed surfaces, select the first pair of contact points and then have a computerised model predict the three dimensional form. This is something that CAD programmes simply cannot do (although some boat building programmes can unfold hulls). John Sharp has included some of the mathematics in his book on D-Form but this is just a start. The limits to D-Forming remain unclear partly because the concept is still developing - if you will pardon the pun. Some D-Forms appear to go beyond a border where the two surfaces resist being joined and I wonder if this has is a three dimensional analogy to negative numbers.

What, I do know for sure is that D-Forms are almost always beautiful, curious shapes that deserve much deeper research and understanding. To repeat Prof. Helmut Pottmann's comment, D-Forms are 'an open problem'.

References

[1] Helmut Pottmann and Johannes Wallner. Computational Line Geometry.Springer-Verlag, 2001.
[2] Paul Bourke has some http://astronomy.swin.edu.au/~pbourke/surfaces/dform/
[3] See the D-Form street furniture on the Wills Watson + Associates web site, streetscape products page.
[8] John Sharp, D-Forms, Tarquin 2006

Links can be found here: www.wills-watson.co.uk

Visualizing Escape Paths in the Mandelbrot Set

Anne M. Burns
Department of Mathematics
Long Island University, C.W. Post Campus
Brookville, NY 11548, USA
aburns@liu.edu

Abstract

This paper describes a method for producing a striking animation of the explosions that take place as the parameter c that defines the Mandelbrot Set is allowed to traverse a path from inside the large cardioid component of the Mandelbrot Set into one of the attached "bulbs" or other regions just outside the set. The presentation will include the animation itself, as well as some of the colorful images obtained by stopping the animation at various points.

1. Introduction

Mathematicians and artists are familiar with the famous orbit diagram for the real-valued function $f_c(x) = x^2+c$ where we see successive period-doubling bifurcations and eventual chaos as the real parameter c decreases along the real number line from 1/4 to -2. In this case the parameter c is actually traveling one of the many paths from c = 1/4 to the outer reaches of the Mandelbrot Set at c = -2. Because we are dealing with real numbers we can plot the fixed or periodic points of f_c on one axis as a function of the parameter c on the other axis. Suppose we want to examine the dynamics by letting c travel along a different path from the origin to some point outside the Mandelbrot Set; in this case both the parameter c and the fixed point w will be complex numbers. How are we to visualize the changing dynamics in this case? By expressing both c and w as functions of a single parameter, r, we can make an animation by plotting several thousand points in the orbit of 0 under $f_c(z)= z^2+c$ for each r, as c travels various escape routes from the Mandelbrot Set, M. By assigning the color of the points in the orbit as a function of r we can create some quite spectacular animations. In [2] I explore the escape routes from M; each route comes with its own unique dynamics and its own amazing graphics illustrating the transition from order to chaos.

 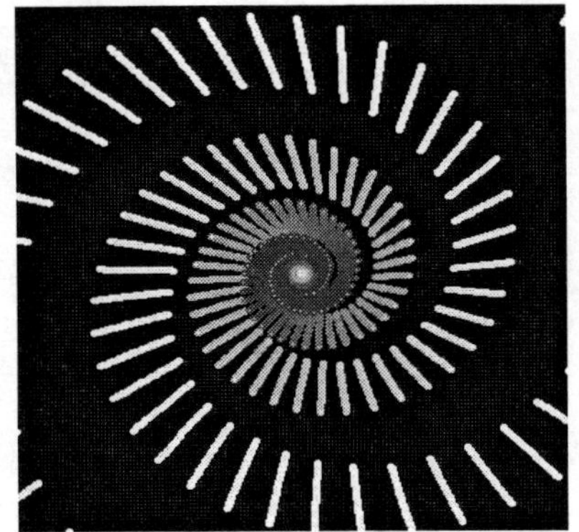 

Figure 1: *Examples of images obtained by stopping the animation*

2. How to find the escape routes

The first step in finding the escape routes is to express both c and the fixed point w as a function of a single parameter r. To find a fixed point, w, of $f_c(z) = z^2+c$, we solve the equation

(1) $w^2+c = w$, and we note that w is attracting if
(2) $|f'(w)| = 2|w| < 1$, or $|w| < 1/2$.

We are going to be interested in the set $M_1=\{c \mid f_c(z)$ has an attracting fixed point$\}$. Solving (1) for c we get $c = w - w^2$ and from (2) letting $w = (r/2)e^{i2\pi\theta}$ gives

(3) $c = \dfrac{re^{i2\pi\theta}}{2} - \dfrac{r^2e^{i4\pi\theta}}{4}$, $0 \leq r < 1$, $0 \leq \theta < 1$.

The boundary of this region ($r = 1$) is the large cardioid (the boundary of M_1) in the Mandelbrot Set. It is well known (see any of the references) that for c on this boundary the dynamics of f_c are determined by the value of θ.

If $\theta = \dfrac{p}{q}$, p and q natural numbers with $p < q$ and $\gcd(p,q) = 1$, then at the point c on the boundary of M_1 a hyperbolic component (or what Devaney calls a "bulb") is attached to M_1. We will call this bulb $M_{p/q}$. The corresponding fixed point is neutral and is a parabolic fixed point. At this parabolic fixed point the attracting fixed point bifurcates into an attracting periodic cycle of period q. For $r < 1$ the corresponding w is attracting and there is a repelling cycle of period q surrounding w.

Figure 2: *Transition from 1/4 to 1/5*: *Orbits keeping r fixed at .975; θ varies from 1/4 to 1/5*

(Note that for the real orbit diagram $\theta = 1/2$ and the bifurcation is period-doubling i.e. $q = 2$.) As r approaches 1 from below, the fixed point w and the repelling cycle coalesce into the neutral fixed point. As r increases beyond 1, c moves into $M_{p/q}$, the fixed point w becomes repelling and the period q cycle becomes attracting. Smaller bulbs are attached to each $M_{p/q}$ and for c in one of these bulbs f_c has an attracting cycle of some finite period (the period will be a multiple of q).

For θ "sufficiently" irrational there is a neighborhood of $w = (1/2)e^{i2\pi\theta}$ called a Siegel disk. In this neighborhood orbits of nearby points look like deformed circles surrounding the fixed point w [4]. There are many books containing beautiful pictures of fractals illustrating the complicated dynamics for values of c near the boundary of the Mandelbrot Set. What we want to do here is to describe how to create an animation of the orbits as c follows the path (3) from inside M_1 across the boundary and into one of the bulbs, or near one of the bulbs.

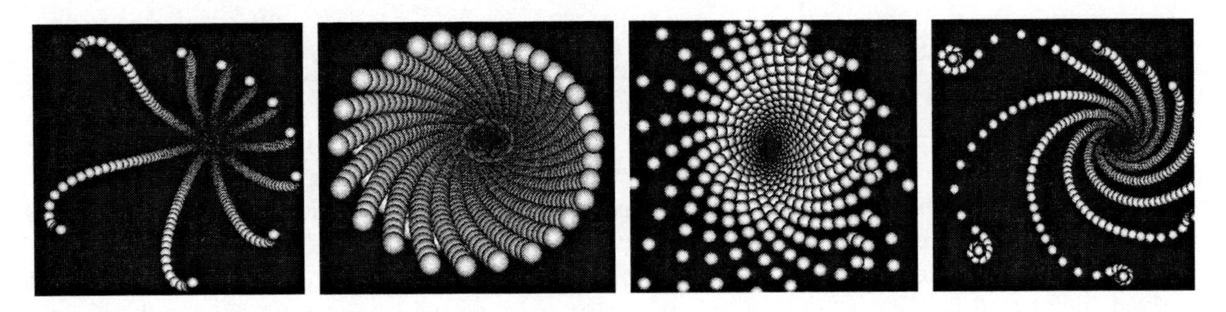

Figure 3. *Transition from 1/10 to 1/9: Orbits keeping r fixed at 1.00438; θ varies from $2\pi/10$ to $2\pi/9$*

3. Orbits for different values of r and θ

In Figures 2 – 4 we show orbits of a single point for various values of r and θ. In Figure 2 we have kept r fixed at .975 and we let θ vary in increments from 1/4 to 1/5. Since $r < 1$, in each case there is an attracting fixed point. In each frame 2000 points in the orbit of 0 were plotted. In the first frame ($p/q = 1/4$) points in the beginning of the orbit surround the fixed point in a "4-pattern". This is very clear when we color every 4th point the same color.

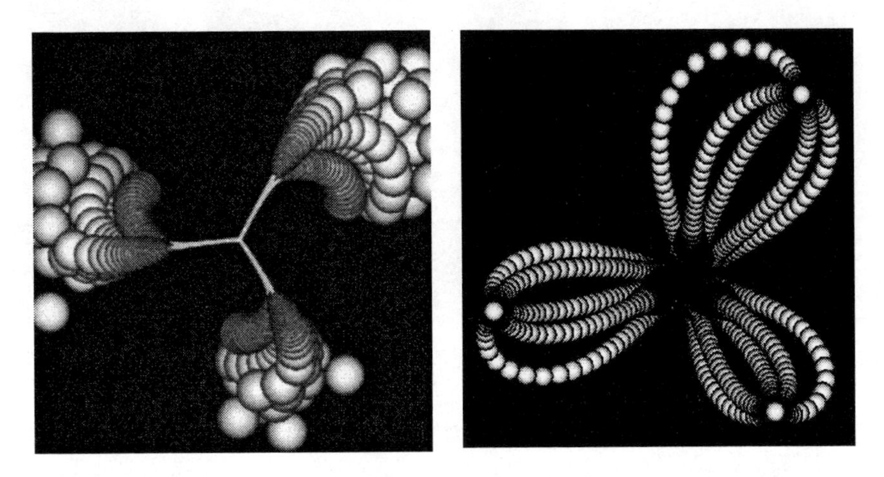

Figure 4. *Transition from attracting fixed point to attracting 3-cycle; θ is fixed at $2\pi/3$; first picture $r = .99$ ($r < 1$), second picture $r = 1.025$ ($r > 1$)*

Similarly in the last frame (p/q = 1/5) we can see the "five"ness of the pattern. In between we see other types of orbits; for example in the third frame p/q = 4/17 and the orbits fall into a "17-pattern". Incredibly, as p/q varies from 1/4 to 1/5, there is an uncountably infinite variety of orbit patterns. In Figure 3 we have let r be slightly greater than 1 and p/q vary from 1/10 to 1/9. In the first frame we see an attracting cycle of period 10 and in the fourth frame we see an attracting cycle of period 9. In between we see an attracting cycle of period 29 in the second frame and another kind of behavior in the third frame

In Figure 4 we see the orbits of 0 where θ is fixed at $2\pi/3$; in the first picture $r < 1$ and there is a fixed point in the center of the screen; in the second frame $r > 1$ and there is an attracting cycle of period 3. Using three colors and coloring every third point in the orbit the same color gives us a better picture of how the orbits behave.

4. The relationship between the fixed and periodic points

To find the cycles of period n for Q_c we have to solve $Q_c^n(w) = w$. There will be 2^n of them. When c = 0, this means $w^n = w$. The solutions will be w = 0 (the attracting fixed point) and the 2^{n-1} th roots of 1. So they will be distributed around the unit circle. For example, if n = 3 there are two fixed points, w = 0 (attracting), w= 1 (repelling) and two 3-cycles: $\{e^{i2\Pi/7}, e^{i4\Pi/7}, e^{i8\Pi/7}\}$ and $\{e^{i6\Pi/7}, e^{i12\Pi/7}, e^{i10\Pi/7}\}$.

If we let $w = (r/2)e^{i2\pi p/q}$, p and q relatively prime natural numbers with $p < q$, be the fixed point, and $c = w - w^2$, when r = 0, w = 0 and there will be at least one repelling q-cycle distributed around the unit circle. In the last paragraph p/q = 1/3 and there were two repelling 3-cycles. As r increases to 1, the fixed point moves along the ray $(r/2)e^{i2\pi p/q}$ and one of the repelling q-cycles surrounds the point w, moving ever closer to w. At r = 1 the fixed point w and the q-cycle coalesce. At $r = 1$ $|Q_c'(w)| = 1$ and the multiplier of the cycle, $\prod_{i=1}^{q} Q'_c(w_i) = 1$. This point is called a parabolic fixed point of Q_c. As r increases beyond 1, the q-cycle becomes attracting and the fixed point becomes repelling. (See Figure 5)

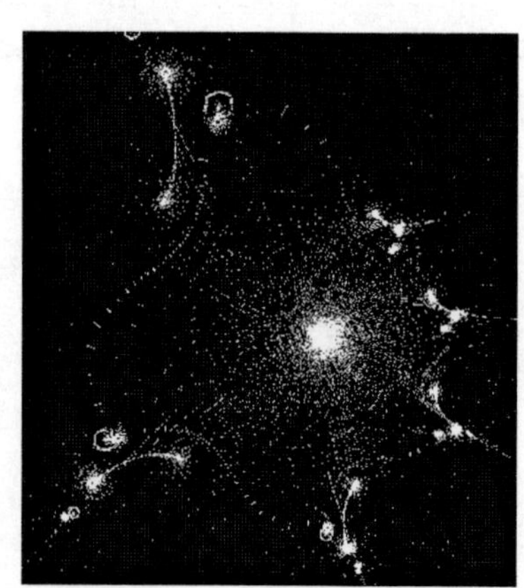

Figure 5: *Three stages in the animation where $\theta = 2\pi(1/6)$. The first frame shows a close-up of the center of the screen after r has increased to about .987; in the second frame r has increased to 1.04. The third frame is a close up of the action near one of the points in the attracting 6-cycle seen in the second frame.*

5. Building the Animation

For each value of θ, where $w = re^{i2\pi\theta}/2$ and $c = w - w^2$, we animate the scene by choosing the center of the computer screen to be the point where the fixed point and cycle coalesce. We allow r to increase in small increments and for each r we plot some number (to be chosen by the user) of points in the orbit of 0. Each orbit is plotted in a different color determined by r, using a continuous ramp of colors. The results are quite spectacular.

After writing the program animating the orbits, we incorporate that program into a larger program where we continuously change the value of θ and observe the amazing changes in dynamics as θ travels around the unit circle. Some of the more spectacular pictures occur when θ is not rational. In this case the parameter c exits M for a brief moment before re-entering in one of the bulbs. Figure 5 shows three stages in the animation where $\theta = 2\pi(1/6)$. In Figure 6 we have illustrated what the screen looks like for $\theta = 2\pi p/q$, where p and q are successive Fibonacci numbers. It is known that the values $\theta = \dfrac{-1+\sqrt{5}}{2}$, $r = 1$ admit a Siegel Disk where orbits behave like deformed rotations about the fixed point w.

Figure 6: *$\theta = 2\pi p/q$, where p and q are successive Fibonacci numbers. In the first frame p = 8, q = 13; in the second frame p = 55, q = 86; in the third frame p = 141, q = 227. In all frames we let r increase to about .99026 before stopping the animation*

Figure 7 shows some stills from the animation for different values of r and θ.

6. Conclusion

Expressing the fixed point and the parameter c as a function of r and θ allows us to produce virtually any kind of orbit that we choose for the function $f_c(z) = z^2+c$. Then by varying the parameters in small increments, we can animate the continuous change that takes place in the nature of the orbits as the parameters change. I have found that this parameterization helps students in a fractals class to find values that yield interesting orbits as in Figure 8. It might suggest other ways of animating different dynamical systems. We might explore using color as a function of the parameters in other ways. For many people visualizing mathematics is vital to understanding it; conversely, understanding the math allows us to create visually compelling images.

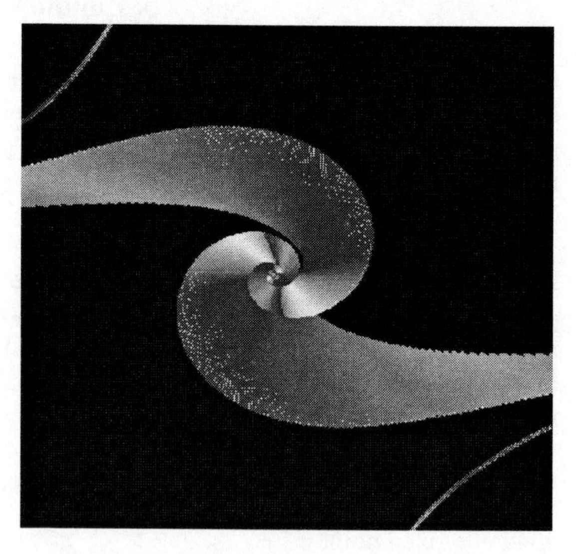

Figure 7. *In the first frame $\theta = 2\pi p/q$ where $p = 1251$ and $q = 5000$; stopped at $r = 1.1987$. In the second frame $\theta = 2\pi p/q$ where $p = 251$ and $q = 500$; stopped at $r = .99459$*

Figure 8. *Orbits of $z^2 + c$*

References

[1] B.Branner, "The Mandelbrot Set", *in R. Devaney and Linda Keen, editors, Chaos and Fractals* (Proceedings of symposia in Applied Mathematics, vol. 39) American Mathematical Society, Providence, RI, 1988

[2] A. Burns, "Plotting the Escape", *Mathematics Magazine,* Vol. 75, No. 2, April, 2002

[3] R. Devaney, *An Introduction to Chaotic Dynamical Systems,* Addison-Wesley, 1987

[4] H.-O. Peitgen and P.H. Richter, *The Beauty of Fractals,* Springer-Verlag, 1986

The Math of Art:
Exploring connections between math and color theory

Amina Buhler-Allen
4240 19th Street
Boulder, Colorado 80304 USA
email: colorfields_@msn.com

Abstract

Simultaneous contrast and extension are fundamental principles in color theory, which directly relate to mathematics. Color study includes study of the proportions of colors and their effects. Using these concepts of the interrelationship of proportions and color can broaden expression much like adding extra colors to a painter's palette.

1. Introduction

With the invention of the camera the role of an artist changed. Representational art became less necessary as a photograph could be taken, instead. This led artists to redefine the parameters of art. The art community was freed from the boundaries of realism. The stage was set for new movements, which created new ways of seeing, teaching, and making art. Discoveries in science and industry advanced the study of what is today color theory.

2. Simultaneous Contrast

2.1. **Chevreul: Contrast of Light Intensity**. In the mid 1800s a chemist with a specialty in dyes, Michael Eugene Chevreul, introduced his book, The Principles of Harmony and Contrast of Colors [1]. Chevreul noticed that a color's character changes; the same color can look lighter or darker depending on its juxtaposition to other colors. It is said that his "Law of Simultaneous Contrasts" was conceived in the tapestry (Gobelin) factory where he noticed that the strength of the black depended on what colors were next to it. This is simultaneous contrast of light intensity or light and dark.

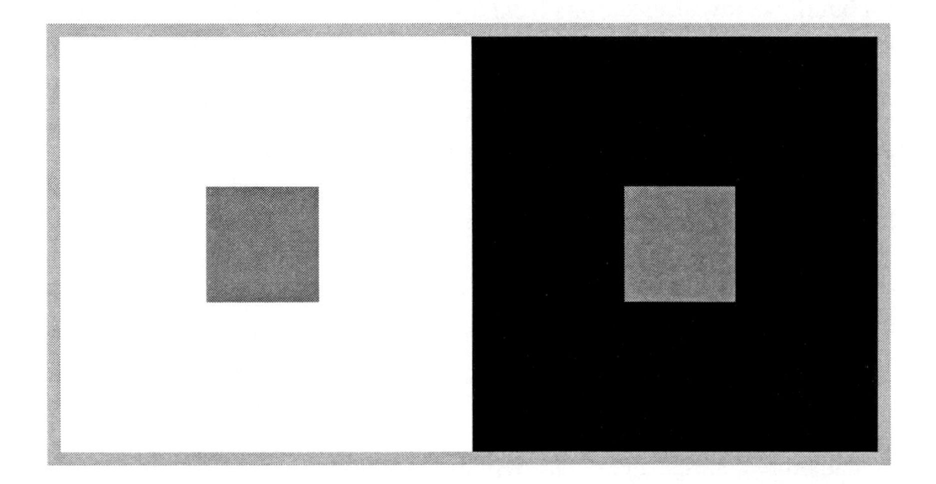

Fig. 1: *Identical gray squares appear differently. The gray square on the white background appears to be darker than the square on the black background.*

Fig 2: *Light/dark simultaneous contrast, using a gray scale. The gray scale on a white background appears differently than the same scale on a black background. When a gray scale is arranged linearly (lightest through darkest), another phenomenon occurs. Look closely at the boundary areas between colors; the same color looks both light and dark.*

Optical mixing, or blending color in the eye, was another phenomenon which Chevreul studied. The Impressionists and Pointillists investigated this technique. Looking close enough to examine the individual strokes or dots on a painting using this style reveals that the color that you see is not the color on the canvas. For example, green areas are made from blues and yellows, which mix in the eye.

Industry started to use this technique in printmaking using a dot matrix of colors to blend colors as we do today. This is referred to as Process color. All colors mix from percentages of four Process colors: cyan (C), magenta (M), yellow (Y) and black (K). The percentage of each Process color breaks down into a dot pattern for that color which ranges between 0-100 percent. Each Process color is printed as a separate color pass. When all four colors are combined there is a series of overlapping dots which create the final print. Color photographs in newspapers are printed in this manner. Pantone colors use the same percentages as process colors, however, rather than a series of dots, the color is mixed as pigment and printed as block color (without dots.) The Pantone colors are labeled using a number system. The standardization of Pantone and Process colors allows for more consistent reproduction of color. Black was added to the primary colors in printmaking to add more contrast and depth. Without black, objects appear "washed out".

2.2. Goethe: Contrast of Hue. The first notations on the "light value" of colors were done by Goethe. Had Chevreul had access to Goethe's work, he would have made use of these light values in his "Law of

Simultaneous Contrast". These "values" or proportions were derived from observation and are still being taught today. In the correct (harmonic) proportions each color will have equal light value. For example, take yellow and purple. Yellow is very bright. Purple is darker. A much larger area of purple is needed to emit the same amount of light that an area of yellow does. This is contrast of hue. The hue is a pure color without white or black. The following proportions were derived from Goethe, elaborated on and taught by Johannes Itten and Josef Albers. Albers taught color theory after Itten at the Bauhaus and later brought these ideas to the United States

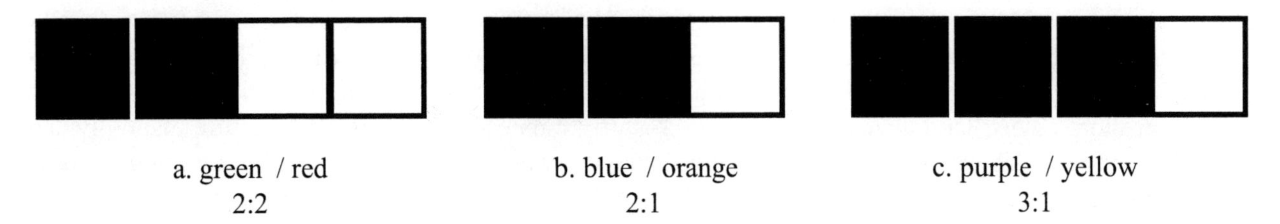

Figure 3: *Simultaneous contrast quantified such that the light intensity of each hue and the area are proportionate. Each color is shown in relation to its complement. 3a represents the light values of green/red, which are equal. For every 2 areas of blue, one area of orange is harmonic in 3b. In 3c, it takes 3 areas of purple for every one of yellow.*

The proportionate areas for the primary and .secondary colors are:

green	:	blue	:	purple	:	red	:	orange	:	yellow
6	:	8	:	9	:	6	:	4	:	3

To use these areas in harmony, the chart is used as follows: for blue / yellow, use 8 areas of blue for every 3 of yellow....8 : 3, etc. There is more detailed reading about these proportions in Johannes Itten's The Art of Color [3].

These proportions were combined with Chevreul's initial observations to make the "Law of Simultaneous Contrasts". Simultaneous contrast includes both contrast of light intensity and contrast of hue.

3. Extension

Using these proportions creates harmony. Deliberately creating ratios that vary from the above chart creates a tension that is called "Extension" or "Proportion of Color". For example, imagine a painting with predominant proportions of blues and purples; add a small orange line and the orange line stands out predominantly. This is the use of the proportion of color. It is a tool to create effects by the limitation or expanse of color by choosing to work with non-proportionate light values. This tension is actually a lack of equilibrium.

Physiologically, the eye wishes to make equilibrium. Creating equilibrium from the imbalance, the eye generates another image called an *afterimage*. The afterimage is always the complementary color of the color viewed. It is strongest with the greatest contrast between the colors. The afterimage is believed to be a result of retinal fatigue. For example: After looking at the color yellow for a protracted time, the color sensors in the eye "burn out", the eye generates the color opposite, purple: this is the afterimage.

Simultaneous contrast is effected by the afterimage even though this image is not a "solid" image. The blending of colors, whether by optical mixing (yellow/ blue areas making green), or by afterimage, are powerful tools that can be manipulated.

4.Shifting Color

"A color has many faces" –Josef Albers

Just as the same middle gray could appear lighter or darker, so can the perception of the same color shift. A color is relative to its surroundings. The most basic shifts have to do with contrast of light intensity or contrast of hue. The secondary (orange, green, purple) colors, as well as other colors which are combinations of two or more colors, lend themselves easily to a shift in appearance. Consider a picture plane divided into two colored backgrounds: one red, one blue. Placing two squares of the same purple color on both background colors makes the purple appear as if it is two different colors. The color has not changed, just its surroundings. Purple is made up of portions of both red and blue. In comparison to red, purple is more blue. On the red ground, the blueness of the purple is apparent. On the blue ground, the redness of the purple is apparent. When juxtaposed next to blue, the quantity of blue in the purple is less than the amount of blue in the blue, therefore, the purple appears more red in contrast.

5. Conclusion

Color can be manipulated the same way a mathematical formula. Using various proportions of color can create harmony or dischord (tension). Until one is familiar with the variables in a formula, the equation is not complete. Knowing the variables of simultaneous contrast and extension facilitates an operative knowledge of connections between the art of color and math.

"Give me mud and I will make the skin of a Venus out of it, if you will allow me to surround it as I please"
- Eugene Delacroix

Bibliography

[1] M.E. Chevreul, *The Principles of Harmony & Contrast of Colors*, Faber Birren, first printed 1839
[2] Goethe (edited by Rupprecht Matthaei), *Goethe's Color Theory*, Van Nostrand Reinhold Company, 1970
[3] Johannes Itten, *The Art of Color*, Van Nostrand Reinhold, 1961
[4] Josef Albers: *The Interaction of Color*, Yale University Press, 1963
[5] Frans Gerritsen, *The Evolution of Color*, Schiffer Publishing LTD, 1982
[6] Stephen J. Sidelinger, *Color Manual*, Prentice-Hall, 1985
[7] Karl Gerstner: *The Spirit of Colors*, MIT Press, 1981

Celtic Knotwork and Knot Theory

Patricia Wackrill
48 Rofant Road
Northwood
Middlesex
HA6 3BE
UK
e-mail: pwackrill@waitrose.com

Abstract

Celtic knotwork is a form of decoration in use for over a thousand years. The designs fill spaces or borders with a pattern derived from plaiting. The designs have no loose ends and may contain more than one closed loop. As in a plait (or braid) of hair, each strand bounces back and forth like a billiard ball to form a pattern of diagonal lines between the edges of the rectangle while crossing over and under others alternately. The dimensions of a rectangular plaited panel can be expressed as the number of bounces there are along the long and short edges. The number of closed loops, referred to as knots by knot theorists, is the greatest common divisor of these two numbers. This paper shows how one can predict the number of loops there will be as a piece of knotwork is created from the panel by removing some of the crossing places and rejoining the loose ends, without crossing to make a gap either looking like) (to make what will be called a horizontal gap, or like this shape turned through 90° to make a vertical gap. As each of the chosen crossing places is removed and the loose ends rejoined in this way a more intricate interlaced design is formed. It is easy enough to trace round the resulting design with coloured pens to find out how many closed loops there are, but the results proved and demonstrated in the paper enable one to predict how the number of loops will change at each stage of the creation of the interlaced design. Such a prediction is not addressed by current knot theory. The first thing to notice is whether a loop crosses itself at the crossing to be removed or whether two different loops cross there. In the first case one or two questions must be answered before the number of loops can be predicted. In the second case the two different loops get combined into one single loop by the rejoining of the loose ends. The difficult part of the research was to devise questions which could be proved to make reliable predictions possible. One's common understanding of how one might take a shortcut back to the start while following a nature trail provides the last link in the chain leading to the prediction. The method will be applied to successive designs produced as each crossing is removed.

1. Introduction

  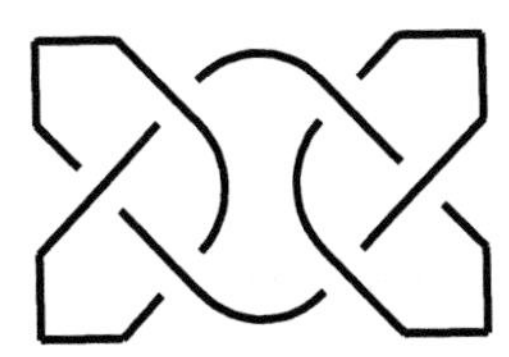

Figure 1: *Two celtic knotwork designs derived from a plaited panel with one loop*

Even a simple plaited panel can be converted to an interlaced design of a type which distinguishes celtic knotwork from the plaits used in Egyptian, Greek or Roman art. Figure 1 demonstrates what happens when the central crossing is removed and the loose ends rejoined to make a gap in the plaiting. The two different ways of rejoining produce different designs, one with two loops and one with one loop. While it is easy enough to determine the number of loops there are even in an intricate interlaced design, with the help of coloured pens, the mathematician seeks an underlying structure which will enable a prediction to be made for the resulting number of loops as each of many crossings is removed. The paper continues with classification of the crossing to be removed and carries on with classifying the ways of rejoining the loose ends to make a gap before proceeding to arguments justifying the predictions.

1. The classification of crossings

A crossing can be:
(a) where a loop crosses itself or
(b) where it crosses another loop.

In case (a) we need to define two types of crossing. To do this we start at the crossing to be replaced and proceed to add arrows along the loop, starting in either direction, until we reach this crossing again. It will then be clear which of the diagrams in Figure 2 below resembles our crossing. The first two resemble part of an upright figure of eight circuit, (not to be confused with what knot theorists call the Figure Eight Knot) while the second two resemble part of a figure of eight lying on its side, ∞; let's call it an infinity circuit. So we can ask whether our crossing to be replaced resembles an 8 crossing or an ∞ crossing, even though the actual loop may have many tangles on either side of the particular crossing to be replaced which disguise its resemblance to one of these two circuits. There is no topological difference between these two circuits; making this distinction will help us to visualise what the different ways of rejoining the loose ends will do to the number of loops.

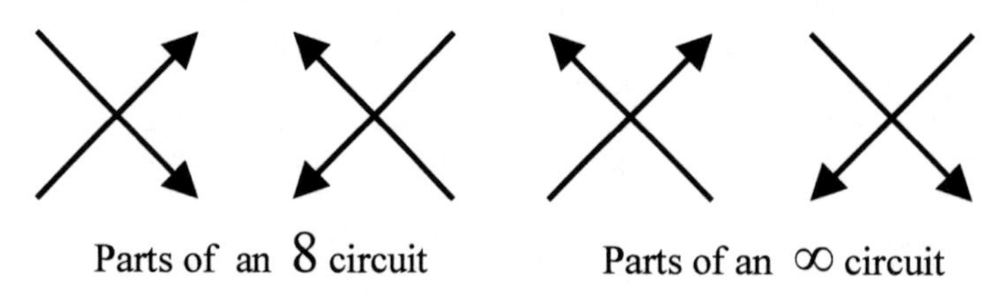

Figure 2: *Different types of crossing.*

2. The classification of gaps

A gap like) (is classified as horizontal because horizontal arrows would indicate its width in a technical drawing, and a gap like this shape turned through 90° as vertical. We can already visualise that a horizontal gap will leave an 8 circuit in tact as one loop and a vertical gap will leave an ∞ circuit in tact as one loop, whereas a vertical gap will cut an 8 circuit into two loops and a horizontal gap will cut an ∞ circuit into two loops. With these classifications of crossing and gaps we can now proceed to arguments supporting a general statement for predicting the number of loops.

3. Arguments justifying the predictions

We start by visualising a simple figure of 8 nature trail shown upright on a map. If one replaced the crossing with a horizontal gap, like) (, one would just follow part of the trail in the reverse direction, whereas a vertical gap would provide a shortcut back to the start. This common understanding can be extended beyond a simple figure of 8.

We will use the plaited panel on the left in Figure 1 to demonstrate such an extension by replacing the central crossing with a gap. The argument, continuing the nature trail analogy, will follow Figure 3 from left to right. Arrows added to the plaited panel show that the central crossing is an 8 crossing. Then, texture added shows the trail as a continuous dotted path followed by a black path. Think of yourself starting at the bottom left hand corner of the trail map and arriving at the central crossing point along the dotted path. If you alter your route so that instead of carrying on straight ahead you turn off to the right, as in the next diagram, then you will find that you have taken a short cut back to the start and left out the black part of the trail. You followed a map with a vertical gap in the middle. If, on the other hand you turn off to the left you will follow the black path, but against the direction of the arrows, as in the right hand diagram, until you reach the central crossing again and can turn left to follow the rest of the dotted path, with the arrows, back to the start. This time you followed a map with a horizontal gap in the middle. We need only join up the second pair of loose ends at the centre of the third diagram and remove all arrows to see that the third and fourth diagrams confirm the effect on the number of loops of replacing the central 8 crossing with vertical and with horizontal gaps, as shown in Figure 1.

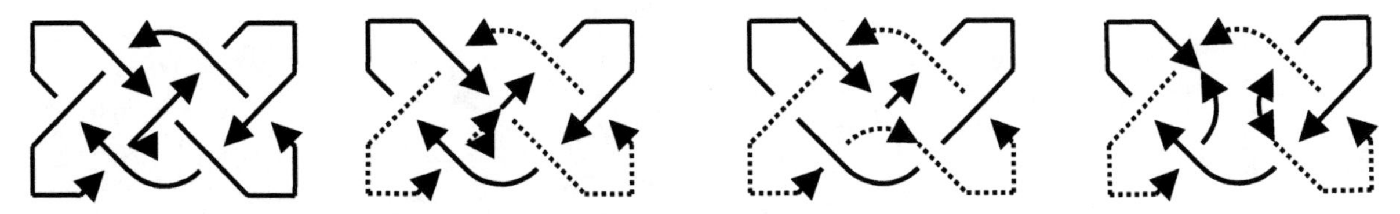

Figure 3: *The nature trail analogy with right and with left turnings instead of crossing ahead.*

We are now ready to summarise our findings in the middle line of Table 1 below. The bottom line arises because an ∞ circuit is merely a rotation of an 8 circuit.

Crossing type	Vertical gap	Horizontal gap
8 crossing	1 extra loop	No extra loops
∞ crossing	No extra loops	1 extra loop

Table 1: *The effects of different types of gap on a loop crossing itself.*

Mathematicians will want a general argument to support this table as a reliable prediction. Consider a nature trail with p consecutively numbered direction posts. Let them be numbered 1 to m as far as the 8 crossing. The original route, on the left in Figure 4, would carry on with numbers $m + 1$ up to n, just before one meets the crossing again, and then from $n + 1$ to p. The effect of a right turn is to miss out posts numbered $m + 1$ up to n; one has kept going in the direction of the arrows, just missing out a piece of the circuit. A trail through these makes one of the two loops, while a trail through those missed out makes the other loop. The effect of the left turn, on the right in Figure 4, is to pass posts 1 up to m in ascending order, then, going against the direction arrows, to post numbered n straight after post numbered m and then posts with decreasing numbers, from $n - 1$ down to $m + 1$, before turning left again to pass post numbered $n + 1$ and then posts with increasing numbers, $n + 2$ up to p. This time all the posts have been passed in a single loop. This generalised argument supports the predictions for the numbers of loops.

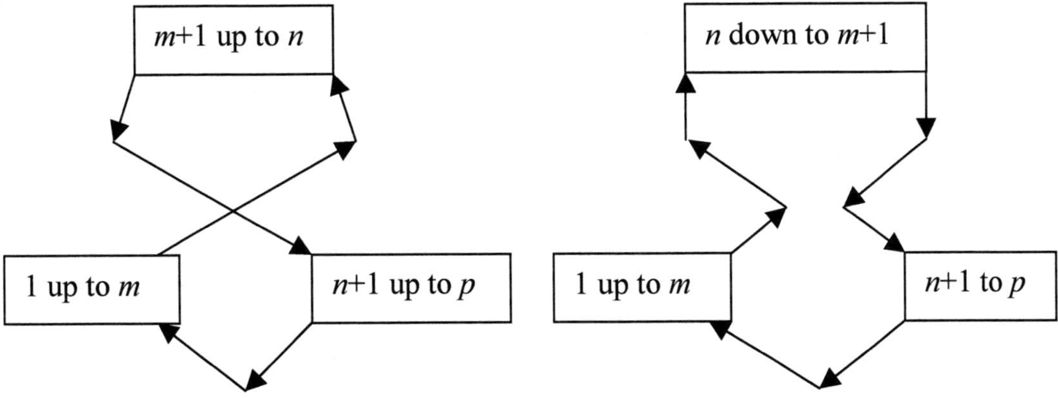

Figure 4: *The nature trail analogy with left turning instead of crossing ahead.*

The generalised argument has focussed on following or going against the direction arrows. Figure 5, with arrows on the gaps enables us to determine the effect on the number of loops with just one question.

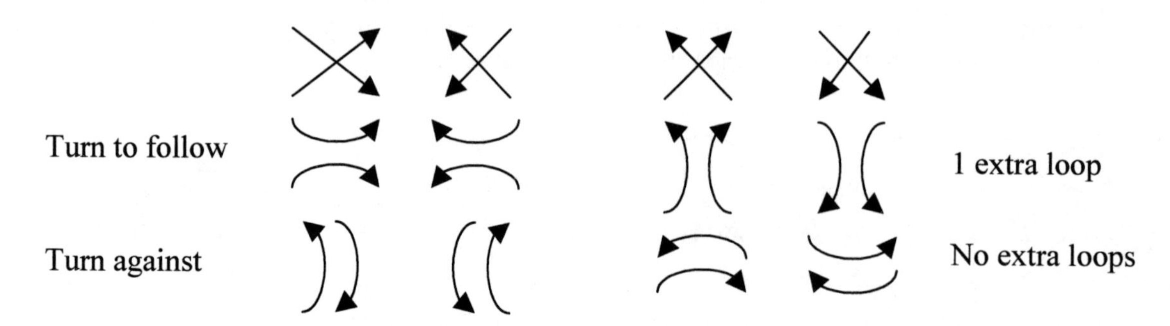

Figure 5: *From crossings to gaps, to number of extra loops.*

So the question is "Does your turn follow the arrow direction or go against it?" Turning to follow makes one extra loop while turning against makes no extra loops.

We now consider the much easier case (b) where a loop crosses another loop as in the middle diagram in Figure 1. The replacement of the left hand crossing with each type of gap is shown in Figure 5.

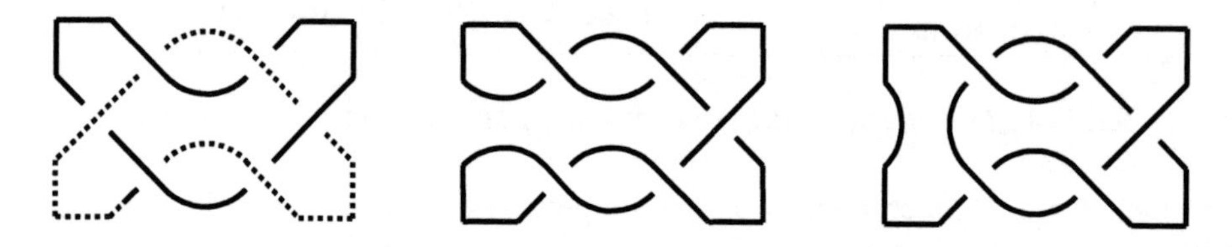

Figure 6: *A place where one loop crosses another replaced with vertical and horizontal gaps.*

In case (b) either type of gap will amalgamate the crossing loops to make one loop because the loose ends consist of two from each loop and if they are to be rejoined without crossing they have to be paired up using one loose end from each loop. So, if the answer to the question of whether a loop crosses itself or another loop at the crossing to be replaced with a gap is that it crosses another loop then one needs no further questions; the replacement will reduce the number of loops in the design by one.

Conclusion

While it is easy to create a knotwork design from a plaited panel by making all the replacements of crossings with gaps at once and then discover how many loops it has by tracing round the thread or threads, the challenge of the research supporting the analysis described in this paper was to find a method of predicting the number of loops there will be at each stage when one replaces one crossing at a time. The mathematician likes to seek out a structure beneath problems which enables many different instances to be analysed with ease. First of all one asks whether the crossing is where a loop crosses itself or another loop. The first case requires matching the crossing to a part of one or other of two simple circuits, and then the orientation of the replacement gap to vertical or horizontal; Table 1 showed the predictions. Figure 5 portrays the same information visually. In the second case the two loops are amalgamated by the replacement gap. This analysis, to predict the number of loops there will be in a Celtic knotwork design brings an area of abstract art born in the Celtic culture into the orbit of knot theory. It provides another bridge linking the joy of the abstract artist to the joy of finding a mathematical structure to predict the answer to a practical question. The analysis can be applied to any knot or link.

Islamic Art: An Exploration of Pattern

Carol Bier
Maryland Institute College of Art
1300 Mount Royal Avenue
Baltimore, Maryland 21217
cbier@mica.edu; cbier@textilemuseum.org; carol.bier@gmail.com

Abstract

As an historian of Islamic art in the Department of Art History at the Maryland Institute College of Art, I am continually learning as I endeavor to teach my students about pattern. Teaching about pattern in Islamic art has facilitated my own exploration of geometry in ways that also benefits my students. This visual presentation explores the results of a single assignment that pertains to coloring a linear plate reproduced in Bourgoin's classic work [6], *Arabic Geometrical Pattern and Design.*

Figure 1. Bourgoin [6], pl. 48.

This visual presentation explores the results of an assignment to students in a class on pattern in Islamic art taught at the Maryland Institute College of Art in Baltimore. The assignment pertains to a single pattern, reproduced as plate 48, among 190 plates showing linear drawings by Jules Bourgoin [6] that document Islamic monuments of Cairo in 1879 (**figure 1**). Bourgoin's book (minus the original French text) has been reprinted many times as one of the Dover Pictorial Archives; it is both inexpensive and readily available.

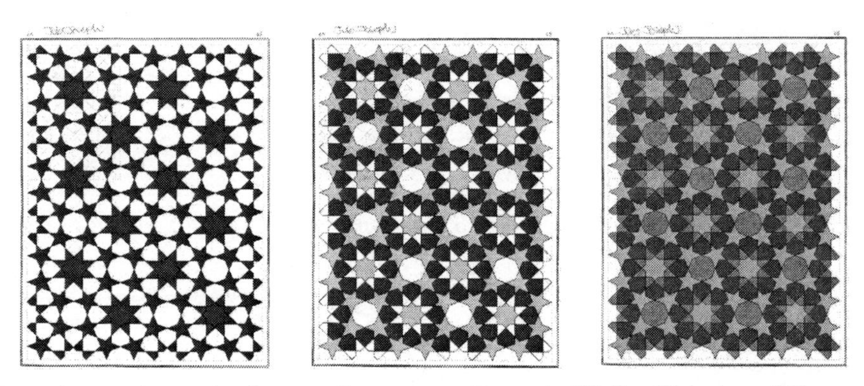

Figure 2. The assignment is to color the same linear pattern (Bourgoin [6], fig. 48) in three different ways: using a single color (*left*), using two colors (*middle*), and using three colors (*right*). Jules Joseph.

The approach I take with the students is intended to introduce them to the manifold possibilities and ambiguities inherent in pattern-making and to the underlying geometric framework for 2-D patterns in Islamic art [2].

The students are asked to reproduce Bourgoin's plate 48 (**figure 1**) three times and color the pattern using one, two, and three or more colors (**figure 2**).

At first, the assignment seems to be restrictive and highly repetitive. But there is no specification as to color or medium. Students tend to respond to the assignment, selecting different colors but using only one medium for each of the colorings. They have used crayons, colored pencils, pastels, water colors, opaque water colors, crayolas, colored markers, and digital paint media.

By restricting the choice of pattern to a single line drawing, with thirty-five students I may receive as many as one hundred and five results, which are usually all different, and significantly so.

The next step is to post the colorings on the wall of the classroom, initially arranged by grouping together each student's work (e.g. **figure 2**). The range of colorings and patterns invariably surprises the entire class, but the implications are immediate. They visually recognize the inherent possibilities for patterns within patterns, rendered apparent by this exercise.

We then explore what I call the attributes of pattern (**figures 2-7; 10-13**). One by one, the students identify such attributes as color, scale, shape, style (that of an individual), medium, grid, combinations of shapes, combinations of colors, and combinations of colors and shapes.

Figure 2, for example, represents the style of one individual's response to the assignment. **Figure 3**, however, isolates different shapes as an attribute distinct from style. Soon the students begin to recognize differences in effects due to the use of an outline, fill techniques, framing, relationships between figure and ground, color juxtapositions, and negative space. **Figures 4 and 5** isolate these other possible attributes of pattern.

Figure 3. Attributes of pattern (shapes). Eight-pointed stars (*left*), Rachel DuVall; hexagons (*center*), DJ William; five-pointed stars (*right*), Geoff Glisson.

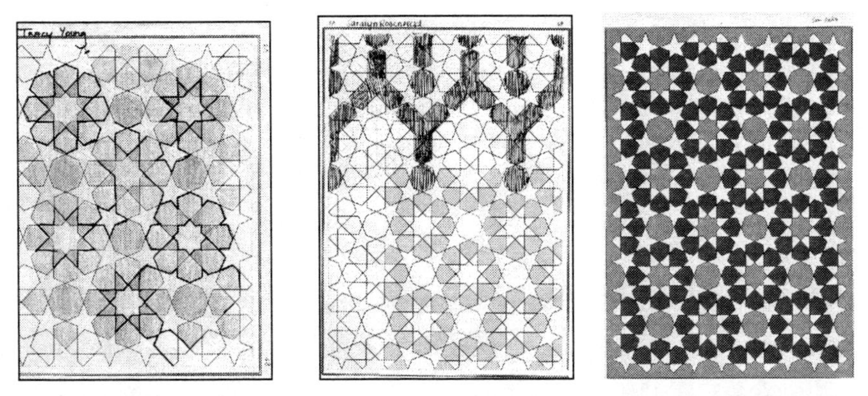

Figure 4. Attributes of pattern: outlines (*left*), Tracy Young; fill techniques (*center*), Saralyn Rosenfield; framing (*right*), Sam Ortiz.

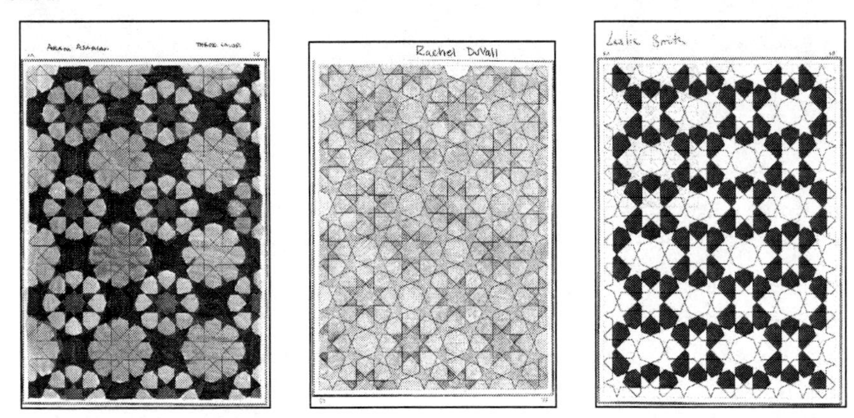

Figure 5. Attributes of pattern: figure/ground relationships (*left*), Aram Asarian; color juxtapositions (*center*), Rachel DuVall; negative space (*right*), Leslie Smith.

Then as the students focus their attention on lines, or axes, and sets of lines (**figure 6**), they begin to ascertain whether the sets are parallel or perpendicular, and if they are orthogonal, whether they are oriented vertically and horizontally or obliquely, or if they cross at other than right angles.

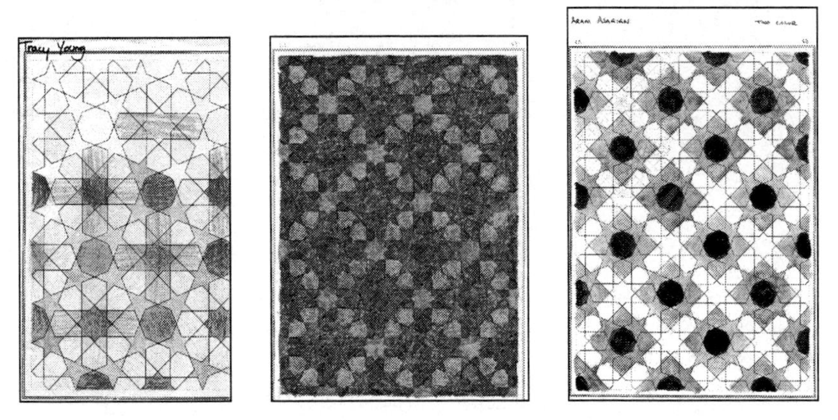

Figure 6. Attributes of pattern (grids): sets of parallel lines and perpendicular lines oriented vertically and horizontally (*left*), Tracy Young; sets of parallel and perpendicular lines oriented horizontally, vertically, and diagonally, with the diagonal axes emphasized by the choice of color, highlighting an oblique grid (*center*), Johanna Regalado; oblique grid (*right*), Aram Asarian.

They begin to recognize orthogonal crossings and oblique grids. They enumerate rows and columns, and notice that some of the rows are offset (**figure 7**). By highlighting individual shapes (octagons) or reserving shapes in negative space (octograms), our perception of the same pattern of form can be affected by the use of color so that we see alternating octograms and octagons (**figure 7**, *left*) or so that we see alternating octograms (**figure** 7, *center*) without referencing the octagons visually. By the further manipulation of color, shape, and negative space, the same linear pattern can again be adjusted so that we see different combinations of stars (**figure** 7, *right*). Through the manipulation of shape in relation to color, one may highlight two forms in alternation (**figure** 7, *left*). By such manipulation of color through alternation, one may highlight the horizontality so that offset forms are not even immediately apparent (**figure 8**, *left*).

Figure 7. Attributes of pattern (alignment of axes): offset rows + columns of octagons and octograms (*left*), DJ William; octograms (*center*), Leslie Smith; multiple stars (*right*), Jules Joseph.

Our focus on alignment and the relationships of rows and columns to each other (**figure 7**) yields to a discussion of symmetry in nature and art [15]. The students learn about translation, rotation, reflection and glide reflection. We examine which rigid motions rely on vectors and axes, and which rely on points (**figure 8**). We look for vectors of translation, axes of reflection, and points of rotation. Sometimes these occur when they are not initially expected. Eventually, the students can recognize local and global symmetries in the plane [14] as well as symmetry-breaking [3].

Figure 8. Playing with pattern: horizontality emphasized through alternation (*left*), Aram Asarian; variations in negative space (*center*), Leslie Smith; four colors using different tonalities and outlines brings out even more relationships (*right*), Johanna Regalado.

Soon enough they realize that our play with colors affects not only the patterns we produce, but that we are also playing with perception; that is, we are doing things with the various attributes of pattern, including color, shape, lines and axes, grids and negative space, which all interact to affect our perception [13]. Dark colors tend to appear as if they recede (**figure 8**, *left*), while bright colors seem to project, playing with our perception of the plane. Different colors affect our perception of different shapes, which relates this exercise also to that of Albers' *Interaction of Color* [1].

For this particular pattern the students identify five shapes: eight-pointed stars, five-pointed stars, hexagons, quadrilaterals, and octagons. They notice that the octagons are regular (equal sides and equal angles), but that the hexagons are not. Then they recognize that the five-pointed stars are effectively negative space, defined by the limits of other shapes, and that the eight-pointed stars (octograms) may be extended by the use of color juxtapositions to form other types of octograms (**figure 9**), emphasizing rotational symmetry. But by looking closely they may recognize that although the rotational symmetries visually dominate, it is the axes of reflection that govern the orientation of the five-pointed stars.

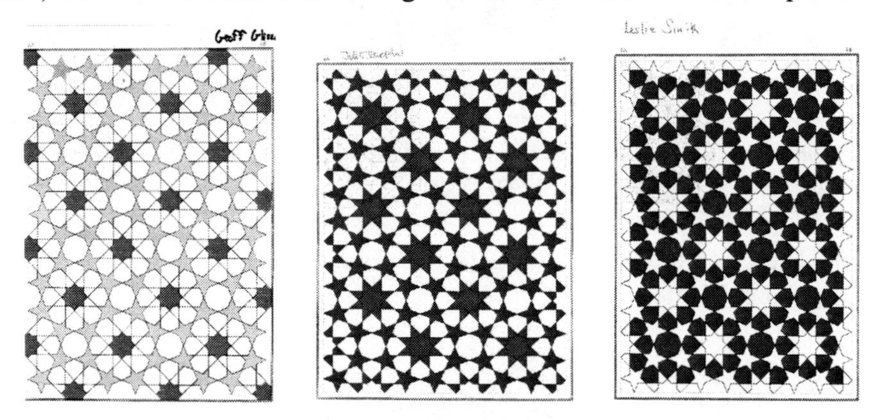

Figure 9. Playing with pattern: eight-pointed stars (octograms) with order 4 rotational symmetry, Geoff Glisson (*left*), Jules Joseph (*center*), Leslie Smith (*right*). These colorings highlight rotational symmetries within the pattern.

Figures 10 – 13 isolate color as an attribute of pattern. The single color exercise accustoms the students' eyes to perceive shapes (**figure 3**) and relationships among shapes. As the students proceed to use two colors, the question arises as to whether white is considered a color. Generally, I then ask if white is a color, and we may discuss differences between pigment and light. Some reply that it is the absence of color, so we begin to discuss the effect of negative space (**figure 10**). What if they really colored in the pattern using only one color? What would happen to the page? What effect would the outlines then have?

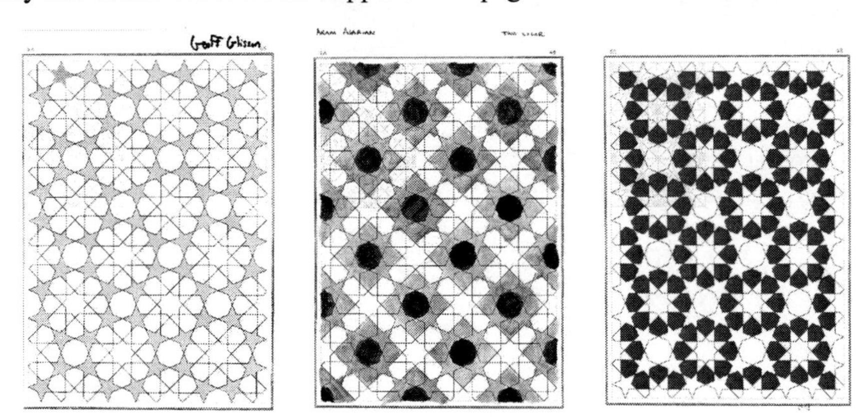

Figure 10. Attributes of pattern: negative space, Geoff Glisson, Aram Asarian, DJ William.

If white is not a color, it may still be considered as negative space, reserved from color. In this case it serves as ground, establishing different relationships between figure and ground (**figure 10**).

The two-color exercise (**figure 11**) emphasizes relationships among shapes. The development of an algorithm, which once established, carries the student through by means of iteration.

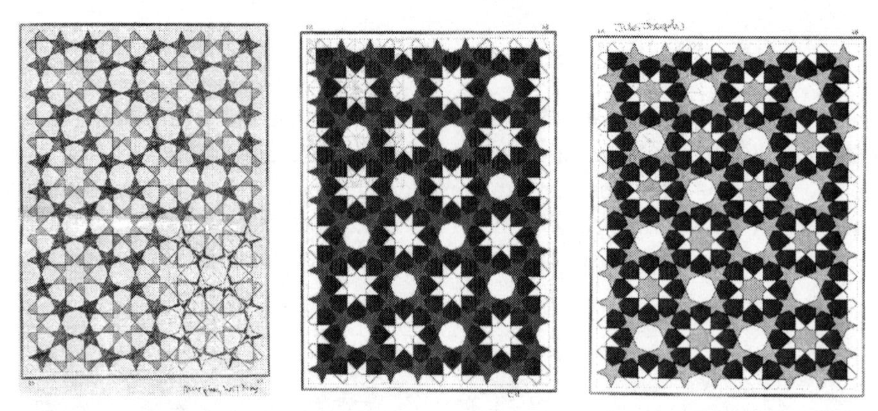

Figure 11. Attributes of pattern: relationships among shapes (using two colors). Eight-pointed stars + five-pointed stars (*left*), Murphey Wilkins; hexagons + five-pointed stars (*center*), DJ William; hexagons + five-pointed stars + eight-pointed stars (*right*), Jules Joseph.

The three color exercise (**figure 12**) enables them to explore other combinations of colors and shapes within a wide range of possibilities.

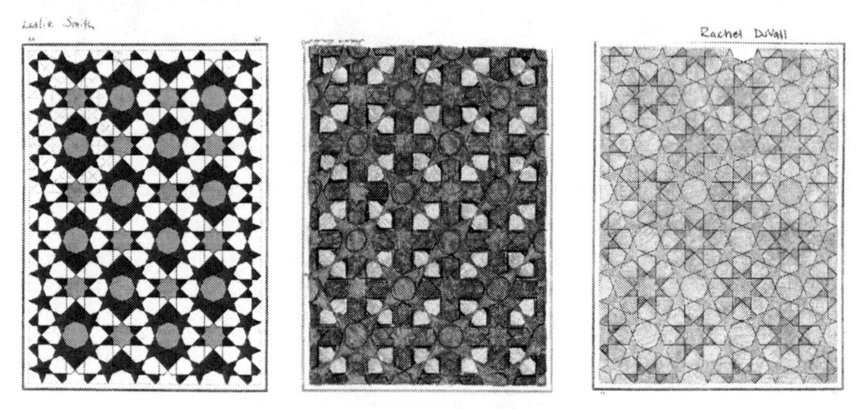

Figure 12. Attributes of pattern: explorations of colors and shapes (using three or more colors), Leslie Smith (*left*), Johanna Regalado (*center*), Rachel DuVall (*right*).

Some students have approached the exercise in an additive manner, selecting a second set of shapes added to the first group (**figure 13**), and a third set added to the second. Other students have approached the exercise in a divisive manner, subdividing areas through the use of color (**figure 14**).

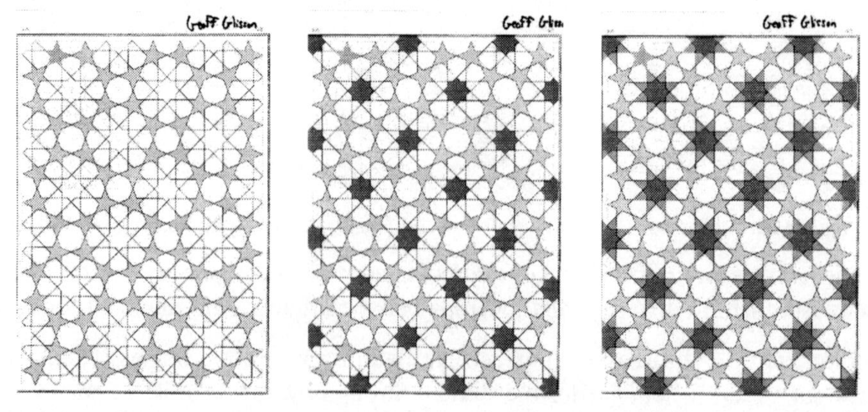

Figure 13. Attributes of pattern (colors). Here the sequence of one, two, and three colors is treated in an additive manner (one color: five-pointed stars; two colors: one color plus eight-pointed stars; three colors: two colors plus quadrilaterals), Geoff Glisson.

What students learn by this simple assignment is profound. The constructivist approach provides each of the students with an individual experience that is eye-opening, mind-bending, insightful, and fun. As a result of this exercise, the students seem to be more ready to take on the challenges of visual analysis when viewing slides of standing monuments of Islamic art, which otherwise may seem so unfamiliar. They are prepared for ambiguity and they take delight in finding patterns within patterns, focusing first on one perceived sequence, then on another, then on a third, and so on.

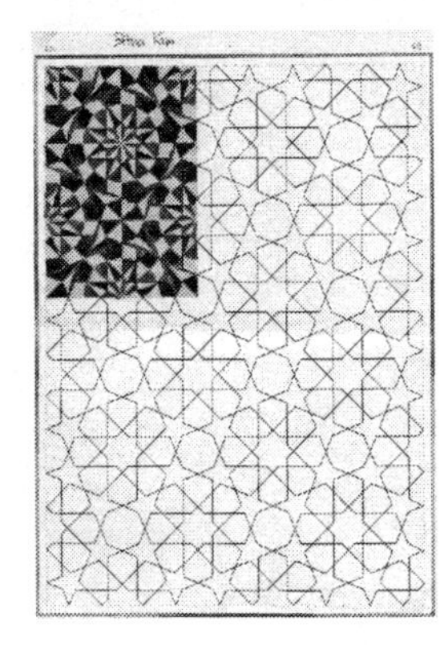

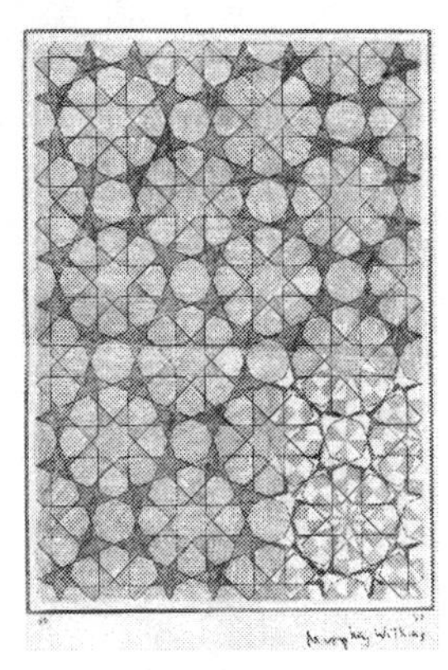

Figure 14. Playing with pattern: subdivisions of space, (*left, center*), Bitna Kim; (*right*), Murphy Wilkins.

The students seem to be more likely to respond to the complexity of Islamic art with a willingness to explore its possibilities, and a willingness to engage their minds with an exploration to realize a visual experience. They are ready for the visual challenges that Islamic art offers [8; 12] and they take pleasure in its perceptual delights (**figure 15**).

They are fascinated that such seemingly complex patterns can be created using only a compass and straight edge, or by paper-folding to create series of intersecting lines and angles with an underlying grid system. They become attuned to seeking out points and lines, calculating angles visually, and recognizing generative units and proportional relationships. As we proceed, as artists, they are introduced to terms that may be unfamiliar – orthogonal, periodic, algorithmic, group theory, set theory, combinatorics, permutations, tilings, tessellations [7; 9; 10; 14]. Although new to their vocabulary, the principles represented through language are already present in their art. Mathematical ideas expressed in art are not so much representational as they are expressive.

As we approach these concepts we attempt to describe what we experienced through this exercise of coloring. Several students mention that they would visit the Islamic galleries at the Walters Art Museum in Baltimore and see objects differently. For mathematics across the curriculum, the next steps can be transferred to other subjects as well. We begin to articulate through words what we see in the effects produced by the use of color: Color may be used to highlight, to frame, to select, to link, to fill, to contrast, to contain, to surround, to provide form, to separate, to pattern, to repeat, to divide, to establish a rhythm, to pulsate, to vibrate, to interrupt, to group together, to tie together, to deny, to imply. Color may be additive or divisive; it may simplify or complicate; it may isolate or provide focus; it may compete or complement; it may make static or it may give a sense of motion. Some other terms that describe what color does, in relation to form: it can confuse, articulate, alternate, outline, obscure, blur, relate.

Figure 15. Wall panels of ceramic tile displayed in the courtyard of the home of Doris Duke at Shangri La, Honolulu, now a museum [5, p. 256, fig. 1]. Similar tile panels were excavated at the site of Takht-i Sulaiman in Iran, dating to the late 13th century [11, p. fig. 94, cat. no. 104]. Note that the two tile panels are identical in linear outline to the pattern analyzed in this exercise, classified by Bourgoin as an "octagon" design [6]. Here, the same pattern is colored in two different ways using the same colors. Photograph by the author.

The value of this exercise is itself manifold. Students become involved in a seemingly simple assignment, which results in an exploration of the attributes of patterns, generally. They become familiarized experientially with pattern-making in a manner fundamental to an understanding of geometric design in Islamic art. They are introduced to mathematical principles inherent to the production of art in many cultures. If they are interested in metaphysics or the contemporary discourse in classical Islamic philosophy or the history of mathematics, they may pursue these subjects through independent study. Of particular relevance is the debate concerning faith and reason, based on interpretations of the works of Plato and Aristotle in the Islamic tradition. The notions of "one and many," "unity and multiplicity" are particularly relevant [4]. And the students' ability to perceive is forever changed, as they begin to recognize that the attributes of pattern interact dynamically. They are drawn into an awareness of the processes of perceiving in a manner that develops an individual visual and perceptual acuity. One student said this exercise put her "in the zone." I asked what that meant, and soon enough we established the notion that it was a zone of deep concentration and focus, one in which mindfulness and mindlessness seem to reside together. The process of repetition seems to facilitate a mindless focus, but at the same time, one must pay very careful attention to the process and nothing else in a meditative manner.

References

[1] Albers, J. *Interaction of Color*. Yale University Press, New Haven. 1975.

[2] Bier, C. *MICA Student Practicums*, http://mathforum.org/geomtery/rugs/resources/practicums, 2005.

[3] Bier, C. "Symmetry and Symmetry-Breaking: An Approach to Understanding Beauty," *Renaissance Banff - Bridges: Mathematical Connections in Art, Music, and Science*, ed. R. Sarhangi and R.V. Moody, pp. 219-26.

[4] Bier, C. "Geometry and the Interpretation of Meaning: Two Monuments in Iran," *Bridges: Mathematical Connections in Art, Music, and Science* (conference proceedings, July 2002), ed. Reza Sarhangi, Winfield, KS. Pp. 67-78. 2002.

[5] Bier, C. and Masunaga, D. "Islamic Art at Doris Duke's Shangri La: Playing with Form and Pattern," *Bridges: Mathematical Connections in Art, Music, and Science*, eds. R. Sarhangi and C. Sequin, pp. 251-58. 2004.

[6] Bourgoin, J. *Arabic Geometrical Pattern and Design*. [Repr. 1879] Dover, New York. 1973.

[7] Beyer, J. *Designing Tessellations: The Secret of Interlocking Pattern*, Contemporary Books, 1999.

[8] Castéra, J-M. *Arabesques: Decorative Art in Morocco*. ACR Éd.Intl. Courbevoie (Paris). 1999.

[9] Devlin, K. *Mathematics: The Science of Patterns*. Scientific American Library. 1994.

[10] Grünbaum, B. and G. C. Shephard. *Tilings and Patterns*. W. H. Freeman, New York. 1987.

[11] Komaroff, L. and Carboni, S. *The Legacy of Genghis Khan: Courtly Art and Culture in Western Asia, 1256-1353*. The Metropolitan Museum of Art, New York, and Yale University Press, New Haven. 2002.

[12] Lee, A.J. "Islamic Star Patterns," *Muqarnas* 4, pp. 182-97. 1987.

[13] Loeb, A.L. *Color and Symmetry*. John Wiley & Sons, New York. 1971.

[14] Schattschneider, D. *M.C. Escher: Visions of Symmetry*. Abrams, New York NY. Revised edition, 2004.

[15] Stevens, P. *Handbook of Regular Patterns: An Introduction to Symmetry in Two Dimensions*. MIT Press, Cambridge, Massachusetts and London. 1981; *Patterns in Nature*.

In Search of Demiregular Tilings

Helmer Aslaksen
Department of Mathematics
National University of Singapore
Singapore 117543
Singapore
aslaksen@math.nus.edu.sg
www.math.nus.edu.sg/aslaksen/

Abstract

Many books on mathematics and art discuss a topic called demiregular tilings and claim that there are 14 such tilings. However, many of them give different lists of 14 tilings! In this paper we will compare the lists from some standard references that give a total of 18 such tilings. We will also show that unless we add further restrictions, there will in fact be infinitely many such tilings. The "fact" that there are 14 demiregular tilings has been repeated by many authors. The goal of this paper is to put an end to the concept of demiregular tilings.

1 Introduction

Several authors, including Ghyka (1946, [5]), Critchlow (1969, [2]), Williams (1979, [13]) and Lundy (2001, [9]), introduce a concept called demiregular tilings They define an edge-to-edge tiling by regular polygons to be demiregular if it has more than one type of vertex, and claim that there are only 14 such tilings. However, they all give different lists of demiregular tilings! In addition, some of them cite Steinhaus (1937, [11]) who says: "[demiregular tilings] are perhaps even more beautiful ... their number is unlimited. (Why?)". Even though he does not explain why there are infinitely many such tilings and just leaves it as an exercise, it is still amazing that some cite him while still claiming that there are only 14 such tilings!

2 N-uniform Tilings

In order to understand Steinhaus's statement that there are infinitely many such tilings, we need to have a precise definition of "type of vertex". We can either say that two vertices are the same if they look the same locally, or we can require that they must be the same globally in the sense that there is a symmetry of the tiling that maps one vertex to the other.

If we consider a tilings consisting of rows of regular triangles, as in the regular tiling of type (3^6), and rows of squares, as in the regular tiling of type (4^4), then there will be vertices of types (4^4) where rows of squares meet, type (3^6) where rows of triangles meet, and type $(3^3 4^2)$ where rows of triangles and squares meet. However, by varying the pattern of rows of triangles and squares, we can get tilings where vertices of the same local type do not belong to the same transitivity class of symmetries, and are therefore not of the same global type.

If we define demiregular to mean more than one local type, then we can easily construct infinitely many different tilings of type $(3^6; 3^3 4^2; 4^4)$, i.e., tilings where all vertices are of these three types, but where there are no mappings taking one tiling onto another. This explains why Steinhaus says there are infinitely many demiregular tilings.

We will define a tiling to be n-Archimedean if it has n different types of vertices, where the type of a vertex refers to the type and order of polygons surrounding the vertex. We will define a tiling to be n-uniform if the group of symmetries of the tiling divide the vertices into n transitivity classes.

	1	2	3	4	5	6	7	8	9	10	11	12	13	14	15	16	17	18	Total
Ghyka	D	D	D	P		D	P	D		D	D	P	D	D	P			D	14
Critchlow	P	P	P	P	P	P	P	P	P	P	P	P	P	P					14
Steinhaus		P						P								P	P		5

Table 1: Eighteen demiregular tilings pictured or described in the literature

It follows from the classification of Archimedean tilings that there are no n-Archimedean tilings for $n > 14$ ([10]), but for $n = 2, ..., 14$ it follows from the arguments above that there are infinitely many n-Archimedean tilings.

It is know that the number of n-uniform tilings is equal to $11, 20, 61, 151, 332, 673$ for $n = 1, ..., 6$. This was proved for $n = 2$ by Krötenheerdt ([8]) in 1969, for $n = 3$ by Chavey ([1]) in 1984 and for $n = 4, 5, 6$ by Galebach ([3]) in 2002 and 2003.

However, if we consider n-Archimedean, n-uniform tilings, known as Krötenheerdt tilings ([8]), we do get a finite family of tilings. The number of Krötenheerdt tilings is $11, 20, 39, 33, 15, 10, 7$ for $n = 1, ..., 7$ and 0 for $n \geq 8$.

This confusion was briefly alluded to by Grünbaum and Shephard in [6]. We hope that the discussion above will put an end to the concept of demiregular tilings. However, we would like to engage in some mathematical archaeology and look at the lists given by Critchlow ([2]), Ghyka ([5]) and Steinhaus ([11]).

3 Lists of Tilings

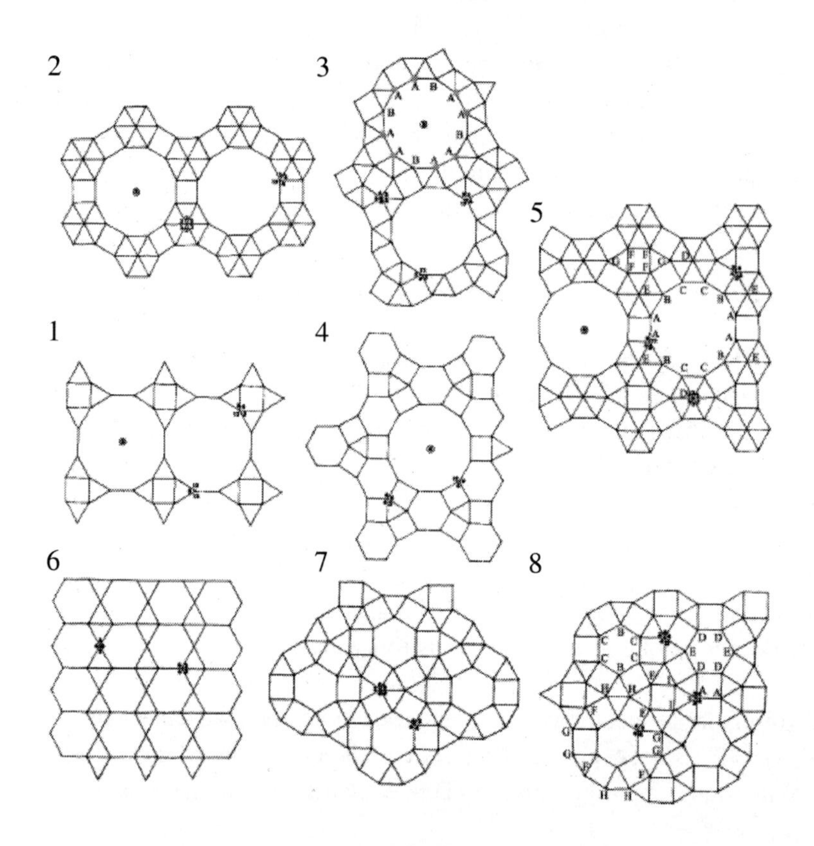

Figure 1: Seventeen demiregular tilings pictured in the literature, part 1

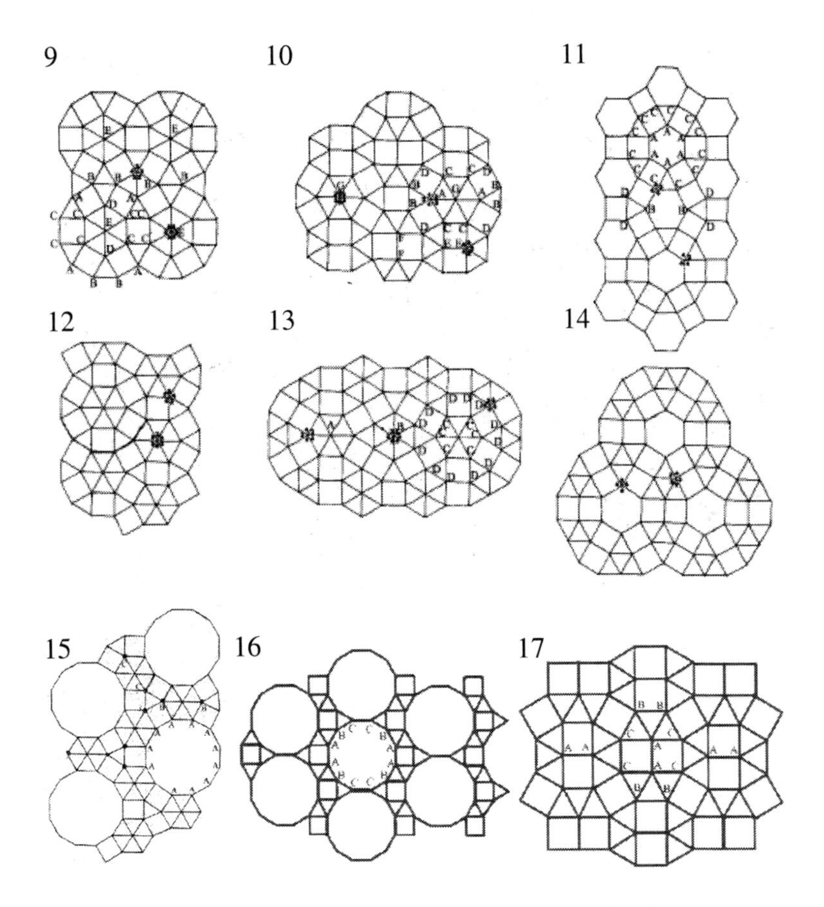

Figure 2: Seventeen demiregular tilings pictured in the literature, part 2

Table 1 lists the 18 tilings described in the three books. Ghyka lists 22 tilings, and claim that seven of them are "regular polymorph" (Archimedean), and 15 are demiregular. He obviously means that there are eight Archimedean and 14 demiregular. He only shows pictures of four of them, but he describes 10 others in a table. Nine of them are of the same type as the tilings in Critchlow that we have listed them above, but number 13 in his list does not correspond to any of Critchlow's pictures and we we have included it as Tiling 18 in Table 1, although it is not in Figures 1 or 2. In Table 1 a P denotes a tiling that is pictured, while a D denotes a tiling that is just described in the list.

Tiling 15 from Ghyka is similar to Tiling 4. The only difference is that the hexagon is broken up into six triangles. Critchlow does not include Tiling 15, but instead adds Tilings 5 and 9 to Ghyka's list.

In various printings of Lundy ([9]), there have been different lists. In one printing the list was identical to Critchlow except that Tilings 8 was replaced by a slightly altered version. In another printing, the heading talks about 14 demiregular tilings, the text says "more than twenty" and the picture shows 24. That's about par for the "theory" of demiregular tilings!

In an early version of [12], the claim about 14 demiregular tilings was repeated, but that has been corrected, and the web page now gives a brief explanation of the problem.

References

[1] D. P. Chavey, *Periodic tilings and tilings by regular polygons*, Ph.D. thesis, Univ. of Wisconsin, Madison, 1984.

[2] Keith Critchlow, *Order in Space, A Design Source Book*, Thames and Hudson, 1969.

[3] Brian L. Galebach, *Number of n-uniform tilings*, The On-Line Encyclopedia of Integer Sequences, http://www.research.att.com/~njas/sequences/A068599.

[4] Brian L. Galebach, *N-uniform Tilings*, http://probabilitysports.com/tilings.html.

[5] Matila Ghyka, *The Geometry of Art and Life*, Sheed and Ward, 1946 (second edition, Dover Publications, 1977).

[6] Branko Grünbaum and G.C. Shephard, Tilings by Regular Polygons, *Mathematics Magazine* **50** (1977), 227–247.

[7] Branko Grünbaum and G.C. Shephard, *Tilings and Patterns*, W. H. Freeman and Company, 1987.

[8] O. Krötenheerdt, Die homogenen Mosaike n-ter Ordnung in der euklidischen Ebene. I, *Wiss. Z. Martin-Luther-Univ. Halle-Wittenberg, Math.-Natur. Reihe* **18** (1969), 273–290.

[9] Miranda Lundy, *Sacred Geometry*, Walker and Company, 2001.

[10] NG Lay Ling, *Tilings and Patterns*, Honours Project, Department of Mathematics, National Univ. of Singapore, 2004, http://www.math.nus.edu.sg/aslaksen/projects/nll.pdf.

[11] Hugo Steinhaus, *Mathematical Snapshots*, Dover Publications, 1999 (originally published in Polish, 1937; first English edition, 1938; Oxford University Press, 1950).

[12] Eric W. Weisstein, *Demiregular Tessellation, from MathWorld — A Wolfram Web Resource*, mathworld.wolfram.com/DemiregularTessellation.html.

[13] Robert Williams, *The Geometrical Foundation of Natural Structure: A Source Book of Design*, Dover Publications, 1979.

Tribute to the Atomium

Samuel Verbiese
Artist
Terholstdreef 46
B – 3090 Overijse
Belgium
E-mail: verbiese@alum.mit.edu

Abstract

This paper describes the project of a sculpture stemming from the pattern of the outside skin layout of the Brussels' Atomium spheres. Two dual polyhedra are considered, the Catalan disdyakis dodecahedron and the Archimedean knotted cuboctahedron. The special projected location and setup of this sculpture makes it a good candidate for celebrating the upcoming 50th anniversary of the Atomium in Brussels built for the World Fair 1958.

1. The pattern of the outside skin layout of the Atomium spheres

The Atomium monument [1] built in 1957 in Brussels, Belgium, for the 1958 World Fair represents a magnification of 165 billion times of the nine atoms defining the cubic centered symmetry of an iron crystal. The sheet metal pieces covering the spheres feature nine large circles, metaphors of the electron orbits, and partition the spheres in 48 equal spherical triangles with a right angle (Fig.1), this layout bearing kind of a triangulated "spherical cube" symmetry naturally aligned in the same orientation of the whole structure.

With the recent replacement of the severely weathered original mirror-hand-polished 1.2 mm thick aluminum skin, with a new chemically polished stainless steel sandwich, and in the framework of my artistic interest in the Zome System [2] already related to the Atomium (Fig. 2), I happened to have my attention drawn on a drawing (Fig. 3) [3] showing the morphing from the Catalan polyhedron called the "disdyakis dodecahedron" to its dual, the Archimedean polyhedron called the "knotted cuboctahedron". The particular symmetry materialised by the triangular facets of the Catalan solid immediately appeared to be closely linked to the symmetry displayed by the spherical triangles on the skin of the Atomium spheres mentioned above. And I could build the Zome model (see Fig. 4) representing possible successive stages of the building of that particular symmetry.

The (magician) story goes as follows : I take a Zome connector from my pocket, and holding it with both hands, I pull out an original "long blue" (white here !) strut ended by two connectors, and use that "antenna" to form the upper part of the tower-shaped model. Then, peeling it like a banana, I generate the 3-D cross representing the unit axes of the Cartesian co-ordinates. Pulling out eight times a central connector along the axis equidistant of the planes in each of the eight three-planes up to the length of a "long blue", itself linked by a triplet of struts to the vertices on the co-ordinate axes, the obtained hollow solid does no longer contain any connector at the center. Next I further pull out these eight connectors to form a cube, which is represented on the lower part of the model. Finally, taking that cube, I just pretend blowing in it to inflate it into a sphere, and there you get precisely the symmetric pattern of the outside skin layout of the Atomium spheres !

Figures : *Assembly of characteristic pictures and images describing the project (see Figure captions in text)*

2. A new sculpture made of a special set of the original Atomium skin pieces.

Now, look at the way the Atomium spheres were originally built. The nine 40 cm wide circular belts, containing series of lights whose firing pattern at night suggests electron movements, cross each other in three different ways : 2 by 2 on spherical squares, 3 by 3 on spherical hexagons and 4 by 4 on spherical octagons each of sides 40 cm (Figs. 1, 6b). If you now imagine yourself placed successively in the center of each of the nine spheres, at 9 meters from the skin, visible from the inside, and asking those little original aluminum squares, hexagons and octagons to virtually migrate towards you, they would have ended-up connecting precisely in nine times a "knotted cuboctahedron" of about 1.4 meter in diameter, and so constituting sort of the essence of the Atomium symmetry ! Bear in mind that the pipes connecting the Atomium spheres do so precisely on the locations of octagons (for the outside pipes forming the large cube) and of hexagons (for the inside pipes connecting the central sphere to the outside ones) : those octagons and hexagons are thus missing facets in the nine new small polyhedral "spheres". Then, you join these nine small polyhedra into the cubic centered disposition of the Atomium, not with long slender cylindrical pipes, like in the Atomium itself, but with short prisms, made of the original plexiglass window panels, which fit in the missing octagons and hexagons. You just realized a 4x4x4 meter sculpture that, instead of standing upright on a corner like the real Atomium, you would set to rest (aren't they tired, those original, "patrimonial" pieces almost 50 years old ?) on one side of the cube. That precious structure could have been positioned some 140 meters away from the Atomium, on the axial lawn down the avenue du Centenaire, its diagonal shooting (Fig. 5a) a laser beam towards the shiny renewed upper sphere of the Atomium to be reflected horizontally, on the precise axis of this avenue (Fig. 2b), over the Brussels' skyline towards impacting a flag (Fig. 2c) on the bronze horses standing on the huge arches of the "Cinquantenaire" built to remember the 50th anniversary of Belgium born in 1830 (see map included, Fig. 5b). As in less than two years the Atomium will have its own 50th anniversary, I called this "active" sculpture, "D'un Cinquantenare à l'Autre" (from one 50th anniversary to another). Unfortunately, the original elements were destroyed and lost during the refurbishing process despite due alarming of the pertaining authorities, so if ever built, this sculpture would be made of new material, probably flat aluminum and plexiglass panels, as metaphors of the original skin, and could include original panels if any still in existence could surface.

3. Figure captions

1: Picture of the Atomium [1] during refurbishing, featuring the old and new skins and part of the supporting structures. The details of the geometry described in the text are apparent.
2: How the Zome System manual [4] inspired me working on the Atomium with Zome : (a) Metazome [5] model, and (b) virtual Zome model developed with Scott Vorthmann's "vZome" software [6].
3: Morphing from the disdyakis dodecahedron to the knotted cuboctahedron [3], inspiring this work.
4: Zome model depicting possible phases of the generation of the symmetry of the Atomium sphere.
5: Setup of the sculpture with a laser beam (a) reflecting on the Atomium upper sphere over Brussels (b) toward the flag on the Cinquantenaire Monument (c). See map "under" picture (b).
6: (a) Zome model of the sculpture, (b) How the new "spheres" came to life.
7: Javaview [7] model of the sculpture.
8: Javaview model rendering of the sculpture, courtesy by Konrad Polthier [8].
9: Other works of mine featuring the Atomium : (a) Looks of Sobieski Park 1 km away from the Atomium, kind of its "shadow" mowed in the prairie by IBGE-BIM, Brussels' Parks authority, which inspired me associating the Atomium (kind of a 3-D labyrinth !) with labyrinths (another theme of interest to me [9]) , (b) early project and (c) realised project; (d) composite knot (yet another theme) based on the Tibetan "infinite loop" [10] morphing into the Atomium (all is within all and vice versa...).

Acknowledgements

I like to dedicate this work both to the late father of the Atomium, ir. André Waterkeyn, due to an aspect of the exceptional beauty of his design that has come to light with this study, and to my own father Jérôme Verbiese who brought me in front of the monument in an early stage of its construction in fall 1957, which certainly helped me becoming an engineer, and now an artist partly engaged in geometric work.
I like to thank my sister ir. Ruth Verbiese, who, knowing the importance of the Atomium in my life, generously offered me which would end up to become the original aluminum triangular plate #823 that sat on the equator and against the west of the main meridian of the central sphere, looking most happily towards the home of my parents, situated at the other side of Brussels, 823 decameters away, aligned on the very axis of the monument and its avenue, past the Cinquantenaire arches ! This gift triggered the study leading to this presentation.
I like also to thank, without naming them, the number of technical individuals in charge of the rehabilitation of the Atomium, now reopened to the public, who supported this project.
Konrad Polthier, the always helpful father of Javaview, provided a nice rendering of the resulting "spheres" and, last but not least, Paul Hildebrandt, co-author and President of Zometool, kindly acknowledges my work (see my Metazome Atomium in his paper on this conference) and contributed material.

References

[1] http://www.atomium.be , official site of the Atomium
[2] http://www.zometool.com , official site of the Zome System
[3] "John" johnarchist@yahoo.com, *"etrunc 2triangulated cube great rhombicuboctaztweb"* , [Actually a morphing from a "disdyakis dodecahedron" to a "knotted cuboctahedron"],
Posted: Dec 20, 2004 on http://ph.groups.yahoo.com/group/ZomeUniverse/photos/view/71b3?b=10.
[4] http://www.zometool.com/products-zomemanual21.html : Zome Manual 2.1
[5] http://www.vorthmann.org/zome , official site of Scott Vorthmann's vZome ("Virtual Zome") SW
[6] http://www.eecs.berkeley.edu/~mihal/zome/metazome.htm , detailed description of Metazome
[7] http://www.javaview.de , official site of "Javaview " imaging SW by Konrad Polthier et al.
[8] Konrad Polthier, personal communication
[9] Samuel Verbiese, *"Amazing Labyrinths"*, poster presented, a.o. at the ISAMA-Bridges 2003 Conference, University of Granada, Granada (Spain), July 23-26, 2003
[10] Samuel Verbiese, *"Merging a Tibetan Endless Loop xith an Ocean Plat"*, poster presented, a.o. at the ISAMA-Bridges 2003 Conference, University of Granada, Granada (Spain), July 23-26, 2003

Copyrights

RHYTHMOS: An Interactive System for Exploring Rhythm from the Mathematical and Musical Points of View

Jakob Teitelbaum and Godfried Toussaint*
School of Computer Science
Center for Interdisciplinary Research in Music Media and Technology
McGill University
Montréal, Québec, Canada
E-mail: godfried@cs.mcgill.ca

Abstract

This paper introduces RHYTHMOS: an interactive software system designed as a tool-kit for the visualization, exploration, understanding, analysis, praxis, and composition of musical notated (symbolic) rhythms. As such it provides user-friendly bridges between art (music composition), performance (praxis), mathematics (cyclic polygons and the distance geometry of point sets on a circle), and science (the psychology of music perception). A description is provided of the system's capability and interactive graphical user interface. Applications to teaching, learning, and practicing rhythms are discussed. Examples are given of the kinds of research that RHYTHMOS facilitates. These include the testing of rhythmic features for the classification, clustering, and phylogenetic analyses of families of rhythms.

1. Introduction

RHYTHMOS is an interactive software system designed as a tool-kit for the visualization, exploration, understanding, analysis, praxis, and composition of musical notated (symbolic) rhythm. It has a graphical user interface (GUI) that allows the user to enter, view and analyze rhythms in a wide variety of geometric representations [12]. Its sound system allows the user to hear the rhythms at different tempos and volume. More than one rhythm may be heard simultaneously, each with a different timbre. One key innovative feature of the system is the ability to view both the rhythm represented as a cyclic necklace on the circle of time, and the full-intervallic content (histogram) of all the inter-onset durations [15]. This allows the user to dynamically control the shape of the histogram and to listen to the resulting rhythms corresponding to the shapes displayed. Another key innovation is the ability to display a rhythm in vertical *braid* notation so that the user can play along following the left-hand and right-hand beats as they occur on the screen [10]. The system contains a library of rhythms, as well as a list of different rhythmic distance (similarity) measures [14] that may be computed to obtain a distance matrix for any family of rhythms selected from its library, thus facilitating the use of phylogenetic analysis programs such as *SplitsTree* [2], [3], [6], [13].

2. The Main Features of RHYTHMOS

2.1. Graphical User Interface. RHYTHMOS is a multi-window system. In the main window, pictured in Figure 1, the user selects one or more rhythms to be explored, and then clicks on the desired view or type of analysis. This, in turn, is displayed in a new window, which may be moved, resized or minimized.

*This research was supported by NSERC.

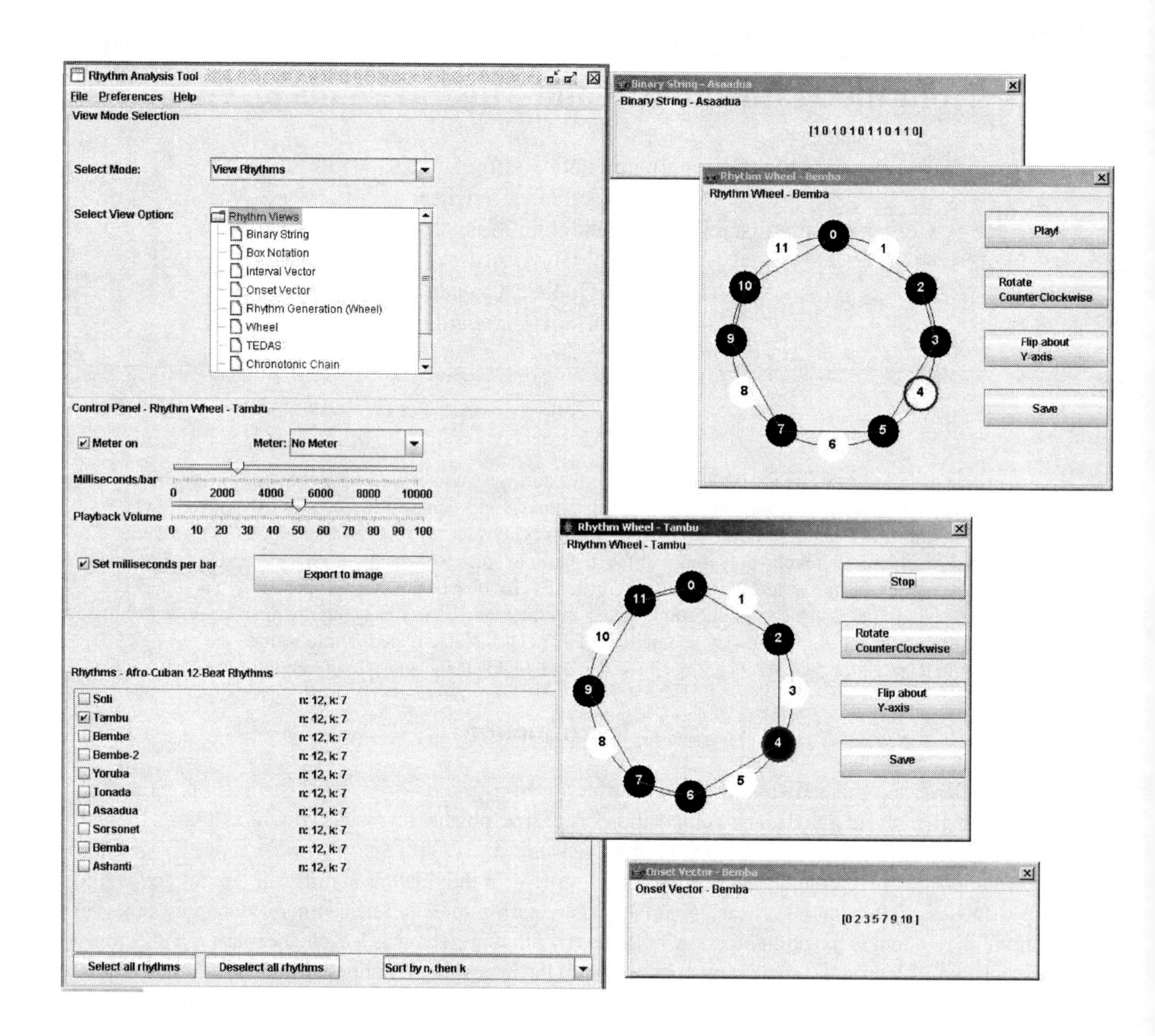

Figure 1: The main window (left) shows view options, a control panel, and a family of rhythms. The control panel has slide bars for tempo and volume adjustment. A variety of underlying meters can be turned on that accompany the rhythm with a different timbre. A window may also be exported as a figure. The four windows on the right show various rhythms represented from top to bottom, respectively, as a binary string, as polygons on circles of time, and as an onset-time-coordinate vector. The highlighted circle on pulse number four of the wheels indicates the pulse being played. Thus the user sees which pulse is being played at the same time it is heard. The position of the windows may be rearranged on the fly as desired.

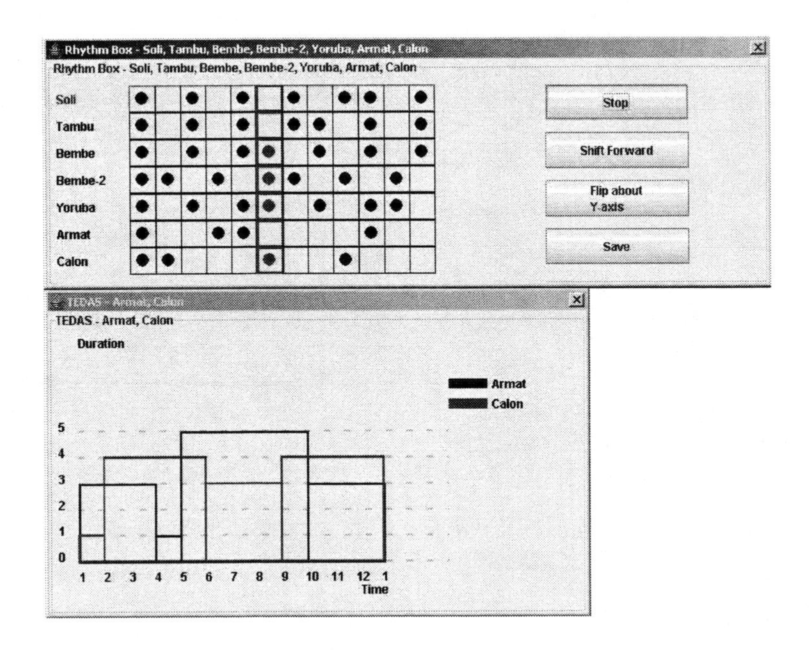

Figure 2: A screen shot of the box and TEDAS representations of some rhythms. The highlighted sixth column in the upper figure indicates that the sixth pulse is being "played" in all the rhythms simultaneously.

There are three basic modes to the program: Input Rhythm, View Rhythms, and Analyse Rhythms. The mode is chosen via the main window, which displays all relevant choices for that mode.

The Input mode allows the user to create rhythms manually, using any of six visual representations. Alternatively, the user may load a predefined library of rhythms from a file, or use existing rhythms to generate new ones, either manually or algorithmically, in the View mode.

Figure 2 is a screen shot illustrating two such visual representations of rhythms: the box notation and the TEDAS representation. The box notation is commonly used by ethnomusicologists. It was popularized in the West by Philip Harland at the University of California in Los Angeles in 1962, and is also known as TUBS (Time Unit Box System). However, it has been used in Korea for hundreds of years [7]. The TEDAS representation was proposed in 1987 by Kjell Gustafson, at the Phonetics Laboratory of the University of Oxford with the aim of visualization [4], [5]. Each temporal element (interval) is a box and both the x and y axes represent the time-durations of the interval.

In the View mode, the user may view one or more rhythms, and a variety of their properties, in any of ten displays. Via these displays, the user may also modify rhythms, generate new rhythms, and listen to rhythms. As mentioned, each display exists in its own window; these windows are dynamically linked. An example of one of the main innovative features of RHYTHMOS is the computation of the inter-onset interval histogram of a rhythm. For example, if the user modifies a rhythm called *Sorsonet*, then the *Sorsonet* histogram will instantaneously be recalculated with the altered rhythm. See the illustration in Figure 3.

The Analysis mode supports a variety of analyses, divided into Rhythm Properties and Distance Measures (dissimilarity measures). These properties may be used to compare rhythms in order to better understand them. Such comparisons help the user to remember rhythms and their inter-relationships, thus serving as a valuable pedagogical tool. RHYTHMOS has the capability of computing a wide variety of standard distances such as the Hamming distance between the rhythm's binary sequences, or the *chronotonic* distance discussed in [14], as well as several new distance measures, such as the *directed swap-distance* analysed in [1]. The chronotonic distance between two rhythms is the total area of the region contained in between their TEDAS functions of time.

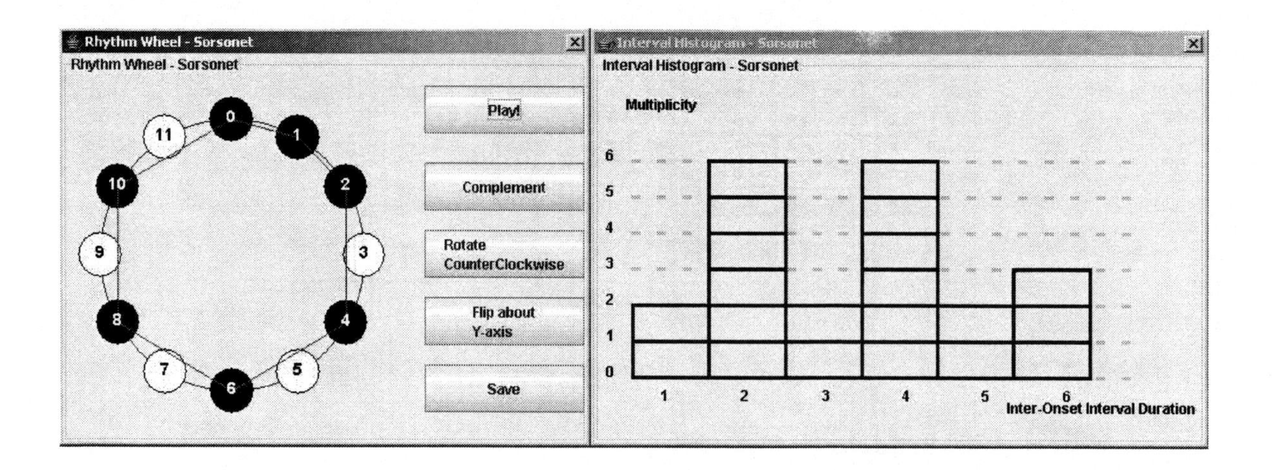

Figure 3: The wheel showing a *Sorsonet* bell-pattern, and its inter-onset-interval histogram. This histogram shows, for each inter-onset interval duration value, the number of times (multiplicity) it occurs in the rhythm.

While some view-specific buttons, such as 'Flip about Y-axis', are contained in the display window, most actions are performed via the main window. Similarly, although the main window shows some of the settings of the display windows (such as tempo), virtually all the information is contained in the display window itself. Thus, the interface is neatly divided into input and output: the user enters information in one window, and receives information from the others.

In addition to the interactive features described in the preceding, RHYTHMOS also helps the user by means of a comprehensive popup system: if the user, while modifying a rhythm, happens to turn it into another rhythm that is already in the library, a popup window will let the user know of the overlap and provide the name of the rhythm in question.

2.2. Audio Playback with Visual Update. RHYTHMOS also supports realtime visual feedback for audio playback of up to eight rhythms, each with an individual timbre and volume, as illustrated in Figure 2. Thus, as a displayed rhythm is played, a highlight visually traces the playback in the display window. Also supported is the simultaneous playback of multiple rhythms, even those having different values of n (number of pulses in the cycle) - this is handled by specifying the time duration of one cycle, regardless of the value of n. In addition to these parameters, the user may also control the tempo, and may choose a background meter.

2.3. System Features. Space limitations allow us to only briefly touch upon three features: Cross-Platform Conformance, Extensibility, and Error Handling.

2.3.1. Cross-Platform Conformance. RHYTHMOS is written entirely in Java. Windows, UNIX, Macintosh, and other computers need only install Java 1.5.

2.3.2. Extensibility. The system's modular, object-oriented design reflects an emphasis on extensibility: new analyses and views may be easily added without disrupting other components. Furthermore extra features, such as the exportation to image and audio files, were implemented with little reprogramming needed for the existing version.

2.3.3. Error Handling. RHYTHMOS contains a robust error handling system, designed to take care of invalid user-input without crashing or damaging existing data. When an error occurs, an informative message appears notifying the user, and describing the problem.

3. The Value of RHYTHMOS to the User

The above features illustrate only some of the capabilities of RHYTHMOS which underpin the true benefit that a wide variety of users will derive from the system. RHYTHMOS introduces new users to an unfamiliar field of study in a visual, practical, gentle, intuitive, exploratory way. In addition, scientists and mathematicians will appreciate its time-saving analytical tools, while users interested in understanding and learning musical rhythm, without learning music theory, will enjoy the multiple visual and aural representations of rhythm that RHYTHMOS offers, as well as its possibilities for playing along with such feedback.

3.1. Wide Accessibility. Much of the value of RHYTHMOS lies in the fact that it is so easy to use; anyone with the most rudimentary of computer skills may, within minutes, create, view, modify, listen to, and play along with rhythms. Furthermore, the user is also exposed to a diverse array of rhythms: on the RHYTHMOS website, rhythmos.cs.mcgill.ca, libraries of rhythms from around the world will be made freely available. Finally, anyone interested in learning about analytical tools, such as rhythmic properties or rhythm similarity measures, may directly apply these tools to rhythms of their choice. This learning-by-doing method provides an engaging and instructive complement to the existing literature in the field.

3.2. Education. RHYTHMOS may also function as a valuable educational tool. Many budding musicians have trouble grasping the abstract nature of musical notation; using the visual displays and audio playback, the rhythms become tangible, helping students learn to visualize and understand rhythms, and link them to Western music notation.

At present there is great interest in teaching mathematics in elementary schools using movement and music, particularly in the Montessori and Waldorf education systems [8]. RHYTHMOS is an ideal tool for accomplishing these educational goals. For example, since polygons represented on the circle of time are in fact rhythms, children may be taught the different types of polygons by listening to how they sound. Although the present version of RHYTHMOS is not directed mainly at young children, by simplifying the user interface to solely emphasize the creation, display, and playback of rhythms, and by disabling the analyses, the system provides an ideal environment for children to explore rhythm on their own. Indeed, a children's version called RHYTHMOS-Junior is also being launched.

3.3. Play-Along Practice. RHYTHMOS offers a variety of possibilities for the user to play along with the audio-visual output, using fingers or hands on a table or other instrument. One important skill for students to learn, is to play a different regular rhythm with each hand, such as 3 beats against 2, 4 beats against 3, etc. For example, in the case of 4 against 3, the left hand could play the 3 = [x . . . x . . . x . . .], and the right hand could play the 4 = [x . . x . . x . . x . .]. Displaying and listening to both rhythms simultaneously on the wheel helps to learn this task more quickly.

Another important rhythm-learning technique is to play *all* the pulses, with both hands alternating, so as to make the rhythm *emerge* from the 'background' of pulses by accentuating the correct onsets. For example, the rhythm [x . x x . x . x x . x .] could be played by starting the pulses with the right hand so that the right hand plays all the odd pulses and the left hand all the even pulses. Then the right hand would accentuate the onsets in [x . x . . . . . x . x .] whereas the left hand would accentuate the onsets in [. . . x . x . x . . . .]. RHYTHMOS can decompose any rhythm into this left-hand-right-hand *braid* notation and display it in a vertical side-by-side fashion so that the user can easily follow along and acquire these skills more easily.

3.4. Mathematical Analysis. RHYTHMOS provides an efficient timesaving tool for intensive mathematical exploration and analysis of rhythms; with a few clicks of the mouse, hours of tedious calculations of

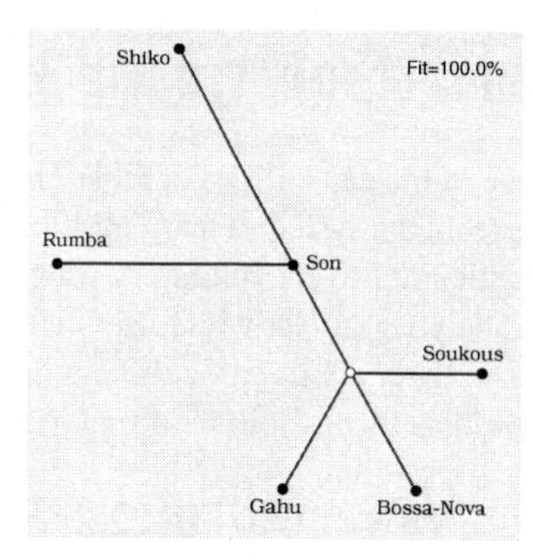

Figure 4: The *SplitsTree* of the six fundamental clave rhythms using the Hamming distance.

a variety of mathematical properies of rhythms, distance measures between rhythms, and histograms of inter-onset durations, are avoided. The dynamic linkage of display windows also encourages exploratory analysis: as a rhythm changes, its displayed properties reflect those changes in real-time. These features allow scientists, for the first time, to thoroughly examine an entire family of rhythms, using multiple tools, in minutes rather than days or weeks.

3.5. Phylogenetic Analysis. RHYTHMOS facilitates the application of techniques for generating phylogenetic trees from the distance matrices it calculates and outputs [6]. One of these methods, called *SplitsTree*, already used in several rhythmic studies, has the desirable property that it produces graphs that are not trees, when the underlying proximity structure is inherently not tree-like [3], [6]. Like the more traditional phylogenetic trees, *SplitsTree* computes a plane graph embedding with the property that the distance in the drawing between any two nodes reflects as closely as possible the true distance between the corresponding two rhythms in the distance matrix. However, if the tree structure does not match the data perfectly then new nodes in the graph may be introduced, as in Figure 4, with the goal of obtaining a better fit. Such nodes may suggest implied "ancestral" rhythms from which their "offspring" may be derived. In addition, edges may be split to form parallelograms, such as in Figure 5. The relative sizes of these parallelograms are proportional to an *isolation index* that indicates how significant the clustering relationships inherent in the distance matrix are. *SplitsTree* also computes the *splitability index*, a measure of the goodness-of-fit of the entire splits graph. This *fit* is obtained by dividing the sum of all the approximate distances in the splits graph by the sum of all the original distances in the distance matrix [3], [6]. The rhythms in Figures 4 and 5 comprise the six fundamental 4/4 time *clave* (and bell) timelines used in traditional African and Afro-Cuban music [12]. RHYTHMOS outputs a distance matrix in a format that may be directly plugged into the *SplitsTree* program.

3.6. Potential Future Applications. In addition to its main function, RHYTHMOS may also be employed in other fields: for instance, as a composition tool for generating new rhythms, or perhaps in cognitive science research on rhythm perception. The extensible design of RHYTHMOS makes any of these applications realistic. Adding a musical-notation function, optimizing the microsecond-accuracy of audio playback, or simplifying the user interface for young children, are all achievable customizations that require little alteration of the main system.

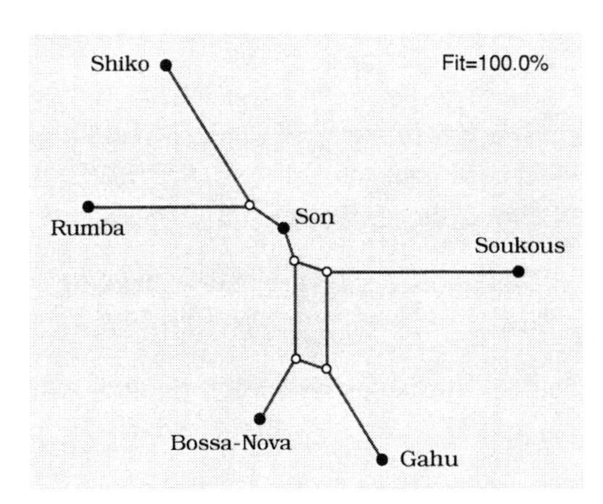

Figure 5: The *SplitsTree* of the six fundamental clave rhythms using the chronotonic distance.

3.7. Advantages over Standard Systems. In terms of visualization capabilities and mathematical analysis of rhythms, RHYTHMOS is the first of its kind: no other software system provides so many analytic tools. As a tool for simply creating rhythms and listening to them, however, RHYTHMOS does have its peers: Notepad [9] and Finale [11]. Notepad and Finale are much more extensive than RHYTHMOS in that they encompass pitch, staves, dynamics, and so on. However, this difference also accounts for the main advantage of RHYTHMOS : it is much less complicated than these systems. For instance, Finale 2004 has 245 different buttons and menu items on the main screen; RHYTHMOS has only 26.

The two other systems on the market that resemble RHYTHMOS (both use a wheel to represent rhythms) are the Flamenco Metronome Compás and the *O-generator*. The Flamenco Metronome Compás is severly limited by being applicable only to the 12-beat patterns of flamenco music, and functions only as a play-along system. The *O-Generator* allows the composition of music with a variety of instruments besides percussion. Unfortunately it is limited to 16-beat patterns, and is designed specifically for ages 10-16. RHYTHMOS on the other hand is useful to both a Ph.D., and an eight-year old child.

4. Conclusion

RHYTHMOS serves as a bridge between the art of music and the science of mathematics; it makes music theory more accessible to the student of music; it carries unacquainted users closer to the realms of structural and mathematical rhythmic analyses. It also serves as a bridge between theory and practice by providing pedagogical tools for playing along with audio-visual feedback. RHYTHMOS has a well-defined function: to aid scientists in the mathematical analysis of rhythms, and to aid all users in exploring and learning about rhythm. Describing *O-Generator*, David Mason, the Content Manager of London Grid for Learning, states "Most music software provides the sentence. *O-Generator* shows how the sentence is put together!" RHYTHMOS goes further by fleshing out the *meaning* of the sentence.

References

[1] Justin Colannino and Godfried Toussaint. An algorithm for computing the restriction scaffold assignment problem in computational biology. *Information Processing Letters*, 95(4):466–471, August 2005.

[2] Miguel Díaz-Bañez, Giovanna Farigu, Francisco Gómez, David Rappaport, and Godfried T. Toussaint. El compás flamenco: a phylogenetic analysis. In *Proc. BRIDGES: Mathematical Connections in Art, Music and Science*, Southwestern College, Kansas, July 30 - August 1 2004.

[3] A. Dress, D. Huson, and V. Moulton. Analysing and visualizing sequence and distance data using SPLITSTREE. *Discrete Applied Mathematics*, 71:95–109, 1996.

[4] Kjell Gustafson. A new method for displaying speech rhythm, with illustrations from some Nordic languages. In K. Gregersen and H. Basbøll, editors, *Nordic Prosody IV*, pages 105–114. Odense University Press, 1987.

[5] Kjell Gustafson. The graphical representation of rhythm. In *(PROPH) Progress Reports from Oxford Phonetics*, volume 3, pages 6–26, University of Oxford, 1988.

[6] Daniel H. Huson. SplitsTree: Analyzing and visualizing evolutionary data. *Bioinformatics*, 14:68–73, 1998.

[7] Lee Hye-Ku. Quintuple meter in Korean instrumental music. *Asian Music*, 13(1):119–129, 1981.

[8] Robert E. Jamison. Rhythm and pattern: discrete mathematics with an artistic connection for elementary school teachers. *DIMACS Series in Discrete Mathematics and Theoretical Computer Science*, 36:203–222, 1997.

[9] Bill Purse. *The Finale Notepad Primer: Learning the Art of Music Notation with Notepad.* Backbeat Books, San Francisco, California, 2003.

[10] Jay Rahn. Turning the analysis around: African-derived rhythms and Europe-derived music theory. *Black Music Research Journal*, 16(1):71–89, 1996.

[11] Thomas E. Rudolph and Vincent A. Jr. Leonard. *Finale: An Easy Guide to Music Notation, Second Edition.* Berklee Press, Boston, Massachusetts, 2005.

[12] Godfried T. Toussaint. A mathematical analysis of African, Brazilian, and Cuban *clave* rhythms. In *Proc. of BRIDGES: Mathematical Connections in Art, Music and Science*, pages 157–168, Towson University, MD, July 27-29 2002.

[13] Godfried T. Toussaint. Classification and phylogenetic analysis of African ternary rhythm timelines. In *Proceedings of BRIDGES: Mathematical Connections in Art, Music and Science*, pages 25–36, Granada, Spain, July 23-27 2003.

[14] Godfried T. Toussaint. A comparison of rhythmic similarity measures. In *Proc. 5th International Conference on Music Information Retrieval*, pages 242–245, Barcelona, Spain, October 10-14 2004. Universitat Pompeu Fabra.

[15] Godfried T. Toussaint. Computational geometric aspects of musical rhythm. In *Abstracts of the 14th Annual Fall Workshop on Computational Geometry*, pages 47–48, Cambridge, Massachusetts, November 19-20 2004. Massachussetts Institute of Technology.

Spidron Domain
The Expanding Spidron Universe

Dániel Erdély & Marc Pelletier
H-1015 Budapest, Battyany str. 31. I./12.
email: edan@spidron.hu www.spidron.hu

Abstract

A number of new discoveries have been made since the last Bridges conference in the area of Spidron research. Shown here are samples of what will be presented in London.

1. Two Dimensions & 2.5 Dimensional Reliefs

Spidron versions of the Penrose-Richert tiles have been discovered, as well as a negative space partner to the classic spidron diamond. Two more reliefs based on semi-regular tilings were also discovered, one based on the tiling of squares and octagons and the other based on tiling of squares, hexagons and dodecagons.

Figures 1 and 2: *Two new semiregular Spidron reliefs*

2. The Splatonic Solids and the Archimedians

In addition to the already known Tetra-Spidro ball and the Octa-Spidro ball, there are three other solids corresponding to the Platonics. Also there are 10 other semiregular Spidron solids. A number of linkage puzzles have been discovered from these shapes.

Figures 3 and 4: *The Splatonics (left), and two Archimedians and eight Quasi-Archemedians (right)*

3. Platonic Disections and Prism Towers

The Platonic solids can be dissected along skew polygons and we've found a family of prisms and towers.

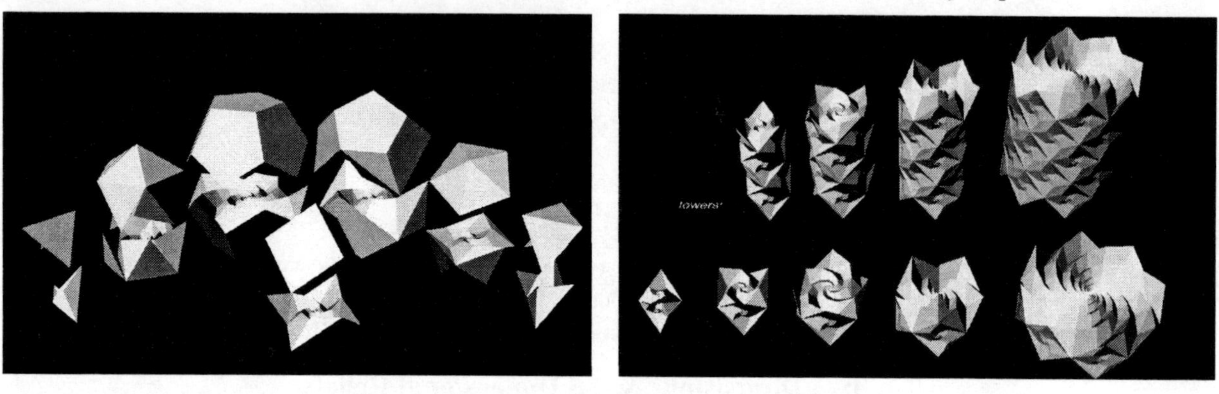

Figures 5 and 6: *Platonic dissections (left) and Prism towers (right)*

4. Non-Periodic Spidron Networks

The A6 and O6 rhombohedra can be used as building blocks for non-periodic arrangements of Spridron nests, with great potential for sculpture.

Figure 7: *The A6 and O6 building blocks*

Special thanks to Amina Allen, Rinus Roelofs, and Walt van Ballegooijen.

For more about Spidrons, refer to D. Erdély, *Some Surprising New Properties of Spidrons*, Renaissance Bridges Proceedings (2005) p. 179-186

An Introduction to Medieval Spherical Geometry for Artists and Artisans

Reza Sarhangi
Department of Mathematics
Towson University
Towson, Maryland, 21252
rsarhangi@towson.edu

Abstract

The main goal of this article is to present some geometric constructions that have been performed on the sphere by a medieval Persian mathematician, Abul Wafa al-Buzjani, which is documented in his treatise *On Those Parts of Geometry Needed by Craftsmen.* These constructions, which have been illustrated as flat images, could be considered the bases of the arts and designs that artists and artisans have created on both the exterior and interior surfaces of a dome. Therefore, such a dome art design is a result of cooperation between mathematicians and artists.

This article also shows that the construction of the icosahedron on a sphere presented in that treatise is not mathematically correct. However, the construction of the spherical dodecahedron is exact. The article also presents flat images of constructions of some Archimedean solids according to the treatise.

1. Introduction

The purpose of this article is to analyze a part of medieval mathematics, a study of the sphere, which has been conducted by a Persian mathematician: Abul Wafa al-Buzjani.

Abul Wafa al-Buzjani was born in Buzjan, near Nishabur, a city in Khorasan, Iran, in 940 A.D. He learned mathematics from his uncles and later on moved to Baghdad when he was in his twenties. He flourished there as a great mathematician and astronomer. He was titled *Mohandes* by the mathematicians, scientists, and artisans of his time, which meant "the most skillful and knowledgeable professional geometer". He died in 997/998 A.D. [1]

Buzjani's important contributions are in the areas of geometry and trigonometry. In geometry he solved problems about compass and straightedge constructions in the plane and on the sphere. Among some other manuscripts which have been disappeared throughout the history of mankind he wrote a treatise: *On Those Parts of Geometry Needed by Craftsmen.*

The treatise by Buzjani was originally written in Arabic, the academic language of the Islamic world of the time, and was translated to the Persian language of its time in two different periods: 10th and 15th centuries. There are only a few known original translations of this book in the world; two of them are kept in two libraries in Iran, *Tehran University Library* and *Astan Ghods Razavi Library*, and another in a library in Paris. The book *Applied Geometry,* which includes a contemporary Persian language translation of Buzjani's treatise appeared first in 1990 and then was republished with some corrections in 1997 [2]. The book also includes another treatise of later centuries: *Interlocks of Similar or Corresponding Figures.* Even though the book introduces a 15th century Persian mathematician as the author of this treatise, there are documents that suggest the possibility of a much earlier time writer, around the 13th century, for this work [3].

To construct shapes and patterns on a sphere, it is essential to have a basic understanding of a type of geometry that does not necessarily follow the foot-steps of Euclidean geometry: spherical geometry. Much formal study of spherical geometry occurred in the nineteenth century. However, some properties of this geometry were known to the Babylonians, Indians, and Greeks more than 2000 years ago. Euclid, in his *Phenomena,* discusses propositions of spherical geometry.

2. Study of Regular Tessellation on a Sphere

In Euclidean geometry of the plane, if p indicates the number of sides of a regular polygon and q the number of copies of the regular polygon about each vertex point, then it is elementary mathematics to show that $(p-2)(q-2) = 4$. Therefore, the number of regular tessellations on the Euclidean plane is three; the equilateral triangle tessellation, $\{3, 6\}$, the square tessellation, $\{4, 4\}$, and the regular hexagonal tessellation, $\{6, 3\}$.

It is interesting to study regular tessellations on a sphere. Since the sum of the angle measures of a spherical triangle is more than 180°, then for the tessellation of a regular p-gon with q copies about each vertex we have $(p-2)(q-2) < 4$.

An important fact about the above formula is the assumption of $p > 2$ for the Euclidean case. However, on the sphere, we may construct regular polygons of only two sides, which are called biangles or lunes. If the angle measure of a biangle is $360°/q$, then we are able to tessellate a sphere with q copies of the biangle.

A surprising fact about the spherical geometry is we don't have the similarity concept of polygons the way that we understand in Euclidean geometry. The following theorem gives a formula for the area of a spherical triangle:

Girard's Theorem: *If ABC is a spherical triangle with interior angles α, β, and γ, then the area of the triangle will be* $\pi r^2 \left(\frac{\alpha+\beta+\gamma-180}{180}\right)$, *where r is the radius of the great circle.*

In the Euclidean plane if two triangles have identical angles then they do not necessarily have the same area. On the sphere, however, if two triangles have congruent corresponding angles, then according to the above theorem they must be congruent. This means that we do not have any non-congruent similar objects on the sphere!

An observation related to this theorem is that there exists infinite number of regular p-gons with different angle sizes. For example, on a sphere, we may construct equilateral triangles with interior angles 70°, 85°, 90°, or any other angle with a value between 60° and 300°.

Going back to our discussion of tessellation, we are interested in determining all of the possible regular tessellations on the sphere. The above few paragraphs have made clear for us that unlike the case for Euclidean geometry, it is not sufficient to say, for example, equilateral triangles tessellate the sphere: It is possible that an appropriate equilateral triangle tessellates the sphere; however, another equilateral triangle with a different angle size does not.

It seems that the problem is much more complicated than its Euclidean version. However, with some information coming from another part of Euclidean geometry, we may be able to find the solution relatively easily.

Suppose that a certain spherical regular p-gon, $p > 2$, with angle measure $360°/q$, can tessellate a sphere. This means a finite number of this regular p-gon can cover the sphere without any gaps or overlaps. Now if all the vertices stay at their places but the sphere becomes flattened on its p-gons, then the result will be a Platonic solid: a polyhedron with identical regular faces, and identical vertices. Therefore, for the case $p > 2$, the problem of the regular spherical tessellation and the Platonic solids are identical.

Platonic solids were known to humans much earlier than the time of Plato. There are carved stones (dated approximately 2000 BC) that have been discovered in Scotland. Some of them are carved with lines corresponding to the edges of regular polyhedra. Specifically among them is a dodecahedral form that shows that the dodecahedron was known to humans much earlier than it appears in any written document. In addition, Icosahedral dice were used by the ancient Egyptians.

Evidence shows that Pythagoreans knew about the regular solids of cube, tetrahedron, and dodecahedron. A later Greek mathematician, Theatetus (415 - 369 BC) has been credited for developing a general theory of regular polyhedra and adding the octahedron and icosahedron to solids that were known earlier.

The name "Platonic solids" for regular polyhedra comes from the Greek philosopher Plato (427 - 347 BC) who associated them with the "elements" and the cosmos in his book *Timaeus*. "Elements," in ancient beliefs, were the four objects that constructed the physical world; these elements are fire, air, earth, and water. Plato suggested that the geometric forms of the smallest particles of these elements are regular polyhedra. Fire is represented by the tetrahedron, earth, cube, air, the octahedron, water the icosahedron, and the almost-spherical dodecahedron, the universe.

The Greeks only considered polyhedra with planar faces. As far as we know from the surviving texts, Buzjani was the first mathematician to consider the projections of polyhedra onto a sphere

3. How to Tessellate a Dome? A Medieval Persian Approach

In the absence of metal beams, domes have been an essential part of the architecture of both official and religious buildings around the world for several centuries. Domes were used to bring the brick structure of the building to conclusion. Based on their spherical constructions, they provided strength to the buildings' foundations and also made the structure more resistant against snow and wind. Besides bringing a sense of strength and protection, the interior designs and decorations resemble sky, heaven, and what a person may expect to see beyond "seven skies" [4].

When we encounter a structure such as a Persian dome and are amazed by the striking beauty and harmony of the varieties of patterns that have been constructed inside and outside of the surface of the dome, a natural question that may come to our minds is if rigorous geometry was involved, either by knowledgeable artisans, or mathematicians of old time, to design the dome. The treatise *On Those Parts of Geometry Needed by Craftsmen* reveals the direct involvement of mathematicians in the study and performance of tiling on the sphere.

Buzjani mentioned in his treatise about the interactions of artists and artisans with mathematicians on topics such as geometric constructions of ornamental patterns and the application of geometry to architectural construction. In Chapter Twelve of his treatise, Buzjani presents the ways that a sphere can be tessellated using properties of Platonic and some Archimedean solids (solids with two or more regular faces with identical vertices). One interesting remark about the original illustrations of this chapter is that all of the spherical constructions have been presented flat, with the hope that the reader uses his imagination to "see" them three-dimensionally.

Figure 1: *Two dome interiors.*

3.1. Tiling with Eight Equilateral Triangles. After explanations of some elementary spherical constructions, in the problem numbered 174 of the treatise, Buzjani illustrates the construction of a spherical octahedron as follows: We want to construct three pair-wise perpendicular great circles. For this, we first construct two perpendicular great circles that meet at *A* (الف) and *C* (ج), and then divide one of the circles into four equal arcs of *AB*, *BC*, *CD*, and *DA*. The great circle that passes through *B* (ب) and *D* (د) — and meets the other circle on *E* (ه) and *F* (ز)—is the desired one. Now we notice that the sphere has been divided into eight spherical equilateral right triangles. This is the spherical octahedron.

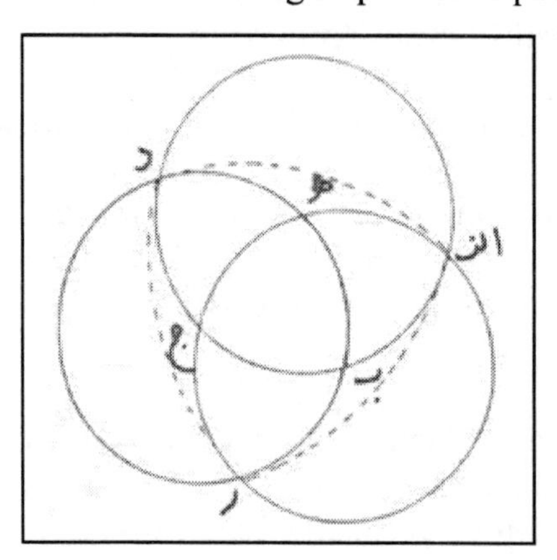
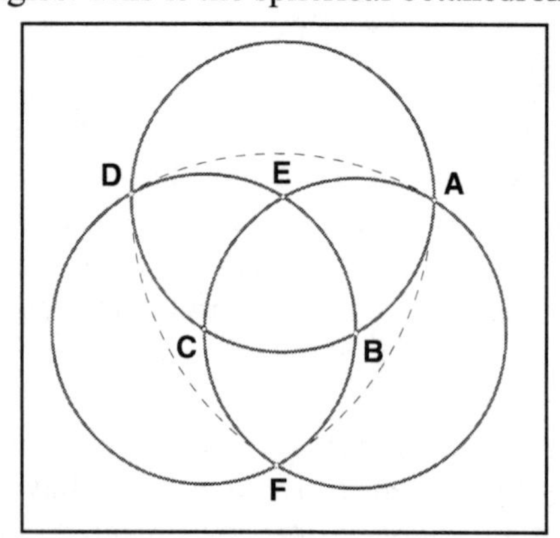

Figure 2: *Tiling of the sphere with 90° equilateral triangles*

3.2. Tiling with four equilateral triangles. Problem numbered 176 is about the construction of a spherical tetrahedron. For this, we need to divide the sphere into four congruent equilateral triangles. The first step is to construct the spherical octahedron as was illustrated in the previous problem.

Figure 3 begins with the construction of Figure 2. Three perpendicular great circles are illustrated that meet pair-wise at *A* and *C*, *B* and *D*, and finally *E* and *F*, and have created eight congruent equilateral triangles. Now let H (ح) be the centroid of the triangle ABE. We then construct three great circles that pass through AH, BH, and EH, respectively. Each circle passes through the centroid of a neighboring triangle, which has only a vertex in common with the original triangle. Name them K (ک), L (ط), and M (ع). Then Δ KLM is one face of our spherical tetrahedron. This process will divide the sphere into four congruent equilateral triangles: a spherical tetrahedron.

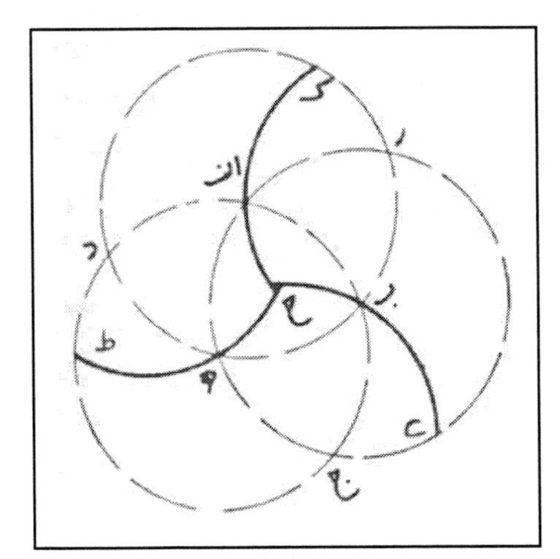

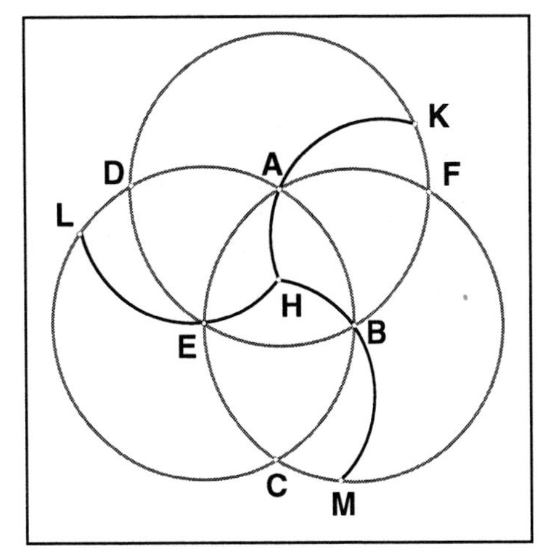

Figure 3 *Tiling of the sphere with 120° equilateral triangles*

3.3. Tiling with six squares. To construct a spherical hexahedron (cube), we first construct the spherical octahedron. Then we connect the centroid of each pair of neighboring triangles that share a common side by an arc from a great circle. This will divide the sphere into six congruent spherical squares.

For a person familiar with the idea of duality of platonic solids, the above construction is an immediate property of the octahedron (or hexahedron), but projected on to a sphere.

Figure 4: *Tiling with six spherical squares*

3.4. Tiling with twenty equilateral triangles. The next spherical construction is for the spherical icosahedron. But as we will see below, it contains an error. In problem numbered 180 the construction has been performed as follows:

Let E and F be the two poles of a sphere (only E (ه) can be seen in the following figure) and the great circle ζ is perpendicular to EF. Divide circle ζ into ten equal arcs C_iC_{i+1}, where $i = 1\ldots9$, and $C_{10}C_1$).

With the centers of C_1 and C_2 and with the radius of C_1C_2, we construct two arcs to meet at A_1 on the E side of the great circle C. With the centers of C_2 and C_3 and with the radius of C_2C_3 we construct two arcs to meet at B_1 on the F side of the great circle ζ. We repeat this procedure to find five points A_i on the E side and five points B_i on the F side of the great circle. Then using great circles passing through $A_1$and B_1, B_1 and A_2, and A_2 and A_1 we will construct the equilateral triangle $A_1B_1A_2$. With this procedure we will construct ten equilateral triangles of $A_iB_iA_{i+1}$, $i = 1\ldots4$, $A_5B_5A_1$, $B_iA_{i+1}B_{i+1}$, $i = 1\ldots4$, $B_5A_1B_1$. Now on the E side we have a spherical regular pentagon of $A_1A_2A_3A_4A_5$ and on the F side another spherical regular pentagon of $B_1B_2B_3B_4B_5$. Using great circles that pass through E and A_i, $i = 1\ldots5$, we are able to divide pentagon $A_1A_2A_3A_4A_5$ into five congruent triangles. We do the same on the other side of the great circle ζ to find five more equilateral triangles. This will conclude our construction of a spherical icosahedron.

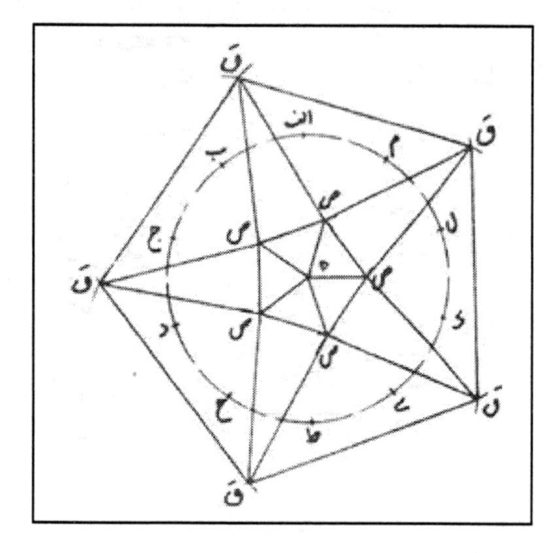

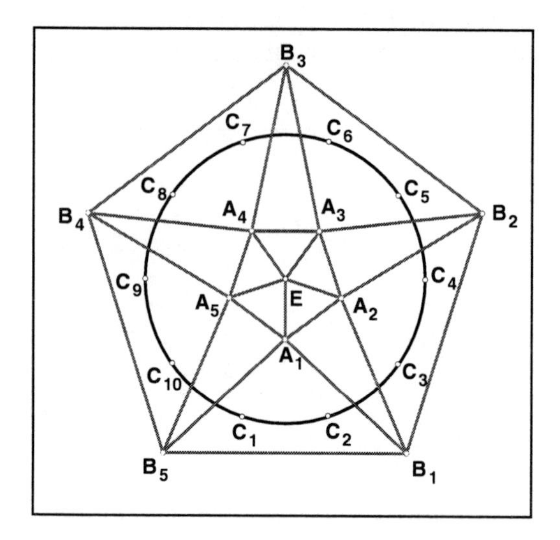

Figure 5: *Tiling with twenty 72° equilateral triangles*

3.5. Tiling with twelve regular pentagon. To construct the spherical dodecahedron, the book gives two methods. The first method uses the duality property of the Platonic solids. In Problem numbered 182 we read that we need first to construct the spherical icosahedron and then find the centroid of each triangle and connect the centroids of the neighboring triangles that have common sides in order to construct a spherical dodecahedron.

The other approach for constructing a spherical dodecahedron that the book illustrates is as follows: We have a sphere S with a given diameter. We first construct segment AB congruent to the diameter and divide it into three congruent segments of AC, CD, and DB. With center D and radius AD we draw a half circle that meets the perpendicular line to AB passing through B at point E. We find H on AB in such a way that $BH = 1/2\ BE$. With the center H and radius HE we find point L on ray AB.

BL is a side of a spherical pentagon that covers the sphere. Now we choose M (here it is ى) on the sphere and draw a circle with radius BL. We divide the circumference into three congruent arcs M_1M_2, M_2M_3, and M_3M_1 (here M_i's are ک, ل, and م).

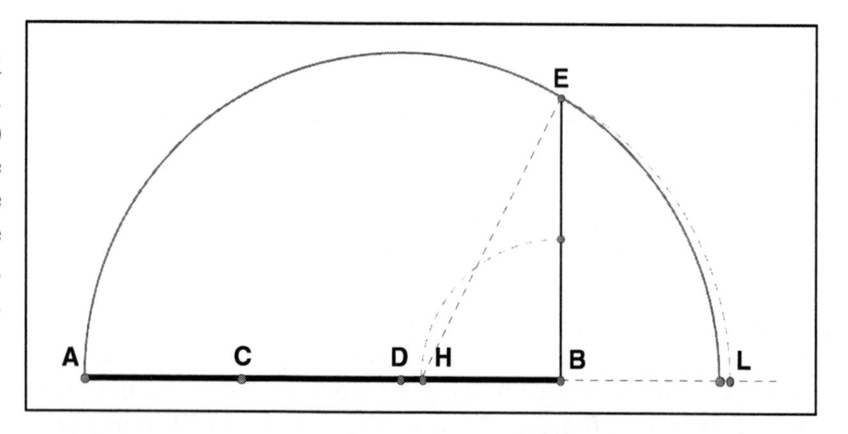

Figure 6: *Geometric construction of a side of spherical pentagon*

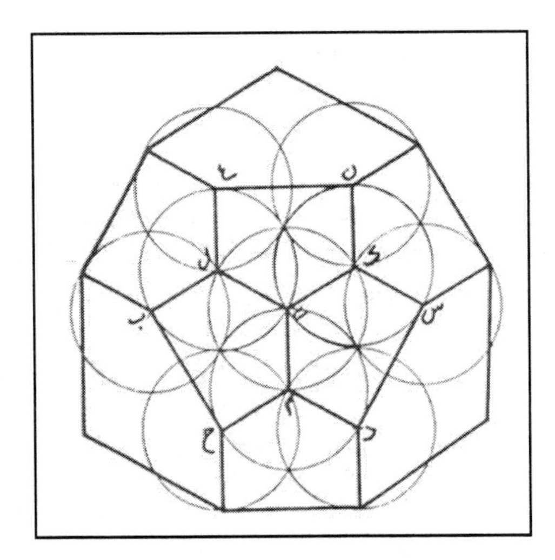

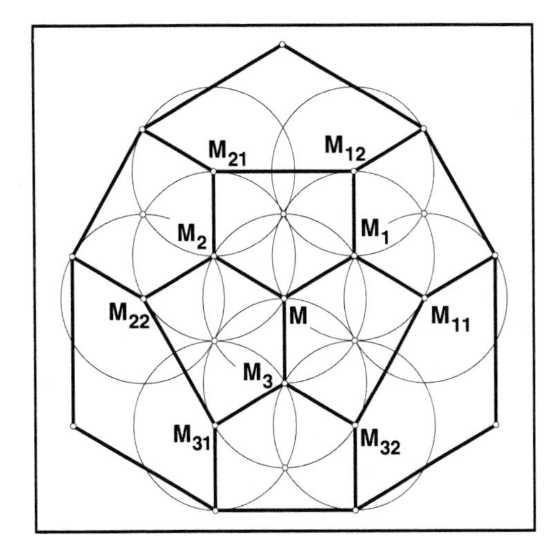

Figure 7: *Tiling of the sphere with twelve 120° regular pentagons*

From each M_i ,i=1,2,3, we make a circle with radius M_iM and divide it into three congruent arcs starting with point M as MM_{i1}, $M_{i1}M_{i2}$, and $M_{i2}M$ (these points are س, ن, ع, ب, ح, and د in the original figure). Now we have three spherical pentagons of $MM_1M_{12}M_{21}M_2$, $MM_2M_{22}M_{31}M_3$, and $MM_3M_{32}M_{11}M_1$. We continue this process to complete the tiling. It is understood by the reader that the sides of each pentagon is a part of a great circle that passes through two vertices.

4. Some Notes about the Buzjani's Constructions

An interesting project, rather than using a software utility, would be to construct these tessellations on a real sphere. This will give us a feeling of how a mathematician of ancient or medieval time could realize an abstract geometric construction on such a surface. Perhaps the Buzjani's tool for construction was a wooden sphere. Nevertheless, we may use a modern sphere, the *Lenart Sphere*, which is manufactured and sold as a pedagogical "spherical blackboard"[5]. The sphere comes with a smooth spherical surface on which we can draw with water-soluble pens, a torus on which to rest the sphere, hemispherical transparencies that fit over the sphere, a spherical ruler/protractor, and a spherical compass and center locator.

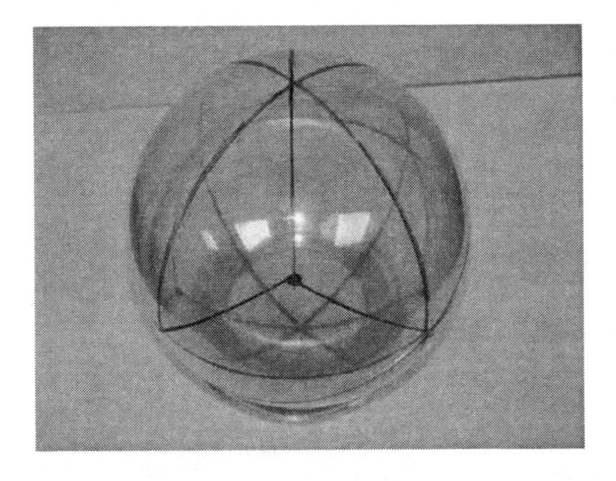

Figure 8: (a) *Tiling with four equilateral triangles*, (b) *Tiling with six squares.*

Figure 8.a presents the spherical tetrahedron and Figure 8.b presents the spherical hexahedron based on the Buzjani's approaches. To understand the logic behind these two constructions, we may take a look at Figure 9. In this figure, the octahedron and its dual, cube, are illustrated. If we start from a vertex of a cube and construct the diagonals of the faces that share that vertex, and connect the other vertices with diagonals of other faces, we will complete a tetrahedron inside the cube. Figure 9 presents this relationship as well and justifies the construction in Figure 3.

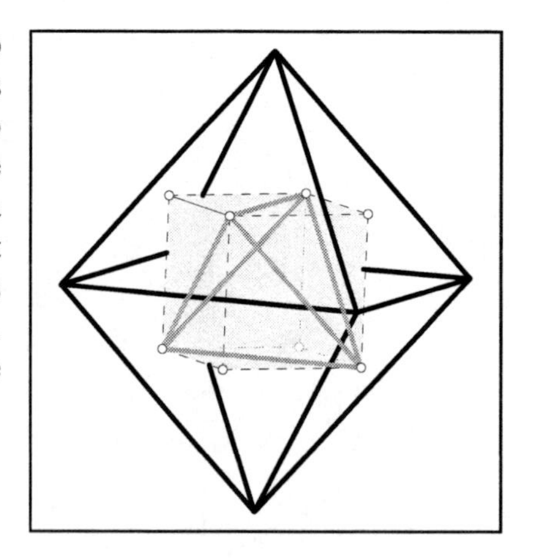

Figure 9: *The octahedron, its dual, the cube, and the tetrahedron.*

To understand the procedure of the mentioned construction of the spherical icosahedron, we need to study Figure 10.a. Points $A_1, A_2, \dots$, $B_1, B_2, \dots$, and $C_1C_2 \dots C_{10}$ are on an icosahedron that seem to be corresponding to the vertices of the Buzjani's constructions of spherical icosahedron in Figure 10.b. In Figure 10.a the triangles $C_1C_2A_1$, and $C_2B_1C_3$ and so on are among the upside-down equilateral triangles that are parts of faces of the icosahedron of $B_5B_1A_1$, $B_1A_2A_1$, and so on.

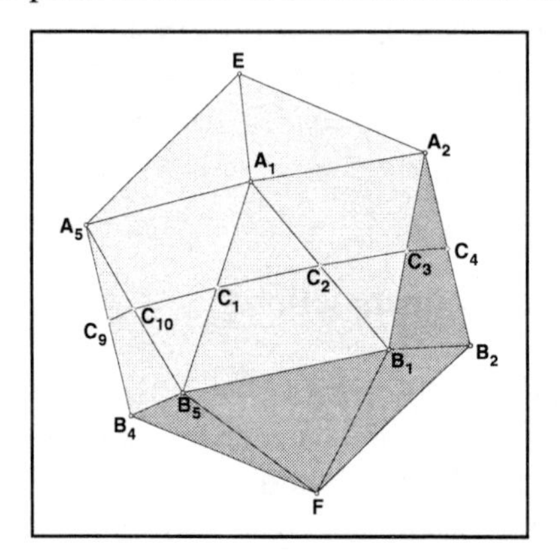

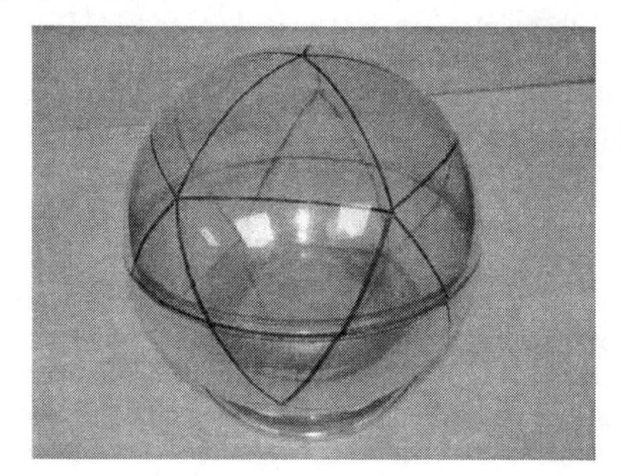

Figure 10: *(a) An icosahedron (b) Buzjani's Tiling with twenty equilateral triangles.*

The logic behind the construction in section 3.4 is since in Figure 10.a, the triangles $C_1C_2A_1$, $C_2B_1C_3$, and so on are equilateral triangles –which are similar to equilateral triangles $B_5B_1A_1$, $B_1A_2A_1$, and so on– by constructing them we are able to find the locations of the vertices A_1, A_2, A_3, A_4, A_5 and B_1, B_2, B_3, B_4, B_5 on the sphere. However, as it was mentioned before, in spherical geometry, the similarity property of the non-congruent triangles fails. This means what we have in Figure 10.a is not an accurate tiling of the sphere with twenty congruent equilateral triangles! In fact the triangles that constitute the antiprism of $A_1A_2A_3A_4A_5B_1B_2B_3B_4B_5$ in Figure 10.b are only isosceles triangles, which are larger than other isosceles triangles that constitute the two pyramids of $EA_1A_2A_3A_4A_5$ and $FB_1B_2B_3B_4B_5$. The problem is when we project the icosahedron in Figure 10.a on the ball that circumscribes it, then the image of $C_1C_2A_1$ on the ball is no longer equilateral. The central angles are the measure of length along the surface. Points C_1 and C_2 are edge midpoints, and are closer to the center of the sphere than the vertex A_1. So different central angles are subtended by equal length segments.

Now we would like to analyze the tiling with twelve regular pentagons presented in section 3.5. Let us assume that the diameter of the sphere is d. Then we find that $EB = \sqrt{3}\,d/3$, $LH = \sqrt{15}\,d/6$, and finally $BL = (\sqrt{5} - 1)\sqrt{3}\,d/3$. This can be expressed as $d = (\sqrt{15} + \sqrt{3})/4\ BL$. But this is the exact value of the circumradius of the dodecahedron with the side congruent to the segment *BL* [6]. Therefore, the Buzjani's construction of the dodecahedron is an exact construction and not an appropriate approximation (He has presented approximations in his treatise for the cases of the non-constructible regular polygons, such as the heptagon and nonagon).

But then there is a question that whether the same mathematician that came up with the exact construction of the spherical dodecahedron—in such an interesting approach that needs a deep investigation to find out how he successfully related the details in Figure 6 to the properties of the dodecahedron—has made such a faulty mistake in the construction of the spherical icosahedron! One possibility would be the addition of new parts to the original manuscript by mathematicians who are supposed to make new copies from the treatise. In fact what we have today are all copies, and perhaps the original document does not exist today. Writing the entire document was the only way of publishing a new book, and, therefore, the correctness of a new copy of a book was totally in the hands of the copy writer and his honesty.

5. Construction of Some Spherical Archimedean Solids

The treatise includes constructions of some spherical Archimedean solids such as cuboctahedron (tiling with eight equilateral triangles and six squares), truncated icosahedron (tiling with twelve pentagons and twenty hexagons), and icosidodecahedron (tiling with twelve pentagons and twenty triangles).

The other Archimedean tilings illustrated by Buzjani include the construction of truncated octahedron—which is done by joining the one-third points of each side to appropriate neighboring one-third points using arcs from great circles and then erasing old vertices of the octahedron—and the truncated tetrahedron—with the same procedure as the former construction. The first construction results in a division of six squares and eight hexagons. The second construction divides the sphere into four triangles and four hexagons.

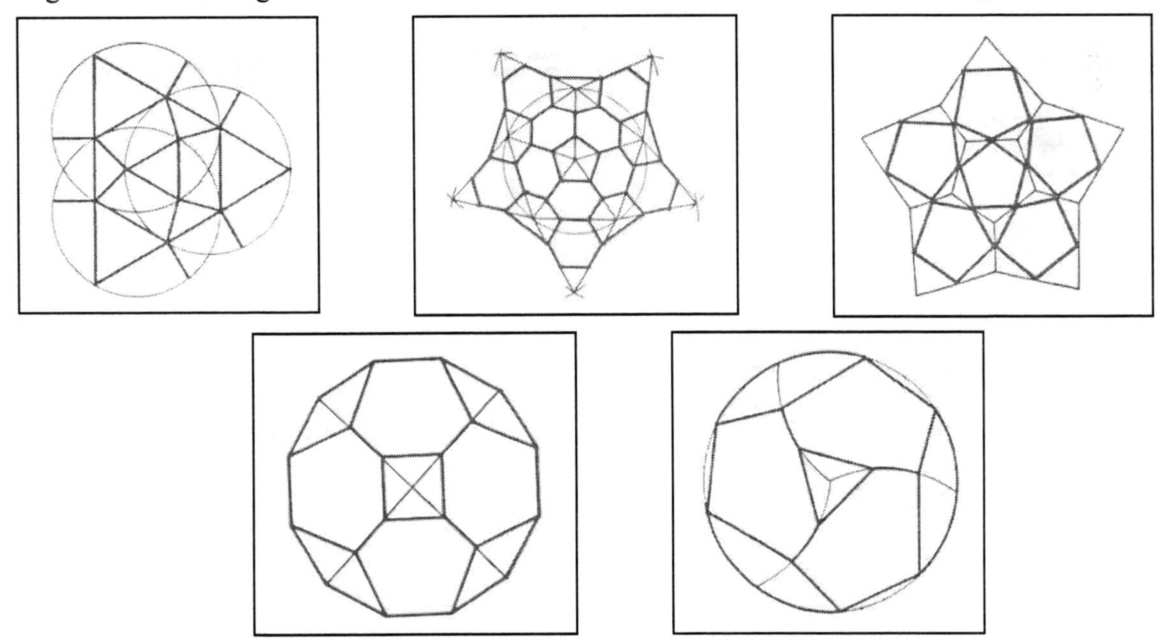

Figure 11: *The flat presentations of the constructions of cuboctahedron, truncated icosahedron, icosidodecahedron, truncated octahedron, and truncated tetrahedron.*

5. Conclusion

In the medieval Persian culture, the structure of objects, such as domes, necessitates the artists and artisans to rely on mathematicians. Such a relationship has occurred during this period as documented in the treatise *On Those Parts of Geometry Needed by Craftsmen.* Even though the observer of such an artwork may fantasize about it by relating the harmony and symmetry of the piece to some sort of revelations, the above explanations present the direct involvement of mathematicians, solely based on the scientific approaches available in those times, for creation of such a work.

References

[1] R. Sarhangi, S. Jablan and R. Sazdanovic, *Modularity in Medieval Persian Mosaics*, Bridges: Mathematical Connections in Art, Music, and Science, Conference Proceedings, Central Plain Book Manufacturing, Winfield, Kansas, 2004.

[2] S. A. Jazbi (translator and editor), هند سه ایرانی: کاربرد هند سه درعمل, *Applied Geometry*, Soroush Press, ISBN 964 435 201 7, Tehran 1997.

[3] A. Özdural, *Mathematics and Arts: Connections between Theory and Practice in the Medieval Islamic World*, Historia Mathematica 27, Academic Press, 2000, pp. 171-301.

[4] R. Sarhangi, *The Sky Within: Mathematical Aesthetics of Persian Dome Interiors,* Nexus, Volume I, Firenze, Italy, 1999, PP 87-97.

[5] I. Lénárt, *Non-Euclidean Adventures on the Lénárt Sphere*, Key Curriculum Press, Berkeley, California, 1996.

[6] Whistler Alley Mathematics, *Platonic Solids*, http://whistleralley.com/polyhedra/platonic.htm

Fabric Sculpture - Jacob's Ladder

Louise Mabbs
Textile artist / teacher / author
12 Alfriston Rd, Battersea, London, SW11 6NN, UK
E-mail: louise.mabbs.textiles@btinternet.com
Website: louisemabbs.co.uk

Abstract

This paper develops ideas from a paper folding idea known as Jacob's Ladder into a fabric sculpture. It shows how, as an artist, I became aware of mathematics in my work. Translating origami concepts into fabric constructions, the nature of fabric affects the form. The opportunities fabric creates suggest possible developments.

I have always loved the pattern making side of maths. I trained in constructed textiles in the mid 80's, specializing as a weaver. After leaving college, I set up my own business teaching quiltmaking, making quilts to commission & originally making soft furnishings for bread & butter. As a textile artist I have always been intrigued with both mathematics and structure, most recently through origami. The work I am describing here combines all of these - using a paperfolding technique called Jacob's Ladder which I have stretched and developed in fabric.

1. Mathematics in my work

Early on in my career, I was asked to submit work to an exhibition called Mathematical Magic. I made two pieces, one based on decimals, "Decimal Rainbows" and the other on primes, "Prime Factors". As a result, I have been developing a mathematical colouring system. I hope my third book will be on this, with the second on Fibonacci inspired textile art.

Later I was commissioned to make such a quilt 'in perspective' (figure 1), although every triangle is straight edged, they appear curved, out of this my spiral quilt series on circular grids developed (figure 2)

Figure 1, Come and see, Creation series no 4, 1997

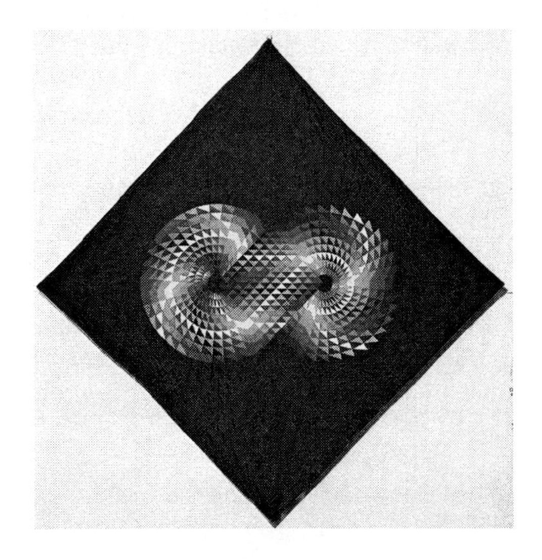

Figure 2, Infinity, Creation series no 7, 2000

In 2002 I was invited to be part of an exhibition in central London, one demonstration day, John Sharp came to see the exhibition. We have been collaborating ever since. He has been showing me the geometrical methods I could envision as structures, but didn't know how to achieve.. For example, I wanted a long, equilateral triangle effect like Toblerone boxes, but circular. John showed me how to work out the grid for a curved structure and coined the term 'Toblercone', a name I adopted. John recognised the shape to be based on a conical construction. The three dimensional fabric sculpture that resulted was coloured according to my mathematical system resulting in spiralling colour bands. It is shown in figure 3.

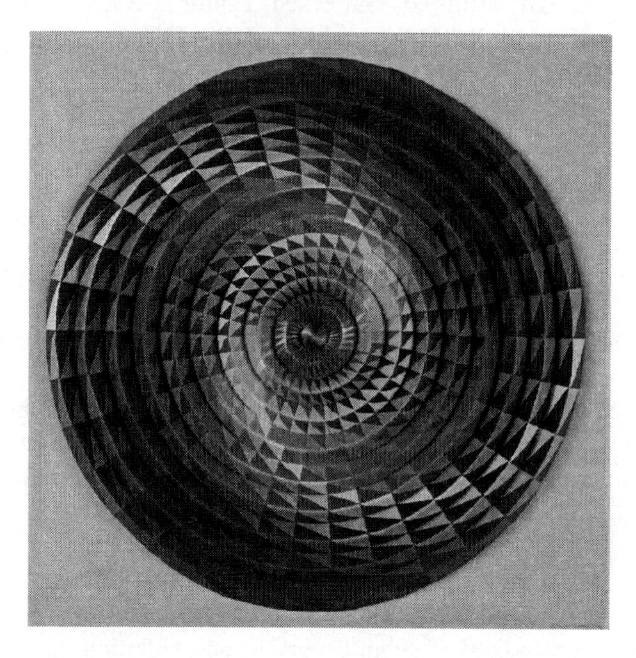

Figure 3, Toblercone Rings 1. 2004

I have recently written a book with another textile artist, Wendy Lowes, called *The Quilter's Guide to Twists and Tucks* which takes basic origami shapes, or existing quilting folds and pushes them a lot further in fabric. The Jacob's Ladder piece I am describing in this paper (see figures 8 - 11) is the final & the most complex of my design projects in the book. Since the book's readership is not particularly mathematical, this paper gives me a chance to describe some of the geometrical properties. Please note that this is **not** the Jacob's Ladder of traditional quilt-making which is a flat pieced patchwork block

2. Origami Jacob's ladder

I played with origami as a child. Then forgot about it. When I met Wendy and her fascinating work through quilting circles, she recommended joining the British Origami Society. I dithered for years until the book commission came, when I subsequently attended two of their biannual conferences. Heinz Stobl was the key guest at my first conference, I was riveted by his work and stretched myself to attend even the advanced lessons. At the time I recognised how amazing his constructions were, but it was the techniques he taught in order to construct his interconnected boxes that interested me. He demonstrated a technique for folding that I know as 'Jacob's Ladder'. I promptly forgot about it, until the deadline for my book proposal.

The structure has many names around the world.

"In Holland this is called a "muizentrapje" (mouses' staircase), in Germany a "hexentrappe" (witches' staircase). Sebastian Kirch calls it in English "witches' staircase" and Heinz told me he heard English and American people call it "witches ladder"." [1]

The Jacob's Ladder is constructed as shown in figure 4. It is made from two strips of paper.

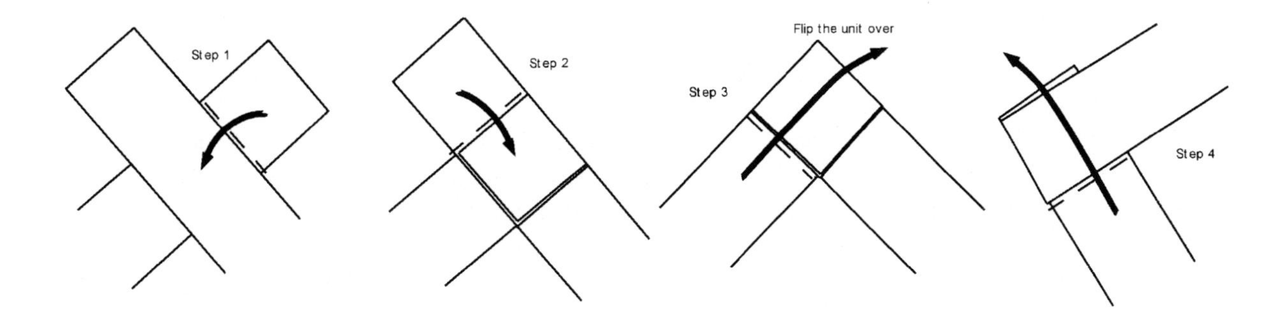

Figure 4, paper construction

Step 1 Lay your two strips on top of each other. Fold the under strip over the top strip so the edges match
Step 2 Fold the top strip back over the first fold so outer edges match
Steps 3 & 4 Turn over the unit & fold the strips alternatively over one another until the whole lengths are folded together. Glue ends together

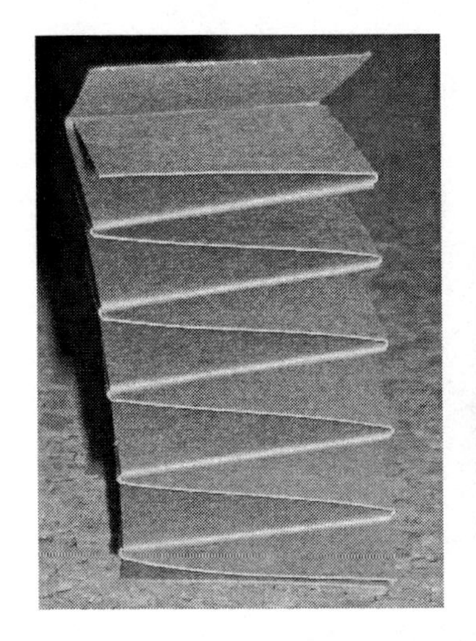

Figure 5a, closed paper Jacob's Ladder **Figure 5b,** open paper ladder

The paper version has specific properties:

- Paper, being stiff, will crease well when you force it to, but it will also crease when you don't want it to and any mistakes in your accuracy of folding exist forever.
- If the paper gets very wet it will rot away completely, if damp it will distort, buckle across the surface and lose it's pristine condition.
- Paper can be slotted together to create sturdy constructions. The Origami purist never glues anything. Everything must be made with interlocking joins.
- Paper ladders are the same colour on both sides. Unless you find a double-laid specialist paper, printed origami paper, or glue two sheets together - which makes folding difficult.

- Paper ladders have a 'natural' extended position. When they are extended to full capacity the outer edges spiral gently down the length. If more stretch is added, you risk tearing the paper and the beauty of their structure.

A fabric version has different properties:

- Fabric on the other hand is drapable. So while it will crease well when intended, between finger nails or under an iron, it will not retain a sturdy shape on it's own. Mistaken creases can mostly be ironed out.
- Fabric is not destroyed by moisture. Indeed it is usually designed to be washed. Though it may loose some qualities obtained by specialist treatments if washed e.g. glazed cotton.
- Fabric will not slot together. It is too slippery. So Fabric Origamists consider it acceptable to use stitch or bonding agents to hold units together. In experimenting with my Jacob's Ladder, I was soon to discover that the stitching was essential to achieving the 'compartments'.
- Stiffening fabrics is essential if you want a rigid structure. The stiffening process depends on the effect you want and whether you need to be able to sew through the stiffening. For my Toblercone pieces I used a plasticised fibre for a very rigid effect since I was mostly only stitching through the multi-fabric pieced quilt top, which was stretched over the structured base. While to quilt it I had to pre punch holes with a large needle. For Jacob's Ladder, I needed to stitch through the stiffening and along the edges, so I needed a softer stiffener. I used fast2fuse, which is an American product. While this does not give me as stiff an effect as I would have liked - the wrinkles on the surface annoy me - because of folding the strips, and hand stitching intersections, I could not use my stiffest support. I am still looking for a better substance for future pieces.
- Fabric ladders can be two coloured. Because you need the middle, stiffening layer, so you may as well use two different fabrics. Besides which it adds to the design possibilities.
- Fabric ladders are comfortable beyond the 'natural' extended position. Since fabric is fluid, other parts of the structure take the strain when distortion occurs and so the new shape remains beautiful.

Folding the fabric version

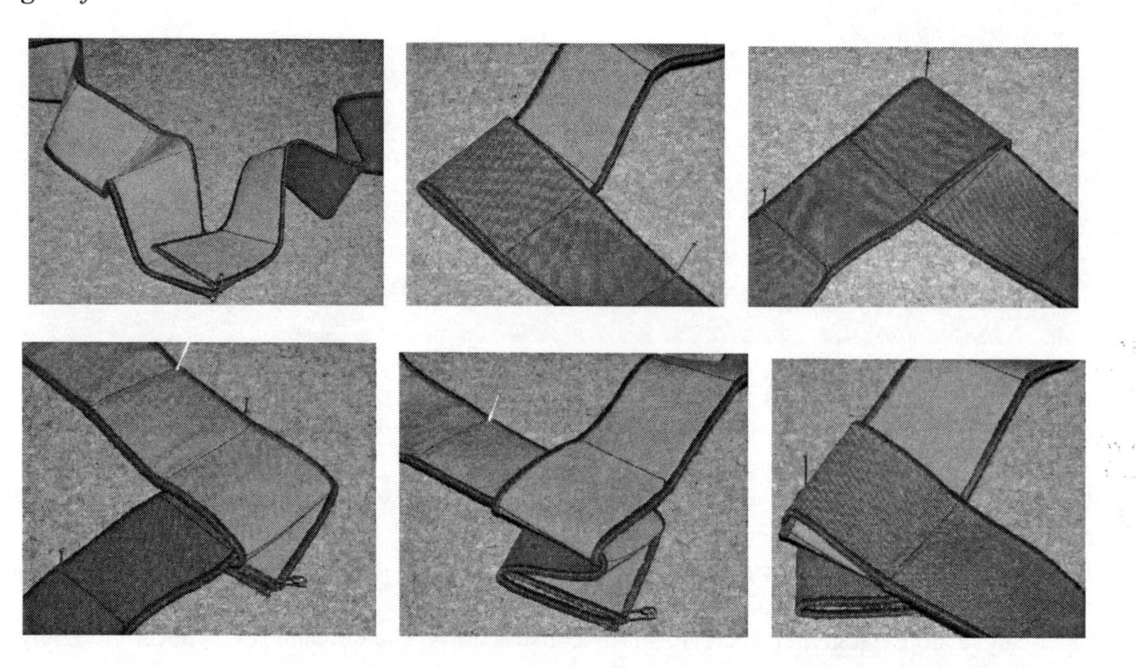

Figure 6a - e, the fabric version of a single strip showing the two colours interchanging

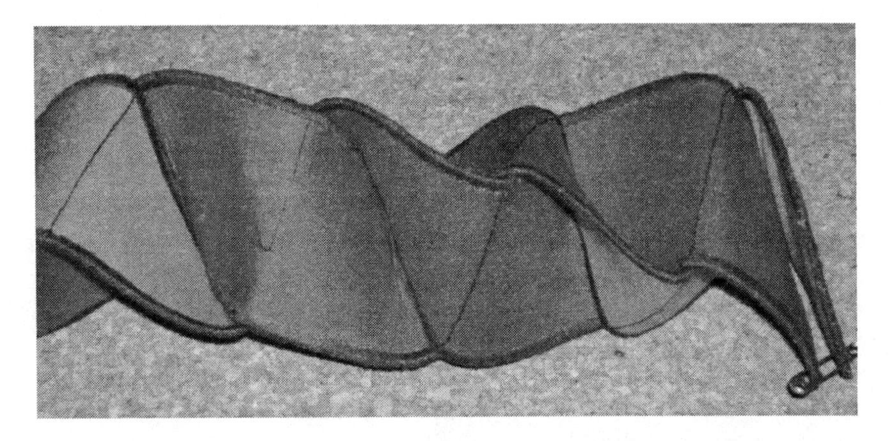

Figure 7 The folded unit before sewing at intersections

3. Making the final piece

Instructions are in my book [6]. These occur as photographic staged instructions. They may be summarised as:

- Cut same width strips of fast2fuse and two pieces of fabric per strip, in several colours.
- Iron each strip of fast2fuse to the two chosen fabric strips.
- Draw regular perpendicular lines the same width as your strip, to create squares.
- Stitch along the lines with a short, straight stitch in contrasting thread.
- Stitch all around the edges of the strips with a close satin stitch on your zig zag settings.
- Fold your strips together along the sewn lines.
- Fold alternate strips clockwise and anticlockwise.
- Sew ends of strips together at top and bottom.
- Join the strips together at the ends of the horizontal straight stitched lines.
- Lay strips in your chosen sequence.
- Join midway along the side of each compartment, as seen in the top section of Jacob's Ladder.
- Alternatively, join compartments several spaces apart to allow for spiralling to occur.
- Hang from small rings attached to the top of the piece.

In reality the instructions are far more intricate in order for it to work.

4. The Mathematics

When I made the piece, I did not initially recognise the mathematical implications of the structure. I had prepared the strips for the first photoshoot for the book and had no idea how to proceed. Several things soon became clear:

a) Using two separate colours on the first strip and two different colours on the second strip of each pair was too chaotic for what I wanted to achieve, which in this case was a sharp rainbow sequence. This was resolved by making two sets in the same colour pairing for each unit.

b) In terms of the smallest number of colours I could use so they would not come up against themselves (4 colour map theory), could I change colours every third strip in this structure?.

c) When the pairs were folded together it was difficult to see how they could be joined together, so I had to alternate clockwise and anticlockwise folding techniques. The piece therefore, involves symmetry.

d) In order to add more interest I used an odd number of colours, in this case 13. Each colour comes up against two other colours.

e) When you fold a paper ladder, the structure appears square, end on, because you fold the strips squarely across each other, though they may spiral above one another if badly folded. When I started to manipulate my fabric ladders, I discovered they appeared triangular in cross section, end on, when stretched beyond the 'natural' extended position of the ladder. Figure 7

f) Fabric ladders have more than one comfortable extended position. Which can be exploited for a variety of effects.

g) At natural comfort position, the outer corners of the unit lie along a line facing 90 degrees from each other

h) At maximum stretch, the corners face opposite directions another 90 degrees. In paper this in uncomfortable but not in fabric, where the outer edges of the compartments take on a curved appearance to accommodate the stretch.

i) When the ladders are extended fully compartments lay flat on the front of the piece, but there are triangular peaks on the reverse (see top of piece).

j) Folded units can also be sewn with the pyramids on alternate sides:

VVVVVVVVVVVVV

k) The links with DNA structure are obvious, if you imagine DNA as a strip of paper twisted where the edges give two helices. The sculpture in Newcastle is a good interpretation (Figure 12).

l) Three helices are generated by my folded structure, similar to the sculpture by John Mayne outside National Theatre, London [5] shown in figure 13

m) Several other questions also arise: Does the number of squares affect the results in terms of rotation? How many strips can be folded into a structure? Or how few? One strip twisted gives two spiralling edges, another interesting structure in itself. Can three strips be successfully plaited? Cords can, but how would flat pieces work?

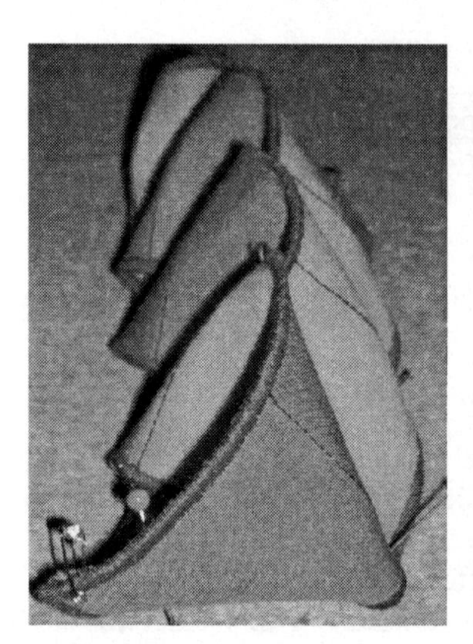

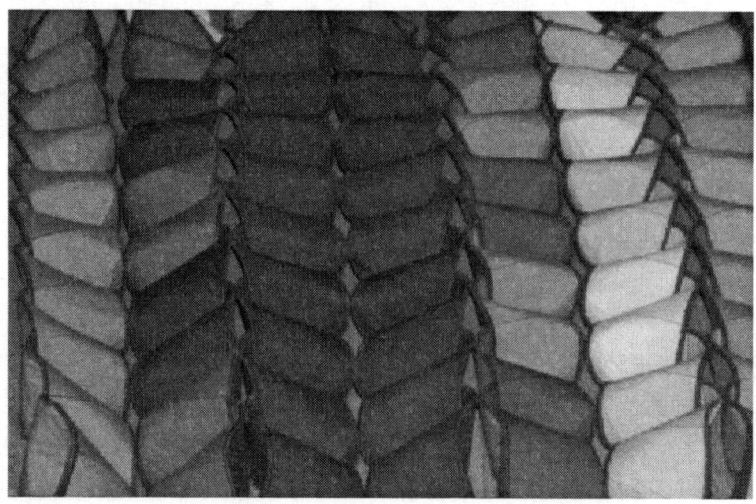

Figure 8a The triangular cross section *and* **8b** The compartments seen on the back

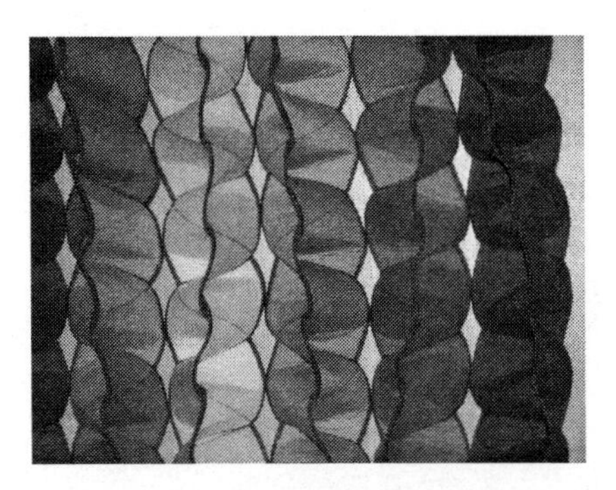

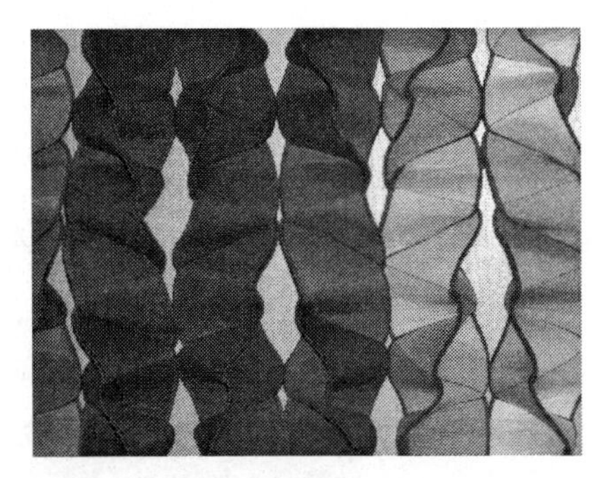

Figure 9 The triangles on the front (fully stretched) **Figure 10** Partially stretched units

Figure 11 Angled view from the top of the sculpture showing the difference in structure in the two halves of the sculpture

Figure 12 The DNA like sculpture in Newcastle

Fig 13 The triple helices section of John Maine's sculpture in London

9. How my structure could be used

There are a number of ways I envisage these sculptures being used:

- Art pieces in indoor public spaces
- Design collaborations with a metalworker?
- Room dividers/screens
- Hanging sculptures within spiral staircases or similar

I will continue to focus on making three dimensional quilts. Along with my Toblercone pieces and a five way picture (i.e. it has five distinct viewing positions like Jaacov Agam's painting technique, which I became aware of in the Pompidou, Paris c 1995 and consolidated in an exhibition with painter George Printezis, from Inverkip Glasgow, who creates similar effects by different methods)

I will be experimenting with the structures. What happens if the units get curved or become circular? What happens if extra strips are added to the unit? How large or small can I make them? I will continue to experiment with stiffening techniques. Or, what happens if the structure is made without stiffening at all?

I will continue to collaborate with other people – John Sharp, other mathematicians and artists in my bid to seek ever more interesting structures and theories. If you do anything with these techniques or ideas, please respect and acknowledge my copyright and send me a copy of your work, for possible inclusion in a future book. Thank you.

11. References

[1] http://home.tiscali.nl/gerard.paula/origami/knotology.html#knippen

[2] www.sermons4kids.com/instructions-ladder-chain.htm

[3] educ.queensu.ca/~fmc/june2002/JacobsLadder.htm

[4] Charles Jencks, architect, sculpture by Life Centre, Newcastle built by local craftsman

[5] John Maine sculpture outside Royal Festival Hall, South Bank, London

[6] Louise Mabbs and Wendy Lowes, The Quilter's Guide to Twists and Tucks, publisher Collins & Brown, ISBN 1843403110, Origami Quilts: 20 folded fabric projects, publisher That Patchwork Place, ISBN 1564776247

[7] For a fascinating collection of DNA inspired items go to www.ncbe.reading.ac.uk/DNA50/ephemera1.html

All photos by Louise Mabbs except figure 3. Shipley Art Gallery

Eva Hild: Topological Sculpture from Life Experience

Nat Friedman
Dept. of Math., Univ. at Albany, Albany, NY 12222.
Artmath@math.albany.edu

Abstract

This is an introduction to the ceramic sculpture of Eva Hild.

Figure 1. Studio.

Eva Hild is a ceramic sculptor who lives and works in the southwest of Sweden. Her works may be described as elegant topological forms that are reminiscent of minimal surfaces since the surfaces basically have a hyperbolic geometry. It is interesting to note that the forms are not mathematically inspired but are inspired by her life experiences, as in the following quote from her web site www.2hild.com/eva_eng.htm

"*Influence, pressure, strain. These words have been the foundation for my current projects that comprise communicating the theme in large hand-built clay forms. Delicate continuously flowing entities in white thin-built clay. They reflect varying degrees of external and internal pressures, and how, as a consequence, perception of inner and outer space is changed or challenged.*

My inspiration is the ever-changing landscape of my own life and environment! I try to relate my work to my life. What is happening and how does it feel? Pressure. Flow. Strain. Ramification. Inside turns outside. As a starting point I put words onto my feelings, and use the vessel form to

translate this into three dimensions. The size of the form relates to my body. The thin walls are pulled and bent in different directions.

Figure 2. Eva Hild.

I feel a great freedom in hand-building. It grows slowly, I have time to reflect, I can change direction, make connections and have a smooth surface with the same thickness. I build big forms, the clay will dry slowly and not collapse. When the form is ready and the clay is dry, I sand away at the surface and then spray it with a slurry of kaolin. The pieces are finally fired in stoneware temperature, about 1250°. "

In the above photograph, one can see that Eva Hild's sculptures are impressively large.

I find them to be amazingly beautiful graceful forms that lovingly enclose space. One can imagine traveling from one enclosed space to another by flying through the narrow passages. Two more sculptures are shown below. The sculptures are completely three-dimensional having no preferred top, bottom, front, or back. Rotate the photos for alternate views. They will also look quite different when viewed from different viewpoints. Each sculpture would also allow one to choose a variety of close-up form-space detail photos with varying light. The sculptures are symphonies of form, space, and light. You are invited to visit her web site above to view additional sculptures.

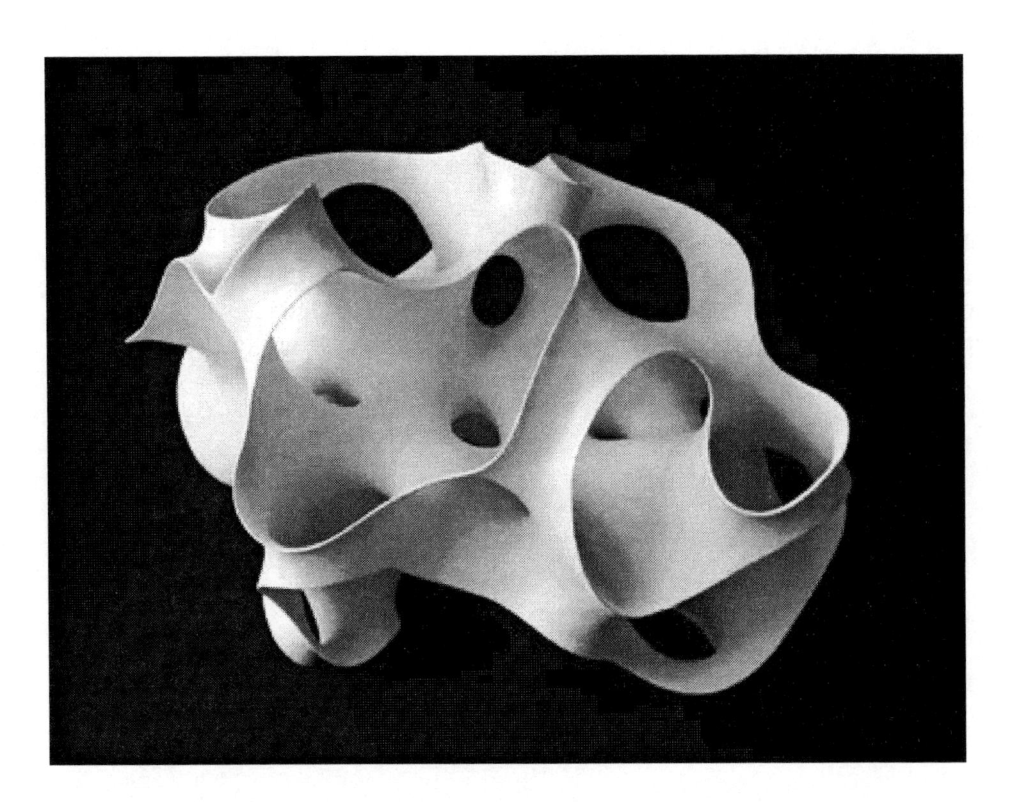

Figure 3.

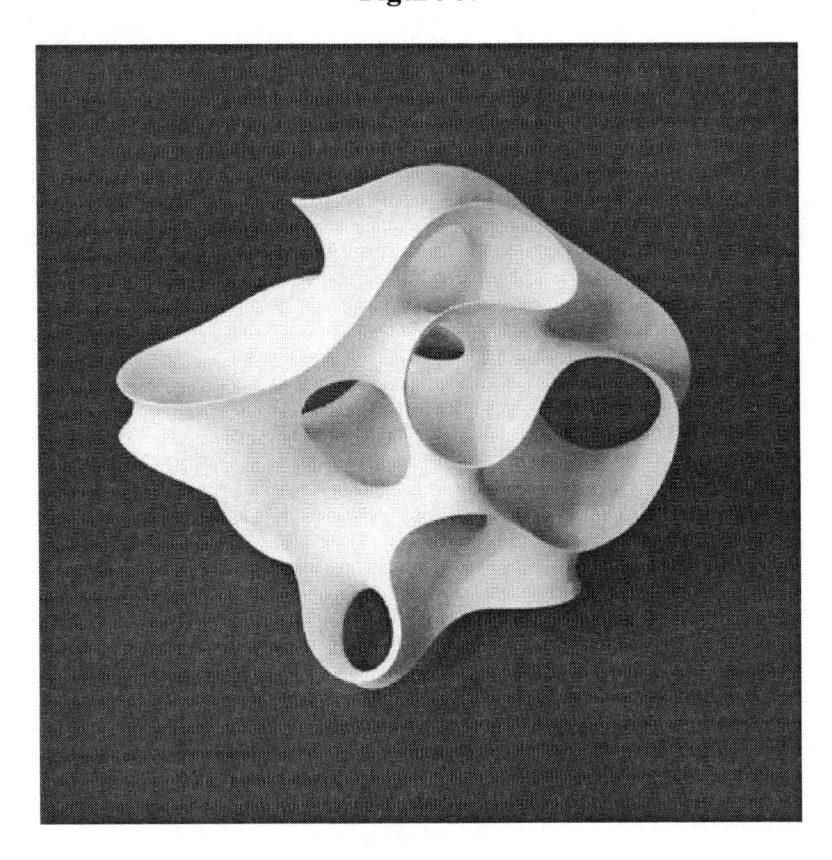

Figure 4.

I have come in contact with many artists while organizing various conferences relating art and mathematics. As in the case of Eva Hild, it always amazes me to find artists whose work looks mathematically inspired but is actually not. They have frequently told me that they do not really understand mathematics. My reply is that "birds fly but they don't understand aerodynamics." Somehow these artists fly on their geometric intuition.
They truly feel the form and shape.

I wish to express my gratitude to Eva Hild for allowing me to include the above images.
Photo credits are: Carl Bengtsson, Andrea Björsell, Ola Kjelbye, Eva Hild.

Interdisciplinary Bridges: A Novel Approach for Teaching Mathematics

Gail Kaplan
Department of Mathematics
Towson University
8000 York Road
Towson, Maryland 21252, USA
gkaplan@towson.edu

Abstract

This paper describes examples of interdisciplinary exploration opportunities that encourage students to use critical and creative thinking skills as they gain understanding and ownership of mathematical ideas. Students enjoy a stimulating journey on a road of discovery.

"All genuine learning comes about through experience. . . Only by taking a hand in the making of knowledge, . . does one ever get a knowledge of the method of knowing."
John Dewey

1. Philosophy of Discovery Based Interdisciplinary Learning

The primary focus of my professional life is the development and dissemination of dynamic approaches to learning mathematics. This focus is built on the belief that the ideal course of study combines the inseparable partners of thought and content, and an essential part of education is to merge the teaching of both. Students need to learn how to think. The thought process must blend analysis with creativity. The goal of both secondary and university classrooms must be to create energetic, engaged learners. To achieve this goal, the ideal faculty member must utilize teaching strategies which necessitate student involvement in the process.

Interdisciplinary discovery projects involve students in a dynamic learning experience. These projects are group oriented; the focus of each project is on exploration, guided by leading questions which encourage mathematical discourse amongst the students. Students much more readily grasp theory that they have "discovered" independently. In the words of one of my students, "You really understand the concept well by the time you finish. In order to finish, . . you need to come up with the concept yourself and this makes it easier for you to use it in the future . . .. The element of discovery involved gives you self confidence . . . " Students taught in this manner experience a unique and extraordinary mathematical journey of learning and become more mathematically powerful.

This paper focuses on two projects. The first is based on *Flatland*, a mathematical novel written by Edwin A. Abbott, and enables students to see connections between mathematics, philosophy, and sociology. The second project is a portfolio assignment providing opportunities for creative writing.

2. Mathematics and Literature Project

2.1. Description of the Project. Consider for a moment a geometry unit including philosophy, sociology, art, theology, and advanced mathematics. In a freshman level high school geometry class, my students read *Flatland*, a mathematical novel written in the 1800s. It is the story of a two dimensional world where woman are straight lines, and men are polygons with various numbers of sides depending on their social status. The initial paragraphs of the novel immediately challenge our perception of reality.

> I CALL our world Flatland, not because we call it so, but to make its nature clearer to you, my happy readers, who are privileged to live in Space.
> Imagine a vast sheet of paper on which straight Lines, Triangles, Squares, Pentagons, Hexagons, and other figures, instead of remaining fixed in their places, move freely about, on or in the surface, but without the power of rising above or sinking below it, very much like shadows - only hard and with luminous edges - and you will then have a pretty correct notion of my country and countrymen. Alas, a few years ago, I should have said "my universe": but now my mind has been opened to higher views of things. . . .
> Place a penny on the middle of one of your tables in Space; and leaning over it, look down upon it. It will appear a circle.
> But now, drawing back to the edge of the table, gradually lower your eye (thus bringing yourself more and more into the condition of the inhabitants of Flatland), and you will find the penny becoming more and more oval to your view; and at last when you have placed your eye exactly on the edge of the table (so that you are, as it were, actually a Flatlander) the penny will then have ceased to appear oval at all, and will have become, so far as you can see, a straight line.

A sphere comes to visit the main character of *Flatland*; "A Square" is his name. Conversations ensue as the Sphere tries to convince Square that there is a world beyond two space.

After the students have read the novel, a colleague from the humanities department leads the class in a discussion of various aspects of the novel, suggesting four categories of inquiry: philosophical, social, mathematical, and cross cultural. The class is divided into groups; each group prepares a presentation explaining the novel in terms of their assigned category. The oral presentation must include specific quotes supporting their ideas as well as a visual aid. The open-ended assignment yields astonishing results. Let us examine the ideas that the students develop.

2.2. Philosophical Aspects. The group focusing on philosophy explains that universal truths in two-space are quite different from universal truths in three-space. In two-space, a sphere is a circle, or a point, or nothing at all. Yet in three-space, a sphere is a ball with which to play. To illustrate their point, the students refer to the following quote. "Were a four-dimensional figure to appear to us, we would only be able to see three of his dimensions. He would likely appear to us as some kind of three dimensional solid, such as a cube or sphere, but he would be able to change his size and form. He would actually consist of an infinite number of cubes or spheres or whatever, just as the sphere consisted of an infinite number of circle." The students link these ideas to the differing beliefs of human beings living in dissimilar cultures.

2.3 Social Aspects. This group focuses on the hierarchy of society in *Flatland.* Women are the lowliest of the low, straight lines. Next are the isosceles triangles. Then there are regular polygons, starting with equilateral triangles, then squares, then pentagons, etc. The priests are circles; they are considered regular polygons with many, many sides. "Circles are the formal names of figures with sides far too numerous to count. This was done out of respect for the highest member of a society in which angularity was despised and smoothness respected." Again, the students explore the notion of a limit. As the number of sides of a regular polygon increases without bound, the object becomes a circle.

2.4 Mathematical Aspects. The group exploring mathematics describes the sphere's visit to two-dimensional space. Initially the sphere touches Flatland only in one point, but it becomes a circle as it moves through the land. The students provide visual tools of this visit for their classmates. A peep box with a slit on the box to represent Flatland and a sphere inside the box which moves up or down by well constructed strings allows the sphere to be seen as a circle until it disappears. See Figure 1. The students describe how the circle "becomes" a point. The students also slice Styrofoam balls to illustrate the idea in another fashion. See Figure 2. Students begin to gain an intuitive idea of a limit. Years later, when studying calculus, these students already possess a general notion of the meaning of limit.

Figure 1: *A student views a sphere in the same manner as an inhabitant of Flatland.*

Figure 2: *Students demonstrate that a sphere is made from infinitely many circles.*

The students explain that "in the two-dimensional world of Flatland the citizens and all inhabitants alike are part of a razor thin existence, less than that of a sheet of paper." They begin to understand that although we represent two dimensional space and think of it as a sheet of paper, this representation is flawed.

2.5 Cross Cultural Aspects. The group dealing with cross-cultural aspects discusses the almost insurmountable difficulty of inhabitants from one land communicating with those from another. In the novel there are four cultures, Pointland, Lineland, Flatland, and Spaceland. See Figure 3. The students describe these cultures and the challenges of communication between and among them. The inhabitants view what they can not understand as having supernatural powers. The square thought sphere was a god because he couldn't understand it. To support their ideas the students cite the following quote. "It seemed that this poor ignorant Monarch – as he called himself – was persuaded that the Straight Line which he called his Kingdom, and in which he passed his existence, constituted the whole world, and indeed the whole of space. Not being able either to move or to see, save in his Straight Line, he had no conception of anything out of it. . Outside his world, or line all was a blank to him; nay, not even a blank, for a blank implies space; say, rather, all was non-existent." When inhabitants travel into other worlds, they experience unexpected differences. "An unspeakable horror seized me. There was a darkness; then a dizzy sickening sensation of sight that was not like seeing; I saw a Line that was no Line; Space that was not Space: I was myself, and not myself. When I could find voice, I shrieked aloud in agony, 'either this is madness or it is Hell'." Again, the students discuss the challenges of exposure to cultures different from one's own.

Figure 3: *Students use each side of a box to represent the four cultures in Flatland.*

The class soon begins to recognize that the boundaries between the mathematical, philosophical, social and cross cultural categories are blurred. Perhaps, these students truly experience the bridges between mathematics and other disciplines!

2.5 Sample Assessment. Creating opportunities for students to explore the realm of philosophical ideas can happen with well structured assessment. Consider the following question and response from the unit test including *Flatland.*

Examination Question: Explain the meaning of this quote in terms of how it relates to *Flatland* and mathematics in general. Include a description of how we can envision a land of four dimensions. "I am not a plane Figure, but a Solid. You call me a Circle; but in reality I am not a Circle, but an infinite number of Circles, of size varying from a Point to a Circle of thirteen inches in diameter, one placed on top of the other."

Student Response 1: "This is an example of how the inhabitants of Flatland see things. They are unable to imagine a solid figure of three dimensions, just as we are unable to imagine a figure of four dimensions. The inhabitant view the world as a thin slice of the third dimension so perhaps we view the world as a thin slice of the forth dimension." [Spelling and grammar as in original.]

Student Response 2: "This quote is the sphere trying to explain his form to the square who is unable to see all three dimensions of the sphere. Because the square can only see two dimensions, he sees the sphere to be a circle able to change his size. This is one of the fundamental ideas in Flatland; how different worlds with different numbers of dimensions can interact and the confusion their interaction causes. This quote also helps us understand how we would see a land of four dimensions. Were a four-dimensional figure to appear to us, we would only be able to see three of his dimensions. He would likely appear to us as some kind of three dimensional solid, such as a cube or sphere, but he would be able to change his size and form. He would actually consist of an infinite number of cubes or spheres or whatever, just as the sphere consisted of an infinite number of circle."

3. Mathematics and Creative Writing

3.1 Introduction. Good teaching orchestrates the learner's interdisciplinary experience so that all aspects of brain operation are addressed, such as analytical thinking, imagination, and emotions. A portfolio can be an excellent vehicle for connections, as well. Consider the following writing assignments from a portfolio project given to a high school algebra class.

3.2. Example 1. The student creates a link between Shakespeare and mathematics. The class had recently completed a unit on graphs and transformations of functions.

Assignment: Explain how you used mathematics or mathematical thinking in another class.

Response:

> In English class we are reading Richard III and are split into groups to discuss the tragedy of Richard. As we were discussing this, I drew a physical representation of Richard's demise. We believed that Richard saw his destiny through his dreams. He was afraid of this fate, which was ultimately his demise. This fear drove him to try to control his fate. What happened was that at the same exact time that his success was apparent as he was becoming a king, he was actually falling farther from being a good man. All of this is represented by the line I graphed, an absolute value, $x = |y|$, that makes its "v" around the x-axis. The positive ray shows his attempts to control his fate by reaching high goals. The negative ray shows where his actions are actually taking him.

3.3. Example 2. One portion of the portfolio assignment is to provide an example of a problem that was totally misunderstood, and the process that eventually led to comprehension. One student wrote,

> When I first stared at this problem it looked like a menacing fire-breathing dragon who could never be conquered, because it looked like it was going to be hard to succeed. Immediately, I ran to Dr. Kaplan for help, hoping she had some kind of magic which would subdue the dragon. After talking with her, the dragon seemed friendly enough, but once out of range of her mathematical magic, the fires returned. Frustrated, I finished the problem to the best of my abilities, and handed it in. When it was given back to me, complete with many helpful comments, the magic was turned over to me and the dragon disappeared in a cloud of smoke.

What a fabulous way to connect mathematics with creative writing!

4. Summary

These interdisciplinary projects are unique learning tools for motivating and encouraging students in the energetic pursuit of mathematical knowledge. Instruction and student involvement are so firmly interwoven that they form inseparable partners in the learning process. Nontraditional interdisciplinary projects serve as genuine learning experiences, allowing students to actively pursue mathematical knowledge in a cooperative setting. Each student is engaged in the learning process; passivity vanishes. Mathematical discourse is an essential ingredient for success. Students explore ideas with one another. In the sharing process, each listens intently to the ideas of others and together they traverse various paths and detours until a successful route is found. The student not only learns the mathematics and its relationship to other disciplines, but also enjoys the satisfaction of playing an active role in an investigative learning process. The journey is challenging, filled with pitfalls of frustration and peaks of excitement until the final "Aha!" moment is achieved.

Reference

Abbott, Edwin A. *Flatland A Romance of Many Dimentions*, Dover Publications, 1992, Unabridged republication of *Flatland*, first publication Seeley & Co., Ltd., London 1884

Concerning the Geometrical in Art

Clifford Singer © 2005
4477 El Campana Way
Las Vegas, Nevada 89121

CliffordhS@aol.com

http://www.cs.unm.edu/~joel/NonEuclid/singer/singer.html
http://www.lastplace.com/EXHIBITS/VIPsuite/CSinger/

Abstract

It is the expanse of thought from earlier twentieth-century Modern art that has been in part an inspiration to my recent painting entitled, *The Blue Rider*. There is History serving as a discipline and the elements that shape the boundaries of style, vision, repetition, method, and constraint in art. This is the role of suggestibility for our perceptions. Expression is not isolated, limited, or confined to a single notion, or arbitrary method. To escape from the perpetual forces of society, tradition, and attitude would be to escape History itself. A process that is introspectively palpable and its individuated, continual, motivated, thematic, imaginative integration of geometrical configuration with color serves as a vehicle to this discipline.

Professor Joel Castellanos, Department of Mathematics and Computer Science, Director of *NonEuclid* at Rice University, has stated, "I am especially drawn to [Clifford Singer's painting entitled] *The Blue Rider*. The lines move me in and around and are suggestive of so many different things and actions, yet never settle anywhere."

Clifford Singer, *The Blue Rider*, 2003©, Acrylic on Plexiglas, 25x42 inches

Clifford Singer, *Cut Space Series*, Composition # 56, ©2005, Acrylic on Lexan, 36x44 inches

Clifford Singer, *Cut Space Series,* Composition # 48, ©2005, Acrylic on Lexan, 36x54 inches

Knot Designs from Snowflake Curves

Paul Gailiunas
25 Hedley Terrace, Gosforth
Newcastle, NE3 1DP, England
email: p-gailiunas@argonet.co.uk

Abstract

The Koch snowflake curve is one of the best-known self-similar fractals. Natural modifications of the polygons that represent the early stages of its generation provide templates for knotwork designs, some of which have been used in bookbindings. The boundary of another well-known self-similar fractal, the Sierpinski gasket, is closely related, and suggests a way to construct fractal knots.

Snowflake Curves

In 1904 Helge von Koch described a curve that has no tangent at any point, encloses a finite area, but the length of the curve between any two of its points is infinite. Start with an equilateral triangle and replace the middle third of every line-segment with two sides of an outward pointing equilateral triangle. Apply this rule to every line-segment, including the ones generated by its application. The curve is generally seen as the limit of a series of polygons, each produced from the previous one by a single application of the replacement rule (figure 1).

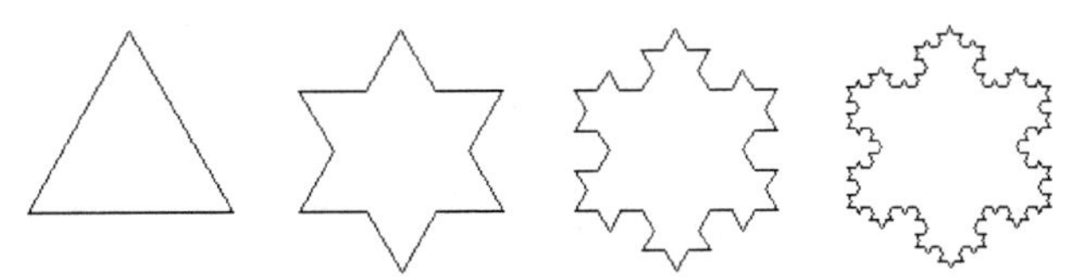

Figure 1: *The first four stages in the development of a Koch snowflake curve.*

The second stage is easily seen as the outline of the polygon (6/2), or Star of David, and this elaboration can extend to further stages in a natural way (figure 2). There are many ways to describe this sequence, for example the Star of David consists of six equilateral triangles joined at their corners to form a ring. At each step make the next in the series by overlaying each small triangle with a copy rotated by 60°.

Another two descriptions of the series are easier to generalize, and will be used later. The first stage consists of n = 6 triangles joined at their corners in a ring. A later stage consists of n = 6 copies of the previous stage joined at their corners in a ring. A scale factor is needed (in this case 1/3) to keep a constant overall size.

Notice that the Star of David has lines parallel in pairs. Consider, for example, the horizontal pair. Make a smaller copy so that its top corner coincides with the top corner of the original, and its *lower* horizontal lies on the *upper* horizontal of the original. Further stages follow the same pattern, but there are many parallels to choose from, so we need to specify corresponding pairs: corresponding parallels are related by a half-turn about the centre of symmetry

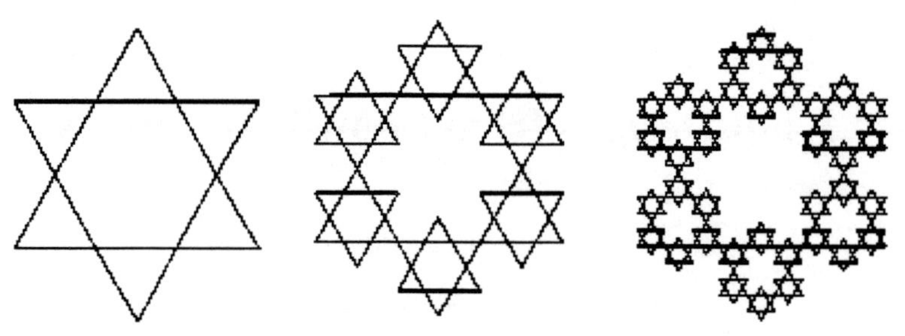

Figure 2: *Elaborating the stages in the development of a Koch snowflake curve.*

Knotwork Designs

Knotwork is a common feature of ornament in almost all cultures, reaching its most elaborate forms in the British Isles in the centuries before the Norman conquest, and in Islamic culture throughout its history. The designs to be derived from snowflake curves are different from traditional patterns, but inevitably they share some characteristics, being generally symmetrical and relying on the illusion of interlacing for their visual impact.

While it is no means an unbreakable rule there seems to be some preference in traditional examples for designs that are unicursal and do not break into disjoint segments. With this restriction, and a requirement that the original pattern in its entirety should form most of the structure of the final design, the number of possibilities becomes manageable, at least in the simpler cases.

The Star of David has three sets of parallel line-segments, and in the first stage of figure 2 they are all of the same type. In the second stage there are three types: a long segment inherited from the first stage, a short segment, its image, and a broken segment, which is nearest the centre. Any symmetric unicursal interlacing must cycle through these three types, although clearly taken from different parallel sets. The end of a cycle must be a rotation by 60° of the start. If it were 120° or 180° then there would be three or two disjoint paths respectively. Changes of direction should not occur too far out, or the original pattern would tend to insignificance, nor within the original line-segments, or some of the original pattern would be lost. Figure 3 shows two possibilities fulfilling these conditions. In both cases an additional, fourth type of line is needed. The next stage has nine types of line, providing a multitude of possibilities.

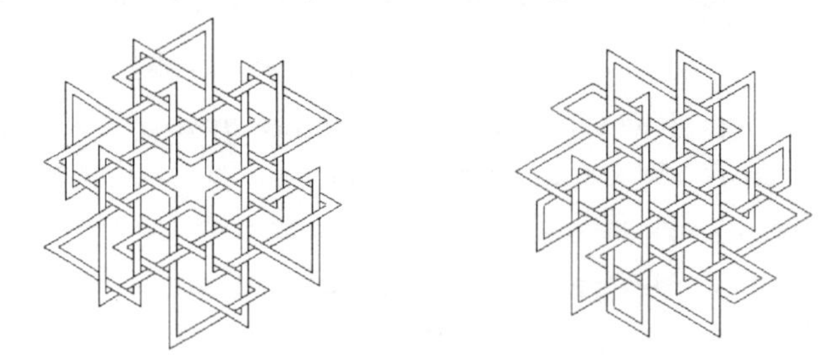

Figure 3: *Two knotwork designs derived from the second stage in figure 2.*

Some Designs Used on Books

Knot designs can be based on other polygons than (6/2) by applying the principles described earlier. Clearly any polygon can be used to form a ring by joining copies at their corners, but the second method, fitting smaller copies into an enclosed region, will only work with star polygons with an even number of sides. Figure 4 shows the first stages if the polygon (8/3) is used.

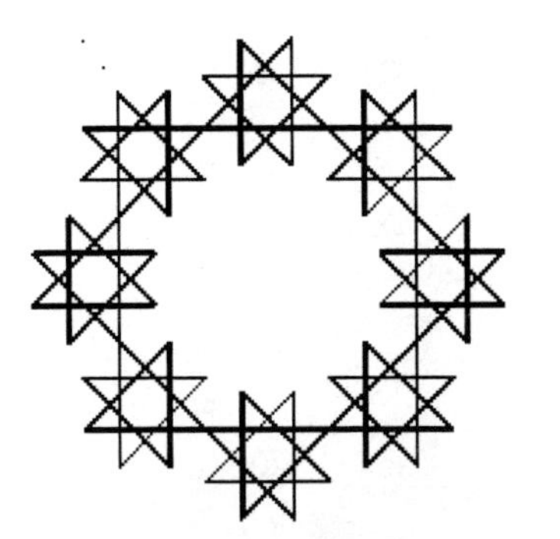

Figure 4: *The two methods give different results with (8/3).*

The second method gives a much more compact pattern, which I have used in the design of a bookbinding (figure 5). I decided that a non-unicursal interlacing is visually more interesting in this case. It makes obvious the relation between the two sizes of (8/3), and I have used colour to emphasise the difference. The book is about Tunisia, so I have chosen a design that relates strongly to traditional Islamic decorations.

Figure 5: *A bookbinding design based on (8/3)*

A more intricate design is based on (10/4). This polygon is a compound of two pentagrams, (5/2), and single copies of (5/2) have been fitted into the (10/4) (figure 6), rather than complete (10/4)s, which would make things unduly complicated.

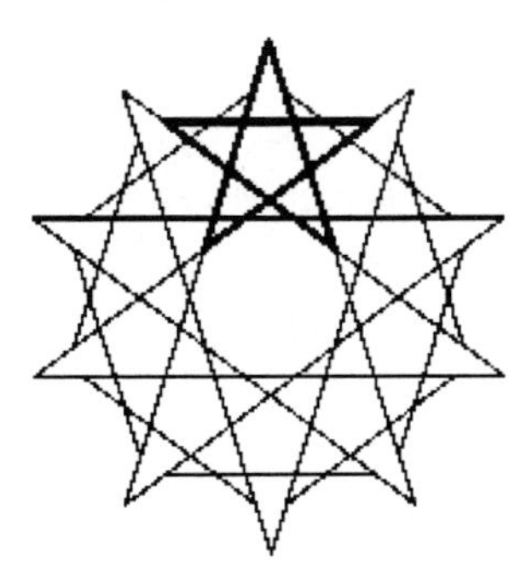

Figure 6: *Pentagrams fitted into (10/4)*

There are only two types of line, making a unicursal pattern almost impossible, but a third type arises naturally by completing small pentagrams at the outer corners. Figure 7 shows the finished design as it appears on the book. It has been adjusted to allow a greater width of straps, so pentagrams such as the one marked in figure 6 have been distorted slightly. The book is *Four Quartets* by T.S.Eliot.

Figure 7: *A bookbinding design based on figure 6.*

A Fractal Knot

All of the examples considered so far have been based on the second stage in the construction of a snowflake curve. Later stages get increasingly complicated, and each one would need a different method to complete a unicursal knotwork design. In order to continue indefinitely each new stage must follow automatically from the previous one, but in all the designs so far the parts added to the underlying structure interfere with each other. The alternative, joining polygons at their corners, provides a workable method, but with an even number of polygons the interlacing cannot be unicursal. The simplest polygon with an odd number of sides is obviously a triangle, and this produces a series that forms the boundary of the series used to generate the Sierpinski gasket: start with a triangle (usually equilateral), join the mid-points of edges to divide into four congruent copies and remove the middle one, repeat with the three remaining, and so on (figure 8).

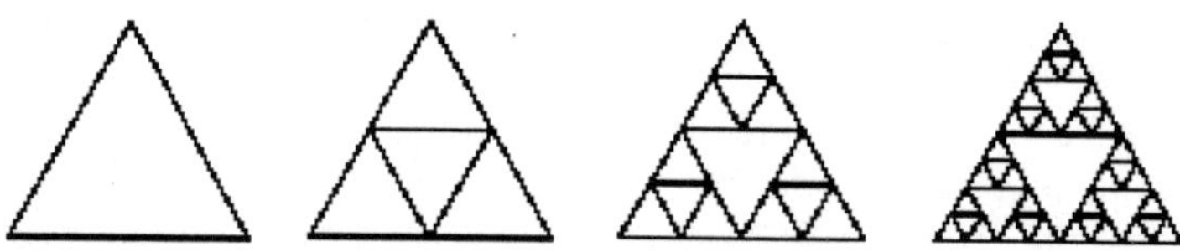

Figure 8: *The first four stages in the development of the boundary of the Sierpinski gasket.*

There are several difficulties in convert this to a knotwork design. Most obviously there is a change of direction at the point where the paths cross. This can be resolved by truncating the triangles to hexagons (figure 9), but considerable care is needed to ensure that each element has the right parity (whether the lacing goes under or over). Nevertheless there is a recursive procedure that ensures that each stage leads correctly to the next.

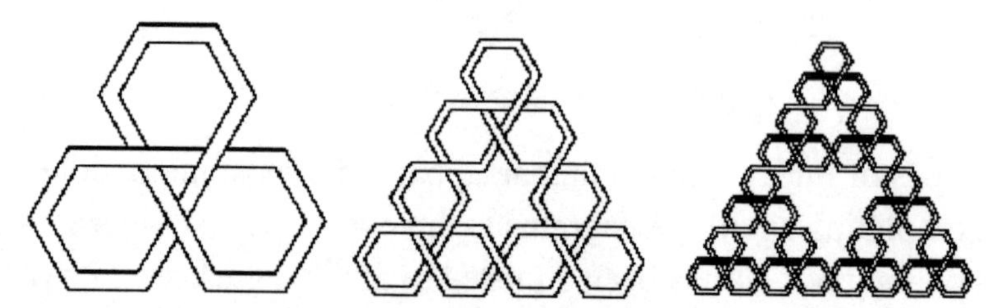

Figure 9: *The first three stages in the generation of a fractal knot.*

Asymmetry in Persian Symmetrical Art and Architecture

Hourieh Mashayekh
Architect and Town Planner
906-257 Lisgar St. Ottawa, Ontario
K2P 0C7 Canada
Email: hourimashayekh@yahoo.ca

Hayedeh Mashayekh
Coordinator and Translator
159-Mostofi Ave, Tehran
14347 Iran
Email: hayedeh_m@yahoo.com

"*Rose petals let us scatter*
And fill the cup with red wine
The firmament let us shatter
And come with new design" (Hafez 14th. century)

Abstract

Since ancient times, the integration of asymmetry in the design of composition has been a common practice in Iranian art and architecture in order to avoid problems such as topography and winds, and/or to comply with cultural and religious believes. This is manifested in mosques where the Mehrabs are[1] turned to the Qebla[2] to face in the direction of Mecca; in some entrances of mosques, public bath houses, or houses, in order to provide more privacy for the users; in town planning of large cities, in order to emphasize the old existing Friday mosques, or to avoid the direct access to a castle or governmental building; in the design of staircases, wind catchers, or in water distribution system; and in decorations such as tiling and miniatures.

The Art of Asymmetry in the Core of Symmetrical Art and Architecture

Islamic architecture derives its characteristic from geometrical symmetry (In this article, Islamic art and architecture refers to the work that was created between 800-1600 AD, and does not necessarily refer to religious architecture and related work).

Designers in Iran have been deeply influenced by the long established traditional rules of symmetry. Nevertheless, when the need arose, these same designers did not hesitate to solve some architectural problems by introducing asymmetrical elements to create a new composition that is harmonious in design.

The city of Isfahan which is located in the center of Iran is an outstanding example of harmony between symmetrical and asymmetrical design.

The *Maydan-e Naghsh-e- Jahan*, an important square in Isfahan with its multifunctional role, is an example of harmonious symmetrical design. As expansion of the square had become necessary, the designers used asymmetrical elements to create a functional design attaching the new buildings to the

1 The Mehrab marks the Quebla,s direction inside the mosques

2 The Qebla is the direction toward Mecca, which is the direction the worshippers face during prayers

Maydan. *Masjideh Shah* (Shah Mosque), being rotated 45 degree to face the Mecca's direction, is attached to the Maydan by using this method.

The Bazaar in Isfahan, located near the Maydan, was originally a symmetrical composition both in plan and section. However, as the city grew around it, the new structures intersected the bazaar modifying the existing symmetry. These structures were carefully designed making the transition a smooth one, and the final composition a pleasing design.

.

Figure 1, *AliGhapoo*
Photographer: Talin Der Georgian, 2000

Figure 2, *Masjideh Shah*
Photographer: Tim Bradley
Aga Khan Archive, 1996

Figure 3, *Masjideh Sheikh Lotfolah*
Photographer: Justin Fitzhugh1996

In the end, these amazing asymmetrical details that modified the symmetrical designs not only added to, but preserved, the beauty of the structures.

We wish to thank Hagit Hadaya and Mehdi Nasrin for their comment and assistance.

Cultural Statistics and Instructional Design

Darius Zahedi
California State University, Fullerton
800 N. State College Blvd.
Fullerton, CA 92831-3599, USA
E-mail: dxzahedi@yahoo.com

Abstract

In online education, a student's first point of contact is the Web interface, a GUI that must induce good feelings and trust. To achieve this, designer needs to be aware of cultural trends shared between members of target group. This requires mathematical formulas and statistical feedbacks so results can be stored, categorized, processed, and retrieved. For example a resulting bar chart can give a designer a vital clue as to what extend a target group tolerates teacher's interference. Efforts towards statistical representation of culture started since late twentieth century. They mostly concentrated on multinational organizations, but now with the table turned and employer being the end-user (students), more sensitivity to the cultural issues must be paid. This paper is a Call to the statisticians inviting them to explore this much-needed young science, with applications that go beyond just commerce and education.

Educational institutions can now reach more people and offer online degrees to more students, who may never ever step on their location's soil. However, when crossing the boundaries of lands, we often cross the boundaries of time as well. A true constructive learning system offers education at the learner's pace and time. The students should be able to login the learning system at their convenience, check their assignments, and take part in discussions by postings their views and giving feedbacks to those of their peers'. After all "learning is characterized not just by the processes within an individual learner but also by the processes shared by and affecting the members of a defined group" (Reiser & Dempsey, 2002). Asynchronous learning environments are blurring the borders that separate learners by lingual, cultural, and time differences. In an English language asynchronous learning system deployed by an institution or organization for a global audience, time difference is a non-issue, and the English language is only an important issue by how it maintains neutral in using references. The instructors and instructional designers, however, need to pay an extra attention to the *cultural issues*.

One of the earliest pioneers and perhaps the most influential psychologist on cultural theories is Lev Vygotsky (1896-1934). He explained that cultural development starts at childhood, first on the social or 'interpsychological' level, and then on the individual or 'intrapsychological' level (Vygotsky, 1978). Le Roi Jones (as cited in Forbes, 2006) describes culture simply as "how one lives and is connected to history by habit". In other words, culture is a learned behavior; it is not a physiological product that is inherited by a newborn genetically. Culture is an evolving discussion and is interpreted differently throughout the history based on the influences of its time and place. Tung (1995) describes culture as "an evolving set of shared beliefs, values, attitudes, and logical processes which provides cognitive maps for people within a given societal group to perceive, think, reason, act, react, and interact". Culture is a variable, not a constant.

"Culture affects who we are, how we think, how we behave, and how we respond to our environment. Above all, it determines how we learn" (Dunn & Marinetti, n.d.). The drop out rates in online courses is considered high and the lack of culturally appropriate learning is a major cause of it. "Students may question the merit in participation, or worse, feel disenfranchised if the course or learning resources do not fit their world view" ("Cross-cultural Issues", 2004).

"The acceptance, use and impact of WWW sites are affected by the cultural backgrounds, values, needs and preferences of learners" (McLoughlin & Oliver, 2000). Web interface is an influential contributor to the users' trust and acceptance of the course material, leading to "e-loyalty" (Cyr, Bonanni, Ilsever, & Bowes, 2005) amongst the participants. In an online learning platform, the Web interface design should be central to the overall delivery media design.

The sayings: "A picture is worth a thousand words" and the "First impression is the last impression" are not enough to emphasize the importance of the audio/visual elements that are used for a culturally diverse audience. An online learning system consists of Web pages displaying data in various colors, shapes, texts, languages, tones, logic, and order. How much will the learner understand and trust these collections will depend largely on the learner's cultural background.

Cultural diversity, however, is not limited to the geographical areas, for example different genders are also considered different cultural groups. Gender is the cultural representative of one's sexual identity. Gender is shaped by its dominant culture, and the expectations about attributes and behaviors appropriate to women or men and their relations (Schalkwyk, 2000). When seeking traits within any given group, we need to evaluate at least two other sub-traits (genders) within the same cultural group to be able to address all the related cross-cultural issues. Hence, when dealing with culture, there are a minimum of '*2 x n*' cultural issues to account for. Where '*2*' represents gender types, and '*n*' is the number of other cultures within the target group. According to Marcus and Gould (2000), even though not everyone may precisely fit within a unique pattern of a given society, there is enough statistical regularity to identify trends and tendencies. Representing culture statistically is an important first step in understanding them.

Geert Hofstede (1997) identified five dimensions of culture and used indexes as their representatives, so cultures can be distinguished and compared statistically. These dimensions are:

1. *Power Distance:* PDI (Power Distance Index) is the amount of tolerance in power distribution within a society. High PDI cultures may tolerate a degree of dictatorship from the leaders. Low PDI cultures on the other hand, see very little gap between the higher-ups and subordinates and expect more equalities. Some of the Web designs reflect this by the importance they give to social and informational hierarchies.

Hi PDI – Malaysia (http://www.unimas.my/)

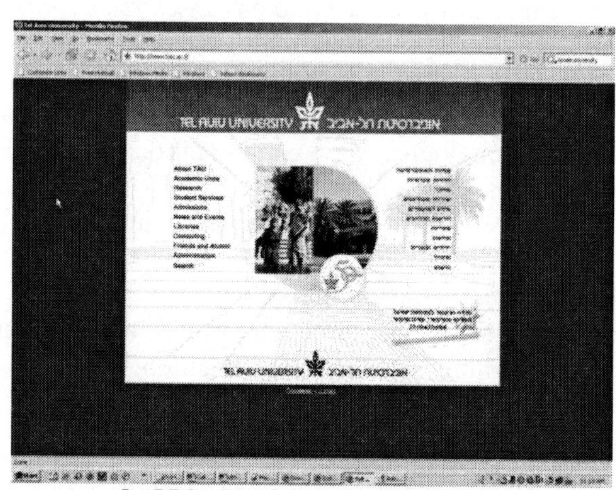

Lo PDI – Israel (http://www.tau.ac.il/)

2. *Individualism vs. Collectivism:* Individualism pertains to cultures with loose ties. Members are expected to look after themselves and their immediate kin. Collectivism believes in strong social ties where the key is trust and loyalty in relationships with other members, and beyond the immediate kin.

IDV (Individualism Index) is used to measure this dimension. Web designs with high IDV usually reflect this by focusing on the youth, new ways, and personal goals.

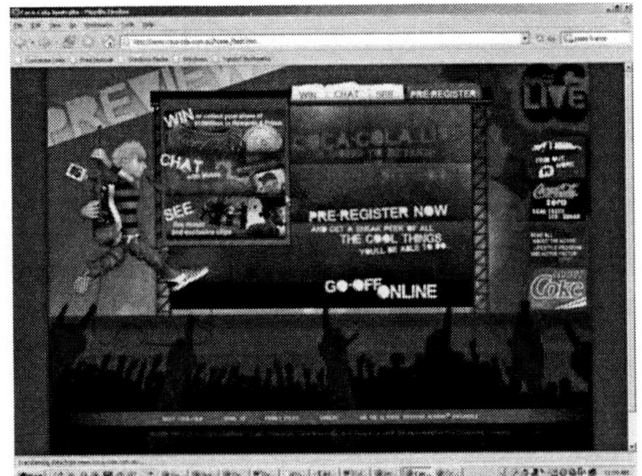

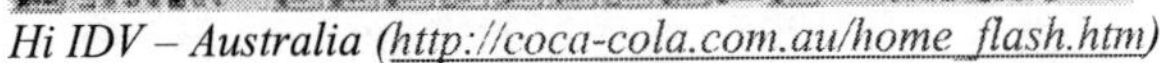
Hi IDV – Australia (http://coca-cola.com.au/home_flash.htm)

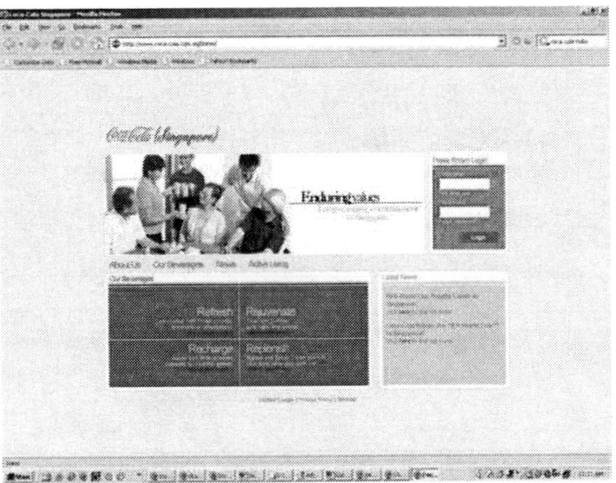
Lo IDV – Singapore (http://www.coca-cola.com.sg/home/)

3. *Masculinity vs. Femininity:* This refers to gender roles within a culture. A culture is masculine when roles are distinctive and such qualities as assertiveness, toughness, and materialism are the dominant qualities. As opposed to the feminine cultures where the roles are overlapping, and qualities such as modesty, softness and family orientation are dominant. MAS (Masculinity Index) is used to measure these gender roles in different cultures. Web designs from countries with high MAS usually have a "mean-to-do-business" appearance. They have direct navigational access, are not very soft, and most of the time gender specific.

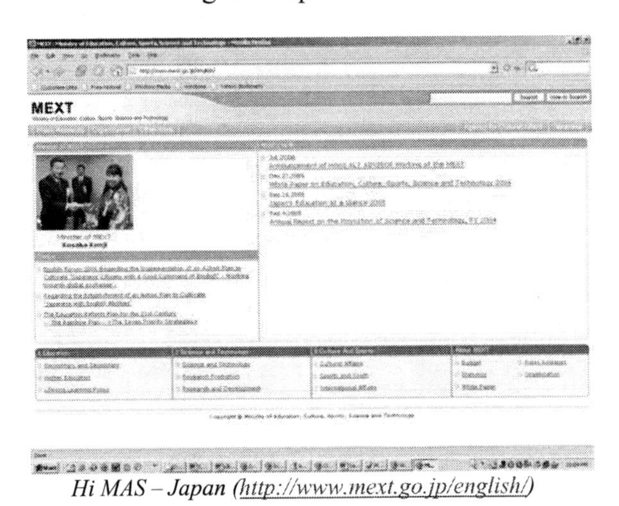

Hi MAS – Japan (http://www.mext.go.jp/english/)

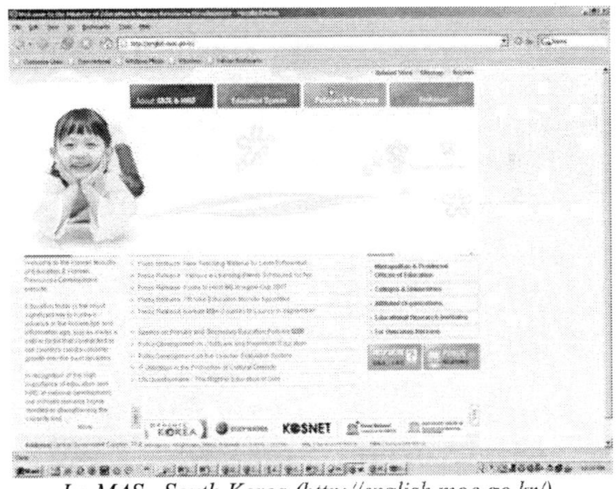

Lo MAS - South Korea (http://english.moe.go.kr/)

4. *Uncertainty Avoidance:* Different cultures react with different levels of anxiety to the unknown. High UAI (Uncertainty Avoidance Index) cultures are suspicious to changes and consider 'different' as dangerous. Low UAI cultures are more open to new ideas. High UAI Web pages offer choices with more expected results and they are usually simpler.

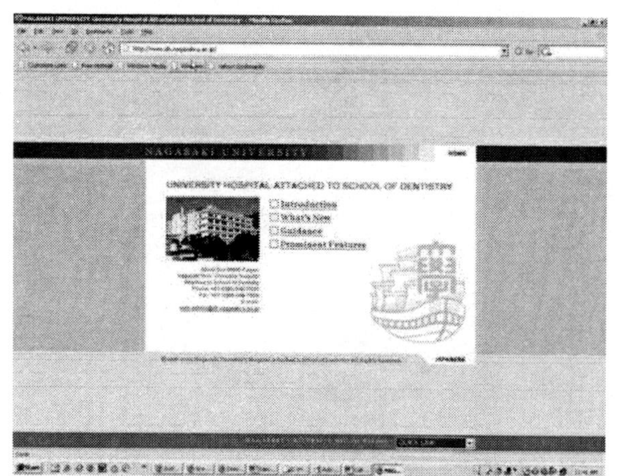

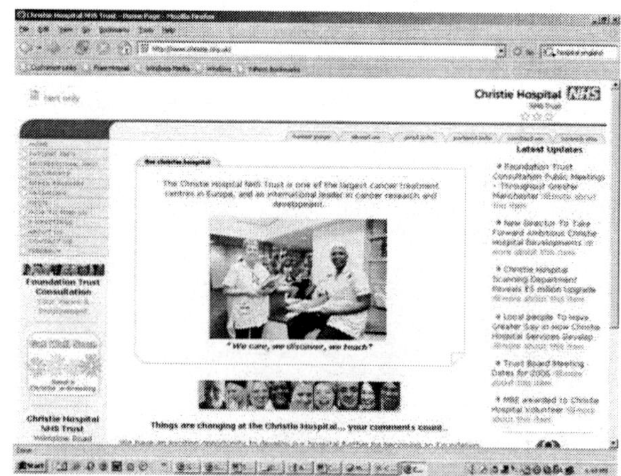

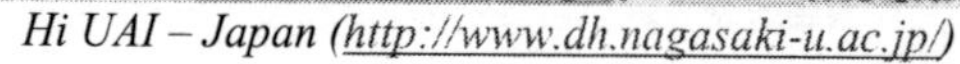

Hi UAI – Japan (http://www.dh.nagasaki-u.ac.jp/) *Lo UAI – England* (http://www.christie.nhs.uk/)

5. *Short-term vs. Long-term Orientation:* This dimension weighs the influence of tradition against progress. A short-term orientated culture is one that has strong traditional bonds, but values are not always measured logically and progress can be comprised for the sake of status and immediate results. In contrast, a long-term orientation sets a practical limit, while respecting traditional and social obligations. This dimension is measured by LTO (Long-Term Orientation) index. High LTO Web pages tend to emphasize on long-term relations.

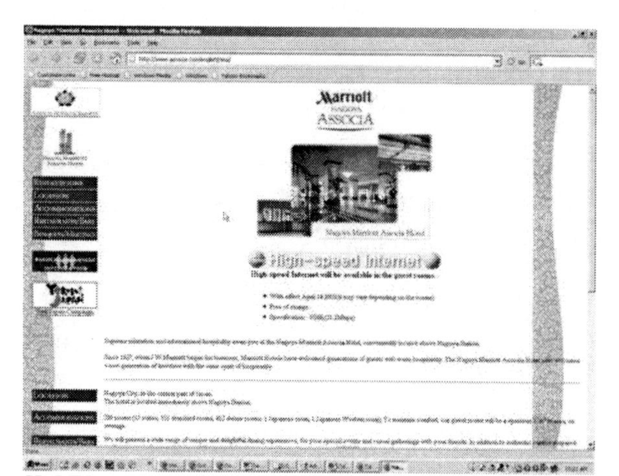

Hi LTO – Japan (http://www.associa.com/english/nma/) *Lo LTO – USA* (http://marriott.com/propertv/propertypage/YYZEC)

	PDI		IDV		MAS		UAI		LTO	
Countries	**rank**	**score**	**rank**	**score**	**rank**	**score**	**rank**	**score**	**rank**	**score**
Arab countries	7	80	26/27	38	23	53	27	68		
Australia	41	36	2	90	16	61	37	51	15	31
Brazil	14	69	26/27	38	27	49	21/22	76	6	65
Canada	39	39	4/5	80	24	52	41/42	48	20	23
China								1	118	
East Africa	21/23	64	33/35	27	39	41	36	52		
France	15/16	63	10/11	71	35/36	43	10/15	86		
Great	42/44	35	3	89	9/10	66	47/48	35	18	25

Britain										
Hong Kong	15/16	68	37	25	18/19	57	49/50	29	2	96
India	10/11	77	21	48	20/21	56	45	40	7	61
Iran	29/30	58	24	41	35/36	43	31/32	59		
Israel	52	13	19	54	29	47	19	81		
Japan	33	54	22/23	46	1	95	7	92	4	80
Malaysia	1	104	36	26	25/26	50	46	36		
Mexico	5/6	81	32	30	6	69	18	82		
Singapore	13	74	39/41	20	28	48	53	8	9	48
South Korea	27/28	60	43	18	41	39	16/17	85	5	75
Taiwan	29/30	58	44	17	32/33	45	26	69	3	87
USA	38	40	1	91	15	62	43	46	17	29

Hofstede's (1991) five cultural dimensions. (adopted from Marcus & Gould, 2000)

Hofstede's cultural model is not the only one, to name a few: 4-dimensional model of Hall and Hall (1990), 14-dimensional model of Reeves (1992), 7-dimensional model of Trompenaars (1993), 19-dimensional model of Collis, Vingerhoets, and Moonen (1997), and 9-dimensional model of Khaslavsky (1998).

In an effort to make available a basic open tool to measure cultural tendencies of target groups, the author has created an instrument called Calculature© which is located at www.InstructionalDesigns.org. Calculature uses Hofstede's cultural indexes, and includes visual elements such as colors and fonts to represent trends. Studies can be created by anyone for any cultural group, be it a country, a village, or a corporation. After all, corporations have their own specific cultures too. The results evolve as database populates with time.

The participants select the cultural 'topic' they wish to participate in, enter their profiles, and answer the questionnaire. The participants' demographic profiles are limited to:

- Gender
- Age
- Country
- Language
- Religion

Questions include those pertaining to the cultural dimensions and responses to a limited variety of Web-friendly colors and fonts:

1. (PDI) To what degree do you accept the power of your higher-ups?
    a. Hi (100) - I accept my supervisors', managers', and leaders' authorities without any questions asked and look up at them as teachers and gurus.
    b. Lo (0) - I see very little gap between the higher-ups and subordinates and expect more equalities.
2. (IDV) How strongly or loose do you define your relation with others in your group?
    a. Hi (100) - I believe that I am an island, an individual responsible only for me and my growth, and my family's well being.
    b. Lo (0) - I believe in strong social ties where the key is trust and loyalty in relationships with other member, and members are expected to contribute into the well-being of the collective.
3. (MAS) How do you describe your nature?

   a. Hi (100) - Assertive, tough, go-getter, materialistic.
   b. Lo (0) - Modest, easy going, soft, family oriented.
4. (UAI) How do you treat uncertainties?
   a. Hi (100) - I do not like changes at work place or at home. I am reluctant to new ideas and look at them with suspicion.
   b. Lo (0) - I am open to changes and look forward to new ideas, even if the outcome is not guaranteed.
5. (LTO) How closely you follow traditional and religious values?
   a. Hi (100) - I follow my religious and traditional values for whatever I do in my day-to-day life and expect the same from others. I trust them completely and follow them blindly.
   b. Lo (0) - I may respect my religious and traditional values but always weigh the consequences, and if in a situation I find them not appropriate or practical, I will ignore them.
6. What color you like best?
7. What color you like the least?
8. What color is the happiest?
9. What color is the saddest?
10. What color is most energizing?
11. What color is most romantic?
12. What color is most spiritual?
13. What font is most appealing?
14. What font is least appealing?

For the first two questions (PDI and IDV), there are two scenarios involved:

- At Work.
- In Politics.

Reason being that the author believes the relations, expectations, and roles are not identical in the two environments. It is popularly believed, however, that cultural values most of the time impact all environments similarly, i.e. family, work, and politics overlap. So by separating these two in our study we have the opportunity to see how far which theory stands true.

Europe (West)

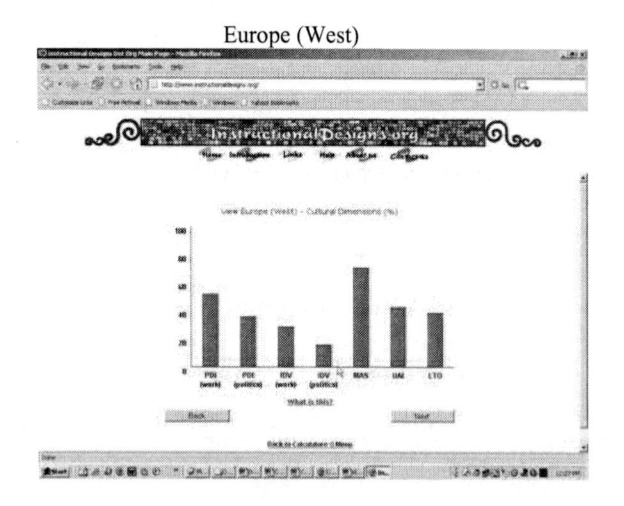

Europe (East)

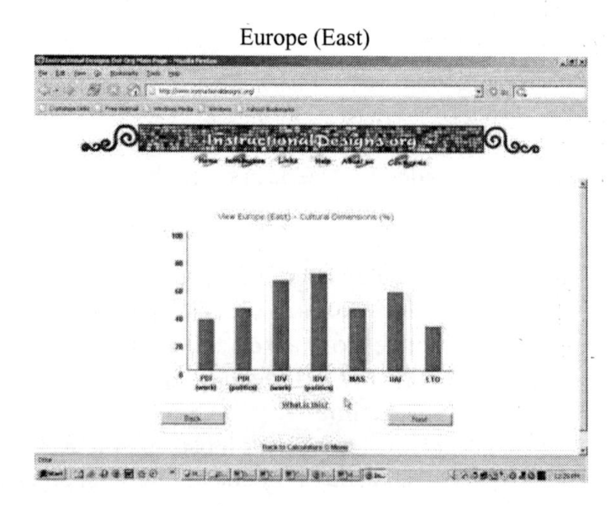

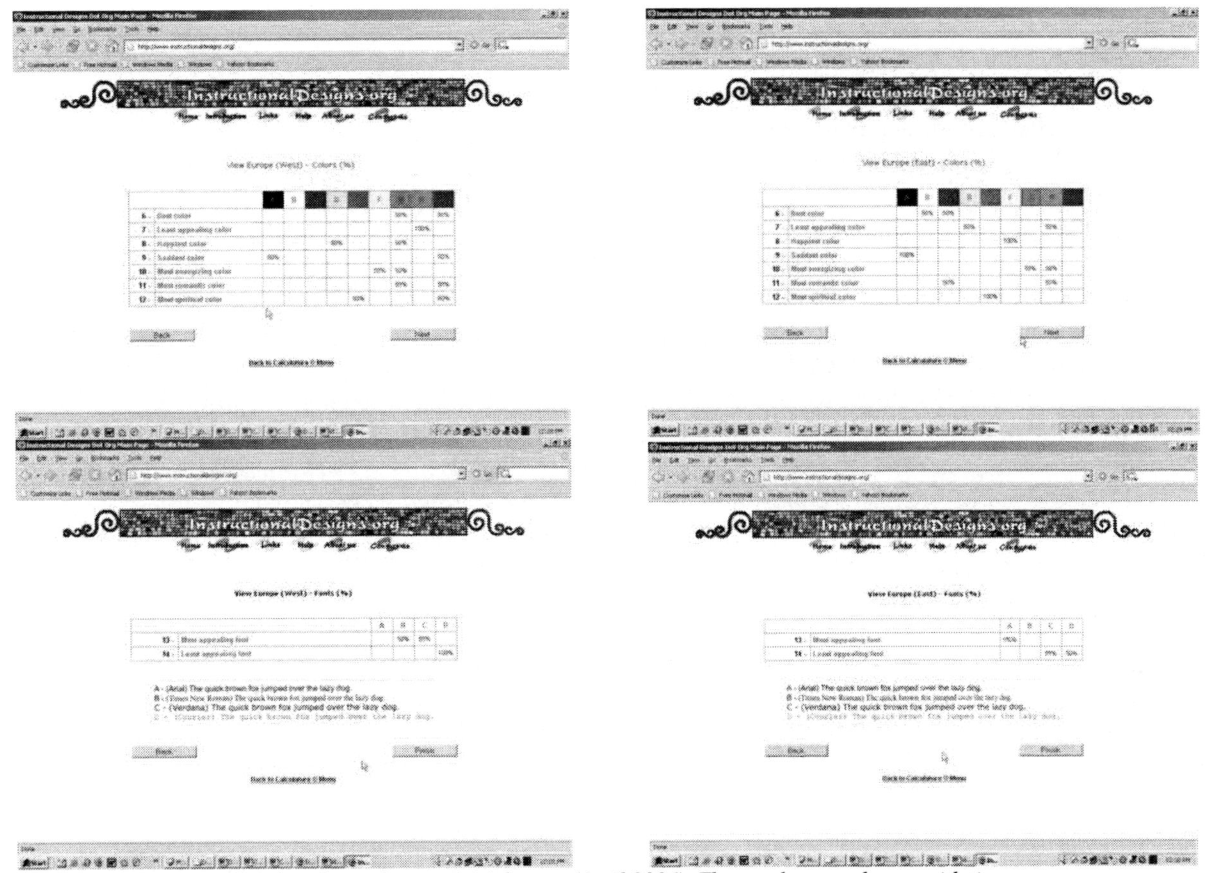

Results of a study at its initial stage (April 2006). The results may change with time.

Faiola and Matei (2005) propose a new theoretical construct that they call cultural cognition theory (CCT), that instead of focusing on the user it explores the source – Web designer's cognitive style. So "by observing the design of graphics, text, and information architecture, we can understand how processes of strategizing by culturally diverse web designers influence their cognitive skills toward a holistic or analytic orientation". McLoughlin and Oliver (2005) warn that "the design of Web based instruction is not culturally neutral, but instead is based on the particular epistemologies, learning theories and goal orientations of the designers themselves", and "one of the limitations in current instructional design models is that they do not fully contexualize the learning experience, and are themselves the product of particular cultures". So in Calculature, the actual interpretations of results are left to the designers because such interpretations need to be handled case by case and within the unique frameworks of both the target groups, and the designers.

Cultural diversity is not just alarms and whistles, it is the salt and pepper of learning, it is in fact a constructive challenge. Diversity can bring richness to the learning, and students who learn in such an environment become better critical thinkers, communicators, problem-solvers and team players (Sugar & Bonk, 1998). Students from different cultures interacting within a structured learning experience, develop greater openness and understand of 'others', and that results in greater productivity (McArthur, n.d.).

Much of the research so far in cultural statistics is inter-country related. The readily available economical and political data make tabulating and formulating comparisons easier. However, with increase in the migration of citizens to virtual communities, there is an immediate need to go beyond countries and focus on the citizens of the world. Cultural statistics today is at its infancy, there is much yet to be discovered beyond our borders.

References

[1] Australian Flexible Learning Framework (2004). *Cross-cultural Issues in Content Development and Teaching Online*. Retrieved October 17, 2005, from http://pre2005.flexiblelearning.net.au/guides/crosscultural.pdf

[2] Collis B., Vingerhoets, J., & Moonen J. (1997). Flexibility as a Key Construct in European Training. *British Journal of Educational Technology*, 28 (3), 199-218.

[3] Cyr, D., Bonanni, C., Bowes, J., & Ilsever, J. (2005). Beyond Trust: Website Design Preferences Across Cultures. *Journal of Global Information Management*, 13(4), 24-52. Retrieved March 8, 2006, from http://www.diannecyr.com/docs/Beyond_Trust.pdf

[4] Dunn, P., Marinetti, A. (n.d.). Cultural Adaptation: Necessity for Global eLearning. *LINE Zine*. Retrieved February 14, 2006, from http://www.linezine.com/7.2/articles/pdamca.htm

[5] Faiola, A., & Matei, S. A. (2005). Cultural Cognitive Style and Web Design: Beyond a Behavioral Inquiry into Computer-mediated Communication. *Journal of Computer-Mediated Communication*, 11(1). Retrieved February 13, 2006, from http://jcmc.indiana.edu/vol11/issue1/faiola.html

[6] Hall, E. T., & Hall, M. R. (1990). *Understanding Cultural Differences: Germans, French, and Americans.* Yarmouth, ME: Intercultural Press.

[7] Hofstede, G. (1997). *Cultures and Organizations: Software of the Mind.* New York, NY: McGraw-Hill.

[8] Jones, R. L. (n.d.). Thoughts: On the Business of Life. *Forbes*, 177(4), 120.

[9] Khaslavsky, J. (1998). Integrating Culture into Interface Design. *CHI 98*, 18(23), 365-366. Retrieved March 6, 2006 from ACM Digital Library.

[11] McArthur, I. (n.d.) Collabor8©: Digital Strategies for Cross-cultural Design Education. Retrieved October 17, 2005, from http://ianmcarthur.net/COLLABOR8/44mca.doc

[12] McLaughlin, C., & Oliver, R. (2000). Designing Learning Environments for Cultural Inclusivity: A Case Study of Indigenous Online Learning at Tertiary Level. *Australian Journal of Educational Technology*, 16(1), 58-72. Retrieved February 10, 2006, from http://www.ascilite.org.au/ajet/ajet16/mcloughlin.html

[13] Marcus, A., & Gould, E. W. (2000). Crosscurrents: Cultural Dimensions and Global Web User-Interface Design. *Interactions*, 7(4), 32-46.

[14] Reeves, T. (1992). Effective Dimensions of Interactive Learning Systems. *Proceedings of the Information Technology for Training and Education Conference (ITTE '92).* Brisbane, Australia: University of Queensland.

[15] Reiser, R. A., & Dempsey, J. V. (2002). *Trends And Issues In: Instructional Design And Technology.* Upper Saddle River, NJ: Merrill Prentice Hall.

[16] Schalkwyk, J. (2000). *CULTURE: a) Culture, Gender Equality and Development Cooperation.* Canadian International Development Agency (June 2000). Retrieved October 17, 2005, from http://www.acdi-cida.gc.ca

[17] Sugar, W., & Bonk, C. (1998). Student Role Play in the World Forum: Analyses of an Arctic Adventure Apprenticeship. In Bonk, C., & King, K. (Eds.), *Electronic Collaborators: Learner-Centered Technologies for Literacy, Apprenticeship, and Discourse.* Mahwah, NJ: Lawrence Erlbaum Associates.

[18] Trompenaars, F. (1993). *Riding the Waves of Culture: Understanding Cultural Diversity in Business.* London, UK: Nicholas Brealey.

[19] Tung, R. L. (1995). International Organizational Behavior. In F. Luthans, (Ed.), *Virtual O.B. Electronic Data Base*, 487-518. New York: McGraw-Hill.

[20] Vygotsky, L. (1978). *Mind in Society: The Development of Higher Psychological Processes.* Cambridge, MA: Harvard University Press.

Musical Scales, Integer Partitions, Necklaces, and Polygons

David Rappaport
School of Computing, Queen's University
Kingston, ON CANADA
DaveR@cs.queensu.ca

Abstract

A musical scale can be viewed as a subsequence of notes taken from a chromatic sequence. Given integers (N, K) $N > K$ we use particular integer partitions of N into K parts to construct distinguished scales. We show that a natural geometric realization of these scales results in maximal polygons.

1 Introduction

In his book on jazz ear training Steve Masakowski [6] speaks of four key scales that form the basis of organization that allows jazz musicians to understand and follow the harmony and melody of a piece. In his paper on seven tone collections Jay Rahn [7] analyzes seven note collections from the perspective of interval structure. With a small exception his seven note collection is identical to the Masakowski collection. In this note we present a combinatorial approach to generating a seven note collection of scales. The collection is not derived from musical consideration but is a collection that satisfies a geometric constraint. Again this collection with some small exceptions matches those of the Masakowski and Rahn collections.

A landmark result in combinatorial music theory is the characterization of the diatonic scale by Clough and Douthett [1]. The characterization can be made in numerous ways. In a subsequent paper Clough et al. [2] enumerate eight different characterizations of the diatonic scale. A more recent mathematical treatment looking at many of these results pertaining to rhythms as well as scales can be found in the paper by Demaine et al. [3] . If we consider a scale as a subsequence of a chromatic circular sequence to be realized by a set of equally spaced points on the circumference of a unit circle, then we can express the distance between notes of the scale as the Euclidean distance between their point representations. Using this representation there is a simple geometric property that encapsulates all of these rules and is unique to the diatonic scale. The diatonic scale is the unique seven note scale (up to rotations) that maximizes the sum of Euclidean distances between every pair of notes.

In this note we take a purely mathematical approach at deriving an interesting group of seven note collections. Rather than base the choice on any deep musical analysis, we will simply define some mathematical objective, solve for it, and report the results. The underlying contribution that is made is a distillation of a plethora of rules and properties into a simple and general framework. Just as the diatonic scale is characterized by a single geometric property, the collection of seven note scales presented are all maximal in a geometric sense.

2 Preliminaries

A musical scale can be viewed as a subsequence of notes taken from a chromatic sequence. Thus we can use the notation (N, K) scales to denote scales of K notes taken from an N note chromatic universe. We develop some notation to describe (N, K) scales in general before we turn our focus to (12,7) scales in particular.

A well known combinatorial object, the "necklace", captures the notion of modelling a scale with a circular sequence. A n-ary necklace is defined as an equivalence class of n-ary strings under rotation, see [10]. More intuitively, think of a string of beads. The string of beads is an implicit sequence that is invariant under rotations. Relating this concept to the diatonic scale consider a string of black and white beads arranged in the same pattern as 12 white and black keys comprising a single octave on the piano. Note that in his well known list of scales Alan Forte [5] considers equivalence classes of strings that are invariant under rotation and reflection. In music theory terminology rotation = transposition and reflection = inversion. The requisite combinatorial structure that can be used to enumerate sequences that are invariant under rotation and reflection is the *bracelet*. Much is known about the cardinality of necklaces and bracelets, and both of these combinatorial objects can be enumerated with very efficient algorithms, see [10, 9].

An additional property that is needed to perform our analysis is to embed the combinatorial necklace onto a concrete geometric surface. Thus we can consider the beads to be placed at equally spaced intervals on the circumference of a circle of radius one.

For an n-ary necklace we can label the beads from $0 \dots n-1$, and by convention always assume that the 0 bead is white. This labelling is useful for describing the distance between any two white beads. We use three distinct distances. Thus for two white beads i, j $i \neq j$, we have:

chromatic The chromatic distance between any two distinct white beads i and j is denoted by $c_{i,j} = |\{k : k \in [i+1 \dots j]\}|$ that is the total number of beads in the substring $[i+1 \dots j]$.

scalar The scalar distance between any two distinct white beads i and j is given by $d_{i,j} = |\{k : k \in [i+1 \dots j]$ and k is white$\}|$ that is the total number of white beads in the substring $[i+1 \dots j]$.

Euclidean The Euclidean distance between any two distinct white beads i and j is denoted by $e_{i,j}$ is the length of the line segment between points representing the beads.

In 1956 Fejes Tóth asked for the configuration of points on a continuous circle of fixed radius that maximizes the sum of inter-point distances. The answer is to place the n points at the vertices of a regular n-gon [4]. Asking a similar question for placing points on a discrete circle, that is a circle with a finite number of equally spaced possible positions with fixed radius, yields the diatonic collection as the unique answer when the number of points is seven and the number of positions on the circle is twelve. In essence the points are spread out as evenly as is possible. Hence the name given by Clough and Douthett *maximally even* [1].

The quantity that is maximized can be written as:

$$\sum_{\text{for all distinct i,j}} e_{i,j}.$$

This sum can be broken down into components, each representing the sum of the inter-point distances for pairs that are at the same scalar distance. Let

$$\Sigma_\ell = \sum_{i,j:d_{i,j}=\ell} e_{i,j}$$

Our distinguished set of scales are defined as those scales whose point representation maximizes individual Σ_ℓ quantities. Noting that $\Sigma_i = \Sigma_{K-i}$ for $i = 1 \dots \lfloor \frac{N}{2} \rfloor$ we consider those scales that maximize $\Sigma_1, \Sigma_2, \dots, \Sigma_{\lfloor \frac{N}{2} \rfloor}$.

We now present a general and efficient way to obtain this collection of scales.

An *integer partition* of a natural number N is a way of writing N as an unordered sum of natural numbers. Consider positive integers N, K $K < N$ and a partition of N using exactly K positive integer summands, that is, $N = a_1 + a_2 + \cdots + a_K$. The values $a_i, 1 \leq i \leq K$ denote the different chromatic distances between notes that are at scalar distance 1, or in musical terminology the length of an interval of a second. In [8] the scales with the property that notes i, j at scalar distance $d_{i,j} = 1$ have their chromatic distance $c_{i,j}$ come from two consecutive values that differ by at most one are examined. It was shown that given integers N, K with $K < N$, there exists a unique m such that $N = m\left\lfloor \frac{N}{K} \right\rfloor + (K-m)\left\lceil \frac{N}{K} \right\rceil$.

Thus let an integer partition of N into K summands $a_i, 1 \leq i \leq k$ such that $a_i \in \{\left\lfloor \frac{N}{K} \right\rfloor, \left\lceil \frac{N}{K} \right\rceil\}$ be called an *even integer partition* of N into K summands, which we denote by EP(N, K).

In [8] it is shown that any scale (N,K) with the property that notes i, j at scalar distance $d_{i,j} = 1$ have chromatic distance $c_{i,j} \in$ EP(N,K) maximizes area. A similar approach can be used to show that the sum of distances between adjacent points is maximized.

As integer partitions are not ordered and our model of a scale is, we need to impose a particular ordering of the summands to obtain a scale. For example consider the values $N = 12$ and $K = 7$ there are three distinct scales resulting from EP(12,7). The method to enumerate these scales is by using combinatorial necklaces. Rather that use necklaces with N beads, K of them white, we use K beads where m are labelled $\left\lfloor \frac{N}{K} \right\rfloor$ and $K - m$ are labelled $\left\lceil \frac{N}{K} \right\rceil$. The distinct number of these necklaces enumerates the different scales with this property. It should be noted that enumerating all two coloured fixed density necklaces (that is we fix the number of beads of each type) can be performed in time that is a linear function of the total number of necklaces enumerated [10].

3 Necklaces and Polygons

For melodic considerations it is desirable to have "smooth" diatonic steps. Thus, as was discussed above, we enumerate the necklaces obtained from EP(12,7).

The triad is the basic building block in Western harmony. A triad consists of three notes and there is an interval of a third between the first and second and the second and third notes. The interval of a third is made up of 3 or 4 chromatic steps. If we think of traversing a scale by leaps of consecutive thirds, that is by skipping over one note, we see that we make two complete revolutions around the scale. This is the intuition that explains why we consider an even integer partition EP(24,7). Observe that $\left\lfloor \frac{24}{7} \right\rfloor = 3$, and that $\left\lceil \frac{24}{7} \right\rceil = 4$, and $24 = 3*4+4*3$. Thus we consider distinct necklaces that can be obtained with seven beads where three are labelled 4, and four are labelled 3.

One of the attributes of the Diatonic scale is that it can be obtained using a generator, see [2]. The next necklace we consider represents the even integer partition EP(36,7). We see that 36 = 6*5+6, or expressed in another way $36 \equiv 1 \pmod 7$. This in turn implies there is a unique necklace corresponding to EP(36,7). Observe that the sequence of diatonic notes taken at chromatic distance 5 corresponds to the familiar *circle of fourths*.

	Interval	Pitch	Notes	Forte No.	M	R
1	1221222	0, 1, 3, 5, 6, 8, 10	Diatonic	7-35	Y	Y
2	1212222	0, 1, 3, 4, 6, 8, 10	Melodic Minor	7-34	Y	Y
3	1122222	0, 1, 2, 4, 6, 8, 10	Neapolitan	7-33	N	Y
4	3343434	0, 1, 3, 5, 6, 8, 10	Diatonic	7-35	Y	Y
5	3343344	0, 1, 3, 4, 6, 8, 10	Melodic Minor	7-34	Y	Y
6	3334434	0, 1, 4, 5, 7, 9, 10	Harmonic Major	7-32	Y	Implicit
7	3334344	0, 2, 3, 5, 7, 8, 11	Harmonic Minor	7-32	Y	Y
8	3333444	0, 1, 3, 5, 6, 9	Aug. Triad & Dim.7	6-Z28	N	N
9	5555556	0, 1, 3, 5, 6, 8, 10	Diatonic	7-35	Y	Y

Table 1: Necklaces and scales corresponding to EP(12,7) [1...3], EP(24,7) [4...8] and EP(36,7) [9].

We enumerate the appropriate necklaces and present the results in Table 1. We have identified each scale by an interval sequence. This corresponds directly to the enumerated necklaces. Each interval sequence is in turn represented by a pitch sequence. This pitch sequence can be interpreted either by fixing 0 to C, or more neutrally to a moveable "Do" system where 0 corresponds to scale degree $\hat{1}$. We then use a familiar name to identify the scales. We obtain five uniquely named 7 note scales, diatonic, melodic minor (ascending), neapolitan (or whole-tone plus a note), harmonic major, and harmonic minor. Note that our analysis produces one 6 note scale, which may be described as an augmented triad superimposed with a symmetric diminished seventh chord. The reason we only get 6 notes is revealed by the interval sequence 3333444. The sequence of four consecutive three's produces a second copy of the 0 pitch. We also use Forte's scale numbering system, [5] as one further standardized representation. Note that the harmonic major and harmonic minor scales share the same Forte number, because one is just a reflection of the other. In the final two columns of the table we denote the scales that are in the Masakowski [6] and Rahn [7] collections respectively. All of our seven note scales appear in the Rahn collection, although the harmonic major is implicit as it is the reflection of the harmonic minor. Only the Neapolitan scale is missing in the Masakowski list.

The necklaces in all but one case lead to a 7-gon inscribed in a regular 12-gon, 24-gon, or 36-gon. Each of the inscribed 7-gons are maximal polygons, that is they maximize either Σ_1, Σ_2, or Σ_3. The diatonic scale is unique as it maximizes all three. Also note that the six note scale that is obtained is an anomaly. Not only does it have less that seven notes, it is not maximal for 6-gons.

4 Discusion

We have shown in one unifying method a way to characterize a collection of seven note scales. The collection is constructed by a simple enumeration of scales that satisfy a geometric property. This collection matches collections that are chosen for conforming to detailed harmonic and melodic considerations. Thus complex properties involving intricate structures are distilled into a simple mathematical formula.

I would like to acknowledge the advice of an anonymous referee who pointed out the connections between these results and those in Jay Rahn's paper.

References

[1] J. Clough and J. Douthett. Maximally even sets. *Journal of Music Theory*, 35:93–173, 1991.

[2] J. Clough, N. Engebretsen, J. Kochavi. Scales, Sets, and Interval Cycles: A Taxonomy. *Music Theory Spectrum*, 21(1):74–104, 1999.

[3] Erik D. Demaine, Francisco Gomez-Martin, Henk Meijer, David Rappaport, Perouz Taslakian, Godfried T. Toussaint, Terry Winograd, and David R. Wood. The Distance Geometry of Music. *submitted to Computational Geometry: Theory and Applications*, 2006.

[4] L. Fejes Tóth. On the sum of distances determined by a pointset. *Acta. Math. Acad. Sci. Hungar.*, 7:397–401, 1956.

[5] Allen Forte. *The Structure of Atonal Music*, Yale University Press, 1973.

[6] Steve Masakowski. *Learning to Hear Your Way Through Music*, Mel Bay Publications, 2004.

[7] Jay Rahn. Coordination of interval sizes in seven-tone collections. *Journal of Music Theory*, 35 (1/2):33–60, 1991.

[8] D. Rappaport. Geometry and Harmony. *Proc. 8th ANNUAL INTERNATIONAL CONFERENCE OF BRIDGES: Mathematical Connections in Art, Music, and Science* Banff, Alberta, 67–72, 2005.

[9] J. Sawada. Generating bracelets in constant amortized time. *SIAM Journal on Computing* 31(1):259–268, 2001.

[10] J. Sawada and F. Ruskey. An efficient algorithm for generating necklaces with fixed density. *SIAM Journal on Computing* 29(2):671-684, 1999.

1927
Two processes of creating form in music

Veryan Weston
25 Meadway
Welwyn Garden City
Herts AL7 4NQ
England
v_weston@hotmail.com

Abstract

It is intended to examine form in two pieces of music written/realised in 1927: Webern's Opus 20 (string trio) and Louis Armstrong's Wild Man Blues (Hot Seven) as a method of evaluating their significance and diametric social relationship. Reference is made to visual art movements and ideas from this period as well as a glance at scientific and mathematical theory which may be seen to have a coincidental relationship with some ideas in art in 1927.

Introduction. Throughout the classical musical cannon within the European tradition, structure can consist of many diverse forms (sonata, ritornello, fugue, canon, rondo etc) built using a particular process of music composition. Over centuries now, written musical notation has been the main system of planning and constructing a piece often before it has ever been heard beyond the composer's own imagination. However, it is considered harder to trace form in improvisation because of it being seen (or not seen!) as shapeless and amorphous due to a popular misconception that there are overriding visceral inclinations with spontaneous invention.

1. Composition

1.1. Background to Anton Webern's string trio - Opus 20. The advent of abstract expressionism through transformation from figurative to abstraction in visual art can be clearly seen to have similarities with musical developments in Germany.

Figure 1: Vasily Kandinsky: *All Saints Day II* 1911 (Munich, Städtishe Galerie)

In Figure 1, similar developments can be found in the music of Schonberg (and Webern) from a later period of time.

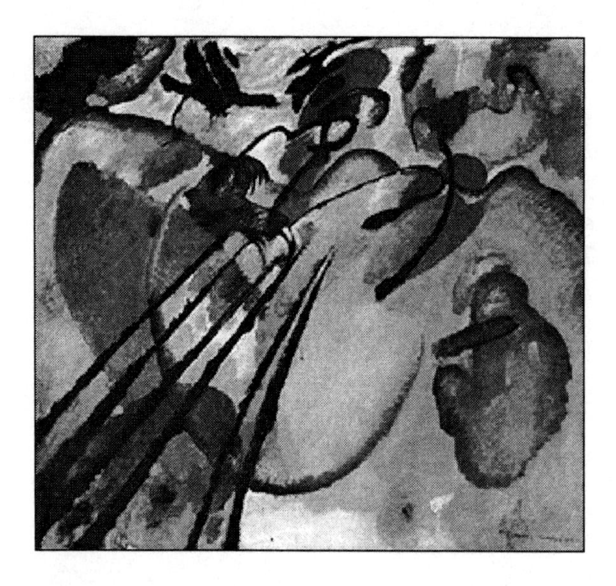

Figure 2: Vasily Kandinsky: *Improvisation 26 (Oars)* 1912 (Munich, Städtishe Galerie)

In Figure 2, Tonality (figurative) is replaced by a freer non-tonal aesthetic and runs a parallel timeline with early abstract expressionism – as heard in Schonberg's works like Five Orchestral Pieces – Opus 16 and Erwartung - Opus 17, the latter taking only 26 days to compose which might suggest a more spontaneous or improvisatory process of composition.

Figure 3: Vasily Kandinsky: *Transversal Line* 1923 (Kunstsammlung Westfalen, Dusseldorf)

In Figure 3, this is later 'cleaned up' once the rules of serialism are applied; expressionism makes way for neater geometric forms which can be seen almost as a more developed and refined form of abstraction, as heard in Webern's Opus 20. Note however the same shapes and placements in all three of Kandinsky's pictures remain related though becoming refined and sparser in the same way that Webern's later compositions became.

1.2. Foreground to Opus 20. Similarly, Webern's music towards the end becomes more transparent culminating in works like Opus 28. So with this refinement, each pitch acquires more importance and value and so is more tightly controlled, leaving no room for any other interpretation. Difference in each performance therefore becomes more negligible. Any character- or personality-trait from the musician's performance of this piece is minimized. The example from the score of Opus 20 below clearly exemplifies the point that every note is more controlled by some form of technical or expressive instruction.

Figure 4: Anton Webern: *Opus 20* for string trio (1927) Bars 4-7

1.3. Dissolving tonal relationships in Opus 20. Also clearly evident from the above example (Figure 4), are very wide intervallic leaps within each instrumental part. This device is used by Webern as a means of disestablishing any trace of tonal gravity. These wide leaps help to dissipate any linear resemblance of scale in order to avoid any accidental tonal inference.

1.4. Relativity in music. Painted in 1910, Kandinsky's picture titled "First Abstract Watercolor", is often considered to be the first piece of non-figurative European art, which runs parallel with Schonberg's previously mentioned middle-period work, and which both coincide with Einstein's Theory of Relativity developed just before this time. The parallel with tonal gravity is obvious. Tonics and dominants are the ground onto which everything returns in the same way that gravitational attraction was recognised and understood for 200 years.

In **1927**, Einstein writes a tribute to the bicentenary of Newton's death. In one part headed "Newton on its limitations", he writes:
"........Space and time were so divested, not of their reality, but of their causal absoluteness (absoluteness-influencing, that is, not -influenced), which Newton was compelled to attribute to them in order to be able to give expression to the laws then known....".
Gravity can be seen to be replaced by relativity – and likewise the listener has the ground taken away from under their feet when the causal absoluteness of tonal gravity is removed from a piece of music. Instead, tone, timbre, texture, and shape become more overriding features in musical and visual composition, the repercussions of which resonate throughout the modernist period over the second half of the 20th Century in the 'West' (as we call it) in general.
By **1927**, theories such as serialism in music as founded by Schonberg and Hauer, and various visual methodologies can be found in theoretical work by visual artists like Kandinsky's "Point and Line to Plane" (1928), Klee's little "Pedagogical Sketchbook" (1923) and Malevich's "Nonobjective World" (1926). These are also reflected in scientific developments of that time. Similarly in theoretical science, Einstein's theories are further developed; one such example would be quantum mechanics.

1.5. The process of composition. In Opus 20, Webern uses for the first time Schonberg's serial system. With this system come Webern's later and far more refined style and its very specific notational instruction. Therefore, when the piece is performed it could be comprehended at best as an eternally endless reproduction of the same accurately played fixed piece which therefore becomes an ossified and empty uncreative process unless we gauge minute differences between each performance as significant.

1.6. Scientific parallels. In **1927** Heisenberg formulates the Uncertainty Principle which expresses the "uncertain relationship" between the position and the momentum (mass times velocity) of a subatomic particle. My research on the internet though would suggest that "........its effect on measurements in the macroscopic world is negligible and can be usually ignored" [2]. But the philosophical relevance to broader issues might well be applied to the second piece of music. In the same way that the differences between a performance of opus 20 are similarly "negligible and can be usually ignored", the wider issue which is thrown up as a result of this theory is "....Heisenberg's result is not a statement about the inability to construct good measuring devices. It is a statement about an intrinsic property of nature... Nature has an essential indeterminacy" [3]

1.7. The principle of uncertainty in music. An alternative to this potential stagnation as a result of too many compositional constraints, might be to have a way of creative music making where the performer is integral in shaping a compositional situation, thus introducing a kind of healthy 'uncertainty principle' to the music. Making music using the spontaneous creation of musical form in improvisation can help both reveal and harness its natural wildness and indeterminacy. An example of this was made in 1927 as follows:

2. Improvisation

2.1. Background to Louis Armstrong's "Hot Seven" version of Wild Man Blues. With the development of sound recording, it was possible to document what were up until the beginning of the 20th century, aural traditions based on their direct relationship with their environment and local community. In **1927**, New Orleans reflected a very culturally cosmopolitan environment significantly different from Viennese life which, at that time, was where most of Webern's Opus 20 was composed.

The Hot Seven recording of Wild Man Blues was made in 1927. It was co-written by Armstrong and Jelly Roll Morton. Its main feature is the continuous series of stop breaks throughout the piece which use the harmonic movement of a quasi-song structure, but the breaks feature instrumental solos by Armstrong and clarinetist Johnny Dodds towards the end.

2.2. Foreground to Wild Man Blues. Unlike Opus 20, very little information can really be gained from any form of notation for this piece however visually explicit, as the vital ingredient is the performance itself by the artists. Nuance and freedom in the form of, for example, rhythmic flexibility, become impossible and pointless to notate, apart from perhaps making as accurate a transcription as possible as a means of evaluating the true complexity of a musician's performance.

Figure 5: Hot 7: *Wild Man Blues* (1927) Opening notation from midi file.

2.3. The symbiosis of composition and improvisation.

There are two creative processes involved in the making of Wild Man Blues, that are inter-dependent and which form the basis for the piece's realisation:

The first process is the composition which creates the foundation from which the second creative process – improvisation/s, can be structured.

"12"

Wild Man Blues, in the same way as Opus 20, has 12 as a key structural number. For Opus 20, Webern uses serialism consciously for the first time, though stretches the standard 12 – tone row to involve a far longer row. The title, Wild Man Blues, might lead us to believe that the harmonic cycle lasts 12 bars, as was/is the norm for blues. In this case though, the harmonic length is likewise stretched far longer.

Form, as in Kandinsky's 2nd example on page 1 is titled *"Improvisation 26 (oars)"*, and as the bracketed word suggests, is obviously derived from something previously figurative.

So the idea of some form of basic frame or structure can be comprehended as providing both a launching point for the process of improvisation as well as an ongoing reference on which to develop and explore ideas using improvisation. In this way, composition becomes only a basis or starting point for something potentially more profound or expressive in the process of improvisation.

It is intended to give a more thorough and in-depth comparative analysis for the conference which will highlight the numerical and formal signifance of each of these pieces of work; attention will be drawn to their contrasting structural relationships which are nevertheless still determined by differences in content.

2.4. References in improvisation. At this point in time, and up until the mid 1960's, harmonic structure in jazz provided a form of restraining mechanism for the player to focus upon and react to. In this way Armstrong's breaks can be seen as a release for the rigidity of the break stops. An example of this might be the sixth break in bar 22 on a dominant chord (a chord that prepares the listener for a tonal resolution, which in this case does not come) where a barrage of sound in the form of scales and arpeggios dovetails beautifully into a sequence of chromatic flurries. So in this instance, tension can be seen as the fixed framework of the harmonic form, and release, the improvisation which is sometimes almost a cry of release from these chains. For this reason, the music has that 'raw edge' which still thankfully provides the listener with an element of exhilarating discomfort, not only offering a glimpse of the social, economic and cultural landscape of New Orleans at that time, but telling a life story of the artists as they perform the music.

2.5. The process of improvisation. Wild Man Blues provides a great example of many aspects of music improvisation. Three such aspects might include:

- **Preparation.** The methodology involved in playing something off the top of your head is ironically founded on substantial preparation. The preparation is based on learning and assimilating useful shapes, ideas, and structures that are relevant to both the performer and their relationship with the community's musical aesthetic. Some of these may be personal discoveries; others are imitations of other admired musicians therefore creating very cohesive structural links between groups of players within environments.

- **Intuition.** The process of applying these above-mentioned ideas to a playing situation often requires a speed of response that relates immediately with sounds being made by other participating musicians. In this way, thought processes are often negligible as there is only time for connecting response to statement, which then is also statement replied to by others. Intuition therefore is a tangible form of defining the relationship with fellow musicians with the direct use of ear/brain/body/instrument/sound.
- **Structure.** Similar to the current statement of "What you eat is what you are", what a musician listens to, practices, preaches, explores at home or in a practice room, will enable the player to create spontaneously both micro and macro form. Often when improvising with others there is either some form of commentary, or even criticism, which does not always need to be sympathetic, e.g. antagonism can also be a useful form of moving and or changing the music's direction or mood. Each consequence therefore helps to define structure and shape in sound.

3. Differences

3.1. Image & Model. Essentially, it must be realised that each artwork here functions in two different ways, in that one is a final and complete recording of an event – similar to a completed picture or image (snapshot). The other is a very specifically notated idea for a performance, so functions as a model. Susanne Langer [1] states "....An image is different from a model, and serves a different purpose. Briefly stated, an image shows how something appears; a model shows how something works….." For this reason perhaps it is unfair to make any specific comparison or value judgment of these as works of art, except to asses the potential efficacy in the two processes of creative music performance.

3.2. Form & Content. Because of the nature of the Bridges Conference's main objective in the study of the relationship between art, mathematics and science, **form** is the principle criterion for observation and analysis. However, it is hoped that the content of a work of art can now be comprehended as directly affecting the form. In this way, the extreme social, economic and cultural differences in the backgrounds between Webern and Armstrong will have some effect on both the process of making as well as the structures discovered in the end product. For this reason, I make the assertion that with every work of art, there is a broader picture to evaluate which is both referential (content) as well as absolute (form).

Bibliography and suggested further readings

[1] **S. Langer.** *Mind: an Essay on Human Feeling,* Vol. 1, 1967, pp. xix, 59

[2] http://en.wikipedia.org/wiki/Uncertainty_principle

[3] **Lightman**, *Great Ideas in Physics*, 210. McGraw-Hill Companies, 2000

[4] **Christopher Small.** *Music Society Education* (John Calder) 1977

[5] **Eddie Prévost.** *No Sound is Innocent.* (Matchless Recordings and Publishing) 1995

[6] **Eddie Prévost.** *Minute Particulars* (Matchless Recordings and Publishing) 2004 – Attention is drawn to p17 where, by strange coincidence, a similar reference is made, but using content as the main focus for examination.

[7] **Derek Bailey.** *Improvisation* (Moorland Publishing) 1980

[8] **Walter Kolneder.** *Anton Webern an introduction to his works* (Faber & Faber) 1968

On Mathematics, Music and Autism

Ioan James
Mathematics Institute
University of Oxford
United Kingdom
E-mail: imj@maths.ox.ac.uk

Abstract

A discussion of research into the psychology of mathematicians, especially in relation to autism, and possible links with the psychology of musicians.

Every science has its own culture and that of mathematics is quite distinctive. As Henri Poincaré said 'mathematics is the activity in which the human mind seems to take least from the outside world, in which it acts or seems to act only of itself and on itself, so that in studying the procedure of geometric thought we may hope to reach what is most essential in the mind of man.' When Bertrand Russell asserted that 'mathematics rightly viewed possesses not only truth but supreme beauty — a beauty cold and austere, like that of good sculpture ... supremely pure, and capable of a stern perfection such as only the greatest art can' he was expressing an extreme view which applies, if at all, only to certain kinds of mathematics. Although Russell was writing a long time ago and expressing his personal feeling for the discipline there are modern mathematicians who would agree with him that mathematics has a special quality, different from that of any other kind of science. In fact many do not regard it as a science at all, rather as one of the arts, and research in the discipline as art for art's sake. For the pure mathematician, the interest seems to lie in the mathematics itself, rather than its application to the real world. Poincaré wrote about 'the feeling of mathematical beauty, of the harmony of numbers and forms, of geometric elegance. This is a true aesthetic feeling which all real mathematicians know.' The contemporary French mathematician Alain Connes explained that 'exploring the geography of mathematics, little by little the mathematician perceives the contours and structure of an incredibly rich world. Gradually he develops a sensitivity to the notion of simplicity that opens up access to new, wholly unsuspected regions of the mathematical landscape.'

Psychologists, especially cognitive psychologists, have long been interested in mathematicians, and. mathematicians have long been interested in cognitive psychology. The Leipzig neurologist Paul Möbius (grandson of the mathematician, August Ferdinand Möbius), an enthusiast for the pseudoscience of phrenology, wrote about mathematicians in his book *Die Anlage zur Mathematik.* in the late nineteenth century. Just a century ago, Henri Poincaré gave a famous lecture on the 'Psychology of Mathematical Invention' at a conference of psychologists in Paris, while about the same time Felix Klein ran a seminar on the subject at Göttingen. Hadamard's well-known monograph *The Psychology of Invention in the Mathematical Field* [1] refers extensively to Poincaré's lecture during which reported extensively on his personal experiences. Poincaré explained that he relied greatly on his unconscious mind. Hadamard's investigations confirmed that many creative mathematicians also rely on intuition although perhaps not to the same extent. Cognitive scientists tell us that most thought is unconscious. They distinguish between verbal thinkers and visual thinkers. My impression is that many mathematicians think in pictures, for example geometers and mathematical physicists, but many others are more verbal thinkers, for example mathematical logicians, but no research has been done on this, as far as I know.

There has been some research into the incidence of psychological disorders among mathematicians. The conclusion seems to be that the incidence of mental illness is much the same as in the general population. Depression is quite common. However, whereas in literature, especially poetry, the incidence of manic-depression is remarkably high, this is not so in mathematics. The same is true of other forms of mental illness. There has also been some research into the incidence of myopia, and one study concludes that this is exceptionally high among mathematicians. Research is being done to try and discover whether there really is a mathematical gene, as has sometimes been suggested, or more plausibly that genetic factors may be involved in the development of a module in the brain which is associated with mathematical activity.

However, whereas the incidence of mental illness is not especially high among mathematicians the situation is quite different in the case of disorders on the autistic spectrum, especially the mild form of autism .known as Asperger's syndrome. Hans Asperger was a Viennese paediatrician who, in 1944, wrote about the disorder to which his name has been attached. The symptoms of Asperger's are generally grouped under six headings, namely impairments of social interaction, all-absorbing narrow interests, repetitive routines, speech and language peculiarities, problems of non-verbal communication, and possibly motor clumsiness. Under each heading there are a bewildering variety of .ways in which the disorder can manifest itself; no individual will exhibit more than some of them, although there should be at least one under most of these broad headings. The disorder usually shows itself in early childhood, and is present throughout life. It is estimated that it affects about 1 in 200 of the general population, males much more than females.

As Asperger himself observed: 'To our own amazement, we have seen that autistic individuals, as long as they are intellectually intact, can almost always achieve professional success, usually in highly specialized academic professions, often in very high positions, with a preference for abstract content. We found a large number of people whose mathematical ability determines their professions.' It is well-established that people with Asperger's syndrome are drawn to mathematics and similar subjects. Mystified by the social world they take refuge in the certainties of mathematics. They tend to enter professions such as computer science, also certain types of engineering. Among creative mathematicians of the past, it is thought that Isaac Newton, Norbert Wiener, Alan Turing, Ronald Fisher, Kurt Gödel and Paul Erdös, amongst others, exhibited Asperger traits, perhaps also Sophie Germain and Emmy Noether. But there is quite a difference between saying that someone exhibited such traits and saying that they had Asperger syndrome, as I shall now explain.

In psychiatry the assessment procedure for a patient with personality problems is fairly standard. Some disorders have obvious physical or behavioural signs but for autism there is no single sign which would uniquely secure the diagnosis. The whole history of the patient has to be considered from birth, the nature of the impairments, their severity and their change over time. Then the person would be observed directly, usually by more than one professional, and would be given a range of standard tests. To facilitate the work a standardised interview procedure is used, so that leading questions are avoided and to ensure that alternative diagnoses are excluded. There are handbooks which contain the currently agreed diagnostic criteria in the form of lists. They are updated from time to time in the light of increasing knowledge. This underlines the fact that a complete set of scientifically objective criteria for the diagnosis of mental disorders has not yet been established. Although a good diagnostician relies as much on experience and intuition as on textbook knowledge nevertheless clinicians need objective procedures or else their conclusions may fail to convince.

Obviously it is impossible to carry out these procedures fully in the case of someone no longer alive. The best that can be done is to search, in the biographical literature, for evidence which is relevant to the standard tests, and then make a judgment based on experience and intuition. Most of the authors of books on autism identify cases of Asperger syndrome, often from far back in history. Uta Frith gives some examples in her book *Autism: Explaining the Enigma,* such as one of the original followers of St Francis of Assisi is an example. Michael Fitzgerald gives some others in his books *Autism and Creativity* and *The Genesis of Artistic Creativity* [2]. Temple Grandin, who has personal experience of the disorder, has also identified people she believes had Asperger syndrome in her *Thinking in Pictures.* Amateurs have also entered the field, for example Norm Ledgin has suggested some Asperger possibles in his *Asperger's and Self-Esteem,* while I have combed the literature for others in my recent book, *Asperger's Syndrome and High Achievement* [3]. I have given Erik Satie and Bela Bartok as examples of composers with the syndrome, Jonathan Swift and Patricia Highsmith as examples of writers, and Vincent van Gogh as an example among painters — there are plenty of others.

What light does this throw on the nature of mathematical creativity, and of creativity generally? Asperger wrote: 'It seems that for success in science or art a dash of autism is essential. For success the necessary ingredient may be an ability to turn away from the everyday world, from the simple practical, an ability to rethink a subject with originality so as to create in new untrodden ways, with all abilities canalised into the one speciality.' Those who have the syndrome live very much in their intellects and certain forms of creativity benefit greatly from this. Certain aspects of the syndrome e.g, workaholism and an extraordinary capacity for persistence can accompany many forms of creativity. When Isaac Newton was asked how he conceived the theory of gravitation he replied that, 'It was through concentration and sheer dedication. I keep the subject constantly before me, till the first dawnings opens slowly, little by little and little into the full and clear light.' The Asperger ability to focus narrowly on a topic and resist distraction is particularly important in mathematics. Mathematicians tend to have strong and narrow interests, and a high degree of focus. An enormous capacity for curiosity and a compulsion to understand are evident in those who have the syndrome, as is a tendency to reject received wisdom and the opinions of experts.

The reader may well wish to be presented with qualitative evidence of the link between science and autism. Fortunately this has recently become available. Simon Baron-Cohen [4] has devised a self-administered questionnaire for measuring the degree to which an adult with normal intelligence has the traits associated with the autistic spectrum. From the answers to the questions a number is obtained, which he calls the autistic-spectrum quotient, and this gives an estimate of where a given individual is situated on the continuum from normality to autism. When the questionnaire was given to 4,175 students at Cambridge University it was completed and returned by over one fifth, with no significant difference in the return rate between disciplines. Natural scientists (including engineers and mathematicians) scored significantly higher than both humanities and social sciences students, confirming the general belief that autistic traits are often associated with scientific skills. Briefly, scientists scored higher than non-scientists; and within the sciences, mathematicians, physical scientists, computer scientists and engineers scored higher than the more human or life-centred sciences of medicine and biology. Further investigation has revealed that a disproportionate number of mathematics students have received a diagnosis of autism, and that autistic individuals tend to have an unusually high proportion of engineers in their families.. Details will be found in a forthcoming article by Baron-Cohen et al [5].

There is far more to be said about this kind of thing than can possibly be included in a short lecture; so may I now refer members of the audience to the forthcoming book *The Mind of the Mathematician* by Michael Fitzgerald and myself [6]. In this there is a section about mathematicians and musicians which I thought might be especially relevant to this conference, and so I will now say something about this. I will also refer to Fitzgerald's *The Genesis of Artistic Creativity* [2] which deals more specifically with possible links between autism and musical talent. Hadamard emphasizes that there are strong resemblances between the psychology of invention in different fields of the arts and sciences and specifically quotes a letter of Mozart's on musical invention.

There seems to be such a contrast between the science of mathematics and the art of music, yet there is a long-standing belief that they are related in some way. As Leibniz wrote to his friend Goldbach, music is a hidden exercise in arithmetic, of a mind unconscious of dealing with numbers. In the words of von Helmholtz 'mathematics and music, the most sharply contrasted fields of intellectual activity which can be found, and yet related, supporting each other, as if to show forth the secret connection which ties together all the activities of our mind.' Some composers seem to have been interested in numerology, if that counts as mathematics. Mozart was one of these and we know that his interests extended into elementary number theory. Stravinsky remarked that [musical form] is at any rate far closer to mathematics than to literature — certainly to something like mathematical thinking and to mathematical relationships. However it is not easy to think of many musicians, whether composers or performers, who were particularly interested in mathematics.

On the other hand, it is easy enough to name people of mathematical ability who were also musical. For example, there was Georg Cantor, who might have followed in the footsteps of his mother's family of musicians and sometimes regretted that he had not. There was James Joseph Sylvester, who believed he had a fine voice and took singing lessons from Gounod. He is quoted as saying 'may not music be described as the mathematics of sense, mathematics as the music of reason?' The many-sided Olinde Rodrigues had some talent as a composer. Richard Dedekind was an accomplished pianist and cellist who composed a chamber opera. Janos Bolyai and A.C. Aitken were exceptionally-fine violinists. Further back we can cite the example of the little-known Hamburg mathematician, Johann Georg Busch, at whose home some of the compositions of Carl Philipp Emmanuel Bach were performed for the first time. Leopold Kronecker was an accomplished pianist and vocalist. Hermann Grassmann was a pianist and composer, some of his arrangements of Pomeranian folk-songs were published. He was also a good singer and conducted a male voice choir for many years. Felix Hausdorff was an excellent pianist and occasionally composed songs; he aspired to be a composer rather than a mathematician. There seems to have been a tendency for mathematicians to marry into musical families, for example Richard Courant and Jacques Hadamard did so, and when they entertained at home there was always music. The famous concerts of chamber music held at the home of the amateur mathematician and musician Emile Lemoine exerted a great influence on the musical life of Paris in the latter part of the nineteenth century.

Albert Einstein had a passion for music, as a way of experiencing and expressing emotion that is impersonal. He was an enthusiastic (but not very good) violinist; Mozart, Bach and Schubert were his favourite composers. Photographs of him playing the violin show a different Einstein from the more familiar images. When he was world-famous as a physicist he is reported to have said that music was as important to him as physics: 'it is a way for me to be independent of people'; on another occasion he described it as the most important thing in his life.

Just after the Second World War the psychologist Geza Revesz [6] conducted a survey of 180 mathematicians, 220 physicists, 206 doctors, and 136 writers, who were asked to complete a short questionnaire. This revealed that 24% of the mathematicians were completely unmusical, as compared with 16% of the physicists, 19% of the doctors, 13% of the writers; 44% of the mathematicians were unmusical, as compared with 33% of the physicists, 41% of the doctors and 29% of the writers; 56% of the mathematicians were musical, as compared with 67% of the physicists, 59% of the doctors, and 71% of the writers; while 9% of the mathematicians were very musical, 9% of the physicists, 6% of the doctors and 11% of the writers. These results do not support the widespread belief that mathematicians generally are particularly musical.

If I had more time I would have liked to describe what has been written about the psychology of musicians, especially the significance of autistic traits in relation to musical talent, but there seems to have been relatively little research in this area. I have already mentioned some well-known composers who are believed to have suffered from Asperger syndrome. Other musicians who may have had the syndrome include the pianist Glenn Gould and the conductor Carlos Schreiber.

Finally a few remarks about savants, which appear from time to time in certain fields; for example there are the lightning calculators. Savant skills seem to appear quite suddenly — they do not need to be consciously learned — and may vanish again equally suddenly. They are most striking when they occur in individuals with otherwise poor intelligence. Savants of various kinds have been much studied, and one conclusion of the research seems to be that they usually, but not always, are in some degree autistic. For further information see Beata Hermelin's book, *Bright Splinters of the Mind* [7].

References

[1] J.Hadamard, *The Psychology of Invention in the Mathematical Field*, Princeton University Press, 1945

[2] M. Fitzgerald, *Autism and Creativity*, Brunner-Routledge, Hove, 2004; *The Genesis of Artistic Creativity*, Jessica Kingsley Publishers, London, 2005.

[3] I. James, *Asperger's Syndrome and High Achievement*, Jessica Kingsley Publishers, London, 2005.

[4] S. Baron-Cohen and S. Wheelwright., The autism-spectrum quotient AQ, *Journal of autism and developmental disorders* 31, 5-17, 2001.

[5] S. Baron-Cohen et al, forthcoming article in a special issue *Mathematical Talent is Linked to Autism* of the the journal *Human Nature.*

[6] G. Revesz, Die Beziehung zwischen Mathematischer und Musicalischer Begabung. Schweizerische Zeitscrift fur Psychologie und ihe Anwendungen 5, 269-81,1946

[7] B. Hermelin, *Bright Splinters of the Mind*, Jessica Kingsley Publishers, London, 2001.

Some related references:

R. C. Archibald, Mathematics and Music, American Mathematical Monthly 31, 1-25, 1924.

H. Asperger, Formen des Autismus bei Kindern, Deutsches Arttzeblatt 14, 4, 1974.

E. T. Bell, *Men of Mathematics*, Victor Gollanz, London, 1937.

D. J. Hershman, and J. Lieb, *Manic Depression and Creativity*, Prometheus, Buffalo, NY, 1998.

I. James, *Remarkable Mathematicians*, Cambridge University Press, 2002.

I. James, *Remarkable Physicists*, Cambridge University Press, 2003.

I. James, Singular Scientists, Journal of the Royal Society of Medicine 96, 36-9, 2000 (Accessed 1 June 2006 at http://www.jrsm.org/cgi/content/full/96/1/36).

I. James, Autism in Mathematicians. Mathematical Intelligencer 25, 61-5, 2003.

K. R. Jamison, *Touched with Fire*. The Free Press, New York, 1996.

A. Storr, *The Dynamics of Creation*, Charles Scribner's Sons, New York, 1972.

• Bridges for Teachers •

• Teachers for Bridges •

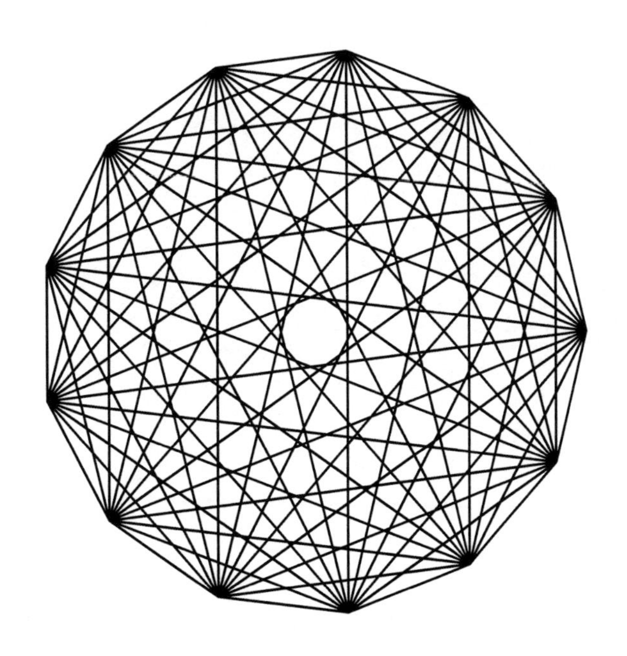

Mathematics Investigations in Art-Based Environments

Mara Alagic
Wichita State University, USA
mara.alagic@wichita.edu

Paul Gailiunas
Gosforth High School, Newcastle, UK
p-gailiunas@argonet.co.uk

Abstract

This paper presents two sources of information about mathematics and art integration. The first source is a brief outline of concepts that will be introduced through the workshop series at this conference. The second source is a collection of insights and resources about mathematics and art integration provided by a group of elementary education teacher candidates.

Introduction

Today's mathematics teachers face many challenges in their efforts to meet requirements of standards-based teaching and high-stakes assessments. They need ongoing support from their professional communities. An emphasis on standards-based teaching, although it appears somewhat limiting, provides formal guidance for developing both conceptual and procedural understandings of mathematical concepts. Educational objectives are focusing more and more on innovative ideas for teaching for understanding [6] which includes integration across curricula, conceptual understanding via multiple representations, contextual learning, and problem-based learning approaches [4, 5]. Teachers need creative ways of accomplishing such demands; they need and like ideas for mathematics education which integrates the arts. In addition to innovative ideas for mathematics education via the arts, teachers need high quality guidance and resources that would model for them how to explicate mathematical concepts that are more or less implicit in art-based contexts. Furthermore, art-based assessments and application of these concepts in new environments may take learning even deeper.

We learn and retain the most from thinking in critical and creative ways. To design an effective, creative, and critical mathematics learning environment teachers must begin with an understanding that students' learning is dependent on the way information is presented. Perception-based representations preserve much of the structure of the original perceptual experience. Meaning-based representations are abstracted from the perceptual details, incorporating the meaning of the experience [2]. Mindfulness in learning requires students to (a) think in meaningful ways to represent what they know and (b) actively engage to create knowledge that reflects their understanding of mathematical ideas [3, 7]. When students represent what they know in their own way, through their own representations, they deepen their understanding of mathematical ideas, expanding their repertoire of representations. Only when individuals can go back and forth between various representations of mathematical concepts (for example, the visual and the analytic) does mathematical understanding occur [1].

Both mathematics and art have their own language, structure, configuration, and means of expression. Creative problem-solving skills, facilitation of both informal and formal approaches, developing expertise in selecting appropriate tools, media and methods are common, in their own contexts, to both mathematics and art apprenticeship. The process and the product of creating and constructing pictures, applets, images, icons even symbols, in order to represent what we perceive to be relevant for an understanding and representing that understanding are in the heart of visualizing [8].

Bridges for Teachers, Teachers for Bridges

Teachers always need good integration ideas. The following is a brief clustering of mathematics and art integration concepts that will be presented at this conference.

- A Mandala is a complex circular design, intended to draw the eye inward to its center. Creating Mandalas provides fascinating bridges between mathematics, art, architecture, history, and science. Techniques used in creating mandalas can be used to demonstrate fractions, angles, trigonometric relations, and fractals. Further, these techniques have the potential to interest and motivate mathematics students (Stang).
- The historical context and reason-result relationship in the creation of dome tessellation by the architect Sinan - designing structures for a specific dimension - evaluating the usage of the dome structure in Sinan's mosques as the result of technology and material brings an understanding of mathematics and art connections in architecture (Sagdic, Vural, & Taygun).
- The creation of math art book forms provides an interesting interplay of 2D and 3D shapes. In the process of paper folding, these basic forms use elementary geometric knowledge, while letting students explore creative ways to use mathematical information in adding content to the pages or faces of these forms (Happersett).
- Hex Signs are circular discs with intricate geometric designs with specific meanings that were hung on barns in the "Pennsylvania Dutch" region of the United States. Construction of Hex Signs involves some interesting mathematics concepts. Common designs include: Rosettes, Birds, and Star Polygons and can be constructed using dynamic geometry tools (Evans).
- Line designs form a basis for mathematical understanding of geometric shapes and relationships of points, segments, and angles. Each of the line segments is really a tangent for each of the curves being formed; however, because of what we focus on, we often see the curves. The Arete of Line Designs explores the historical (including the George Boole connection), philosophical, and pedagogical nature of line designs (Round).
- The Plato Bead, a Bead Dodecahedron, is an example of a polyhedron and is another way to represent and learn the properties of regular and semi-regular solids. With a history dating back several hundred years in China, a bead polyhedron can be used with various sizes, colors, and types of beads (Shea).
- Sliceform models are three-dimensional objects created by slicing a solid many times in two directions. Modeling and physically constructing mathematical models may be accomplished using a computer, a printer, craft knife, glue, and paperboard (Luecking).
- A game called "zellij multipuzzle" is a set of 669 zellij-style tiles, of which one side is white and the other a different color so that each tile can be utilized in a positive or negative configuration. Designed by using the technique of laser cutting, this game is an introduction to the art of geometrical arabesque (Castera).

- Topological Mesh Modeling with hands-on experiments can be accomplished using topological modeler, TopMod. TopMod provides a wide variety of interactive techniques that allow the creation of unusual and interesting shapes by changing the topology of 2-manifold meshes (Akleman & Srinivasan).
- Paper Sculptures with Vertex Deflection are mathematically motivated developable surfaces (examples: sculptures by Ilhan Koman); a variety of shapes creating saddle, maxima, and minima using nip and tuck can be constructed and can provide an introduction to mathematical ideas, such as the Gauss-Bonet theorem. (Akleman, Koman, & Akgün).
- Stick models based on Platonic polyhedra convey some geometry concepts. These physical models are applicable to students from elementary school through graduate school, as with hands-on experiences, they build on their existing level of knowledge (McDermott)
- Reflecting on Vermeer's painting of The Music Lesson, the basic use of the RMS90 Modular Scale can be utilized to directly deduce all the elements of the scene in perspective, essentially recreating the perspective outline of the painting (García-Salgado).

References

[1] Duval, R. (1999). *Representation, vision, and visualization: Cognitive functions in mathematical thinking.* Paper presented at the annual meeting of the Annual Meeting of the North American Chapter of the International Group for the Psychology of Mathematics Education, Morelos, Mexico.

[2] Fuys, D., Geddes, D., & Tischler, R. (1988). *The van Hiele model of geometric thinking among adolescents*: Journal of Research in Mathematics Education, Monograph 3.

[3] Langer, E. J. (2000). Mindful learning. *Current Directions in Psychological Science, 9*(6), 220-223.

[4] National Council of Teachers of Mathematics. (2000). *Principles and standards for school mathematics*. Reston, VA: National Council of Teachers of Mathematics.

[5] National Research Council. (2000). *How people learn: Brain, mind, experience, and school.* Washington, D.C.: National Academy Press.

[6] Perkins, D. N. (1993). Teaching for understanding [Electronic version]. *American Educator: The Professional Journal of the American Federation of Teachers, 17*(3), 28-35.

[7] Salomon, G., & Globerson, T. (1987). Skill may not be enough: The role of mindfulness in learning and transfer. *International Journal of Educational Research, 11*, 623-637.

[8] Shaffer, D. W. (1995). Symmetric Intuitions: Dynamic Geometry/Dynamic Art. *Symmetry: Culture and Science, 6*(3), 476-479.

Moving Beyond Geometric Shapes: Other Connections Between Mathematics and the Arts for Elementary-grade Teachers

Virginia Usnick
Marilyn Sue Ford
Department of Curriculum and Instruction
University of Nevada, Las Vegas
4505 Maryland Parkway
PO Box 453005
Las Vegas, NV 89154-3005 USA
E-mails: vusnick@unlv.nevada.edu and fordm@unlv.nevada.edu

Abstract

When classroom teachers are asked to identify connections between mathematics and art, they typically refer to geometric concepts. In an attempt to broaden their understanding of potential connections, this paper presents activities that involve common vocabulary, probability, and imagery.

Introduction

In 1989, the National Council of Teachers of Mathematics published its *Curriculum and Evaluation Standards for School Mathematics* [1]. The Council proposed ten standards. Five focused on content and five presented processes. Since the publication of this document as well as its revised edition in 2000, many states have rewritten their own state documents to reflect the vision of the Council. Some states have chosen to include all five process standards while others have chosen to embed processes within the content strands.

The process that is often embedded within the content listings is "connections." This standard encourages teachers to develop curriculum that enables students to make connections among mathematical representations (e.g., concrete, pictorial, and abstract levels of knowledge), between mathematical domains (e.g., algebra and geometry), and between mathematics and other content domains.

Since the *Standards* were originally published over 15 years ago, one could hope that students entering teacher education programs today would be aware of numerous connections within mathematics and between mathematics and areas outside of mathematics. The hope might be even higher when considering inservice teachers who have experienced teacher professional development either through university-level methods courses or conferences sponsored by mathematics-related professional organizations. Unfortunately, this hope is easily dashed when one interviews both preservice and inservice teachers.

Two groups of teachers (one inservice and one preservice) were asked to identify how mathematics could be connected to five areas common to an elementary or middle-school curriculum: science, social studies, art, language arts, and physical education. Many of both the inservice and preservice teachers indicated they'd experienced more difficulty thinking of connections between mathematics and art than with any of the other content areas. One of the preservice teachers, who has a bachelor's degree in music, stated he'd had a "tough time thinking about a connection between math and art" as his degree was in music. He pointed out that the data collection sheet had indicated "art" and not "the arts," so he'd thought about "paintings, not music or dance."

A quick perusal of their responses showed that the variety of connections between mathematics and art was smaller for the inservice teachers than the preservice teachers. Both groups tended to focus on geometric connections but one or two within each group mentioned patterns, golden ratio, and music. Only one student (a preservice teacher) mentioned anything other than content (e.g., angles, proportions, shapes, color, timelines). She stated that both mathematicians and artists use spacial [sic] reasoning.

In an effort to broaden teachers' perspectives of how mathematics may be connected to "the arts," several classroom-based activities will be discussed in this paper. While some will result in products with geometric bases, the focus will be on how other mathematical content is used to generate the end product.

Activities

It's All in the Name. Investigating patterns generated by children's names can be an introduction to algebraic thinking. While this activity does not necessarily produce a "work of art," it provides a visual image for concepts that aggravate many middle-grade teachers and students: common multiples and least common multiples.

Give students a sheet of centimetre grid paper. Have students outline a 5x5 section. Then, beginning in the upper-left corner of the section, have students lightly print their names, one letter per cell, wrapping to the next row if their name is longer than 5 letters. Continue printing until all 25 cells are filled. Using two colors of markers, students color vowels in one color and consonants in another. When complete, students look for and describe patterns in their design. If they cannot see a pattern in the grid, have them add rows and continue their name and coloring. Then students look for other students who have the same design. Samples for the names of Ginny and Marilyn are shown below with consonants shaded and vowels left white.

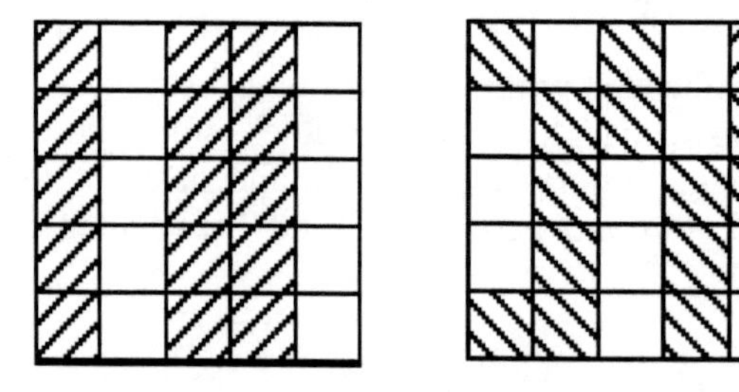

Figure 1: *5x5 grid for the names Ginny and Marilyn.*

Once students have discussed patterns generated on a 5x5 grid, students can be asked a variety of questions such as:

a) What might you pattern look like if the grid were 6x6 rather than 5x5?
b) How is the pattern of a name with an even number of letters the same as (or different from) the pattern of a name with an odd number of letters?
c) What would the pattern of a 9-letter name look like on a 5x5 grid?

Variation of It's All in the Name. Again provide students with cm grid paper. Have students outline a 10x10 section. After printing their names as in the previous activity, have them use a black (or other dark color) to shade in the cells that have the last letter of their name. Have students find a partner whose design is different from their own. Then, numbering the cells from 1 to 100 from the upper-left corner and moving across the grid, have them find the numbers of the darkened cells. For example, for the name Ginny, the darkened cells would be 5, 10, 15...100 while Marilyn's would be 7, 14, 21...98. Then have students find all the darkened cells they have in common (i.e., common multiples) as well as the first darkened cell in common (least common multiple).

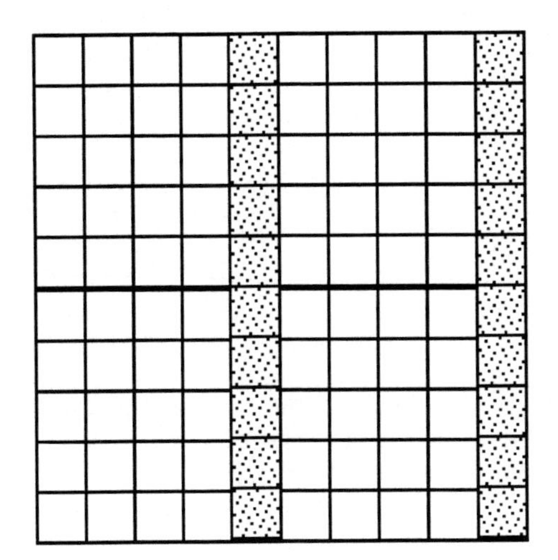

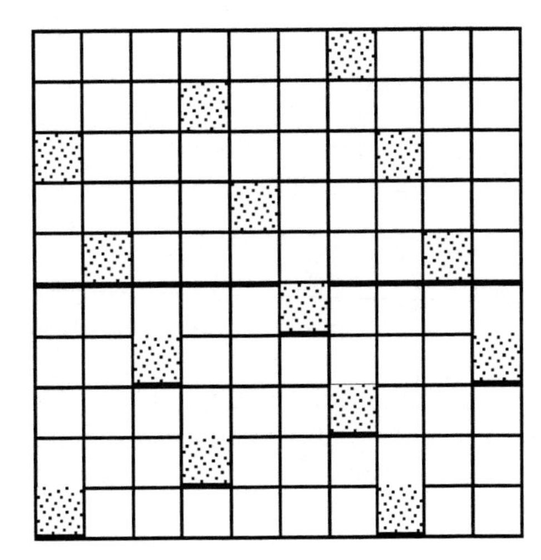

Figure 2: *10x10 grids for last letters of Ginny and Marilyn.*

Clapping Names. A similar activity for finding common multiples has a musical foundation and may be more appropriate for auditory learners. This activity is taken from the work of Schaffer, Stern, and Kim [2].

Students determine which letters in their names are consonants and which are vowels. While quietly spelling their names, they slap their thighs for consonants and clap their hands for vowels. For example, *Ginny* would be Slap, Clap, Slap, Slap, Clap. Have students make the final sound of their name louder than the rest of the word. Also, have students practice their pattern several times, being sure not to make a break at the end of the name. So, if *Ginny* was done four time, the sequence would be Slap, Clap, Slap, Slap, **Clap**, Slap, Clap, Slap, Slap, **Clap**, Slap, Clap, Slap, Slap, **Clap**, Slap, Clap, Slap, Slap, **Clap**. Then, working in partners, have students do their slap-clap patterns multiple times, using the same rhythm. Have them determine when their final sounds meet. Whether the final sound is a vowel or consonant is immaterial as the common multiples appear as a result of the louder sound at the end of the names. For *Ginny* and *Marilyn*, the first loud sound together would be the 35th sound. This activity may need a third person to count the number of sounds as children (and adults) often get lost in the slap-clap sequence.

S C S S **C** S C S S **C** S C S S **C** S C S S **C** S C S S **C** S C S S **C** S C S S **C**
S C S C S C **S** S C S C S C **S** S C S C S C **S** S C S C S C **S** S C S C S C **S**

Figure 3: *Ginny and Marilyn as slaps and claps.*

Conga Rhythms. A variation of the Clapping Names activity provides a connection between mathematics and rhythms and symbols in music. In *Conga Drumming: A Beginner's Guide to Playing with Time*, Dworsky and Sansby introduce a form of representation that uses both numbers and symbols. Again, multiples and common multiples are found when sounds converge. [3]

Chancey Art. While many teachers quickly identify geometric connections between mathematics and art, few would recognize the potential of using probability to generate artistic products. However, that is the basis for many of the wall paintings of Sol LeWitt, a contemporary minimalist. Some activities result in an almost Picasso-looking picture.

Collect pictures from magazines (one for each student) and trim them into squares. Have students fold a picture into 16 squares (fold twice horizontally and twice vertically) and cut along the fold lines. Then, onto a backing paper, have them piece the 16 cells back into the original picture (do not glue yet). Using a regular die, students roll the die, perform the associated movement on the first cell (upper-left corner), and glue the piece in place. Students repeat the roll, move, and glue sequence on each of the other 15 squares. The following moves might be used:

1 one-quarter turn clockwise
2 one-quarter turn counter-clockwise
3 half turn
4 flip so backside is facing upwards
5 trade piece with one in some other cell
6 do not change the piece's position

An extension of this activity will require two copies of each original picture. After students have folded, cut, and moved the cells of the original picture, have the extra copy of each picture available for display. Students need to match the uncut picture with the transformed one. Having pictures with similar colors makes the activity slightly harder.

Another extension focuses on the mathematical foundation of the activity. Provide each student with a 1x4 strip similar to the one below.

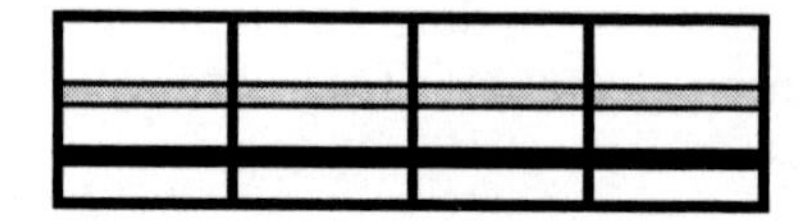

Figure 4: *1x4 strip used in Chancey Art follow-up lesson*

Have students separate the strip into 4 square cells and put them back into a 1x4 strip that looks like the original (i.e., lines match). Then each student rolls a die and performs the move on the first cell (left-most cell). Have students determine how many other students have a 1x4 strip that looks exactly like theirs. What is each student's chance of finding a matching strip? One might think that, since there are 6 potential actions, the chance of finding a potential match is 1 in 6. However, this is only true if each move results in a unique "look." If the strip shown above had only one line across the middle, both quarter turns would end up with the line in the same position. Therefore, it is critical that each move result in a different look.

Then have students roll the die and move the next piece. Again, have them find matching strips and discuss why there might be fewer matches. Continue with the third and fourth cells.

Since this activity quickly results in few matching strips, it may be easier for students to find patterns and apprehend the influence of combinations if a modified set of moves is used. For example, rather than having each number on the die produce a different movement, changes in the position of the cell could be simply based on whether the result of the die is odd or even. Then, there would be 2 outcomes for the first cell, 2 for the second, 2 for the third, and 2 for the fourth, or 16 combinations. The length of the strip can also be shortened for younger children.

Circle Art. Roll a regular die. On a 5x5 grid, use any point as the center of a circle that will have a radius of 1. If the number on the die is 1 or 4, draw a quarter circle. If the number is 2 or 5, draw a half circle. If the number is 3 or 6, draw three-quarters of a circle.

Roll again. Draw another arc ("unfinished" circle) using the second end-point of the previous arc as the beginning point of the new arc. Repeat 8 more times. Each curve must stay within the limits of the 5x5 grid. If using the last end-point will force you out of the grid, you may choose another point. You may retrace previously drawn curves and reuse end-points.

When you have rolled the die 10 times, transfer your design to the upper-left quadrant on a 9x9 grid. Complete the 9x9 grid by flipping the design over the vertical and horizontal lines. Shade your final product in a symmetrical fashion.

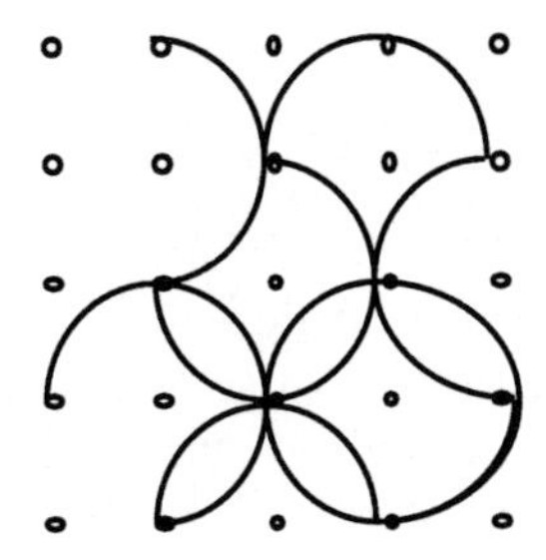

Figure 5: *Result of rolling 2, 3, 1, 1, 5, 6, 2, 1, 4, 3*

Figure 6: *Design flipped over horizontal and vertical axes, grid points removed.*

Thistle Quilts. Quilts are often used to demonstrate the use of transformational geometry. However, they can also be used to teach about sequencing. Provide students with square pieces of paper 8 or 10 cm on a side. Have them find the midpoints of the sides. If they use a folding method, the edges of the paper should just be pinched.

Beginning at either a vertex or a midpoint, have them draw a line segment to a non-adjacent vertex or midpoint. They should continue drawing segments beginning at a vertex or a midpoint, keeping the following rules in mind:

1. They may not cross a segment already drawn. That is, they must stop when a new line segment reaches a previously-drawn segment, and
2. At least one of the regions created by the new line segment must be a triangle.

They may use the same beginning point more than once. But the only time an ending point may be used more than once is if it is on the perimeter of the square. Draw segments until 8 regions have been created. Quilt squares are then traded with a partner. Students determine the order in which the line segments were drawn. Encourage students to determine whether the quilt square could have resulted from more than one sequence.

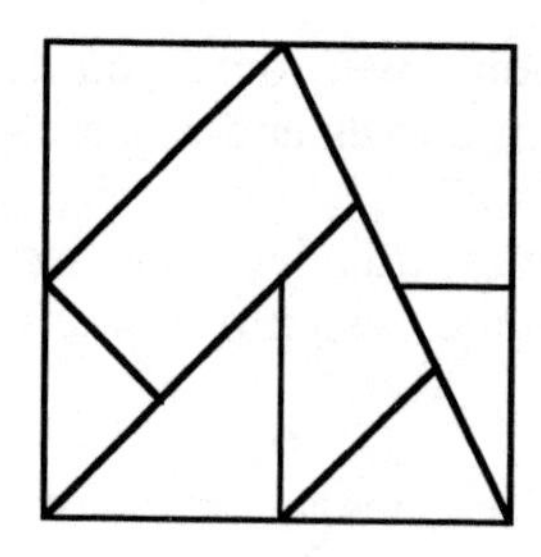

Figure 8: *Sample of a thistle quilt block.*

Taxi-cab Art. Students can investigate geometric concepts through taxi-cab geometry. The co-ordinate plane is likened to city streets that run parallel or perpendicular to each other. To travel from one point to another, a taxi-cab must stay on the streets (grid lines) and may not move into a cell as that would be viewed as driving into or through buildings.

First, have students choose a number between 4 and 8. Later in the activity, this number will be used to determine when their travels are finished. Students roll a regular die. On a 5x5 grid, they begin at any point and draw a line that shows a taxi-cab's trip. The trip covers as many "blocks" as the number on the die. During the trip, the cab can turn corners but cannot make U-turns at or between intersections.

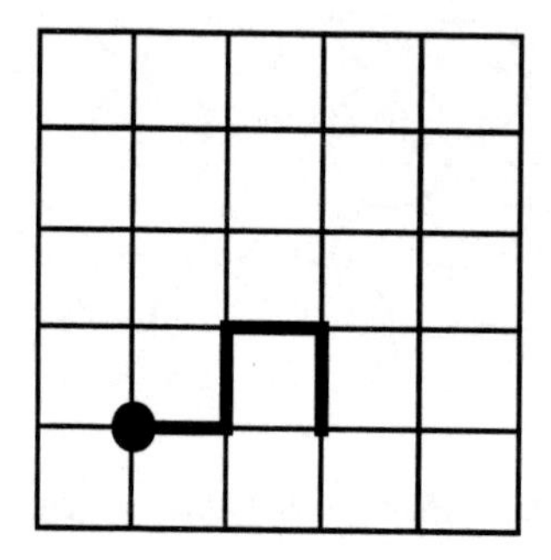

Figure 9: *Sample of a four-block trip beginning at the circle.*

Students roll again and draw another cab ride. The new trip starts where the last one ended. During this and subsequent trips, students may go "around blocks," thus creating a closed figure. They may also create a closed figure using parts of previous trips. Each trip must stay within the limits of the 5x5 grid. They continue rolling and drawing cab rides until they have generated as many closed figures as the number they choose at the start of this activity.

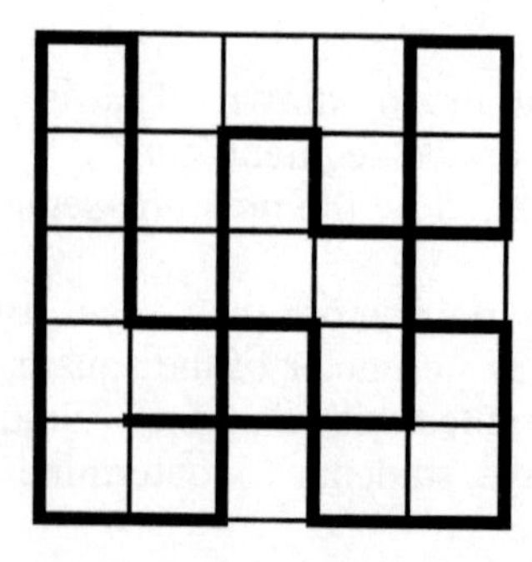

Figure 10: *Taxi-cab trips resulting in 5 closed figures.*

When they have made the appropriate number of closed figures, they transfer their design to the upper-left quadrant on a 9x9 grid and complete the 9x9 grid by flipping the design over the vertical and horizontal axes. The final step of this activity is to shade the end product in a symmetrical fashion.

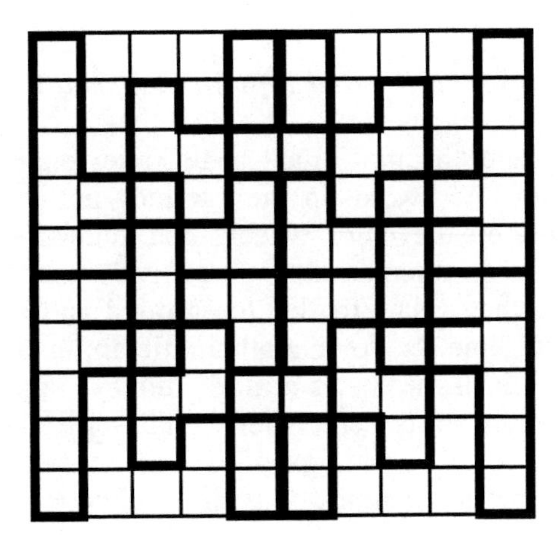

Figure 11: *Taxi-cab trips resulting in 5 closed figures, reflected across horizontal and vertical axes.*

Imaginings. While many activities produce a tangible artistic product, some classroom activities develop mental skills, such as visualization. Below are several exercises that are purported to develop one's ability to visualize. More activities can be found at <http://math.ucsd.edu/~doyle/docs/mpls/handouts/handouts.html>.

1. Picture your first name. Read off the letters backwards. Try other words.
2. Imagine a square. Picture the midpoints of each side. Cut off each corner of the square, cutting from midpoint to midpoint. Sketch the result of the leftover shape. Rearrange the cut-off points to form a new square.
3. Visualize 2 squares. Place the second square centered over the first square but at a forty-five degree angle to the first. Sketch the intersection of the 2 squares.
4. Mark the sides of a square into thirds. Cut off each of its corners back to the marks. Sketch the result.
5. How many colors are required to color the faces of a cube if no two adjacent faces have the same color?

This activity is rich with ambiquity. Results of the imaginings often depend upon the original images created in the student's mind. A debriefing session held at the end of each exercise allows students to describe their results and provides opportunities to enrich vocabulary and develop precise meanings and definitions. An alternative closure to each activity is to have students instructor a partner to sketch the image they describe and discuss words and phrases which were helpful or un-helpful.

Some "imaginings" have more than one possible answer. Have students choose one and reword it so there is only one possible answer. Also have students reword an imagining so there is more than one possible answer.

Common Terminology. In addition to sharing certain processes such as visualization, mathematics and the arts share terminology. However, their definitions and uses are not always consistent. Provide students with a list of words similar to that given below. Have them discuss how the term is used in both mathematics and the arts and decide whether the meanings are similar or not.

Additive/Subtractive
Complement/Complementary
Line
Positive/Negative
Proportion

Figure 7. *Sample words used in both mathematics and the arts.*

While "additive/subtractive" and "proportional" may have relatively similar meanings to both a mathematician and an artist, the other words in the list may not. To an artist, "complementary" refers to the hue directly opposite on the color wheel; to a mathematician it could mean two angles whose sum is 90 degrees. To a mathematician, "positive" and "negative" refer to a number's relationship to zero; to an artist, they relate to the foreground and background. Possibly the biggest difference occurs with the word "line." To a mathematician, a line has no beginning or end, is straight, and has no width; to an artist a line is a mark left by a dot or point moving continuously through space or over a surface. It starts somewhere, ends somewhere, and may have texture. It may be thick or thin; straight, curvy, or zigzagged; horizontal, vertical, or diagonal. When artistically-talented students express frustration with mathematics, it may be rooted in their confusion about terminology.

Each of the described activities focuses on connections between mathematics and art as a visual, and typically 2-dimensional product. Other connections could involve dance or music and symmetry or fractions. In addition to the numerous connections involving concepts, teachers should also investigate how processes such as problem solving and communication are common to both mathematics and the arts.

References

[1] National Council of Teachers of Mathematics, *Curriculum and Evaluation Standards for School Mathematics.* Reston, VA, 1989.

[2] K. Schaffer, E. Stern, & S. Kim. *Math Dance*. URL: http://www.mathdance.org

[3] A. Dworsky & B. Sansby. *Conga Drumming: A Beginner's Guide to Playing with Time.* Dancing Hands Music: Minneapolis, MN, 1994.

A Geometric Inspection of Pennsylvanian Dutch Hex Signs

Evan G. Evans and Reza Sarhangi
Department of Mathematics
Towson University
8000 York Road
Towson, MD 21252, U.S.A.
eevans3@towson.edu and rsarhangi@towson.edu

Abstract

This paper discusses the mathematics that is involved in the construction of "Hex Signs" and describes the construction of such signs. Hex Signs are circular discs with intricate geometric designs with specific meanings that were hung on barns in the "Pennsylvania Dutch" region of the United States. Common designs include: Rosettes, Birds, and Star Polygons.

1. History of the "Hex Sign"

Since the beginning of time man has created designs to portray his feelings. He used these designs to communicate his feelings, and at times, to protect him from what he did not understand or feared. Some common cave drawings are now believed to have been created for good luck for the upcoming harvest. In the middle ages, Europeans were using decorative symbols and motifs on everyday objects to brighten up their lives. These symbols also had a dual purpose, one of religious and ritual protection. Some symbols were used to ward off supposed evil spirits, while others were considered to be good luck or a key for a long and healthy life. Often these symbols took on specific geometric designs which were representative of their purpose. Others included everyday objects such as birds, tulips, hearts, and stars with each representing a specific and different meaning [6]. Over time these symbols became widely known and used among the Europeans.

Figure 1: *Single Distlefink, Double Eagle, and the Welcome Bird*

In the early 19th century, a large wave of German farmers immigrated to America and a significant portion settled in the plush valley lands between Harrisburg and Philadelphia, Pennsylvania. This valley came to be known as "Pennsylvania Dutch" country. Although the name is a misnomer, these early settlers weren't Dutch, they were German; however, "Duetch" is the German name for someone from Germany. These farmers were mainly German Lutheran and Reformed settlers that commonly decorated everyday items with colorful designs. These designs generally used the common motifs carried over from the European middle-ages. In the middle of the 19th century folk artistes began enlarging these old geometric designs and transforming them onto barns. Whether these signs, later to be known as HEX SIGNS, were made to display ones cultural heritage, simple decorative pleasure, encourage prosperity, or to ward off evil spirits is widely debated.

Figure 2: *Two barns with Hex signs.*

One theory is that the geometric circles are representative of a family heritage, such as a coat-of-arms commonly used in England [1]. A farmer would commission a folk artist to paint a geo-circle design with various symbols and motifs that represented a particular lineage. A good portion of hex signs found today have the same picture and design, thus ruling out the theory that a hex sign shows one's lineage.

Another theory is that a hex sign is simply for decorative purposes. This arises from the fact that the Pennsylvania Dutch people have a history of painting everyday items with colorful designs. However, the act of painting a hex sign on one's barn in these times took a lot of time and money, neither of which the average farmer had much of. Most hex signs created in the last 50 years are identical or similar in design, leading one to believe that the purpose for the creation of hex signs now involves the need to feed the growing tourism industry in that area of Pennsylvania.

The last and most widely accepted theory is that these hex signs are exactly that, HEX SIGNS. In German "hex" means "witch" and it is believed that these beautiful geometric designs were created to ward off evil spirits or used to beckon the spirits for prosperity in the upcoming years [4].

Despite all the various theories one thing is certain, these simple looking geometric designs are far from easy to create. They are each composed of circles that surround various animal, flower, or star designs arranged to convey a specific motif or meaning. These symbolic meanings include: a heart representing love, a tulip for faith, an eagle for strength, the color white for purity, a black circle for the belief in Christ, and a brown border for long life.

We shall dissect some commonly found hex signs, discuss their meanings, and how they were constructed. Upon closer scrutiny, one will see the incredible geometric relevance of these hex signs, as well as a greater appreciation for their makers. Remember, these constructions were made from simple tools (nail, rope, hammer, straight edge) and paint.

We will first examine the most common geometric design found in the Pennsylvania Dutch Hex Signs, the six pointed Rosette. The single rosette is a basic motif of hex signs and it is one of the most ancient designs in the world. The rosette appears on buildings, furniture, gravestones and pottery. A green scalloped border symbolizes smooth sailing through life and the red colored rosette is used to symbolize strength. This sign would be considered a potent safeguard against harm and portrayed a strong sense of good fortune. This design is based on seven congruent circles. One circle as the main central circle and six circles that pass through the center of the original circle with their centers located on the circumference of the main central circle. One can see the relationship between this figure and the "Kissing Number of a Circle" which is also six.

Figure 3: *The six pointed Rosette, its geometric construction, and the Kissing Number of a circle .*

This single rosette has a D_6 dihedral symmetry with order 12 and is found abundantly in the Pennsylvanian Dutch countryside. It has appeared painted on barns in the mid 19th century and continues to be prominent to this day. It is now more common to find these hex signs already painted on circular cuts of plywood ready for installation.

2. Other Hex Signs and their Meaning

Other hex signs have animal figures and common objects that carry different meanings. The Distlefink was the good luck bird of the Pennsylvania Dutch (Figure 1). It was actually a stylized version of the goldfinch. The goldfinch eats thistle seed and uses thistle down for its nest and was called a thistlefinch from which comes the Pennsylvania Dutch "Distelfink". This double distlefink sign is also a common hex sign. The two birds give a double measure of good fortune and the Trinity tulips stand for faith in yourself, faith in what you do and faith in your fellow man. The scalloped border symbolizes ocean waves for smooth sailing through life. This particular hex sign has no rotational properties but does have a one fold reflection down the center (an m1 or pm11) [5]. These designs were generally made by a local folk artist who was commissioned by a Pennsylvanian Dutch Farmer to paint such a hex sign on his barn. Some folk artists had created stencils of particular hex signs and would simply trace the hex sign onto the barns, while other talented folk artists would paint by free hand creating a distinct and unique hex sign each time.

Figure 4: *Double Distlefink*

Some of the more complex geometric hex signs make use of star polygons. A star polygon is defined by the following theorem.

> Theorem: Let n be the number of equally spaced points on a circle. Begin at a point and, going around in one direction, join every kth point. Then (n, k) star polygon—a star based on meeting all the vertices in a single stroke— exists if and only if $k \neq 1$, $k \neq (n-1)$, and n and k are relatively prime [3].

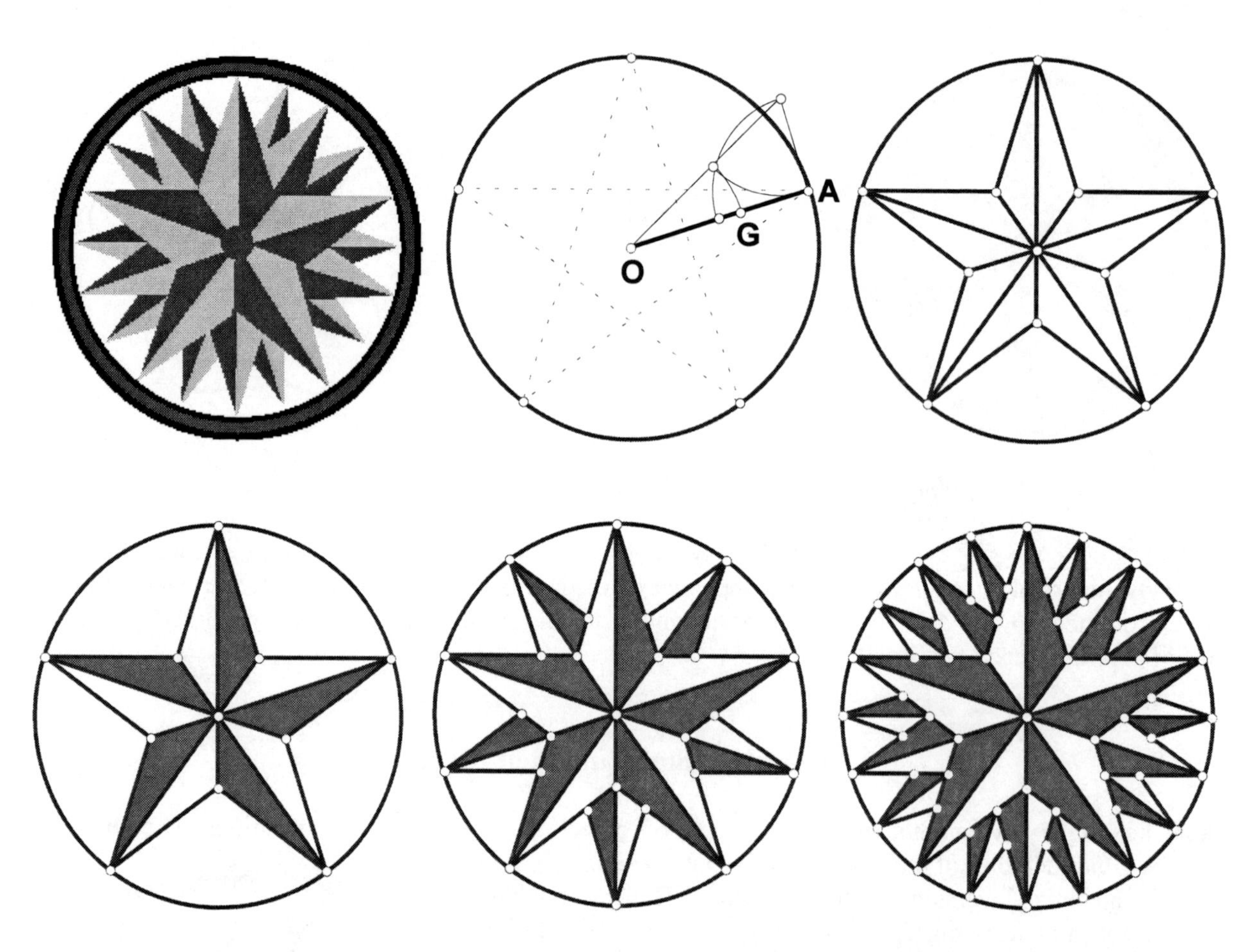

Figure 5: *The Triple Star, a (5, 2) Star Polygon, Pentagram, that has been constructed based on a circle with radius OA divided into ten congruent arcs to OG, the large part of the Golden Cut, and the process of making the Triple Star from a (5, 2) Star Polygon.*

The Pennsylvanian Dutch folk artist used this simple concept to create the spectacular triple star hex sign. The Triple Star motif symbolizes good luck, success and happiness. The ring of brown is associated with the cycle of life making this particular sign a wish for a lifetime of happiness. Although this hex sign appears to have a very basic construction, upon further study one will see the complexity involved in creating such a design. On close inspection we can see the hex sign is made up of four overlapping pentagonal star polygons (5, 2), *Pentagram*. One can now see the complexity in design of this simple looking hex sign and its C_5 cyclic symmetry of order 5 [5]. The relationship between the golden cut and the production of a decagonal figure on a circle within this segment is another interesting point about this design.

Some hex sign designs include the construction of a smaller circle within the bigger circle to bring out more artistic value and meaning. The eight pointed star is another fairly common hex design; with the blue star symbolizing goodwill, the tulips representing faith and trust in man, and the sheaves of wheat symbolizing abundance. This design was often used to decorate large buildings and it proclaims abundance and goodwill for all. This hex sign makes use of two (8, 3) star polygons; one on the outer ring (blue star) and one superimposed in the center circle (red star). This design has a D_8 dihedral symmetry with order 16.

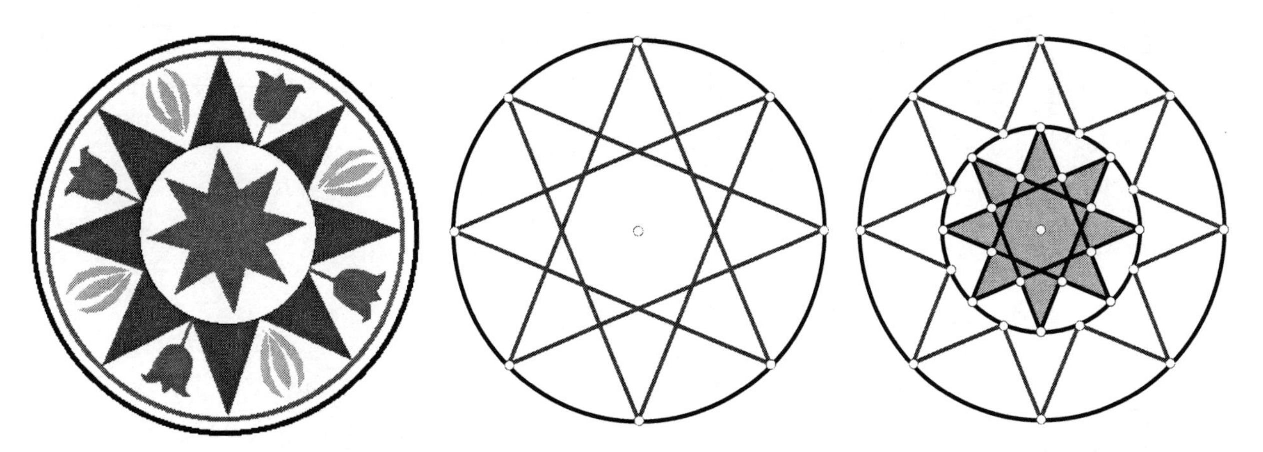

Figure 6: *A hex design based on the construction of a (8, 3) Star Polygon.*

The leaf designs below are the *Maple Leaf* (lower left) and the *Mighty Oak Leaf* (lower right) hex signs. The first depicts five large maple leaves radiating from a design center, sporting an array of colorful earth tones. The leaves portray the diversity and beauty of life here on earth. A good luck eight pointed star (in the center) completes the design. The design represents appreciation of life's beauty and the sweetness and purity of life. This design is based on the division of the outer ring into five equal parts (the five maple leaves). The Oak Leaf or Mighty Oak is made up of four oak leaves in bold colors radiating from the center. The Oak symbolizes strength in body, mind and character. The four colors of the leaves symbolize the seasons of life and the wavy border in the outer ring symbolizes smooth sailing through life [6]. The Oak Leaf hex sign has an outer ring that is based on division of a circle into eight equal arcs with an inner circle having a six petal rosette as we presented in Figure 3.

Figure 7: *The Maple Leaf hex design with an inner (8, 3) Star polygon, and The Mighty Oak Leaf with an inner six pointed Rosette.*

The stylized tulip with its three petals is a dominate feature in Pennsylvania Dutch folk art. It is referred to as the *Double Trinity Tulip* and it symbolizes the trinity as well as faith, hope and charity. The heart in this sign (as well as other Pennsylvanian German folk art) is not the heart of sentimental "Victorian" valentines. Rather, it is religious in its representation of the heart of God, the source of all love, and hope for a future life. The colors in this heart are used to give it additional meanings. Red symbolizes strong emotion and blue is used to indicate strength, especially spiritual strength. The white background adds purity and the solid black circle conveys unity in Christ. This hex sign has a one fold reflection down the vertical center while the inner blue rosette has a D_6 dihedral symmetry of order 12.

Figure 8: *Double Trinity Tulip*

The next hex sign, known as the *Daddy Hex*, has an outer ring divided into twelve equally spaced petals and an inner ring divided into eight equally spaced petals [4]. The outer rosette provides twelve months of good luck, while the smaller rosette provides an added measure of good luck during difficult times of the year. The outer ring of this particular hex sign looks like a D_4 symmetry but closer inspections reveals that if we consider the colors of the pedals, then the design is only a C_4 cyclic group since no mirror reflection is admitted. The inner circle then would be a C_2 cyclic group.

We must give credit to those Pennsylvanian Dutch folk artists for creating such magnificent signs. The geometry and mathematical properties that these particular creations have is too profound to discuss in this paper, but further investigation is encouraged.

Figure 9: *Daddy Hex*

Reference

[1] Ensminger, R. (1992). *The Pennsylvania Barn: Its Origin, Evolution, and Distribution in North America*. The Johns Hopkins University Press.

[2] Igou, B. (2001, October). *The Story of the Hex Sign.* Amish Country News, cover article. Retrieved November 3, 2005, from http://www.amishnews.com/featurearticles/Storyof hexsigns.htm

[3] Sarhangi R. (2004). *Elements of Geometry for Teachers*, 2nd Edition, Pearson Education, Boston, Massachusetts

[4] Smith, E. & Horst, M.(1993). *Hex Signs and other Barn Decorations*. Lebanon PA. Applied Arts Publishers.

[5] Washburn, D. & Crowe, D. (1992). *Symmetries of Culture: Theory and Practice of Plane Pattern Analysis.* University of Washington Press.

[6] Yoder, D. & Graves, T.(2000). *Hex Signs: Pennsylvania Dutch Barn Signs and their Meaning.* Mechanicsburg, PA. Stackpole Books.

Creating Sliceforms with 3D Modelers

Stephen Luecking
School of Computer Science, Telecommunications and Information Systems
DePaul University
243 South Wabash
Chicago, IL 60604
sluecking@cs.depaul.edu

Abstract

3D or CAD modeling programs can provide tools for the beginner to quickly create mathematical models known as sliceforms, or, in the terminology of computer graphics, raster surfaces. This tutorial and workshop provides the novice with the tools and procedures for modeling and physically constructing these models using their PC, a printer, craft knife, glue and paperboard.

Sliceforms – to borrow the term coined by John Sharp – is a technique utilized by Alexander von Brill and Felix Klein in the construction of mathematical surfaces from cardboard. These forms, produced in Munich circa 1870, sold throughout the world to mathematics educators. Later the technique was used to construct extremely strong wooden surfaces for press-molding or vacuum-forming sheet material into contoured surfaces.

The construction method uses cross-section curves from the surface laid out along an orthogonal, x,y grid so that the curves intersect at each grid crossing. The curves present as the edges of planar slices of the volume enclosed by the surface. A series of x slices then intersect a series of y slices to promulgate the surface and its volume. CAD modeling programs permit the creation of the surface and tools for generating cross-sections. These cross-sections may then be printed onto a cover weight stock, cut out and assembled in the proper order so as to replicate the cross-sections produced by the modeling program.

Figure 1: *John Sharp, Sliceforms.* **Figure 2**: *Charles and Rae Eames, press mold form for aircraft fuselage.*
Figure 2: *Charles and Rae Eames, fuselage segment pressed from mold.*

This workshop/tutorial uses Rhino v.3, available for free download at www.rhino3d.com. This is a fully functional demo copy with no time limit, but limited to 25 saves. The tutorial assumes familiarity with using the Windows PC interface.

Creating the Surface/Solid

This workshop will use geometric solids already available for insertion into the graphic. These are known as geometric primitives and their normal function is to combine with other such primitives to represent more complex objects. Rhino features, among its primitives, spheres, ellipsoids and paraboloids. This tutorial will use the paraboloid.

1. Open Rhino and the default window should appear with a number of menus and tool icons.

   Click on the Box icon in the left toolbar and right click on the parabolic surface in the fly-out toolbar. This will give the option of creating the parabolic solid by first clicking on a vertex point and then on a focus point

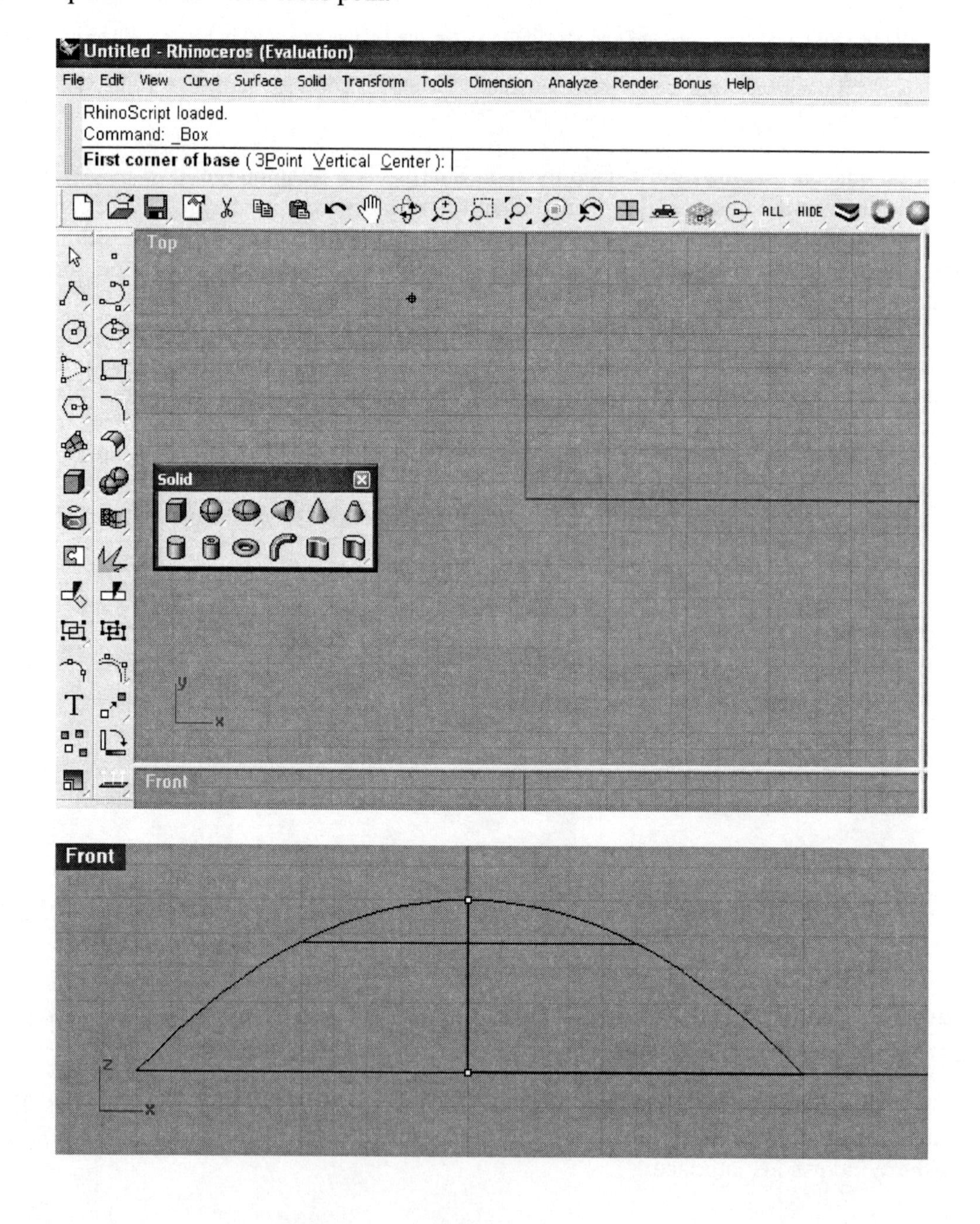

2. There will be four viewports on the screen labeled Top, Front, Right and Perspective. Click in the Front view to set the vertex and then drop down to set the focus. Now slide the mouse to on side and click to set the rim of the paraboloid.

Snap Ortho Planar Osnap (Note that it is easiest to click on the grid points and to set the vertex on the green ***y*** axis and the focus on ***0,0*** point of the grid. The third click to set the rim should then be on the red ***x*** axis. To aid in this precise placement activate the **Snap** button on the Snap Menu at the bottom of the screen.)

3. The paraboloid will appear in wireframe mode in all four windows. To ensure that it fits well into all windows right click on the Zoom Extents icon in the top toolbar. To aid in visualizing this solid click on the Shaded Viewport icon in the top toolbar. Shade each viewport by first clicking in the viewport and then clicking on the icon. Right clicking this icon will return the viewport to wireframe.

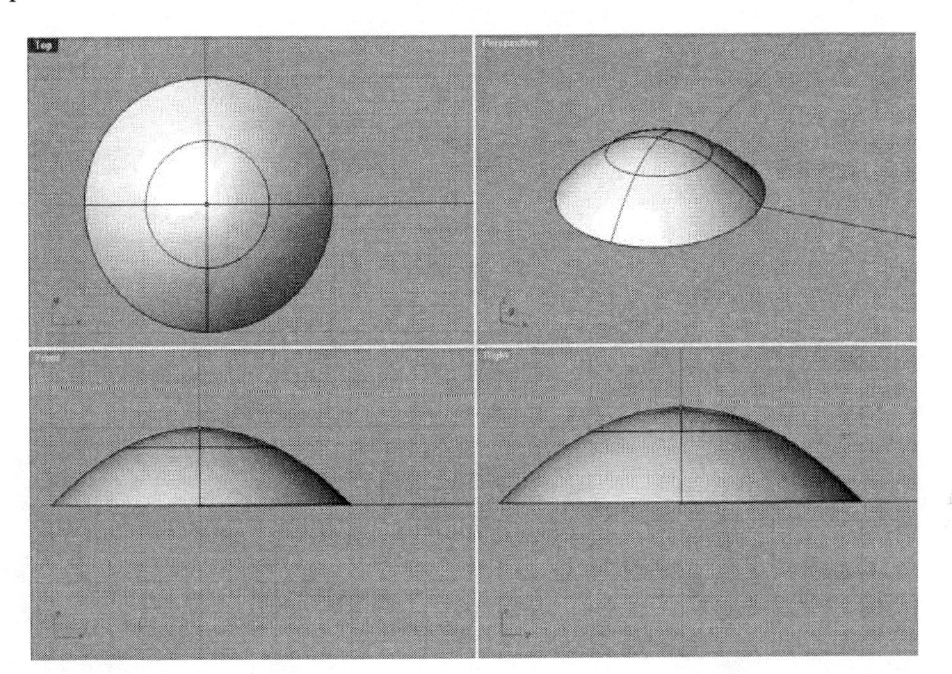

4. Click on the surface to select it. (Note: when selected the wireframe turns yellow.) Now click on the Solid Tools icon in the left toolbar and in the fly-out window click on the Cap Planar Holes icon. This closes the base of the parabolic surface and it becomes a solid.

Slicing the Surface

5. The next step is to generate the cross-sections of the paraboloid. Select the paraboloid and then click on the Curve from Object icon and click on the Contour icon in the fly-out window.

6. There will be a prompt in the Command line above the toolbars: “Contour plane base point.” Working in the top window, click on the left end of the paraboloid. A second prompt reads: “Direction perpendicular to contour planes.” Now click at a horizontal (along the x axis) to the right. The final prompt reads: “Distance between contours.” Type in the desired spacing between contours. In the example below that distance was 2, which equals 2 grid units.

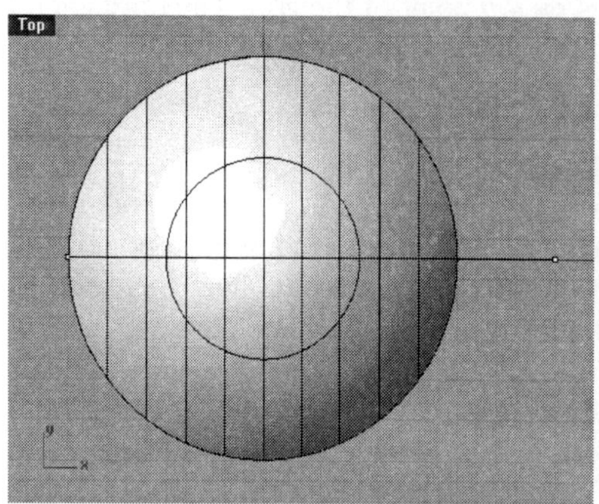

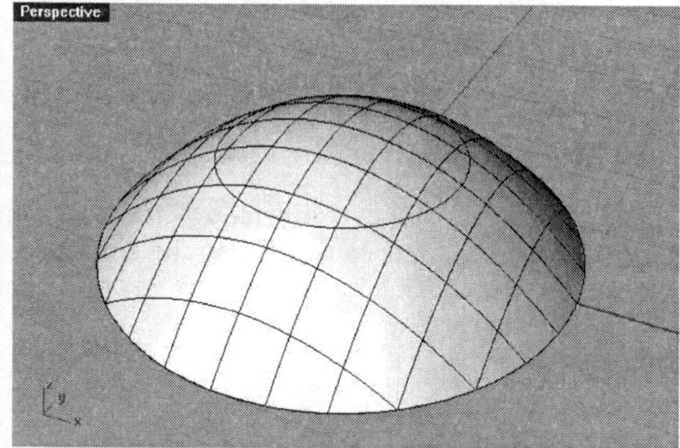

The program will automatically draw in the contours. Repeat this process in the y direction.

7. Select the paraboloid and delete by hitting the delete key. This will leave the contour curves. Select all of these curves. Now click on the Surface icon in the left toolbar and choose the Surface from Planar Curves icon.

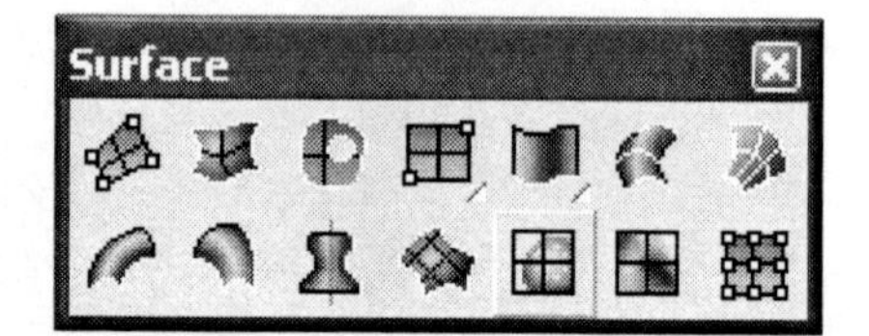

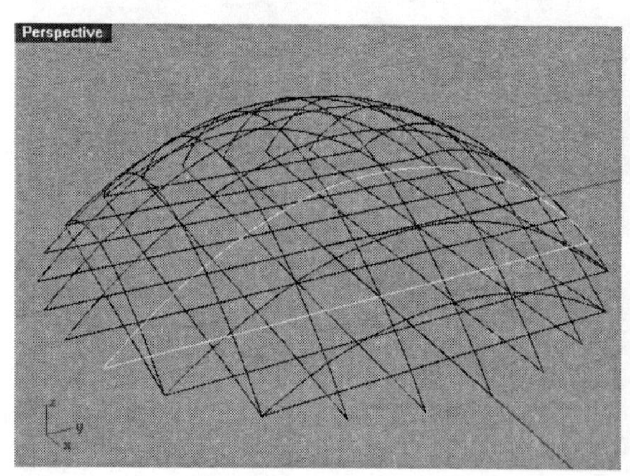

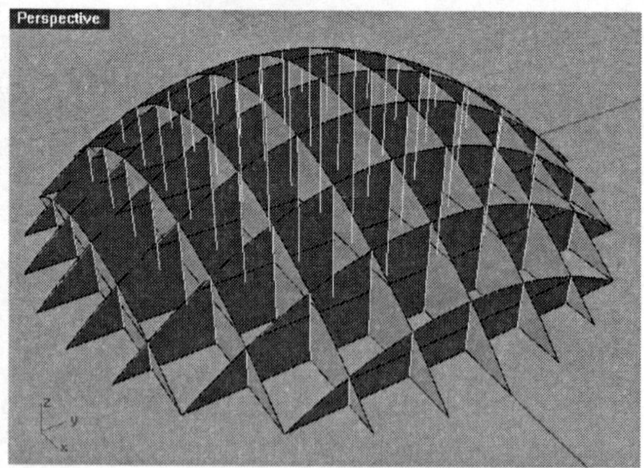

8. The contour curves are now section planes and the result is a preview of the sliceform to come. Select all of these planes and from the Curves from Objects menu choose the Intersection icon. This command draws a line through each intersection.

9. ALL Click on the All icon in the top toolbar and choose select Select Surfaces from the fly-out window. This will select each planar section, but not the curves. Delete to leave only the curve sets.

10. For the next few steps work in the Top and Front viewports. Drag the selection arrow to form a box moving from left to right. Dragging from this direction will select only those items entirely within the selection box. (Dragging right to left will select any item the box touches.)

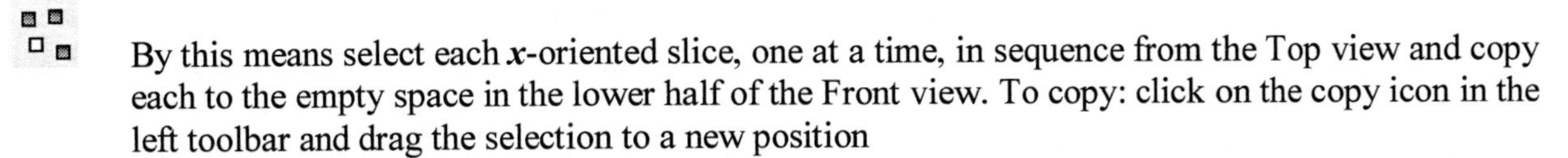

By this means select each *x*-oriented slice, one at a time, in sequence from the Top view and copy each to the empty space in the lower half of the Front view. To copy: click on the copy icon in the left toolbar and drag the selection to a new position

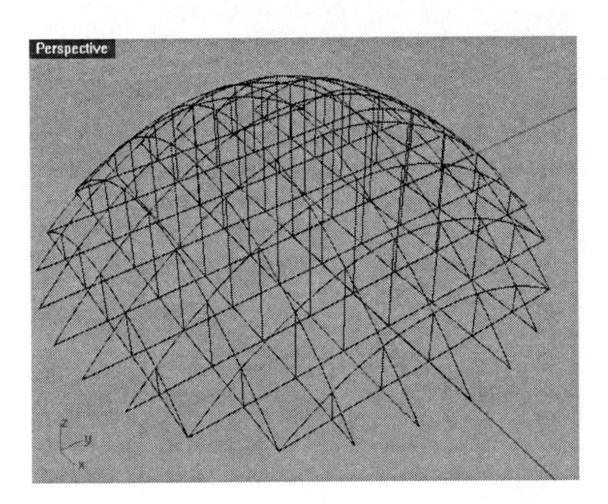

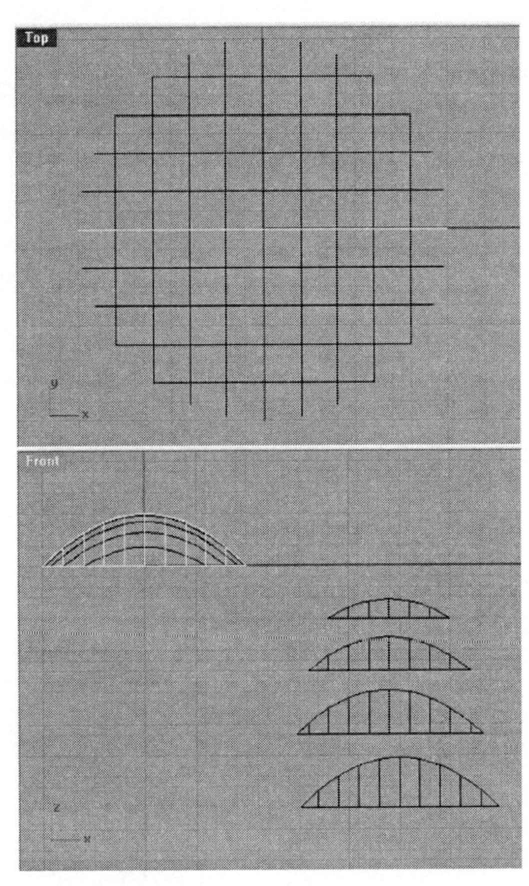

Repeat this procedure with the *y*-oriented slices, dragging their copies into the lower half of the Right view port.

11. The Front view port now displays the ***x*** slices and the Right view port displays the ***y*** slices. In addition there are lines to mark each intersection of slices. To ease the manufacture of the physical sliceform will require one more step.

Go to the Snap bar at the bottom of the screen and activate **Osnap** (object snap). Then choose the midpoint object snap only. Click on the Point icon in the left toolbar and choose the Multiple Points tool. This tool permits the placement of points with each click of the mouse. Click on each intersection line and a point will snap to its midpoint.

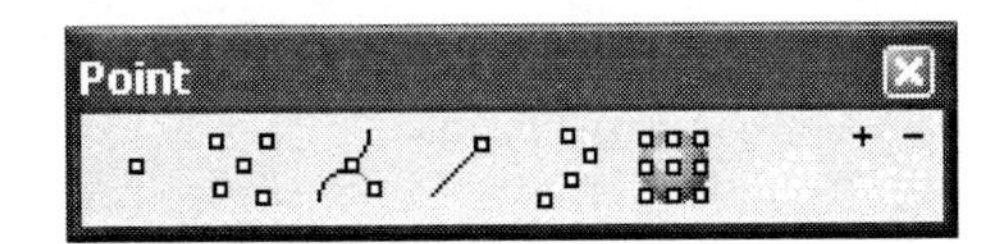

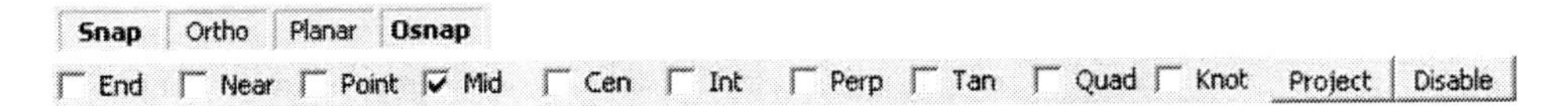

Printing the Slices

The slices are now ready to print, using a heavy paper stock, a cover weight or higher, and an ink jet printer (laser printers cannot manage heavy papers).

12. To prepare for printing, select and drag the individual slices into vertical groups and draw a rectangle around them, similar in proportion to a standard sheet of paper. The Rectangle tool resides in the left tool bar. Click on the icon and then click and drag and click again to draw the framing rectangle. Keep the Snap tool active. The exact same rectangle should be used to frame each grouping of slices! Use the copy tool if necessary to ensure this congruency of rectangles.

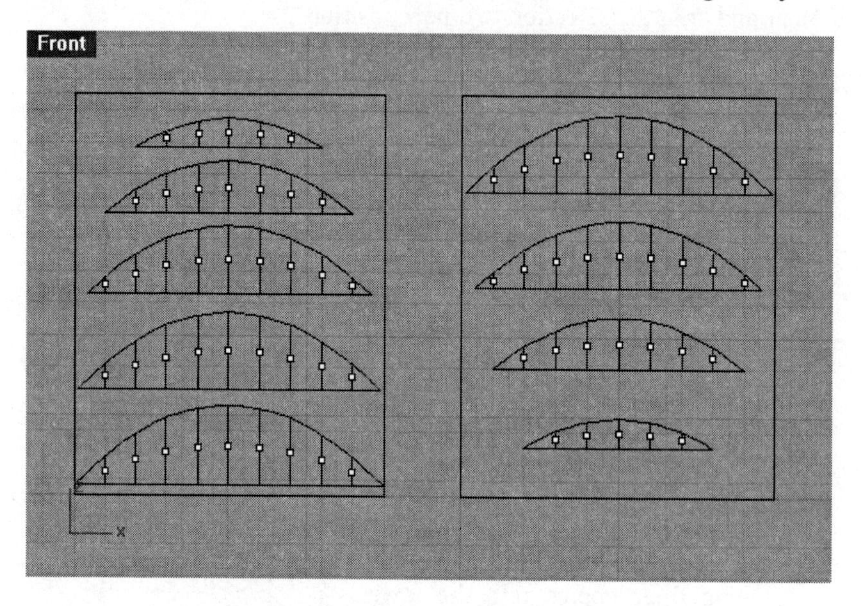

13. Go to the File menu and choose Print. The following window will appear:

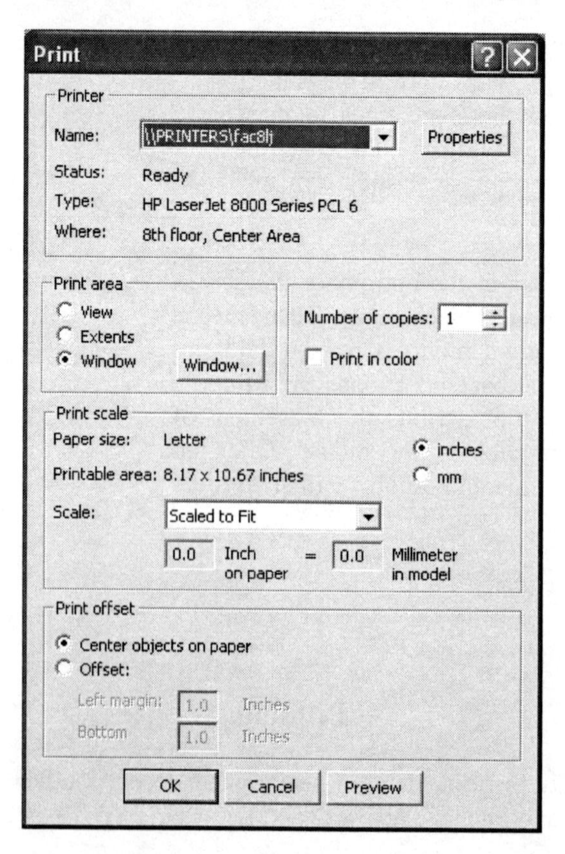

In the Print area sub-menu select Window then click on the Window… button. The Print menu will disappear and a cursor will appear that is to be clicked and dragged to match the framing rectangles. The Print menu now reappears.

In the Print scale sub-menu make sure the Scale box reads, "Scaled to Fit" and click OK. The image within the rectangle will print to fit the page. Repeat this to print each group of slices. The fact that the framing rectangles determine the print area and that they are all the same will ensure that the slices are all printed to the same scale. A typical slice would appear as below:

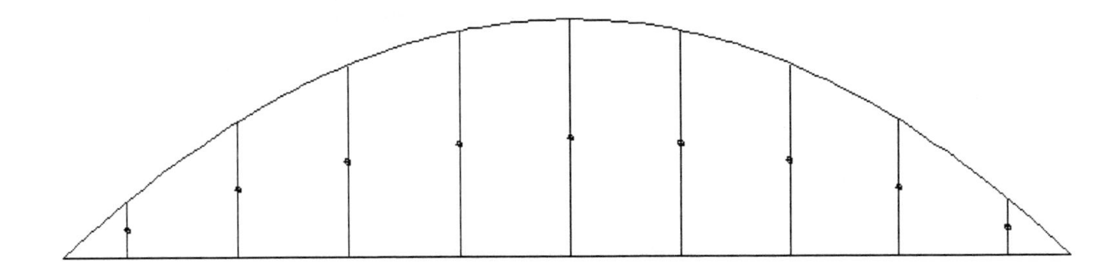

Assembling the Sliceform

14. It is a good idea to label the direction – ***x*** or ***y*** – of the slices, as well as their numbers in sequence, by lightly penciling these in on each. Cut out the slices with a craft knife. The slices must now have slots cut at each intersection with the ***x*** slices slotted from below and the ***y*** slices slotted from above (or vice verse). These slots must be cut as thinly as possible, but with a double cut to assure some breadth to the slot. The sliver of paper curling up from the slot should look like a thread. The points at the middle of each line of intersection marks the extent of these slots.

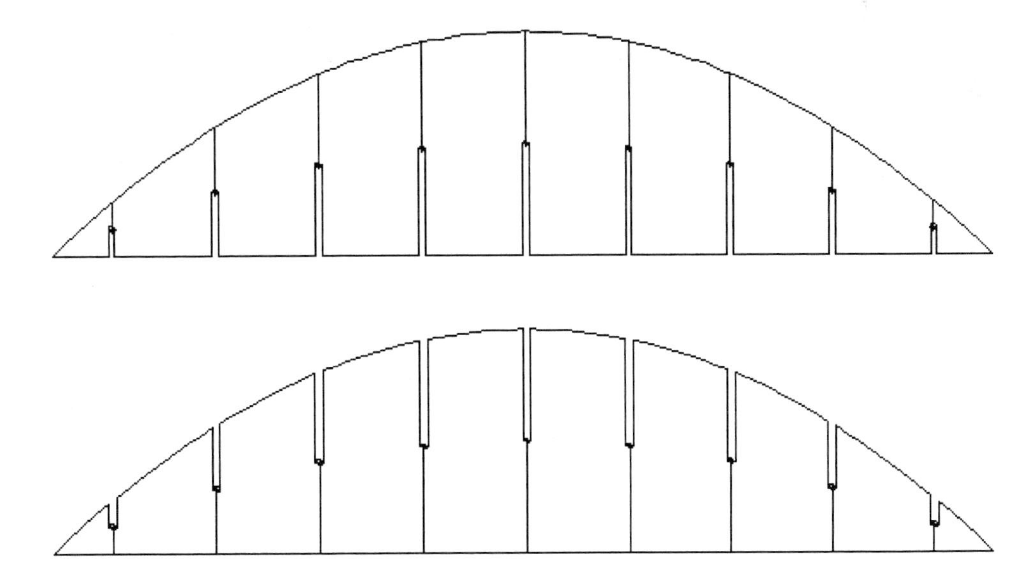

15. Assemble by inserting the bottom-slotted slices into the top-slotted slices being careful to maintain the position of the slice within the model. Like an accordion this arrangement will collapse flat. To make the structure rigid glue the bottom edge of the slices to a board. In the case of the sample paraboloid, a circular base of the diameter of the paraboloid would work well, too.

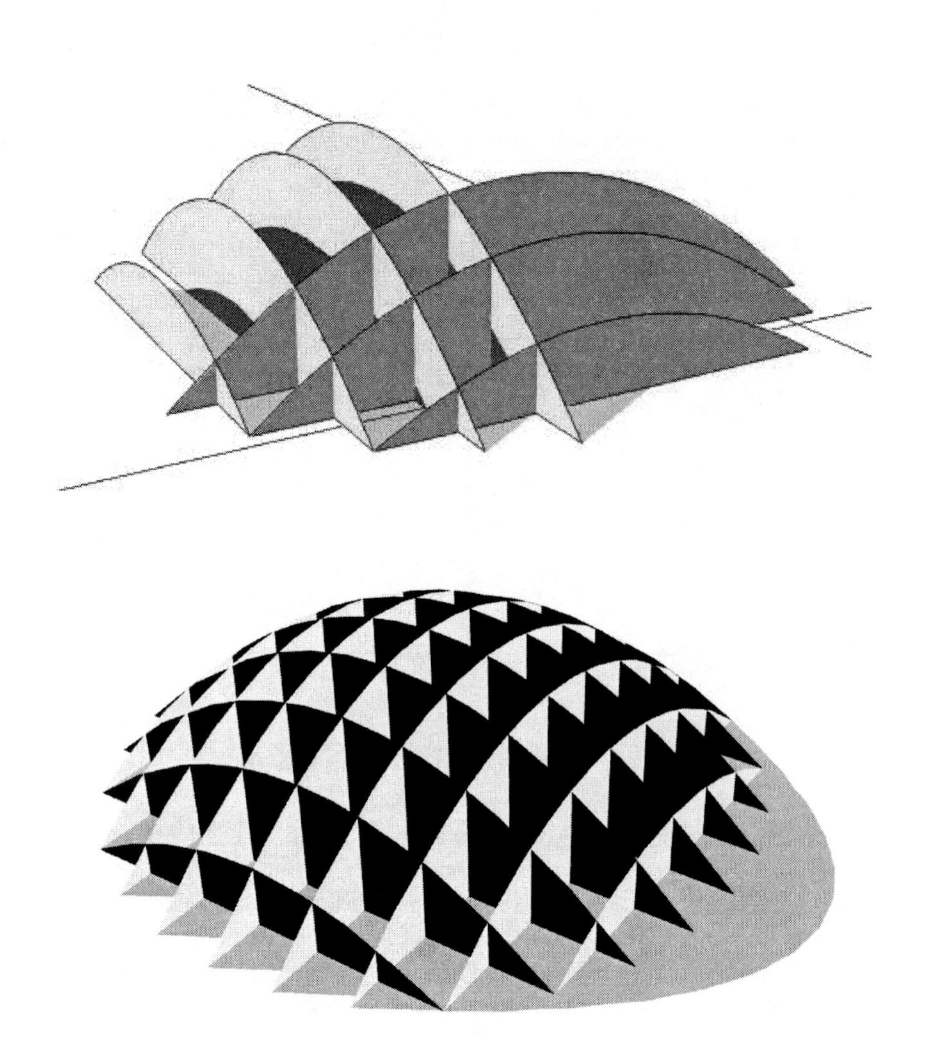

Paper Sculptures with Vertex Deflection

Tevfik Akgün
Faculty of Art and Design
Design Communication
Department
Yildiz Technical University
Istanbul, Turkey
akgunbt@yildiz.edu.tr

Ahmet Koman
Molecular Biology and
Genetics Department
Boğaziçi University
Istanbul, Turkey
akoman@boun.edu.tr

Ergun Akleman
Visualization Sciences Program,
Department of Architecture
Texas A&M University
College Station, Texas, USA
ergun@viz.tamu.edu

Abstract

This workshop presents mathematical concepts vertex deflection and Gauss-Bonnet Theorem with hands on experiences using paper, plastic, stapler and glue. We show how to create sculptor Ilhan Koman's mathematically motivated developable surfaces [1, 3, 4]. We also present how one can construct a variety of shapes creating saddle, maxima and minima using nip and tuck.

1 Workshop Overview

This workshop shows how a mathematical concept called vertex deflections [2] can be used to intuitively construct developable surfaces using paper, plastic, stapler and glue [1]. We will provide an intuitive introduction to vertex deflection using Ilhan Koman's sculptures as shown in Figure 1 [1, 3]. The sculptures of Koman discussed in [1] visually provides information about vertex deflection and can help anybody to understand local behavior around a extreme point such as saddle or maxima as shown in Figure 2. We also show how to create other Koman developable sculptures such as hyperforms (see [1].)

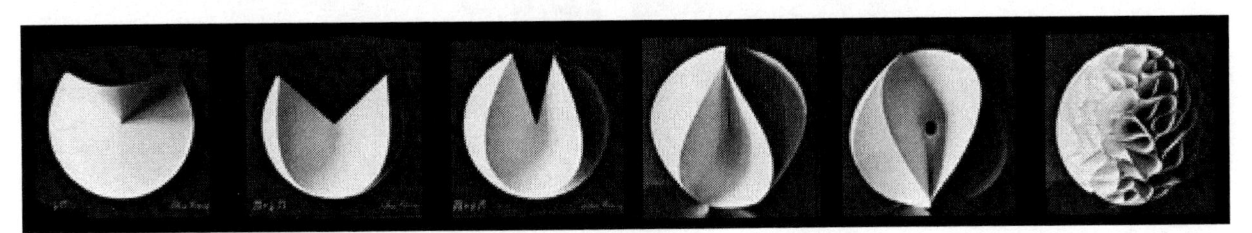

Figure 1: Ilhan Koman's Developable Sculptures. These were constructed with sheet metal in the 1980's. In these sculptures, the connections are almost invisible (photos by Tayfun Tunçelli).

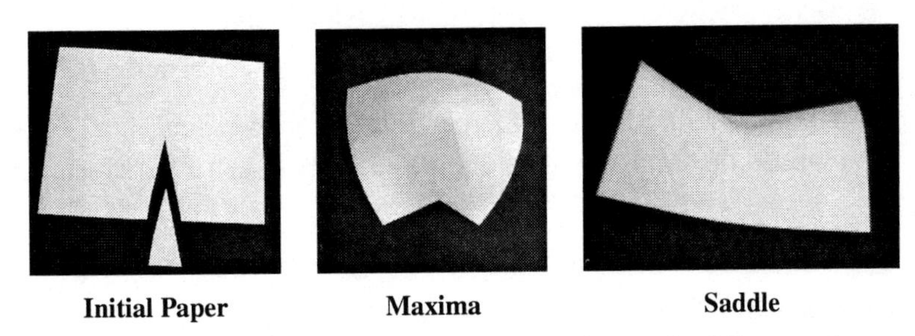

Figure 2: Creation of maxima and saddle with subtracting and adding angles to a flat surface.

The workshop is intended for all Bridges attendees who are interested in the creation of interesting shapes and sculptures; and teaching how to construct those sculptures. With the papers, staplers and glue in the room all participants will have an hands-on experience with vertex deflections.

The beginning part of the workshop covers construction of Ilhan Koman's developable sculptures after a short overview. Ilhan Koman's sculptures such as hyperforms will be created by participants with hands-on experiments. The concepts of saddle, minimum and maximum will be introduced by nipping and tucking circular pieces of papers. In the second part, we show to intuitively construct paper sculptures with hands-on experiences, such as the ones shown in Figure 3. We will also create Architectural forms together with participants. Figure 3 shows two examples of paper sculptures we have created based on the insight coming from vertex deflections [2] and Gauss-Bonnet theorem [5].

 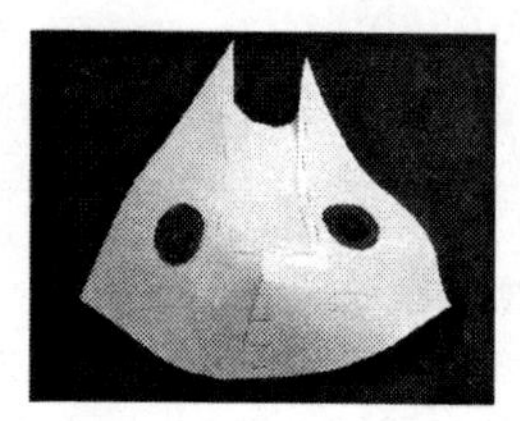  

Figure 3: Paper sculptures. Cat mask is created from a 8.5×11 paper that is cut into one square and one rectangular pieces. These pieces are later stapled together to create the cat mask. Swan is created from a shaped and folded paper using only three staples.

Figure 4: Architectural forms that are created using vertex deflections.

References

[1] Tevfk Akgun, Ahmet Koman and Ergun Akleman, "Developable Sculptural Forms of Ilhan Koman" Proceedings of Bridges 2006, Mathematical Connections between Art, Music and Science, London, August 2006.

[2] C. R. Calladine, "Theory of Shell Structures", Cambridge University Press, Cambridge, 1983.

[3] Koman Foundation web-site; http://www.koman.org

[4] John Sharp, D-forms and Developable Surfaces, Bridges 2005, pp. 121-128, 2005.

[5] "Eric W. Weisstein", "Gauss-Bonnet Formula", 2005, From MathWorld–A Wolfram Web Resource. http://mathworld.wolfram.com/Gauss-BonnetFormula.html

Understanding the Mathematics Based Formulation on Dome Tessellation in Architect Sinan's Mosques Design

Zafer Sagdic
Mujdem Vural
Gokce Tuna Taygun
Architecture Department
Yildiz Technical University
80630 Besiktas
Istanbul, Turkiye
zafersagdic@hotmail.com
mujdemvural@hotmail.com
gokcetunataygun@hotmail.com

The K-12 education is a comprehensive learning program, which includes curriculum, tools, materials and an innovative lesson delivering system. In college education, it is not always easy for teachers to have students' attention in history lessons. According to the description of K-12 education, some new teaching methods should be suggested to students on their history learning program for helping them to understand the history easily.

In general history lessons, the world history is told by giving the dates of events and mostly using some "spectacles", which are looking through "war history". The reasons and the results are important on history and they are always related with each other. Sometimes for better understanding the reasons and results of socio-cultural based relations should be researched on history education. The history of architecture gives lots of clues on that point of view. Thus, the Ottoman Empire can be chosen as an example for this workshop. The aim of this workshop is to understand the reason-result relationship on the creation of dome tessellation of architect Sinan -while designing structure for a specific dimension- by interactive education. To fulfil this aim the workshop will evaluate the usage of the dome structure in Sinan's mosques as the result of technology and material.

The Ottoman Empire was one of the biggest empire of the world history. And Sultan Suleiman the Magnificent period was the Golden Age, not only because of the treasury, but also because of developments on the socio-cultural life; and mostly on the architecture by Architect Sinan's designs.

During the Ottoman history, monumental sized buildings have been constructed by sultans to show their power. Mostly these buildings were mosques. The size of the mosque has been created according to the tradition of the praying lines facing to the Kibble[1] while namaz[2]. Thus, the main aim is to cover the "x" distance of praying medium (downwards-to-upwards).

All over the history the monumental mosque has not only been the symbol of the sultan ruled the Empire, but also the symbol of the Imperial Power in that age. Using this monumentality of structure was the priority of Architect Sinan's designs, who was one of the superior architect-building engineers of his age with Michelangelo. The proof of this is his stone masonry structures are still "alive"/ survived within last

[1] Kible: the direction of Mecca.
[2] Namaz: the ritual worship of muslims, five times a day.
[3] Pendant: circular triangle; a structural element of roofs or ceilings

400 years. Structure is the most important point of Architect Sinan's designs. Instead of using downwards-to-upwards space organization in mosques, Architect Sinan created the medium from "x" distance dome (upwards-to-downwards). He created the main dome first; extend over an area, using the maximum limits permitted by the material and the structure. Thus, his method is "dome tessellation". Definition of tessellation is a collection of shapes that fit together to cover a surface without overlapping or leaving gaps. Often a repeating geometric pattern, may of which may also be referred it to as tiling. Types of tessellation include translation, rotation and reflection. The study of tessellation can integrate many disciplines across any full curriculum-in art, math, language arts, social studies, etc. (www.artlex.com, lexicon of visual art terminology).

The organization of dome tessellation in Architect Sinan's designs are various in types, however, the theory is the same. The unity in his designs is to generate the space organization from the dome tessellation. There is always a main dome at the centre. Force coming from the main dome is supported by semi-domes at sides, quarter-domes at corners and pendants[3] at the transition to the base of the building. And finally, at the façades, arches are used to carry the force coming from the structural elements.

Geometric definitions of structural elements of tessellation are given below.
Main dome = approximately half sphere,
Semi-dome = either half sphere or quarter of a sphere,
Quarter-dome=either half sphere or quarter of a sphere,
Pendant = triangular sphere

Selected examples are shown on the following pages

1. Suleymaniye Mosque
Construction period : 1550-1557, under the reign of : Sultan Suleiman the Magnificent
2. Selimiye Mosque
Construction period : 1568-1575, under the reign of : Sultan Selim II
3. Shehzade Mehmed Mosque
Construction period : 1543-1548, under the reign of : Sultan Suleiman the Magnificent
4. Mihrimah Sultan Mosque
Construction period : 1540-1548, supported by : Mihrimah Sultan
5. Kilic Ali Pasha Mosque
Construction period : 1580, supported by : Kilic Ali Pasha

References

[1] A. Kuran, Mimar Sinan, Hürriyet Vakfı Yayınları, Istanbul, 1986.
[2] N. Camlıbel, Sinan Mimarliginda Yapı Strukturunun Analitik Incelenmesi, YTÜ Yayinlari, Istanbul, 1998.
[3] R.Gunay, Mimar Sinan ve Eserleri, YEM Yayin, Yapi-Endustri Merkezi Yayinlari, Istanbul, 2002.

*Images number 3-4-7-9-11-14 are from N. Camlıbel, Sinan Mimarliginda Yapı Strukturunun Analitik Incelenmesi, YTÜ Yayinlari, Istanbul, 1998.

Image number 1 is from R.Gunay, Mimar Sinan ve Eserleri, YEM Yayin, Yapi-Endustri Merkezi Yayinlari, Istanbul, 2002.

All of the other images are from A. Kuran, Mimar Sinan, Hürriyet Vakfı Yayınları, Istanbul, 1986.

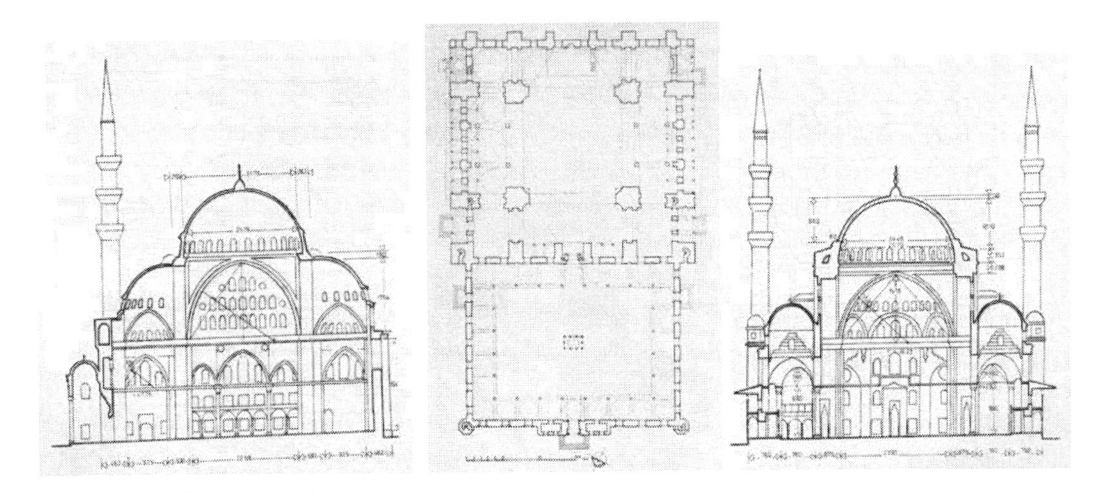

Figure 1, *Plan and cross-sections of the Suleymaniye Mosque, Istanbul-Turkiye*

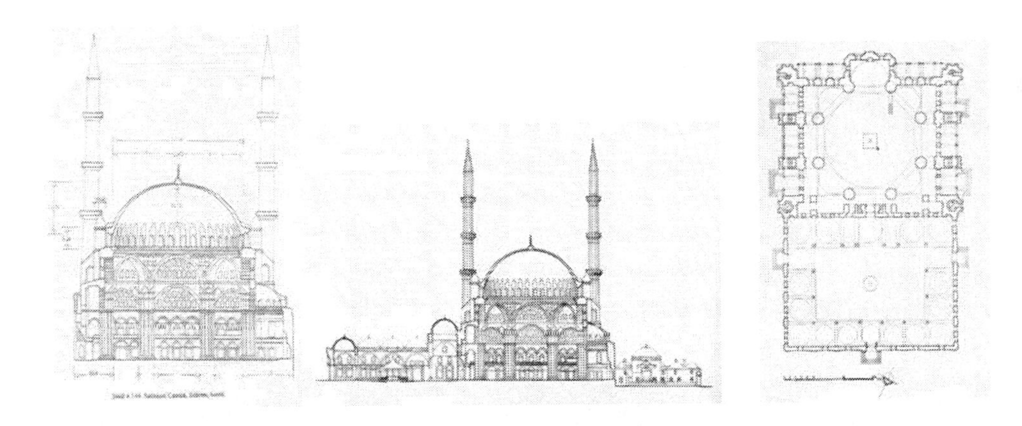

Figure 2: *Plan and cross-sections of the Selimiye Mosque, Edirne-Turkiye*

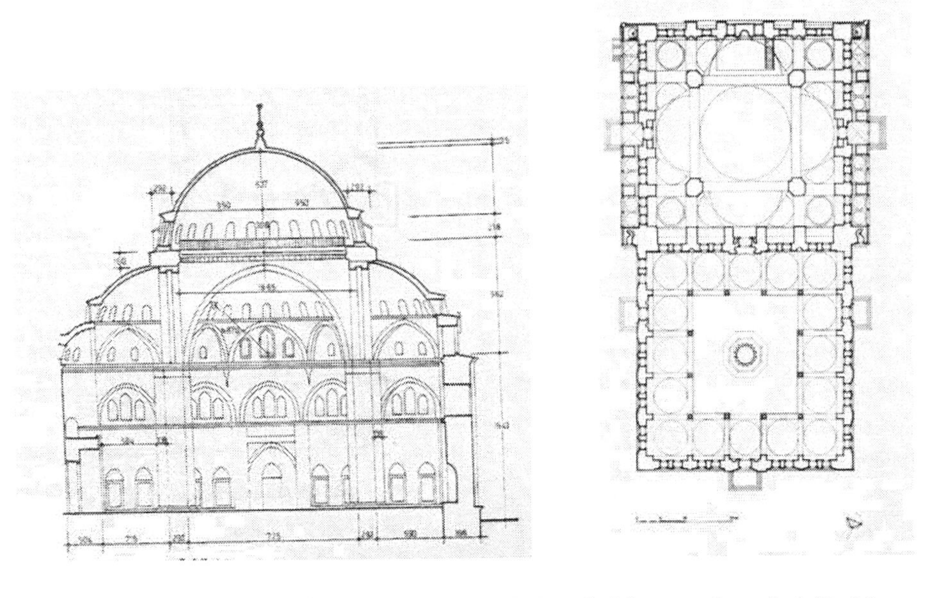

Figure 3: C*ross-section and plan of the Shehzade Mosque, Istanbul-Turkiye*

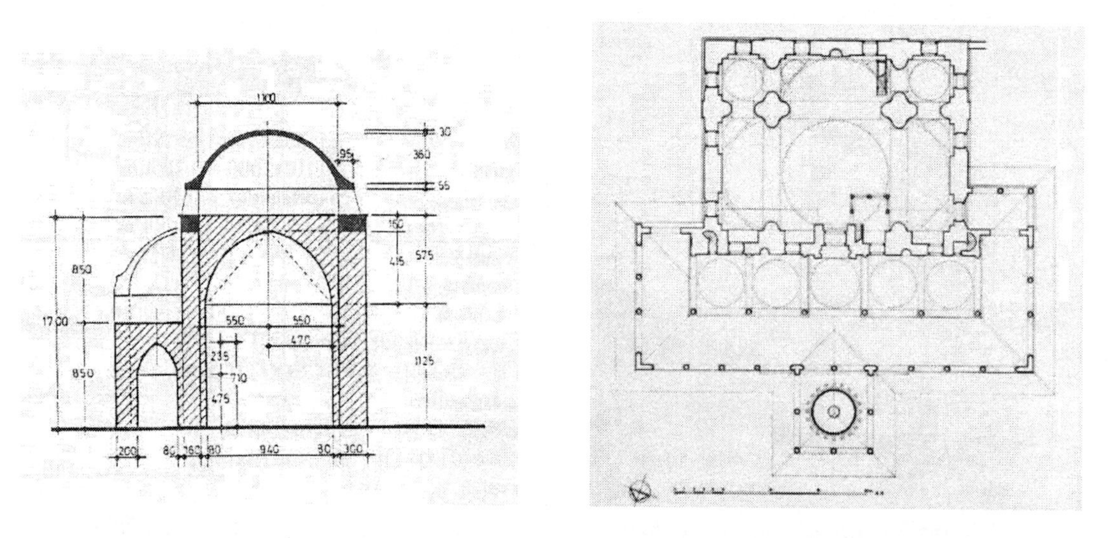

Figure 4: *Cross-section and plan of the Uskudar Mihrimah Mosque, Istanbul-Turkiye*

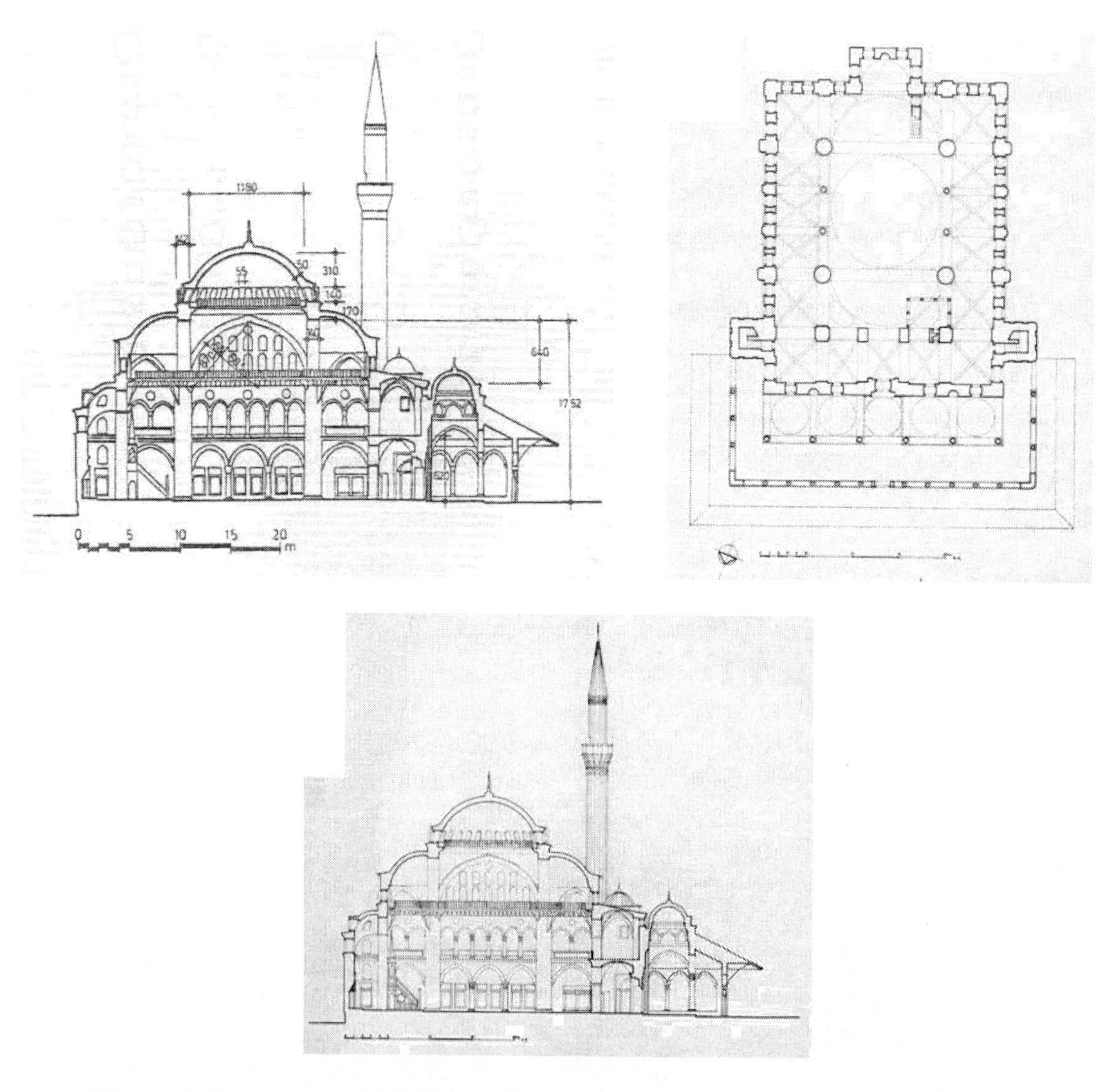

Figure 5: *Cross-sections and plan of theKilic Ali Pasha Mosque, Istanbul-Turkiye.*

Mandala and 5, 6 and 7 fold Division of the Circle

Paul F. Stang
Townshend International School
188 Dolni Trebonin
Czech Republik 382 01
Paul.Stang@Townshend.cz

Abstract

The Compass is perhaps oldest of all the math and drawing tools. When did someone think to put two sticks together, hold one in place and twirl the other, or link two pegs with a rope, pound one in the ground and use the other to draw circles in the dirt? It is commonly known that with only compass, ruler and pencil, a six-fold division of the circle can be made. An amazing array of 2 and 3 dimensional possibilities then follow, to form bridges between Math, Art, History, Culture and Science and even Mythology and Magic! Mathematics is learned through the hands, creativity and social interaction. Further, the compass, when coupled with the phi proportion, can be used to obtain 5 and 7 fold division of the circle. The Initiate, interested in mastering the compass, must begin this journey of exploration by ensuring precision. Often, the compass user grips the device too firmly, pressing harder in an effort to ensure quality. The result of this 'muscling' is often that the point makes an overly large hole in the paper, the compass opens from the pressure, making a spiral, and the paper slips. The proper way to grasp the compass is to twirl the upper post between thumb and index finger, so that it pirouettes. In this way it makes a crisp circle. The image may be faint but we can twirl the compass more times for better definition, rather than pressing harder. With brief explanations, we will now proceed rapidly through a multitude of forms.

The Path to Mandala

Workshop participants will learn easy geometric drawings, to ensure familiarity and accuracy with the compass (so that they can similarly enable this with their students). They will create basic art forms, known as Mandalas. They will continue, learning how to create flat, three dimensional figures and build platonic solids from flat patterns. They will finish with more challenging techniques, by developing the phi proportion and using it to generate 5 and 7 pointed stars. This final work will expose for them the minds of Da Vinci and the builders of Stonehenge and the Great Pyramid of Giza, and more!

Participants should come away with fascinating material that can really interest and hence motivate mathematics students, while making strong connections between art and history.

We begin with a compass and ruler to draw mathematical, artistic imagery very useful at developing new interest in students who normally don't like mathematics, and new challenges for those who do. Next will be seen how to make equilateral hexagons with compass and ruler. This will give participants a basic understanding of how to make regular polygons.

We will draw basic Mandalas, and relate them to historical usage in Eastern - Western cultures, providing fascinating bridges between mathematics and art and architecture, history and science. It will be shown how these techniques can be used to demonstrate fractions, angles, trigonometric relations and fractals. Seeing how to ensure precision, we will continue then with more complex drawings.

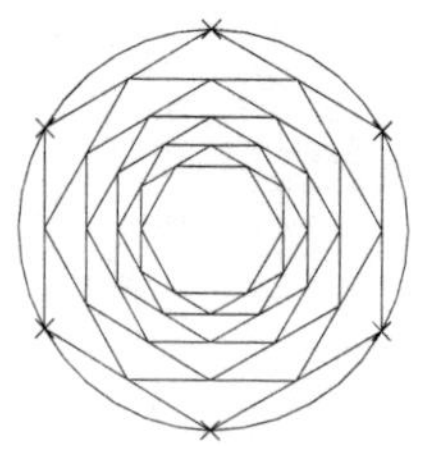

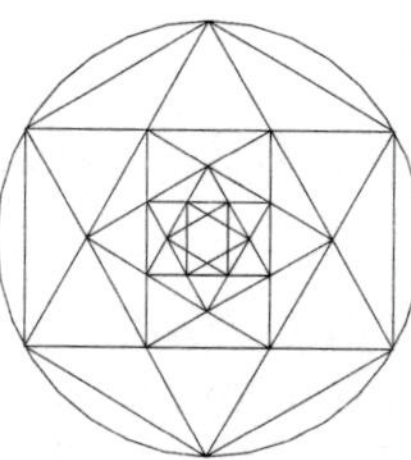

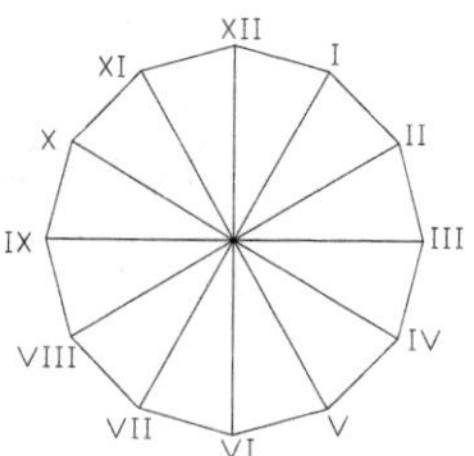

Next, a variety of images will be created to show how 3-D can be made on a flat paper. Participants will continue with related techniques on flat patterns that they will then be able to cut out and create three of the platonic solids from. We will see that many common, practical, commercial, religious and mystical imagery can also be generated by these same methods.

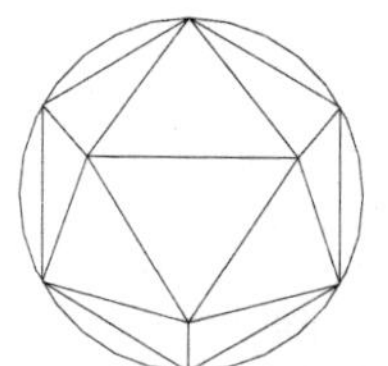
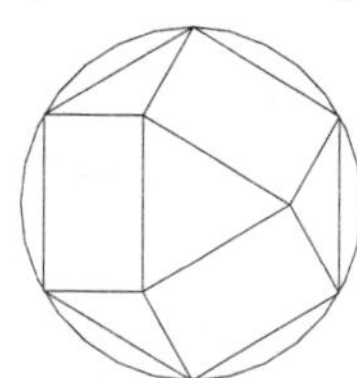
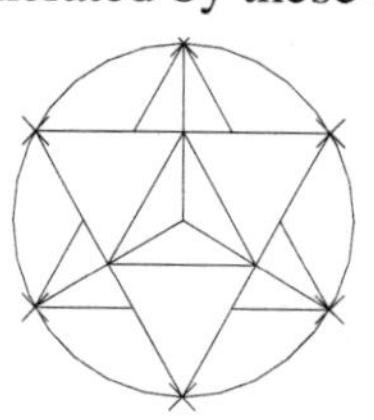
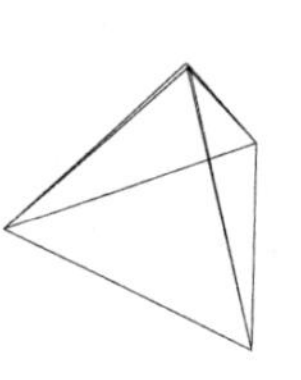
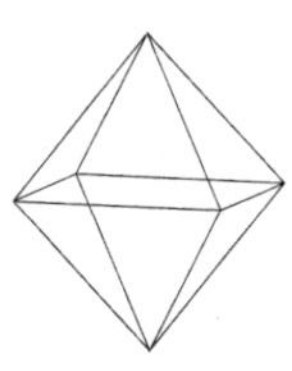
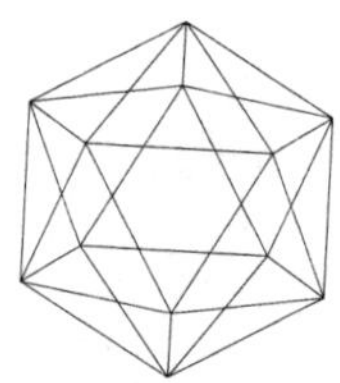

The 5-pointed star; a common symbol in society, can also be made with just compass and ruler. Not surprisingly, given all of the phi relationships in the five pointed star, we will use phi for it's construction. More interesting is that the elusive 7-pointed star can also be constructed using phi!

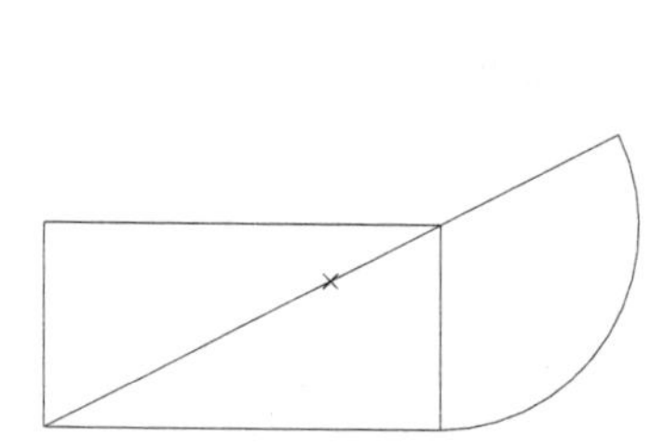
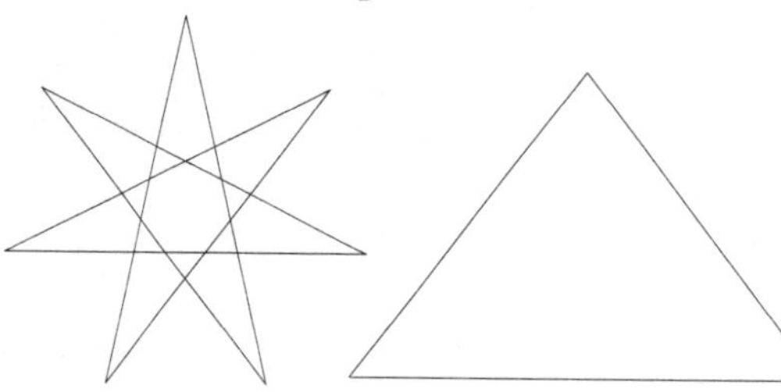
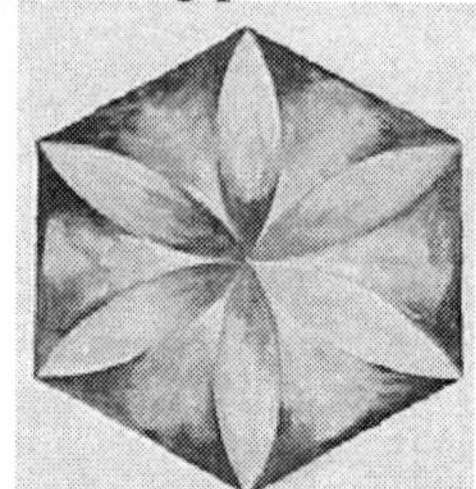

The ease with which all of this work can be done, and the accurate, beautiful pictures are a great surprise for Math students and a good return from long vacations, when the winter doesn't seem to want to end, or a lot of people are absent and it is difficult to maintain continuity.

Conclusion

We have worked through key steps companion to trigonometry in the second or third year of High School. We can pull in students with historical, artistic, social and athletic intelligence who don't normally do well in mathematics. Improvements in tests and overall effort normally exceed one letter grade. Students usually gifted in mathematics, have their characters broadened by such work. Still to be done; develop the 9-division of the circle!

References

[1]Plato, The *dialogues of Timaeus and Critias.*
[2]Michael Schneider, *A beginner's guide to constructing the Universe*, HarperCollins, New York, 1994
[3]Alexander Thom, *Megalithic Sites in Britain*, Oxford University Press, London, 1967
[4]Gerald S. Hawkins, *Stonehenge Decoded*, Doubleday and Company, New York, 1965
[5]W. M. Flinders Petrie, *The Pyramids and Temples of Gizeh*, 1883
[6]Robert Lawlor, *Sacred Geometry*, Philosophy and Practise, 1982

Mathematical Book Forms for Teachers

Susan Happersett
249 4th street
Jersey City, NJ 07302, USA
E-mail: fibonaccisusan@yahoo.com

The sequential properties of basic mathematics facilitate the creation of math art book forms. This workshop presents three artists book forms with mathematical significance for school teachers. Scissors, glue stick, protractor, straight edge and pre-cut paper are the only required equipment to make all three forms.

Book Form 1: Single Sheet Instant Books
Book Form 2: Star Books
Book Form 3: Trihexaflexagons

These basic forms use elementary geometric knowledge, while letting Student explore creative ways to use mathematical information in adding the content to the pages or faces of these forms.

Book Form 1: Single Sheet Instant Books

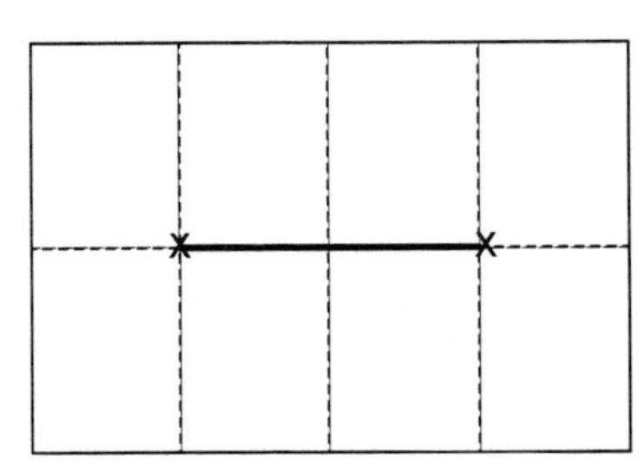

Figure 1

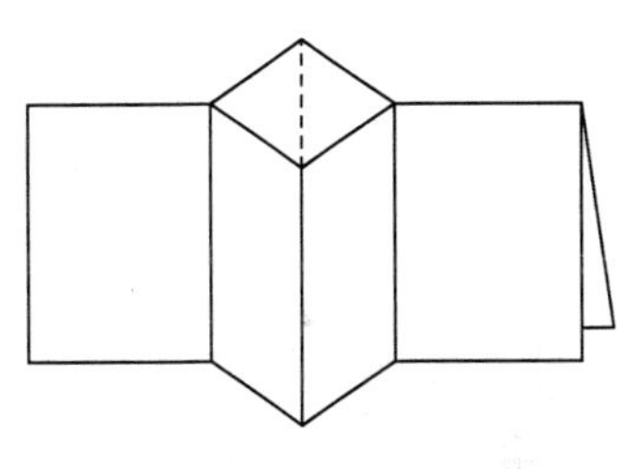

Figure 2

The basis for this book is a single sheet of photocopying paper (8.5" x 11"). The great thing about these books is that you only need to have images on one side of the paper to create a book with 2-sided pages. Start by folding the paper in half, then each half in half again, like you were making a fan. Then unfold and fold in half the other way, lengthwise. See figure 1. Unfold again and make a slit in the last fold between the two X's in figure 1. To assemble the book, fold lengthwise again, open up the slit and pinch closed, perpendicular to the original slit. See figure 2. Now, fold all the pages together, and you will end up with an 8-page book. This form works well for books about counting, fractions and other basic concepts. Since it is one-sided, it is perfect for making large editions using photocopied images.

Book form 2: Star Books

Each book requires at least 4 sheets of square paper. Take one sheet and fold it in half. Unfold, and then fold in half in the other direction. Make sure both folds are going the same way (both "hills"). See figure 3. Unfold again. Now fold diagonally, creating a valley fold. Unfold your square. It should now look like figure 4.

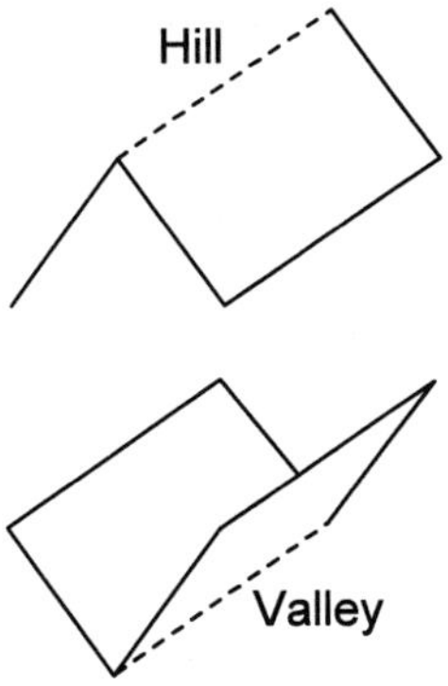

Hill

Valley

Figure 3

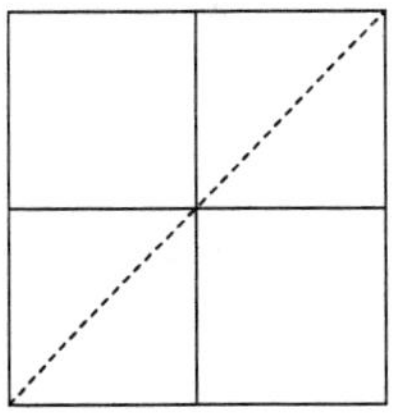

Figure 4

Pinch in at the diagonal valley fold and flatten to form a small square. This is the basic component of the book. Repeat this process for all other sheets of paper. When all of the square components are complete, they are glued together, folded point to folded point, one on top of the other to make a stack. When the glue is dry, the outside corners can be opened to reveal a star. This type of pop-up book has great visual impact, because there are multiple elements each student can make one square that can be assembled into a group book as a class project. With a slight variation in gluing and more sheets of paper, this form can also be made into a snake book that can be displayed open along a wall.

Book Form 3: Trihexaflexagons

Martin Gardner defined Flexagons in Scientific American in 1956: 'Flexagons are paper polygons, folded from straight or crooked strips of paper, which have the fascinating property of changing their faces when they are "flexed".' [1] Garner attributes the discovery of Flexagons to Arthur Stone, an English graduate student at Princeton in 1939. Trihexaflexagons are shaped like hexagons and have three distinct faces. The construction of Trihexaflexagons requires more accurate folding and can be frustrating for someone too young. I start with a strip of paper in a 1 to 7 ratio, and then draw a line with 30 degree angle from the corner. See figure 5. Cut off the little right triangle. Now you are ready to fold: bring the corner point up to the top edge and create the first fold to make an equilateral triangle. See figure 6.

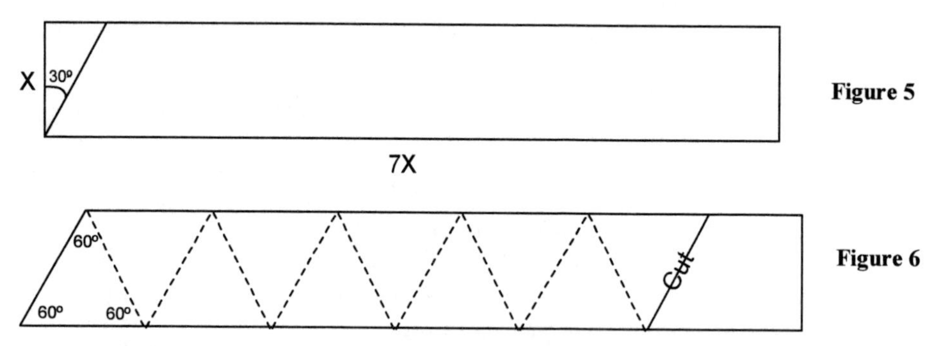

Figure 5

Figure 6

Continue folding, under and over, like a fan. Bring the corner to the edge, folding carefully, so that each time you make a new equilateral triangle. When you have 10 equilateral triangles, stop folding. You should have some paper left over, cut that off. Right now you should have a stack of folded triangles. Take the last triangle, wrap it around, and glue it to the first triangle. Once the glue dries you can flatten out your Trihexaflexagon. You can change the face on your Trihexaflexagon by pinching in towards the center and then re-opening on another fold. Decorating the three faces of the Trihexaflexagon is the fun part. Seeing how the faces change can be an interesting way to explore equilateral triangles and to discuss symmetry.

Assembling these forms corresponds with Van Hiele's levels of understanding geometry [2]. The 2nd level, analysis, relates to identifying the properties of rectangles within rectangles in Form 1. The 3rd level, abstraction, relates to the relationship between the square and the triangles in Form 2 and the interaction between the equilateral triangles in Form 3. Folding flat sheets of paper into 3D shapes helps visualize the relationship between 2D and 3D geometry. Several topics could be addressed: the study of scale and surface area (Form 1) and the study of angles (Forms 2 and 3).

References

[1] Martin Garner, *Hexaflexagons and Other Mathematical Diversions*, University of Chicago Press, Chicago, 1959

[2] P.M. van Hiele, *Structure and Insight. A Theory of Mathematics Education*, Academic Press, Orlando, 1986

The Arête of Line Designs

Michael Round
educator: Center for *auto*SOCRATIC Excellence
USA Director: Theory of Constraints for Education
13234 Long Street, Overland Park, Kansas USA 66213
round@rationalsys.com

Abstract

This workshop will explore the historical, philosophical, and pedagogical nature of line designs, with a focus on good designs and what constitutes the proper context and good environment ensuring "joy in work" is realized, now and in the future.

Math curriculum frequently includes line design lesson plans, a careful structure on how to achieve a desired result, and a method to grade results. To ensure relevance to a topic, many line designs are crafted to mimic some picture, such as a face or a valentine. Further "enhancements" include many java-enabled web sites affording ease of creation of these designs. As this integration of line designs increases in math curriculum, it's instructive to consider what it is about these designs we consider "good".

Mary Everest Boole, wife of famed mathematician George Boole and known by many as the origin of this type of activity, said the following regarding this process:

> *"The beauty of some of the designs is unquestionable; and there can be no second opinion about the value of the method, as training, from the point of view of geometry as well as from that of art. What is not quite so obvious at first sight is its bearing on the training of the unconscious mind for science. Without the slightest intellectual strain it puts the children through that normal sequence of orderly attention to classification and detail, interspersed with nodal points of synthesis, which may be called the very breathing-rhythm of the scientific discoverer.*
>
> *But to make this exercise of any use there must be no copying from diagrams; the value of it depends on the child evoking a curve, watching it growing, under his fingers, from mere obedience to a law ... and beauty has resulted, not from understanding but from obedience ... the act of evoking a curve 'out of the everywhere into here', by simple obedience to a rhythmic law, lodges an impression on the unconscious mind which will be ready to surge up in ten years' time." [1]*

Clearly, what we as adults consider "good" qualities are necessary conditions for a good activity, but are they sufficient? Ms. Boole addresses 'orderly attention' and 'classification of detail', leading to 'beauty' as the result, with a particular benefit the training of the mind for future excellence. How is this *process* captured in a rubric concentrating on *product*? To emphasize this point, Edith Somervell said the following about the "process versus product" dilemma:

> *"Beautiful curves are produced by a process so simple and automatic that the most inartistic child can succeed in generating beauty by mere conscientious accuracy; and the habit of doing this tends to produce a keen feeling for line. It has also been noticed in*

some cases, where clean, pure, and strong colour has been used, that a remarkable sensitiveness to colour relation has grown." "The results obtained by a child, of exquisite curved and flower forms on the 'back' of his card, by faithful obedience to a dull little rule in making straight stitches on the 'front', is of the nature a miracle. It should, therefore, be hardly necessary to insist that the less said the better, when the little worker produces anything especially beautiful or unexpected." [2]

What as adults can we do regarding the working environment to ensure a good process *with* beautiful results? What types of designs *are* appropriate for children? How do we translate "dull little rule" into practice? And finally, if the method is so powerful we can see the impression years down the road, how can we ensure the children continue creating such designs *outside* the classroom?

Figure 1: Results of the Prepared Environment

To ensure the integrity of the environment for quality work, certain conditions must be met. For example, "mental perfection" in the design-layout process requires exact calculations. A circle consisting of 16 equally spaced points requires angle measures differing by 22.5 degrees. Not seeing that *exact* measure on the protractor, the child knows the design to be mentally flawed. A necessary condition for the good environment, then, is the creation of "mentally perfect" designs. What are other necessary conditions?

This teacher workshop addresses many of these issues as the result of a number of after-school clubs I've conducted over the past couple of years, integrating the "prepared environment" philosophy of Maria Montessori [3] with the idea of flow [4]. Current experiments in the working environment, including the introduction of background percussion music to highlight the rhythmic nature of the activities, and various materials demonstrating the dynamic quality of these patterns, will be explored as well.

[1] Mary Everest Boole, *Preparation of the Child for Sciences*. University of Oxford. 1904
[2] Edith L. Somervell, *A Rhythmic Approach to Mathematics*. George Philip & Son, Ltd, London. 1906
[3] E.M. Standing, *Maria Montessori: Her Life and Work*. Penguin Books USA. 1984
[4] Mihaly Csikszentmihalyi, *Flow: The Psychology of Optimal Experience*. Harper & Row Publishers, New York. 1990
[5] Robert M. Pirsig, *Zen and the Art of Motorcycle Maintenance*. William Morrow and Company, Inc. 1974
[6] W. Edwards Deming, *The New Economics*. MIT. 1993

The Plato Bead—A Bead Dodecahedron

Laura Shea
7682 E. Windcrest Row
Parker, Colorado 80134
E-mail: dancingrainbow@comcast.net

Abstract

Creating polyhedra with beads is another way to learn the properties of regular and semi-regular solids. The instructions given below are for the dodecahedron (The Plato Bead). In a bead polyhedron each face becomes open space; each edge becomes one bead; each vertex becomes a thread void. The structure is light and open. The size of the overall bead changes with the size of beads used.

One of the five Platonic solids, the dodecahedron consists of 12 equilateral pentagon faces, 30 edges, and 20 vertices. Considered the Platonic symbol for the universe, this twelve-sided form may also represent time with each face being a month of the year or one of the twelve signs of the zodiac. In the bead polyhedron a bead stands in for each of the 30 edges. The 12 faces of the form become open spaces. Each of the vertices becomes a void surrounded by three beads and thread. Because our support medium is thread, each polygon face of the polyhedron becomes a softer shape, approximating a circle. The bead polyhedron more closely resembles a sphere than the geometrically angular dodecahedron but still retains many properties of the dodecahedron.

Beaded beads based on regular and semi-regular polyhedrons have been part of the largely undocumented bead lexicon in China for several hundred years. According to Valerie Hector's research, Chinese beadworkers call the dodecahedron *mei* or "plum blossoms". The Chinese often use a two strand thread path. The Plato Bead uses a one strand thread path.

This pattern uses two colors of beads "A" (white) and "B" (blue) to create the impression of alternating rows of color. (Two colors make it easier for the new beader to work and follow the pattern.) The text lists Row, Step, Bead # and Colors (A, B). Each circle or loop of beads completed in each step will contain a total of 5 beads (five edges of each pentagon face). Each bead will be shared by another circle. Three beads will surround each thread void. Take care to pull the monofilament through completely as it can kink inside a bead hole and release later, weakening the integrity of the piece. The bead polyhedron's stability depends on the bead holes being as full of thread as possible and reasonably tight tension.

Variations

Other transformations of the dodecahedral shape happen when more than one size of bead is used within one structure. Examples and patterns will be available in class. These transformations are not the traditional ones achieved with paper models.

Figure 1:
The Plato Bead.

Figure 2:
Dodecahedron Variations, bead size and color.

The Plato Bead

Row 1

Step 1: String 5 beads AAAAA. (Beads 1, 2, 3, 4, 5). Go through Bead 1 forming a circle. Leave at least a six-inch tail. The tail will be useful later to help hold onto the bead as it forms.

(**Tip:** Form a circle not a teardrop, the thread must be coming out the opposite side of Bead 1 from the tail.)

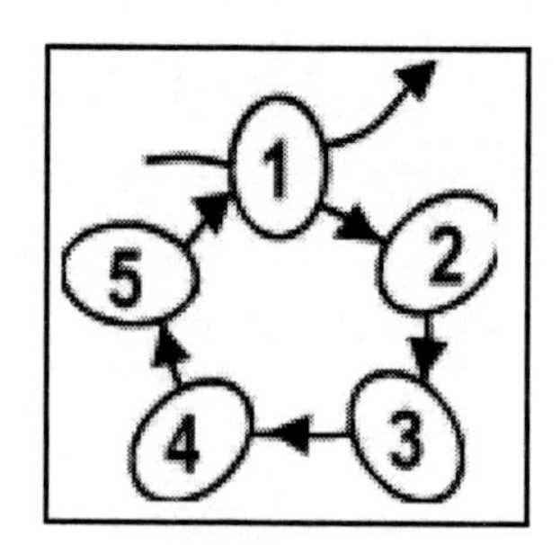

Figure 3:
Row 1 completed. Note the star shape.

Row 2

Step 2: String 4 beads BAAB.(Beads 6, 7, 8, 9). Go through Bead 1 (Row 1). Thread is in a void. Continue through Bead 2 (Row 1).

(**Tip:** A tendency of students is stop after going through Bead 1 and then add beads. Understanding where each forming void is, helps tremendously. TIIAV=Thread is in a void.)

Step 3: String 3 beads BAA. (Beads 10, 11, 12).Go through Bead 6 (mystery/ Elijah bead, Row 2) and Bead 2 (Row 1). TIIAV. Continue through Bead 3 (Row 1).

(**Tip:** When first learning this technique, students tend to forget to include the beads I call "mystery" or "Elijah" beads. At Passover it is custom to lay a place at the table for the prophet Elijah, the unexpected guest. MEB=mystery/Elijah bead.)

Step 4: String 3 beads BAA. (Beads 13, 14, 15). Go through Bead 10 (MEB, Row 2) and Bead 3 (Row 1). TIIAV. Continue through Bead 4 (Row 1).

Step 5: String 3 beads BAA. (Beads 16, 17, 18). Go through Bead 13 (MEB, Row 3) and Bead 4 (Row 1). TIIAV. Continue through Bead 5 (Row 1) and Bead 9 (Row 1).

Step 6: String 2 beads AA. (Beads 19, 20). Go through Bead 16 (MEB, Row 2), Bead 5 (Row 1), and Bead 9 (Row 2). TIIAV. Continue through Bead 8 (Row 2).

(**Tip:** at the end of end completed Row the partially completed bead will have a five-pointed shape.)

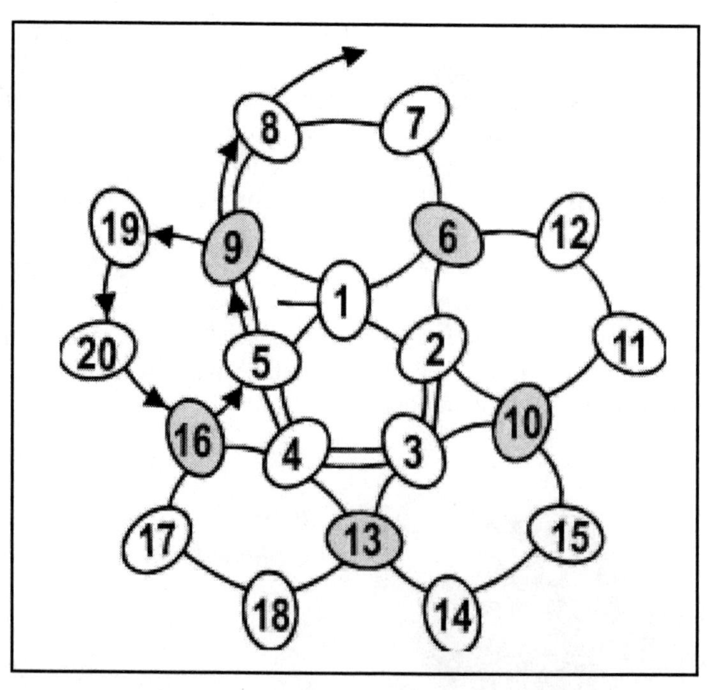

Figure 4: *Rows 1 & 2 completed. The star shape continues to repeat.*

Row 3

Step 7: String 3 beads BAB. (Beads 21, 22, 23). Go through Bead 19 (MEB, Row 2) and Bead 8 (Row 2). TIIAV. Continue through Bead 7 (Row 2) and Bead 12 (Row 2).

Step 8: String 2 beads BA. (Beads 24, 25). Go through Bead 21 (MEB, Row 3), Bead 7 (Row 2) and Bead 12 (Row 2). TIIAV. Continue through Bead 11 (Row 2) and Bead 15 (Row 2).

Step 9: String 2 beads BA. (Beads 26, 27). Go through Bead 24 (MEB, Row 3), Bead 11 (Row 2) and Bead 15 (Row 2). TIIAV. Continue through Bead 14 (Row 2) and Bead 18 (Row 2).

Step 10: String 2 beads BA. (Beads 28, 29). Go through Bead 26 (MEB, Row 3), Bead 14 (Row 2) and Bead 18 (Row 2). TIIAV. Continue through Bead 17 (Row 2) and Bead 20 (Row 2) and Bead 23 (Row 3).

Step 11: String 1 bead A (Bead 30, the final bead). Go through Bead 28 (MEB, Row 3), Bead 17 (Row 2), Bead 20 (Row 2) and Bead 23 (Row 3). TIIAV.

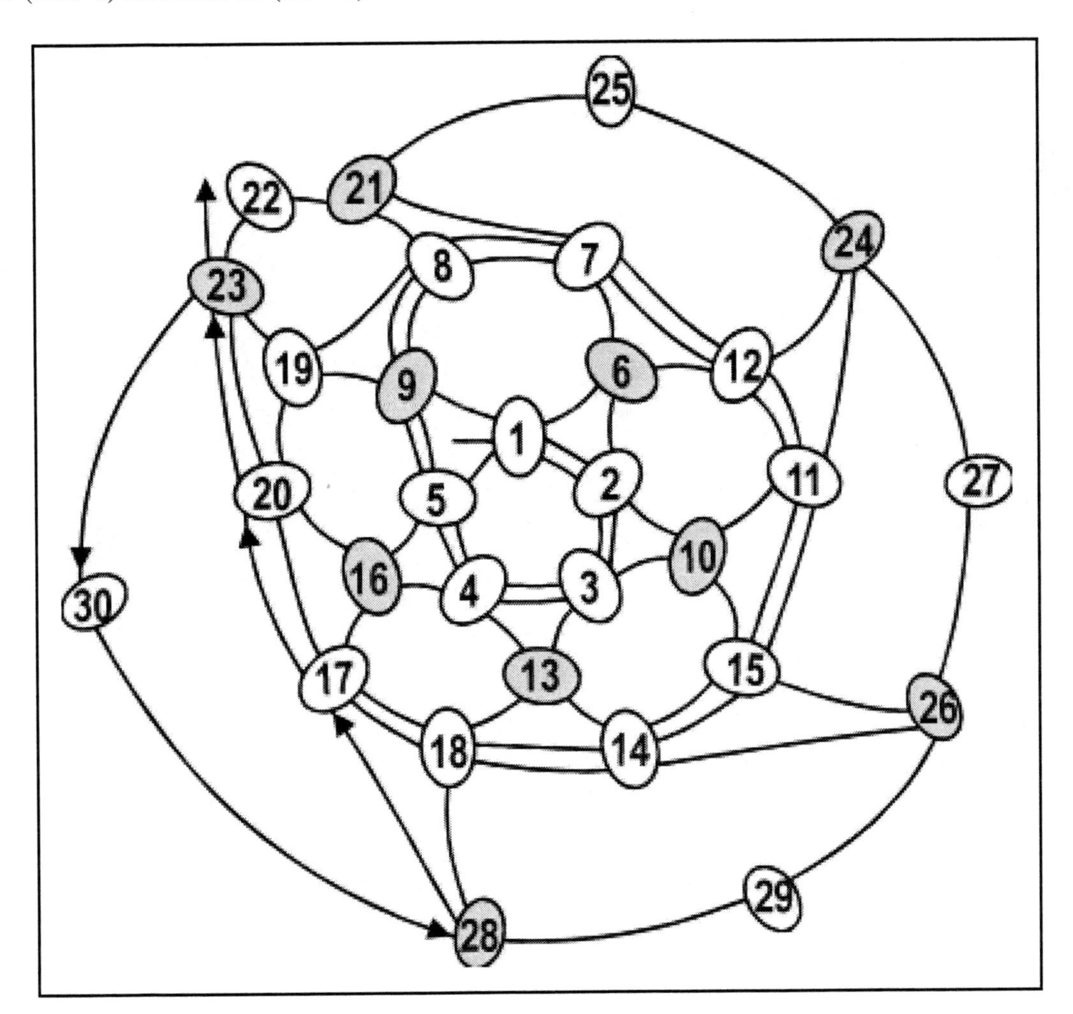

Figure 5: *Rows 1 & 2 & 3 completed.*

Row 4

Step 12: There are five remaining beads to connect. They should all be "A" beads. Continue through Beads 22, 25, 27, 29, and 30. Secure the thread by passing back through several circles, being careful not to cross any voids. Bury the beginning tail in the same way. The thread will lock itself if it passes through enough circles. Optional: knotting or gluing to secure the thread.

Secure the thread by passing back through several circles being careful not to cross a void.

Optional: additional knotting or gluing to secure thread.

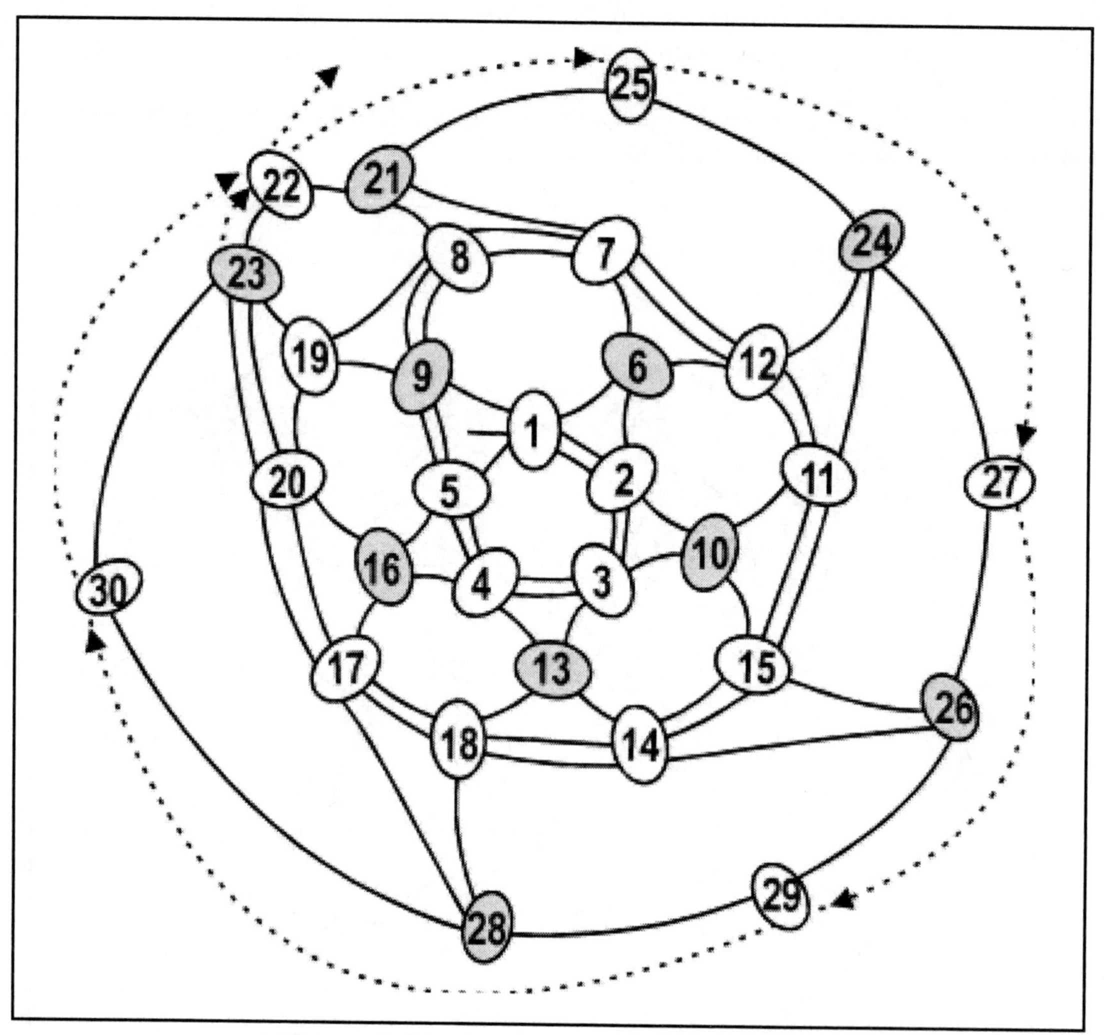

Figure 6: *The finished bead dodecohedron.*

Materials

This pattern for BTTB requires 30 faceted plastic beads (8 mm) and 2 yards of 15# test monofilament. In a class room situation or at home the beader will also need a small dish for the beads, scissors and sharp-pointed tweezers. Contact me at the above email for information about other bead/ thread combinations.

Building Simple and Not So Simple Stick Models

Robert McDermott
Center for High Performance Computing
University of Utah
Salt Lake City, Utah, 84112, USA
E-mail: mcdermott@chpc.utah.edu

Abstract

Physical models are invaluable for conveying concepts in geometry. In this paper, I explain how to build stick models based on the Platonic polyhedra. Supplies for these models were thin bamboo shish kebab sticks from a grocery store, and vinyl tubing from a hardware store; both supplies are inexpensive and readily available. The tools used were a ruler for measuring the length of sticks, a clipper to cut the sticks, a scissor to cut the tubing, and a punch to make holes in the tubing. These tools are also reasonably inexpensive and readily available. Grade school, high school, undergraduate and graduate level students have made models with these supplies and tools and all of them have taken away something meaningful related to their existing level of knowledge.

1. Introduction

Inspiration to produce these simple stick models came from Ron Resch, my Ph.D. thesis advisor, with his intriguing film entitled "The Paper and Stick Thing Film" [1]. My investigation into building stick models includes an interest in using them in the class room. With some preparation of materials before the class, these models can be built in class. The supplies are widely available and inexpensive, and the tools exist in most home tool boxes.

It is important to think of a physical model of a regular polyhedron as a "sketch" of the concept for that polyhedron. There is perfection associated with a regular polyhedron which makes it impossible to achieve with any physical materials. Failure to meet perfection is true no matter how precise the physical pieces are for building the model. Consequently, the less exact the pieces become, the more of a "sketch" the model becomes. However, these "sketches" support the concept and are useful examples for discussing many geometric relationships.

My models are all based on the Platonic solids. They serve as a significant foundation for the study of three-dimensional space [2]. The better a student understands these models, the better that student is able to understand and imagine more sophisticated concepts about three-dimensional space. Additional understanding of these models is achieved by adding the sense of touch to the sense of sight. A student who is able to handle these models, reinforces their knowledge gained through sight. The size of the models is also important. It has been studied by J.J. Gibson, [3] that a model which is held in the hands provides insight. A model that is approximately a handful, is better than one that is considerably smaller or considerably larger [3]. A model the size of a students hand can be quickly rotated providing many views. A geometric relationship that exists in many places in a model can be seen nearly simultaneously. This multiplicity of views reinforces both the local and global nature of geometric relationships.

Regular polyhedron models have a very simple definition. A regular model consists entirely of a single-sized regular polygon, whose edges are all the same length. All vertices of a regular solid have the same number of regular polygons or edges meeting at that vertex. The number of edges meeting at a vertex is the ***degree*** of that vertex. Each vertex looks the same as every other vertex of a regular solid. Therefore, once a single vertex is made for a model the remaining vertices of that model are all the same. This condition implies that such vertices have equal solid angles. In Coxeter's Regular Polytopes [4], page 15, it states that solids with regular faces and regular solid angles are regular solids.

The remainder of this paper provides a list of supplies, tools, and directions for constructing stick models related to Platonic solids. The models serve to highlight the Platonic solids, their mid-edges, their mid-faces and their mid-cell, as well as, provide views of their dual polyhedron.

2. Supplies and Tools

Supplies for these models consist of thin bamboo shish kebab skewers and three sizes of vinyl tubing. Sticks were purchased from a variety of supermarkets, and come in a standard length of approximately 10 inches. Any length of stick in this vicinity can serve to build these models. The vinyl tubing can be purchased from a variety of hardware stores. The size of the tubing used relates to the ***degree*** of the vertices for the model. Tubing measuring ¼" is used for ***degree*** three vertices, with ½" tubing used for ***degree*** four vertices and ¾" tubing used for ***degree*** five and ***degree*** six vertices.

A ruler is used for measuring the sticks. A clipper is used to cut the sticks. A scissor is use to cut the tubing and a punch is used to make holes in the tubing. Once a stick is cut to length it becomes an ***edge*** in a model and is referred to as an ***edge*** for the remainder of this paper. The strength of your grip comes into play when cutting the sticks. Different people may well decide on different tools for this repetitive task. A scissor with blades of approximately 4" will successfully cut the vinyl tubing. The tubing is cut producing a ring of tubing approximately ¼" in length. For the remainder of this paper this ring of tubing will be referred to as a ***vertex***. A leather punch with a single #6 size punch can be found at a crafts store. It is used to make holes in the vinyl tubing, so that ***edges*** easily insert into the holes to create a ***vertex***. It is best if the size of the hole and the size of the stick are closely matched so that the tubing will serve to hold the sticks firmly, and do not easily slide out of the hole. Also the size matching allows the stick to be easily inserted into the ring.

3. Platonic Solids, High-Lighting Their Mid-Edges, Mid-Faces, Mid-Cells and their Duals.

3.1. Tetrahedron. I started my stick models with the most basic of regular Platonic solids, the tetrahedron. I cut 6 sticks 8 3/4" long for the six ***edges*** and cut four pieces of ¼" tubing for the 4 ***degree*** 3 ***vertices***. I used the punch to make three evenly spaced holes around the ring of tubing at approximate 120 ***degree*** separation between the holes in each of these 3 ***vertices***. With these ***edges*** and ***vertices***, I assembled a model. This model, although very simple, served to establish an approach for building more complex models. I started by inserting three ***edges*** into one ***vertex*** **Figure 1a**. On the end of each of these three ***edges*** I placed a ***vertex*** **Figure 1b**. An ***edge*** was inserted between two of these three ***vertices***, **Figure 1c**. I inserted another ***edge*** into the remaining hole of one of those ***vertices***, and placed the other end of that ***edge*** into a hole of the remaining ***vertex*** **Figure 1d**. Now only one ***edge*** remained and two holes remained in two ***vertices***. When this final ***edge*** was inserted into the final two holes of those ***vertices*** a tetrahedron was formed **Figure 1e**.

Note: A finished tetrahedron can be adjusted to more closely approximate the ideal of a regular tetrahedron. Take a close look at each of the ***vertices*** to adjust the position of the ***edges*** within the hole of the ***vertex*** so that the ends of these ***edges*** closely resemble edges that meet at a theoretical point in the center of this ***vertex*** **Figure 1a**. Repeat this observing and adjusting of the remaining ***vertices***, and your model will have an appearance closer to the ideal.

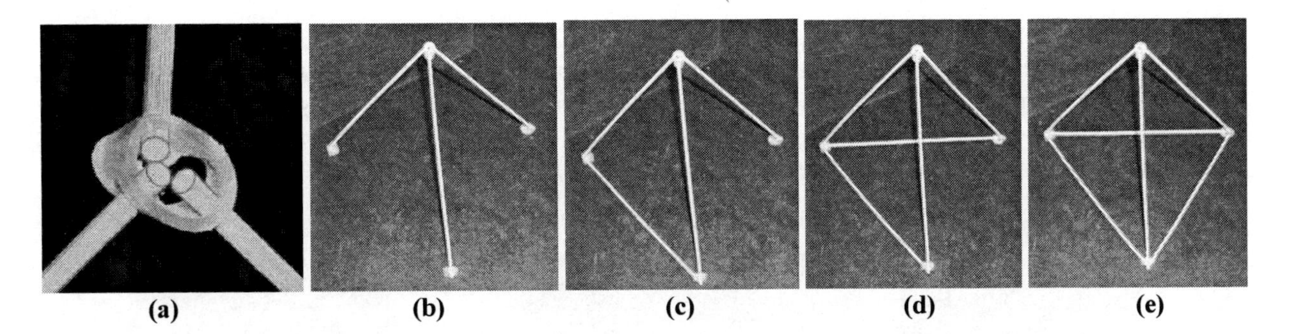
(a) (b) (c) (d) (e)

Figure 1: 1 ***vertex*** **(a)**, 3 ***edges*** **(b)**, 4 ***edges*** **(c)**, 5 ***edges*** **(d)**, a tetrahedron **(e)**.

3.2. Mid-Edge Tetrahedron. The next model built drew the viewer's attention to the mid-edge of the tetrahedron. I cut 6 11½" ***edges*** and 4 ***degree*** 3 ***vertices***. In addition, 12 5¾" ***edges*** were cut, as well 6 ***degree*** 6 ***vertices*** were cut and punched from ¾' tubing. Before assembling the tetrahedron, each ***edge*** of the tetrahedron had one of the ***degree*** 6 ***vertices*** slide onto each ***edge*** so that the ***vertex*** was positioned near the middle of the ***edge***, **Figure 2a**. Holes on opposite sides of each ***vertex*** were used with an 11½" ***edge***. Once each ***edge*** had a ***vertex*** positioned near its mid-point a tetrahedron was assembled. The ***half-edges*** were inserted into the holes of the mid-edge ***vertices*** **Figure 2b.** An octahedron is formed with these 12 ***edges*** **Figure 2c.**

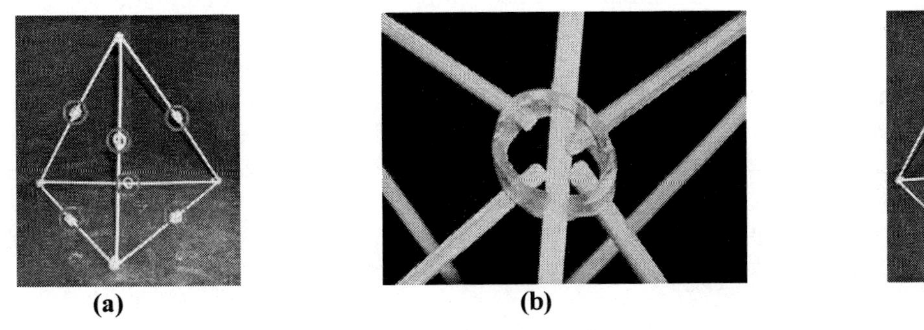
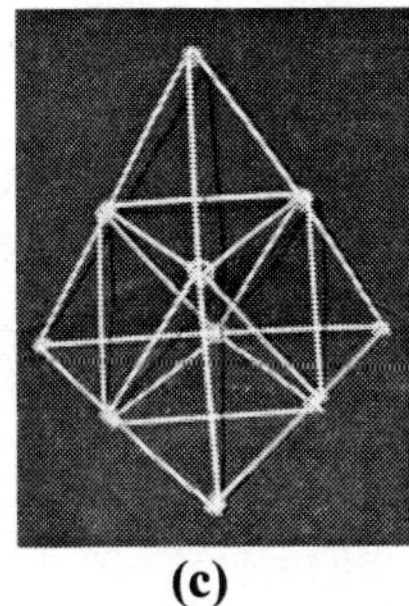
(a) (b) (c)

Figure 2: vertex at mid-edge **(a)**, ***vertex***, full-***edge*** & half-***edges*** **(b)**, tetrahedron & octahedron **(c)**

3.3. Mid-Face Tetrahedron. The next model drew attention to the ***mid-face*** of the tetrahedron. This model consisted of 6 11¼" ***edges***, 12 6½" ***edges***, 6 3¾" ***edges*** and 8 ***degree*** 6 ***vertices***. Four of the 8 ***vertices*** were combined with the 6 shorter ***edges*** to form a tetrahedron as seen in the center of **Figure 3a**. These ***edges*** will use every other hole in the ***degree*** 6 ***vertices***. Each of the 12 mid-length ***edges*** will have one of their ends inserted into the open holes in the 4 ***degree*** 6 ***vertices*** of the recently assembled smaller tetrahedron. These ***edges*** will be collected into groups of three and then inserted into the remaining ***degree*** 6 ***vertices***. Finally, the long ***edges*** were inserted into the open holes of the 4 ***degree*** 6 vertices to form an outside tetrahedron **Figure 3a.**

3.4. Mid-Cell Tetrahedron. The next model drew the viewer's attention to the mid-cell of the tetrahedron. This model had 6 9 ¼" ***edges***, 4 6" ***edges***, 5 ***degree*** 4 ***vertices***. The 6 longer ***edges*** and 4 ***degree*** 4 ***vertices*** were assembled into a tetrahedron **Figure 3b**. The remaining 4 shorter ***edges*** would have one end inserted into each of the 4 ***vertices*** of this tetrahedron. The 4 free ends of these shorter ***edges*** were inserted into the remaining ***degree*** 4 ***vertex.***

3.5. Dual Tetrahedron. For this model 6 11" ***edges*** and 24 5½" ***edges*** were cut with 8 ***degree*** 3 ***vertices*** and 6 ***degree*** 8 ***vertices***. This assembly is very similar to the Mid-Edge tetrahedron assembly previously.

On 6 of the longer ***edges*** place a ***degree*** 8 ***vertex*** near the mid-edge **Figure 3c**. These 6 ***edges*** were assembled into a tetrahedron with 6 ***degree*** 3 ***vertices***. With 12 of the shorter ***edges*** were inserted into the ***degree*** 8 ***vertices*** as in ***Figure*** *2c*. The remaining 12 shorter ***edges*** will be inserted into the 4 ***degree*** 3 ***vertices*** in groups of three. Their free ends were inserted into the ***degree*** 8 ***vertices*** to form a second tetrahedron that was the same size as the initial tetrahedron for this model.

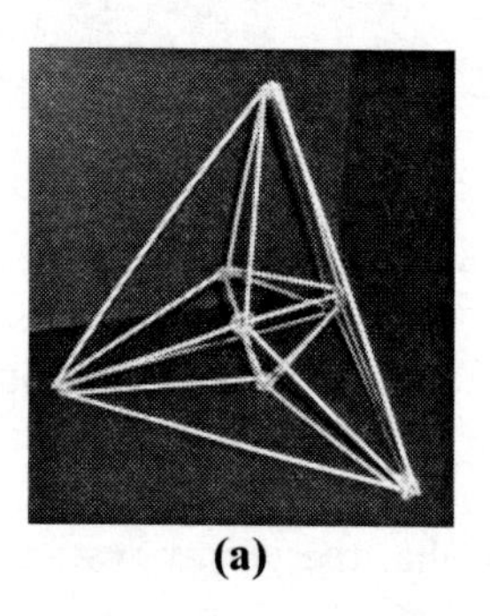
(a)

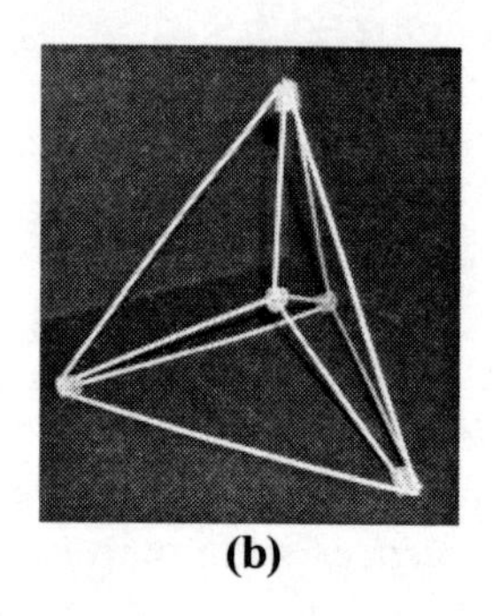
(b)

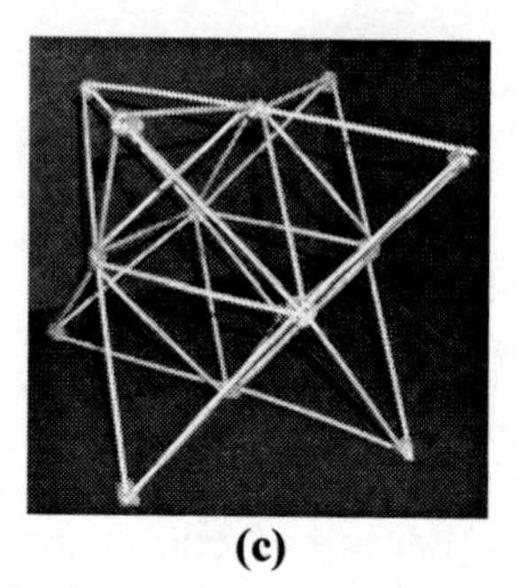
(c)

Figure 3: mid-points of faces **(a)**, mid-point of fcell **(b)**, stella octangula **(c)**

When a single tetrahedron is joined together with a second tetrahedron that is identical in size in the manner described above a stella octangula is formed. This model is also the stellation of a regular octahedron. A stellation is formed when non-adjacent faces are extended so that they intersect each other and the edges resulting form star polygons as in Coxeter [4] Section 6.2 Page 96.

3.6. A Tetrahedron Inside a Cube and that Cube Inside a Dodecahedron. For this model there were 6 6 7/8" ***edges*** with 4 ***degree*** 9 ***vertices*** to build an initial tetrahedron. Use every third hole in these 4 ***vertices.*** Subsequently, 12 5 1/8" ***edges*** were cut with 4 ***degree*** 6 ***vertices.*** Three of the shorter ***edges*** were inserted into each of the ***degree*** 3 ***vertices*** forming a corner **Figure 4a**. Each of the three free ends of this corner were inserted into every third open holes of the 4 ***degree*** 9 ***vertices*** on one face of the tetrahedron **Figure 4b.** When this model is completed there is a cube outside of a tetrahedron.

To form a dodecahedron outside this cube 30 2 7/8" ***edges*** were cut along with 12 ***degree*** 3 ***vertices.*** Five of these shorter ***edges*** and two ***degree*** 3 ***vertices*** formed a ***winged edge*** assembly **Figure 4c**. The four free ends of this ***winged edge*** assembly were inserted into open holes on the four ***vertices*** on one ***face*** of the cube. The central ***edge*** of the ***winged edge*** assembly needed to alternate its direction so that there are five of these shorter ***edges*** surrounding a single ***edge*** of the cube **Figure 4d.** These five shorter ***edges*** combine to form a pentagon and the twelve ***edges*** of the cube are inside twelve pentagons that form a regular dodecahedron outside the cube **Figure 4d.**

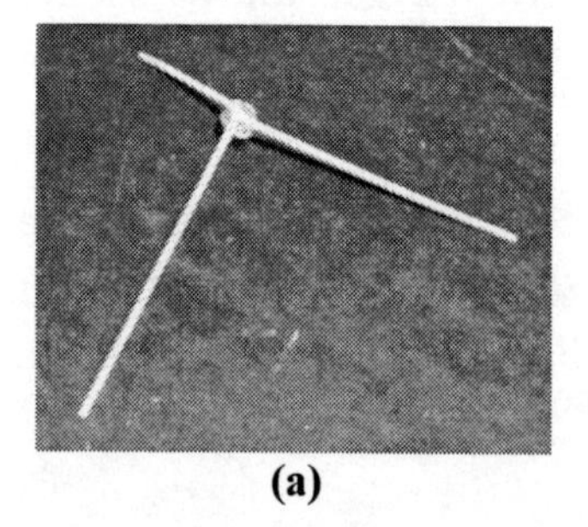
(a)

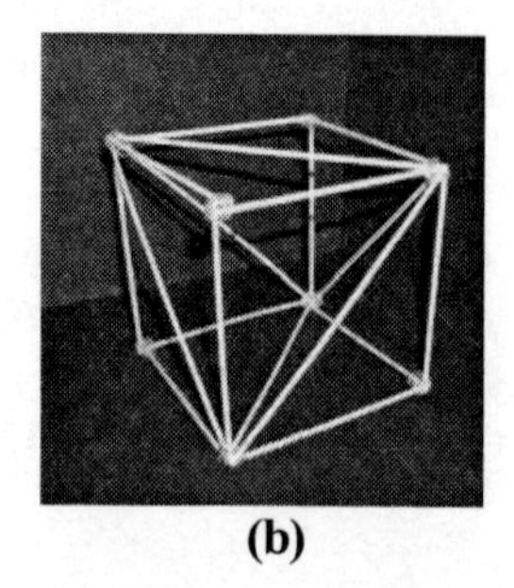
(b)

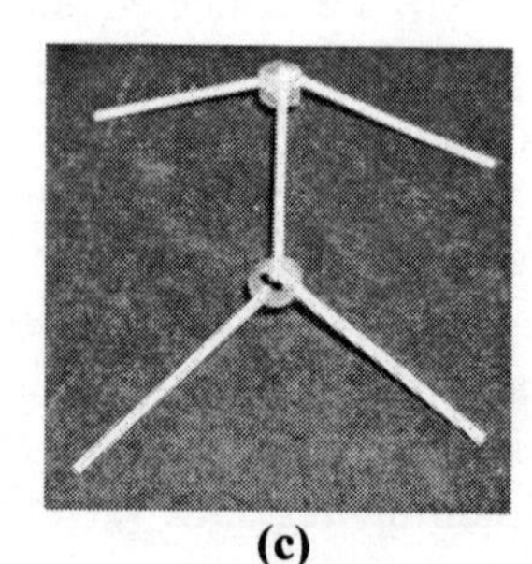
(c)

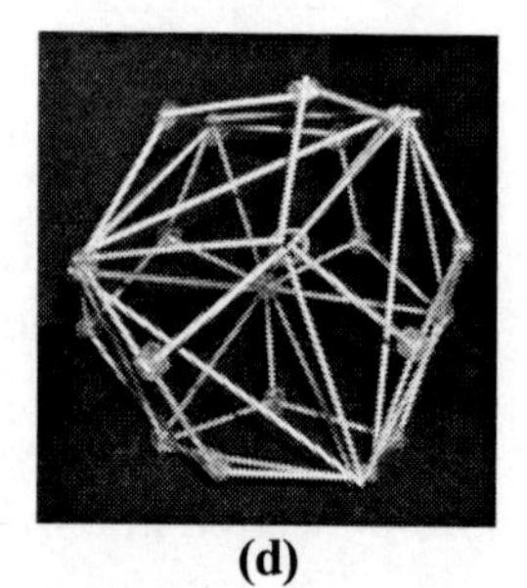
(d)

Figure 4: cube corner **(a)**, cube outside **(b)**, winged edged **(c)**, dodecahedron outside **(d)**

3.7. Octahedron, Icosahedron, Tetrahedron with octahedron and icosahedron inside. Building an octahedron with 12 ***edges*** and 6 ***drgree 4 vertices Figure* 5a** and an icosahedron with 30 ***edges*** and 12 ***degree*** 5 ***vertices*** **Figure 5b** were very similar to building the initial tetrahedron. Taking the model from **Figure 2c** with the octahedron inside the tetrahedron an additional polyhedron was built inside this octahedron. For the icosahedron 12 ***degree*** 6 ***vertices*** and 30 3 5/16" ***edges*** cut. For this model the ***degree*** 6 ***vertices*** will be placed on an edge of the octahedron but not at its mid-edge. This ***vertex*** will be placed at a location that divides the ***edge*** into the golden ratio (approximately 3½" and 2¼" for the 5¾" ***edge***). Within each of the 8 ***faces*** of the octahedron a ***face*** of an icosahedron was formed. Near each of the 6 ***vertices*** of the octahedron 2 ***faces*** of the icosahedron were formed for a 20 face icosahedron **Figure 5c.**

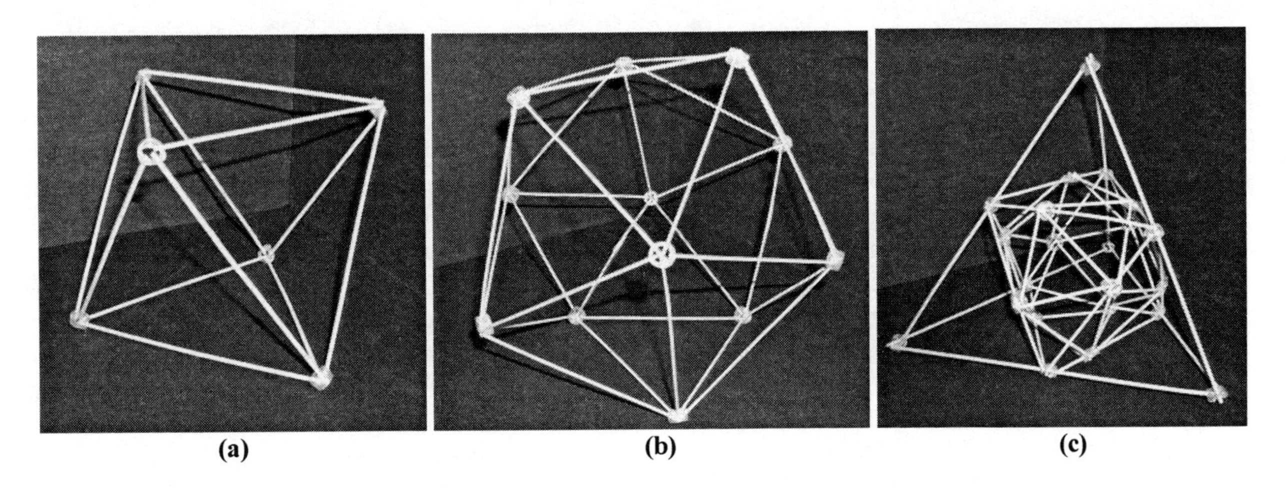

(a) **(b)** **(c)**

Figure 5: octahedron **(a)**, icosahedron **(b)**, tetrahedron , octahedron, icosahedron **(c)**

3.8. Stellated Dodecahedron & Stellated Icosahedron. Just for fun two stellations were built for a dodecahedron and an icosahedron **Figure 6**. These models were spatially and visually quite interesting as well as being less simple to build from shish kebob stick for ***edges*** and holes punched in vinyl tubing rings for ***vertices.*** Each has 30 ***edges,*** where one has 12 ***degree*** 5 ***vertices*** and the other has 20 ***degree*** 3 ***vertices.*** These models were assembled starting with a single ***vertex*** and observing star polygons.

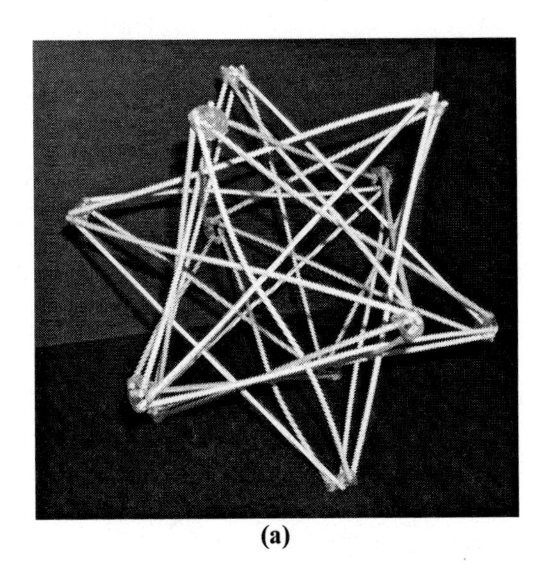

(a)

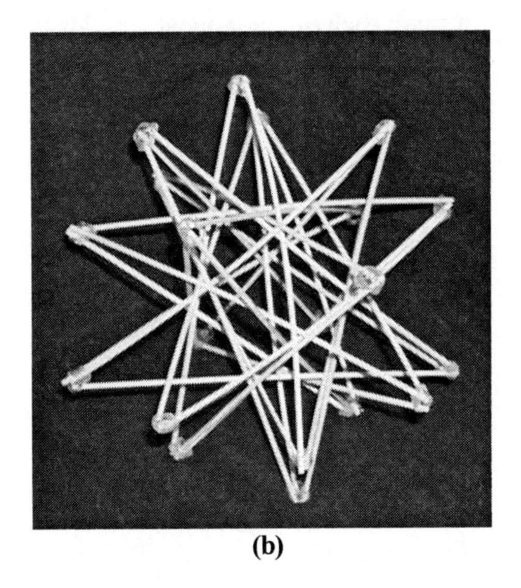

(b)

Figure 6: stellated dodecahedron **(a)**, stellated icosahedron **(b)**

4. Handling Stick Models

Each of the models in this paper has multiple axes of symmetry. The presence of so much symmetry leads the students into spending a considerable amount of time with the handling of these models. Using index fingers to help orient these models aligns the different axes of symmetry for vertices, edges and faces of each model to more clearly observe the symmetry. These models can have a vertex, an edge or a face set on a flat surface to study them individually. Some models can have a set of vertices set on a flat surface and studied from this orientation. All in all, a long time can be spent with this set of models observing them individually and using them as examples for other concepts to be studied by students at all levels of their education.

5. Conclusion

I have presented these simple stick models to students ranging from elementary school to graduate school for a period of more than twenty years. These presentations have always been well received by both the students and their teachers. The teachers have shown their appreciation by wanting to keep the models to hang in their classrooms and serve as constant examples for future study. Recently, I presented the models to eighth grade students at the McGillis School in Salt Lake City, Utah. This class has fifteen students interested in learning and enthusiastic about a presentation that includes models. They listened to my words about each model, but they were most interested in their opportunity to handle the models. I have been giving presentations to this group of students since they were in pre-school as our daughter is a member of this class.

Each group of students, depending on their individual maturity and their educational experience, takes away something different, even though the materials are exactly the same and the words used are very much the same. Emphasizing the tactile experience of handling each of the models provides the students with a recognizably different experience than merely viewing three-dimensional images of the regular solids on a piece of paper or on a computer screen

With the stick models, as opposed to the paper models [2], the students think of themselves as having x-ray vision, seeing from the near side to the far side of the models. Also, students can visually and physically align geometric relationships from both the near side of the model and the far side as they are study each of the models individually.

Acknowledgements

I would like to thank, my son, Colin, for photographing the models that appear in this paper. I would also like to thank, my spouse, Deborah, for proof reading drafts of this paper.

References

[1] Resch, Ronald D., Paper and Stick Thing Film, on a DVD available at RonResch.com, 2004.
[2] McDermott, R. J., A Physical Proof for Five and Only Five Regular Solids, Bridges 2005.
[3] Gibson, J.J., The Senses Considered as a Perceptual System, 1966, Reprinted in 1983 by Greenwood Press, West Port Conn 1983.
[4] Coxeter, H.S.M., Regular Polytopes, Third Edition, Dover Publications, Inc., NY, 1973.

Topological Mesh Modeling

Ergun Akleman
Visualization Sciences Program,
Department of Architecture
Texas A&M University
College Station, Texas, USA

Vinod Srinivasan
Visualization Sciences Program,
Department of Architecture
Texas A&M University
College Station, Texas, USA

Abstract

This workshop presents Topological Mesh Modeling with hand-on experiments using our topological modeler, TopMod. Our modeler provides a wide variety of interactive techniques that allow to create unusual and interesting shapes by changing the topology of 2-manifold meshes.

1. Introduction

This workshop presents Topological Mesh Modeling with hands-on experiences using TopMod, a manifold mesh modeling system that includes most of the work presented in our Topological Mesh Modeling papers. We provide TopMod to all the audience and we have interactive demonstrations.

TopMod allows interactively changing the topology of 2-manifold polygonal meshes. It guarantees topological consistency of polygonal meshes. TopMod is based on minimal and complete operations: insert/delete edge and create/remove vertex. Using these operations, handles can be created and deleted, holes can be opened or closed, polygonal meshes can be connected or disconnected. These minimal operations are also highly consistent with subdivision algorithms. In particular, these operations can easily be included into a subdivision modeling system such that the topological changes and subdivision operations can be performed alternatively during model construction.

TopMod provides wide variety of ways to create high genus shapes such as rind modeling, curved handles, connected and manifold Sierpinsky polyhedra, wire and column modeling. It also provides probably the most complete set of subdivision schemes available. It's subdivision categorization helps to understand mesh topological properties of remeshing schemes of subdivision surfaces. Figure 1 shows some models created using TopMod and printed with a Fused Deposition Machine in 3D. For more information about TopMod see Topological Mesh Modeling page:
http://www-viz.tamu.edu/faculty/ergun/research/topology/

2. Intended Audience and Presentation Requirements

The course is intended for all Bridges attendees who are interested in the creation of interesting shapes and sculptures; and teaching how to create those sculptures. The software is also useful to teach topology to students. We provide software to all participants and the participants who bring their laptops have hands-on experience. We encourage the participants to bring their laptop. They can also download TopMod from our web-page before workshop: http://www-viz.tamu.edu/faculty/ergun/research/topology/index.html

Figure 1: 3D prints of the models created in TopMod. For each sculpture, we took two photographs from slightly different point of views. The sculptures are photographed on a mirror. Background is eliminated. These shapes are made from ABS plastic and printed using a Fused Deposition Machine (FDM). They are later painted using an acrilic paint. The ribbon shape on the left is really a genus-1 surface, but, during the construction of this surface, genus changed several times.

3. Workshop Schedule

The workshop will cover how to use TopMod while intuitively teaching Topological Mesh Modeling with hands-on experiences in two hours. Artistic aspects of topological mesh modeling will also be covered. The following is the schedule of the workshop.

1. **Introduction:** In this section, we introduce speakers and present a short overview. The attendees who bring their laptop will install software.
2. **Operations:** In this section, we introduce minimal and complete operations for mesh modeling which are insert/delete edge & create/remove vertex operations. We also cover splice operation and Euler operators. We also show cubical handles.
3. **High Genus Modeling:** This section introduces high genus modeling operators, namely multi-segment curved handles, rind modeling, wire modeling, column modeling.
4. **Subdivision & Remeshing:** This section introduces subdivision schemes and provides a taxonomy for subdivision scheme, namely primary conversion schemes, dual conversion schemes, primary preservation schemes, dual preservation schemes.
5. **Others:** In this section, we cover subjects such as generalized extrusions, generalized handles, generalized & manifold fractals.

Vermeer's the Music Lesson in Modular Perspective

Tomás García-Salgado
National Autonomous University of Mexico
tgsalgado@perspectivegeometry.com

Abstract

This workshop has the aim to recreate the perspective outline of the Music Lesson. The reader may notice in Figure 1, how the painting's image formation clearly fits in my *RMS90 Modular Scale*. We will learn the basic use of this *scale* to directly deduce all the elements of the scene in perspective. Therefore, neither a plan nor an elevation is required for the practice, just a good photograph copy of the painting is needed. I will provide these copies and the *scales* as well, while the participants should bring some A4 sheets, a portable drawing board, squares, eraser, and pencils (gray and yellow). It would take about 75 minutes to perform it. You will remember those high school days.

Theoretical Basis for the Practice

A fine proof of the use of a camera obscura (a darkened chamber in which the real image of an object is received through a small opening or lens and focused in natural color onto a facing surface rather than recorded on a film or plate.) in Vermeer's paintings is supported by the same image formation that several of them seem to have. I have found this feature in: The Music Lesson, Lady Standing at the Virginals, the Concert, The Girl with a Wineglass, Lady Written a Letter, The Glass of Wine, the Allegory of Faith, and the Art of Painting. In my opinion, such a feature was unattainable by using any type of perspective method available at the time. To fully capture the Music Lesson's floor in perspective, Vermeer could have removed the furniture to catch it with the camera obscura, and later go back to recompose the scene. Otherwise, the accurate rendering of the floor would have been difficult to achieve. Let us suppose that he put marks at the canvas' border to retain the vanishing lines of the floor, or even better, using a removable frame to put them over it. What advantage could he have gained? Simply, he could verify the outline of the floor at any time during the painting's process without the aid of the camera obscura. Therefore, it is doubtful that he used both lateral *distance vanishing points* (*dvp*) for this task since they fall outside of the canvas.

The Workshop

To begin, we will draw the floor tiles with the aid of the *RMS90 scale*. Place the P scale at the left border of your drawing to measure and trace all the transversal lines of the floor. Then deduce the *dvp* through the bottom modulation of the floor —although Vermeer most likely did not use them— since they are essential to recreate the perspective outlining of the floor. Second, all the architectural measures are determined by using the floor tiles modulation as a spatial reference. In Figure 1, for instance, the column width can be related to the floor by carrying vertical lines to it, finding its position between the P = 3.5 *m* and P = 5 *m*. Hence, by using the floor modulation at either of these depths, the column height is determined as well. Third, the characters and furniture are deduced in the same manner. Finally, once the *vdp'* are determined, at both sides of the drawing board, the 'real' observer's distance can be measured modularly, and prove that the painting's image formation corresponds to an approximate angle of 90 degrees.

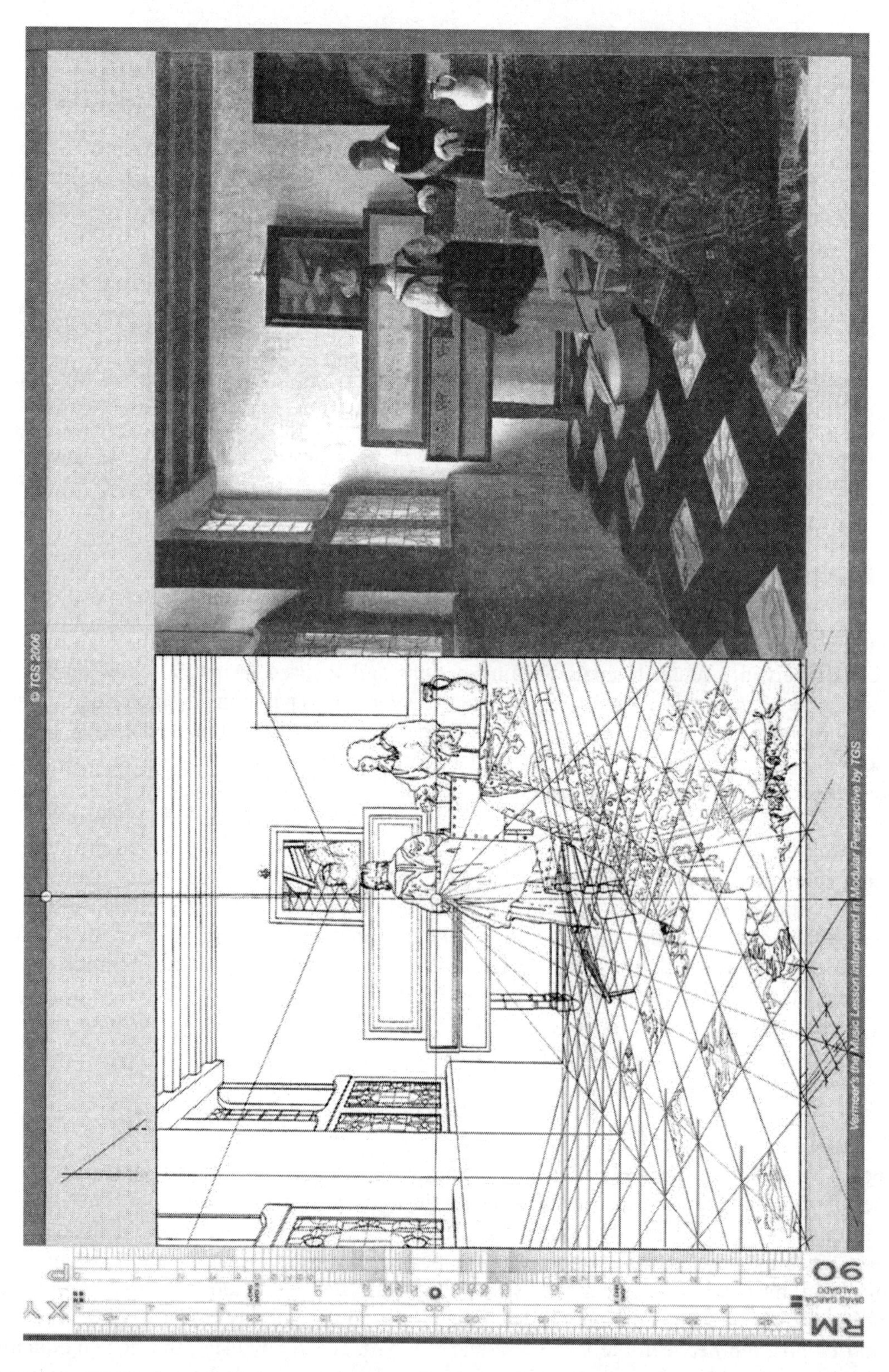

Figure 1. *Vermeer's the Music Lesson perspective outlining according to my Modular Perspective method. You may notice how the RMS90 scale perfectly matches with the floor-tiled depths one by one. The main problem in perspective is to solve depths, which is exactly what the RMS was invented for.*

Zellij Multipuzzle

Jean Marc Castera
6 rue Alphand, 75013 Paris
jm@castera.net

Using the technique of laser cutting I recently made a game I named "zellij Multipuzzle". It is a set of 669 zellij-style tiles. One side colored in white, the other in a different color, so each tile can be used in a "positive" or "negative" configuration, according to the necessary alternation of colors. There is also a set of units with which you can make frames of different sizes. This make the game easier for beginners.
This is the most efficient way I have experimented for an introduction to the art of geometrical arabesque : direct immersion in the galaxy of zellij !

1. Description of the game. The shape and the number of the tiles have been chosen to provide the largest possibilities of solutions in each situation (**fig. 2** is an example of two different solutions for the same beginning). Those solutions are standard motives that people are rediscovering, but they also can find original variations.
There is only one shape that does not have any symmetry axes (**Fig. 1**, left column of tiles, bottom), so it requires its chiral shape .
The frames are a help for beginners. They are made of different pieces, which allows different and fast configurations (beginning with a simple configuration, then removing parts of the frame to go to a more difficult level), and also makes the game easy to transport.

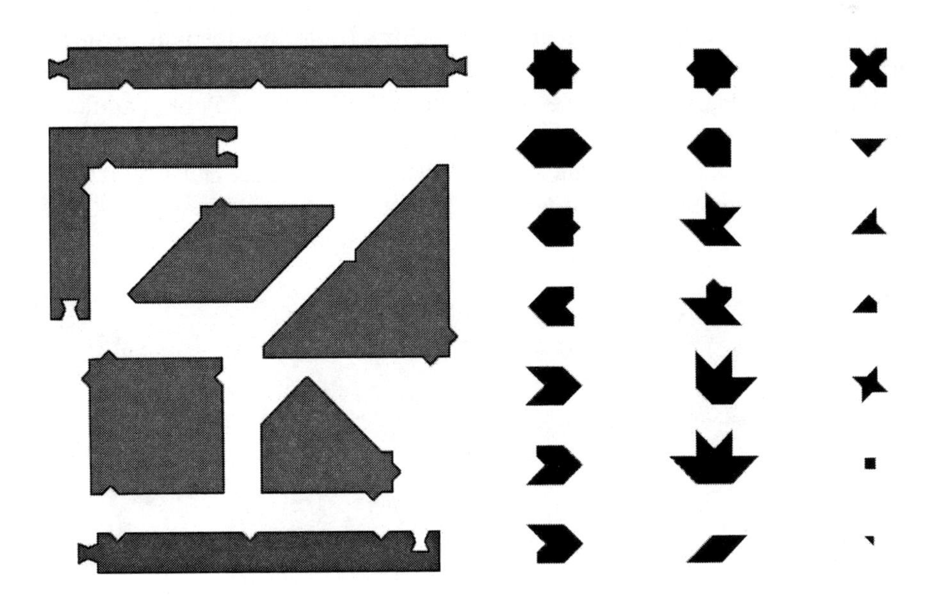

Figure1 : *Left : the different pieces of the frames. Right : the different shapes of the 669 tiles.*

2. Basic rules.

1. Alternation of the colors: adjacent tiles have to be of different colors (like a checkerboard). Two colors are available.
2. Continuity of the line: the line never stops running (except at the limit of the composition). That means that in each crossing you have an even number of lines. I say that the line have to keep alive, and everyone understands.
3. Recommendation 1: avoid having more than two lines crossing at the same point.
4. Recommendation 2: avoid having a line that changes of direction at a crossing.

An interesting way is to work according to the symmetries of the frame, and then break some of these symmetries.

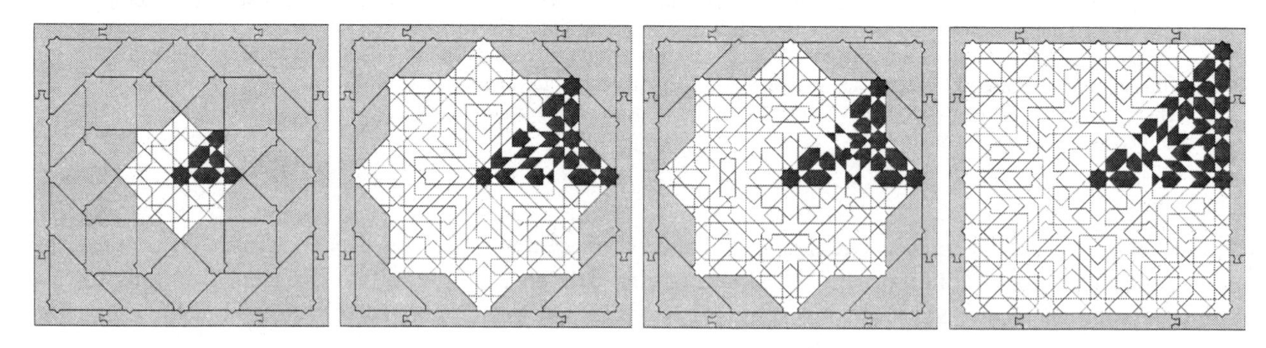

Figure 2: *Examples with different arrangements of the frames, or different solutions with the same frame.*

3. Previous experimentations (2005).

- Arabic World Institute (IMA), Paris. Public: high school students.
  The IMA uses the puzzle activity regularly (with different groups of people) as an introduction to geometric art.
- "La Cité des Sciences", Paris (Paris science museum). Public: mixed, adults and children.
- "Salon des Jeux et de la Culture Mathématique", Paris (a 4-days festival of mathematic games in Paris). Public: young people.
- International interdisciplinary conference "Science and Art", Athens. Public: adults.

In the first workshop I gave the participants some models to reproduce. But I realized that many people prefer to make their own models. Now I do not give models anymore. I explain with few words the very simple rules, and I let people the pleasure of (re)discovering patterns.

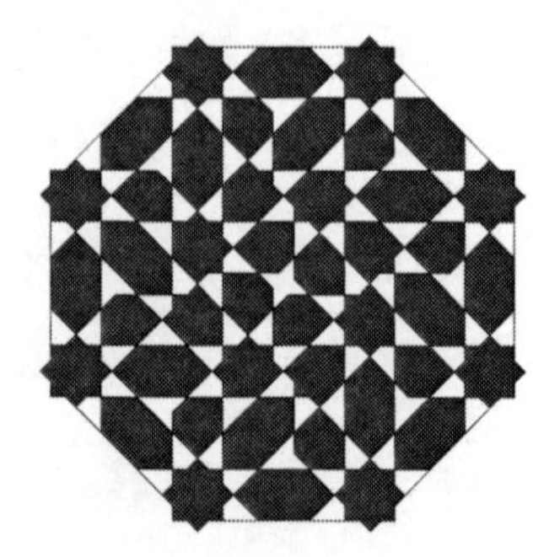

Figure 3 : *left and center: two patterns made by children without the use of models in a 45 minutes session. Sometimes you need a shape that is not available in the game, but you can often make it assembling small ones: this is the "tangram aspect" of the game. In the second example (with the small frame) they first found a symmetric solution, then I encouraged them to break some symmetries. Right: an example made without frames, with an octagonal limit, and no axes of symmetries but a center of rotational symmetry.*

Bridges London

INDEX

INDEX